Infosys Science Foundation Series

Infosys Science Foundation Series in Mathematical Sciences

The **Infosys Science Foundation Series in Mathematical Sciences**, a Scopus-indexed book series, focuses on high-quality content in the domain of pure and applied mathematics, biomathematics, financial mathematics, operations research and theoretical computer science. With this series, Springer hopes to provide readers with monographs, textbooks, edited volume, handbooks and professional books of the highest academic quality on current topics in relevant disciplines. Literature in this series will appeal to a wide audience of researchers, students, educators and professionals across the world.

Elisabetta Barletta · Sorin Dragomir ·
Mohammad Hasan Shahid · Falleh R. Al-Solamy

Differential Geometry

Riemannian Geometry and Isometric Immersions (Book I-B)

Springer

Elisabetta Barletta
Dipartimento di Matematica, Informatica ed Economia
Università degli Studi della Basilicata
Potenza, Potenza, Italy

Sorin Dragomir
Dipartimento di Matematica, Informatica ed Economia
Università degli Studi della Basilicata
Potenza, Potenza, Italy

Mohammad Hasan Shahid
Department of Mathematics
Jamia Millia Islamia
New Delhi, Delhi, India

Falleh R. Al-Solamy
Department of Mathematics
King Abdul Aziz University
Jeddah, Saudi Arabia

ISSN 2363-6149 ISSN 2363-6157 (electronic)
Infosys Science Foundation Series
ISSN 2364-4036 ISSN 2364-4044 (electronic)
Infosys Science Foundation Series in Mathematical Sciences
ISBN 978-981-96-1630-5 ISBN 978-981-96-1631-2 (eBook)
https://doi.org/10.1007/978-981-96-1631-2

Mathematics Subject Classification: 53C05, 53C07, 53C12, 53C17, 53C18, 53C40, 53C43, 53C56, 53C60, 53Z05

This Springer imprint is published by the registered company Springer Nature Singapore Pte Ltd.
The registered company address is: 152 Beach Road, #21-01/04 Gateway East, Singapore 189721, Singapore

Preface to Book I-B

The present book is the second volume of a series of four, aimed to the investigation of the relationship between a branch of complex analysis of functions of several complex variables, which is the theory of tangential Cauchy–Riemann equations, and a branch of mathematical physics, which is general relativity and gravitation theory, including quantum mechanical aspects. The theory of Cauchy–Riemann manifolds itself exhibits features belonging to mathematical disciplines culturally distinct from complex analysis, i.e., differential geometry and the theory of partial differential equations, both linear and quasi-linear. The announced volumes are

Book I-A. *Differential Geometry: Manifolds, Bundles, Characteristic Classes*
Chapter 1. Manifolds and Tensor Calculus
Chapter 2. Differential Actions and Principal Bundles
Chapter 3. Infinite Dimensional Differential Geometry
Chapter 4. Parallel Displacement and Exponential Mappings
Chapter 5. Holonomy
Chapter 6. Complex Line Bundles
Chapter 7. Characteristic Classes
Appendix A. Algebraic Topology Methods

Book I-B. *Differential Geometry: Riemannian Geometry and Isometric Immersions*
Chapter 1. Riemannian Geometry
Chapter 2. Finslerian Geometry
Chapter 3. Isometric Immersions
Chapter 4. Riemannian Foliations
Chapter 5. Sobolev Inequalities on Submanifolds

Book I-C. *Differential Geometry: Foundations of Cauchy–Riemann and Pseudohermitian Geometry*
Chapter 1. Cauchy–Riemann Manifolds
Chapter 2. Pseudohermitian Geometry
Chapter 3. Tangential Cauchy–Riemann Complex

Book I-D. *Differential Geometry: Advanced Topics in CR and Pseudohermitian Geometry.*

Book I-B consists of five chapters, whose contents are described as follows:

Chapter 1 is introductory to Riemannian geometry, with an emphasis on the geometry of geodesics in Riemannian manifolds. Smooth 1-parameter variations of a given geodesic lead to Jacobi fields and the Jacobi equation. We give the needed algebraic preparation on the Riemann–Christoffel tensor field and corresponding sectional curvatures of 2-planes tangent to a Riemannian manifold. We discuss Ricci curvature and Einstein manifolds, i.e., Riemannian manifolds whose Ricci tensor is proportional to the metric. The natural geometric structures on the total space of the tangent bundle over a Riemannian manifold (a Riemannian metric G, the so-called Sasaki metric, and an almost complex structure compatible to G, i.e., making G into an almost Hermitian metric) are briefly described. Some space is devoted to a technique of integration on a Riemannian manifold, i.e., to the calculation if integrals of functions on a Riemannian manifold by exploiting the exponential mapping and therefore the geometry of geodesics on that Riemannian manifold. The Laplace–Beltrami operator (on functions and on differential forms) on a Riemannian manifold is discussed vis-a-vis to the geometry of geodesics, and in particular, one recalls an important mathematical analysis tool, which is the Bochner–Lichnerowicz formula. The familiar expression of the Laplace–Beltrami operator in radial coordinates is prepared for applications considered later on in this book. Chapter 1 ends with a review of Hodge–de Rham's theory and its applications to the study of the topology of compact semi-simple Lie groups.

Chapter 2 is a brief introduction to Finslerian geometry, focusing on the theory of geodesics in a Finslerian manifold. The reason for the presence of Chap. 2 in a book essentially devoted to Riemannian geometry and isometric immersions between Riemannian manifolds is to be searched for in Shiing–Shen Chern's manifest paper[1] pleading in favor of the "return" *en vogue* of the study of the theory of Finsler spaces.

Chapter 3 starts with a detailed presentation of the fundamentals of the theory of isometric immersions between Riemannian manifolds (the first and second fundamental forms of an isometric immersion, the normal bundle, the derivation of the Gauss and Weingarten formulas and the Gauss–Ricci–Codazzi equations). We discuss classical results on totally geodesic and minimal isometric immersions,

[1] S-S. Chern, "Finsler geometry is just Riemannian geometry without the quadratic restriction", Notices of Amer. Math. Soc., 43(1996), 959–963.

including Cartan's theorem (on the axiom of r-planes, belonging to the rather opposite[2] philosophy at work in submanifold theory, i.e., regarding the manner in which requirements on the geometry of the second fundamental form of the submanifolds of a given Riemannian manifold, influence on the geometry of that ambient space), Nomizu's theorem relating the totally geodesic submanifolds of a Riemannian manifold M to the structure of the fixed point set of an isometry $[F \in \text{Isom}(M)]$ leading to the determination of certain totally geodesic submanifolds of a Brieskorn sphere, the study of minimal submanifolds of spheres embracing classical examples such as Clifford tori, Chern's theorem (on minimal submanifolds of a Riemannian manifold of constant sectional curvature), and the classification of totally umbilical submanifolds of Euclidean space. Minimal surfaces, their Weierstrass representations, and (normal) Gauss maps are also treated in some detail. As a natural *seguito* of our presentation of the geometry of minimal submanifolds, we introduce the basics of the theory of harmonic maps. Examples abound. Chapter 3 also presents an array of modern results known colloquially as "characterizations of the sphere", e.g., descriptions, in terms of their extrinsic and intrinsic geometry, of submanifolds of Euclidean space that are, under requirements on their second fundamental form and (intrinsic) curvature, isometric to a sphere. Submanifolds of spheres are discussed also vis-a-vis to Chen's conjecture. Some room is also given to the newer topic of biharmonic isometric immersions.

Chapter 3 ends up with the detailed exposition of a beautiful result by Eugenio Calabi on weak elliptic inequalities in Riemannian geometry. The reviewed work[3] by Calabi is perhaps better known to experts in PDEs, yet the arguments in the proof maintain a solid bond to the geometry of submanifolds in Riemannian manifolds, and Riemannian geometry in general, hence authors' choice to include Calabi's result in the present volume.[4] The ideas and techniques in Chap. 3 are perhaps the red thread through the four volumes of the present book.

Chapter 4 is devoted to the theory of Riemannian foliations, to the geometry of their second fundamental form and mean curvature, to the practical use of the associated spectral sequences, and to a brief presentation of de Rham–Hodge theory (for Riemannian foliations of compact Riemannian manifolds). Chapter 4 ends with two more specialized topics, i.e., the study of I. Vaisman's curvature groups of a Riemannian foliation, and a ramification of the theory of harmonic maps within foliation theory (the theory of foliated harmonic maps).

Chapter 5 aims to introduce the reader familiar with the differential geometric aspects of the theory of isometric immersions, such as presented in Chap. 3, to certain

---

[2] The first, direct, purpose of submanifold theory is perhaps to describe the intrinsic geometry of a given submanifold, from information on the second fundamental form of the given immersion and the geometry (mainly curvature properties) of the ambient space.

[3] Cf. E. Calabi, "An extension of E. Hopf's maximum principle with applications to Riemannian geometry", Duke Math. J., 25(1958), 45–56.

[4] The PDEs preparation needed to fully understand the proof to Theorem 3.233 is relegated to Volume VI, Functional Analysis and Partial Differential Equations: *Basics of Partial Differential Equations Theory*, of our 11 volumes book project (the plan of the forthcoming volumes V–XI is given in the Preface to Volume I).

mathematical analysis themes within geometric analysis on submanifolds of Riemannian manifolds, such as the extension of Sobolev and isoperimetric inequalities from the Euclidean case to the realm of Riemannian geometry.

The approach to most proofs is frankly tailored for didactical purposes,[5] and in few cases, the approach is but "monographic". The bibliographic references through Book I-B embrace one of the following two forms: there is of course a list of references at the end of the volume, yet a number of bibliographic items are indicated in a footnote, placed in an ϵ-neighborhood of the specific result discussed.[6]

Potenza, Italy — Elisabetta Barletta
Potenza, Italy — Sorin Dragomir
New Delhi, India — Mohammad Hasan Shahid
Jeddah, Saudi Arabia — Falleh R. Al-Solamy

[5] Dedicated to authors' many students over the years, and with the precise aim to furnish to the reader the means for the differential geometric preparation that appears to the authors of this book appropriate for ultimately studying differential geometric (and mathematical analysis) aspects of the theory of tangential Cauchy–Riemann equations and its application to general relativity and gravitation theory.

[6] A choice also dictated by didactical purposes.

Contents

About the Authors

Elisabetta Barletta is a Professor of Mathematical Analysis at the Department of Mathematics, Computer Science, and Economy, Università degli Studi della Basilicata, Potenza, Italy. Her research interests regard complex analysis of functions of several complex variables, reproducing kernel Hilbert spaces, geometry of Levi flat CR manifolds, and proper holomorphic maps of pseudoconvex domains.

Sorin Dragomir is a Professor of Mathematical Analysis at the Department of Mathematics, Computer Science, and Economy, Università degli Studi della Basilicata, Potenza, Italy. His research interests regard the applications of subelliptic theory, Lorentzian geometry, and Cauchy–Riemann geometry, to the theory of singularities of spacetime.

Mohammad Hasan Shahid is a Professor at the Department of Mathematics, Jamia Millia Islamia, New Delhi, India. He earned his Ph.D. in Mathematics from Aligarh Muslim University, India, on the topic "On geometry of submanifolds" in 1988 under (Late) Prof. Izhar Husain. Earlier, he served as Associate Professor at King Abdul Aziz University, Jeddah, Saudi Arabia, from 2001 to 2006. He was a recipient of the postdoctoral fellowship from the University of Patras, Greece, from October 1997 to April 1998. Recently, he was awarded the Sultana Nahar Distinguished Teacher Award of the Year 2017–2018 for his outstanding contribution to research.

Falleh R. Al-Solamy is President at King Khalid University, Abha, Saudi Arabia. Earlier, he was a Professor of Differential Geometry at King Abdulaziz University, Jeddah, Saudi Arabia. He studied Mathematics at King Abdulaziz University, Jeddah, Saudi Arabia, and earned his Ph.D. in Mathematics from the University of Wales Swansea, Swansea, UK, in 1998, under Prof. Edwin Beggs. His research interests concern the study of the geometry of submanifolds in Riemannian and semi-Riemannian manifolds, Einstein manifolds, and applications of differential geometry in physics.

Chapter 1
Riemannian Geometry

1.1 Riemannian and Semi-Riemannian Metrics

1.1.1 Historical Note

Modern Riemannian geometry arises essentially from the differential geometry of surfaces[1] in Euclidean space $\mathbb{R}^3$. A surface $S \subset \mathbb{R}^3$ may be described as the locus of all points $(x, y, z) \in \mathbb{R}^3$ which satisfy an equation of the form

$$\phi(x, y, z) = 0. \tag{1.1}$$

By the implicit function theorem, if ϕ is of class C^1 and $\phi_z \neq 0$ along (1.1) then (1.1) may be written as

$$z = h(x, y). \tag{1.2}$$

Equation (1.2) emphasizes on the bidimensional character of the surface. One may argue that, with the passage from (1.1) to (1.2), a lack of symmetry[2] arises, as to the role played by the Cartesian coordinates (x, y, z). The removal of this inadequacy, allowing in particular for weakening the regularity conditions imposed on the defining function ϕ [conditions sufficient to guarantee the representation (1.2)], is due to

---

[1] Cf. e.g., H.W. Guggenheimer, *Differential Geometry* (Dover Publications, Inc., New York, 1977), pp. 206–356.

[2] By definition a surface S of Eq. (1.1) satisfies $D\phi(\mathbf{x}) \neq 0$ fro any point $\mathbf{x} \in S$. If $D\phi(\mathbf{x}_0) = (\phi_x(\mathbf{x}_0), \phi_y(\mathbf{x}_0), \phi_z(\mathbf{x}_0)) \neq 0$ then at least one of the components of the gradient of ϕ evaluated at $\mathbf{x}_0$ is nonzero, e.g., $\phi_z(\mathbf{x}_0) \neq 0$ and then, by the permanence of sign of continuous functions, $\phi_z(\mathbf{x}) \neq 0$ at any point $\mathbf{x}$ of a neighborhood Ω of $\mathbf{x}_0$ in $\mathbb{R}^3$; then one arrives to the representation (1.2) of the portion of surface $S \cap \Omega$. In this approach, the variable z appears as privileged. The question is clearly formal and springs merely from the way notations have been fixed (a change of basis in $\mathbb{R}^3$ followed by relabelling of variables, shows that there are no "favorite" variables).

E. Barletta et al., *Differential Geometry*, Infosys Science Foundation Series,
https://doi.org/10.1007/978-981-96-1631-2_1

Fig. 1.1 Carl Friedrich Gauss (1777–1855): German mathematician, astronomer and physicist. **G.** contributed in a decisive manner to the development of several fields of mathematics and physics, such as mathematical analysis, number theory, statistics, numerical calculus, differential geometry, geodesy, geophysics, magnetism, electrostatics, astronomy, and optics

Gauss. Gauss represented (open pieces of) surfaces as images of *parametrizations* (Fig. 1.1)[3]

$$\varphi = (\varphi_1, \varphi_2, \varphi_3) : U \subset \mathbb{R}^2 \to \mathbb{R}^3, \quad \begin{cases} x = \varphi_1(u, v), \\ y = \varphi_2(u, v), \\ z = \varphi_3(u, v), \end{cases} \quad (u, v) \in U.$$

The parameters (u, v) may be thought of as coordinates of the point $p = \varphi(u, v) \in \mathbb{R}^3$ (the *Gaussian coordinates*, or *curvilinear coordinates*, of the point p). For the same surface, there exists an infinity of distinct parametrizations. Indeed if $F = (F_1, F_2) : U \to V \subset \mathbb{R}^2$ is a C^1 diffeomorphism (a *transformation of Gaussian coordinates*) $u' = F_1(u, v)$, $v' = F_2(u, v)$, $(u, v) \in U$, then the (parametrized) surfaces

$$\varphi_1 = \varphi : U \to \mathbb{R}^3, \quad \varphi_2 = \varphi \circ F^{-1} : V \to \mathbb{R}^3,$$

have the same support in $\mathbb{R}^3$ [i.e., $\varphi_1(U) = \varphi_2(V)$]. A fundamental question in differential geometry is then to express the laws possessing an intrinsic geometric meaning in a form not depending of the Gaussian coordinates of the given surface. To give a simple example let us consider a surface $\varphi : U \to \mathbb{R}^3$ and a curve $\gamma : [0, 1] \to \mathbb{R}^3$ lying on $S = \varphi(U)$, i.e., $\gamma(t) = \varphi(a(t))$ for any $0 \le t \le 1$ and for some $a : [0, 1] \to U$. The scalar parametric equations of the curve γ are

$$x = \varphi_1(a(t)), \quad y = \varphi_2(a(t)), \quad z = \varphi_3(a(t)), \qquad 0 \le t \le 1.$$

Moreover let us assume that the maps $\varphi : U \to \mathbb{R}^3$ and $a : [0, 1] \to U$ are respectively of class C^1 and of class C^1 up to the boundary. The length $L(\gamma)$ of the curve γ is

[3] A moment's thought shows that a parametrization isn't merely the inverse of a local chart (thinking the given surface as a 2-dimensional differentiable manifold) but the local representation (with respect to a given local chart) of the inclusion, assumed to be an immersion (of the given surface in $\mathbb{R}^3$).

$$L(\gamma) = \int_0^1 \|\dot{\gamma}(t)\| \, dt = \int_0^1 \left[\left(\frac{dx}{dt}\right)^2 + \left(\frac{dy}{dt}\right)^2 + \left(\frac{dz}{dt}\right)^2 \right]^{1/2} dt$$
$$= \int_0^1 \left[\sum_{i,j=1}^2 g_{ij} \frac{da^i}{dt} \frac{da^j}{dt} \right]^{1/2} dt$$

where we have set $g_{ij} = \sum_{A=1}^{3} \frac{\partial \varphi_A}{\partial u^i} \frac{\partial \varphi_A}{\partial u^j}$ with $u^1 = u$ and $u^2 = v$ [the coefficients of the *first fundamental form* of the surface with respect to the Gaussian coordinates (u^1, u^2)]. If $(u'^1, u'^2) = F(u^1, u^2)$ is a coordinate transformation then

$$g'_{k\ell} = \frac{\partial \left(F^{-1}\right)^i}{\partial u'^k} \frac{\partial \left(F^{-1}\right)^j}{\partial u'^\ell} \left(g_{ij} \circ F^{-1}\right). \tag{1.3}$$

As an elementary consequence of the transformation law (1.3) of the coefficients g_{ij} [with respect to a transformation of Gaussian coordinates $(u^1, u^2) \mapsto (u'^1, u'^2)$], for any point $p \in S = \varphi(U)$ the inner product g_p on the tangent space at p to S

$$g_p(X, Y) = \sum_{i,j=1}^{2} g_{ij}(u) X^i Y^j, \quad X, Y \in T_p(S), \quad p = \varphi(u),$$

$$X = \sum_{i=1}^{2} X^i \left(\frac{\partial}{\partial u^i}\right)_p, \quad Y = \sum_{j=1}^{2} Y^j \left(\frac{\partial}{\partial u^j}\right)_p, \quad u = (u^1, u^2) \in U,$$

is well defined, i.e., the definition doesn't depend upon the choice of curvilinear coordinates of the point p. As Gauss has shown, several geometric properties of the surface S may be deduced[4] from the study of the first fundamental form (or *induced metric*) $g_p : T_p(S) \times T_p(S) \to \mathbb{R}$, $p \in S$. These considerations were the starting point of the studies performed by Bernhard Riemann[5] in 1892. Riemann observed that the confinement at the dimension two cases (surfaces in $\mathbb{R}^3$) as in Gauss' studies is incidental and proposed, in the language of modern differential geometry, the study of n-dimensional Riemannian manifolds. Therefore one commonly attributes to Riemann the "discovery" of Riemannian manifolds, although Riemann considered only open subsets of $\mathbb{R}^n$ endowed with non-flat Riemannian metrics (metrics distinct from the Euclidean metric on $\mathbb{R}^n$). Indeed the notion of an abstract differentiable manifold, as operated within nowadays mathematics, belongs

[4] Cf. e.g., M.P. Do Carmo, *Differential Geometry of Curves and Surfaces* (Prentice Hall, Inc., Englewood Cliffs, New Jersey, 1976), pp. 92–99.

[5] Cf. B. Riemann, *On the Hypotheses which lie at the Bases of Geometry*. Translated by William Kingdon Clifford (Nature **VIII**(183, 184), 14–17, 36, 37).

Fig. 1.2 Bernhard Riemann (1826–1866): German mathematician. **R.** has contributed to the development of mathematical analysis and differential geometry. As an infant, **R.** is remembered as shy and fearful yet in possession of exceptional mathematical abilities. **R.** was interested both in mathematics and formal biblical studies, with an intention of embracing the ecclesiastical career. **R.** started the systematic study of mathematics at the University of Göttingen, under the supervision of Gauss, who also persuaded **R.** to give up theological studies and devote his efforts entirely to mathematics. In his first lectures (in 1854) **R.** sets the basis of what today one calls Riemannian geometry which may be thought of as the geometric basis (even if in a different signature) of GRG theory. While the observation that Riemannian geometry (in its Lorentzian version) may be associated - form a historical point of view—to the successive scientific creation by Einstein is a common occurrence in nowadays textbooks, by far less emphasis is set on the suggestion by **R.** that dimensions higher than 3 and 4 should be used in the description of physical reality (a suggestion proposing **R.** as a precursor of Kaluza-Klein type theories)

to H. Poincaré[6] and it became clear only after the work by John Nash[7] that a Riemannian manifold may be isometrically immersed in an Euclidean space (of sufficiently high dimension). At any rate, it is a fact, even in the original approach by Riemann, that the intrinsic concepts introduced by Gauss (e.g., the curvature of a surface) may be easily extended to the case of (abstract, n-dimensional) Riemannian manifolds and the so-called non-Euclidean geometry (discovered by Lobacevski and Bolyai) is but a particular case of Riemannian geometry, as well as the familiar Euclidean geometry (Figs. 1.2, 1.3 and 1.4).

1.1.2 Bilinear Symmetric Forms

We start with a preparation of linear algebra. Let V be a n-dimensional and let $g : V \to V \to \mathbb{R}$ be a $\mathbb{R}$-bilinear and symmetric form.

Definition 1.1 (1) We say g is *nondegenerate* if

$$g(u, v) = 0 \text{ for any } v \in V \Longleftrightarrow u = 0.$$

---

[6] Jules Henri Poincaré (1854–1912), French mathematician. Poincaré's work is fundamental for an array of mathematical disciplines. The book *On the theory of Fuchsian functions* (1881) is commonly regarded as one of Poincaré's major contributions.

[7] Cf. J. Nash, The imbedding problem for Riemannian manifolds. Ann. Math. **63**(1), 20–63 (1956).

Fig. 1.3 Nikolai Ivanovici Lobacevski (1792–1856), a Russian mathematician. **L.** discovered, independently from J. Bolyai, the *non-Euclidean geometry* in the context of which the parallels axiom is replaced by the prescription that by a given point non-incident to a given line one may conduct two parallel lines (parallel to the given line). Although Lobacevki's non-Euclidean geometry is but a particular example of Riemannian metric (the *hyperbolic metric*) on $\mathbb{R}^2$, of negative constant sectional curvature (and then the efforts to "perceive" the new metric and its properties by considerations of synthetic geometry, in the plane of Euclidean geometry and in the presence of the former, become somewhat questionable), the discovery by **L.** may be looked at historically as a revolution in the understanding of the concept of space and its influences substantially on the successive development of mathematical thought. The main work by **L.** is *Foundations of geometry* (1830)

Fig. 1.4 Janos Bolyai (Cluj-Napoca 1802–Târgu-Mureş 1860), Hungarian mathematician native of Transylvania. **B**. discovered, independently from N.I. Lobacevski, the non-Euclidean geometry. **B.**'s main work is *Appendix* (1832), written in Latin as an appendix to the book *Tentamen juventutem studiosam in elementa matheseos purae introducendi* by Farkas Bolyai (1775–1856), father of **B**. Due to financial difficulties **B**. was forced to present his arguments in a short manuscript (24 pages, the only ever published by **B.**) and developed an ad hoc formal language aimed to give maximum concision to the manuscript yet resulting in a text rather difficult to read. As N.I. Lobacevski had already published (in 1829) a similar work (also containing a generalization of the theory in question) the historical importance of **B.**'s contribution appears as questionable. One should nevertheless stress the great respect shown by Gauss to *Appendix* and the fact that **B.**'s mathematical legacy, even if in manuscript form only, counts over 20.000 pages. These may be found in the library Bolyai-Teleki at Târgu Mureş, in Romania, where **B.** passed away. For the purposes of this book, it is noteworthy that **B**. suggested that a new (with respect to Newton's universal attraction law) mathematics/physics model would be necessary should space geometry be non-Euclidean

(2) We say g is *positive definite* (respectively *negative definite*) if $g(u,u) \geq 0$ (respectively $g(u,u) \leq 0$) for any $u \in V$ and $g(u,u) = 0 \Longleftrightarrow u = 0$.

(3) We say g is *positive semi-definite* (respectively *negative semi-definite*) if $g(v,v) \geq 0$ (respectively $g(v,v) \leq 0$) for any $v \in V$.

(4) We say g is *definite* (respectively *semi-definite*) if g is positive or negative definite (respectively if g is positive or negative semi-definite). □

Any definite bilinear symmetric form is in particular nondegenerate.

Let g be a bilinear and symmetric form on V and let $W \subset V$ be a subspace. The restriction of g to $W \times W$, i.e., the map $g : W \times W \to \mathbb{R}$, is bilinear and symmetric, too. Said restriction is denoted by $g|_W$ (rather than $g|_{W\times W}$). If g is (semi) definite then $g|_W$ is (semi) definite, too.

Definition 1.2 The *index* $\nu = \text{ind}(g)$ of a bilinear symmetric form $g \in V^{0,2}$ is the highest real dimension of a real subspace $W \subset V$ such that $g|_W$ be negative definite. □

Hence $0 \leq \nu \leq n$ and $\nu = 0$ if and only if g is positive semi-definite.

Definition 1.3 The *quadratic form* associated to the bilinear symmetric form g is the function $q : V \to \mathbb{R}$ defined by $q(v) = g(v, v)$ for any $v \in V$. □

It is often easier to work with q rather than g. On the other hand, by passing from g to q no information is lost, for g may be recovered from q by the so-called *polarization formula*

$$g(u, v) = \frac{1}{2}\{q(u+v) - q(u) - q(v)\}, \quad u, v \in V.$$

Let $\{e_i : 1 \leq i \leq n\} \subset V$ be a linear basis and let $\{e^i : 1 \leq i \leq n\} \subset V^*$ be the dual linear basis. The matrix of g with respect to the basis $\{e_i : 1 \leq i \leq n\}$ is

$$g_{ij} = g(e_i, e_j) \in \mathbb{R}, \quad 1 \leq i, j \leq n.$$

Since g is symmetric $[g_{ij}]_{1\leq i,j\leq n} \in \text{Mat}_{n\times n}(\mathbb{R})$ is symmetric.

Proposition 1.4 *A symmetric bilinear form g is nondegenerate if and only if the matrix of g with respect to some basis of V, and then with respect to any basis of V, is nonsingular.*

As g is also a $(0, 2)$-tensor, i.e., $g \in V^{0,2}$, one has $g = g_{ij}\, e^i \otimes e^j$. With respect to the basis $\{e_i : 1 \leq i \leq n\}$ the quadratic form associated to the bilinear symmetric form g reads

$$q(v) = g(v, v) = v^i v^j g(e_i, e_j) = g_{ij} v^i v^j$$

for any $v = v^i e_i \in V$.

1.1.3 Inner Products

Let $g : V \times V \to \mathbb{R}$ be a bilinear symmetric form.

Definition 1.5 If g is nondegenerate then g is an *inner product*. An inner product g is *indefinite* if g is not definite. □

The reader is certainly familiar with definite inner products (from standard courses on linear algebra, taught to both physics and mathematics students). Several properties of definite inner products do survive for indefinite inner products. However indefinite inner products also enjoy new and surprising properties.

Example 1.6 (*Minkowski* inner product on $\mathbb{R}^2$) Let $V = \mathbb{R}^2$ and let $g : \mathbb{R}^2 \times \mathbb{R}^2 \to \mathbb{R}$ be given by $g(u, v) = u^1 v^1 - u^2 v^2$ for any $u = (u^1, u^2)$ and $v = (v^1, v^2)$. g is nondegenerate. Indeed if $g(u, v) = 0$ for any $v \in \mathbb{R}^2$ then $g(u, e_i) = 0$ for any $i \in \{1, 2\}$. Here $\{e_1, e_2\} \subset \mathbb{R}^2$ is the canonical basis. Consequently

$$0 = g(u, e_i) = u^i, \quad i \in \{1, 2\},$$

i.e., $u = 0$ or g is an inner product. On the other hand $g(v, v) = (v^1)^2 - (v^2)^2$ so that $g(e_1, e_1) > 0$ and $g(e_2, e_2) < 0$, i.e., g is indefinite.

Definition 1.7 A pair (V, g) consisting of a real vector space of finite dimension, and an inner product g on V is called an *inner product space*. □

Definition 1.8 Let (V, g) be an inner product space. A vector $v \in V$ is *null* if $g(v, v) = 0$ yet $v \neq 0$. □

In Example 1.6 the null vectors form a degenerate quadric

$$\mathcal{C} = q^{-1}(0) = \{(v^1, v^2) \neq 0 : (v^1)^2 - (v^2)^2 = 0\}$$

(the *null cone*) which is the union of two *null lines*

$$\mathcal{C} = L_+ \cup L_-, \quad L_\pm = \{(v^1, v^2) \neq 0 : v^1 \pm v^2 = 0\}.$$

If in turn $c \in \mathbb{R} \setminus \{0\}$ then the set $q^{-1}(c)$ is a hyperbola whose asymptotes are the null lines $L_\pm$.

Definition 1.9 Let (V, g) be an inner product space.

(i) Two vectors $u, v \in V$ are *orthogonal*, and one adopts the notation $u \perp v$, if $g(u, v) = 0$.

(ii) A system of vectors $\{v_1, \ldots, v_p\} \subset V$ is *orthogonal* if it consists of mutually orthogonal vectors.

(iii) An orthogonal system $\{v_1, \ldots, v_p\}$ is *orthonormal* if $g(v_i, v_i) \in \{\pm 1\}$ for any $1 \leq i \leq p$. □

Exercise 1.10 Any orthonormal system in an inner product space is free.

Solution. Let $\mathcal{S} = \{v_1, \ldots, v_k\}$ be an orthonormal system. If

$$\sum_{i=1}^{k} \lambda_i v_i = 0$$

then

$$0 = g\left(\sum_{i=1} \lambda_i v_i,\ v_j\right) = \sum_{i=1}^{k} \lambda_k\, g(v_k,\ v_j) = \lambda_j\, g(v_j,\ v_j) = \pm\lambda_j$$

and hence $\lambda_j = 0$, for any $1 \leq j \leq k$. Q.e.d.

Proposition 1.11 *Let* (V, g) *be an inner product space. Then* g *is indefinite if and only if* g *admits null vectors.*

Proof The implication "$\Longleftarrow$" is obvious. The opposite implication is more difficult to prove and requires a few classical ingredients, such as the *Gramm-Schmidt process* and the *canonical form* of the quadratic form associated with the given inner product.

Lemma 1.12 *Let* g *be an inner product on* V *possessing no null vectors, i.e.,* $g(v, v) \neq 0$ *for any* $v \in V \setminus \{0\}$. *Then* V *admits an orthonormal basis.*

Proof Let $\{v_1, \ldots, v_n\}$ be just any basis of V. We build, recurrently, a new system of vectors $\{e_i : 1 \leq i \leq n\}$ by setting

$$\tilde{e}_1 = v_1,\ \ e_1 = \frac{1}{|g(\tilde{e}_1,\ \tilde{e}_1)|^{1/2}}\,\tilde{e}_1,\ \ \tilde{e}_i = v_i + \sum_{j=1}^{i-1} \lambda_{ij} e_j,\ \ e_i = \frac{1}{|g(\tilde{e}_i,\ \tilde{e}_i)|^{1/2}}\,\tilde{e}_i,\ \ 2 \leq i \leq n,$$

where the scalars $\lambda_{ij} \in \mathbb{R}$ are chosen such that

$$\tilde{e}_i \perp e_j,\quad 2 \leq i \leq n,\quad 1 \leq j \leq i - 1,$$

i.e., at every step i the new vector $\tilde{e}_i$ be orthogonal to the vectors precedingly built $\{e_j : 1 \leq j \leq i - 1\}$. Moreover each of the vectors $\{e_j : 1 \leq j \leq i - 1\}$ possesses the property $g(e_j, e_j) \in \{\pm 1\}$. That is to say

$$0 = g(\tilde{e}_i,\ e_j) = g(v_i,\ e_j) + \sum_{k=1}^{i-1} \lambda_{ik}\, g(e_k,\ e_j) = g(v_i,\ e_j) + \lambda_{ij}\, g(e_j,\ e_j) \Longrightarrow$$

$$\Longrightarrow \lambda_{ij} = -\frac{g(v_i,\ e_j)}{g(e_j,\ e_j)},\quad 2 \leq i \leq n,\ \ 1 \leq j \leq i - 1.$$

Hence

$$\tilde{e}_i = v_i - \sum_{j=1}^{i-1} \frac{g(v_i,\ e_j)}{g(e_j,\ e_j)}\, e_j,\quad 2 \leq i \leq n. \tag{1.4}$$

Clearly $\{e_1, \ldots, e_n\}$ is a basis in V and an orthonormal system. Lemma 1.12 is proved.

The adopted method of proof is known as the *Gramm-Schmidt process* and may obviously be used to produce, starting from a free system of vectors, an orthonormal system having the same number of vectors.

The second ingredient needed in the proof of Proposition 1.11 is the representation of g, and of the associated quadratic form q, with respect to an orthonormal basis, provided such a basis exists. If $\{e_i : 1 \le i \le n\} \subset V$ is an arbitrary basis and $\{e^i : 1 \le i \le n\} \subset V^*$ is the corresponding dual basis then (by $g \in V^* \otimes V^* = V^{0,2}$)

$$g = g_{ij}\, e^i \otimes e^j, \quad q(v) = g_{ij} v^i v^j, \quad v = v^i e_i \in V.$$

When $\{e_i : 1 \le i \le n\}$ is orthonormal one has $\epsilon_i = g(e_i, e_i) \in \{\pm 1\}$ and hence

$$g = \epsilon_i \delta_{ij}\, e^i \otimes e^j = \sum_{i=1}^{n} \epsilon_i\, e^i \otimes e^i, \quad q(v) = \epsilon_i \delta_{ij} v^i v^j = \sum_{i=1}^{n} \epsilon_i (v^i)^2 . \tag{1.5}$$

Formula (1.5) gives the *canonical form* of q. The process of choosing an orthonormal basis of V and representing q as in (1.5) is commonly known as *bringing q to the canonical form.*

Lemma 1.13 *Let g be a bilinear symmetric form on V. Let us assume that there is an orthonormal basis $\{e_i : 1 \le i \le n\}$. Then $\epsilon_j = 1$ (respectively $\epsilon_i = -1$) for any $1 \le i \le n$ if and only if g is positive definite (respectively negative definite).*

Proof Let us assume that $\epsilon_i = 1$ for any $1 \le i \le n$. Then $g(v, v) = \sum_{i=1}^{n} (v^i)^2 \ge 0$ and $g(v, v) = 0 \iff v^i = 0, \quad (1 \le i \le n) \iff v = 0$, i.e., g is positive definite.

Exercise 1.14 Complete the proof of Lemma 1.13.

Let us prove now Proposition 1.11. Let us assume that g is indefinite. We wish to show that g possesses null vectors. We argue by contradiction, i.e., we assume that $g(v, v) \ne 0$ for any $v \in V \setminus \{0\}$. Then we are under the hypothesis of Lemma 1.12 and hence we can choose an orthonormal basis $\{e_i : 1 \le i \le n\}$. In particular

$$g(v, v) = \epsilon_1 (v^1)^2 + \cdots + \epsilon_n (v^n)^2, \quad \epsilon_i \in \{\pm 1\}.$$

The ϵ_i cannot have the same sign, for in that case (by Lemma 1.13) g would be definite. Hence, by eventually relabelling the vectors $\{e_1, \ldots, e_n\}$, there is an integer $0 < \nu < n$ such that $\epsilon_1 = \cdots = \epsilon_\nu = -1 = -\epsilon_{\nu+1} = \cdots = -\epsilon_n$, i.e.,

$$q(v) = -\sum_{i=1}^{\nu} (v^i)^2 + \sum_{i=\nu+1}^{n} (v^i)^2 .$$

At this point one may easily produce examples of vectors $v \in V \setminus \{0\}$ such that $q(v) = 0$. For instance, if $s = \min\{\nu, n - \nu\}$ then let us set

$$v = v^i e_i, \quad v^i = \begin{cases} 1 & \text{if } 1 \le i \le s, \\ -1 & \text{if } \nu + 1 \le i \le \nu + s, \\ 0 & \text{otherwise}, \end{cases}$$

so that v is null. Q.e.d.

When g is an indefinite inner product, one may not think of two orthogonal vectors as forming a right angle. Indeed, any null vector is orthogonal to itself. Let (V, g) be an inner product space. If $W \subset V$ is a subspace of V then one sets

$$W^\perp = \{v \in V : v \perp W\}, \quad v \perp W \Longleftrightarrow \forall\, w \in W \;:\; v \perp w.$$

Clearly $W^\perp$ is a vector subspace of V.

Definition 1.15 $W^\perp$ is called W *perp*. □

As seen in the preceding definition, one gives up calling $W^\perp$ the orthogonal complement of W since in general the sum $W + W^\perp$ may fail to coincide with V. For instance

Example 1.16 If (V, g) is the 2-dimensional Minkowski space $(\mathbb{R}^2,\ e^1 \otimes e^1 - e^2 \otimes e^2)$ and $W \subset \mathbb{R}^2$ is the space spanned by the vector $(1, 1)$ then $W^\perp = W$. Indeed

$$W^\perp = \{v \in \mathbb{R}^2 : v \perp W\} = \{v \in \mathbb{R}^2 : v \perp (1, 1)\} = \{v \in \mathbb{R}^2 : v^1 - v^2 = 0\} = \{(\lambda, \lambda) : \lambda \in \mathbb{R}\} = W.$$

Nevertheless $W^\perp$ possesses the following familiar properties.

Proposition 1.17 *Let W be a subspace of an inner product space V. Then*

$$\dim_{\mathbb{R}}(W) + \dim_{\mathbb{R}} W^\perp = \dim_{\mathbb{R}} V, \tag{1.6}$$

$$\left(W^\perp\right)^\perp = W. \tag{1.7}$$

Proof Let us assume that $k = \dim_{\mathbb{R}} W$, $0 \le k \le n$. If $k = 0$, and hence $W = (0)$, then $W^\perp = V$ so that (1.6) holds trivially. Also $V^\perp = (0)$ (as g is nondegenerate) hence (1.7) holds good, too. The case $k = n$ may be dealt with in a similar manner. From now on we assume that $0 < k < n$. Let $\{e_1, \ldots, e_k\}$ be a basis of W. By standard results of linear algebra there exists a system of vectors $\{e_{k+1}, \ldots, e_n\} \subset V$ such that $\{e_i : 1 \le i \le n\}$ is a basis of V (one "completed" the system $\{e_1, \ldots, e_k\}$ to get a basis of V). Now

$$W^\perp = \{v \in V : g(v, e_\alpha) = 0,\ \ 1 \le \alpha \le k\} = \{v = v^i e_i \in V : g_{i\alpha} v^i = 0,\ \ 1 \le \alpha \le k\},$$

i.e., $W^\perp$ is isomorphic to the space of solutions $\mathcal{S}$ of the linear system

$$\sum_{i=1}^{n} g_{i\alpha} x^i = 0, \quad 1 \le \alpha \le k.$$

As g is nondegenerate $A = [g_{ij}]_{1 \le i,j \le n} \in \mathrm{GL}(n, \mathbb{R})$. It follows that the columns of A are linearly independent. Any subsystem of a free system is free, too, hence the columns

$$\left\{ \left(g_{1\alpha}, \ldots, g_{n\alpha} \right)^T \; : \; 1 \le \alpha \le k \right\}$$

are linearly independent. Hence the rectangular matrix

$$[g_{i\alpha}]_{\substack{1 \le i \le n \\ 1 \le \alpha \le k}} \in \mathrm{Mat}_{n \times k}(\mathbb{R})$$

has rank k so that $\dim_{\mathbb{R}} \mathcal{S} = n - k$. Finally $W^\perp \approx \mathcal{S}$ yields $\dim_{\mathbb{R}} W^\perp = n - k$ and (1.6) is proved.

One has $(W^\perp)^\perp = \{v \in V : v \perp W^\perp\}$ hence $W \subset (W^\perp)^\perp$. Indeed if $w \in W$ then $g(w, v) = 0$ for any $v \in V$, and in particular for any $v \in W^\perp$, hence $w \perp W^\perp$, i.e., $w \in (W^\perp)^\perp$. On the other hand [by (1.6) applied twice]

$$\dim_{\mathbb{R}} (W^\perp)^\perp = \dim_{\mathbb{R}}(V) - \dim_{\mathbb{R}} W^\perp = n - (n - k) = k = \dim_{\mathbb{R}} W$$

and then it must be that $(W^\perp)^\perp = W$. Q.e.d.

Definition 1.18 Let (V, g) be an inner product space. A subspace $W \subset V$ is said to be *nondegenerate* if the bilinear symmetric form $g|_W$ is nondegenerate. Otherwise W is *degenerate*. □

If (V, g) is an inner product space with g definite, for any subspace $W \subset V$ the restriction $g|_W$ of g to $W \times W$ is a definite bilinear symmetric form, and in particular $g|_W$ is nondegenerate. If in turn g is an indefinite inner product then V admits degenerate subspaces. For instance, the 1-dimensional space generated by a null vector $v \in V \setminus \{0\}$ is clearly a degenerate subspace. Hence, in general, a subspace of an inner product space may fail to be an inner product space. This difficulty adds up to the one mentioned above, that in general $W^\perp$ fails to be a complement to W.

Proposition 1.19 *A subspace $W \subset V$ of an inner product space is nondegenerate if and only if $V = W \oplus W^\perp$.*

Proof By a basic result in linear algebra

$$\dim_{\mathbb{R}} W + \dim_{\mathbb{R}} W^\perp = \dim_{\mathbb{R}}(W + W^\perp) + \dim_{\mathbb{R}}(W \cap W^\perp)$$

(which is *Grassmann*'s formula). By Proposition 1.17 one has $\dim_{\mathbb{R}} W + \dim_{\mathbb{R}} W^\perp = n$ and hence

$$\dim_{\mathbb{R}}(W + W^\perp) = n - \dim_{\mathbb{R}}(W \cap W^\perp). \tag{1.8}$$

Now let us assume that W is a nondegenerate subspace. Then if $v \in W^\perp \cap W^\perp$ it follows that $v \in W$ and $v \perp W$ so that (by the nondegeneracy of $g|_W$) it must be $v = 0$, i.e., $W \cap W^\perp = (0)$ or equivalently the sum $W + W^\perp$ is direct (and we may write $W \oplus W^\perp$). In particular formula (1.8) becomes $\dim_{\mathbb{R}}(W \oplus W^\perp) = n$, a fact that, together with $W \oplus W^\perp \subset V$, yields $W \oplus W^\perp = V$. Q.e.d.

Exercise 1.20 Prove the opposite implication.

At this point, once the main properties of the space $W^\perp$ (for a given subspace W of an inner product space) were studied, we may prove a more general version of Lemma 1.12. Precisely

Proposition 1.21 *Any inner product space $V \neq (0)$ admits an orthonormal basis.*

Proof In any inner product space there exist non-null vectors. Indeed if $g(v, v) = 0$ for any $v \in V$ then

$$0 = g(v + w, v + w) = g(v, v) + 2g(v, w) + g(w, w) = 2g(v, w)$$

for any $v, w \in V$. Finally the assumption that $g(v, w) = 0$ for any $w \in V$ yields, as g is nondegenerate, that $v = 0$ and then $V = (0)$, a contradiction.

Hence there is a vector $v_1 \in V$ such that $g(v_1, v_1) \neq 0$. Let us set $e_1 = |g(v_1, v_1)|^{-1/2} v_1$ so that $g(e_1, e_1) \in \{\pm 1\}$. If $n = 1$ then the proof of the proposition is complete. If in turn $n \geq 2$ then let us show that one may add a second vector e_2 to the system $\mathcal{S}_1 = \{e_1\}$ in such a manner that e_2 be orthogonal to e_1 and such that $g(e_2, e_2) \in \{\pm 1\}$. Let $W_1 = \{\lambda e_1 : \lambda \in \mathbb{R}\}$ be the space spanned by $\mathcal{S}_1$. That space is certainly nondegenerate and then (by Proposition 1.11) $W_1 \oplus W_1^\perp = V$. As $\dim_{\mathbb{R}} W_1 = 1$ and $\dim_{\mathbb{R}} V = n \geq 2$ it follows that $W_1^\perp \neq (0)$. Then there is $v_2 \in W_1^\perp$ such that $g(v_2, v_2) \neq 0$ and we may set $e_2 = |g(v_2, v_2)|^{-1/2} v_2 \in W_1^\perp$ so that $e_2 \perp e_1$ e $g(e_2, e_2) \in \{\pm 1\}$ and then $\mathcal{S}_2 = \{e_1, e_2\}$ is an orthonormal system in V. For any $k \in \{1, \dots, n-1\}$ we consider the proposition $P(k)$ asserting that given an orthonormal system $\mathcal{S}_k = \{e_1, \dots, e_k\}$ there is a vector $e_{k+1} \in V$ such that $\mathcal{S}_k \cup \{e_{k+1}\}$ be an orthonormal system. As previously shown $P(1)$ holds true. We wish to show that the proposition $P(k-1) \Longrightarrow P(k)$ is true. Let us assume that $P(k-1)$ is true. Let $\mathcal{S}_k = \{e_1, \dots, e_k\}$ be an orthonormal system and let

$$W_k = \{\lambda_1 e_1 + \cdots + \lambda_k e_k : \lambda_1, \dots, \lambda_k \in \mathbb{R}\}$$

be the subspace of V generated by $\mathcal{S}_k$. The system $\mathcal{S}_k$ is orthonormal and hence $\mathcal{S}_k$ is free. Then $\dim_{\mathbb{R}} W_k = k$. Moreover let us observe that W_k is nondegenerate. Indeed if $v \in W_k$ and $g(v, w) = 0$ for any $w \in W_k$ then $g(v, e_j) = 0$ for any $1 \leq j \leq k$ or

$$0 = g(v, e_j) = \sum_{i=1}^{k} \lambda_i g(e_i, e_j) = \pm \lambda_j$$

so that $v = 0$. Next, by Proposition 1.11

$$W_k \oplus W_k^{\perp} = V$$

and hence (by $k < n$) $W_k^{\perp} \neq (0)$. There is a vector $v_{k+1} \in W_k^{\perp}$ such that $g(v_{k+1}, v_{k+1}) \in \{\pm 1\}$ hence the system

$$\mathcal{S}_k \cup \{e_{k+1}\}, \quad e_{k+1} = \frac{1}{|g(v_{k+1}, v_{k+1})|^{1/2}} v_{k+1},$$

is orthonormal and then $P(k+1)$ is true. Q.e.d.

Proposition 1.22 *Let (V, g) be an inner product space and let $\{e_i : 1 \leq i \leq n\}$ be an orthonormal basis. Let us set $\epsilon_i = g(e_i, e_i) \in \{\pm 1\}$. Then*

$$v = \sum_{i=1}^{n} \epsilon_i \, g(v, e_i) \, e_i \tag{1.9}$$

for any $v \in V$.

The numbers $g(v, e_i) \in \mathbb{R}$ in the decomposition (1.9) are the *Fourier coefficients* of v. The decomposition (1.9) is the *orthonormal development* of v.

Exercise 1.23 Prove (1.9).

Let (V, g) be an inner product space and $W \subset V$ a nondegenerate subspace. Let $\pi : V \to W$ be the map defined as follows. Let $v \in V$ and let us set $\pi(v) = w$ where $v = w + w^{\perp}$, $w \in W$, $w^{\perp} \in W^{\perp}$, is the unique decomposition of v with respect to the direct sum decomposition $V = W \oplus W^{\perp}$.

Definition 1.24 $\pi : V \to W$ is the *orthogonal projection* on the nondegenerate subspace W. □

Exercise 1.25 Let $W \neq (0)$ be a nondegenerate subspace of the inner product space (V, g). Let $\{e_\alpha : 1 \leq \alpha \leq k\}$ be an orthonormal basis of W. If $\pi : V \to W$ is the orthogonal projection on W, show that

$$\pi(v) = \sum_{\alpha=1}^{k} \epsilon_\alpha \, g(v, e_\alpha) \, e_\alpha, \quad v \in V.$$

Definition 1.26 Let (V, g) be an inner product space and let ν be the index of g as a bilinear symmetric form. The number ν is called the *index* of (V, g) and we write $\nu = \mathrm{ind}(V, g)$ or merely $\nu = \mathrm{ind}(V)$. □

Proposition 1.27 *Let (V, g) be an inner product space of index ν. Then for any orthonormal basis $\{e_i : 1 \leq i \leq n\}$ the index ν equals the number of minus signs in the canonical form of the associated quadratic form, i.e.,*

$$q(v) = -\sum_{i=1}^{\nu}(v^i)^2 + \sum_{i=\nu+1}^{n}(v^i)^2$$

for any $v \in V$.

Proof Let $\{e_i : 1 \le i \le n\}$ be an orthonormal basis and let us set $\epsilon_i = g(e_i, e_i) \in \{\pm 1\}$. Let $0 \le \mu \le n$ such that

$$\epsilon_1 = \cdots = \epsilon_\mu = -1 = -\epsilon_{\mu+1} = \cdots = -\epsilon_n .$$

We must show that $\nu = \mu$. We only discuss the case $1 < \mu < n$ (while the cases $\mu \in \{1, n\}$ are left as an exercise to the reader). Let $W_\mu \subset V$ be the subspace generated by the vectors $\{e_i : 1 \le i \le \mu\}$. Then $\dim_{\mathbb{R}}(W_\mu) = \mu$ and $g|_{W_\mu}$ is negative definite, so that $\mu \le \nu$.

Viceversa, let $W \subset V$ such that $g|_W$ be negative definite. Let us consider the map

$$f : W \to W_\mu, \quad f(w) = -\sum_{i=1}^{\mu} g(w, e_i)\, e_i, \quad w \in W.$$

Clearly $f : W \to W_\mu$ is linear. Let $w \in \operatorname{Ker}(f)$ so that $w \in W_\mu^\perp$, i.e.,

$$w = \sum_{i=\mu+1}^{n} g(w, e_i)e_i .$$

Moreover $g|_W$ is negative definite hence

$$0 \ge g(w, w) = \sum_{i,j=\mu+1}^{n} g(w, e_i)g(w, e_j)g(e_i, e_j) = \sum_{i=\mu+1}^{n} g(w, e_i)^2$$

and then $g(w, e_i) = 0$ for any $\mu + 1 \le i \le n$, i.e., $w = 0$. We have shown that $\operatorname{Ker}(f) = (0)$ and hence that f is injective. It follows that $\dim_{\mathbb{R}}(W) \le \dim_{\mathbb{R}}(W_\mu) = \mu$ and then, by the arbitrariness of the choice of W, it follows that $\nu \le \mu$. Q.e.d.

We have shown that for any nondegenerate subspace $W \subset V$ of an inner product space (V, g), one may build an orthonormal basis of V by completing an orthonormal basis of W. This yields

Corollary 1.28 *Let $W \subset V$ be a nondegenerate subspace of the inner product space (V, g). Then*

$$\operatorname{ind}(V) = \operatorname{ind}(W) + \operatorname{ind}(W^\perp).$$

Definition 1.29 Let (V, g) and (V', g') be two inner product spaces. A linear map $T : V \to V'$ *preserves the inner products* if

$$g'(Tu, Tv) = g(u, v), \quad u, v \in V.$$

□

Let us observe that any linear map $T : V \to V'$ preserving the inner products in injective. Indeed if $Tu = 0$ then for any $v \in V$

$$g(u, v) = g'(Tu, Tv) = 0$$

and hence (as g is nondegenerate) $u = 0$. Moreover let us observe that a linear map $T : V \to V'$ preserves the inner products if and only if T preserves the quadratic forms associated to the inner products g e g', i.e., $q'(Tv) = q(v)$ per ogni $v \in V$. One of the implications is obvious. The other follows from the polarization formula.

Definition 1.30 A linear isomorphism $T : V \to V'$ is a *linear isometry* if T preserves the inner products g e g'. □

As any linear application preserving the inner products is already a monomorphism (of finite dimensional vector spaces) it follows that:

Corollary 1.31 *Let $T : V \to V'$ be a linear map. Then T is a linear isometry if and only if T preserves the inner products and* $\dim_{\mathbb{R}}(V) = \dim_{\mathbb{R}}(V')$.

One proves easily that

Proposition 1.32 *Two inner product spaces V and V' are equi-dimensional and have the same index if and only if there is a linear isometry of V into V'.*

Exercise 1.33 1. Let $b \in V^{0,2}$ be a bilinear symmetric form on the vector space V. By definition the *null space* of B is the set

$$\mathrm{Null}(b) = \{v \in V : b(v, w) = 0, \quad \forall\, w \in V\}.$$

Also the *null cone* of b is

$$\lambda = \{v \in V \setminus \{0\} : b(v, v) = 0\}.$$

Let us set $\lambda_0 = \lambda \cup \{0\}$ so that $\lambda_0 \supset \mathrm{Null}(b)$. Show that

(i) $\mathrm{Null}(b)$ is a subspace of V.
(ii) λ_0 is a subspace of V if and only if $\lambda_0 = \{0\}$ or $\lambda_0 = V$.
(iii) b nondegenerate $\Longleftrightarrow \mathrm{Null}(b) = \{0\}$.
(iv) b definite $\Longleftrightarrow \lambda_0 = \{0\}$.
(v) b semi-definite $\Longleftrightarrow \mathrm{Null}(b) = \lambda_0$.

2. Let (V, g) be an inner product space of index $\nu = \text{ind}(V)$. Show that there is a subspace $W \subset V$ of dimension $\dim_{\mathbb{R}}(W) = \min\{\nu, n - \nu\}$ such that $g|_W = 0$.

3. Let $O(n)$ be the orthogonal group, i.e., $O(n) = \{A \in GL(n, \mathbb{R}) : AA^t = I_n\}$ where A^t is the transpose of A. Let us consider the inner products g and g_1 on $\mathbb{R}^n$ such that

$$g(x, y) = g(Ax, Ay), \quad g_1(x, y) = g_1(Ax, Ay),$$

for any $x, y \in \mathbb{R}^n$. Show that there is a constant $c \in \mathbb{R} \setminus \{0\}$ such that $g_1 = cg$.

1.1.4 Semi-Riemannian Manifolds

Let M be a differentiable manifold, of class C^∞ and dimension n.

Definition 1.34 A *semi-Riemannian metric* on M is a tensor field $g \in \mathcal{T}^{0,2}(M) = C^\infty(\tau_M^{0,2})$ such that for any point $p \in M$ the bilinear map $g_p \in T_p^{0,2}(M) = T_p^*(M) \otimes T_p^*(M)$ is an inner product of constant index $\nu = \text{ind}\left(T_p(M), g_p\right)$.

A pair (M, g) consisting of a differentiable manifold M and a semi-Riemannian metric g on M is a *semi-Riemannian manifold.*

The number $\nu \in \{0, 1, \dots, n\}$ is the *index* of the semi-Riemannian manifold (M, g).

A semi-Riemannian manifold (M, g) of index $\nu = 0$ is a *Riemannian manifold* and g is a *Riemannian metric*.

A semi-Riemannian manifold (M, g) of index $\nu = 1$ is a *Lorentzian manifold* and g is a *Lorentzian metric*. □

Given a semi-Riemannian metric g on M, for any point $p \in M$ the map $g_p : T_p(M) \times T_p(M) \to \mathbb{R}$ is a nondegenerate bilinear symmetric form on $T_p(M)$. Clearly one may consider the function

$$\nu : M \to \mathbb{Z}, \quad \nu(p) = \text{ind}(T_p(M), g_p), \quad p \in M.$$

In Definition 1.34 one requests that the function $\nu : M \to \mathbb{Z}$ be constant. If g is a Riemannian metric then g_p is a positive definite inner product for any $p \in M$. If (U, x^i) is a local coordinate system on M and g is a semi-Riemannian metric on M then

$$g_p = g_{ij}(p)(dx^i)_p \otimes (dx^j)_p, \quad p \in U.$$

The functions $g_{ij} : U \to \mathbb{R}$ are differentiable of class C^∞. One may also write $g = g_{ij}\, dx^i \otimes dx^j$ on U. As each g_p is nondegenerate one has $\det[g_{ij}(p)] \neq 0$ for any $p \in U$ while for a Riemannian metric g one has $\det[g_{ij}(p)] > 0$ for any $p \in U$.

Proposition 1.35 *Any differentiable manifold admits a Riemannian metric.*

Proof Let $\{U_\alpha : \alpha \in I\}$ be a locally finite open cover such that $\overline{U}_\alpha$ is a compact subset of the domain of a local chart (V_α, x_α^i). Let $\{\varphi_\alpha : \alpha \in I\}$ be a smooth partition of unity subordinated to the cover $\{U_\alpha : \alpha \in I\}$. On each open set V_α we consider the Riemannian metric g_α given by

$$g_{\alpha,p}(u,v) = \sum_{i=1}^{n} u^i v^i, \quad p \in V_\alpha, \quad u, v \in T_p(M), \quad u = u^i \left(\frac{\partial}{\partial x_\alpha^i}\right)_p, \quad v = v^i \left(\frac{\partial}{\partial x_\alpha^j}\right)_p.$$

As the sets $A_\alpha = \overline{U}_\alpha$ are closed one may smoothly extend any $g_\alpha|_{A_\alpha} \in C^\infty(A_\alpha, \tau_M^{0,2})$ to a globally defined tensor field $\tilde{g}_\alpha \in \mathcal{T}^{0,2}(M)$ such that $\tilde{g}_\alpha|_{A_\alpha} = g_\alpha|_{A_\alpha}$, and in particular $\tilde{g}_\alpha|_{U_\alpha} = g_\alpha|_{U_\alpha}$. Now let us set

$$g = \sum_{\alpha \in I} \varphi_\alpha \tilde{g}_\alpha \in \mathcal{T}^{0,2}(M).$$

One may easily check that g is a Riemannian metric on M. Q.e.d.

Not every n-dimensional manifold admits a semi-Riemannian metric of index $0 < \nu < n$. For instance, if $n = 2$, the only compact differentiable manifolds admitting semi-Riemannian metrics of index $\nu = 1$ are the Klein bottle and the torus, cf. e.g., N. Steenrod.[8]

Let (M, g) be a Riemannian manifold. The *length* of the tangent vector $v \in T_p(M)$ is $\|v\| = g_p(v, v)^{1/2}$. If $u, v \in T_p(M) \setminus \{0\}$ then the *angle* between u and v is the number $\theta \in [0, \pi]$ determined by

$$\cos\theta = \frac{g_p(u,v)}{\|u\| \, \|v\|}.$$

The angle θ is well defined because of

$$-1 \le \frac{g_p(u,v)}{\|u\| \, \|v\|} \le 1$$

[as following from the Cauchy-Schwartz inequality for the positive definite inner product g_p] and $\cos : [0, \pi] \to [-1, 1]$ is a one-to-one and on-to function.

Definition 1.36 Let (M, g) be a Riemannian manifold and $\gamma : [a, b] \to M$ differentiable curve. The *length* of γ is

$$L(\gamma) = \int_a^b \|\dot{c}(t)\| \, dt.$$

□

Definition 1.37 A *change of parameter* is a C^∞ diffeomorphism

$$\phi : [c, d] \to [a, b].$$

[8] Cf. N. Steenrod, *Topology of Fiber Bundles* (Princeton University Press, 1951).

Given a differentiable curve $\gamma : [a, b] \to M$ we set $\overline{\gamma}(\tau) = \gamma(\phi^{-1}(\tau))$ for any $\tau \in [c, d]$. The differentiable curve $\overline{\gamma} : [c, d] \to M$ is got from $\gamma : [a, b] \to M$ by *reparametrization*. □

One checks easily that

Proposition 1.38 *Let $\gamma : [a, b] \to M$ be a differentiable curve and $\overline{\gamma} = \gamma \circ \phi^{-1} : [c, d] \to M$ a reparametrization of γ. Then $L(\gamma) = L(\overline{\gamma})$, i.e., γ and $\overline{\gamma}$ have the same length.*

In other words, the length of a differentiable curve doesn't depend upon its parametrization.

A differentiable curve $\gamma : [a, b] \to M$ is *regular* if $\dot{c}(t) \neq 0$ for any $t \in [a, b]$. For regular curves on Riemannian manifolds, one may use a canonical parametrization, as furnished by the following:

Proposition 1.39 *Let $\gamma : [a, b] \to M$ be a regular curve in the Riemannian manifold (M, g) and let us set*

$$\phi : [a, b] \to [0, L(\gamma)], \quad \phi(t) = \int_a^t \|\dot{\gamma}(\tau)\| \, d\tau, \quad a \leq t \leq b.$$

Then ϕ is a change of parameter and the curve $\overline{\gamma} = \gamma \circ \phi^{-1} : [0, L(\gamma)] \to M$ is canonically parametrized, i.e., $\|\dot{\overline{\gamma}}(s)\| = 1$ for any $s \in [0, L(\gamma)]$.

The reparametrization $\overline{\gamma} : [0, L(\gamma)] \to M$ built within Proposition 1.39 is commonly referred to as the reparametrization of $\gamma : [a, b] \to M$ by *arc length*. Clearly $s = \int_a^t \|\dot{\gamma}(\tau)\| \, d\tau$ is the length of the curve $\gamma : [a, t] \to M$ for any $t \in [a, b]$.

1.1.5 Canonical Measure

Let (M, g) be a n-dimensional Riemannian manifold. Let $\phi = (\xi^1, \ldots, \xi^n) : U \to \mathbb{R}^n$ be a local chart on M and let us set

$$g_{U,\phi} := \left[g_{ij}(\xi^1, \ldots, \xi^n) \right], \quad \left| g_{U,\phi} \right| := \det g_{U,\phi} > 0.$$

Note that $\phi(U)$ is equipped with the positive measure

$$\mu_{U,\phi} = \left| g_{U,\phi} \right|^{1/2} d\xi^1 \cdots d\xi^n .$$

We endow U with the measure

$$\nu_{U,\phi} := \phi^{-1}\left(\mu_{U,\phi}\right).$$

If φ is a continuous function of compact support $\mathrm{supp}(\varphi) \subset U$ then

$$\nu_{U,\phi}(\varphi) = \int_{\phi(U)} (\varphi \circ \phi^{-1}) \left|g_{U,\phi}\right|^{1/2} d\xi^1 \cdots d\xi^n .$$

If (U', ϕ') is another local chart on M then (by a change of variable under the integral sign)

$$\nu_{U,\phi}\big|_{U\cap U'} = \nu_{U',\phi'}\big|_{U\cap U'}$$

so that (by a partition of unity argument) there is a unique measure v_g on M such that

$$\mathrm{v}_g\big|_U = \nu_{U,\phi}$$

for any local chart (U, ϕ). Said v_g is a positive measure.[9]

Definition 1.40 The positive measure v_g is called the *canonical measure* of the Riemannian manifold (M, g). □

Therefore, for any Riemannian manifold, the notions of a measurable subset and of an integrable function are well defined. If M is compact, the total mass of M is denoted by $\mathrm{Vol}(M, g)$ and called the *volume* of (M, g).

The canonical measure is invariant by isometries.

Let us assume that M is oriented. Then integration of a top degree differential form makes sense on M, and so does the notion of a measure associated to that form.[10] One may build,[11] in a natural manner, a volume form η_g on (M, g) such that v_g is the measure associated to η_g.

Definition 1.41 The *volume form* η_g on M is the differential n-form on M given by[12]

$$\left(\eta_g\right)_x := \eta\big(T_x(M)\big), \quad x \in M.$$

□

[9] Cf. Theorem 17, Chap. 4, in L. Schwartz, *Cours d'Analuse de l'École Polytechnique* (Harmann, 1967).

[10] Cf. L. Scwartz, *op. cit.*, Chap. 6, § 6.

[11] A bit of linear algebra is needed, as follows. Let E be a n-dimensional oriented Euclidean vector space. Let $\eta(E) \in \Lambda^n E^*$ be the exterior n-form taking the value 1 on any right-handed orthonormal basis of E. Let α and β be two bases of E, with α right-handed and orthonormal, and let $\{u^1, \ldots, u^n\}$ and $\{v^1, \ldots, v^n\}$ be the corresponding dual bases. Let P be the passage matrix from α to β. Let $g_\alpha = I_n$ and g_β be the matrices associated to α and β, *vis-a-vis* to the Euclidean structure of E. Then

$$g_\beta = P^T g_\alpha P = P^T P \Longrightarrow \left|g_\beta\right| = |P|^2, \quad \eta(E) = u^1 \wedge \cdots \wedge u^n = \left|P\right| v^1 \wedge \cdots \wedge v^n,$$

so that $\eta(E) = \varepsilon \left|g_\beta\right|^{1/2} v^1 \wedge \cdots \wedge v^n$ where $\varepsilon \in \{\pm 1\}$ (depending on whether β is right handed, or left handed).

[12] With respect to the notations in the previous footnote the Euclidean space E is $\big(T_x(M), g_x\big)$.

With respect to a local chart $\phi = (\xi^1, \ldots, \xi^n) : U \to \mathbb{R}^n$

$$\eta_g \circ \phi^{-1} = \varepsilon \left|g_{U,\phi}\right|^{1/2} d\xi^1 \wedge \cdots \wedge d\xi^n$$

with $\varepsilon \in \{\pm 1\}$ (according to whether (U, ϕ) is positively oriented). Therefore v_g is the measure associated to the volume form η_g.

1.1.6 Levi–Civita Connection

Let (M, g) be a semi-Riemannian manifold.

Definition 1.42 A linear connection ∇ on M is a *metric connection* if g is parallel with respect to ∇, i.e., $\nabla_X g = 0$ for any $X \in \mathfrak{X}(M)$. □

A linear connection ∇ su M is a metric connection if and only if

$$X(g(Y, Z)) = g(\nabla_X Y, Z) + g(Y, \nabla_X Z), \quad X, Y, Z \in \mathfrak{X}(M),$$

an identity known as *Ricci's identity*.

Exercise 1.43 Let us denote by $\mathcal{C}(\tau_M, g)$ the set of all metric connections on (M, g), i.e., $\mathcal{C}(\tau_M, g) = \{\nabla \in \mathcal{C}(\tau_M) : \nabla g = 0\}$. Show that $\mathcal{C}(\tau_M, g)$ is an affine subspace of $\mathcal{C}(\tau_M)$.

The property of a linear connection of being "metric" admits the following geometric interpretation.

Proposition 1.44 *Let $\gamma : [a, b] \to M$ be a differentiable curve and $\mathcal{E}^t_\gamma : T_{\gamma(a)}(M) \to T_{\gamma(t)}(M)$ the parallel displacement operator along γ, for any $t \in [a, b]$, with respect to the linear connection ∇. Then ∇ is a metric connection if and only if $\mathcal{E}^t_g$ is a linear isometry of $(T_{\gamma(a)}(M), g_{\gamma(0)})$ into $(T_{\gamma(t)}(M), g_{\gamma(t)})$ for any $t \in [a, b]$.*

Proof Let us assume that $\nabla g = 0$. Let us observe that for any function $f \in C^\infty(\Gamma, \mathbb{R})$ one has $\dot\gamma(t)(f) = (f \circ \gamma)'(t)$ for any $t \in [a, b]$. Let $v \in T_{\gamma(0)}(M)$ be a fixed tangent vector and let $X \in C^\infty(\Gamma, \tau_M)$ be the solution to the Cauchy problem

$$\left(\nabla_{\dot\gamma} X\right)_{\gamma(t)} = 0, \quad a \leq t \leq b,$$

$$X_{\gamma(0)} = v.$$

By definition $\mathcal{E}^t_\gamma(v) = X_{\gamma(t)}$. Let us define $f \in C^\infty(\Gamma, \mathbb{R})$ by setting $f = g(X, X)$. Let us observe that

$$f(\gamma(0)) = \left\{g_{\gamma(t)}(X_{\gamma(t)}, X_{\gamma(t)})\right\}_{t=0} = g_{\gamma(0)}(v, v).$$

On the other hand

$$(f \circ \gamma)'(t) = \dot{\gamma}(t)(f) = \dot{\gamma}(t)\,[g(X, X)] = 2\,g(\nabla_{\dot{\gamma}} X,\ X)_{\gamma(t)} = 0$$

hence $f \circ \gamma$ is the constant function, i.e., $f(\gamma(t)) = f(\gamma(0))$, $a \leq t \leq b$. Finally

$$g_{\gamma(t)}(\mathcal{E}^t_\gamma(v),\ \mathcal{E}^t_\gamma(v)) = g(X, X)_{\gamma(t)} = f(\gamma(t)) = f(\gamma(0)) = g_{\gamma(0)}(v, v),$$

i.e., $\mathcal{E}^t_\gamma$ preserves the quadratic forms associated to the inner products $g_{\gamma(0)}$ and $g_{\gamma(t)}$.

Viceversa, let us assume that for any differentiable curve $\gamma : [a, b] \to M$ the parallel displacement operator $\mathcal{E}^t_\gamma : T_{\gamma(0)}(M) \to T_{\gamma(t)}(M)$ with respect to ∇ is a linear isometry of $(T_{\gamma(0)}(M),\ g_{\gamma(0)})$ onto $(T_{\gamma(t)}(M),\ g_{\gamma(t)})$ for any $t \in [a, b]$. Let us consider a $g_{\gamma(0)}$-orthonormal basis $\{e_1, \ldots, e_n\} \subset T_{\gamma(0)}(M)$ and let us set

$$E_i(t) = \mathcal{E}^t_\gamma(e_i) \in T_{\gamma(t)}(M), \quad 1 \leq i \leq n, \quad t \in [a, b].$$

As $\mathcal{E}^t_\gamma$ is a linear isometry it follows that $\{E_i(t) : 1 \leq i \leq n\}$ is a $g_{\gamma(t)}$-orthonormal basis in $T_{\gamma(t)}(M)$. Let us consider the arbitrary vector fields $X, Y, Z \in \mathfrak{X}(M)$ and the point $p \in M$. If $v = X_p \in T_p(M)$ then let us consider a differentiable curve $\gamma : [a, b] \to M$ such that $\gamma(t_0) = p$ and $\dot{\gamma}(t_0) = v$ for some $t_0 \in [a, b]$. We may then represent the vector fields Y and Z along γ in the following manner

$$Y_{\gamma(t)} = Y^i(\gamma(t))E_i(t), \quad Z_{\gamma(t)} = Z^i(\gamma(t))E_i(t),$$

where $Y^i, Z^i : [a, b] \to \mathbb{R}$ are differentiable functions uniquely determined by that representation. Moreover

$$g(Y, Z)_{\gamma(t)} = Y^i(t)Z^j(t)g_{\gamma(t)}(E_i(t),\ E_j(t)) = Y^i(t)Z^j(t)\epsilon_i\delta_{ij} = \epsilon_i Y^i(t)Z^i(t).$$

On the other hand

$$g(\nabla_X Y, Z)_p + g(Y, \nabla_X Z)_p = g_p((\nabla_X Y)_p,\ Z_p) + g_p(Y_p,\ (\nabla_X Z)_p) =$$

[as the tangent vector $(\nabla_X Y)_p$ is determined by the value of X at p, and the same is true for $(\nabla_X Z)_p$]

$$= g_{\gamma(0)}((\nabla_{\dot{\gamma}} Y)_{\gamma(0)},\ Z_{\gamma(0)}) + g_{\gamma(0)}(Y_{\gamma(0)},\ (\nabla_{\dot{\gamma}} Z)_{\gamma(0)}) =$$

$$= \left\{g_{\gamma(t)}((\nabla_{\dot{\gamma}} Y)_{\gamma(t)},\ Z_{\gamma(t)}) + g_{\gamma(t)}(Y_{\gamma(t)},\ (\nabla_{\dot{\gamma}} Z)_{\gamma(t)})\right\}_{t=0}$$

and if $f^i \in \mathcal{F}(M)$ and $V_i \in \mathfrak{X}(M)$ are respectively differentiable functions and vector fields such that $f^i(\gamma(t)) = Y^i(t)$ e $V_{i,\gamma(t)} = E_i(t) = \mathcal{E}^t_\gamma(e_i)$ per ogni $a \leq t \leq b$ then $(\nabla_{\dot{\gamma}} V_i)_{\gamma(t)} = 0$ and hence

$$(\nabla_{\dot\gamma} Y)_{\gamma(t)} = (\nabla_{\dot\gamma}(f^i V_i))_{\gamma(t)} = \left\{\dot\gamma(f^i)\, V_i + f^i \nabla_{\dot\gamma} V_i\right\}_{\gamma(t)} =$$

$$= \dot\gamma(f^i)_{\gamma(t)} V_{i,\gamma(t)} = (f^i \circ \gamma)'(t) E_i(t) = \frac{dY^i}{dt}(t) E_i(t)$$

and the same manner

$$(\nabla_{\dot\gamma} Z)_{\gamma(t)} = \frac{dZ^i}{dt}(t) E_i(t)$$

for any $t \in [a, b]$. Finally

$$g(\nabla_X Y, Z)_p + g(Y, \nabla_X Z)_p =$$

$$= \left\{\left[\frac{dY^i}{dt}(t) Z^j(t) + Y^i(t)\frac{dZ^j}{dt}(t)\right] g_{\gamma(t)}(E_i(t), E_j(t))\right\}_{t=0} =$$

$$= \left\{\left[\frac{dY^i}{dt}(t) Z^j(t) + Y^i(t)\frac{dZ^j}{dt}(t)\right] \epsilon_i \delta_{ij}\right\}_{t=0} = \epsilon_i \left(Y^i Z^i\right)'(0)$$

and

$$v(g(Y, Z)) = \dot\gamma(0)(g(Y, Z)) = (g(Y, Z) \circ \gamma)'(0) = \left(\epsilon_i Y^i Z^i\right)'(0)$$

hence

$$\{X(g(Y, Z)) - g(\nabla_X Y, Z) - g(Y, \nabla_X Z)\}_p = 0$$

for any point $p \in M$. Q.e.d.

Theorem 1.45 *For any semi-Riemannian manifold* (M, g) *there is a unique linear connection* ∇ *such that* (i) $\nabla g = 0$ *and* (ii) $T = 0$.

Theorem 1.45 is the *fundamental theorem of semi-Riemannian geometry*.

Definition 1.46 The linear connection furnished by Theorem 1.45 is called the *Levi-Civita connection* of (M, g). □

Therefore Levi-Civita connection is a metric and symmetric linear connection. In this sense it is canonical, i.e., equations (i)–(ii) in Theorem 1.45 allow us to choose, in a manner which may be thought of as canonical, a linear connection in the affine space $\mathcal{C}(\tau_M, g)$. By solving a certain tensorial system, one may determine all elements of $\mathcal{C}(\tau_M, g)$.

Proof of Theorem 1.45 We first prove uniqueness, i.e., the set $\{\nabla \in \mathcal{C}(\tau_M, g) : T = 0\}$ has cardinality ≤ 1. If a metric and symmetric linear connection ∇ exists then [by $\nabla g = 0$] for any $X, Y, Z \in \mathfrak{X}(M)$

$$X(g(Y, Z)) = g(\nabla_X Y, Z) + g(Y, \nabla_X Z), \tag{1.10}$$

$$Y(g(Z,X)) = g(\nabla_Y Z, X) + g(Z, \nabla_Y X), \tag{1.11}$$

$$Z(g(X,Y)) = g(\nabla_Z X, Y) + g(X, \nabla_Z Y), \tag{1.12}$$

where Eqs. (1.11)–(1.12) are respectively derived from (1.10) and from (1.11) by a circular permutation of the arguments $X \to Y \to Z \to X$. Let us sum up Eqs. (1.10)–(1.11) and subtract (1.12) from the resulting equation. We obtain

$$X(g(Y,Z)) + Y(g(Z,X)) - Z(g(X,Y)) =$$

$$= g(\nabla_X Y, Z) + g(Y, \nabla_X Z) + g(\nabla_Y Z, X) + g(Z, \nabla_Y X) -$$

$$- g(\nabla_Z X, Y) - g(X, \nabla_Z Y)$$

that is (by $T = 0$)

$$X(g(Y,Z)) + Y(g(Z,X)) - Z(g(X,Y)) =$$

$$= 2\,g(\nabla_X Y, Z) + g(Z, [Y,X]) + g(Y, [X,Z]) + g([Y,Z], X)$$

a relation which is commonly written in the form

$$2\,g(\nabla_X Y, Z) = X(g(Y,Z)) + Y(g(Z,X)) - Z(g(X,Y)) + \tag{1.13}$$

$$+ g([X,Y], Z) - g([Y,Z], X) + g([Z,X], Y)$$

and certainly (1.13) asserts that if a metric and symmetric linear connection ∇ exists, then it is unique, as implicitly given by the right-hand side of Eq. (1.13). The last statement clearly relies on the fact that g is nondegenerate [and then $g(X,Y) = 0$ for any $Y \in \mathfrak{X}(M)$ yields $X = 0$].

Now we prove existence, i.e., that $\{\nabla \in \mathcal{C}(\tau_M, g) : T_\nabla = 0\} \neq \emptyset$. Let us consider the map

$$\nabla : \mathfrak{X}(M) \times \mathfrak{X}(M) \to \mathfrak{X}(M)$$

defined as follows. Let $X, Y \in \mathfrak{X}(M)$ be two arbitrary vector fields and let $\nabla_X Y \in \mathfrak{X}(M)$ be the vector field implicitly defined by Eq. (1.13) for any $Z \in \mathfrak{X}(M)$. To start with we must show that $\nabla \in \mathcal{C}(\tau_M)$. The fact that ∇ is a derivation in the second argument may be proved as follows. For any $f \in \mathcal{F}(M)$ and any $X, Y, Z \in \mathfrak{X}(M)$ one has [by (1.13)]

$$2\,g(\nabla_X(fY), Z) = X(g(fY, Z)) + (fY)(g(Z,X)) - Z(g(X, fY)) +$$

$$+ g([X, fY], Z) - g([fY, Z], X) + g([Z,X], fY)$$

or

$$2\,g(\nabla_X(fY), Z) =$$

$$= X(f)g(Y, Z) + f\,X(g(Y, Z)) + f\,Y(g(Z, X)) -$$

$$- Z(f)g(X, Y) - f\,Z(g(X, Y)) +$$

$$+ X(f)g(Y, Z) + f\,g([X, Y], Z) + Z(f)g(Y, X) +$$

$$+ f\,g([Z, Y], X) + f\,g([Z, X], Y)$$

and then, after a few simplifications

$$2\,g(\nabla_X(fY), Z) = 2\,X(f)g(Y, Z) +$$

$$+ f\,\{X(g(Y, Z)) + Y(g(Z, X)) - Z(g(X, Y)) +$$

$$+ g([X, Y], Z) - g([Y, Z], X) + g([Z, X], Y)\}$$

hence [once again by (1.13)]

$$g(\nabla_X(fY), Z) = X(f)g(Y, Z) + f\,g(\nabla_X Y, Z)$$

and finally [as g is nondegenerate and $Z \in \mathfrak{X}(M)$ is arbitrary]

$$\nabla_X(fY) = X(f)Y + f\nabla_X Y.$$

Exercise 1.47 The reader should check, in a manner similar to the above, that (1.13) also implies

$$\nabla_{f_1 X_1 + f_2 X_2} Y = f_1 \nabla_{X_1} Y + f_2 \nabla_{X_2} Y,$$

$$\nabla_X(\lambda_1 Y_1 + \lambda_2 Y_2) = \lambda_1 \nabla_X Y_1 + \lambda_2 \nabla_X Y_2,$$

$$\forall\ f_1, f_2 \in \mathcal{F}(M), \quad \forall\ X_1, X_2, X, Y_1, Y_2, Y \in \mathfrak{X}(M), \quad \forall\ \lambda_1, \lambda_2 \in \mathbb{R}.$$

Next we show that the linear connection given by (1.13) has zero torsion, i.e., $T(X, Y) = 0$ for any $X, Y \in \mathfrak{X}(M)$. It suffices to check that $g(T(X, Y), Z) = 0$ for any $Z \in \mathfrak{X}(M)$. By (1.13)

$$2\, g(T(X, Y), Z) = 2\, g(\nabla_X Y, Z) - 2\, g(\nabla_Y X, Z) - 2\, g([X, Y], Z) =$$

$$= X(g(Y, Z)) + Y(g(Z, X)) - Z(g(X, Y)) +$$

$$+ g([X, Y], Z) - g([Y, Z], X) + g([Z, X], Y) -$$

$$- Y(g(X, Z)) - X(g(Z, Y)) + Z(g(Y, X)) -$$

$$- g([Y, X], Z) + g([X, Z], Y) - g([Z, Y], X) - 2\, g([X, Y], Z) = 0$$

[by exploiting the symmetry of g and the skew-symmetry of the Lie product].

Exercise 1.48 Check that $(\nabla_X g)(Y, Z) = 0$ per ogni $X, Y, Z \in \mathfrak{X}(M)$, once again as a consequence of (1.13).

We devote the remainder of the section to certain considerations of local character. Let ∇ be the Levi-Civita connection of the semi-Riemannian manifold (M, g). Let (U, x^i) be a local coordinate system on M. Let us consider the local coefficients $\Gamma^i_{jk} \in \mathcal{F}(U)$ of ∇ with respect to (U, x^i), i.e.,

$$\Gamma^i_{jk} \partial_i = \nabla_{\partial_j} \partial_k \,.$$

It is a classical matter that the functions Γ^i_{jk} may be explicitly computed in terms of the local coefficients $g_{ij} = g(\partial_i, \partial_j)$ of the metric, and their partial derivatives (thus providing an alternative proof of the uniqueness of the Levi-Civita connection). Let us set as customary

$$g_{ij||k} = (\nabla_{\partial_k} g)(\partial_i,\, \partial_j) \in \mathcal{F}(U).$$

Then

$$g_{ij||k} = \partial_k \left(g\left(\partial_i,\, \partial_j\right)\right) - g\left(\nabla_{\partial_k}\partial_i,\, \partial_j\right) - g\left(\partial_i,\, \nabla_{\partial_k}\partial_j\right) =$$

$$= g_{ij|k} - \Gamma^\ell_{ki} g_{\ell j} - g_{i\ell}\Gamma^\ell_{kj}$$

and hence (as $\nabla g = 0 \Longrightarrow g_{ij||k} = 0$)

$$g_{ij|k} = \Gamma^\ell_{ki} g_{\ell j} + g_{i\ell}\Gamma^\ell_{kj} \,. \tag{1.14}$$

Let us derive a similar equation by a circular permutation of indices $i \to j \to k \to i$ in (1.14)

$$g_{jk|i} = \Gamma^\ell_{ij} g_{\ell k} + g_{j\ell}\Gamma^\ell_{ik} \,. \tag{1.15}$$

Again by a circular permutation of indices $i \to j \to k \to i$ in (1.15) one obtains

$$g_{ki|j} = \Gamma^\ell_{jk} g_{\ell i} + g_{k\ell}\Gamma^\ell_{ji} \,. \tag{1.16}$$

Let us sum up Eqs. (1.14)–(1.15) and subtract Eq. (1.16). In classical local differential geometry this approach is known as the *Christoffel process*. One obtains

$$g_{ij|k} + g_{jk|i} - g_{ki|j} = \Gamma^{\ell}_{ki} g_{\ell j} + g_{i\ell}\Gamma^{\ell}_{kj} + \Gamma^{\ell}_{ij} g_{\ell k} + g_{j\ell}\Gamma^{\ell}_{ik} - \Gamma^{\ell}_{jk} g_{\ell i} - g_{k\ell}\Gamma^{\ell}_{ji}$$

or [by $g_{ij} = g_{ji}$ and $\Gamma^i_{jk} = \Gamma^i_{kj}$]

$$2\, g_{j\ell}\Gamma^{\ell}_{ik} = g_{ij|k} + g_{jk|i} - g_{ki|j}\,. \tag{1.17}$$

If we set $[g^{ij}] = [g_{ij}]^{-1}$ then $g_{j\ell} g^{jm} = \delta^m_\ell$ hence contraction by g^{jm} in (1.17) leads to

$$2\,\Gamma^m_{ik} = g^{jm}\left(g_{ij|k} + g_{jk|i} - g_{ki|j}\right)$$

or

$$\Gamma^m_{ik} = g^{jm}\,\{ik, j\} \tag{1.18}$$

where we have set by definition

$$\{ij, k\} = \frac{1}{2}\left(g_{ik|j} + g_{jk|i} - g_{ij|k}\right).$$

Definition 1.49 The functions $\{ij, k\} \in \mathcal{F}(U)$ are the *Christoffel symbols of the first kind.* □

It is customary to set

$$\begin{Bmatrix} i \\ jk \end{Bmatrix} = g^{i\ell}\,\{jk, \ell\}$$

hence formulas (1.18) may also be written

$$\Gamma^m_{ik} = \begin{Bmatrix} m \\ ik \end{Bmatrix}.$$

Definition 1.50 The functions $\begin{Bmatrix} i \\ jk \end{Bmatrix} \in \mathcal{F}(U)$ are the *Christoffel symbols of the second kind.* □

Let $\{E_i : 1 \le i \le n\}$ be an arbitrary local frame of the tangent bundle, defined on the open set $U \subset M$. Let us set $g_{ij} = g(E_i, E_j) \in \mathcal{F}(U)$ and let us consider the local differential 1-forms $\omega^i_j \in \Omega^1(U)$ determined by

$$\nabla E_j = \omega^i_j \otimes E_i$$

where ∇ is the Levi-Civita connection of (M, g). If $\{\omega^i : 1 \leq i \leq n\}$ are the local 1-forms dual to $\{E_i : 1 \leq i \leq n\}$, i.e., $\omega^i(E_j) = \delta^i_j$ then (by the symmetry of ∇)

$$0 = T(X, Y) = \nabla_X Y - \nabla_Y X - [X, Y] = \nabla_X \left\{\omega^i(Y)E_i\right\} - \nabla_Y \left\{\omega^i(X)E_i\right\} - [X, Y] =$$

$$= X(\omega^i(Y))E_i + \omega^j(Y)\nabla_X E_j - Y(\omega^i(X))E_i - \omega^j(X)\nabla_Y E_j - [X, Y] =$$

$$= \{X(\omega^i(Y)) - Y(\omega^i(X)) - \omega^i([X, Y]) + \omega^j(Y)\omega^i_j(X) - \omega_j(X)\omega^i_j(Y)\}E_i =$$

$$= 2\{(d\omega^i)(X, Y) + (\omega^i_j \wedge \omega^j)(X, Y)\}E_i$$

hence

$$d\omega^i + \omega^i_j \wedge \omega^j = 0. \tag{1.19}$$

Moreover (by $\nabla g = 0$)

$$\begin{aligned} X(g_{ij}) = X(g(E_i, E_j)) &= g(\nabla_X E_i,\ E_j) + g(E_i,\ \nabla_X E_j) = \\ &= \omega^k_i(X)g(E_k,\ E_j) + g(E_i,\ E_k)\omega^k_j(X) \end{aligned}$$

hence

$$dg_{ij} = g_{kj}\ \omega^k_i + g_{ik}\ \omega^k_j\,. \tag{1.20}$$

In particular if the local fields $\{E_i : 1 \leq i \leq n\} \subset \mathfrak{X}(U)$ are orthonormal, i.e., $g_{ij} = \epsilon_i \delta_{ij}$ with $\epsilon_i \in \{\pm 1\}$ then Eq. (1.20) become

$$0 = \epsilon_k \delta_{kj} \omega^k_i + \epsilon_i \delta_{ik} \omega^k_j$$

or

$$\epsilon_j \omega^j_i + \epsilon_i \omega^i_j = 0. \tag{1.21}$$

Note that in formula (1.21) the indices i, j are free (aren't summation indices).

Formulas written with respect to local orthonormal frames violate Einstein's summation conventions. They are however simpler than formulas written with respect to an arbitrary local frame.

If (M, g) is a Riemannian manifold (and then $\epsilon_i = 1$ per ogni $1 \leq i \leq n$) formulas (1.21) become

$$\omega^i_j = -\omega^j_i\,. \tag{1.22}$$

Let us examine a standard, yet important, example. Let $M = \mathbb{R}^n$ and let ∇ be the connection associated to the full parallelism structure of $\mathbb{R}^n$, i.e.,

$$\nabla_X Y = X(Y^i)\frac{\partial}{\partial x^i}, \quad Y = Y^i \frac{\partial}{\partial x^i}, \tag{1.23}$$

where $(x^1, \dots, x^n)$ are the Cartesian coordinates on $\mathbb{R}^n$. On the other hand $\mathbb{R}^n$ admits a natural Riemannian metric g given by

$$g(X, Y) = \sum_{i=1}^{n} X^i Y^i, \tag{1.24}$$

$$X = X^i \frac{\partial}{\partial x^i}, \quad Y = Y^i \frac{\partial}{\partial x^i}.$$

Definition 1.51 The Riemannian metric (1.24) is called the *Euclidean metric* su $\mathbb{R}^n$. □

Note that the linear connection (1.23) is symmetric. Indeed

$$T(X, Y) = \nabla_X Y - \nabla_Y X - [X, Y] = X(Y^i)\partial_i - Y(X^i)\partial_i - [X, Y] = 0.$$

On the other hand (1.23) is also a metric connection. Indeed

$$(\nabla_X g)(Y, Z) = X(g(Y, Z)) - g(\nabla_X Y, Z) - g(Y, \nabla_X Z) =$$

$$= X\left(\sum_{i=1}^{n} Y^i Z^i\right) - X(Y^i) g(\partial_i, Z) - X(Z^i) g(Y, \partial_i) =$$

$$= \sum_{i=1}^{n} \{X(Y^i)Z^i + Y^i X(Z^i)\} - X(Y^i) \sum_{j=1}^{n} \delta_i^j Z^j - X(Z^i) \sum_{j=1}^{n} Y^j \delta_i^j = 0.$$

Here we set $\partial_i \equiv \partial/\partial x^i$ for the sake of simplicity. The uniqueness statement in the fundamental theorem of semi-Riemannian geometry implies that (1.23) is the Levi-Civita connection of the Riemannian manifold $(\mathbb{R}^n, g)$ where g is the Euclidean metric.

Let $\nu \in \mathbb{Z}$ such that $0 \le \nu \le n$. Let us observe that $\mathbb{R}^n$ admits also the semi-Riemannian metric g given by

$$g(X, Y) = -\sum_{i=1}^{\nu} X^i Y^i + \sum_{j=\nu+1}^{n} X^j Y^j. \tag{1.25}$$

The linear connection (1.23) is the Levi-Civita connection of $\mathbb{R}^n$ endowed with the semi-Riemannian metric (1.25).

It is customary to denote by $\mathbb{E}^n$ the n-dimensional *Euclidean space*, i.e., the pair consisting of $\mathbb{R}^n$ and the Euclidean metric (1.24). Also one denotes by $\mathbb{E}^n_\nu$ the *semi-Euclidean space*, i.e., the pair consisting of $\mathbb{R}^n$ and the semi-Riemannian metric (1.25). Then $\mathbb{E}^n_0 = \mathbb{E}^n$. Both semi-Riemannian manifolds got for $\nu = 1$ or $\nu = n - 1$

(i.e., $\mathbb{E}^n_1$ and $\mathbb{E}^n_{n-1}$) are commonly referred to as *Minkowski spaces*. Other pieces of terminology of frequent use indicate each $\mathbb{E}^n_\nu$ with $0 < \nu < n$ as *pseudo-Euclidean spaces* and $\mathbb{E}^n_1$ and $\mathbb{E}^n_{n-1}$ as *Lorentz spaces*.

Another example of a Riemannian metric, frequently met in differential geometry practice, is that induced by the Euclidean metric on any submanifold $M \subset \mathbb{E}^n$. The properties of these metrics, referred to as *induced metrics* or *first fundamental forms*, will be studied in detail in subsequent subsections of the present section.

Exercise 1.52 1. Let M be a n-dimensional differentiable manifold. Let us assume that M is parallelizable and let us consider a full parallelism structure $\{X_1, \dots, X_n\}$ on M. Show that the formulas

$$g(X_i, X_j) = \delta_{ij}, \quad 1 \le i, j \le n,$$

determine a Riemannian metric g on M and compute its Levi-Civita connection.

2. Show that for any Riemannian manifold (M, g) there is a canonical isomorphism of $\mathcal{F}(M)$-modules $\mathfrak{X}(M) \approx \Omega^1(M)$.

3. Let (M, g) be a Riemannian manifold and let $T \in \mathcal{T}^{1,2}(M)$. Show that there is a unique linear connection $\nabla \in \mathcal{C}(\tau_M)$ such that $\nabla g = 0$ and the torsion of ∇ is T.

4. Let (M_i, g_i), $i \in \{1, 2\}$ be two Riemannian manifolds. Moreover let $\pi_i : M_1 \times M_2 \to M_i$ be the canonical projections ($i \in \{1, 2\}$). Show that

$$g = \pi_1^* g_1 + \pi_2^* g_2$$

is a Riemannian metric on the product manifold $M = M_1 \times M_2$. Compute the Christoffel symbols of g with respect to a local coordinate system of M got as the direct product of two local charts (U_i, φ_i) of M_i, $i \in \{1, 2\}$.

5. Let (M, g) be a n-dimensional Riemannian manifold and $\nabla \in \mathcal{C}(\tau_M, g)$ a metric linear connection. Show that the holonomy group of ∇ at an arbitrary point is a subgroup of the orthogonal group $\mathrm{O}(n)$.

1.2 Geodesics and Jacobi Fields

Let (M, g) be a n-dimensional semi-Riemannian manifold. The auto-parallel curves of the Levi-Civita connection ∇ of (M, g) are the *geodesics* of (M, g). Hence $\gamma(t) \in M$, $t \in [a, b]$, is a geodesic of (M, g) if

$$\left(\nabla_{\dot\gamma}\dot\gamma\right)_{\gamma(t)} = 0, \quad a \le t \le b.$$

Ricci's identity yields $\|\dot\gamma(t)\| = \text{constant}$. Indeed

$$\frac{d}{dt}\left\{g_{\gamma(t)}(\dot\gamma(t), \dot\gamma(t))\right\} = 2\, g_{\gamma(t)}\left(\left(\nabla_{\dot\gamma}\dot\gamma\right)_{\gamma(t)}, \dot\gamma(t)\right) = 0$$

hence $g_{\gamma(t)}(\dot\gamma(t), \dot\gamma(t)) = g_{\gamma(a)}(\dot\gamma(a), \dot\gamma(a))$. Q.e.d. The geodesics equations may be locally written

$$\frac{d^2\gamma^i}{dt^2}(t) + \left\{ \begin{matrix} i \\ jk \end{matrix} \right\} (\gamma(t)) \frac{d\gamma^j}{dt}(t) \frac{d\gamma^k}{dt}(t) = 0$$

where $\gamma^i(t) = x^i(\gamma(t))$, $1 \le i \le n$, while $\left\{ \begin{matrix} i \\ jk \end{matrix} \right\}$ are the Christoffel coefficients of the second kind. Of course all properties of auto-parallel curves of a given linear connection hold good for the geodesics of a given Riemannian manifold.

Definition 1.53 A semi-Riemannian manifold is *geodesically complete* if any geodesic of (M, g) is contained in a geodesic of (M, g) which is defined on the whole real line $\mathbb{R}$. □

Another way to state Definition 1.53 is to say that (M, g) is geodesically complete when every geodesic $\gamma : [a, b] \to M$ may be extended to a geodesic defined on the whole of $\mathbb{R}$, i.e., there is a geodesic $\tilde\gamma : \mathbb{R} \to M$ such that $\tilde\gamma|_{[a,b]} = \gamma$. Let us observe that $\mathbb{E}^n$ e $\mathbb{E}^n_\nu$ are geodesically complete semi-Riemannian manifolds.

Exercise 1.54 1. Let $U = \{(x, y) \in \mathbb{R}^2 : y > 0\}$. Compute the geodesics of U knowing that the components of the Riemannian metric g prescribed on U are

$$g_{11} = g_{22} = \frac{1}{y}, \qquad g_{12} = 0.$$

2. Let (M, g) be a Riemannian manifold and let ∇ be its Levi-Civita connection. Let $\overline{\nabla} \in \mathcal{C}(\tau_M, g)$ be another metric connection on M. Let us assume that the torsion $\overline{T}$ of $\overline{\nabla}$ obeys to

$$g(\overline{T}(X, Y), Y) = 0, \quad X, Y \in \mathfrak{X}(M).$$

Show that ∇ and $\overline{\nabla}$ have the same geodesics.

Let (M, g) be a n-dimensional semi-Riemannian manifold. Let $U \subset \mathbb{R}^2$ be an open set, and let $\varphi : U \to M$ be a parametrized surface.

Definition 1.55 A *vector field along* φ is a differentiable map

$$V : (t, s) \in U \longmapsto V_{(t,s)} \in T_{\varphi(t,s)}(M).$$

□

Let us assume that for every fixed s_0 with (t, s_0) the curve $t \mapsto \varphi(t, s_0)$ is a geodesic parametrized by arc length. Let us set

$$E = \varphi_*(\partial/\partial t), \quad V = \varphi_*(\partial/\partial s).$$

Then

$$\nabla_E E = 0. \tag{1.26}$$

Moreover, note that $[E, V] = 0$. Then (as ∇ is torsion-free)

$$\nabla_E V = \nabla_V E. \tag{1.27}$$

Consequently

$$\nabla_E^2 V = \nabla_E(\nabla_E V) = \nabla_E(\nabla_V \nabla E). \tag{1.28}$$

Then [by (1.26) and the very definition of the curvature tensor R]

$$\nabla_E^2 V - R(E, V)E = 0. \tag{1.29}$$

Definition 1.56 $V = \varphi_* \left(\frac{\partial}{\partial s}\right) = \frac{\partial \varphi}{\partial s}$ is the *infinitesimal variation* induced by the family of geodesics $s =$ constant. □

Definition 1.57 A tangent vector field $X(t)$ along a geodesic $\gamma(t)$ is said to be a *Jacobi field* if

$$\frac{\nabla^2 X(t)}{dt^2} - R(\dot{\gamma}(t), X(t))\dot{\gamma}(t) = 0. \tag{1.30}$$

□

Proposition 1.58 *A Jacobi field $X(t)$ along the geodesic $\gamma(t)$ is determined by $X(t_0)$ and $\frac{\nabla X}{dt}(t_0)$ for some value t_0 of the parameter.*

The set of all Jacobi vector fields along a geodesic $\gamma(t)$ is a $2n$-dimensional real vector space J_γ. Also, the subspace $\{X \in J_\gamma : X(t_0) = 0\}$ of all Jacobi fields along γ that vanish at a fixed point on the geodesic γ, is n-dimensional.

Jacobi fields on M may be built by the following simple procedure. Let $v \in T_p(M)$ and let $\overline{v}$ be the vector field on $T_p(M)$ associated to v by the canonical identification of the tangent spaces at points of $T_p(M)$ with $T_0(T_p(M))$. Let $v_1, v_2 \in T_p(M)$ and let $U \subset \mathbb{R}^2$ be the open set consisting of all $(t, s) \in \mathbb{R}^2$ such that $\exp_p$ is defined at $t(v_2 + s\, v_1)$.

Proposition 1.59 *The map*

$$\varphi : U \to M, \quad \varphi(t, s) = \exp_p\left[t(v_2 + s v_1)\right],$$

is a 1*-parameter family of geodesics in M and the vector field*

$$X = (\exp_p)_*(t\, \overline{v}_1)$$

is a Jacobi field along every geodesic of the family.

Proposition 1.60 *Let $X(t)$ be a Jacobi field along the geodesic $\gamma(t)$ such that*

$$X(0) = 0, \quad \frac{\nabla X}{dt}(0) = v \neq 0.$$

Then

$$X(t) = \left(\exp_{\gamma(0)}\right)_*(t\,\overline{v}).$$

1.3 Riemann-Christoffel Tensor Field

Let (M, g) be a semi-Riemannian manifold and ∇ the Levi-Civita connection of (M, g). Let $R = R^{\nabla} \in \mathcal{T}^{1,3}(M)$ be the curvature tensor field of the Levi-Civita connection, i.e.,

$$R(X, Y)Z = \nabla_X \nabla_Y Z - \nabla_Y \nabla_X Z - \nabla_{[X,Y]} Z$$

per ogni $X, Y, Z \in \mathfrak{X}(M)$.

Definition 1.61 R is called the *curvature* of the semi-Riemannian manifold (M, g). □

Clearly $R(X, Y) = -R(Y, X)$ as for any linear connection. In addition the curvature of (M, g) satisfies the so-called *Bianchi identities*.

Theorem 1.62 *For any $X, Y, Z \in \mathfrak{X}(M)$*

$$R(X, Y)Z + R(Y, Z)X + R(Z, X)Y = 0, \tag{1.31}$$

$$(\nabla_X R)(Y, Z) + (\nabla_Y R)(Z, X) + (\nabla_Z R)(X, Y) = 0. \tag{1.32}$$

Proof Taking into account the symmetry of ∇ and Jacobi's identity

$$R(X, Y)Z + R(Y, Z)X + R(Z, X)Y =$$

$$= \nabla_X \nabla_Y Z - \nabla_Y \nabla_X Z - \nabla_{[X,Y]} Z + \nabla_Y \nabla_Z X - \nabla_Z \nabla_Y X - \nabla_{[Y,Z]} X +$$

$$+ \nabla_Z \nabla_X Y - \nabla_X \nabla_Z Y - \nabla_{[Z,X]} Y =$$

$$= \nabla_X (\nabla_Y Z - \nabla_Z Y) + \nabla_Y (\nabla_Z X - \nabla_X Z) + \nabla_Z (\nabla_X Y - \nabla_Y X) -$$

$$- \nabla_{[X,Y]} Z - \nabla_{[Y,Z]} X - \nabla_{[Z,X]} Y =$$

$$= \nabla_X [Y, Z] + \nabla_Y [Z, X] + \nabla_Z [X, Y] - \nabla_{[X,Y]} Z - \nabla_{[Y,Z]} X - \nabla_{[Z,X]} Y =$$

$$= [X, [Y, Z]] + [Y, [Z, X]] + [Z, [X, Y]] = 0.$$

The covariant derivative of R is given by

$$(\nabla_X R)(Y, Z)W = \nabla_X(R(Y, Z)W) - R(\nabla_X Y, Z)W - R(Y, \nabla_X Z)W - R(Y, Z)\nabla_X W$$

for any $X, Y, Z, W \in \mathfrak{X}(M)$. A straightforward yet rather involved calculation shows that the symmetry of ∇ together with Jacobi identity implies (1.32). Q.e.d.

Definition 1.63 The identities (1.31)–(1.32) are called respectively the *first* and *second Bianchi identities*. □

The first Bianchi identity may be written as

$$\sum_{XYZ} R(X, Y)Z = 0 \tag{1.33}$$

where $\sum_{XYZ}$ is the cyclic sum over all permutations of the arguments X, Y, Z. Similarly, the second Bianchi identity may be written as

$$\sum_{XYZ} (\nabla_X R)(Y, Z) = 0. \tag{1.34}$$

The writings (1.33)–(1.34) have only the advantage of shortness.

Theorem 1.64 *For any* $X, Y, Z, W \in \mathfrak{X}(M)$

$$g(R(X, Y)Z, W) = -g(R(X, Y)W, Z). \tag{1.35}$$

Proof The identity (1.35) follows from $\nabla g = 0$ by a straightforward calculation. Q.e.d.

Definition 1.65 The *Riemann-Christoffel tensor field* of the semi-Riemannian manifold (M, g) is

$$R : \mathfrak{X}(M) \times \mathfrak{X}(M) \times \mathfrak{X}(M) \times \mathfrak{X}(M) \to \mathcal{F}(M),$$

$$R(X, Y, Z, W) = g(R(Z, W)Y, X), \quad X, Y, Z, W \in \mathfrak{X}(M).$$

□

Let us observe that the Riemann-Christoffel tensor field $R \in \mathcal{T}^{0,4}(M)$ is commonly denoted by the same symbol as the curvature tensor field.

Theorem 1.66 *The Riemann-Christoffel R is skew-symmetric in the first two arguments and the last two arguments. Moreover*

$$R(X, Y, Z, W) = R(Z, W, X, Y) \tag{1.36}$$

for any $X, Y, Z, W \in \mathfrak{X}(M)$.

Follows from the first Bianchi identity.

1.4 Sectional Curvature

We confine ourselves to the Riemannian category. The case of semi-Riemannian manifolds of index $\nu > 0$ is briefly discussed later on in this monograph. Let then (M, g) be a Riemannian manifold.

Definition 1.67 A 2-*plane* is a 2-dimensional vector subspace of the tangent space $T_p(M)$ at some $p \in M$. The *sectional curvature* of a 2-plane $\sigma \subset T_p(M)$ is the number

$$k(\sigma) = R_p(u, v, u, v)$$

where $\{u, v\} \subset \sigma$ is a g_p-orthonormal basis of σ. □

To check that the number $k(\sigma) \in \mathbb{R}$ is well defined we need to show that $R_p(u, v, u, v)$ doesn't depend upon the choice of orthonormal basis in σ. Indeed if $\{u', v'\} \subset \sigma$ is another orthonormal basis

$$g_p(u', u') = g_p(v', v') = 1, \quad g_p(u', v') = 0,$$

then

$$\begin{pmatrix} u' \\ v' \end{pmatrix} = \begin{pmatrix} a & b \\ c & d \end{pmatrix} \begin{pmatrix} u \\ v \end{pmatrix}, \quad A = \begin{pmatrix} a & b \\ c & d \end{pmatrix} \in \mathrm{GL}(2, \mathbb{R}).$$

Next

$$g_p(u, u) = g_p(v, v) = 1, \quad g_p(u, v) = 0,$$

yields

$$1 = g_p(u', u') = a^2 + b^2, \quad 1 = g_p(v', v') = c^2 + d^2, \quad 0 = g_p(u', v') = ac + bd,$$

hence $A \in \mathrm{O}(2)$.

$$R_p(u', v', u', v') = a R_p(u, v', u', v') + b R_p(v, v', u', v') =$$

$$= ac R_p(u, u, u', v') + ad R_p(u, v, u', v') + bc R_p(v, u, u', v') + bd R_p(v, v, u', v') =$$

(as R_p is skew-symmetric in the first two variables)

$$= (ad - bc)R_p(u, v, u', v') = (ad - bc)\left\{aR_p(u, v, u, v') + bR_p(u, v, v, v')\right\} =$$

(by the skew-symmetry of R_p in the last two variables)

$$= (ad - bc)\left\{adR_p(u, v, u, v) + bcR_p(u, v, v, u)\right\} =$$

$$= (ad - bc)^2 R_p(u, v, u, v) = \det(A)^2 R_p(u, v, u, v) = R_p(u, v, u, v)$$

as $\det(A) \in \{\pm 1\}$. Q.e.d.

If $\{u, v\} \subset \sigma$ is a g_p-orthonormal basis and $\{u', v'\} \subset \sigma$ is an arbitrary basis then

$$g_p(u', u')g_p(v', v') - g_p(u', v')^2 = (ad - bc)^2$$

hence

$$k(\sigma) = \frac{R_p(u', v', u', v')}{g_p(u', u')g_p(v', v') - g_p(u', v')^2}. \tag{1.37}$$

Given positive integers $k, n \in \mathbb{N}$ such that $k \leq n$ and a real n-dimensional vector space V we denote by $G_k(V)$ the set of all vector subspaces $\sigma \subset V$ such that $\dim_{\mathbb{R}} \sigma = k$. One may organize $G_k(V)$ as a differentiable manifold (the *Grassmann manifold* of k-subspaces in V). Let us set as customary

$$G_2(\tau_M) = \bigcup_{p \in M} G_2(T_p(M))$$

so that the *sectional curvature* of (M, g) is a function $k : G_2(\tau_M) \to \mathbb{R}$.

Theorem 1.68 *Let $R, R' \in \mathcal{T}^{0,4}(M)$ be two tensor fields of type $(0, 4)$ on M possessing the symmetries of the Riemann-Christoffel tensor field, i.e.,*

$$R(X, Y, Z, W) = -R(Y, X, Z, W), \quad R(X, Y, Z, W) = -R(X, Y, W, Z),$$

$$R(W, Z, X, Y) + R(W, X, Y, Z) + R(W, Y, Z, X) = 0,$$

$$R(X, Y, Z, W) = R(Z, W, X, Y),$$

for any $X, Y, Z, W \in \mathfrak{X}(M)$ and the same for R'. If

$$R(X, Y, X, Y) = R'(X, Y, X, Y) \tag{1.38}$$

for any $X, Y \in \mathfrak{X}(M)$ then $R = R'$.

Consequently, if at a point $p \in M$ the sectional curvature of every 2-plane $\sigma \subset T_p(M)$ is known then the Riemann-Christoffel tensor field evaluated at p is determined. In other words, if two Riemannian metrics on M have the same sectional curvature then their Riemann-Christoffel tensor fields coincide, as well.

Proof of Theorem 1.68 Let us set $S = R - R'$ so that [by the assumption (1.38)]

$$S(X, Y, X, Y) = 0, \quad X, Y \in \mathfrak{X}(M).$$

We wish to show that $S = 0$. Let $\{e_i : 1 \le i \le n\} \subset T_p(M)$ be a basis and let us represent $u, v \in T_p(M)$ as

$$u = \sum_{i=1}^{n} u^i e_i, \quad v = \sum_{i=1}^{n} v^i e_i .$$

Then

$$0 = S_p(u, v, u, v) = u^i v^j u^k v^\ell S_{ijkl}(p) \tag{1.39}$$

where we have set $S_{ijk\ell}(p) = S_p(e_i, e_j, e_k, e_\ell)$. As u and v are arbitrary we may choose their components such that for any fixed pair of indices $i_0, j_0 \in \{1, \ldots, n\}$

$$u^{i_0} = 1, \quad i \in \{1, \ldots, n\} \setminus \{i_0\} \Longrightarrow u^i = 0,$$

$$v^{j_0} = 1, \quad j \in \{1, \ldots, n\} \setminus \{j_0\} \Longrightarrow v^j = 0,$$

so that [by (1.39)] $S_{i_0 j_0 i_0 j_0} = 0$. As the indices i_0 and j_0 are arbitrary it follows that

$$S_{ijij}(p) = 0. \tag{1.40}$$

If in turn we consider three distinct indices $\{i_0, j_0, k_0\} \subset \{1, \ldots, n\}$ and the tangent vectors u and v with the components

$$u^{i_0} = 1, \quad i \in \{1, \ldots, n\} \setminus \{i_0\} \Longrightarrow u^i = 0,$$

$$v^{j_0} = v^{k_0} = 1, \quad j \in \{1, \ldots, n\} \setminus \{j_0, k_0\} \Longrightarrow v^j = 0,$$

then [by (1.39)]

$$0 = S_{ijk\ell}(p) u^i v^j u^k v^\ell = S_{i_0 j i_0 \ell}(p) v^j v^\ell = v^\ell \left\{ S_{i_0 j_0 i_0 \ell} + S_{i_0 k_0 i_0 \ell} \right\} =$$

$$= S_{i_0 j_0 i_0 j_0}(p) + S_{i_0 j_0 i_0 k_0}(p) + S_{i_0 k_0 i_0 j_0}(p) + S_{i_0 k_0 i_0 k_0}(p) =$$

[by (1.40)]

$$= 2 S_{i_0 j_0 i_0 k_0}(p)$$

due to (1.40) and to

$$S(X, Y, Z, W) = S(Z, W, X, Y), \quad X, Y, Z, W \in \mathfrak{X}(M).$$

We have proved that

$$S_{ijik}(p) = 0. \tag{1.41}$$

Finally we consider the tangent vectors u and v such that each of them has but two nonzero components (u^{i_0}, u^{k_0}) and (v^{j_0}, v^{k_0}) so that identity (1.39) yields

$$0 = S_{ijk\ell}(p)u^i v^j u^k v^\ell = \left\{S_{i_0 jk\ell}(p) + S_{k_0 jk\ell}(p)\right\} v^j u^k v^\ell =$$

$$= \left\{S_{i_0 ji_0\ell}(p) + S_{k_0 ji_0\ell}(p) + S_{i_0 jk_0\ell}(p) + S_{k_0 jk_0\ell}(p)\right\} v^j v^\ell =$$

[by (1.41)]

$$= \left\{S_{k_0 j_0 i_0\ell}(p) + S_{i_0 j_0 k_0\ell}(p) + S_{k_0\ell_0 i_0\ell}(p) + S_{i_0\ell_0 k_0\ell}(p)\right\} v^\ell =$$

$$= S_{k_0 j_0 i_0 j_0}(p) + S_{i_0 j_0 k_0 j_0}(p) + S_{k_0\ell_0 i_0 j_0}(p) + S_{i_0\ell_0 k_0 j_0}(p) +$$

$$+ S_{k_0 j_0 i_0\ell_0}(p) + S_{i_0 j_0 k_0\ell_0}(p) + S_{k_0\ell_0 i_0\ell_0}(p) + S_{i_0\ell_0 k_0\ell_0}(p)$$

and all terms with two repeated indices are zero, e.g.,

$$S_{k_0 j_0 i_0 j_0}(p) = S_{j_0 k_0 j_0 i_0}(p) = 0$$

by the skew-symmetry of S in the first two variables and the last two variables

$$S(X, Y, Z, W) = -S(Y, X, Z, W), \quad S(X, Y, Z, W) = -S(X, Y, W, Z),$$

and by (1.41). It follows that

$$S_{k\ell ij}(p) + S_{i\ell kj}(p) + S_{kji\ell}(p) + S_{ijk\ell}(p) = 0. \tag{1.42}$$

On the other hand

$$S(W, Z, X, Y) + S(W, X, Y, Z) + S(W, Y, Z, X) = 0$$

implies

$$S_{\ell kij}(p) + S_{\ell ijk}(p) + S_{\ell jki}(p) = 0$$

which together with (1.42) yields $S_{ijk\ell}(p) = 0$. Q.e.d.

Definition 1.69 A Riemannian manifold (M, g) is said to be *isotropic* if the sectional curvature of any 2-plane $\sigma \subset T_p(M)$ depends only on the point p, i.e., there is a function $f : M \to \mathbb{R}$ such that $k = f \circ \pi$ where $\pi : G_2(\tau_M) \to M$ is the canonical projection. □

By the very definition of sectional curvature, if M is isotropic then

$$R(X, Y)Z = f\left[g(Y, Z9X - g(X, Z)Y\right]. \tag{1.43}$$

Definition 1.70 A Riemannian manifold (M, g) is a *space of constant curvature* if its sectional curvature $k : G_2(M) \to \mathbb{R}$ is a constant function. □

Theorem 1.71 *Every isotropic connected Riemannian manifold of dimension $n \geq 3$ is a space of constant curvature.*

Theorem 1.71 is the classical theorem of F. Schur. The proof is a mere exercise relying on the second Bianchi identity.

The sectional curvature of indefinite semi-Riemannian manifolds was studied by L. Graves and K. Nomizu,[13] and by J.K. Beem.[14] In particular the paper by L. Graves and K. Nomizu (cf. *op. cit.*) generalizes (from the Riemannian to the semi-Riemannnian case) Cartan's theorem on manifolds satisfying the axiom of r-planes (Theorem 3.31).

1.5 Ricci Curvature, Einstein Manifolds

Let (M, g) be a n-dimensional semi-Riemannian manifold, of index ν. Let ∇ be the Levi-Civita connection of (M, g) and let $K = K^\nabla$ be its Riemann-Christoffel tensor field. The *Ricci curvature* of (M, g) is the $(0, 2)$-tensor field

$$\text{Ric} : \mathfrak{X}(M) \times \mathfrak{X}(M) \to C^\infty(M)$$

defined as follows. Let $x_0 \in M$ and let $\{E_j : 1 \leq j \leq n\}$ be a local orthonormal frame

$$g(E_j, E_k) = \epsilon_j \, \delta_{jk}, \quad \epsilon_j = -1, \quad 1 \leq j \leq \nu, \quad \epsilon_k = 1, \quad \nu + 1 \leq k \leq n,$$

on $T(M)$, defined on an open neighborhood $U \subset M$ of x_0. We set

$$\text{Ric}_{x_0}(v, w) = \sum_{j=1}^{n} K_{x_0} \epsilon_j \left(E_{j, x_0}, \, v, \, E_{j, x_0}, \, w\right) \tag{1.44}$$

for any $v, w \in T_{x_0}(M)$. The definition of $S_{x_0}(v, w)$ doesn't depend upon the choice of local orthonormal frame about x_0. The *scalar curvature* of (M, g) is the function

[13] Cf. L. Graves, K. Nomizu, On sectional curvature of indefinite metrics. Math. Ann. **232**, 267–272 (1979).

[14] Cf. J.K. Beem, Pseudo-Riemannian manifolds with totally geodesic bisectors. Proc. Amer. Math. Soc. **49**, 212–215 (1975).

$\rho : M \to \mathbb{R}$ defined by

$$\rho(x_0) = \sum_{j=1}^{n} \epsilon_j \operatorname{Ric}(E_j, E_j)_{x_0} .$$

Theorem 1.72 *Let (M, g) be a connected semi-Riemannian manifold of dimension $n \geq 3$ such that* $\operatorname{Ric}(X, Y) = \lambda g(X, Y)$ *for some $\lambda \in C^\infty(M)$. Then λ is constant.*

Follows easily from the second Bianchi identity.

Definition 1.73 A n-dimensional ($n \geq 3$) Riemannian manifold such that

$$R(X, Y) = \lambda g(X, Y) \tag{1.45}$$

for some $\lambda \in \mathbb{R}$ and any $X, Y \in \mathfrak{X}(M)$, is called an *Einstein manifold.* □

Einstein manifolds are rich in geometrical properties, and are widely studied within Riemannian geometry, cf. e.g., A.L. Besse.[15] Several examples belonging to the semi-Riemannian[16] category will be exhibited[17] in the volume of this book devoted to General Relativity Theory.

Proposition 1.74 *Any Riemannian manifold of constant sectional curvature is an Einstein manifold.*

Proof Follows from (1.43) and (1.44).

Proposition 1.75 *Every Einstein manifold of dimension $n = 3$ has constant sectional curvature.*

The proof is a mere exercise, relying on the Schur theorem.

Example 1.76 (1) Let $S^n(r) \subset \mathbb{E}^{n+1}$ be the standard sphere of radius $r > 0$. The induced metric on $S^n(r)$ has (positive) constant sectional curvature $1/r^2$.

(2) Let L^{n+1} denote $\mathbb{R}^{n+1}$ equipped with the semi-Riemannian metric

$$\sum_{j=1}^{n} dx^j \otimes dx^j - dx^{n+1} \otimes dx^{n+1} .$$

Then L^{n+1} is a Lorentzian manifold. The equation

$$-(x^{n+1})^2 + \sum_{j=1}^{n} (x^j)^2 = -r^2$$

---

[15] A.L. Besse, *Einstein Manifolds* (Springer, Berlin-Heidelberg, 1987).

[16] Actually, Lorentzian.

[17] As solutions to the Einstein field equations.

defines a hypersurface H^n in L^{n+1}. The manifold H^n equipped with the induced metric is the n-dimensional *hyperbolic space*. H^n has (negative) constant sectional curvature $-1/r^2$. For $r = 1$, the hyperbolic space $H_n = H^n(-1)$ is isometric to $\mathbb{R}^n_+ = \{y \in \mathbb{R}^n \; : \; y^n > 0\}$ equipped with the Riemannian metric

$$\frac{1}{(y^n)^2} \sum_{j=1}^{n} dy^j \otimes dy^j \, .$$

Note first that $\mathbb{R}^n_+$ is itself a space of constant curvature -1. Indeed, let us set

$$E_j = y^n \, \frac{\partial}{\partial y^j}$$

so that $\{E_j \; : \; 1 \le j \le n\}$ is a (globally defined) orthonormal frame on $T(\mathbb{R}^n_+)$. The sectional curvatures determined by $\{E_j \; : \; 1 \le j \le n\}$ equal -1. To build the announced isometry, one considers first the map

$$h : H^n \to \mathbb{R}^n, \quad h\big(x^1, \ldots, x^{n+1}\big) = \big(u^1, \ldots, u^n\big), \quad u^j = \frac{x^j}{1 + x^{n+1}}, \quad 1 \le j \le n,$$

and then the inversion h' in $\mathbb{R}^n$ of center $p = (0, 0, \ldots, -1)$ given by the vector equation

$$\vec{py} = \frac{2}{\|\vec{pu}\|} \, \vec{pu} \, .$$

This yields a diffeomorphism $F : H^n \to \mathbb{R}^n$ preserving the metrics. □

Definition 1.77 A n-dimensional Riemannian manifold $M = M^n(k)$ of constant sectional curvature k is called a *space form* if M is simply connected and complete. □

$\mathbb{E}^n$, S^n and H^n are examples of space forms. For a detailed study of space forms one may see the classical monograph by J. Wolf.[18]

1.6 Canonical Structures on the Tangent Bundle

Let (M, g) be a Riemannian manifold, and let ∇ be its Levi-Civita connection. The manifold $T(M)$ [the total space of the tangent bundle over M] carries a natural Riemannian metric G given by

$$G(A, B) = g^{\pi}(KA, \; KB) + g^{\pi}(LA, \; LB) \tag{1.46}$$

[18] J.A. Wolf, *Spaces of Constant Curvature* (Mc Graw-Hill, New York, 1967).

for any $A, B \in \mathfrak{X}(T(M))$. To explain the notations adopted in (1.46), we need a few preparatory elements. Let $\pi : T(M) \to M$ be the canonical projection, and let $\pi^{-1}T(M) \to T(M)$ be the pullback of $T(M) \to M$ via π. The fiber in the pullback bundle over $v \in T(M)$ is

$$\left(\pi^{-1}TM\right)_v = T_{\pi(v)}(M).$$

Let L be the vector bundle morphism

$$L : T\big(T(M)\big) \to \pi^{-1}T(M),$$

$$L_v A := (d_v \pi)A, \quad A \in T_v\big(T(M)\big), \quad v \in T(M).$$

Let $x_0 \in M$ be a point, and let $(U, \tilde{x}^i)$ be a local coordinate system about x_0. Let X_i be the sections in the pullback bundle given by

$$X_i \in C^\infty\big(\pi^{-1}(U), \pi^{-1}T(M)\big), \quad X_i(v) = \left(\frac{\partial}{\partial \tilde{x}^i}\right)_{\pi(v)}, \quad v \in \pi^{-1}(U),$$

so that $\{X_i : 1 \le i \le n\}$ is a local frame in $\pi^{-1}T(M)$, defined on the open set $\pi^{-1}(U)$. Let $\left\{ {i \atop jk} \right\}$ be the Christoffel symbols of the second kind, i.e.,

$$\nabla_{\partial_j} \partial_k = \left\{ {i \atop jk} \right\} \partial_i, \quad \partial_i \equiv \frac{\partial}{\partial \tilde{x}^i}.$$

Let $\left(\pi^{-1}(U), x^i, \dot{x}^i\right)$ be the induced local coordinates on $T(M)$, i.e.,

$$x^i(v) = \tilde{x}^i\big(\pi(v)\big), \quad v = \dot{x}^i(v)\left(\frac{\partial}{\partial \tilde{x}^i}\right)_{\pi(v)}, \quad v \in \pi^{-1}(U).$$

Let us consider the tangent vector fields

$$\frac{\delta}{\delta x^i} \in \mathfrak{X}\big(\pi^{-1}(U)\big), \quad 1 \le i \le n,$$

$$\frac{\delta}{\delta x^i} = \frac{\partial}{\partial x^i} + N_i^j \frac{\partial}{\partial \dot{x}^j}, \quad N_j^i := \left\{ {i \atop jk} \right\} \dot{x}^k \in C^\infty\big(\pi^{-1}(U)\big).$$

Next, let us consider the assignment

$$N : v \in T(M) \longmapsto N_v \subset T_v\big(T(M)\big)$$

defined as follows. Let $v \in T(M)$ and let $x := \pi(v) \in M$. Let $(U, \tilde{x}^i)$ be a local coordinate system on M about x. We set

$$N_v := \left\{\lambda^j \left(\frac{\delta}{\delta x^j}\right)_v \ : \ \lambda_j \in \mathbb{R}, \ \ 1 \le j \le n\right\}.$$

The definition of N_v doesn't depend upon the choice of local coordinates about $\pi(v)$. By the very definition N is a C^∞ distribution on $T(M)$. Next, let us consider

$$V \ : \ v \in T(M) \longmapsto V_v := \operatorname{Ker}\big(d_v \pi\big) \subset T_v\big(V(M)\big).$$

Locally V is the span of $\{\partial/\partial \dot{x}^j \ : \ 1 \le j \le n\}$, so V is a C^∞ distribution on $T(M)$, as well.

Definition 1.78 N and V are referred to as the *horizontal* and *vertical* distributions on $T(M)$, respectively. □

Lemma 1.79 *For every $v \in T(M)$*

$$T_v\big(V(M)\big) = N_v \oplus V_v \, . \tag{1.47}$$

The proof is left as an exercise to the reader. We shall need the vector bundle morphisms

$$\gamma : \pi^{-1}T(M) \to V, \quad \beta : \pi^{-1}T(M) \to N,$$

$$\gamma_v X_i(v) = \left(\frac{\partial}{\partial \dot{x}^i}\right)_v, \quad \beta_v X_i(v) = \left(\frac{\delta}{\delta x^i}\right)_v, \quad 1 \le i \le n,$$

The definitions of γ_v and β_v do not depend on the choice of local coordinates about $\pi(v)$.

Proposition 1.80 *$\gamma : \pi^{-1}T(M) \to V$ and $\beta : \pi^{-1}T(M) \to N$ are vector bundle isomorphisms.*

Proof Exercise.

Definition 1.81 γ and β are respectively called the *vertical lift* and *horizontal lift*. □

Exercise 1.82 (1) Show that

$$\gamma_v X = \frac{dC}{dt}(0), \quad X \in T_{\pi(v)}(M), \quad v \in T(M),$$

where

$$C : (-\epsilon, \, \epsilon) \to T(M), \quad C(t) = v + t\, X, \quad |t| < \epsilon \, .$$

(2) Show that the restriction of L to N

$$L : N \to \pi^{-1}T(M)$$

is a vector bundle isomorphism whose inverse in the horizontal lift.

(3) Show that the following sequence of vector bundles and vector bundle morphisms

$$0 \to \pi^{-1}T(M) \xrightarrow{\gamma} T\big(T(M)\big) \xrightarrow{L} \pi^{-1}T(M) \to 0$$

is exact. □

Let $Q_v : T_v(T(M)) \to V_v$ be the projection, with respect to the direct sum decomposition (1.47). Let us consider the vector bundle morphism

$$K : T\big(T(M)\big) \to \pi^{-1}T(M), \quad K_v A = Q_v A, \quad A \in T_v(T(M)), \quad v \in T(M).$$

Definition 1.83 K is referred to as the *Dombrowski map*. □

Exercise 1.84 (1) Show that

$$K\,\frac{\delta}{\delta x^i} = 0, \quad K\,\frac{\partial}{\partial \dot{x}^i} = X_i, \quad 1 \le i \le n.$$

(2) Show that the following sequence of vector bundles and vector bundle morphisms

$$0 \to \pi^{-1}T(M) \xrightarrow{\beta} T(T(M)) \xrightarrow{K} \pi^{-1}T(M) \to 0$$

is exact. □

We end our explanation of the notations in (1.46) by recalling the bundle metric g^π. By definition, $g^\pi = \pi^{-1}g$ is the pullback of g via π, i.e., locally

$$g^\pi\big(X_i,\, X_j\big) = g_{ij} \circ \pi, \quad g_{ij} = g\Big(\frac{\partial}{\partial \tilde{x}^i},\, \frac{\partial}{\partial \tilde{x}^j}\Big).$$

g^π is a Riemannian bundle metric on the pullback bundle $\pi^{-1}T(M)$.

Lemma 1.85 *The* (0, 2)*-tensor field G on $T(M)$ given by* (1.46) *is a Riemannian metric on $T(M)$.*

Proof Exercise.

Definition 1.86 G is called the *Sasaki metric* on $T(M)$. □

Exercise 1.87 (1) Show that the local components of the Sasaki metric with respect to the local coordinate system $\big(\pi^{-1}(U),\, x^i,\, \dot{x}^i\big)$ are

$$G_{ij} = g_{ij} \circ \pi + \big(g_{rs} \circ \pi\big)\Big(\Big\{ {r \atop pi} \Big\} \circ \pi\Big)\Big(\Big\{ {s \atop qj} \Big\} \circ \pi\Big)\, \dot{x}^p\, \dot{x}^q,$$

$$G_{j,\,k+n} = G_{j+n,\,k} = \Big(\big\{rj,\, k\big\} \circ \pi\Big)\, \dot{x}^r, \quad G_{j+n,\,k+n} = g_{jk} \circ \pi\,.$$

(2) Show that the vertical and horizontal distributions on $T(M)$ are orthogonal with respect to the Sasaki metric. □

Every tangent vector field $X \in \mathfrak{X}(M)$ determines a section $\hat{X} \in C^\infty(\pi^{-1}TM)$ (the *natural lift* of X) given by

$$\hat{X}(v) = X_{\pi(v)}, \quad v \in T(M).$$

For instance, every X_i is the natural lift of $\partial/\partial\tilde{x}^i$.

Definition 1.88 The *Liouville vector* is the section $\mathfrak{L} \in C^\infty(\pi^{-1}TM)$ defined by

$$\mathfrak{L}(v) = v, \quad v \in T(M).$$

□

Not every section in $\pi^{-1}T(M)$ is[19] the natural lift of a tangent vector field on M. For every $X \in \mathfrak{X}(M)$ we adopt the notations

$$X^H = \beta\,\hat{X}, \quad X^V = \gamma\,\hat{X}.$$

Definition 1.89 X^H, $X^V \in \mathfrak{X}(T(M))$ are respectively called the *horizontal lift* and the *vertical lift* of X. □

The curvature tensor field

$$R : \mathfrak{X}(M) \times \mathfrak{X}(M) \times \mathfrak{X}(M) \to \mathfrak{X}(M)$$

induces a $C^\infty(T(M))$-multi-linear map

$$R^\pi : C^\infty(\pi^{-1}TM) \times C^\infty(\pi^{-1}TM) \times C^\infty(\pi^{-1}TM) \to C^\infty(\pi^{-1}TM)$$

locally given by

$$\{R^\pi(X, Y)Z\}(v) := X^i\, Y^j\, Z^k \left\{R\Big(\frac{\partial}{\partial\tilde{x}^i}, \frac{\partial}{\partial\tilde{x}^j}\Big)\frac{\partial}{\partial\tilde{x}^k}\right\} \circ \pi\Big\}_{\pi(v)},$$

$$X, Y, Z \in C^\infty(\pi^{-1}TM), \quad v \in T(M),$$

$$X\big|_{\pi^{-1}(U)} = X^i\, X_i, \quad Y\big|_{\pi^{-1}(U)} = Y^j\, X_j, \quad Z\big|_{\pi^{-1}(U)} = Z^k\, X_k\,.$$

[19] For instance, the Liouville vector $\mathfrak{L}$ isn't the natural lift of any tangent vector field on M.

The definition of $\{R^\pi(X, Y)Z\}(v)$ doesn't depend upon the choice of local coordinates $(U, \tilde{x}^i)$ about $\pi(v)$.

Exercise 1.90 Show that

$$R^\pi(\hat{X}, \hat{Y})\hat{Z} = \{R(X, Y)Z\} \circ \pi$$

for any $X, Y, Z \in \mathfrak{X}(M)$. □

Proposition 1.91 *The Levi-Civita connection $\tilde{\nabla}$ of $T(M)$ equipped with the Sasaki metric G satisfies*

(i) $\tilde{\nabla}_{X^H} Y^H = (\nabla_X Y)^H - \frac{1}{2}\gamma\, R^\pi(\hat{X}, \hat{Y})\mathfrak{L}$,
(ii) $\tilde{\nabla}_{X^V} Y^H = \frac{1}{2}\beta\, R^\pi(\mathfrak{L}, \hat{X})\hat{Y}$,
(iii) $\tilde{\nabla}_{X^H} Y^V = (\nabla_X Y)^V + \frac{1}{2}\beta\, R^\pi(\mathfrak{L}, \hat{Y})\hat{X}$,
(iv) $\tilde{\nabla}_{X^V} Y^V = 0$,

for any $X, Y \in \mathfrak{X}(M)$.

Proof The proof is standard, cf. e.g., Blair [Bla].

The geometry of the tangent bundle[20] over a Riemannian manifold, including the technique of horizontal, vertical, and complete lifts[21] of vector fields, was a very popular subject in the 1960s and 1970s, yet faded soon after into a collection of more or less elementary examples. The topic was essentially abandoned[22] as soon as it became clear that most differential geometric objects (naturally associated to the Riemannian structure g on M) on $T(M)$ enjoy no particular geometric properties, unless g is flat. For instance

Corollary 1.92 *The vertical and horizontal distributions V and H on $T(M)$ are parallel with respect to the Levi-Civita connection $\tilde{\nabla}$ of $(T(M), G)$ if and only if (M, g) is flat (i.e., $R = 0$).*

Corollary 1.93 *The horizontal distribution H on $T(M)$ is integrable if and only if (M, g) is flat.*

Corollary 1.94 *If g is flat then the Sasaki metric G is flat.*

---

[20] Cf. K. Yano, *Tangent and Cotangent Bundles* (M. Dekker, Inc., New York, 1973).

[21] Cf. K. Yano, S. Ishihara, Horizontal lifts of tensor fields and connections to tangent bundles. J. Math. Mech. **16**, 1015–1030 (1967).

[22] A fauna of examples of Riemannian metrics on $T(M)$, other than the Sasaki metric, were introduced in an effort to produce examples of Riemannian structures which are less "rigid" than the Sasaki metric, i.e., additional requirements, such as integrability, parallelism, etc., do not yield severe limitations on the base metric g, cf. e.g., S. Dragomir, D. Perrone, *Harmonic Vector Fields. Variational Principles and Differential Geometry* (Elsevier Inc., New York, 2012), § 5.1.

Theorem 1.95 *If the curvature tensor field $\tilde{R}$ of the Sasaki metric G is covariant constant (i.e., $\tilde{\nabla}\tilde{R} = 0$) then (M, g) is flat.*

Proof See O. Kowalski.[23]

$T(M)$ admits the almost complex structure J defined by

$$J X^H = X^V, \quad J X^V = -X^H,$$

for every $X \in \mathfrak{X}(M)$.

Proposition 1.96 *The Sasaki metric G is almost Hermitian, with respect to the almost complex structure $\tilde{J}$. If the almost complex structure $\tilde{J}$ is integrable then (M, g) is flat.*

1.7 Geodesics and Integration

We start with recalling a bit of linear algebra. Let E and F be two Euclidean spaces, of the same dimension n, and let $f : E \to F$ be a linear map. Let $\beta_E \subset E$ and $\beta_F \subset F$ be orthonormal bases, and let M_β be the matrix associated to f, relative to the bases β_E and β_F. Let $\beta'_E \subset E$ and $\beta'_F \subset F$ be another pair of orthonormal bases, and $M_{\beta'}$ the matrix of f with respect and β'_E and β'_F. Let P be the matrix of passage from β_E to β'_E, and Q the matrix of passage from β_F to β'_F. Then

$$M_{\beta'} = Q^{-1}\, M_\beta\, P = Q^T\, M_\beta\, P$$

[because of $Q \in \mathrm{O}(n)$] yielding

$$\det M_{\beta'} = (\det Q)\,(\det M_\beta)\,(\det P) \tag{1.48}$$

and then

$$\left|\det M_{\beta'}\right| = \left|\det M_\beta\right|$$

[again, as $P,\ Q \in \mathrm{O}(n) \Longrightarrow \left|\det P\right| = \left|\det Q\right| = 1$]. The meaning of (1.48) is that $\left|\det M_\beta\right|$ doesn't depend on the choice of orthonormal bases $\beta_E \subset E$ and $\beta_F \subset F$.

Definition 1.97 The number $\left|\det M_\beta\right|$ is called the *modulus of the determinant* of f, and is denoted by $\left|\det f\right|$. □

---

[23] Cf. O. Kowalski, Curvature of the Riemannian metric on the tangent bundle of a Riemannian manifold. J. reine und angew. Math. **220**, 124–129 (1971).

Let $b \subset E$ be an arbitrary basis. Let $\Phi : E \to F$ be an isometry. Let M_b be the matrix of f relative to the bases $b \subset E$ and $\Phi(b) \subset F$. Then

$$\left|\det M_b\right| = \left|\det f\right|.$$

We apply the previous consideration to Riemannian geometry, as follows. Let (M, g) be a Riemannian manifold, and let $x \in M$ be a point. Let $V \subset T_x(M)$ and $U \subset M$ be respectively neighborhoods of the zero tangent vector $0_x \in T_x(M)$ and the point $x \in M$, such that the exponential map $\exp_x$ gives a diffeomorphism

$$\exp_x \big|_V : V \simeq U.$$

For every $v \in V$ we set[24]

$$\theta(v) := \left|\det\left(d_v \exp_x\right)\right|.$$

Let v_x be the Lebesgue measure on the Euclidean space $\left(T_x(M),\, g_x\right)$.

Proposition 1.98 *For any continuous function on U*

$$\int_U \varphi\, \mathrm{v}_g = \int_V \left(\varphi \circ \exp_x\right) \theta\, \mathrm{v}_x\,.$$

The proof is well known.[25]

Let $S_x(M)$ be the unit sphere in $T_x(M)$, i.e.,

$$S_x(M) = \{v \in T_x(M)\ :\ g_x(v, v) = 1\}.$$

Let $\rho : S_x(M) \to (0, +\infty)$ be a continuous function. Let

$$V = \left\{v \in T_x(M)\ :\ |v| < \rho\left(\frac{1}{|v|}\, v\right)\right\}.$$

Let $W \subset S_x(M) \times \mathbb{R}_+^*$ be the open set

$$W = \{(u, r)\ :\ r < \rho(u)\}.$$

Then $(u, r) \longmapsto ru$ is a diffeomorphism $W \simeq V \setminus \{0\}$.

---

[24] With respect to the notations in Definition 1.97 one has

$$E = T_x(M) \simeq T_v\left(T_x(M)\right), \quad F = T_{\exp_x v}(M), \quad f = d_v \exp_x\,.$$

[25] Cf. e.g., Proposition C.III.2, p. 55, in M. Berger, P. Gauduchon, E. Mazet, *Le Spectre d'une Variété Riemannienne*, Lecture Notes in Mathematics, vol. 194 (1971).

Definition 1.99 Let $v \in V \setminus \{0\}$. The unique $(u, r) \in W$ such that $v = r\,u$ are the *polar coordinates* of v. □

Let us consider the function $\theta : W \to \mathbb{R}_+$ given by

$$\theta(r,\, u) = \theta(ru)r^{n-1}\,. \tag{1.49}$$

Proposition 1.100 *For every continuous function φ on U*

$$\int_U \varphi\, \mathrm{v}_g = \int_W \varphi(\exp_x ru)\,\theta(u,\, r)\, du\, dr.$$

Proof Follows by Proposition 1.98 and a coordinate change under the integral sign. This is but a particular instance of a general method for computing integrals on compact manifolds.

Let (M, g) be a compact Riemannian manifold and let $x \in M$. For every $u \in S_x(M)$ let C_u be the geodesic of initial data (x, u), parametrized by arc length. For sufficiently small values of the parameter s the curve $C_u\big|_{[0,s]}$ is minimizing. Let A_u consist of all s such that $C_u\big|_{[0,s]}$ is minimizing. If $s \in A_u$ then $[0, s] \subset A_u$. Also A_u is bounded. Indeed, let

$$\delta(M, g) := \max_{x,x' \in M} d(x,\, x')$$

be the diameter of M. Then, if $s > \delta(M, g)$ then certainly $s \notin A_u$. Let us set

$$\rho(u) := \sup A_u\,.$$

Then $\rho(u) \in A_u$. Indeed, let γ be a curve joining x to $C_u(\rho(u))$ and let $\{s_k\}_{k \geq 1} \subset A_u$ be a sequence tending to $\rho(u)$ as $k \to \infty$. For every k

$$s_k \leq \ell(\gamma) + \rho(u) - s_k$$

and tending with $k \to \infty$

$$\rho(u) \leq \ell(\gamma).$$

So actually $A_u = [0,\, \rho(u)]$. One may show[26] that ρ is continuous on $S_x(M)$. Thus ρ is uniformly continuous, and one may deduce that the set

$$\dot{U}_x := \left\{ v \in T_x(M)\ :\ |v| = \rho\Big(\frac{1}{|v|}\, v\Big) \right\}$$

[26] Cf. [KoNo], vol. II, p. 98.

has measure zero. On the other hand, let

$$U_x := \left\{ v \in T_x(M) \ : \ |v| < \rho\left(\frac{1}{|v|}\, v\right) \right\}.$$

U_x is an open neighborhood of the origin 0_x and $\dot{U}_x = \partial U_x$. The exponential mapping $\exp_x$ is a diffeomorphism of U_x onto its image, and U_x is the largest neighborhood of $0_x \in T_x(M)$ possessing this property. Let us set

$$\mathcal{C}_x := \exp_x\left(\dot{U}_x\right).$$

Definition 1.101 $\mathcal{C}_x$ is called the *cut locus* of the point x in (M, g). □

Proposition 1.102 *$\mathcal{C}_x$ has measure zero and*

$$M = \exp_x\left(U_x\right) \cup \mathcal{C}_x\,.$$

Corollary 1.103 *Let (M, g) be a compact Riemannnian manifold, and let $x \in M$. Let $U_x \subset T_x(M)$ be the largest open neighborhood of 0_x such that* $\exp_x$ *is a diffeomorphism of U_x on its image. Let $f \in L^1(M, \mathrm{v}_g)$ be a v_g-integrable function on M. Then*

$$\int_M f\, \mathrm{v}_g = \int_{U_x} (f \circ \exp_x)\, \theta\, \mathrm{v}_x\,.$$

1.8 Laplace–Beltrami Operator

Let (M, g) be a n-dimensional oriented Riemannian manifold.

Definition 1.104 Let $\xi \in \mathfrak{X}(M)$. The *divergence* of ξ is the function $\mathrm{div}(\xi) : M \to \mathbb{R}$ determined by

$$d\left(\xi \rfloor \omega\right) = \mathrm{div}(\xi)\, \omega$$

where $\omega = d\, \mathrm{v}_g$. □

We shall need the operator

$$\delta : \Omega^1(M) \to C^\infty(M), \quad \delta\, \alpha = -\mathrm{div}\left(\alpha^\sharp\right), \quad \alpha \in \Omega^1(M).$$

Let us assume M is compact. We endow $\Omega^p(M)$ with the scalar product

$$\langle \alpha, \beta \rangle = \int_M g^*(\alpha, \beta)\, d\, \mathrm{v}_g, \quad \alpha, \beta \in \Omega^p(M). \tag{1.50}$$

Proposition 1.105 *For any* $f \in \Omega^0(M)$ *and* $\alpha \in \Omega^1(M)$

$$\langle df,\ \alpha\rangle = \langle f, \delta\alpha\rangle$$

i.e., δ *is the formal adjoint of* d.

The identity in Proposition 1.105 may also be written

$$\int_M g^*(df,\ \alpha)\, d\,\mathrm{v}_g = \int_M f\ \delta\alpha\ d\,\mathrm{v}_g\,.$$

The proof of Proposition 1.105 follows by Green's lemma.

If $d : \Omega^p(M) \to \Omega^{p+1}(M)$ is the exterior differentiation operator on p-forms then $\delta : \Omega^{p+1}(M) \to \Omega^p(M)$ is defined by

$$\langle d\alpha,\ \beta\rangle = \langle \alpha,\ \delta\,\beta\rangle,$$

$$\alpha \in \Omega^p(M), \quad \beta \in \Omega^{p+1}(M).$$

The operators div and δ may be explicitly expressed in local coordinates, as follows. Let $(U,\ x^i)$ be a local coordinate system on M and let us set

$$\theta = \sqrt{\mathfrak{g}}, \quad \mathfrak{g} = \det\left[g_{ij}\right],$$

so that

$$\omega = d\,\mathrm{v}_g = \theta\, dx^1 \wedge \cdots \wedge dx^n \tag{1.51}$$

on U. Then

$$\operatorname{div}(\xi) = \frac{1}{\theta}\frac{\partial(\theta\,\xi^i)}{\partial x^i}\,. \tag{1.52}$$

Let us set $X_i \equiv \partial/\partial x^i$ for simplicity. Let $\alpha = \alpha_i\, dx^i$. Then

$$\alpha^\sharp = g^{ij}\alpha_j\, X_i\,. \tag{1.53}$$

Consequently [by (1.52)–(1.53)]

$$\delta\alpha = -\frac{1}{\theta}\,\frac{\partial(\theta\, g^{ij}\,\alpha_j)}{\partial x^i}\,. \tag{1.54}$$

The operator δ may be described as the trace of the covariant derivative. Let $\alpha \in \Omega^1(M)$. Its covariant derivative $\nabla\alpha$ is the $C^\infty(M)$-bilinear form

$$(\nabla\alpha)(X, Y) = (\nabla_X\alpha)Y = X(\alpha(Y)) - \alpha(\nabla_X Y).$$

If $\{E_i \,:\, 1 \leq i \leq n\}$ is a local g-orthonormal frame, the trace of $\nabla\alpha$ is

$$\operatorname{trace}(\nabla\alpha) = \sum_{i=1}^{n} (\nabla\alpha)(E_i,\, E_i). \tag{1.55}$$

Note that

$$g\big(\nabla_X \alpha^\sharp,\, Y\big) = (\nabla\alpha)(X, Y). \tag{1.56}$$

Consequently trace $\nabla\alpha$ is the trace of the endomorphism $X \mapsto \nabla_X \alpha^\sharp$.

Proposition 1.106 *For any differential* 1*-form* $\alpha \in \Omega^1(M)$ *on the Riemannian manifold* (M, g)

$$\delta\alpha = -\operatorname{trace} \nabla\alpha.$$

Definition 1.107 The *Laplacian* is the operator

$$\Delta : \Omega^0(M) \to \Omega^0(M), \quad \Delta f = \delta\, df.$$

□

In local coordinates

$$\Delta f = -\frac{1}{\theta}\frac{\partial}{\partial x^i}\Big(\theta\, g^{ij}\,\frac{\partial f}{\partial x^j}\Big). \tag{1.57}$$

Proposition 1.108 *For any* $u, v \in \Omega^0(M)$

$$\Delta(uv) = u\Delta v - 2g^*(du,\, dv) + v\Delta u.$$

By Proposition 1.106

$$\Delta f = -\operatorname{trace}\big(\operatorname{Hess}(f)\big) \tag{1.58}$$

where

$$\operatorname{Hess}(f) := \nabla\, df.$$

With respect to a local g-orthonormal frame $\{E_i \,:\, 1 \leq i \leq n\}$, defined on an open set $U \subset M$

$$\Delta f = -\sum_{i=1}^{n} \operatorname{Hess}(f)(E_i,\, E_i). \tag{1.59}$$

The expression (1.59) allows for the following geometric interpretation. Let $x \in U$ and let us set $v_i = E_{i,x}$. Let $\gamma_i(t)$ be the geodesic of initial data

$$\gamma_i(0) = x, \quad \frac{d\gamma_i}{dt}(0) = v_i\,.$$

Then[27]

$$(\Delta f)(x) = -\sum_{i=1}^{n} \frac{d^2(f \circ \gamma_i)}{dt^2}(0). \tag{1.60}$$

The Laplacian is a Riemannian invariant, i.e., for every isometry $\phi : M \to N$ of two Riemannian manifolds M and N

$$\Delta_M(f \circ \phi) = (\Delta_N f) \circ \phi, \quad f \in \Omega^0(N). \tag{1.61}$$

Proposition 1.109 *If $p : M \to N$ is a Riemannian submersion with totally geodesic[28] fibers then*

$$\Delta_M(f \circ \phi) = (\Delta_N f) \circ \phi \tag{1.62}$$

for any $f \in \Omega^0(N)$.

1.8.1 Bochner–Lichnerowicz Formula

Let (M, g) be a compact, orientable, Riemannian manifold. As d and δ are one the formal adjoint of the other

$$\langle \Delta u,\ v\rangle = \langle u,\ \Delta v\rangle, \tag{1.63}$$

$$\langle \Delta u,\ u\rangle = \|du\|^2. \tag{1.64}$$

[27] If $\gamma(t)$ is a geodesic with $\gamma(0) = x$ and $\dot{\gamma}(0) = v$ then

$$\frac{d(f \circ \gamma)}{dt}(t) = \frac{\partial f}{\partial x^i}(\gamma(t))\frac{d\gamma^i}{dt}(t),$$

$$\frac{d^2(f \circ \gamma)}{dt^2}(t) = \frac{\partial^2 f}{\partial x^i\,\partial x^j}(\gamma(t))\,\frac{d\gamma^i}{dt}(t)\,\frac{d\gamma^j}{dt}(t) + \frac{\partial f}{\partial x^i}(\gamma(t))\frac{d^2\gamma^i}{dt^2}(t) =$$

$$= \left[\frac{\partial^2 f}{\partial x^j\,\partial x^k}(\gamma(t)) - \frac{\partial f}{\partial x^i}(\gamma(t))\left\{ \begin{matrix} i \\ jk \end{matrix} \right\}(\gamma(t))\right]\frac{d\gamma^j}{dt}(t)\,\frac{d\gamma^k}{dt}(t) =$$

$$= \text{Hess}(f)_{\gamma(t)}\big(\dot{\gamma}(t),\ \dot{\gamma}(t)\big),$$

$$\frac{d^2(f \circ \gamma)}{dt^2}(0) = \text{Hess}(f)_x\big(v,\ v\big).$$

[28] Cf. the next section.

Proposition 1.110 *For every* $f \in \Omega^0(M)$

$$-\frac{1}{2}\,\Delta\big(|df|^2\big) = \big|\mathrm{Hess}(f)\big|^2 - |\Delta f|^2 + \mathrm{Ric}\big((df)^\sharp,\ (df)^\sharp\big). \tag{1.65}$$

The proof of (1.65) follows from:

Lemma 1.111 *For every* $\alpha \in \Omega^1(M)$ *and any* $X, Y \in \mathfrak{X}(M)$

$$\nabla_X \nabla_Y \alpha - \nabla_Y \nabla_X \alpha - \nabla_{[X,Y]}\alpha = \big(R(X,Y)\alpha^\sharp\big)^\flat .$$

It is convenient to prove (1.65) on a neighborhood $U \subset M$ of every point $x_0 \in M$ endowed with a g-orthonormal frame $\{E_i\ :\ 1 \le i \le n\}$ build by parallel translation of a given g_{x_0}-orthonormal basis $\{v_i\ :\ 1 \le i \le n\} \subset T_{x_0}(M)$. Parallel translation is meant along the geodesics $\gamma_i(t)$ of initial data

$$\gamma_i(0) = x_0, \quad \dot\gamma_i(0) = v_i .$$

Said local frame has the advantage that

$$\big(\nabla_{E_i} E_j\big)_{x_0} = 0, \tag{1.66}$$

resulting in the customary simplification of calculations.

Definition 1.112 The *Laplacian* $\Delta : \Omega^p(M) \to \Omega^p(M)$ is defined by

$$\Delta\alpha = d\delta\alpha + \delta d\alpha \tag{1.67}$$

for any $\alpha \in \Omega^p(M)$. □

The Bochner formula (1.65) generalizes to differential forms of arbitrary degree[29]

$$-\frac{1}{2}\Delta\big(|\alpha|^2\big) = \big|\nabla\alpha\big|^2 - g^*\big(\alpha,\ \Delta\alpha\big) + \mathrm{WBL}(R,\ \alpha) \tag{1.68}$$

for every $\alpha \in \Omega^p(M)$. Here $\mathrm{WBL}(R,\ \alpha)$ is quadratic in α and linear in the curvature tensor R.

1.8.2 Laplacian in Radial Coordinates

Let $B_\epsilon(x_0) \subset M$ be a metric ball of center x_0 and radius $\epsilon > 0$ sufficiently small, such that the exponential map $\exp_{x_0}$ is a diffeomorphism onto $B_\epsilon(x_0)$. Let f be a

[29] Cf. formulas (2)–(3), p. 4, by A. Lichnerowicz, *Géométrie des groups de transformations* (Dunod, Paris, 1958).

C^∞ function on $B_\epsilon(x_0)$, depending only on the distance from x_0, i.e., there is a C^∞ function $\varphi : [0, \epsilon) \to \mathbb{R}$ such that

$$f(x) = \varphi\big(r(x)\big), \quad r(x) = d(x_0, x), \quad x \in B_\epsilon(x_0). \tag{1.69}$$

Let θ be the function in (1.51). Let us set

$$\theta' = \frac{\partial\theta}{\partial r} := g^*\big(dr, d\theta\big).$$

Proposition 1.113

$$\Delta f = -\frac{d^2\varphi}{dr^2} - \frac{d\varphi}{dr}\Big(\frac{\theta'}{\theta} + \frac{n-1}{r}\Big)$$

at every point $x \in B_\epsilon(x_0)$, $x \neq x_0$.

Proof Let $x \in B_\epsilon(x_0)$. There is a unique geodesic γ, parametrized by arc length s, joining x_0 to x. Let $v_1 \in T_x(M)$ be the tangent vector to γ at its point x. By the very definition of r

$$\gamma\big(r(x)\big) = x, \quad \frac{d\gamma}{ds}\big(r(x)\big) = v_1 .$$

Let us complete v_1 to a g_x-orthonormal basis

$$\big\{v_1, v_2, \dots, v_n\big\} \subset T_x(M). \tag{1.70}$$

Let γ_i, $2 \le i \le n$ be the $n-1$ geodesics, parametrized by arc length α, of initial data

$$\gamma_i(0) = x, \quad \frac{d\gamma_i}{d\alpha}(0) = v_i, \quad 2 \le i \le n.$$

By (1.60)

$$(\Delta f)(x) = -\frac{d^2(f \circ \gamma)}{ds^2}(r(x)) - \sum_{i=2}^{n} \frac{d^2(f \circ \gamma_i)}{d\alpha^2}(0). \tag{1.71}$$

The first term on the right-hand side of (1.71) is

$$-\frac{d^2\varphi}{dr^2}(r(x)).$$

We are left with the calculation of

$$\frac{d^2(f \circ \gamma_i)}{d\alpha^2}(0).$$

For every fixed $2 \leq i \leq n$ and α sufficiently small, let C_α be the geodesic joining x_0 and $\gamma_i(\alpha)$. Then[30]

$$(f \circ \gamma_i)(\alpha) = \varphi\big(\ell(C_\alpha)\big) \tag{1.72}$$

where $\ell(C_\alpha)$ is the length of C_α. Then [by (1.72)]

$$\frac{d^2(f \circ \gamma_i)}{d\alpha^2}(0) = \frac{d^2\varphi}{ds^2}\big(r(x)\big)\left[\frac{d}{d\alpha}\Big\{\ell(C_\alpha)\Big\}_{\alpha=0}\right]^2 + \tag{1.73}$$

$$+\frac{d\varphi}{dr}\big(r(x)\big)\,\frac{d^2}{d\alpha^2}\Big\{\ell(C_\alpha)\Big\}_{\alpha=0}.$$

As v_i is orthogonal to γ

$$\frac{d}{d\alpha}\Big\{\ell(C_\alpha)\Big\}_{\alpha=0} = 0.$$

As to the term $\frac{d^2}{d\alpha^2}\Big\{\ell(C_\alpha)\Big\}_{\alpha=0}$ [by the second variation formula for the length functional]

$$\frac{d^2}{d\alpha^2}\Big\{\ell(C_\alpha)\Big\}_{\alpha=0} = g\big(Y_i(r(x)),\ Y_i'(r(x))\big) \tag{1.74}$$

where $Y_i \in J_\gamma$ is the Jacobi field along γ determined by the 1-parameter family of geodesics C_α, that is

$$Y_i(0) = 0, \quad Y_i(r(x)) = v_i\ .$$

where $Y_i' = \nabla_{\dot\gamma} Y_i$. Then

$$(\Delta f)(x) = -\frac{d^2\varphi}{dr^2}\big(r(x)\big) + \tag{1.75}$$

$$-\frac{d\varphi}{dr}\big(r(x)\big)\sum_{i=2}^{n} g\big(Y_i\big(r(x)\big),\ Y_i'\big(r(x)\big)\big).$$

Next, we compute $\sum_{i=2}^{n} g\big(Y_i\big(r(x)\big),\ Y_i'\big(r(x)\big)\big)$ in terms of θ and its first derivative. Recall that θ is the determinant of the exponential map at x_0, hence θ^{-1} is the determinant of $\big(\exp_{x_0}\big)^{-1}$. Thus (by also applying the Gauss lemma[31])

---

[30] One has

$$(f \circ \gamma_i)(\alpha) = f\big(\gamma_i(\alpha)\big) = \varphi\big(r\big(\gamma_i(\alpha)\big)\big) = \varphi\big(\ell(C_\alpha)\big)$$

as C_α is the minimizing geodesic joining x_0 and $\gamma_i(\alpha)$ [that achieves the distance $r\big(\gamma_i(\alpha)\big)$ from x_0 to $\gamma_i(\alpha)$].

[31] By the Gauss lemma in Riemannian geometry, the exponential is radially an isometry.

$$\theta^{-1}(x) = \left|(d_x \exp_{x_0}^{-1})v_2 \wedge \cdots \wedge (d_x \exp_{x_0}^{-1})v_n\right|. \tag{1.76}$$

Thus, if x is the point $p = \gamma(s)$ on the curve γ

$$\theta^{-1}(p) = \frac{\left|(d_p \exp_{x_0}^{-1})Y_2(s) \wedge \cdots \wedge (d_p \exp_{x_0}^{-1})Y_n(s)\right|}{\left|Y_2(s) \wedge \cdots \wedge Y_n(s)\right|}. \tag{1.77}$$

Then $(d_p \exp_{x_0}^{-1})Y_i(s) \in T_{x_0}(M)$ is realized as transverse vector of the family of geodesics $\left\{\exp_{x_0}^{-1} \circ C_\alpha\right\}$, corresponding to the segments joining in $T_{x_0}(M)$ the origin 0_{x_0} to

$$r\,\dot\gamma(0) + \frac{\alpha}{\left|(d_x \exp_{x_0}^{-1})v_i\right|}\,(d_x \exp_{x_0}^{-1})v_i\,.$$

It follows that

$$(d_p \exp_{x_0}^{-1})Y_i(s) = \frac{s}{r}\,(d_x \exp_{x_0}^{-1})v_i \tag{1.78}$$

and hence

$$\theta^{-1}(p) = \frac{s^{n-1}}{r^{n-1}} \frac{\left|(d_x \exp_{x_0}^{-1})v_2 \wedge \cdots \wedge (d_x \exp_{x_0}^{-1})v_n\right|}{\left|Y_2(s) \wedge \cdots \wedge Y_n(s)\right|}. \tag{1.79}$$

Finally

$$\theta(p) = \frac{r^{n-1}}{s^{n-1}}\left|Y_2(s) \wedge \cdots \wedge Y_n(s)\right|\theta(x). \tag{1.80}$$

The derivative with respect to s of the determinant $\left|Y_2(s) \wedge \cdots \wedge Y_n(s)\right|$ equals $\sum_{i=1}^n g(Y_i(s),\ Y_i'(s))$. Hence

$$\frac{\partial\theta}{\partial s}(p) = -(n-1)\frac{r^{n-1}}{s^n}\left|Y_2(s) \wedge \cdots \wedge Y_n(s)\right|\theta(x) + \tag{1.81}$$

$$+\frac{r^{n-1}}{s^{n-1}}\Big[\sum_{i=2}^n g(Y_i(s),\ Y_i'(s))\Big]\theta(x).$$

Therefore, at the point x

$$\frac{\partial\theta}{\partial s}(x) = -\frac{n-1}{r}\,\theta(x) + \Big[\sum_{i=2}^n g(Y_i(s),\ Y_i'(s))\Big]\theta(x). \tag{1.82}$$

Then

$$\frac{\theta'}{\theta}(x) = \sum_{i=2}^{n} g(Y_i(s),\ Y_i'(s)) - \frac{n-1}{r}. \tag{1.83}$$

Consequently, (1.75) may be rewritten as

$$(\Delta f)(x) = -\frac{d^2\varphi}{ds^2}(r(x)) - \frac{d\varphi}{ds}(r(x))\Big[\frac{\theta'}{\theta}(x) + \frac{n-1}{r(x)}\Big]$$

or briefly

$$\Delta f = -\frac{d^2\varphi}{dr^2} - \frac{d\varphi}{dr}\Big(\frac{\theta'}{\theta} + \frac{n-1}{r}\Big).$$

Proposition 1.113 is proved.

1.9 Hodge–de Rham Theory

1.9.1 Star Isomorphism

Let $(M,\ g)$ be a n-dimensional, orientable, Riemannian manifold with the metric tensor g. Let $\big(U,\ x^i\big)$ be a local coordinate system on M. Every differential p-form $\alpha \in \Omega^p(M)$ may be locally represented as

$$\alpha = \frac{1}{p!} a_{i_1\cdots i_p}\, dx^{i_1} \wedge \cdots \wedge dx^{i_p} = a_{(i_1\cdots i_p)}\, dx^{i_1} \wedge \cdots \wedge dx^{i_p}, \quad a_{i_1\cdots i_p} \in C^\infty(U,\ \mathbb{R}),$$

on U. Writing $(i_1 \cdots i_p)$ implies that $i_1 < \cdots < i_p$. We shall need the (*generalized*) *Kronecker symbol*

$$\delta^{j_1\cdots j_p}_{i_1\cdots i_p} = \det \begin{pmatrix} \delta^{j_1}_{i_1} & \cdots & \delta^{j_1}_{i_p} \\ \vdots & & \vdots \\ \delta^{j_p}_{i_1} & \cdots & \delta^{j_p}_{i_p} \end{pmatrix}.$$

The Kronecker symbol enjoys the following properties[32]

(i) $\delta^{j_1\cdots j_p}_{i_1\cdots i_p}$ is skew-symetric both in the upper indices j_k and the lower indices i_k.
(ii) $\delta^{(j_1\cdots j_p)}_{(i_1\cdots i_p)} = \delta^{j_1}_{i_1} \ldots \delta^{j_p}_{i_p}$.

[32] It should be observed that (ii) $\Longleftrightarrow$ (iii) $\Longleftrightarrow$ (iv).

(iii) For every system of numbers $\{a_{(i_1\cdots i_p)} \in \mathbb{R} \,:\, 1 \le i_1 < \cdots < i_p \le n\}$

$$a_{(i_1\cdots i_p)} = \delta^{j_1\cdots j_p}_{i_1\cdots i_p}\, a_{(j_1\cdots j_p)}\,.$$

(iv) For every p-vector $a_{i_1\cdots i_p}$

$$a_{i_1\cdots i_p} = \frac{1}{p!}\,\delta^{j_1\cdots j_p}_{i_1\cdots i_p}\, a_{j_1\cdots j_p}\,.$$

If $\alpha = a_{(i_1\cdots i_p)}\, dx^{i_1} \wedge \cdots \wedge dx^{i_1}$ and $\beta = b_{(j_1\cdots j_q)}\, dx^{j_1} \wedge \cdots \wedge dx^{j_q}$ then

$$\alpha \wedge \beta = c_{i_1\cdots i_{p+q}}\, dx^{i_1} \wedge \cdots \wedge dx^{i_{p+q}}, \tag{1.84}$$

$$(p+q)!\, c_{i_1\cdots i_{p+q}} = \delta^{(j_1\cdots j_p)(k_1\cdots k_q)}_{i_1\cdots i_{p+q}}\, a_{j_1\cdots j_p}\, b_{k_1\cdots k_q},$$

$$(p+1)!\, d\alpha = \delta^{j\,(i_1\cdots i_p)}_{k_1\cdots k_{p+1}}\, \frac{\partial a_{(i_1\cdots i_p)}}{\partial u^j}\, dx^{k_1} \wedge \cdots \wedge dx^{k_{p+1}}\,. \tag{1.85}$$

The definition (1.84) readily implies

$$\alpha \wedge \beta = (-1)^{pq}\, \beta \wedge \alpha\,.$$

The *star isomorphism* $* : \Omega^p(M) \to \Omega^{n-p}(M)$ is

$$*\ \alpha = a^*_{(j_1\cdots j_{n-p})}\, dx^{j_1} \wedge \cdots \wedge dx^{j_{n-p}},$$

$$a^*_{j_1\cdots j_{n-p}} = \epsilon_{(i_1\cdots i_p) j_1\cdots j_{n-p}}\, a^{(i_1\cdots i_p)},$$

$$\epsilon_{i_1 i_2 \ldots i_n} = \delta^{1\,2\ldots n}_{i_1 i_2 \ldots i_n}\, \sqrt{G}, \quad G = \det\big[g_{ij}\big],$$

$$a^{i_1\cdots i_p} = g^{i_1 j_1} \cdots g^{i_p j_p}\, a_{j_1\cdots j_p}, \quad \big[g^{ij}\big] = \big[g_{ij}\big]^{-1}.$$

Then

$$*\ 1 = \epsilon_{1\cdots n}\, dx^1 \wedge \cdots \wedge dx^n = \sqrt{G}\, dx^1 \wedge \cdots \wedge dx^n\,.$$

The operator $*$ enjoys the following properties

(i) $*(\alpha+\beta) = *\alpha + *\beta$, $*(f\,\alpha) = f\ *\alpha$,
(ii) $*\ *\ \alpha = *\,(*\,\alpha) = (-1)^{pn+p}\,\alpha$,
(iii) $\alpha \wedge *\beta = \beta \wedge *\alpha$,
(iv) $\alpha \wedge *\alpha = 0 \Longleftrightarrow \alpha = 0$,

for any $\alpha,\ \beta \in \Omega^p(M)$ and any $f \in C^\infty(M)$. Also, if $\alpha = a_{(i_1\cdots i_p)}\, dx^{i_1} \wedge \cdots \wedge dx^{i_p}$ and $\beta = b_{(j_1\cdots j_p)}\, dx^{j_1} \wedge \cdots \wedge dx^{j_p}$ then

$$\alpha \wedge *\beta = a^{(i_1\cdots i_p)}\, b_{(i_1\cdots i_p)}\ *1. \tag{1.86}$$

The L^2 *inner product* on $\Omega^p(M)$ is defined by

$$(\alpha,\ \beta)_{L^2} = \int_M \alpha \wedge *\beta$$

whenever the integral is convergent. This is of course the case when we assume that M is compact. The L^2 inner product enjoys the following properties

(i) $(\alpha,\ \alpha)_{L^2} \geq 0$ and $= 0 \iff \alpha = 0$,
(ii) $(\alpha,\ \beta)_{L^2} = (\beta,\ \alpha)_{L^2}$,
(iii) The L^2 inner product is $\mathbb{R}$-bilinear,
(iv) $(*\alpha,\ *\beta)_{L^2} = (\alpha,\ \beta)_{L^2}$,

for any $\alpha,\ \beta \in \Omega^p(M)$. The forms α and β are *orthogonal* $(\alpha,\ \beta)_{L^2} = 0$.

1.9.2 Harmonic Forms

The operator $\delta : \Omega^p(M) \to \Omega^{p-1}(M)$ is defined by

$$\delta\alpha = (-1)^{np+n+1}\ * d * \alpha\,.$$

The *codifferential* $\delta\alpha$ of α enjoys the following properties

(i) δ is $\mathbb{R}$-linear,
(ii) $\delta\,\delta\,\alpha = 0$,
(iii) $*\,\delta\,\alpha = (-1)^p\, d * \alpha$ and $*d\alpha = (-1)^{p+1}\,\delta * \alpha$.

Definition 1.114 The form α is *coclosed* if $\delta\alpha = 0$ and *coexact* if $\alpha = \delta\beta$ for some $\beta \in \Omega^{p+1}(M)$ (one also says that α *cobounds* β). □

Definition 1.115 The form α is *harmonic*[33] if it is closed (i.e., $d\alpha = 0$) and coclosed ($\delta\alpha = 0$). □

Let $\Delta \equiv \delta\, d + d\,\delta : \Omega^p(M) \to \Omega^p(M)$ be the *Laplace-Beltrami* operator. K. Kodaira calls[34] α harmonic if $\Delta\alpha = 0$. Let $\Omega^p_H(M)$ be the set of all harmonic p-forms. Then $\Omega^p_H(M)$ is a subspace of $\Omega^p(M)$ yet not an algebra.

---

[33] The definition belongs to W.V.D. Hodge, *The Theory and Applications of Harmonic Integrals* (Cambridge University Press, London, New York, 1952).

[34] Cf. K. Kodaira, Harmonic fields in Riemannian manifolds (generalized potential theory). Ann. Math. **50**, 587–665 (1949).

1.9.3 Orthogonality Relations

Let us assume M to be compact and orientable. If $\alpha \in \Omega^p(M)$ and $\beta \in \Omega^{p+1}(M)$ then (by the Stokes theorem[35])

$$\int_M d\big(\alpha \wedge *\beta\big) = 0. \tag{1.87}$$

As

$$d\big(\alpha \wedge *\beta\big) = d\alpha \wedge *\beta + (-1)^p\, \alpha \wedge d * \beta$$

formula (1.87) becomes

$$\int_M d\alpha \wedge *\beta = (-1)^{p-1} \int_M \alpha \wedge d * \beta$$

or[36]

$$(d\alpha,\ \beta)_{L^2} = (\alpha,\ \delta\beta)_{L^2}, \tag{1.88}$$

i.e., δ is the formal adjoint of d. Equation (1.88) may also be written

$$(\alpha,\ d\beta)_{L^2} = (\delta\alpha,\ \beta)_{L^2}$$

for any $\alpha \in \Omega^p(M)$ and $\beta \in \Omega^{p-1}(M)$. As a consequence of (1.88).

Corollary 1.116 *A differential p-form α is closed if and only if α is orthogonal to every coexact p-form.*

Proof Let $\alpha \in \Omega^p(M)$ be closed, i.e., $d\alpha = 0$. Let $\gamma \in \Omega^p(M)$ be a coexact p-form, i.e., $\gamma = \delta\beta$ for some $\beta \in \Omega^{p+1}(M)$. Then [by (1.88)]

$$0 = (d\alpha,\ \beta)_{L^2} = (\alpha,\ \delta\beta)_{L^2} = (\alpha,\ \gamma)_{L^2},$$

i.e., α is orthogonal to γ. Viceversa, let $\alpha \in \Omega^p(M)$ be orthogonal to $\delta\, \Omega^{p+1}(M)$. Then, for every $\beta \in \Omega^{p+1}(M)$

$$0 = (\alpha,\ \delta\,\beta)_{L^2} = (d\alpha,\ \beta)_{L^2}$$

---

35 Cf. e.g., S.I. Goldberg, *op. cit.*, pp. 62–63.

36 Indeed (by the very definition of δ)

$$\delta\beta = (-1)^{n(p+1)+n+1} * d * \beta,$$

$$d * \beta \in \Omega^{n-p}(M) \stackrel{*}{\longrightarrow} \Omega^p(M) \stackrel{*}{\longrightarrow} \Omega^{n-p}(M), \quad * * d * \beta = (-1)^{(n-p)n+(n-p)}\, d * \beta,$$

$$(-1)^{p-1}\, d * \beta = (-1)^{p-1+(n-p)(n+1)} * * d * \beta = (-1)^{n(n+1)+2n} * \delta\,\beta = * \delta\,\beta.$$

and in particular for $\beta = d\alpha$

$$0 = (d\alpha,\, d\alpha)_{L^2} = \|d\alpha\|^2_{L^2} \Longrightarrow d\alpha = 0.$$

Q.e.d.

Similar to the above.

Corollary 1.117 *A differential form is coclosed if and only if is orthogonal to all exact forms.*

Next, we may easily show that on a compact manifold the definitions of harmonicity as given by Hodge and Kodaira are equivalent, i.e.,

Corollary 1.118 *Let M be a compact orientable Riemannian manifold. Let $\alpha \in \Omega^p(M)$. Then*

$$\alpha \in \Omega^p_H(M) \Longleftrightarrow \Delta\alpha = 0.$$

Proof Let α such that $\Delta\alpha = 0$. Then

$$0 = (\Delta\alpha,\, \alpha)_{L^2} = (\delta d\alpha, \alpha)_{L^2} + (d\delta\alpha,\, \alpha)_{L^2} = (d\alpha,\, d\alpha)_{L^2} + (\delta\alpha,\, \delta\alpha)_{L^2} = \|d\alpha\|^2_{L^2} + \|\delta\alpha\|^2_{L^2}$$

yielding $d\alpha = 0$ and $\delta\alpha = 0$. Q.e.d. The converse is trivial.

1.9.4 Hodge–de Rham Decomposition Theorem

Let M be a compact orientable Riemannian manifold, and let $\Omega^p_d(M)$ and $\Omega^p_\delta(M)$ denote the spaces of d-exact, respectively δ-exact, p-forms on M.

Theorem 1.119 *Let $\beta \in \Omega^p(M)$. The following statements are equivalent*

(i) *There is $\alpha \in \Omega^p(M)$ such that $\Delta\alpha = \beta$.*
(ii) *β is orthogonal to $\Omega^p_H(M)$.*

Proof (i)$\Longrightarrow$(ii). Let α such that $\Delta\alpha = \beta$. Then for any harmonic form $\gamma \in \Omega^p_H(M)$ [i.e., $\Delta\gamma = 0$]

$$(\beta,\, \gamma)_{L^2} = (\Delta\alpha,\, \gamma)_{L^2} = (\alpha,\, \Delta\gamma)_{L^2} = 0.$$

Q.e.d.

The opposite implication (ii) $\Longrightarrow$ (i) is highly nontrivial[37] and was first provided by W.V.D. Hodge.[38] Cf. also S.I. Goldberg, *op. cit.*, Appendix C, pp. 296–300.

---

[37] Relies on Fredholm's theory of integral equations.

[38] Cf. *op. cit.*

As $\Omega^p_H(M)$ is finite dimensional,[39] say $\dim_{\mathbb{R}} \Omega^p_H(M) = m$, one may consider a basis $\{\varphi_1, \ldots, \varphi_m\}$ of harmonic forms that is orthonormal

$$\big(\varphi_i, \varphi_j\big)_{L^2} = \delta_{ij} .$$

For every $\alpha \in \Omega^p(M)$ we set

$$\alpha_H := \sum_{i=1}^{m} \big(\alpha, \varphi_i\big)_{L^2} \varphi_i \in \Omega^p_H(M).$$

We claim that $\alpha - \alpha_H$ is orthogonal to $\Omega^p_H(M)$. Indeed

$$\big(\alpha - \alpha_H, \varphi_j\big)_{L^2} = \big(\alpha, \varphi_j\big)_{L^2} - \Big(\sum_{i=1}^{m} \big(\alpha, \varphi_i\big)_{L^2}, \varphi_j\Big)_{L^2} = 0.$$

Q.e.d. Consequently (by Theorem 1.119) there is a form γ such that

$$\Delta\gamma = \alpha - \alpha_H .$$

Let us set

$$\alpha_d := d\,\delta\,\gamma \in \Omega^p_d(M), \quad \alpha_\delta := \delta\, d\,\gamma \in \Omega^p_\delta(M),$$

so that

$$\alpha = \alpha_d + \alpha_\delta + \alpha_H . \tag{1.89}$$

The decomposition (1.89) is easily seen to be also unique. Formula (1.89) is the *Hodge-de Rham Decomposition Theorem*.

1.9.5 The Operators d, δ and Δ in Terms of Covariant Derivatives

Let α be a p-form locally expressed as

$$\alpha = \frac{1}{p!} a_{i_1 \cdots i_p} dx^{i_1} \wedge \cdots \wedge dx^{i_p} = a_{(i_1 \cdots i_p)} dx^{i_1} \wedge \cdots \wedge dx^{i_p} .$$

[39] Cf. again S.I. Goldberg, *op. cit.*, Appendix C, p. 299.

The covariant derivative of α [with respect to the Levi-Civita connection ∇^g of $(M,\ g)$] is

$$D_i\, a_{i_1\cdots i_p} := \left(\nabla^g_{\partial_i}\alpha\right)\left(\partial_{i_1},\ \ldots,\ \partial_{i_p}\right) =$$

$$= \frac{\partial a_{i_1\cdots i_p}}{\partial x^i} - \sum_{\sigma=1}^{p} \left\{ {j \atop i_\sigma\, i} \right\} \alpha_{i_1\cdots i_{\sigma-1}\, j\, i_{\sigma+1}\cdots i_p}\,. \tag{1.90}$$

The exterior derivative of α is [by (1.85)]

$$d\alpha = (d\,a)_{(i_1\cdots i_{p+1})}\, dx^{i_1}\wedge\cdots\wedge dx^{i_{p+1}}, \quad (d\,a)_{i_1\cdots i_{p+1}} = \delta^{j\,(j_1\cdots j_p)}_{i_1\cdots i_{p+1}}\, \frac{\partial\, a_{(j_1\cdots j_p)}}{\partial x^j}\,.$$

Let us substitute the ordinary (partial) derivatives in terms of covariant derivatives [from (1.90)] and observe the cancelation of the Christoffel symbols

$$(d\,a)_{i_1\cdots i_{p+1}}\, dx^{i_1}\wedge\cdots\wedge dx^{i_{p+1}} =$$

$$= \delta^{j\,(j_1\cdots j_p)}_{i_1\cdots i_{p+1}} \left[D_j\, a_{(j_1\cdots j_p)} + \sum_{\sigma=1}^{p} \left\{ {k \atop j_\sigma\, j} \right\} \alpha_{j_1\cdots j_{\sigma-1}\, k\, j_{\sigma+1}\cdots j_p} \right] dx^{i_1}\wedge\cdots\wedge dx^{i_{p+1}} =$$

$$= \delta^{j\,(j_1\cdots j_p)}_{i_1\cdots i_{p+1}}\, D_j\, a_{(j_1\cdots j_p)}\, dx^{i_1}\wedge\cdots\wedge dx^{i_{p+1}}\,.$$

The codifferential of α is

$$\delta\alpha = (\delta\alpha)_{(i_1\cdots i_{p-1})}\, dx^{i_1}\wedge\cdots\wedge dx^{i_{p-1}},$$

$$(\delta\alpha)_{i_1\cdots i_{p-1}} = (-1)^{np+n+1}\left(*\, d\, *\, \alpha\right)_{i_1\cdots i_{p-1}} = -g^{ij}\, \delta^{(j_1\cdots j_p)}_{i\, i_1\cdots i_{p-1}}\, D_j\, a_{(j_1\cdots j_p)}\,.$$

The Laplace-Beltrami operator $\Delta\alpha = \delta\, d\, \alpha + d\, \delta\, \alpha$ is

$$\left(\Delta\alpha\right)_{i_1\cdots i_p} = -g^{ij}\, D_j\, D_i\, a_{i_1\cdots i_p} + \sum_{\sigma=1}^{p} a_{i_1\cdots i_{\sigma-1}\, j\, i_{\sigma+1}\cdots i_p}\, R^j_{i_\sigma} +$$

$$+ \frac{1}{2}\sum_{\sigma=1}^{p}\sum_{\rho=1}^{p} a_{i_1\cdots i_{\rho-1}\, j\, i_{\rho+1}\cdots i_{\sigma-1}\, i\, i_{\sigma+1}\cdots i_p}\, R^{ij}{}_{i_\rho\, i_\sigma},$$

$$R^{ij}{}_{k\ell} = g^{jm}\, R^i{}_{mk\ell},$$

and in particular for $p = 0$

$$\Delta f = -g^{ij}\, D_j\, D_i\, f.$$

1.9.6 Green's Operator

Let M be a compact orientable Riemannian manifold.

Theorem 1.120 (i) *For every $\alpha \in \Omega^p(M)$ there is a unique harmonic p-form $H[\alpha] \in \Omega^p_H(M)$ such that*

$$(\alpha, \beta)_{L^2} = (H[\alpha], \beta)_{L^2}$$

for any $\beta \in \Omega^p_H(M)$.

(ii) *For any $\alpha \in \Omega^p(M)$*

$$H\big[H[\alpha]\big] = H[\alpha].$$

(iii) *For every $\alpha \in \Omega^p(M)$ there is a solution $\beta \in \Omega^p(M)$ to the equation*

$$\Delta\beta = \alpha - H[\alpha]. \tag{1.91}$$

Any two solutions differ by a harmonic function, hence there is a unique solution β to (1.91) *that is orthogonal to $\Omega^p_H(M)$.*

Proof (i) Let $\alpha \in \Omega^p(M)$. Then (by the Hodge-de Rham decomposition theorem) α decomposes uniquely as

$$\alpha = \alpha_d + \alpha_\delta + \alpha_H \in \Omega^p_d(M) \oplus \Omega^p_\delta(M) \oplus \Omega^p_H(M)$$

hence we may define $H : \Omega^p(M) \to \Omega^p(M)$ by $H[\alpha] := \alpha_H$. Then $\alpha - H[\alpha]$ is orthogonal of $\Omega^p_H(M)$, so that for any $\beta \in \Omega^p_H(M)$

$$0 = (\alpha - H[\alpha], \beta)_{L^2} = (\alpha, \beta)_{L^2} - (H[\alpha], \beta)_{L^2}.$$

Q.e.d.

(ii) Let $\{\varphi_i : 1 \le i \le m\}$ be a L^2 orthonormal basis in $\Omega^p_H(M)$. For every $\alpha \in \Omega^p(M)$

$$H\big[H[\alpha]\big] = \sum_{j=1}^{m} \big(H[\alpha], \varphi_j\big)_{L^2} \varphi_j =$$

$$= \sum_{j=1}^{m} \left(\sum_{i=1}^{m} (\alpha, \varphi_i)_{L^2} \varphi_i, \varphi_j \right)_{L^2} = \sum_{i,j=1}^{m} \delta_{ij} (\alpha, \varphi_i)_{L^2} \varphi_j = \alpha.$$

Q.e.d.

(iii) Let $\alpha \in \Omega^p(M)$ and let $\beta, \beta' \in \Omega^p(M)$ such that

$$\Delta\beta = \alpha - H[\alpha], \quad \Delta\beta' = \alpha - H[\alpha], \tag{1.92}$$

hence $\Delta(\beta - \beta') = 0$, i.e., $\beta - \beta'$ is a harmonic form. In other words, β and β' differ by a harmonic form. Next, let β and β' be two forms satisfying (1.92) and in addition to that both β and β' are orthogonal to $\Omega^p_H(M)$. In particular β and β' are orthogonal to $\beta - \beta'$. Then

$$\|\beta - \beta'\|^2 = (\beta,\ \beta - \beta')_{L^2} - (\beta',\ \beta - \beta')_{L^2} = 0 \Longrightarrow \beta = \beta'.$$

Q.e.d.

The unique solution β to (1.91), orthogonal to $\Omega^p_H(M)$, furnished by (iii) in Theorem 1.120 is denoted by $\beta =: G\,\alpha$. Summing up

$$\alpha = \Delta(G\,\alpha) + H[\alpha], \quad (G\,\alpha,\ \gamma)_{L^2} = 0, \quad \gamma \in \Omega^p_H(M).$$

Definition 1.121 $G : \Omega^p(M) \to \Omega^p(M)$ is the *Green operator*. □

Theorem 1.122 (i) $H\big[G\,\alpha\big] = 0$ *for any* $\alpha \in \Omega^p(M)$.

(ii) *The operators* H *and* G *commute with* d, δ, Δ *and* $*$.

(iii) G *is self-adjoint, i.e.,*

$$\big(G\,\alpha,\ \beta\big)_{L^2} = \big(\alpha,\ G\,\beta\big)_{L^2}$$

for any $\alpha,\ \beta \in \Omega^p(M)$.

(iv) $\big(G\,\alpha,\ \alpha\big)_{L^2} \geq 0$ *with equality if and only if* α *is harmonic, i.e.,*

$$\big(G\,\alpha,\ \alpha\big)_{L^2} = 0 \Longleftrightarrow \alpha \in \Omega^p_H(M).$$

Exercise 1.123 Prove Theorem 1.122. □

1.9.7 Killing Vector Fields

Let $(M,\ g)$ be a compact orientable Riemannian manifold, and let $X \in \mathfrak{X}(M)$ be a tangent vector field. Let $\overline{\mathscr{L}}_X : \Omega^p(M) \to \Omega^p(M)$ be the formal adjoint of $\mathscr{L}_X$ [the Lie derivative at the direction X], i.e.,

$$\big(\overline{\mathscr{L}}_X\alpha,\ \beta\big)_{L^2} = \big(\alpha,\ \mathscr{L}_X\beta\big)_{L^2}$$

for any $\alpha,\ \beta \in \Omega^p(M)$. On the other hand [by (1.86)]

$$\big(\mathscr{L}_X\alpha,\ \beta\big)_{L^2} = \int_M g^*\big(\mathscr{L}_X\alpha,\ \beta\big) \ *1 = \int_M \big(\mathscr{L}_X\alpha\big) \wedge *\beta =$$

[as $\mathscr{L}_X$ is a derivation]

$$= \int_M \left\{ \mathscr{L}_X(\alpha \wedge *\beta) - \alpha \wedge \mathscr{L}_X(*\,\beta) \right\}.$$

Note that $\alpha \wedge *\beta$ is a top degree form on M, hence $d\,\alpha \wedge *\beta = 0$. Then [by Cartan's formula $\mathscr{L}_X = d \circ i_X + i_X \circ d$ and the Stokes theorem]

$$\int_M \mathscr{L}_X(\alpha \wedge *\beta) = \int_M d\, i_X\, \alpha \wedge *\beta = 0.$$

Finally[40]

$$\left(\alpha,\, \overline{\mathscr{L}}_X\beta\right)_{L^2} = -\int_M \alpha \wedge \mathscr{L}_X(*\,\beta) = \int_M \alpha \wedge (-1)^{(n-1)(n-p)+1} \;*\,*\mathscr{L}_X\,*\,\beta =$$

$$= (-1)^{(n+1)(n-p)+1} \int_M h^*\left(\alpha,\, *\mathscr{L}_X\,*\,\beta\right)\; *1$$

yielding

$$\overline{\mathscr{L}}_X\beta = (-1)^{np+p+1}\; *\,\mathscr{L}_X\,*\,\beta\,. \tag{1.93}$$

Another useful expression of $\overline{\mathscr{L}}_X$ is got from Cartan's formula[41]

$$\left(\alpha,\, \overline{\mathscr{L}}_X\beta\right)_{L^2} = \left(\mathscr{L}_X\alpha,\, \beta\right)_{L^2} = \left(d\, i_X\alpha + i_X\, d\,\alpha,\, \beta\right)_{L^2} =$$

$$= \left(i_X\,\alpha,\, \delta\,\beta\right)_{L^2} + \left(d\,\alpha,\, \epsilon(X^\flat)\,\beta\right)_{L^2} = \left(\alpha,\, \epsilon(X^\flat)\,\delta\,\beta + \delta\,\epsilon(X^\flat)\,\beta\right)_{L^2}$$

so that

$$\overline{\mathscr{L}}_X = \epsilon(\eta)\,\delta + \delta\,\epsilon(\eta) \tag{1.94}$$

where $\eta = X^\flat$ (or $X = \eta^\sharp$). It follows that[42]

$$\mathscr{L}_X\alpha + \overline{\mathscr{L}}_X\alpha = \tag{1.95}$$

$$= (\delta\eta)\,\alpha + \sum_{\rho=1}^{p} g^{jk}\left(\mathscr{L}_X g\right)_{k\, i_\rho}\, \alpha_{i_1\cdots i_{\rho-1}\, j\, i_{\rho+1}\cdots i_p}\, dx^{i_1} \wedge \cdots \wedge dx^{i_p}\,.$$

---

[40] Because of

$$\mathscr{L}_X * \beta \in \Omega^{n-p}(M) \xrightarrow{*} \Omega^p(M) \xrightarrow{*} \Omega^{n-p}(M), \quad ** = (-1)^{(n-p)n+(n-p)}\,\mathrm{id}.$$

[41] Also $\epsilon(\eta)\,\beta = \eta \wedge \beta$ and $\left(i_X\alpha,\, \beta\right)_{L^2} = \left(\alpha,\, \epsilon(\eta)\,\beta\right)_{L^2}$ with $X = \eta^\sharp$.

[42] Cf. also (3.7.11) in S.I. Goldberg, *op. cit.*, p. 109.

Definition 1.124 Let $L \subset \mathfrak{X}(M)$ be a Lie subalgebra. A p-form $\alpha \in \Omega^p(M)$ is said to be L-*invariant* if $\mathscr{L}_X \alpha = 0$ for every $X \in L$. □

Definition 1.125 A tangent vector field $X \in \mathfrak{X}(M)$ is a *Killing vector field* if $\mathscr{L}_X g = 0$. □

Let $\mathfrak{i}(M, g)$ be the Lie algebra of all Killing vector fields (infinitesimal motions) of (M, g). As a corollary to (1.95).

Theorem 1.126 *Let M be a compact orientable Riemannian manifold. The harmonic forms on M are $\mathfrak{i}(M, g)$-invariant differential forms, i.e., $\mathscr{L}_X \alpha = 0$ for any $\alpha \in \Omega^p_H(M)$ and any Killing vector field $X \in \mathfrak{i}(M, g)$.*

1.10 Compact Lie Groups

1.10.1 Grassman Algebra of a Lie Group

Let G be a compact, connected, n-dimensional Lie group, and let $\mathfrak{g}$ be its Lie algebra. Let $\mathfrak{g}^*$ be the dual of $\mathfrak{g}$. Let $\Omega^1(G)_{\text{l.i.}}$ be the space of all left invariant differential 1-forms on G. For every $a \in G$ let $j_a : \mathfrak{g} \simeq T_a(G)$ be the isomorphism given by $j_a(A) = A_a$, for every left invariant vector field $A \in \mathfrak{g}$.

Lemma 1.127 *The map*

$$\alpha \in \mathfrak{g}^* \longmapsto \tilde{\alpha} \in \Omega^1(G), \quad \tilde{\alpha}_a(v) := \alpha\big(j_a^{-1} v\big), \quad v \in T_a(G), \quad a \in G,$$

is $\Omega^1(G)_{\text{l.i.}}$-valued, and determines an isomorphism

$$\mathfrak{g}^* \simeq \Omega^1(G)_{\text{l.i.}} \,. \tag{1.96}$$

Proof For any $a, b \in G$ the diagram

$$\begin{array}{ccc} T_b(G) & \xrightarrow{d_b L_a} & T_{ab}(G) \\ j_b \uparrow & & \uparrow j_{ab} \\ \mathfrak{g} & = & \mathfrak{g} \end{array}$$

is commutative, where $L_a : G \to G$ is the left translation with $a \in G$. Hence, for any $v \in T_b(G)$

$$\big(L_a^* \tilde{\alpha}\big)_b \, v = \tilde{\alpha}_{ab} \, (d_b L_a) \, v = \alpha\big(j_{ab}^{-1} \, (d_b L_a) \, v\big) = \alpha\big(j_b^{-1} \, v\big) = \tilde{\alpha}_b \, v,$$

i.e., $\tilde{\alpha}$ is left invariant. Q.e.d.

Lemma 1.128 *For any $\alpha \in \mathfrak{g}^*$ and any $A \in \mathfrak{g}$ the function*

$$a \in G \longmapsto \tilde{\alpha}_a(A_a) \in \mathbb{R} \tag{1.97}$$

is constant.

Proof Indeed the function (1.97) equals identically $\alpha(A) \in \mathbb{R}$. Q.e.d.

By taking into account the isomorphism (1.96) the left invariant differential 1-form $\tilde{\alpha}$ is often denoted by the same symbol α.

Corollary 1.129 *For any $\alpha \in \mathfrak{g}^*$ and any $X, Y \in \mathfrak{g}$*

$$(d\alpha)(X, Y) = -\frac{1}{2}\alpha([X, Y]). \tag{1.98}$$

Let $\{X_\alpha : 1 \le \alpha \le n\} \subset \mathfrak{g}$ be a linear basis, and let $\{\omega^\alpha : 1 \le \alpha \le n\} \subset \mathfrak{g}^*$ be the dual basis, i.e., $\omega^i(X_j) = \delta^i_j$. The forms α in $\mathfrak{g}^*$ or $\Omega^1(G)_{\text{l.i.}}$ are referred[43] to as the *Maurer-Cartan forms*. As $\mathfrak{g}$ is a Lie algebra, there exist constants ${C_{\beta\gamma}}^\alpha \in \mathbb{R}$, uniquely determined by

$$[X_\beta, X_\gamma] = {C_{\beta\gamma}}^\alpha X_\alpha$$

[the *structure constants* of $\mathfrak{g}$ with respect to the basis $\{X_\alpha\}$]. Clearly

$$C^\alpha_{\beta\gamma} = -C^\alpha_{\gamma\beta} \tag{1.99}$$

[as the Lie product is skew-symmetric].

Corollary 1.130

$$d\omega^\alpha = -\frac{1}{2} C^\alpha_{\beta\gamma}\, \omega^\beta \wedge \omega^\gamma . \tag{1.100}$$

Proof By (1.98)

$$(d\omega^\alpha)(X_\beta, X_\gamma) = -\frac{1}{2}\omega^\alpha([X_\beta, X_\gamma]) = -\frac{1}{2}{C_{\beta\gamma}}^\alpha,$$

$$C^\alpha_{\rho\sigma}(\omega^\rho \wedge \omega^\sigma)(X_\beta, X_\gamma) = \frac{1}{2} C^\alpha_{\rho\sigma}(\omega^\rho \otimes \omega^\sigma - \omega^\sigma \otimes \omega^\rho)(X_\beta, X_\gamma) =$$

$$= \frac{1}{2} C^i_{\rho\sigma}\left\{\delta^\rho_\beta \delta^\sigma_\gamma - \delta^\sigma_\gamma \delta^\rho_\beta\right\} = C^\alpha_{\beta\gamma},$$

yielding (1.100). Q.e.d.

Definition 1.131 The equations (1.100) are called the *Maurer-Cartan equations.* □

[43] According to the classical terminology, cf. e.g., S.I. Goldberg, *Curvature and Homology* (Dover Publications, Inc., New York, 1982), p. 133.

The Jacobi identity in $\mathfrak{g}$

$$[X_\alpha, [X_\beta, X_\gamma]] + [X_\beta, [X_\gamma, X_\alpha]] + [X_\gamma, [X_\alpha, X_\beta]] = 0$$

yields further relations between the structure constants

$$C^\rho_{\beta\gamma} C^\sigma_{\alpha\rho} + C^\rho_{\gamma\alpha} C^\sigma_{\beta\rho} + C^\rho_{\alpha\beta} C^\sigma_{\gamma\rho} = 0.$$

Pullback by a C^1 map and exterior differentiation commute, hence for every Maurer-Cartan form α

$$L_a^* d\,\alpha = d\big(L_a^*\,\alpha\big) = d\,\alpha,$$

i.e., if α is left invariant then so does $d\alpha$. In other words

$$d\Big[\Omega^1(G)_{\text{l.i.}}\Big] \subset \Omega^2(G)_{\text{l.i.}}\,.$$

For the same reason

$$d\Big[\Omega^p(G)_{\text{l.i.}}\Big] \subset \Omega^{p+1}(G)_{\text{l.i.}}, \quad p \geq 1.$$

Lemma 1.132 (i) *For every $p \in \mathbb{N}$ and any set of constants*

$$\Big\{A_{\alpha_1 \dots \alpha_p} \in \mathbb{R} \;:\; 1 \leq \alpha_1, \,\dots,\, \alpha_p \leq n\Big\}$$

the differential p-form

$$A_{\alpha_1 \dots \alpha_p}\, \omega^{\alpha_1} \wedge \cdots \wedge \omega^{\alpha_p} \tag{1.101}$$

is left invariant, and (ii) *every left invariant p-form may be expressed in the form* (1.101).

Lemma 1.133 (i) *The map*

$$\omega \in \Lambda^p\big(\mathfrak{g}^*\big) \longmapsto \tilde{\omega} \in \Omega^p(G),$$

$$\tilde{\omega}\big(A_1, \,\dots,\, A_p\big)_a = \omega\big(A_1, \,\dots,\, A_p\big), \quad A_1, \,\dots,\, A_p \in \mathfrak{g}, \quad a \in G,$$

is $\Omega^p(G)_{\text{l.i.}}$-valued and determines an isomorphism

$$\Lambda^p\big(\mathfrak{g}^*\big) \simeq \Omega^p(G)_{\text{l.i.}}\,.$$

In particular $\dim_{\mathbb{R}} \Omega^p(G)_{\text{l.i.}} = \binom{n}{p}$.

(ii) *The exterior differential* $d : \Omega^p(G)_{\text{l.i.}} \to \Omega^{p+1}(G)_{\text{l.i.}}$ *induces a* $\mathbb{R}$*-linear map* $d : \Lambda^p(\mathfrak{g}^*) \to \Lambda^{p+1}(\mathfrak{g}^*)$ *such that the diagram*

$$\begin{array}{ccc} \Omega^p(G)_{\text{l.i.}} & \xrightarrow{d} & \Omega^{p+1}(G)_{\text{l.i.}} \\ \uparrow & & \uparrow \\ \Lambda^p(\mathfrak{g}^*) & \xrightarrow{d} & \Lambda^{p+1}(\mathfrak{g}^*) \end{array}$$

is commutative.

Lemma 1.134 *Every Lie group is an orientable manifold.*

Indeed $\omega^1 \wedge \cdots \wedge \omega^n$ is a top degree nowhere zero form on G.

1.10.2 Invariant Differential Forms

For every $X \in \mathfrak{g}$ we consider the map

$$\operatorname{ad}(X) : \mathfrak{g} \to \mathfrak{g}, \quad \operatorname{ad}(X)\, Y := [X, Y], \quad Y \in \mathfrak{g}.$$

Clearly $\operatorname{ad}(X)$ is a $\mathbb{R}$-linear map.

Lemma 1.135 *The map*

$$\operatorname{ad} : \mathfrak{g} \to \operatorname{Hom}_{\mathbb{R}}(\mathfrak{g}), \quad X \in \mathfrak{g} \longmapsto \operatorname{ad}(X),$$

is a representation.[44]

Proof $\operatorname{Hom}_{\mathbb{R}}(\mathfrak{g})$ is organized as an algebra with the ordinary commutator of two linear maps $\mathscr{A}, \mathscr{B} \in \operatorname{Hom}_{\mathbb{R}}(\mathfrak{g})$

$$[\mathscr{A}, \mathscr{B}] := \mathscr{A} \circ \mathscr{B} - \mathscr{B} \circ \mathscr{A}.$$

Then for any $X, Y, Z \in \mathfrak{g}$ (by the Jacobi identity)

$$\operatorname{ad}([X, Y])\, Z = [[X, Y], Z] = -[[Y, Z], X] - [[Z, X], Y] =$$

$$= \operatorname{ad}(X)\,[Y, Z] + \operatorname{ad}(Y)\,[Z, X] = \operatorname{ad}(X) \circ \operatorname{ad}(Y)\, Z - \operatorname{ad}(Y)\operatorname{ad}(X)\, Z =$$

$$= [\operatorname{ad}(X), \operatorname{ad}(Y)]\, Z\,.$$

Q.e.d.

[44] That is ad is a morphism of Lie algebras.

Definition 1.136 ad : $\mathfrak{g} \to \mathrm{Hom}_{\mathbb{R}}(\mathfrak{g})$ is called the *adjoint representation* of $\mathfrak{g}$. □

For every $X \in \mathfrak{g}$ let

$$\mathscr{L}_X : \Omega^p(G) \to \Omega^p(G)$$

be the Lie derivative at the direction X.

Lemma 1.137 *For every $X \in \mathfrak{g}$ the Lie derivative $\mathscr{L}_X\omega$ of a left invariant p-form ω is a left invariant p-form, i.e.,*

$$\mathscr{L}_X\Big[\Omega^p(G)_{\text{l.i.}}\Big] \subset \Omega^p(G)_{\text{l.i.}}\,.$$

Proof Let $\omega \in \Omega^p(G)_{\text{l.i.}}$ and $a \in G$. For any $v_1, \ldots, v_p \in T_b(G)$ let $A_1, \ldots, A_p \in \mathfrak{g}$ such that $j_b(A_\alpha) = A_{\alpha,\,b} = v_\alpha$ for any $1 \leq \alpha \leq p$. Then

$$\Big[L_a^*(\mathscr{L}_X\omega)\Big]_b\big(v_1, \ldots, v_p\big) = \big(\mathscr{L}_X\omega\big)_{ab}\big((d_bL_a)v_1, \ldots, (d_bL_a)v_p\big) =$$

$$= \big(\mathscr{L}_X\omega\big)_{ab}\big((d_bL_a)A_{1,\,b}, \ldots, (d_bL_a)A_{p,\,b}\big) =$$

[as each A_α is left invariant]

$$= \big(\mathscr{L}_X\omega\big)_{ab}\big(A_{1,\,ab}, \ldots, A_{p,\,ab}\big) = \big(\mathscr{L}_X\omega\big)\big(A_1, \ldots, A_p\big)_{ab} =$$

$$= \Big\{X\Big(\omega(A_1, \ldots, A_p)\Big) - \sum_{\alpha=1}^{p} \omega(A_1, \ldots, A_{\alpha-1}, \big[X, A_\alpha\big], A_{\alpha+1}, \cdots, A_p)\Big\}_{ab} =$$

[as $\omega\big(A_1, \ldots, A_p\big)$ is a constant function on G]

$$= -\sum_{\alpha=1}^{p} \omega_{ab}\big(A_{1,\,ab}, \ldots, A_{\alpha-1,\,ab}, \big[X, A_\alpha\big]_{ab}, A_{\alpha+1,\,ab}, \cdots, A_{p,\,ab}\big) =$$

(as $[X, A_\alpha]$ is a left invariant vector field)

$$= -\sum_{\alpha=1}^{p} \omega_{ab}\big((d_bL_a)v_1, \ldots, (d_bL_a)v_{\alpha-1}, (d_bL_a)\big[X, A_\alpha\big]_b, (d_bL_a)v_{\alpha+1}, \cdots, (d_bL_a)v_p\big) =$$

$$= -\sum_{\alpha=1}^{p} \big(\mathscr{L}_a^*\omega\big)_b\big(v_1, \ldots, v_{\alpha-1}, \big[X, A_\alpha\big]_b, v_{\alpha+1}, \ldots, v_p\big) =$$

[as ω is left invariant, and once again $\omega(A_1, \ldots, A_p) \in \mathbb{R}$ and $v_\alpha = A_{\alpha, b}$]

$$= X\big(\omega(A_1, \ldots, A_p)\big)_b - \sum_{\alpha=1}^{p} \omega\big(A_1, \ldots, A_{\alpha-1}, [X, A_\alpha], A_{\alpha+1}, \ldots, A_p\big)_b =$$

$$= \big(\mathscr{L}_X \omega\big)\big(A_1, \ldots, A_p\big)_b = \big(\mathscr{L}_X \omega\big)_b\big(v_1, \ldots, v_p\big).$$

Q.e.d.

Lemma 1.138

$$\mathscr{L}_{X_j}\omega^i = C_{kj}{}^i\, \omega^k\,.$$

Proof Let $a \in G$, $v \in T_a(G)$ and $A := j_a^{-1}(v) \in \mathfrak{g}$. Then

$$\big(\mathscr{L}_{X_j}\omega^i\big)_a\, v = \Big\{\big(\mathscr{L}_{X_j}\omega^i\big)\, A\Big\}_a = \Big\{X_j\big(\omega^i(A)\big) - \omega^i\big([X_j, A]\big)\Big\}_a =$$

[as $\omega^i(A) \in \mathbb{R}$ and $A = \omega^k(A)\, X_k$ with $\omega^k(A) \in \mathbb{R}$]

$$= -\omega^k(A)_a \Big\{\omega^i\big([X_j, X_k]\big)\Big\}_a = -\omega^k(A)_a\, C_{jk}{}^i = C_{kj}{}^i\, \omega_a^k(v).$$

Q.e.d.

Let us recall Cartan's formula

$$\mathscr{L}_X = d \circ i_X + i_X \circ d, \quad X \in \mathfrak{X}(G).$$

Cf. e.g., Lemma 3.5.1 in S.I. Goldberg, *op. cit.*[45]

Corollary 1.139

$$\mathscr{L}_X \circ d = d \circ \mathscr{L}_X\,.$$

For every differential form $\omega \in \Omega^1(G)$ we adopt the notation

$$\epsilon(\omega) : \Omega^p(G) \to \Omega^{p+1}(G), \quad p \geq 0,$$

$$\epsilon(\omega)\, \beta = \omega \wedge \beta, \quad \beta \in \Omega^p(G).$$

[45] That is formula (3.5.1) in S.I. Goldberg, *op. cit.*, p. 102.

Lemma 1.140

$$d = \frac{1}{2}\,\epsilon(\omega^\alpha)\,\mathscr{L}_{X_\alpha} \tag{1.102}$$

everywhere on $\Omega^p(G)_{\text{l.i.}}$, $p \geq 0$.

Proof We only need to check (1.102) for $p = 0$ and $p = 1$. As $\Omega^0(G)_{\text{l.i.}} \simeq \mathbb{R}$ the equality (1.102) holds trivially (i.e., both sides are identically zero) for $p = 0$. Next

$$\frac{1}{2}\,\epsilon(\omega^\alpha)\,\mathscr{L}_{X_\alpha}\,\omega^\beta = \qquad \text{[by Lemma 1.138]}$$

$$= \frac{1}{2}\,\epsilon(\omega^\alpha)\,C_{\gamma\alpha}{}^{\beta}\,\omega^\gamma = -\frac{1}{2}\,C_{\alpha\gamma}{}^{\beta}\,\omega^\alpha \wedge \omega^\gamma = d\,\omega^\beta$$

[by (1.100)]. Q.e.d.

Let $\beta \in \Omega^p(G)_{\text{l.i.}}$ so that

$$\beta = B_{\alpha_1\cdots\alpha_p}\,\omega^{\alpha_1} \wedge \cdots \wedge \omega^{\alpha_p},$$

with $B_{\alpha_1\cdots\alpha_p} \in \mathbb{R}$. Then (as the Lie derivative acts as a derivation on exterior monomials)

$$\mathscr{L}_{X_\gamma}\beta = B_{\alpha_1\cdots\alpha_p}\sum_{k=1}^{p}\omega^{\alpha_1} \wedge \cdots \wedge \omega^{\alpha_{k-1}} \wedge \mathscr{L}_{X_\gamma}\omega^{\alpha_k} \wedge \omega^{\alpha_{k+1}} \wedge \cdots \wedge \omega^{\alpha_p} =$$

$$= B_{\alpha_1\cdots\alpha_p}\sum_{k=1}^{p} C_{\sigma\gamma}{}^{\alpha_k}\,\omega^{\alpha_1} \wedge \cdots \wedge \omega^{\alpha_{k-1}} \wedge \omega^\sigma \wedge \omega^{\alpha_{k+1}} \wedge \cdots \wedge \omega^{\alpha_p} =$$

$$= \sum_{k=1}^{p} B_{\alpha_1\cdots\alpha_{k-1}\rho\alpha_{k+1}\cdots\alpha_p}\,C_{\alpha_k\gamma}{}^{\rho}\,\omega^{\alpha_1} \wedge \cdots \wedge \omega^{\alpha_p}$$

hence [by (1.102)]

$$d\beta = \frac{1}{2}\,\epsilon\big(\omega^\gamma\big)\,\mathscr{L}_{X_\gamma}\beta =$$

$$= \frac{1}{2}\,\epsilon\big(\omega^\gamma\big)\sum_{k=1}^{p} B_{\alpha_1\cdots\alpha_{k-1}\rho\alpha_{k+1}\cdots\alpha_p}\,C_{\alpha_k\gamma}{}^{\rho}\,\omega^{\alpha_1} \wedge \cdots \wedge \omega^{\alpha_p}$$

or

$$d\beta = \frac{1}{2}\sum_{k=1}^{p} B_{\alpha_1\cdots\alpha_{k-1}\rho\alpha_{k+1}\cdots\alpha_p}\,C_{\alpha_k\gamma}{}^{\rho}\,\omega^\gamma \wedge \omega^{\alpha_1} \wedge \cdots \wedge \omega^{\alpha_p}\,. \tag{1.103}$$

Definition 1.141 A differential form $\beta \in \Omega^\bullet(G)$ is said to be $\mathfrak{g}$-*invariant* if $\mathscr{L}_X \beta = 0$ for every $X \in \mathfrak{g}$. □

Corollary 1.142 *Every* $\mathfrak{g}$-*invariant form is closed.*

Proof Follows from (1.102).

1.10.3 Compact Semi-simple Lie Groups

Definition 1.143 A Lie group is *semi-simple* if its Killing form

$$B(X, Y) = -\text{trace}\left\{\text{ad}(X) \circ \text{ad}(Y)\right\}, \qquad X,\ Y \in \mathfrak{g},$$

is nondegenerate. □

Let us set

$$g_{\alpha\beta} := C_{\alpha\sigma}{}^{\rho}\, C_{\rho\beta}{}^{\sigma}\,.$$

As the structure constants are skew-symmetric, $\left[g_{\alpha\beta}\right]$ is symmetric. It may be shown that G is semi-simple if and only if the matrix $\left[g_{\alpha\beta}\right]$ has rank n. When G is a compact semi-simple Lie group then $\left[g_{\alpha\beta}\right]$ is positive definite.[46] The section is devoted to a few local differential geometric properties of a compact semi-simple Lie group. Let us set

$$\left[g^{\alpha\beta}\right] := \left[g_{\alpha\beta}\right]^{-1}.$$

Also, let us set

$$C_{\alpha\beta\gamma} := g_{\gamma\alpha}\, C_{\alpha\beta}{}^{\sigma}\,. \tag{1.104}$$

As a consequence of the Jacobi identity for $\mathfrak{g}$ [and the resulting identities satisfied by the structure constants $C_{\alpha\beta}{}^{\gamma}$] $C_{\alpha\beta\gamma}$ is skew-symmetric in any pair of indices.

Let $\left(U,\ u^1,\ \ldots,\ u^n\right)$ be a local coordinate system on G. Then

$$X_\alpha = \xi^i_\alpha\, \frac{\partial}{\partial u^i}$$

for some $\left[\xi^i_\alpha\right] : U \to \text{GL}(n,\ \mathbb{R})$. Next, let us set

$$h^{ij} := \xi^i_\alpha\, \xi^j_\beta\, g^{\alpha\beta}$$

---

[46] Cf. S.I. Goldberg, *op. cit.*, p. 136.

so that $\left[h^{ij}(a)\right]$ is positive definite for any $a \in U$. Let us set

$$h := h_{ij}\, du^i \otimes du^j, \quad \left[h_{ij}\right] := \left[h^{ij}\right]^{-1},$$

so that h is a Riemannian metric on G.

Exercise 1.144 Show that (i) h is a globally[47] defined Riemannian metric on G, (ii) h is completely determined by G, i.e., the definition of h doesn't depend upon the choice of linear basis $\{X_\alpha\} \subset \mathfrak{g}$. □

Solution. (i) Let $f = \{X_\alpha : 1 \leq \alpha \leq n\}$ be a basis in $\mathfrak{g}$. Let $\left(U', u'^i\right)$ be another local coordinate system on G with $U \cap U' \neq \emptyset$. Let us consider the Riemannian metrics $h(f)$ and $h'(f)$ [respectively on U and U']

$$h(f) = h_{ij}\, du^i \otimes du^j, \quad h'(f) = h'_{ij}\, du'^i \otimes du'^j,$$

$$\left[h'_{ij}\right] = \left[h'^{ij}\right]^{-1}, \quad h'^{ij} = \xi'^i_{\ \alpha}\, \xi'^j_{\ \beta}\, g^{\alpha\beta},$$

$$\left[g^{\alpha\beta}\right] = \left[g_{\alpha\beta}\right]^{-1}, \quad g_{\alpha\beta} = C_{\alpha\sigma}{}^{\rho}\, C_{\rho\beta}{}^{\sigma}.$$

If

$$X_\alpha = \xi'^i_{\ \alpha} \frac{\partial}{\partial u'}, \quad \frac{\partial}{\partial u^j} = \frac{\partial u'^i}{\partial u^j} \frac{\partial}{\partial u'^i}, \quad \det\left[\frac{\partial u'^i}{\partial u^j}\right] \neq 0 \ \text{ on } \ U \cap U',$$

then

$$\xi'^i_{\ \alpha} = \frac{\partial u'^i}{\partial u^j}\, \xi^j_\alpha$$

so that

$$h'^{ij} = \frac{\partial u'}{\partial u^k}\, \xi^k_\alpha\, \frac{\partial u'^j}{\partial u^\ell}\, \xi^\ell_\beta\, g^{\alpha\beta}$$

or

$$h'^{ij} = h^{k\ell} \frac{\partial u'}{\partial u^k} \frac{\partial u'^j}{\partial u^\ell}$$

yielding

$$h'_{ij} = h_{k\ell} \frac{\partial u^k}{\partial u'^i} \frac{\partial u^\ell}{\partial u'^j}$$

and then $h(f)\big|_{U \cap U'} = h'(f)\big|_{U \cap U'}$. Q.e.d.

[47] The common misunderstanding that h is a but locally defined Riemannian metric (on U) is perhaps fueled by Sect. 4.3 in S.I. Goldberg's popular monograph [4.3. ***Local*** *geometry of a compact semi-simple Lie group*, S.I. Goldberg, *op. cit*, p. 136].

(ii) Let $\hat{f} \equiv \{\hat{X}_\alpha \;:\; 1 \le \alpha \le n\} \subset \mathfrak{g}$ be another basis, so that

$$\hat{X}_\beta = a^\alpha_\beta \, X_\alpha$$

for some $\left[a^\alpha_\beta\right] \in \mathrm{GL}(n,\, \mathbb{R})$. Then

$$\hat{X}_\beta = \hat{\xi}^i_\beta \, \frac{\partial}{\partial u^i}, \quad \hat{\xi}^i_\beta \in C^\infty(U,\, \mathbb{R}), \quad \hat{\xi}^i_\beta = a^\alpha_\beta \, \xi^i_\alpha \,.$$

Let $h(\hat{f})$ be the Riemannian metric on G defined by

$$h(\hat{f}) = \hat{h}_{ij} \, du^i \otimes du^j, \quad \left[\hat{h}_{ij}\right] = \left[\hat{h}^{ij}\right]^{-1}, \quad \hat{h}^{ij} = \hat{\xi}^i_\alpha \, \hat{\xi}^j_\beta \, \hat{g}^{\alpha\beta}, \quad \left[\hat{g}^{\alpha\beta}\right] = \left[\hat{g}_{\alpha\beta}\right]^{-1},$$

$$\hat{g}_{\alpha\beta} = \hat{C}_{\alpha\sigma}{}^{\rho} \, \hat{C}_{\rho\beta}{}^{\sigma}, \quad \hat{C}_{\alpha\beta}{}^{\gamma} \, \hat{X}_\gamma = \left[\hat{X}_\alpha,\, \hat{X}_\beta\right].$$

Then

$$a^\sigma_\alpha \, a^\rho_\beta \, C_{\sigma\rho}{}^{\lambda} = \hat{C}_{\alpha\beta}{}^{\gamma} \, a^\lambda_\gamma$$

or

$$\hat{C}_{\alpha\beta}{}^{\mu} = a^\sigma_\alpha \, a^\rho_\beta \, C_{\sigma\rho}{}^{\lambda} \, b^\mu_\lambda, \quad \left[b^\alpha_\beta\right] := \left[a^\alpha_\beta\right]^{-1}.$$

Consequently

$$\hat{g}_{\alpha\beta} = a^\epsilon_\alpha \, a^\gamma_\beta \, g_{\epsilon\gamma}$$

or

$$\hat{g}^{\alpha\sigma} = g^{\mu\rho} \, b^\sigma_\mu \, b^\alpha_\rho$$

yielding

$$\hat{h}^{ij} = g^{\gamma\rho} \, \xi^i_\gamma \, \xi^j_\beta$$

or $\hat{h}^{ij} = h^{ij}$ and then $\hat{h}_{ij} = h_{ij}$ or $h(\hat{f}) = h(f)$. Q.e.d.

Let us consider the differential 1-forms

$$\left\{v^\alpha \;:\; 1 \le \alpha \le n\right\} \subset \Omega^1(U), \quad v^\alpha = \eta^\alpha_i \, du^i, \quad \eta^\alpha_i := g^{\alpha\beta} \, \xi^j_\beta \, h_{ij} \,.$$

Lemma 1.145

$$\xi^i_\alpha \, \eta^\alpha_j = \delta^i_j, \quad \xi^i_\alpha \, \eta^\beta_i = \delta^\beta_\alpha \,. \tag{1.105}$$

Next, let us set

$$\Gamma^i_{jk} = \xi^i_\alpha \, \frac{\partial \eta^\alpha_j}{\partial u^k} \,. \tag{1.106}$$

Lemma 1.146 (i) *There is a unique globally defined linear connection* ∇ *on* G *such that*

$$\nabla_{\partial_j} \partial_k = \Gamma^i_{jk}\, \partial_i\,, \quad \partial_j \equiv \frac{\partial}{\partial u^j} \tag{1.107}$$

for any local coordinate system $(U,\, u^i)$ *on* G*. Moreover* (ii) ∇ *is completely determined by* G*, i.e., the definition of* ∇ *doesn't depend upon the choice of basis* $\{X_\gamma\} \subset \mathfrak{g}$.

Proof (i) Let $\left(U',\, u'^i\right)$ be another local coordinate system with $U \cap U' \neq \emptyset$, and let us set

$$\eta'^{\alpha}_j = g^{\alpha\beta}\, \xi'^k_\beta\, h'_{jk}\,, \qquad \Gamma'^i_{jk} = \xi'^i_\alpha\, \frac{\partial \eta'^\alpha_j}{\partial u'^k}\,.$$

The transformation laws

$$\xi'^i_\alpha = \frac{\partial u'^i}{\partial u^j}\, \xi^j_\alpha\,, \quad h'_{jk} = \frac{\partial u^r}{\partial u'^j}\, \frac{\partial u^s}{\partial u'^k}\, h_{rs}\,,$$

yield

$$\eta'^\alpha_j = \eta^\alpha_r\, \frac{\partial u^r}{\partial u'^j}$$

and then

$$\Gamma'^i_{jk} = \frac{\partial u'^i}{\partial u^\ell}\, \frac{\partial^2 u^\ell}{\partial u'^k\, \partial u'^j} + \frac{\partial u'^i}{\partial u^\ell}\, \frac{\partial u^r}{\partial u'^j}\, \frac{\partial u^s}{\partial u'^k}\, \Gamma^\ell_{rs}$$

which is the ordinary transformation law for the coefficients of a linear connection. Therefore (1.107) determines a globally defined linear connection on G. Q.e.d.

(ii) Let $\hat{X}_\beta = a^\alpha_\beta\, X_\alpha$ be another basis in $\mathfrak{g}$. Then

$$\hat{g}^{\alpha\beta} = b^\alpha_\rho\, b^\beta_\mu\, g^{\rho\mu}\,, \quad \hat{\xi}^i_\beta = a^\alpha_\beta\, \xi^i_\alpha\,, \quad \hat{h}_{ij} = h_{ij}\,,$$

hence $\hat{\eta}^\alpha_j = \hat{g}^{\alpha\beta}\, \hat{\xi}^k_\beta\, \hat{h}_{jk}$ becomes

$$\hat{\eta}^\alpha_j = b^\alpha_\rho\, \eta^\rho_j\,.$$

Finally $\hat{\Gamma}^i_{jk} = \hat{\xi}^i_\alpha\, \dfrac{\partial \hat{\eta}^\alpha_j}{\partial u^k}$ becomes $\hat{\Gamma}^i_{jk} = \Gamma^i_{jk}$. Q.e.d.

By $\xi^i_\alpha\, \eta^\alpha_j = \delta^i_j$ the definition (1.106) may also be written as

$$\Gamma^i_{jk} = -\frac{\partial \xi^i_\alpha}{\partial u^k}\, \eta^\alpha_j\,. \tag{1.108}$$

Therefore, the torsion tensor field $T_{jk}{}^{i} = \Gamma^{i}_{jk} - \Gamma^{i}_{kj}$ of the linear connection furnished by Lemma 1.146 is

$$T_{jk}{}^{i} = \xi^{i}_{\alpha}\Big(\frac{\partial \eta^{\alpha}_{j}}{\partial u^{k}} - \frac{\partial \eta^{\alpha}_{k}}{\partial u^{j}}\Big). \tag{1.109}$$

The equations $\big[X_{\beta},\, X_{\gamma}\big] = C_{\beta\gamma}{}^{\alpha}\, X_{\alpha}$ may be written in local coordinates as

$$\xi^{j}_{\beta}\,\frac{\partial \xi^{i}_{\gamma}}{\partial u^{j}} - \xi^{j}_{\gamma}\,\frac{\partial \xi^{i}_{\beta}}{\partial u^{j}} = C_{\beta\gamma}{}^{\alpha}\,\xi^{i}_{\alpha}\,.$$

Let us contract with η^{β}_{k} so that to obtain

$$\frac{\partial \xi^{i}_{\gamma}}{\partial u^{k}} - \eta^{\beta}_{k}\,\xi^{\ell}_{\gamma}\,\frac{\partial \xi^{i}_{\beta}}{\partial u^{\ell}} = \eta^{\beta}_{k}\,C_{\beta\gamma}{}^{\alpha}\,\xi^{i}_{\alpha}\,.$$

Next, contraction with η^{γ}_{j} leads to

$$\eta^{\gamma}_{j}\,\frac{\partial \xi^{i}_{\gamma}}{\partial u^{k}} - \eta^{\beta}_{k}\,\frac{\partial \xi^{i}_{\beta}}{\partial u^{j}} = \eta^{\gamma}_{j}\,\eta^{\beta}_{k}\,C_{\beta\gamma}{}^{\alpha}\,\xi^{i}_{\alpha}\,. \tag{1.110}$$

Consequently [by (1.108) and (1.110)]

$$T_{jk}{}^{i} = \Gamma^{i}_{jk} - \Gamma^{i}_{kj} = -\frac{\partial \xi^{i}_{\alpha}}{\partial u^{k}}\,\eta^{\alpha}_{j} + \frac{\partial \xi^{i}_{\alpha}}{\partial u^{j}}\,\eta^{\alpha}_{k} =$$

$$= C_{\beta\gamma}{}^{\alpha}\,\xi^{i}_{\alpha}\,\eta^{\beta}_{j}\,\eta^{\gamma}_{k}\,. \tag{1.111}$$

Equation (1.111) implies.

Lemma 1.147 *$T_{jk\ell} := h_{i\ell}\, T_{jk}{}^{i}$ is skew-symmetric in each pair of indices.*

Corollary 1.148 (i) *The metric tensor h is parallel with respect to the linear connection ∇, i.e., $\nabla h = 0$.*

(ii) *In particular, the connection coefficients Γ^{i}_{jk} and the Christoffel symbols $\left\{ {i \atop jk} \right\}$ [of the Riemannian manifold $(G,\, h)$] are related by*

$$\Gamma^{i}_{jk} = \left\{ {i \atop jk} \right\} + \frac{1}{2}\,T_{jk}{}^{i}\,. \tag{1.112}$$

(iii) *Consequently*

$$\left\{ {i \atop jk} \right\} = \frac{1}{2}\,\xi^{i}_{\alpha}\Big(\frac{\partial \eta^{\alpha}_{j}}{\partial u^{k}} + \frac{\partial \eta^{\alpha}_{k}}{\partial u^{j}}\Big). \tag{1.113}$$

Proof (i) Let us contract with η_i^β in $\Gamma^i_{jk} = \xi^i_\alpha \dfrac{\partial \eta^\alpha_j}{\partial u^k}$ so that to obtain

$$\frac{\partial \eta^\beta_j}{\partial u^k} = \eta^\beta_i \, \Gamma^i_{jk} \,. \tag{1.114}$$

Let us take the partial derivative of

$$h_{ij} = \eta^\alpha_i \, \eta^\beta_j \, g_{\alpha\beta}$$

with respect to u^k so that to obtain

$$\frac{\partial h_{ij}}{\partial u^k} = g_{\alpha\beta}\Big(\frac{\partial \eta^\alpha_i}{\partial u^k}\, \eta^\beta_j + \eta^\alpha_i \, \frac{\partial \eta^\beta_j}{\partial u^k}\Big) =$$

[by (1.114)]

$$= g_{\alpha\beta}\Big\{\eta^\beta_j \, \eta^\alpha_\ell \, \Gamma^\ell_{ik} + \eta^\alpha_i \, \eta^\beta_\ell \, \Gamma^\ell_{jk}\Big\} =$$

[by $g_{\alpha\beta}\,\eta^\beta_j = \xi^k_\alpha \, h_{jk}$, a consequence of the very definition of η^β_j]

$$= \xi^s_\alpha \, h_{js} \, \eta^\alpha_\ell \, \Gamma^\ell_{ik} + \eta^\alpha_i \, \xi^s_\alpha \, h_{\ell s} \, \Gamma^\ell_{jk}$$

that is [by (1.105)]

$$\frac{\partial h_{ij}}{\partial u^k} = h_{j\ell} \, \Gamma^\ell_{ik} + h_{\ell i} \, \Gamma^\ell_{jk} \,. \tag{1.115}$$

Next, we compute the covariant derivative of h with respect to ∇, i.e.,

$$\big(\nabla_{\partial_k} h\big)\big(\partial_i, \, \partial_j\big) = \frac{\partial h_{ij}}{\partial u^k} - \Gamma^\ell_{ki} \, h_{\ell j} - \Gamma^\ell_{kj} \, h_{i\,\ell} =$$

[by (1.115)]

$$= h_{j\ell}\left(\Gamma^\ell_{ik} - \Gamma^\ell_{ki}\right) + h_{i\ell}\left(\Gamma^\ell_{jk} - \Gamma^\ell_{kj}\right) = h_{j\ell} \, T_{ik}{}^\ell + h_{i\ell} \, T_{jk}{}^\ell = T_{ikj} + T_{jki} = 0$$

[by the skew-symmetry of T_{ijk}]. We may conclude that $\nabla h = 0$. Q.e.d.

(ii) As $\nabla h = 0$

$$\frac{\partial h_{ij}}{\partial u^k} = \Gamma^\ell_{ki} \, h_{\ell j} + \Gamma^\ell_{kj} \, h_{i\,\ell},$$

$$\frac{\partial h_{jk}}{\partial u^i} = \Gamma^\ell_{ij} \, h_{\ell k} + \Gamma^\ell_{ik} \, h_{j\,\ell},$$

$$\frac{\partial h_{ki}}{\partial u^j} = \Gamma^\ell_{jk} \, h_{\ell i} + \Gamma^\ell_{ji} \, h_{k\,\ell},$$

hence [by adding the first two equations, and subtracting the third]

$$2\,\{ki,\ j\} = h_{\ell j}\Big(\Gamma^{\ell}_{ki} + \Gamma^{\ell}_{ik}\Big) + h_{i\ell}\,T_{kj}{}^{\ell} + h_{\ell k}\,T_{ij}{}^{\ell}$$

or

$$2\,\{ki,\ j\} = 2\,h_{j\ell}\,\Gamma^{\ell}_{ik} + T_{kij} + T_{kji} + T_{ijk}$$

where $\{ij,\ k\}$ are the Christoffel symbols of the first kind, i.e.,

$$\{ij,\ k\} = \frac{1}{2}\Big(\frac{\partial h_{ik}}{\partial u^j} + \frac{\partial h_{jk}}{\partial u^i} - \frac{\partial h_{ij}}{\partial u^k}\Big).$$

Moreover [by the skew-symmetry of T_{ijk}]

$$2\,\{ki,\ j\} = 2\,h_{j\ell}\,\Gamma^{\ell}_{ik} + T_{ijk}$$

and contraction with h^{js} leads to

$$\left\{ {s \atop ki} \right\} = \Gamma^{s}_{ik} + \frac{1}{2}\,T_{ki}{}^{s}$$

which is (1.112). Q.e.d.

(iii) By statement (ii) in Corollary 1.148

$$\left\{ {i \atop jk} \right\} = \Gamma^{i}_{jk} - \frac{1}{2}\,T_{jk}{}^{i} =$$

[by the very definition of Γ^{i}_{jk} together with (1.109)]

$$= \frac{1}{2}\,\xi^{i}_{\alpha}\left(\frac{\partial \eta^{\alpha}_{j}}{\partial u^k} + \frac{\partial \eta^{\alpha}_{k}}{\partial u^j}\right).$$

Q.e.d.

Corollary 1.149 *The elements of the Lie algebra $\mathfrak{g}$ of a compact semi-simple Lie group G are infinitesimal isometries of $(G,\ h)$, i.e., $\mathfrak{g} \subset \mathfrak{i}(G,\ h)$.*

Proof The covariant derivative of the differential 1-form υ^{α} with respect to the Levi-Civita connection ∇^{h} of $(G,\ h)$ is

$$\eta^{\alpha}_{j\,\|k} = \big(\nabla^{h}_{\partial_k}\upsilon^{\alpha}\big)\,\partial_j = \eta^{\alpha}_{j\,|k} - \left\{ {i \atop kj} \right\}\eta^{\alpha}_{i},$$

$$\partial_i \equiv \frac{\partial}{\partial u^i}, \quad \varphi_{|i} = \frac{\partial \varphi}{\partial u^i}, \quad \varphi \in C^1(G).$$

Then [by (1.113) and $\xi^i_\beta \, \eta^\alpha_i = \delta^\alpha_\beta$]

$$\eta^\alpha_{j\,||\,k} = \eta^\alpha_{j\,|\,k} - \frac{1}{2}\,\xi^i_\beta \left(\eta^\beta_{j\,|\,k} + \eta^\beta_{k\,|\,j}\right)\eta^\alpha_i = \frac{1}{2}\left(\eta^\alpha_{j\,|\,k} - \eta^\alpha_{k\,|\,j}\right)$$

or [as (1.109) implies $\eta^\beta_i \, T_{jk}{}^i = \eta^\beta_{j\,|\,k} - \eta^\beta_{k\,|\,j}$]

$$\eta^\alpha_{j\,||\,k} = \frac{1}{2}\,\eta^\alpha_i \, T_{jk}{}^i \, .$$

Let us contract with ξ^ℓ_α so that to obtain

$$\xi^\ell_\alpha \, \eta^\alpha_{j\,||\,k} = \frac{1}{2}\,T_{jk}{}^\ell \, . \tag{1.116}$$

The covariant derivative of a tangent vector field $X = X^i \, \partial_i \in \mathfrak{X}(U)$ with respect to ∇^h is

$$X^i{}_{||\,j} = (\nabla^h_{\partial_j} X)\partial_i = \left[X^i{}_{|\,j} + X^k \left\{ {i \atop jk} \right\}\right]\partial_i \, .$$

As $\xi^\ell_\alpha \, \eta^\alpha_j = \delta^\ell_j$

$$\xi^\ell_\alpha \, \eta^\alpha_{j\,||\,k} = -\eta^\alpha_j \, \xi^\ell_{\alpha\,||\,k}$$

hence (1.116) becomes

$$-\,\eta^\alpha_j \, \xi^\ell_{\alpha\,||\,k} = \frac{1}{2}\,T_{jk}{}^\ell$$

and contraction with ξ^j_β leads to

$$\xi^\ell_{\beta\,||\,k} = \frac{1}{2}\,\xi^j_\beta \, T_{kj}{}^\ell \, . \tag{1.117}$$

As $\nabla^h h = 0$

$$\big(\mathscr{L}_X h\big)(Y,\, Z) = h\big(\nabla^h_Y X,\; Z\big) + h\big(Y,\; \nabla^h_Z X\big)$$

for any $X,\; Y,\; Z \in \mathfrak{X}(G)$. Then

$$\big(\mathscr{L}_{X_\beta} h\big)(\partial_j,\; \partial_k) = h_{\ell k}\, \xi^\ell_{\beta\,||\,j} + h_{j\ell}\, \xi^\ell_{\beta\,||\,k} =$$

[by (1.117) and the skew-symmetry of T_{ijk}]

$$= \frac{1}{2}\,\xi^i_\beta \left\{T_{ji}{}^\ell \, h_{\ell k} + T_{ki}{}^\ell \, h_{j\ell}\right\} = \frac{1}{2}\,\xi^i_\beta \left\{T_{jik} + T_{kij}\right\} = 0,$$

that is X_β is a Killing vector field on $(G,\; h)$. Q.e.d.

1.10.4 Harmonic Forms on Compact Semi-simple Lie Groups

Let G be a compact semi-simple Lie group.

Theorem 1.150 *Let $\alpha \in \Omega^p(G)$ be a $\mathfrak{g}$-invariant differential form. Then*

(i) *$d\alpha$ is $\mathfrak{g}$-invariant.*
(ii) *$* \alpha$ is $\mathfrak{g}$-invariant.*
(iii) *If $\alpha = d\beta$ then β is $\mathfrak{g}$-invariant.*

Proof (i) Let $X \in \mathfrak{g}$ hence (by hypothesis) $\mathscr{L}_X\alpha = 0$. Then

$$\mathscr{L}_X\, d\,\alpha = d\,\mathscr{L}\,\alpha = 0,$$

i.e., $d\alpha$ is $\mathfrak{g}$-invariant. Q.e.d.

(ii) By (1.95) for every $X \in \mathfrak{X}(G)$

$$\mathscr{L}_X\alpha + \overline{\mathscr{L}}_X\alpha = (\delta\,\eta)\,\alpha +$$

$$+\sum_{r=1}^{p} h^{jk}\left(\mathscr{L}_X\, h\right)_{ki_r}\,\alpha_{i_1\cdots i_{r-1}\,j\,i_{r+1}\cdots i_p}\,du^{i_1}\wedge\cdots\wedge du^{i_p}$$

where $\eta = X^\flat$. If $X \in \mathfrak{g} \subset \mathfrak{i}(G,\, h)$ the previous formula simplifies accordingly, i.e.,

$$(\delta\,\eta)\,\alpha = \overline{\mathscr{L}}_X\alpha = (-1)^{np+p+1}\, *\,\mathscr{L}_X\, *\alpha\,.$$

Finally, we need to recall[48] that X is an infinitesimal conformal transformation [i.e., $\mathscr{L}_X h = \lambda\, h$ for some $\lambda \in C^\infty(G)$] if and only if $\delta\, X^\flat = 0$. Consequently $*\mathscr{L}_X\, *\,\alpha = 0$, and hence $\mathscr{L}_X\, *\,\alpha = 0$ (as $*$ is an isomorphism). Q.e.d.

(iii) Let $\alpha \in \Omega^p(G)$ be an arbitrary p-form on G. Then (by the Hodge-de Rham decomposition theorem)

$$\alpha = \alpha_d + \alpha_\delta + H[\alpha] \in \Omega^p_d(G) \oplus \Omega^p_\delta(G) \oplus \Omega^p_H(G)$$

hence

$$\alpha_d + \alpha_\delta = \alpha - H[\alpha] = \Delta\big(\mathbf{G}\,\alpha\big)$$

where $\mathbf{G} : \Omega^p(G) \to \Omega^(G)$ is the Green operator. Summing up, we have obtained the following (tautologous yet useful) identity[49] (for the Hodge-de Rham decomposition of α)

---

[48] Cf. S.I. Goldberg, *op. cit.*, p. 107.

[49] Certainly holding true should one replace G by any compact orientable Riemannian manifold.

$$\alpha = \Delta(\mathbf{G}\,\alpha) + H[\alpha],$$

$$\alpha = d\,\delta\,\mathbf{G}\,\alpha + \delta\,d\,\mathbf{G}\,\alpha + H[\alpha].$$

Let us assume that $\alpha = d\beta$ for some $\beta \in \Omega^{p-1}(G)$. Then $\delta\,d\,\mathbf{G}\,\alpha = 0$ and $H[\alpha] = 0$, so that

$$\alpha = d\,\delta\,\mathbf{G}\,\alpha\,.$$

Let us recall that $\delta = (-1)^{np+n+1} * d\,*$, hence

$$\alpha = d * d\,\gamma, \tag{1.118}$$

$$\gamma := (-1)^{np+n+1} * \mathbf{G}\,\alpha.$$

Next [as α is $\mathfrak{g}$-invariant, and by (1.118)] for any $X \in \mathfrak{g}$

$$0 = \mathscr{L}_X\alpha = \mathscr{L}_X\big(d * d\,\gamma\big) = d\,\mathscr{L}_X * d\,\gamma = -d * \overline{\mathscr{L}}_X\,d\,\gamma =$$

[by $\mathscr{L}_X = -\overline{\mathscr{L}}_X$ for any $X \in \mathfrak{g} \subset \mathfrak{i}(G, h)$]

$$= d * \mathscr{L}_X\,d\,\gamma = -d * d\,\mathscr{L}_X\,\gamma$$

or

$$d * d\,\mathscr{L}_X\,\gamma = 0. \tag{1.119}$$

Then [as $\mathscr{L}_X\,\gamma$ is a $(n-p+1)$-form]

$$\delta\,d\,\mathscr{L}_X\,\gamma = (-1)^{n(n-p+1)+n+1} * d * \big(d\,\mathscr{L}_X\,\gamma\big) = 0$$

[by using (1.119) at the last step] or

$$\delta\,d\,\mathscr{L}_X\,\gamma = 0. \tag{1.120}$$

Next [by (1.120)]

$$0 = \big(\delta\,d\,\mathscr{L}_X\,\gamma,\ \mathscr{L}_X\gamma\big)_{L^2} = \|d\,\mathscr{L}_X\,\gamma\|^2 \Longrightarrow d\,\mathscr{L}_X\,\gamma = 0$$

so that

$$0 = \delta\,d\,\mathscr{L}_X\,\gamma = \mathscr{L}_X\,d\,\gamma$$

implying that $d\gamma$ is $\mathfrak{g}$-invariant. Therefore [by statement (ii) in Theorem 1.150] $*\,d\,\gamma$ is $\mathfrak{g}$-invariant. Finally

$$\alpha = d\big(*\,d\,\gamma\big) = d\,\beta$$

so that β is cohomologous to a $\mathfrak{g}$-invariant form. Q.e.d.

Theorem 1.151 *On any compact semi-simple Lie group G a differential form is harmonic if and only if it is $\mathfrak{g}$-invariant.*

Proof By Theorem 1.126 every harmonic form $\alpha \in \Omega^p_H(G)$ is $\mathfrak{i}(G, h)$-invariant. On the other hand $\mathfrak{i}(G, h) \supset \mathfrak{g}$, hence every harmonic form is $\mathfrak{g}$-invariant.

Viceversa, let β be a $\mathfrak{g}$-invariant p-form on G. Any $\mathfrak{g}$-invariant form is closed, so that $d\beta = 0$. We claim that β is also coclosed, so it is harmonic. Indeed

$$\delta\beta = (-1)^{np+n+1} * d * \beta =$$

[as $d = (1/2)\, \epsilon(\omega^\alpha)\, \mathscr{L}_{X_\alpha}$]

$$= \frac{1}{2}(-1)^{np+n+1} * \epsilon(\omega^\alpha)\, \mathscr{L}_{X_\alpha} * \beta = \frac{1}{2}(-1)^{np+n+1} * \epsilon(\omega^\alpha) * *^{-1} \mathscr{L}_{X_\alpha} * \beta =$$

[as $i_X = (-1)^{np+n} * \epsilon(\eta) *$ on $\Omega^p(G)$ with $\eta = X^\sharp$, and $\left(\omega^\alpha\right)^* = X_\alpha$]

$$= -\frac{1}{2} i_{X_\alpha} *^{-1} \mathscr{L}_X * \beta =$$

[as $\mathscr{L}_{X_\alpha} * = * \mathscr{L}_{X_\alpha}$]

$$= -\frac{1}{2} i_{X_\alpha} \mathscr{L}_{X_\alpha} \beta = 0$$

(as β is $\mathfrak{g}$-invariant). Q.e.d.

As a corollary of the proof

$$\delta = -\frac{1}{2} \sum_{\alpha=1}^{n} i_{X_\alpha} \circ \mathscr{L}_{X_\alpha}$$

showing that δ is an anti-derivation in $\Lambda^\bullet\left(\mathfrak{g}^*\right)$. Consequently

Corollary 1.152 *On any compact semi-simple Lie group the exterior product of two harmonic forms is harmonic.*

Let us denote by $\Omega^p_{\mathfrak{g}}(G)$ the space of all $\mathfrak{g}$-invariant p-forms on G. The statement in Theorem 1.151 is briefly that

$$\Omega^p_{\mathfrak{g}}(G) = \Omega^p_H(G)$$

for any compact semi-simple Lie group G.

Theorem 1.153 *The first and second Betti numbers of a compact semi-simple Lie group vanish.*

Proof Let $\beta \in \Omega^p_H(G)$ be a harmonic p-form. By Theorem 1.151 the form β is $\mathfrak{g}$-invariant

$$\beta = B_{\alpha_1 \cdots \alpha_p}\, \omega^{\alpha_1} \wedge \cdots \wedge \omega^{\alpha_p}, \quad B_{\alpha_1 \cdots \alpha_p} \in \mathbb{R},$$

and hence closed, i.e.,

$$0 = d\beta = \frac{1}{2} \sum_{r=1}^{p} B_{\alpha_1 \cdots \alpha_{r-1}\, \rho\, \alpha_{r+1} \ldots \alpha_r}\, C_{\alpha_r\, \alpha}{}^{\rho}\, \omega^{\alpha} \wedge \omega^{\alpha_1} \wedge \cdots \wedge \omega^{\alpha_p}$$

yielding

$$\sum_{r=1}^{p} B_{\alpha_1 \cdots \alpha_{r-1}\, \rho\, \alpha_{r+1} \ldots \alpha_r}\, C_{\alpha_r\, \alpha}{}^{\rho} = 0. \tag{1.121}$$

Let $p = 1$, i.e., let $\beta = B_{\alpha_1}\, \omega^{\alpha_1} \in \Omega^1_H(G)$. Then [by (1.121)]

$$B_{\rho}\, C_{\alpha_1 \alpha}{}^{\rho} = 0.$$

Let us contract with $C_{\gamma}{}^{\alpha\alpha_1}$ [which is $g^{\alpha\rho}\, C_{\gamma\rho}{}^{\alpha_1}$] and observe[50] that

$$0 = B_{\rho}\, C_{\alpha_1 \alpha}{}^{\rho}\, C_{\gamma}{}^{\alpha\alpha_1} = -B_{\rho},$$

i.e., $\beta = 0$.

Next let $p = 2$, i.e., let

$$\alpha = A_{\alpha\beta}\, \omega^{\alpha} \wedge \omega^{\beta} \in \Omega^2_H(G) = \Omega^2_{\mathfrak{g}}(G)$$

so that [by $d\alpha = 0$]

$$A_{\rho\beta}\, C_{\alpha\gamma}{}^{\rho} + A_{\alpha\rho}\, C_{\beta\gamma}{}^{\rho} = 0. \tag{1.122}$$

By cyclic permutation of the indices $\alpha \to \beta \to \gamma \to \alpha$ in (1.122) we obtain another two identities of the sort

$$A_{\rho\gamma}\, C_{\beta\alpha}{}^{\rho} + A_{\beta\rho}\, C_{\gamma\alpha}{}^{\rho} = 0,$$

$$A_{\rho\alpha}\, C_{\gamma\beta}{}^{\rho} + A_{\gamma\rho}\, C_{\alpha\beta}{}^{\rho} = 0.$$

Adding the three resulting relations leads to

$$2\, A_{\rho\beta}\, C_{\alpha\gamma}{}^{\rho} + 2\, A_{\alpha\rho}\, C_{\beta\gamma}{}^{\rho} + 2\, A_{\rho\gamma}\, C_{\beta\alpha}{}^{\rho} = 0$$

---

[50] Indeed

$$C_{\alpha_1\alpha}{}^{\rho}\, C_{\gamma}{}^{\alpha\alpha_1} = C_{\alpha_1}{}^{\alpha\rho}\, C_{\gamma\alpha}{}^{\alpha_1} = C_{\gamma\alpha}{}^{\alpha_1}\, C_{\alpha_1\, \beta\sigma}\, g^{\alpha\beta}\, g^{\sigma\rho} =$$
$$= -C_{\gamma\alpha}{}^{\alpha_1}\, C_{\alpha_1\, \sigma\beta}\, g^{\alpha\beta}\, g^{\sigma\rho} = -C_{\gamma\alpha}{}^{\alpha_1}\, C_{\alpha_1\, \sigma}{}^{\alpha}\, g^{\sigma\rho} = -g_{\gamma\sigma}\, g^{\sigma\rho} = -\delta^{\rho}_{\gamma}\,.$$

or [by using (1.122) once more]

$$A_{\rho\gamma}\, C_{\beta\alpha}{}^{\rho} = 0.$$

Finally, contraction with $C_{\sigma}{}^{\alpha\beta}$ gives[51] $A_{\rho\gamma} = 0$, i.e., $\alpha = 0$. Q.e.d.

If one sets as customary $b_p(G) = \dim_{\mathbb{R}} H^p_{\text{dR}}(G)$ (the *Betti numbers* of G) then the brief statement of Theorem 1.153 is

$$b_1(G) = 0, \quad b_2(G) = 0,$$

for any compact semi-simple Lie group G. The proof of Theorem 1.153 relied on the classical result that the pth Betti number $b_p(M)$ of a compact orientable manifold equals[52] the maximum number of linearly independent harmonic forms.

Theorem 1.154 *Let G be a compact n-dimensional Lie group*[53] *such that $b_p(G) = \binom{n}{p}$ for any $0 \leq p \leq n$. Then the Euler characteristic of G is zero, i.e., $\chi(G) = 0$.*

Proof Indeed

$$\chi(G) = \sum_{p=0}^{n} (-1)^p\, b_p(G) = \sum_{p=0}^{n} (-1)^p \binom{n}{p} = 0.$$

Q.e.d.

The rather strong assumption in Theorem 1.154 is realized in the abelian case. Precisely

Theorem 1.155 *Let G be a n-dimensional compact, connected, abelian Lie group. Then*

$$b_p(G) = \binom{n}{p}, \quad 0 \leq p \leq n.$$

Proof As G is abelian, so does its Lie algebra $\mathfrak{g}$. Hence the structure constants vanish

$$0 = \left[X_\alpha,\, X_\beta\right] = C_{\alpha\beta}{}^{\gamma}\, X_\gamma \Longrightarrow C_{\alpha\beta}{}^{\gamma} = 0.$$

If $X_\alpha = \xi^i_\alpha\, \partial/\partial u^i$ we endow G with the Riemannian metric h given by

$$h^{ij} = \sum_{\alpha=1}^{n} \xi^i_\alpha\, \xi^j_\alpha\,. \tag{1.123}$$

[51] Cf. the calculation in the previous footnote.

[52] Cf. e.g., S.I. Goldberg, *op. cit.*, Appendix C.

[53] Not necessarily semi-simple.

Note that

$$\mathscr{L}_{X_\beta}\, \omega^\alpha = C_{\gamma\beta}{}^{\alpha}\, \omega^\gamma = 0$$

so that $\omega^\alpha \in \Omega^1_{\mathfrak{g}}(G)$. Then (by the proof of Theorem 1.151) ω^α is a harmonic 1-form with respect to the metric h [defined by (1.123)]. Yet $\mathscr{L}_X$ is a derivation, hence $\mathscr{L}_X \alpha = 0$ for any $X \in \mathfrak{g}$ and any left invariant form $\alpha \in \Omega^p(G)_{\text{l.i.}}$ yielding $b_p(G) = \dim_{\mathbb{R}} \Lambda^p\big(\mathfrak{g}^*\big)$. Q.e.d.

1.11 Curvature and Fundamental Groups

Let Γ be a finitely generated group, and let $\{g_1, \ldots, g_p\} \subset \Gamma$ be a system of generators of Γ. For every positive integer $s \in \mathbb{N}$ let $\gamma(s)$ be the number[54] of distinct group elements that can be expressed as words of length $\leq s$ in the generators $\{g_1, \ldots, g_p\}$ and their inverses.

Definition 1.156 The function $\gamma : \mathbb{N} \to \mathbb{N}$ is called the *growth function* associated to Γ and the system of generators $\{g_1, \ldots, g_p\}$. □

The inequality $\gamma(s+t) \leq \gamma(s)\,\gamma(t)$ implies an exponential upper bound on the growth function, i.e., $\gamma(s) \leq \gamma(1)^s$.

Exercise 1.157 Show that the limit $\lim_{s\to\infty} \gamma(s)^{1/s}$ exists and is finite.

Solution. For every fixed t, let us set[55] $k := [s/t] + 1$. Then

$$\gamma(s) \leq \gamma(kt) \leq \gamma(t)^k \leq \gamma(t)^{1+s/t} \Longrightarrow \limsup \gamma(s)^{1/s} \leq \gamma(t)^{1/t}$$

and hence

$$\limsup \gamma(s)^{1/s} \leq \inf\left\{\gamma(t)^{1/t}\right\} \leq \liminf\, \gamma(s)^{1/s}\,.$$

Q.e.d. □

The following lemma shows to what extent the asymptotic behavior of $\gamma(s)$ as $s \to \infty$ is independent of the choice of generators.

Lemma 1.158 *Let $A = \{g_1, \ldots, g_p\}$ and $B = \{h_1, \ldots, h_q\}$ be two different generator systems for the same finitely generated group Γ. Let γ_A and γ_B be the corresponding growth functions. Then*

$$\gamma_B(t) \leq \gamma_A(k_A\, t), \quad \gamma_A(s) \leq \gamma_B(k_B\, s), \quad t,\, s \in \mathbb{N}, \tag{1.124}$$

for some constants $k_A,\, k_B \in \mathbb{N}$.

---

[54] For instance, if Γ is a free abelian group of rank 2 with generators $\{x,\, y\}$ then $\gamma(s) = 2s^2 + 2s + 1$.

[55] If $a \in \mathbb{R}$ then $[a]$ denotes the integer part of a.

Proof For sufficiently large k_A, every $h_j \in B$ can be written as a word of length $\leq k_A$ in the generators $A \cup A^{-1}$. Hence $\gamma_B(t) \leq \gamma_B\big(k_A\, t\big)$. Q.e.d.

Definition 1.159 Γ has *exponential growth* if $\lim_{s\to\infty} \gamma(s)^{1/s} > 1$. □

Let M be a Riemannian manifold and let $\pi_1(M)$ be the fundamental group of M. Classical results by Milnor (cf. [Mil]) study the behavior of the growth function of a finitely generated subgroup $\Gamma \subset \pi_1(M)$.

Theorem 1.160 *Let M be a complete n-dimensional Riemannian manifold whose Ricci curvature tensor is positive semi-definite. Then the growth function γ of every finitely generated subgroup $\Gamma \subset \pi_1(M)$ satisfies*

$$\gamma(s) \leq C\, s^n \tag{1.125}$$

for some constant $C > 0$ depending on the system of generators entering the definition of γ.

Theorem 1.161 *Let M be a compact Riemannian manifold, of negative sectional curvature. Then the growth of the fundamental group $\pi_1(M)$ is at least exponential, i.e.,*

$$\gamma(s) \geq a^s$$

for some constant $a > 1$.

Cf. Theorems 1 and 2 in [Mil], p. 1. Theorems 1.160 and 1.161 should be paralleled with the classical results by Myers[56] (cf. [Mye]), Preissmann[57] (cf. [Pre]), and to the Hadamard-Cartan theorem.[58]

Proof of Theorem 1.160 Let $\tilde{M}$ be the universal covering space of M and let $x_0 \in \tilde{M}$ be a base point. Let $V(r) := \text{vol}\big[\tilde{B}_r(x_0)\big]$ be the volume of the closed ball of center x_0 and radius r in $\big(\tilde{M}, \tilde{d}\big)$, where $\tilde{d} = d_{\tilde{g}}$ is the distance function associated to the Riemannian metric $\tilde{g}$ on $\tilde{M}$ (the pullback of the Riemannian metric g on M). By a classical result of Bishop (cf. [Bis]) if the Ricci curvature is non-negative $\text{Ric}_{\tilde{g}}(X, X) \geq 0$ then

$$V(r) \leq \omega_n\, r^n \tag{1.126}$$

where ω_n is the volume of the unit ball in the Euclidean space $\mathbb{R}^n$. Let us identify the fundamental group $\pi_1(M)$ to the group of all covering transformations of $\tilde{M}$ over M. Let $\Gamma \subset \pi_1(M)$ be a finitely generated subgroup, and let $g_1, \ldots, g_p : \tilde{M} \to \tilde{M}$ be a finite generator system of Γ. Let us set

---

[56] If M is a compact Riemannian manifold whose Ricci tensor is positive definite then $\pi_1(M)$ is finite.

[57] If M is a compact Riemannian manifold of sectional curvature < 0 then every abelian subgroup of $\pi_1(M)$ is cyclic.

[58] If M is a complete Riemannian manifold of sectional curvature ≤ 0 then $\pi_i(M) = 0$ for every $i \geq 2$.

$$\mu := \max\left\{\tilde{d}\big(x_0,\ g_i(x_0)\big)\ :\ 1 \le i \le p\right\}.$$

Let us observe that the ball $\tilde{B}_{\mu s}(x_0)$ contains at least $\gamma(s)$ distinct points of the set $\{g(x_0)\ :\ g \in \pi_1(M)\}$. For sufficiently small $\epsilon > 0$

$$\tilde{B}_\epsilon(x_0) \cap g\big[\tilde{B}_\epsilon(x_0)\big] = \emptyset, \quad g \in \pi_1(M) \setminus \{\mathrm{id}\}.$$

Then $\tilde{B}_{\mu s+\epsilon}(x_0)$ contains at least $\gamma(s)$ disjoint sets of the family $\{g\big[B_\epsilon(x_0)\big]\ :\ g \in \pi_1(M)\}$, thus yielding

$$\gamma(s)\, V(\epsilon) \le V(\mu s + \epsilon). \tag{1.127}$$

Finally [by (1.126)–(1.127)]

$$\gamma(s) \le c\, s^n, \quad s \ge 1, \quad c := \frac{\omega_n\, (\mu + \epsilon)^n}{V(\epsilon)}.$$

Q.e.d.

Proof of Theorem 1.161 Let $-\alpha^2 < 0$ be an upper bound on the sectional curvature of M. By a result of Günter (cf. [Gun])

$$V(r) \ge n\, \omega_n \int_0^r \left(\frac{\sinh\, \alpha x}{\alpha}\right)^{n-1} d\,x. \tag{1.128}$$

The number in the right-hand side of the inequality (1.128) is the volume of a ball of radius r in a space form of negative sectional curvature, and asymptotically

$$\omega_n \int_0^r \left(\frac{\sinh\, \alpha x}{\alpha}\right)^{n-1} d\,x \approx 2\,c\, \exp(\lambda\, r), \quad r \to \infty,$$

$$c := \frac{n\, \omega_n}{(n-1)\,(2\alpha)^n}, \quad \lambda := (n-1)\,\alpha.$$

Consequently

$$V(r) > c\, \exp(\lambda\, r) \tag{1.129}$$

for sufficiently large r. Let δ be the diameter of M. Let $\Pi\ :\ \tilde{M} \to M$ be the projection. Then $\Pi\big[\tilde{B}_\delta(x_0)\big] = M$. Therefore, the family

$$\{g\, \tilde{B}_\delta(x_0)\ :\ g \in \pi_1(M)\}$$

is a locally finite[59] covering of $\tilde{M}$. Let

$$F := \{g \in \pi_1(M) \ : \ \tilde{B}_\delta(x_0) \cap g\, \tilde{B}_\delta(x_0) \neq \emptyset\}.$$

Also, let

$$\nu := \min \left\{ d\big(g\, \tilde{B}_\delta(x_0),\ \tilde{B}_\delta(x_0)\big) \ : \ g \in \pi_1(M) \setminus F \right\} > 0.$$

We need

Lemma 1.162 *Let $B := \tilde{B}_\delta(x_0)$. If $d\big(x_0,\ g\,B\big) < \nu t + \delta$ for some $t \in \mathbb{N}$, then there exist $f_1, \ \ldots,\ f_t \in F$ such that $g = f_1 \cdots f_t$.*

Cf. Lemma 2 in [Mil], p. 3. To prove Lemma 1.162, let $y \in g\, B$ such that

$$d\big(x_0,\ y\big) < \nu t + \delta$$

and let $y_1,\ y_2,\ \cdots,\ y_{t+1} = y$ be points on the minimizing geodesic β that joins x and y. Then

$$d\big(x_0,\ y_1\big) \leq \delta, \quad d\big(y_i,\ y_{i+1}\big) > \nu,$$

and

$$y_i \in h_i\, B, \quad 1 \leq i \leq t,$$

for some $h_i \in \pi_1(M)$, with $h_1 = 1$ and $h_{t+1} = g$. Let us set $f_i := h_t^{-1} h_{i+1}$, so that $f_1\, f_2\, \ldots\, f_t = g$. As

$$h_i^{-1}\, y_i \in B, \quad h_i^{-1}\, y_{i+1} \in f_i\, B, \quad d\big(h_i^{-1}\, y_i,\ h_i^{-1}\, y_i\big) < \nu,$$

it follows[60] that $f_i \in F$. Q.e.d.

In particular F is a system of generators of $\pi_1(M)$. Let γ_F be the growth function of $\pi_1(M)$, defined by using F as a set of generators. By Lemma 1.162 the family

$$\big\{ f_1\, f_2\, \ldots\, f_t\, B \ : \ f_i \in F, \quad 1 \leq i \leq t \big\} \tag{1.130}$$

is a cover of the interior of $\tilde{B}_{\nu t+\delta}(x_0)$. The family (1.138) has cardinality $\gamma_F(t)$, hence

$$\gamma_F(t)\, V(\delta) \geq V(\nu t + \delta). \tag{1.131}$$

---

[59] Otherwise, for some $r > 0$ the (compact) neighborhood $\tilde{B}_r(x_0)$ would intersect infinitely many of the sets $g\, \tilde{B}_\delta(x_0)$, and hence $\tilde{B}_{r+\delta+\epsilon}(x_0)$ would contain infinitely many disjoint sets $g\, \tilde{B}_\epsilon(x_0)$, each of which having volume $V(\epsilon) > 0$, in contradiction with $V(r + \delta + \epsilon) < \infty$.

[60] By the very definitions of ν and F.

By Lemma 1.158 there is $k \in \mathbb{N}$ such that $\gamma(k\,t) \geq \gamma_F(t)$ or

$$\gamma(s) \geq \gamma_F\left(\left[\frac{s}{k}\right]\right), \quad s \geq k. \tag{1.132}$$

By (1.129) and (1.131)–(1.132)

$$\gamma(s) \geq c_1 \exp\left(\lambda_1 s\right)$$

$$c_1 := c\,\frac{\exp\left(\lambda(\delta - \nu)\right)}{V(\delta)} > 0, \quad \lambda_1 := \frac{\lambda\,\nu}{k} > 0.$$

for sufficiently large s. As $\gamma(s) > 1$ for any $s \in \mathbb{N}$, there is $a > 1$ such that $\gamma(s) \geq a^s$ for every s. Q.e.d.

1.12 Kanai Graphs

The exposition in this section relies on work by M. Kanai,[61] P. Soardi,[62] and P. Soardi and M. Yamasaki.[63] The notion of a *rough isometry* is however implicit in the work by M. Gromov[64] while Kanai's approach to the discretization, or combinatorial approximation, of Riemannian manifolds was suggested by J. Milnor[65], who also provided examples of pairs of Riemannian manifolds that are roughly isometric to each other.

1.12.1 Quasi-isometries, Pseudo-isometries, and Rough Isometries

Let M and N be two n-dimensional Riemannian manifolds, equipped with the Riemannian metrics g and h.

Definition 1.163 A C^∞ diffeomorphism $\phi : M \to N$ is a *quasi-isometry* if there is a constant $a \geq 1$ such that

---

[61] Cf. M. Kanai, Rough isometries, and combinatorial approximations of geometries of non-compact manifolds. J. Math. Soc. Japan **37**(3), 391–413 (1985); Rough isometries and the parabolicity of Riemannian manifolds. J. Math. Soc. Japan **38**(2), 227–238 (1986).

[62] Cf. P. Soardi, Rough isometries and Dirichlet finite harmonic functions on graphs. Proc. Amer. Math. Soc. **119**(4), 1239–1248 (1993).

[63] Cf. P. Soardi, M. Yamasaki, Parabolic index and rough isometries. Hiroshima Math. J. **23**, 333–342 (1993).

[64] M. Gromov, *Hyperbolic Manifolds, Groups, and Actions*. Annals of Mathematics Studies 97 (Princeton University Press, Princeton, 1981).

[65] Cf. J. Milnor, A note on curvature and fundamental group. J. Diff. Geom. **2**, 1–7 (1968).

$$a^{-1}\, g_x\big(v,\, v\big)^{1/2} \le h_{\phi(x)}\big((d_x\phi)v,\, (d_x\phi)v\big)^{1/2} \le a\, g_x(v,\, v)^{1/2} \tag{1.133}$$

for any $v \in T_x(M)$ and any $x \in M$. If there is a quasi-isometry $\phi : M \to N$ then M and N are *quasi-isometric* Riemannian manifolds. □

Definition 1.164 A continuous map $\phi : M \to N$ is a *pseudo-isometry* if there exist constants $a \ge 1$ and $b > 0$ such that

$$a^{-1}\, d_g(x,\, y) - b \le d_h\big(\phi(x),\, \phi(y)\big) \le a\, d_g(x,\, y) \tag{1.134}$$

for any $x,\ y \in M$. □

Here d_g denotes the distance function on M associated to the Riemannian metric g. Pseudo-isometries were introduced by G.D. Mostow[66] as a tool for studying discontinuous subgroups of semi-simple Lie groups. Every quasi-isometry is a pseudo-isometry. The converse is false, as the relation associated to (1.133) is an equivalence relation, while the relation associated to (1.134) is not. Indeed examples of complete Riemannian manifolds M and N exist such that there is a pseudo-isometry $M \to N$ while there is no isometry $N \to M$, i.e., the relation associated to (1.134) is not symmetric.

Let M be a metric space, equipped with the distance function $d : M \times M \to [0,\, +\infty)$. For every $x \in M$ let $B_r(x) \subset M$ be the metric ball of center x and radius $r > 0$. If $S \subset M$ is a subset then $B_r(S)$ denotes r-neighborhood of S, i.e., $B_r(S) = \big\{x \in M\ :\ d(x,\, S) < r\big\}$.

Definition 1.165 Let $\epsilon > 0$. A subset $S \subset M$ is *ϵ-full* if $M = B_\epsilon(S)$. Also S is *full* if it is ϵ-full for some $\epsilon > 0$. □

Definition 1.166 Let M and N be two metric spaces, with the distance functions d and ρ, respectively. A map[67] $\phi : M \to N$ is said to be a *rough isometry* if

(i) The set $\phi(M)$ is full in N,
(ii) There exist constants $a \ge 1$ and $b > 0$ such that

$$a^{-1}\, d(x,\, y) - b \le \rho\big(\phi(x),\, \phi(y)\big) \le a\, d(x,\, y) + b \tag{1.135}$$

for any $x,\ y \in M$. □

The requirement (1.135) is weaker than (1.134), and of course weaker than (1.133). The notion is essentially due to M. Gromov[68] and the relation associated to (1.135) is that of *coarse equivalence*.

---

[66] Cf. G.D. Mostow, *Strong Rigidity of Locally Symmetric Spaces*. Annals of Mathematics Studies, 78 (Princeton University Press, Princeton, 1973).

[67] Not necessarily continuous.

[68] Cf. M. Gromov, *op. cit.*, pp. 183–213.

Every quasi-isometry (respectively pseudo-isometry) with a full image is a rough isometry. Composition of two rough isometries is a rough isometry.

Lemma 1.167 *For every rough isometry $\phi : M \to N$, there is a rough isometry $\phi^- : N \to M$ such that*

$$x \in M \longmapsto d\big(\phi^- \circ \phi(x),\, x\big) \in [0,\, +\infty), \quad y \in N \longmapsto \rho\big(\phi \circ \phi^-(y),\, y\big) \in [0,\, +\infty), \tag{1.136}$$

are bounded functions.

Proof By definition the image of ϕ is full in N, i.e., there is $\epsilon > 0$ such that $B_\epsilon\big[\phi(M)\big] = N$. Let $y \in N$ be an arbitrary point. Then

$$\epsilon > \rho\big(y,\, \phi(M)\big) = \inf\Big\{\rho\big(y,\, \phi(x)\big) \ : \ x \in M\Big\} = \text{g.l.b. of } \Big\{\rho\big(y,\, \phi(x)\big) \ : \ x \in M\Big\} \Longrightarrow$$

$$\Longrightarrow \epsilon \quad \text{is not a l.b.} \quad \Longrightarrow \quad \exists\, x_\epsilon \in M \quad \text{such that } \epsilon > \rho\big(y,\, \phi(x_\epsilon)\big).$$

Let $A(y) \subset M$ be the subset consisting of all $x \in M$ such that $\rho\big(y,\, \phi(x)\big) < \epsilon$. Then $x_\epsilon \in A(y)$ so that $A(y) \neq \emptyset$. For every $y \in N$, let us pick up exactly one $x \in A(y)$ and set[69] by definition $\phi^-(y) := x$. To end the proof of Lemma 1.167 we should check (1.136). For instance, let us show that the set

$$B = \Big\{d\big(\phi^-\big[\phi(x)\big],\, x\big) \in [0,\, +\infty) \ : \ x \in M\Big\}$$

is bounded. For every $x \in M$, let us set $y := \phi(x) \in N$. By the very definition of ϕ^- one has $\phi^-(y) \in A(y)$ hence

$$\rho\big(y,\, \phi\big[\phi^-(y)\big]\big) < \epsilon$$

[where $\epsilon > 0$ was previously chosen such that $B_\epsilon\big[\phi(M)\big] = N$]. Then (as ϕ is a rough isometry)

$$\epsilon > \rho\big(y,\, \phi\big[\phi^-(y)\big]\big) = \rho\big(\phi(x),\, \phi\big[\phi^-(y)\big]\big) \geq a\, d\big(x,\, \phi^-(y)\big) - b \Longrightarrow$$

$$\Longrightarrow b + \epsilon > d_g\big(x,\, \phi^-(y)\big) = d\big(x,\, \phi^-\big[\phi(x)\big]\big).$$

Q.e.d.

Definition 1.168 A rough isometry $\phi^- : N \to M$ such as furnished by Lemma 1.167 is called a *rough inverse* of ϕ. □

Definition 1.169 Two metric spaces M and N are *roughly isometric* if there is a rough isometry $\phi : M \to N$. □

---

[69] A construction clearly *not* independent from the axiom of choice.

The relation in Definition 1.169 is an equivalence relation. This is one of the advantages of weakening the requirement (1.134) to the requirement (1.135).

Any two metric spaces of finite diameter are roughly isometric. In particular, the information that two metric spaces are roughly isometric is only interesting in the non-compact case. Rough isometries "neglect" compact factors: for any roughly isometric metric spaces M and N, and any compact metric space K, M is roughly isometric to $N \times K$.

The work by Milnor (cf. [Mil]) is thought of (by Kanai [Kan1]) as motivational[70] for the discrete, or combinatorial, approach to Riemannian geometry. Let Γ be a finitely generated group, and let $A \subset \Gamma$ be a finite system of generators. For any element $x \in \Gamma \setminus \{1\}$ let $|x|_A \in \mathbb{N}$ be the smallest positive integer k such that x is a product of k elements of $A \cup A^{-1}$, i.e.,

$$|x|_A = \min \Big\{ k \in \mathbb{N} \;:\; \exists\, a_1, \ldots, a_k \in A \cup A^{-1} \;\text{ such that }\; x = a_1 \cdots a_k \Big\}.$$

Here $A^{-1} := \big\{ a^{-1} \;:\; a \in A \big\}$. Also we set $|1|_A := 0$.

Definition 1.170 The function $|\cdot|_A : \Gamma \to \mathbb{Z}_+$ is the *word norm* of Γ with respect to the system of generators A. □

Lemma 1.171 *For any* $x,\ y \in \Gamma$

(i) $|x|_A \geq 0$ *and* $|x|_A = 0 \Longleftrightarrow x = 1$,
(ii) $|x^{-1}|_A = |x|_A$,
(iii) $|x\, y|_A \leq |x|_A + |y|_A$.

Also

(iv) *The word norms corresponding to two systems of generators A and B are equivalent, i.e., there is a constant* $a \geq 1$ *such that*

$$\frac{1}{a}\, |x|_A \leq |x|_B \leq |x|_A, \quad x \in \Gamma.$$

The proof is straightforward. The word norm is of course formally related to the growth function $\gamma = \gamma_A : \Gamma \to \mathbb{N}$. Indeed, let us set

$$E(s) := \big\{ x \in \Gamma \;:\; |x|_A \leq s \big\}, \quad s \in \mathbb{N}.$$

Then $\gamma(s) = \sharp \big[E(s) \big]$.

[70] Examples of pairs of Riemannian manifolds that are but roughly isometric to each other, are furnished in [Mil].

Lemma 1.172 *Let M be a complete Riemannian manifold, and let Γ be a finitely generated group acting freely on M as a properly discontinuous group of isometries of M such that M/Γ is compact. Let $o \in M$ be a fixed point, and let us set*

$$\|x\| = d\big(o,\, x\,o\big), \quad x \in \Gamma.$$

Then for any $x,\ y \in \Gamma$

(i) $\|x\| \geq 0$ *and* $\|x\| = 0 \Longleftrightarrow x = 1$,
(ii) $\|x^{-1}\| = \|x\|$,
(iii) $\|xy\| \leq \|x\| + \|y\|$.

The proof is straightforward. By a result of Milnor (cf. [Mil]) under the assumptions of Lemma 1.172 there exist constants $a \geq 1$ and $b \geq 0$ such that

$$\frac{1}{a}\,|x|_A - b \leq \|x\| \leq a\,|x|_A, \quad x \in \Gamma\,. \tag{1.137}$$

Next, let us set

$$\delta_A(x,\, y) := \left|x^{-1}y\right|_A, \quad x,\, y \in \Gamma.$$

Lemma 1.173 (i) *δ_A is a left invariant metric on Γ,*
(ii) *The map*

$$\varphi : \Gamma \to M, \quad \varphi(x) = x\,o, \quad x \in \Gamma$$

is a rough isometry of $(\Gamma,\, \delta_A)$ into $(M,\, d)$ and

$$\frac{1}{a}\,\delta_A(x,\, y) - b \leq d\big(\varphi(x),\, \varphi(y)\big) \leq a\,\delta_A(x,\, y), \quad x,\, y \in \Gamma.$$

In particular, as being roughly isometric is an equivalence relation, if Γ is a discrete group acting freely as a properly discontinuous group of isometries on the complete Riemannian manifolds M and N such that the quotients M/Γ and N/Γ are compact, then M is roughly isometric to N. Once again as a consequence of (1.137), J. Milnor has shown (cf. [Mil]) that the volume growth rate of M is dominated by that of Γ, i.e., there are constants $a' \geq 1$, $b' \geq 0$ and $c \geq 1$ such that

$$\frac{1}{c}\cdot \sharp\left\{x \in \Gamma \,:\, |x|_A \leq \frac{r}{a'} - b'\right\} \leq \mathrm{vol}\big[B_r(o)\big] \leq c\cdot \sharp\left\{x \in \Gamma \,:\, |x|_A \leq a'r + b'\right\}. \tag{1.138}$$

The inequalities (1.138) offer a clear indication that the geometry of the Riemannian manifold M may be approximated by the combinatorial geometry of the discrete group Γ. The merit of the work by Kanai (cf. [Kan1,Kan2]) is to show that the geometric properties of M (such as the volume growth rate of M, or the validity of an isoperimetric inequality on M) are mapped to N by any rough isometry

$\varphi : M \to N$. As rough isometries aren't assumed to be continuous maps, the local geometry of M will not be mapped to that of N, in general. To compensate for that, one confines oneself to n-dimensional Riemannian manifolds M obeying to the following condition.

Fundamental Assumption. (i) *The Ricci curvature of M is bounded from below by a constant*

$$\mathrm{Ric}(X, X) \geq -(n-1)\, K^2, \quad X \in \mathfrak{X}(M), \tag{1.139}$$

with $K > 0$, *and* (ii) *the injectivity radius of M is positive*

$$\mathrm{inj}(M) > 0. \tag{1.140}$$

As well known,[71] on every complete Riemannian manifold M satisfying the fundamental assumption (1.139)–(1.140)

$$\mathrm{vol}\left[B_r(p) \leq V(r),\right. \tag{1.141}$$

$$\mathrm{area}\left[\partial\, B_r(p)\right] \leq A(r), \tag{1.142}$$

for any $p \in M$ and any $r > 0$. Also, the function

$$r > 0 \longmapsto \frac{\mathrm{vol}\left[B_r(p)\right]}{V(r)} \tag{1.143}$$

is monotone non-increasing for any $p \in M$, and

$$\mathrm{vol}\left[B_r(p)\right] \geq V_0\, r^n, \quad p \in M, \quad r \in \left(0, \ \frac{1}{2}\,\mathrm{inj}(M)\right] \subset \mathbb{R}, \tag{1.144}$$

Here $V(r)$ is the volume of a geodesic ball in the simply connected complete n-dimensional Riemannian manifold of constant sectional curvature $-K^2$, and $A(r)$ is the "area" of the corresponding geodesic sphere. Also V_0 is a constant depending only on the dimension n and $\mathrm{inj}(M)$ denotes the injectivity radius of M.

1.12.2 Nets in Riemannian Geometry

M. Kanai's results on the invariance under rough isometries of (certain) geometric properties of a Riemannian manifold rely (cf. [Kan1, Kan2]) on the approximation of said Riemannian manifold by a combinatorial structure called a *net*. J. Milnor's

[71] The inequalities (1.141)–(1.142) and the monotonicity of the function (1.143) follow from comparison theorems in Riemannian geometry, solely under the assumption that the Ricci curvature is bounded from below. The inequality (1.144) is a result by Croke [Cro].

work shows (cf. [Mil]) that the orbit $\Gamma\, o$ of the action of Γ on M is a net in the sense of Definitions 1.174 and 1.176 below, for every Riemannian manifold M obeying to the fundamental assumptions (1.139)–(1.140) above, and by stating (1.138) we emphasized that the geometry of the discrete group is indeed intimately tied to that of the Riemannian manifold M. As it will be shown shortly, a net in a complete Riemannian manifold has a canonical combinatorial metric, corresponding to the word metric in the case of a finitely generated group. Also a net in a complete Riemannian manifold with Ricci curvature bounded from below will obey to (1.137), and hence the net will appear to be roughly isometric to the manifold.

1.12.2.1 Net Structures, Nets, Combinatorial Metrics

Let P be a countable set.

Definition 1.174 A family $N = \{N(p) \, : \, p \in P\}$ of parts $N(p)$ of P is called a *net structure* on P if

(i) $N(p)$ is a finite subset of P,
(ii) For any $p, q \in P$

$$q \in N(p) \Longleftrightarrow p \in N(q).$$

□

Definition 1.175 Let $N = \{N(p) \, : \, p \in P\}$ be a net structure on P. Every point $q \in N(p)$ is called a *neighbor* of p. □

Definition 1.176 A pair (P, N) consisting of a countable set P and a net structure N of P is called a *net*. □

Every net is a locally finite countable graph.

Definition 1.177 Let (P, N) be a net, and let $\{p_j \, : \, 0 \le j \le \ell\}$ be a system of points in P. Then $\mathbf{p} = (p_0, \dots, p_\ell)$ is called a *path* from p_0 to p_ℓ if $p_j \in N(p_{j-1})$ for any $1 \le j \le \ell$. The number $\ell(\mathbf{p}) := \ell \in \mathbb{Z}_+$ is the *length* of the path $\mathbf{p}$. □

Definition 1.178 A net (P, N) is *connected* if for any $p, q \in P$ there is a path $\mathbf{p} = (p_0, \dots, p_\ell)$ from $p_0 = p$ to $p_\ell = q$. □

Let $\Omega_P(p, q)$ be the set of all paths in the net $\{P, N\}$ connecting p to q.

Definition 1.179 Let (P, N) be a connected net, the distance function

$$\delta(p, q) = \min \{\ell(\mathbf{p}) \, : \, \mathbf{p} \in \Omega_P(p, q)\}$$

is the *combinatorial metric* of the net (P, N). □

Let $\sharp(A)$ denote the cardinality of the set A.

Definition 1.180 A net (P, N) is *uniform* if $\sup\{\sharp[N(p)] : p \in P\} < \infty$. □

Lemma 1.181 (i) *Let (P, N) be a uniform, connected net. Then there is $\lambda \geq 1$ such that*

$$\sharp\Big\{p \in P : \delta(p, S) \leq r\Big\} \leq \lambda^r \, \sharp(S)$$

for any $r \geq 0$ and any finite subset $S \subset P$.

(ii) *Let P and Q be two connected nets, with P uniform, and let $\varphi : P \to Q$ be a rough isometry with respect to the combinatorial metrics of P and Q. Then there is $\mu > 0$ such that*

$$\sharp(S) \leq \mu \, \sharp\big[\varphi(S)\big]$$

for any finite subset $S \subset P$.

1.12.2.2 Nets on Riemannian Manifolds

Let (M, g) be a complete Riemannian manifold, and let $d = d_g$ be the distance function on M springing from the given Riemannian metric g.

Definition 1.182 Let $P \subset M$ be a subset and $\epsilon > 0$ a positive number. The set P is a said to be *ϵ-separated* if

$$d(p, q) \geq \epsilon$$

for any pair of distinct points $p, q \in M$, $p \neq q$. □

Definition 1.183 Let $P \subset M$ be an ϵ-separated set. P is said to be *maximal* if it is maximal with respect to inclusion, i.e., for every ϵ-separated set $Q \subset M$, if $P \subset Q$ then $P = Q$. □

Exercise 1.184 Show that any maximal ϵ-separated subset $P \subset M$ is ϵ-full in M. □

Solution. We must show that $B_\epsilon(P) = M$. We argue by contradiction, i.e., let us assume that $M \setminus B_\epsilon(P) \neq \emptyset$, and let $x_0 \in M \setminus B_\epsilon(P)$. Then $x_0 \notin B_\epsilon(P) = \big\{x \in M : d(x, P) < \epsilon\big\}$ hence

$$\epsilon \leq d(x_0, P) = \inf\big\{d(x_0, x) : x \in P\big\}$$

so that $x_0 \notin P$ and $d(x_0, x) \geq \epsilon$ for every $x \in P$. This observation, together with the fact that P is ϵ separated, implies that $Q := P \cup \{x_0\}$ is ϵ-separated, and $Q \supset P$ (strict inclusion), in contradiction with the maximality of P. Q.e.d.

Let (M, g) be a complete Riemannian manifold, and let d be the distance function on M associated with g. Let $\epsilon > 0$ and let $P \subset M$ be a maximal ϵ-separated subset. For every $p \in P$ let us set

$$N(p) := \{q \in P \ : \ 0 < d(p, q) \leq 2\epsilon\}, \quad N := \{N(p) \ : \ p \in P\}. \tag{1.145}$$

Exercise 1.185 Show that N is a net structure of P. □

Exercise 1.186 Let $P \subset M$ be a maximal ϵ-separated subset of a complete Riemannian manifold M. Let N be the net structure of P given by (1.145). Show that the net (P, N) is connected provided that M is connected. □

Definition 1.187 A maximal ϵ-separated subset $P \subset M$ of a complete Riemannian manifold, together with the net structure given by (1.145), is called an *ϵ-net* in M. □

Lemma 1.188 *Let M be a complete n-dimensional Riemannian manifold with Ricci curvature bounded by below, i.e.,*

$$\operatorname{Ric}(X, X) \geq -(n-1)K^2$$

for some $K > 0$ and any $X \in \mathfrak{X}(M)$. Let $P \subset M$ be an ϵ-separated set. Then for any $r > 0$ and any $x \in M$ there is $\nu = \nu(n, K, \epsilon, r) > 0$ such that

$$\sharp\left\{p \in P : x \in B_r(p)\right\} \leq \nu.$$

In particular, every ϵ-net in a complete Riemannian manifold with Ricci curvature bounded from below is uniform.

Cf. Lemma 2.3 in [Kan1], p. 397.

Proof of Lemma 1.188 Let $r > 0$ and $x \in M$ and let us set

$$P_x := \{p \in P \ : \ x \in B_r(p)\}.$$

Note that

$$B_{\epsilon/2}(p) \subset B_{r+\epsilon/2}(x) \subset B_{2r+\epsilon/2}(p) \tag{1.146}$$

for any $p \in P_x$. Indeed, let $y \in B_{\epsilon/2}(p)$ so that $d(p, y) < \epsilon/2$. Yet $p \in P_x \Longrightarrow x \in B_r(p) \Longrightarrow d(p, x) < r$ and hence

$$d(x, y) \leq d(x, p) + d(p, y) < r + \frac{\epsilon}{2} \Longrightarrow y \in B_{r+\epsilon/2}(p),$$

i.e., $B_{\epsilon/2}(p) \subset B_{r+\epsilon/2}(x)$. Q.e.d. Similarly, let $y \in B_{r+\epsilon/2}(x)$ so that

$$y \in B_{r+\epsilon/2}(x) \Longrightarrow d(x, y) < r + \frac{\epsilon}{2},$$

$$d(p, y) \leq d(p, x) + d(x, y) < r + r + \frac{\epsilon}{2} \Longrightarrow y \in B_{2r+\epsilon/2}(p)$$

yielding $B_{r+\epsilon/2}(x) \subset B_{2r+\epsilon/2}(p)$. Q.e.d. Next, let us observe that

$$\text{vol}\big[B_{\epsilon/2}(p)\big] \geq \frac{V\left(\frac{\epsilon}{2}\right)}{V\left(2r + \frac{\epsilon}{2}\right)} \text{vol}\big[B_{2r+\epsilon/2}(p)\big]. \tag{1.147}$$

To check (1.147) let us consider the function

$$f : M \times (0, +\infty) \to \mathbb{R}, \quad f(x, r) = \frac{\text{vol}\left[B_r(x)\right]}{V(r)}.$$

As $r_1 \leq r_2 \Longrightarrow f(x, r_1) \geq f(x, r_2)$ it follows that

$$\frac{\epsilon}{2} < 2r + \frac{\epsilon}{2} \Longrightarrow f\left(p, \frac{\epsilon}{2}\right) \geq f\left(p, 2r + \frac{\epsilon}{2}\right)$$

which is (1.147). Moreover, let us observe that the family

$$\left\{B_{\epsilon/2}(p) \ : \ p \in P\right\}$$

consists of mutually disjoint balls. We argue by contradiction. If there are $p_1, \ p_2 \in P$ such that $p_1 \neq p_2$ and $B_{\epsilon/2}(p_1) \cap B_{\epsilon/2}(p_2) \neq \emptyset$, let $y \in B_{\epsilon/2}(p_1) \cap B_{\epsilon/2}(p_2)$. Then

$$d\big(p_j, y\big) < \frac{\epsilon}{2}, \quad j \in \{1, 2\},$$

so that [as P is ϵ-separated and $p_1 \neq p_2$]

$$\epsilon \leq d(p_1, p_2) \leq d(p_1, y) + d(y, p_2) < \frac{\epsilon}{2} + \frac{\epsilon}{2} = \epsilon,$$

a contradiction. In particular the balls $\{B_{\epsilon/2}(p) \ : \ p \in P_x\}$ are mutually disjoint, and [by (1.146)] are contained in $B_{r+\epsilon/2}(x)$, hence

$$\text{vol}\left[B_{r+\epsilon/2}(x)\right] \geq \sum_{p\in P_x} \text{vol}\left[B_{\epsilon/2}(p)\right] \geq \quad \text{[by (1.147)]}$$

$$\geq \sum_{p\in P_x} \frac{V\left(\frac{\epsilon}{2}\right)}{V\left(2r + \frac{\epsilon}{2}\right)} \text{vol}\big[B_{2r+\epsilon/2}(p)\big] \geq$$

[of course $B_{2r+\epsilon/2}(p) \supset B_{r+\epsilon/2}(x)$ implies $\text{vol}\big[B_{2r+\epsilon/2}(p)\big] \geq \text{vol}\big[B_{r+\epsilon/2}(x)\big]$]

$$\geq \frac{V\left(\frac{\epsilon}{2}\right)}{V\left(2r+\frac{\epsilon}{2}\right)} \text{vol}\big[B_{r+\epsilon/2}(x)\big] \,\sharp\,(P_x)$$

that is

$$\sharp\,(P_x) \leq \frac{V\left(2r+\frac{\epsilon}{2}\right)}{V\left(\frac{\epsilon}{2}\right)} =: \nu\,.$$

Q.e.d.

Lemma 1.189 *Let M be a n-dimensional complete Riemannian manifold with Ricci curvature bounded from below by $-(n-1)\,K^2$, and let $P \subset M$ be an ϵ-net. Let δ be the combinatorial metric of P. Then $(P,\,\delta) \simeq (M,\,d)$ (a rough isometry), i.e.,*

$$\frac{1}{2\epsilon}\,d\big(p_1,\,p_2\big) \leq \delta\big(p_1,\,p_2\big) \leq a\,d\big(p_1,\,p_2\big) + b, \quad p_1,\,p_2 \in P \tag{1.148}$$

for some constants $a \geq 1$ and $b \geq 0$ depending only on n, K and ϵ. Consequently the inclusion $P \hookrightarrow M$ is a rough isometry.

Cf. Lemma 2.5 in [Kan1], p. 397. Lemma 1.189 generalizes J. Milnor's inequality (1.137) and provides a basic tool for understanding the geometry of a Riemannian manifold M in terms of the combinatorial geometry of a ϵ-net in M.

Proof of Lemma 1.189 Let $p_1,\,p_2 \in P$. If $p_1 = p_2$ there is nothing to check. Let $p_1 \neq p_2$. To check the first inequality in (1.148) let $\mathbf{q} \in \Omega_P(p_1,\,p_2)$ be a path connecting p_1 and p_2, i.e., $\mathbf{q} = (q_0,\,\ldots,\,q_\ell)$ with $q_j \in P$ for any $0 \leq j \leq \ell$, and $q_0 = p_1, q_\ell = p_2$, and

$$q_j \in N(q_{j-1}) = \big\{q \in P \;:\; 0 < d(q_{j-1},\,q) \leq 2\,\epsilon\big\}$$

for any $1 \leq j \leq \ell$, and $\ell(\mathbf{q}) = \ell$. Then

$$d\big(p_1,\,p_2\big) \leq \sum_{j=0}^{\ell-1} d\big(q_j,\,q_{j+1}\big) \leq 2\,\epsilon\,\ell = 2\,\epsilon\,\ell(\mathbf{q})$$

so that $(1/(2\epsilon))\,d\big(p_1,\,p_2\big)$ is a lower bound of the set $\Delta(p_1,\,p_2) := \big\{\ell(\mathbf{q}) \;:\; \mathbf{q} \in \Omega_P(p_1,\,p_2)\big\}$. Consequently

$$\delta\big(p_1,\,p_2\big) = \inf \Delta(p_1,\,p_2) = \text{g. l. b. of } \Delta(p_1,\,p_2) \geq \frac{1}{2\epsilon}\,d\big(p_1,\,p_2\big).$$

Q.e.d. To check the second inequality in (1.148) let γ be a minimizing geodesic in M, parametrized by arc length, joining p_1 and p_2. Let us consider the set

$$P_\gamma := \{q \in P \ : \ B_\epsilon(q) \cap \gamma \neq \emptyset\}.$$

Claim 1. $\{B_\epsilon(q) \ : \ q \in P_\gamma\}$ *covers* γ.

Proof The proof relies on the fact that any maximal ϵ-separated subset $P \subset M$ is ϵ-full. Indeed, let x be a point of γ, so that

$$x \in \gamma \subset M = B_\epsilon(P) \Longrightarrow \epsilon > d(x, P) = \inf\{d(x, p) \ : \ p \in P\} \Longrightarrow$$

$$\Longrightarrow \ \epsilon \text{ is not a l. b. of the set } \{d(x, p) \ : \ p \in P\} \Longrightarrow \ \exists\, p \in P \ : \ d(x, p) < \epsilon \Longrightarrow x \in B_\epsilon(p).$$

Q.e.d.

Claim 2. $\delta(p_1, p_2) \leq \sharp(P_\gamma)$.

Proof

1.12.3 Volume Growth Rate

The strategy of the proofs of M. Kanai's results (cf. *op. cit.*) on the invariance under rough isometries of Riemannian geometric properties may be briefly described as follows. Let M and N be complete Riemannian manifolds obeying to the fundamental assumptions (1.139)–(1.140), and let us assume that $M \simeq N$ (a rough isometry). Any rough isometry $M \simeq N$ will induce a rough isometry $P \simeq Q$ between nets $P \subset M$ and $Q \subset N$ [as a consequence of an inequality similar to (1.137)]. On the other hand, an inequality such as (1.138) should imply that the geometries of P and Q are the same as those of M and N. At this point, all one needs to show is that the roughly isometric nets $P \simeq Q$ have the same geometry, which is commonly not a difficult task. Therefore, M. Kanai's work focuses on establishing discrete approximation results such as (1.138).

Let M be a Riemannian manifold and $o \in M$.

Definition 1.190 Let $m \in \mathbb{Z}_+$. We say M is a Riemannian manifold of *polynomial growth* of *order* m if

$$\inf\left\{k \in \mathbb{N} \ : \ \limsup_{r\to\infty} \frac{\text{vol}\left[B_r(0)\right]}{r^k}\right\} = m. \tag{1.149}$$

□

Definition 1.191 M is said to be a Riemannian manifold of *exponential growth* if

$$\liminf_{r\to\infty} \frac{\log \operatorname{vol}\left[B_r(o)\right]}{r} > 0. \tag{1.150}$$

□

Definitions 1.190 and 1.191 do not depend upon the choice of point $o \in M$.

Theorem 1.192 *Let $V \in \{M, N\}$ be two complete Riemannian manifolds of Ricci curvature bounded from below by a constant $-\{\dim(V) - 1\} K^2$, $K > 0$, and of positive injectivity radius* $\operatorname{inj}(V) > 0$. *Assume that $M \simeq N$ (a rough isometry). Then*

$$M \text{ is of polynomial growth of order } m \Longleftrightarrow N \text{ is of polynomial growth of order } m,$$

$$M \text{ is of exponential growth} \Longleftrightarrow N \text{ is of exponential growth.}$$

Cf. Theorem 3.3. in [Kan1], p. 398. The proof of Theorem 1.192 is to show that the growth rate of a Riemannian manifold is approximated by that of a ϵ-net in that manifold, where the growth rate of a net is defined as follows.

Definition 1.193 Let P be a net and $o \in P$ a point. P is said to be of *polynomial growth* of *order m* if

$$\inf \left\{k > 0 \;:\; \limsup_{r\to\infty} \frac{\sharp\left\{p \in P \;:\; \delta(o, p) < r\right\}}{r^k} = m.\right.$$

P is said to be of *exponential growth* if

$$\liminf_{r\to\infty} \frac{\log \sharp\left\{p \in P \;:\; \delta(o, p) < r\right\}}{r} > 0.$$

□

The reader should observe that the definition of growth rate for a net is defined by formally replacing $\operatorname{vol}\left[B_r(o)\right]$ in (1.149)–(1.150) by $\sharp\left\{p \in P \;:\; \delta(o, p) < r\right\}$. The meaning of M. Kanai's result is that the volume growth rates of geodesic balls in Riemannian manifolds are invariant under rough isometries. The proof of Theorem 1.192 requires.

Lemma 1.194 *Let P and Q be two roughly isometric uniform nets. Then*

$$P \text{ is of polynomial growth of order } m \Longleftrightarrow Q \text{ is of polynomial growth of order } m,$$

$$P \text{ is of exponential growth} \Longleftrightarrow Q \text{ is of exponential growth.}$$

Lemma 1.195 *Let M be a complete Riemannian manifold of Ricci curvature bounded from below by $-(n-1)K^2$ and with* $\operatorname{inj}(M) > 0$*. Let $P \subset M$ be a ϵ-net. Then*

$$M \text{ is of polynomial growth of order } m \iff P \text{ is of polynomial growth of order } m,$$

$$M \text{ is of exponential growth} \iff P \text{ is of exponential growth.}$$

Cf. Lemmas 3.4 and 3.6 in [Kan1], p. 399.

Proof of Theorem 1.192 Let M and N be two Riemannian manifolds obeying to the assumptions of Theorem 1.192, and let $P \subset M$ and $Q \subset N$ be nets. Note that both P and Q are uniform. By Lemma 1.189 any rough isometry $M \simeq N$ induces a rough isometry $P \simeq Q$. Hence P and Q have the same growth rate. Yet (by Lemma 1.195) the growth rates of P and Q coincide with those of M and N, respectively. We may conclude that geodesic balls in M and N have the same volume growth rate. Q.e.d.

1.12.4 Isoperimetric Constants

Let M be a n-dimensional complete Riemannian manifold, and let $0 < m \le \infty$. Let us set

$$I_m(M) = \begin{cases} \inf\limits_{\Omega} \dfrac{\operatorname{area}\left(\partial\Omega\right)}{\left(\operatorname{vol}\Omega\right)^{(m-1)/m}} & \text{if } 0 < m < \infty, \\[2ex] \inf\limits_{\Omega} \dfrac{\operatorname{area}\left(\partial\Omega\right)}{\operatorname{vol}\left(\Omega\right)} & \text{if } m = \infty. \end{cases}$$

where Ω ranges over all nonempty smoothly bounded domains $\Omega \subset M$ such that area $\left(\partial\Omega\right)$ is well defined.

Definition 1.196 $I_m(M)$ is the *m-dimensional isoperimetric constant* $I_m(M)$ of the Riemannian manifold M. □

The isoperimetric inequality for the Euclidean space $\mathbb{R}^n$ implies that

$$I_m(\mathbb{R}^n) = \begin{cases} \dfrac{\omega_{n-1}^{1/n}}{n^{(n-1)n}} & \text{if } m = n, \\[2ex] 0 & \text{if } m \neq n. \end{cases}$$

The isoperimetric constant vanishes if M is compact. If M is a complete Riemannian manifold then

$$I_n(M) \leq I_n(\mathbb{R}^n),$$

and if $m < n$ then[72] $I_m(M) = 0$. If M is a simply connected complete Riemannian manifold M whose sectional curvature is bounded from above by a negative constant, then $I_\infty(M) > 0$. Also, if M is a n-dimensional complete Riemannian manifold of non-negative Ricci curvature then $I_m(M) = 0$ for any $m > n$. Cf. Buser [Bus], Osserman [Oss], and Yau [Yau], for an array of results on isoparametric constants and ramifications.

Theorem 1.197 *Let $V \in \{M, N\}$ be two complete Riemannian manifolds of Ricci curvature bounded from below by a constant $-\big(\dim(V) - 1\big) K^2$ with $K > 0$, and such that* $\operatorname{inj}(V) > 0$. *Let us assume that M and N are roughly isometric and that* $\dim(V) \leq m \leq \infty$. *Then*

$$I_m(M) > 0 \Longleftrightarrow I_m(N) > 0.$$

Cf. Theorem 4.1 in [Kan1], p. 400. The strategy of the proof of Theorem 1.197 is the same as that of Theorem 1.192, that is one first introduces "isoperimetric constants" for nets, and proves that the statement holds in the discrete approximation setting.

Let P be a net.

Definition 1.198 The *boundary* ∂S of a subset $S \subset P$ is

$$\partial S := \big\{p \in P \ : \ \delta(p, S) = 1\big\}.$$

□

Definition 1.199 The *m-dimensional isoperimetric constant* of P is

$$I_m(P) := \inf_S \frac{\sharp(\partial S)}{(\sharp\, S)^{(m-1)/m}}$$

where S ranges over all nonempty finite subsets of P. □

The proof of Theorem 1.197 relies on the following lemmas.

Lemma 1.200 *Let P and Q be roughly isometric uniform nets. Then*

$$I_m(P) > 0 \Longleftrightarrow I_m(Q) > 0.$$

[72] By looking at $\lim_{r \to 0} \operatorname{area}\big[\partial B_r(o)\big] / \big\{\operatorname{vol} B_r(o)\big\}^{(m-1)/m}$.

Lemma 1.201 *Let M be a n-dimensional complete Riemannian manifold of Ricci curvature bounded from below by $-(n-1)K^2$ and of positive injectivity radius* $\operatorname{inj}(M) > 0$. *Let $P \subset M$ be an ϵ-net. Then*

$$I_m(M) > 0 \Longleftrightarrow I_m(P) > 0$$

for every $m \in [n, \infty]$.

1.12.5 Liouville's Theorem

By the classical theorem of Liouville, every positive harmonic function on the Euclidean space is a constant. By a result of Yau (cf. [Yau]) the statement holds[73] on any Riemannian manifold M of non-negative Ricci curvature. The work[74] by Moser [Mos], admits a geometric interpretation (cf. Greene and Wu [GrWu]) implying that on any $M \simeq \mathbb{R}^n$ (a quasi-isometry), all positive harmonic functions are constants. By Theorem 1.197 certain isoperimetric inequalities are invariant under rough isometries. This suggests that J. Moser's result (cf. *op. cit.*) should extend to Riemannnian manifolds M that are roughly isometric to the Euclidean space. As shown by M. Kanai (cf. *op. cit.*) this is indeed the case under the additional assumption that M has Ricci curvature bounded from below by $-(n-1)K^2$ and $\operatorname{inj}(M) > 0$. Precisely

Theorem 1.202 *Let M be a n-dimensional complete Riemannian manifold with* $\operatorname{Ric}(X, X) \geq -(n-1)K^2$ *and* $\operatorname{inj}(M) > 0$. *Assume that M is roughly isometric to $\mathbb{R}^m$ for some $m \geq n$. Then every positive harmonic function on M is a constant.*

Cf. Theorem 5.1 in [Kan1], p. 404. Theorem 1.202 follows immediately from the following lemmas.

Lemma 1.203 *Let M be a complete Riemannian manifold with* $\operatorname{Ric}(X, X) \geq -(n-1)K^2$ *and* $\operatorname{inj}(M) > 0$, *and such that $M \simeq \mathbb{R}^2$ (a rough isometry). Then every positive superharmonic function is a constant.*

Lemma 1.204 *Let M be a complete Riemannian manifold with* $\operatorname{Ric}(X, X) \geq -(n-1)K^2$ *and* $\operatorname{inj}(M) > 0$, *and such that $M \simeq \mathbb{R}^m$ (a rough isometry) for some* $m \geq \max\{n, 3\}$. *Then there are bounded domains $\Omega_k \subset M$, $k \in \mathbb{N}$, such that*

---

[73] The proof relies on a gradient estimate for positive harmonic functions.

[74] Relying on the Sobolev inequalities, which are known to be closely related to the isometrimetric inequalities.

(i) $\overline{\Omega}_k \subset \Omega_{k+1}$ *and* $M = \bigcup_{k=1}^{\infty} \Omega_k$,

(ii) *For every* $k \geq 1$ *and every positive harmonic function* u *on* Ω_k, *the Harnack inequality*

$$\sup_{\Omega_k} u \leq h \inf_{\Omega_k} u \tag{1.151}$$

holds for some constant $h > 0$ *independent of* u *and* k.

Cf. Propositions 5.2 and 5.3 in [Kan1], p. 404.

1.12.6 Rough Isometries and Parabolicity of Riemannian Manifolds

By a result of Kanai [Kan2], parabolicity is also preserved by rough isometries.

Definition 1.205 A complete Riemannian manifold M is *parabolic* if all positive superharmonic functions on M are constants. □

Parabolicity may be characterized as follows. Let $p_t(x, y)$, with $t > 0$ and $x, y \in M$, be the minimal positive fundamental solution to the heat equation $\partial u/\partial t = \Delta u$ [with u defined on $(0, +\infty) \times M$], cf. e.g., Itô [Ito1], Dodziuk [Dod1]. Then M is non-parabolic if and only if the Green function

$$g(x, y) = \int_0^{\infty} p_t(x, y)\, dt$$

exists (cf. Itô [Ito2]). The Euclidean space $\mathbb{R}^n$ is non-parabolic if and only if $n \geq 3$.

Theorem 1.206 *Let* $V \in \{M, N\}$ *be two complete Riemannian manifolds of Ricci curvature bounded from below by* $-\big(\dim(V) - 1\big)K^2$, *of positive injectivity radius* $\mathrm{inj}(V) > 0$, *and such that* $M \simeq N$ *(a rough isometry). Then*

$$M \text{ is parabolic} \iff N \text{ is parabolic}.$$

Cf. Theorem 1 in [Kan2], p. 228.

Corollary 1.207 *Let* M *be a normal covering of a compact Riemannian manifold whose deck transformation group is abelian of rank* m. *Then*

$$M \text{ is parabolic} \iff m \leq 2.$$

This is a result by Lyons and Sullivan [LySu]. Corollary 1.207 follows from Theorem 1.206 because M is roughly isometric to a Euclidean space, unless M is compact (cf. [Kan1] or our preceding section). The proof of Theorem 1.206 relies again on Kanai's approximation of the "continuous" geometry of a Riemannian manifold M by the combinatorial geometry of a ϵ-net in M. Let P be a net, and let us consider the linear operator L acting on functions u defined on P, given by

$$Lu(p) = \frac{1}{\sharp N(p)} \sum_{q \in N(p)} u(q) - u(p), \quad p \in P.$$

As well known L enjoys[75] properties similar to those of the Laplace-Beltrami operator on a Riemannian manifold.

Definition 1.208 A function u on P is *superharmonic* if $Lu \leq 0$. □

Definition 1.209 A net P is *parabolic* if every positive superharmonic function on P is a constant. □

Let M be a complete Riemannian manifold, and let $P \subset M$ be a maximal ϵ-separated subset. Let us set $N(p) := \{q \in P \,:\, 0 < d(p, q) \leq 3\epsilon\}$. Then $\{N(p) \,:\, p \in P\}$ is a net structure of P, and the resulting net will still[76] be called a ϵ-net.

Theorem 1.210 *Let M be a n-dimensional complete Riemannian manifold of Ricci curvature bounded from below by a constant $-(n-1)K^2$ with $K > 0$, and of positive injectivity radius* $\mathrm{inj}(M) > 0$. *Let $\epsilon > 0$ and let $P \subset M$ be a ϵ-net. Then*

$$M \ \textit{is parabolic} \iff P \ \textit{is parabolic}.$$

Cf. Theorem 2 in [Kan2], p. 229. Theorem 1.206 follows from Theorem 1.210. Results similar to Theorem 1.210 were obtained by Varopoulos [Var], and Lyons and Sullivan [LySu], by constructing a discrete random walk approximating the Brownian motion on a Riemannian manifold. Instead, the approach in [Kan2] relies on a criterion of parabolicity in terms of the capacity of a bounded domain in the given Riemannian manifold, and said criterion is established by a variational argument.

---

[75] Cf. e.g., Dodziuk [Dod1], Dynkin and Yushkevic [DyYu].

[76] There is a slight discrepancy (with respect to the previous sections) in the definition of the set $N(p)$ of all neighbors of p [previously $N(p) = \{q \in P \,:\, 0 < d(p, q) \leq 2\epsilon\}$] that is required in [Kan2] for technical reasons.

1.12.6.1 Capacity and Parabolicity

Let (M, g) be a complete Riemannian manifold, and let $\Omega \subset M$ be a smoothly bounded domain.

Definition 1.211 The *capacity* of Ω is

$$\operatorname{cap}(\Omega) = \inf\left\{ \int_M |\nabla u|^2 \, d\,\mathrm{v}_g \; : \; u \in C_0^\infty(M), \;\; u\big|_\Omega = 1 \right\}.$$

□

Lemma 1.212 *M is non-parabolic* $\Longleftrightarrow$ cap $(\Omega) > 0$.

Cf. Proposition 3 in [Kan2], p. 230.

Proof of Lemma 1.212 We follow the arguments by Fischer-Colbrie and Schoen [FCSc].

"$\Longrightarrow$" Let M be non-parabolic. Let $p \in \Omega$ be a point, and let us set $v(x) = \log g(p, x)$, where g is the Green function. Then, for any $u \in C_0^\infty(M)$ with $u = 1$ on Ω (by Green's lemma)

$$\int_{M\setminus\Omega} u^2 \, |\nabla v|^2 \, d\,\mathrm{v}_g = -\int_{M\setminus\Omega} u^2 \, \Delta v \, d\,\mathrm{v}_g =$$

$$= -\int_{\partial\Omega} \frac{\partial v}{\partial \nu} \, d\,\sigma + 2\int_{M\setminus\Omega} u \, g\big(\nabla u, \, \nabla u\big) \, d\,\mathrm{v}_g \leq$$

$$\leq -\int_{\partial\Omega} \frac{\partial v}{\partial \nu} \, d\,\sigma + 2\int_{M\setminus\Omega} |u| \, |\nabla u| \, |\nabla v| \, d\,\mathrm{v}_g \leq$$

$$\leq -\int_{\partial\Omega} \frac{\partial v}{\partial \nu} \, d\,x + \int_{M\setminus\Omega} |\nabla u|^2 \, d\,x + \int_{M\setminus\Omega} u^2 \, |\nabla v|^2 \, d\,\mathrm{v}_g$$

yielding

$$\int_{\partial\Omega} \frac{\partial v}{\partial \nu} \, d\,\sigma \leq \int_M |\nabla u|^2 \, d\,\mathrm{v}_g \, .$$

Here ν is the inward normal vector field on $\partial\Omega$. Consequently

$$\operatorname{cap}(\Omega) \geq \int_{\partial\Omega} \frac{\partial v}{\partial \nu} \, d\,\sigma > 0.$$

Q.e.d.

"$\Longleftarrow$" Let $\operatorname{cap}(\Omega) > 0$. Let $\Omega_k \subset M$ be a sequence of smoothly bounded domains such that

$$\overline{\Omega} \subset \Omega_1, \quad \overline{\Omega}_k \subset\subset \Omega_{k+1}, \quad M = \bigcup_{k \geq 1} \Omega_k .$$

For each $k \geq 1$ there is a function $u_k \in C^\infty\big(\overline{\Omega}_k \setminus \Omega\big)$ which is harmonic in $\Omega_k \setminus \overline{\Omega}$ and $u_k = 1$ o $\partial\Omega$ and $u_k = 0$ on $\partial\Omega_k$. Let us observe that

$$\operatorname{cap}(\Omega) = \lim_{k \to \infty} \int_{\Omega_k \setminus \Omega} \left|\nabla u_k\right|^2 d\, \mathrm{v}_g .$$

By Harnack's inequality and using Schauder estimates, one may pick[77] up a subsequence $\{u_{k_j}\}_{j \geq 1}$ of $\{u_k\}_{k \geq 1}$ that converges in the Hölder $C^{2,\alpha}$-norm, on any compact subset of $M \setminus \Omega$, to a positive function $u \in C^\infty(M \setminus \Omega)$ that is harmonic in $M \setminus \overline{\Omega}$, and satisfies $u = 1$ on $\partial\Omega$. The extension of u by $u = 1$ on Ω is a positive superharmonic function on M. Hence, to prove non-parabolicity of M, it suffices to show that u is not constant. Indeed (by Green's lemma)

$$\int_{\partial\Omega} \frac{\partial u}{\partial \nu}\, d\sigma = \lim_{j \to \infty} \int_{\partial\Omega} \frac{\partial u_{k_j}}{\partial \nu}\, d\sigma = \lim_{j \to \infty} \left|\nabla u_{k_j}\right|^2 d\, \mathrm{v}_g = \operatorname{cap}(\Omega) > 0$$

implying that u is not constant. Q.e.d.

1.12.6.2 Discrete Potential Theory

A few potential theoretic aspects of nets are briefly recalled, by following Dodziuk [Dod2]. Let P be a net, and let us set

$$\nu(p) = \sharp\big[N(p)\big], \quad \pi(p, q) = \begin{cases} \dfrac{1}{\nu(p)} & \text{if } q \in N(p), \\ 0 & \text{otherwise.} \end{cases} ,$$

so that the combinatorial Laplacian reads

$$Lu(p) = \sum_{q \in N(p)} \pi(p, q)\, u(q) - u(p), \quad p \in P,$$

[77] Cf. e.g., Gilbarg and Trudinger [GiTr].

for any function $u : P \to \mathbb{R}$. Recall that a function $u : P \to \mathbb{R}$ is superharmonic if $Lu \leq 0$, and that P is parabolic when[78] there are no positive superharmonic functions on P excepts for (positive) constants. Next, let us consider the functions

$$\pi_k : P \times P \to \mathbb{R}, \quad k \in \mathbb{Z}_+,$$

defined recurrently by

$$\pi_0 : P \times P \to \mathbb{R}, \quad \pi(p, q) := \begin{cases} 1 & \text{if } p = q, \\ 0 & \text{if } p \neq q, \end{cases}$$

$$\pi_{k+1}(p, q) = \sum_{r \in P} \pi_k(p, r)\, \pi(r, q), \quad k \geq 0.$$

π_k is a discrete analog to the heat kernel on a Riemannian manifold.

Definition 1.213 The *Green function* of P is

$$g(p, q) = \sum_{k=0}^{\infty} \pi_k(p, q).$$

□

Exercise 1.214 Show that

(1) If P is connected and $g(p_0, q_0) < \infty$ for some $p_0, q_0 \in P$, then $g(p, q) < \infty$ for any $p, q \in P$.

(2) If $g < \infty$ then for any $p, q \in P$

$$Lg_q(p) = \begin{cases} -1 & \text{if } p = q, \\ 0 & \text{if } p \neq q, \end{cases}$$

where

$$g_q : P \to \mathbb{R}, \quad g_q(p) = g(p, q).$$

□

The following result is a discrete analog to a result by Itô [Ito2] (and will be referred to as *Itô's Lemma*).

[78] A known characterization of parabolicity within probability theory is that P is parabolic $\iff$ the Markov chain on P with transition probability π is recurrent.

Lemma 1.215 *P is non-parabolic* $\Longleftrightarrow g < \infty$.

Cf. Lemma 4 in [Kan2], p. 232. The proof (a mere algebraic manipulation) is left as an exercise to the reader.

Definition 1.216 For every pair of functions $u, v : P \to \mathbb{R}$ we consider the functions

$$\langle Du, Dv \rangle, \quad |Du| : P \to \mathbb{R},$$

$$\langle Du, Dv \rangle(p) := \sum_{q \in N(p)} \{u(q) - u(p)\} \{v(q) - v(p)\},$$

$$|Du|(p) := \sqrt{\langle Du, Dv \rangle(p)}, \quad p \in P.$$

□

Definition 1.217 The *support* of a function $u : P \to \mathbb{R}$ is $\text{Supp}(u) = \{p \in P : u(p) \neq 0\}$. □

Lemma 1.218 *Let $u, v : P \to \mathbb{R}$ be two functions, at least one of finite support. Then*

$$\sum_{p \in P} \Big\{2 \, \nu \, u \, Lv + \big\langle Du, Dv \big\rangle \Big\}(p) = 0. \tag{1.152}$$

The proof of Lemma 1.218 is straightforward.

Definition 1.219 Let $S \subset P$ be a finite set. The *capacity* of S is

$$\text{cap}(S) := \inf \Big\{ \sum_{p \in P} |Du|^2(p) \ : \ u \in \mathscr{F}_S(P) \Big\}$$

where $\mathscr{F}_S(P)$ consists of all functions $u : P \to \mathbb{R}$ such that $\text{Supp}(u) < \infty$ and $u = 1$ on S. □

Lemma 1.220 *Let P be a net, and let $S \subset P$ be a finite subset, $S \neq \emptyset$. Then*

$$P \ \text{ is non-parabolic } \iff \text{cap}(S) > 0.$$

Cf. Proposition 6 in [Kan2], p. 233. Lemma 1.220 is a discrete analog to Lemma 1.212. Given a net P, let us recall the combinatorial metric $\delta(p, q) = \inf \{\ell(\mathbf{p}) \ : \ \mathbf{p} \in \Omega_P(p, q)\}$ on P, and that P is uniform when $\sup_{p \in P} \sharp \big[N(p)\big] < \infty$.

Corollary 1.221 *Let P and Q be two roughly isometric uniform nets. Then P is parabolic* $\Longleftrightarrow$ *Q is parabolic.*

Cf. Corollary 7 in [Kan2], p. 234.

Lemma 1.222 (Local Poincaré inequality) *Let M be a n-dimensional complete Riemannian manifold with Ricci curvature bounded from below by a constant $-(n-1)K^2$ with $K \geq 0$. Then for any geodesic ball $B_r(p) \subset M$ there is a constant $\beta = \beta(n, K, r) > 0$ such that*

$$\int_{B_r(p)} |\nabla u| \, d\,\mathrm{v}_g \geq \beta \int_{B_r(p)} \left| u - u^* \right| \, d\,\mathrm{v}_g, \quad u \in C^\infty\left(\overline{B_r(p)}\right).$$

Here

$$u^* := \frac{1}{\mathrm{vol}\left[B_r(p)\right]} \int_{B_r(p)} u \, d\,\mathrm{v}_g \, .$$

***Proof of Lemma* 1.222** Let us set $B := B_r(p)$ for brevity. It suffices to prove the lemma for $u \in C^\infty\left(\overline{B}\right)$ obeying to

(i) $u^* = 0$,
(ii) u has but finitely many critical points,
(iii) $\mathrm{vol}\left\{x \in B \, : \, u(x) > 0\right\} \leq \mathrm{vol}\left\{x \in B \, : \, u(x) < 0\right\}$.

Let us set $D_t = \left\{x \in B \, : \, u(x) > t\right\}$ with $t \geq 0$. Then

$$\mathrm{vol}\left(D_t\right) \leq \frac{1}{2} \mathrm{vol}(B), \quad t \geq 0.$$

By Buser's local isoperimetric inequality (cf. Buser [Bus])

$$\mathrm{area}\left(B \cap \partial D_t\right) \geq c_1 \, \mathrm{vol}\left(D_t\right), \quad t \geq 0,$$

for some constant $c_1 = c_1(n, K, r) > 0$. Hence

$$\int_B |\nabla u| \, d\,\mathrm{v}_g \geq \int_{D_0} |\nabla u| \, d\,\mathrm{v}_g = \int_0^\infty \left(B \cap \partial D_t\right) d\,t \geq$$

$$\geq c_1 \int_0^\infty \mathrm{vol}\,(D_t) \, d\,t = c_1 \int_{D_0} u \, d\,\mathrm{v}_g = \frac{c_1}{2} \int_B |u| \, d\,\mathrm{v}_g \, .$$

Q.e.d.

Let $P \subset M$ be a ϵ-net in M. Let us recall that the balls $\{B_{\epsilon/2}(p) \, : \, p \in P\}$ are mutually disjoint, and the balls $\{B_\epsilon(p) \, : \, p \in P\}$ cover M, merely as a consequence of the fact that P is maximal and ϵ separated. Also, let us recall that, as a consequence of the assumption on the Ricci curvature [bounded by below by $-(n-1)K^2$]

(i) for every $r > 0$, there is a constant $\mu(r) > 0$, such that for any $x \in M$

$$\sharp \left[P \cap B_r(x)\right] \leq \mu(r),$$

and in particular P is uniform.

(ii) $(P, \delta) \simeq (M, d)$ (a rough isometry).

If in addition [to the assumption on the Ricci curvature] $\text{inj}(M) > 0$ then

(iii) $\text{vol}\, B_r(p) \geq V_0(r)$ for any $p \in M$ and any $r \in \left(0, \frac{1}{2}\text{inj}(M)\right]$,

(iv) $\text{vol}\, B_r(p) \leq V_1(r)$ for any $p \in M$ and any $r > 0$,

where $V_0(r)$ and $V_1(r)$ are positive constants independent of $p \in M$. Also let us recall that (iii) is a result by Croke [Cro], while (iv) follows from the Comparison Theorem in Riemannian geometry.

***Proof of Theorem* 1.210** Let M be a complete Riemannian manifold of Ricci curvature bounded from below by $-(n-1)K^2$ and of positive injectivity radius. By statements (i)–(ii) above any two nets in M are uniform and roughly isometric. Hence (by Corollary 1.221) one of the two nets is parabolic if and only if so does the other. Therefore, it suffices to prove Theorem 1.210 for a ϵ-net $P \subset M$ with $0 < \epsilon \leq \frac{1}{2}\text{inj}(M)$. We illustrate the ideas of the proof by checking the implication "$\Longrightarrow$". Let us assume that M is not parabolic, and let $\Omega \subset M$ be a smoothly bounded domain. Then (by Lemma 1.212) $\text{cap}(\Omega) > 0$.

Claim 1. *The finite subset $S = \{p \in P \,:\, B_{2\epsilon}(p) \cap \Omega \neq \emptyset\}$ has positive capacity.*

Claim 1 clearly implies (by Lemma 1.220) that P is not parabolic.

Proof of Claim 1 Let $p \in P$ and let us consider a function $\eta_p \in C_0^\infty(M)$ such that

$$0 \leq \eta_p \leq 1, \quad \eta_p = 1 \text{ on } B_\epsilon(p), \quad \eta_p = 0 \text{ on } M \setminus B_{2\epsilon}(p),$$

$$\left|\nabla \eta_p\right| \leq c_1 \text{ on } M,$$

where $c_1 > 0$ is a constant independent of p. Net, let us set

$$\xi_p(x) := \frac{\eta_p(x)}{\sum_{q \in P} \eta_q(x)}, \qquad x \in M.$$

Let us observe that there is a constant $c_2 > 0$ such that $\left|\nabla \xi_p\right| \leq c_2$. Indeed

$$\left|\nabla \xi_p\right| \leq \frac{\left|\nabla \eta_p\right|}{\sum_q \eta_q} + \eta_p \frac{\sum_q \left|\nabla \eta_q\right|}{\left(\sum_q \eta_q\right)^2} \leq \left|\nabla \eta_p\right| + \sum_q \left|\nabla \eta_q\right| \leq (\nu_0 + 2)\, c_1$$

where $\nu_0 := \sup_{p\in P} \sharp\left[N(p)\right] < \infty$. Let $u^* : P \to \mathbb{R}$ be an arbitrary function such that $\mathrm{Supp}(u^*) < \infty$ and $u^* = 1$ on S, and let us set

$$u(x) := \sum_{p\in P} u^*(p)\,\xi_p(x), \quad x \in M.$$

Then, for any $x \in B_\epsilon(p)$

$$\nabla u(x) = \sum_{q\in N(p)\cup\{p\}} u^*(q)\,\nabla\xi_q(x) = \quad [\text{by} \sum_{q\in N(p)\cup\{p\}} \nabla\xi_q(x) = 0]$$

$$= \sum_{q\in N(p)\setminus\{p\}} \left\{u^*(q) - u^*(p)\right\}\nabla\xi_q(x) \Longrightarrow$$

$$\Longrightarrow \left|\nabla u(x)\right|^2 \le c_2^2\,\nu_0\left|Du^*\right|^2(p).$$

At this point [by statement (iv) above]

$$\int_M \left|\nabla u\right|^2 d\,\mathrm{v}_g \le \sum_{p\in P}\int_{B_\epsilon(p)} \left|\nabla u\right|^2 d\,\mathrm{v}_g \le c_3 \sum_{p\in P}\left|Du^*\right|^2(p)$$

where $c_3 := c_2^2\,\nu_0\,V_1(\epsilon)$. Finally [by $u = 1$ on Ω]

$$c_3 \sum_{p\in P}\left|Du^*\right|^2(p) \ge \mathrm{cap}(\Omega) > 0$$

yielding $\mathrm{cap}(S) > 0$. Q.e.d.

1.12.7 Rough Isometries and Harmonic Functions on Graphs

Let $G = \{V(G),\ E(G)\}$ be a graph with vertex set $V(G)$, and edge set $E(G)$. Here $E(G) \subset \left\{E \in \mathscr{P}\big(V(G)\big)\ :\ \sharp(E) = 2\right\}$, an assumption that excludes "multiple edges" and "self-loops". For every vertex $x \in V(G)$, a *neighbor* of x is a vertex $y \in V(G)$ such that $\{x,\ y\} \in E(G)$. Let $N_G(x)$ denote the set of all neighbors of x. Graphs in this section are infinite [i.e., $V(G)$ is an infinite set] and every vertex $x \in V(G)$ has neighbors, i.e., $N_G(x) \neq \emptyset$. A *path* in G is a set $\mathbf{p} = \left\{x_0,\ x_1,\ \dots,\ x_\ell\right\}$ with $\ell \in \mathbb{N}$, $x_j \in V(G)$, $0 \le j \le \ell$, and $x_{j+1} \in N_G(x_j)$ for any $0 \le j \le \ell - 1$. The *length* of the path $\mathbf{p} = \left\{x_0,\ x_1,\ \dots,\ x_\ell\right\}$ is the number $\ell(\mathbf{p}) := \ell$. Given two vertices $x,\ y \in V(G)$, let $\Omega_G(x,\ y)$ be the set of all paths $\mathbf{p} = \{x_0,\ x_1,\ \dots,\ x_\ell\}$ *joining* x and

y, i.e., $x_0 = x$ and $x_\ell = y$. G is *connected* if $\Omega_G(x, y) \neq \emptyset$ for any $x, y \in V(G)$. The *degree* of a vertex $x \in V(G)$ is

$$\deg(x) := \sharp\big[N_G(x)\big].$$

Graphs in this section are connected and of *bounded degree*, i.e., there is a positive constant $M > 0$ such that

$$\sup_{x \in V(G)} \deg(x) \leq M. \tag{1.153}$$

The *distance* between two vertices $x, y \in V(G)$, $x \neq y$, is

$$d_G(x, y) := \inf\big\{\ell(\mathbf{p}) \ : \ \mathbf{p} \in \Omega_G(x, y)\big\}.$$

If $x = y$ we set by definition $d_G(x, x) = 0$. Two vertices $x, y \in V(G)$ are *adjacent* (and one occasionally writes $x \sim y$) if

$$x \in N_G(y) \Longleftrightarrow y \in N_G(x) \Longleftrightarrow d_G(x, y) = 1.$$

An *orientation* of G is a couple of functions $\mathrm{i}, \mathrm{t} : E(G) \to V(G)$ such that

$$e = \big\{\mathrm{i}(e), \ \mathrm{t}(e)\big\}, \quad \forall \, e \in E(G).$$

G is *directed* is an orientation (i, t) is prescribed on G. For every $k \in \mathbb{N}$ the *k-fuzz* of G is the graph $G^k = \{V(G^k), \ E(G^k)\}$ with $V(G^k) = V(G)$ and

$$E(G^k) = \big\{\{x, y\} \ : \ x, y \in V(G), \ \ 1 \leq d_G(x, y) \leq k\big\}.$$

In particular $E(G) \subset E(G^k)$. A *morphism* of graphs $\psi : G_1 \to G_2$ is a map of vertices $\psi : V(G_1) \to V(G_2)$ such that

$$d_G(x, y) = 1 \Longrightarrow d_{G_2}\big(\psi(x), \ \psi(y)\big) \leq 1.$$

For every morphism $\psi : G_1 \to G_2$ the *image* of G_1 in G_2 is the graph G' defined by

$$V\big(G'\big) := \psi\big(V(G_1)\big) \subset V(G_2),$$

$$E\big(G'\big) := \Big\{\big\{\psi(x), \ \psi(y)\big\} \in E(G_2) \ : \ \big\{x, y\big\} \in E(G) \ \text{ with } \ \psi(x) \neq \psi(y)\Big\}.$$

The function $d_G : V(G) \times V(G) \to \mathbb{Z}_+$ is a distance function, or metric, on $V(G)$, so that $\big(V(G), \ d_G\big)$ is a metric space. Two graphs G_1 and G_2 are *roughly isometric* if the corresponding metric spaces $\big(V(G_a), \ d_{G_a}\big)$, $a \in \{1, 2\}$, are roughly isometric, i.e., there is a map $\phi : V(G_1) \to V(G_2)$ such that there are constants $a > 0$ and

$b > 0$ with the property

$$\frac{1}{a} d_{G_1}(x, y) - b \le d_{G_2}\big(\phi(x), \phi(y)\big) \le a\, d_{G_1}(x, y) + b \tag{1.154}$$

for any $x, y \in V(G_1)$.

Lemma 1.223 *Let G_a, $a \in \{1, 2\}$, be two roughly isometric graphs, and let $\phi : V(G_1) \to V(G_2)$ be a rough isometry of $(V(G_1), d_{G_1})$ into $(V(G_2), d_{G_2})$, i.e., ϕ satisfies* (1.154) *for some $a > 0$ and $b > 0$. Then for every $k \in \mathbb{N}$, $k \ge a + b$, the map ϕ determines a morphism $G_1 \to G_2^k$.*

The proof is left as an exercise to the reader.

1.12.7.1 Yamasaki's Potential Theory

We recall a few elements of discrete potential theory, as created for the needs of the theory of infinite networks. The results in this section belong to Yamasaki [Yam1, Yam3], and Kayano and Yamasaki [KaYa]. Let $f : V(G) \to \mathbb{R}$ be a function. The *Dirichlet sum* of f is

$$D(f) := \sum_{x \sim y} \left| f(x) - f(y) \right|^2 . \tag{1.155}$$

f is *Dirichlet finite* if $D(f) < \infty$. Let $o \in V(G)$ be a fixed reference vertex. Let $\mathbf{D}(G)$ denote the space of all Dirichlet finite functions $f : V(G) \to \mathbb{R}$. For every $f \in \mathbf{D}(G)$ we set

$$\| f \|_{\mathbf{D}(G)} := \Big(|f(o)|^2 + D(f) \Big)^{1/2} . \tag{1.156}$$

Equation (1.156) is a norm on $\mathbf{D}(G)$. Actually $\mathbf{D}(G)$ is a Hilbert space with the inner product

$$(f, g)_{\mathbf{D}(G)} := f(o)\, g(o) + \sum_{x \sim y} \{ f(x) - f(y) \} \{ g(x) - g(y) \}$$

and (1.156) is the corresponding norm. Let $\ell_0(G)$ be the space of all functions $f : V(G) \to \mathbb{R}$ of finite support, i.e., $\mathrm{Supp}(f) < \infty$. As $(V(G), d_G)$ is a metric spaces, and in particular a Hausdorff space, one may adopt the ordinary notion of support, i.e.,

$$\mathrm{Supp}(f) := \overline{V(G) \setminus Z(f)}, \quad Z(f) := \{ x \in V(G) \; : \; f(x) = 0 \}$$

[here $\overline{A}$ denotes the closure of the subset $A \subset V(G)$]. However, a moment's thought shows[79] that $\text{Supp}(f) = V(G) \setminus Z(f)$. Let $\mathbf{D}_0(G)$ denote the closure of $\ell_0(G)$ in $\mathbf{D}(G)$.

Definition 1.224 A graph G is said to be *recurrent*[80] if $\mathbf{D}_0(G) = \mathbf{D}(G)$. □

Exercise 1.225 Show that G is recurrent $\iff$ $\mathbf{1} \in \mathbf{D}_0(G)$, where $\mathbf{1} : V(G) \to \mathbb{R}$ is the constant function $\mathbf{1}(x) = 1$. □

Cf. Theorem 3.2 in [Yam1]. Let G be a directed graph, with the orientation (i, t). For every vertex $v \in V(G)$, let $O(v)$ be the set of neighbors $u \in N_G(v)$ such that $\mathrm{i}(e) = v$ and $\mathrm{t}(e) = u$, for some edge $e \in E(G)$. That is $\big\{\{v, u\} \in E(G) : u \in O(v)\big\}$ is the set of all *outgoing* edges of G issuing at v. Similarly, let $I(v)$ be the set of all neighbors u of v such that $\mathrm{i}(e) = u$ and $\mathrm{t}(e) = v$, for some $e \in E(G)$, i.e., $\big\{\{u, v\} \in E(G) : u \in I(v)\big\}$ is the set of all *ingoing* edges terminating at v. A *flow* f from v is an assignment of a non-negative real number to every edge such that, at every vertex distinct from v the incoming flow equals the outgoing flow, i.e.,

$$\sum_{u \in O(v)} f\big(\{v, u\}\big) = \sum_{u \in I(v)} f\big(\{u, v\}\big).$$

The net flow leaving v

$$\sum_{u \in O(v)} f\big(\{v, u\}\big)$$

is the *value* of f. The *energy* of the flow is

$$\mathscr{E}(f) = \sum_{e \in E(G)} f(e)^2 .$$

The graph G is *transient* from the vertex v if G admits a flow from v of positive value and finite energy. When G is connected the definition doesn't depend upon the choice of v, and G is referred to merely as *transient*. Cf. also Markvorsen, McGuinness, and Thomassen [MGT].

Exercise 1.226 Let G be a connected graph. Show that G is recurrent $\iff$ G is not transient. □

---

[79] Indeed, let $x \in \text{Supp}(f)$, hence there is a sequence $\{x_\nu\}_{\nu \geq 1} \subset V(G) \setminus Z(f)$ such that $d_G(x_\nu, x) \to 0$ as $\nu \to \infty$. Let us assume that $f(x) = 0$. Then $x_\nu \neq x$ for any $\nu \geq 1$ hence $d_G(x_\nu, x) \geq 1$, a contradiction.

[80] The ordinary definition of the notion is probabilistic, i.e., G is *recurrent* if the simple random walk on G assigning equal probability of passing from a vertex to any of its neighbors, is recurrent. For the time being we avoid the probabilistic treatment of the topic, by adopting as a definition (cf. Definition 1.224) an equivalent description of recurrent graphs (cf. Theorem 3.2 in [Yam1]).

More generally, the *energy* of a function $\psi : E(G) \to \mathbb{R}$ is

$$\mathscr{E}(\psi) = \sum_{e \in E(G)} \psi(e)^2$$

and the *energy* of a function $\phi : V(G) \to \mathbb{R}$ is the energy of $\psi(e) = \left|\phi(\mathrm{t}(e)) - \phi(\mathrm{i}(e))\right|$, i.e., the energy of ϕ is the Dirichlet sum $D(\phi)$.

Definition 1.227 A function $h : V(G) \to \mathbb{R}$ is *harmonic* if

$$h(x) = \frac{1}{\deg(x)} \sum_{y \sim x} h(y), \quad x \in V(G). \tag{1.157}$$

□

Let $\mathbf{HD}(G)$ denote the space of harmonic functions $h \in \mathbf{D}(G)$. By a result of Yamasaki (cf. [Yam2]).

Theorem 1.228 *Let G be a transient graph. Then for every $f \in \mathbf{D}(G)$ there is a unique $g \in \mathbf{D}_0(G)$ and a unique $h \in \mathbf{HD}(G)$ such that*

$$f = g + h. \tag{1.158}$$

Formula (1.158) in Theorem 1.228 is referred to as *Royden's decomposition* of $f \in \mathbf{D}(G)$.

1.12.7.2 Extremal Length of a Set of Paths in a Graph

Let $\mathbf{P}$ be a set of non-self-intersecting infinite paths in a graph G.

Definition 1.229 The *extremal length* $\lambda_G(\mathbf{P})$ is defined by

$$\frac{1}{\lambda_G(\mathbf{P})} = \inf_{w} \mathscr{E}(w) \tag{1.159}$$

where the infimum is take over all functions $w : E(G) \to [0, +\infty)$ such that

$$\mathscr{E}(w) < \infty, \quad \sum_{e \in E(P)} w(e) \geq 1, \quad P \in \mathbf{P},$$

where $E(P)$ and $V(P)$ are the edges and vertices of P. □

Definition 1.230 A property is said to hold for *almost all paths* in $\mathbf{P}$ if said property holds for all $P \in \mathbf{P} \setminus \mathbf{S}$, for some subset $\mathbf{S} \subset \mathbf{P}$ with $\lambda_G(\mathbf{S}) = \infty$. □

Let $o \in V(G)$, and let $\mathbf{P}_o$ be the set of all non-self-intersecting infinite paths whose first vertex is o. Let $f \in \mathbf{D}(G)$ and $P \in \mathbf{P}_o$. Then

$$P = \{o = x_0, x_1, \ldots, x_\nu, \ldots\},$$

$$x_{\nu+1} \in N_G(x_\nu).$$

Let us assume that $\lim_{\nu\to\infty} f\left(x_\nu\right)$ exists and is finite. Said limit is denoted by

$$f(P) = \lim_{\substack{x \to \infty \\ x \in V(P)}} f(x) \in \mathbb{R}.$$

We recall

Theorem 1.231 *Let $f \in \mathbf{D}(G)$. Then*

(1) $f(P) = \lim_{\substack{x \to \infty \\ x \in V(P)}} f(x)$ *exists and is finite for almost every* $P \in \mathbf{P}_0$.

(2) *If* $f \in \mathbf{D}_0(G)$ *then* $f(P) = 0$ *for almost every* $P \in \mathbf{P}_0$.

(3) *If* $f \in \mathbf{HD}(G)$ *is not a constant, then there is no constant* $\gamma \in \mathbb{R}$ *such that* $f(P) = \gamma$ *for almost every* $P \in \mathbf{P}_0$.

Cf. Theorems 3.1 and 3.3 in Kayano and Yamasaki [KaYa], and Theorem 3.1 in Yamasaki [Yam3].

1.12.7.3 Dirichlet Finite Harmonic Functions

Let $\mathcal{O}_{\mathbf{HD}}$ denote the set of all graphs G which admit no nonconstant harmonic function of finite Dirichlet sum.

Definition 1.232 An element $G \in \mathcal{O}_{\mathbf{HD}}$ is said to be a *graph of class* $\mathcal{O}_{\mathbf{HD}}$. □

The main result in Soardi [Soa] is

Theorem 1.233 *Let G_1 and G_2 be roughly isometric infinite connected graphs of bounded degree. Then*

$$G_1 \in \mathcal{O}_{\mathbf{HD}} \Longrightarrow G_2 \in \mathcal{O}_{\mathbf{HD}}.$$

Cf. [Soa], p. 1242. The proof of Theorem 1.233 follows from the three lemmas below.

Lemma 1.234 *Let G be a transient connected graph of bounded degree. If $G \in \mathcal{O}_{\mathbf{HD}}$ then $G^k \in \mathcal{O}_{\mathbf{HD}}$ for every $k \in \mathbb{N}$.*

Cf. Lemma 1 in [Soa], p. 1242.

Lemma 1.235 *Let G and G′ be transient connected graphs of bounded degree. Let us assume that*

(i) *There is a morphism* $\psi : G \to G'$ *such that* G' *is the image of G by* ψ,
(ii) *There is a constant* $m \in \mathbb{R}$ *such that for any* $x, y \in V(G)$

$$\psi(x) = \psi(y) \Longrightarrow d_G(x, y) \leq m.$$

Then $G \in \mathscr{O}_{\mathbf{HD}} \Longrightarrow G' \in \mathscr{O}_{\mathbf{HD}}$.

Cf. Lemma 2 in [Soa], p. 1243.

Lemma 1.236 *Let G be a transient connected subgraph of a connected graph H of bounded degree, and let us assume that the inclusion* $G \hookrightarrow H$ *is a rough isometry. Then* $G \in \mathscr{O}_{\mathbf{HD}} \Longrightarrow H \in \mathscr{O}_{\mathbf{HD}}$.

Cf. Lemma 3 in [Soa], p. 1244.

Proof of Theorem 1.233 If G_1 and G_2 are recurrent, then $G_1, G_2 \in \mathscr{O}_{\mathbf{HD}}$ by Theorem 6.1 in [Yam2]. Let us assume that G_1 and G_2 are transient, and let $\phi : V(G_1) \to V(G_2)$ be a rough isometry, i.e., ϕ obeys to (1.154) for some constants $a > 0$ and $b > 0$. Let $k > a + b$. Then ϕ is a morphism of G_1 into G_2^k. Let G_1' be the image of G_1 in G_2^k by ϕ. Then [by Lemma 1.235 with $m = ab$, together with (1.154)] $G_1' \in \mathscr{O}_{\mathbf{HD}}$. Therefore [by Lemma 1.236] $G_2^k \in \mathscr{O}_{\mathbf{HD}}$. Finally [by Lemma 1.234] $G_2 \in \mathscr{O}_{\mathbf{HD}}$. Q.e.d.

1.12.8 Parabolic and Hyperbolic Infinite Networks

Through the present section we follow mainly the work by Yamasaki [Yam1]. Let X and Y be two infinite countable sets, and let $K : X \times Y \to \{0, \pm 1\}$ be a function such that

(i) For every $y \in Y$ the set $e(y) := \{x \in X : K(x, y) \neq 0\}$ consists of exactly two points $e(y) = \{x_1, x_2\}$ such that $K(x_1, y) K(x_2, y) = -1$,
(ii) For every $x \in X$ the set $Y(x) := \{y \in Y : K(x, y) \neq 0\}$ is nonempty and finite,
(iii) For any $x, x' \in X$ there exist $x_1, \ldots, x_n \in X$ and $y_1, \ldots, y_{n+1} \in Y$ such that $e(y_j) = \{x_{j-1}, x_j\}$, $1 \leq j \leq n + 1$, with $x_0 := x$ and $x_{n+1} := x'$.

Elements of X are referred to as *nodes*, while elements in Y are called *arcs*. The function K is the *node-arc incidence function*. A node $x \in X$ is *incident* with an arc $y \in Y$ if $K(x, y) \neq 0$. If $x \in X$ is incident with $y \in Y$ one also says that the arc y is *incident* with the node x. Once this terminology is adopted, $e(y)$ is the set of all nodes incident with the arc y, while $Y(x)$ is the set of all arcs incident with the node x. By

Axiom (i) above every edge y is incident precisely with two nodes $e(y) = \{x_1, x_2\}$, and one distinguishes between the two nodes as, up to eventually relabeling x_1 and x_2, one has

$$K\big(x_1, y\big) = -1, \quad K\big(x_2, y\big) = +1.$$

With this distinction in mind, one may give the two nodes the new names $x_- := x_1$ and $x_+ := x_2$, and speak about the *initial* and *terminal* nodes $x_\pm$ of the edge y. By Axiom (ii) for every node $x \in X$ there exist finitely many arcs incident with x, and in particular a node is never *isolated*, i.e., $Y(x) \neq \emptyset$.

Definition 1.237 A synthetic object $N = \{X, Y, K, r\}$ consisting of a triple $\{X, Y, K\}$ as above, and a function $r : Y \to (0, +\infty)$, is called an *infinite network*, and r is the *resistance* of N. □

A graph G may be associated to an infinite network $N = \{X, Y, K, r\}$ by setting

$$V(G) := X, \quad E(G) := \big\{e(y) \ : \ y \in Y\big\}.$$

By Axiom (i) there are no loops. By Axiom (iii) the graph G is connected.

1.12.8.1 Function Spaces on an Infinite Network

Let $N = \{X, Y, K, r\}$ be an infinite network. Let $L(X)$ be the set of all functions $u : X \to \mathbb{R}$. The *support* of $u \in L(X)$ is $\mathrm{Supp}(u) := \{x \in X \ : \ u(x) \neq 0\}$. The *Dirichlet sum* of *order* p (with $1 < p \leq \infty$) of $u \in L(X)$ is

$$D_p(u) := \sum_{y \in Y} \frac{\Big|\sum_{x \in X} K(x, y)\, u(x)\Big|^p}{r(y)^{p-1}}, \qquad 1 < p < \infty,$$

$$D_\infty(u) := \sup_{y \in Y} \frac{\Big|\sum_{x \in X} K(x, y)\, u(x)\Big|}{r(y)}, \qquad p = \infty.$$

Next, let us set

$$L_0(X) := \big\{u \in L(X) \ : \ \sharp\,(\mathrm{Supp}(u)) < \infty\big\},$$

$$\mathbf{D}^{(p)}(N) := \big\{u \in L(X) \ : \ D_p(u) < \infty\big\}.$$

Also, for any subset $A \subset X$

$$\mathbf{D}^{(p)}(N; A) = \big\{u \in \mathbf{D}^{(p)}(N) \ : \ u = 0 \ \text{ on } \ A\big\},$$

$$L_0^A(X) = \big\{u \in L_0(X) \ : \ u = 0 \ \text{ on } \ A\big\}.$$

Then $\mathbf{D}^{(p)}(N;\ A)$, $1 < p < \infty$, is a reflexive Banach space with the norm $u \in \mathbf{D}^{(p)}(N;\ A) \mapsto \left[D_p(u)\right]^{1/p}$. Also $\mathbf{D}^{(\infty)}(N;\ A)$ is a Banach space with the norm $u \in \mathbf{D}^{(\infty)}(N;\ A) \mapsto D_\infty(u)$. Let $\mathbf{D}_0^{(p)}(N;\ A)$ be the closure of $L_0^A(X)$ in $\mathbf{D}^{(p)}(N;\ A)$. Let $x_0 \in X$ be a fixed node, and let us set

$$\|u\|_p = \left[D_p(u) + \left|u(x_0)\right|^p\right]^{1/p}, \quad 1 < p < \infty,$$

$$\|u\|_\infty = D_\infty(u) + \left|u(x_0)\right|.$$

Lemma 1.238 *For every finite subset $F \subset X$, there is a constant $M(F) > 0$ such that*

$$\sum_{x \in F} \left|u(x)\right| \leq M(F)\, \|u\|_p$$

for any $u \in \mathbf{D}^{(p)}(N)$. Consequently, if $u_\nu \in \mathbf{D}^{(p)}(N)$ and $\|u_\nu - u\|_p \to 0$ then $u_\nu(x) \to u(x)$ as $\nu \to \infty$, for any $x \in X$.

Cf. Lemma 2.1 and its corollary in [Yam1], p. 137. The proof is left as an exercise to the reader. Cf. also Lemma 1 in [Yam4].

Lemma 1.239 *$\mathbf{D}^{(p)}(N)$ is a Banach space with the norm $u \mapsto \|u\|_p$. If $1 < p < \infty$ then $\mathbf{D}^{(p)}(N)$ is reflexive.*

Cf. Proposition 2.1 in [Yam1], p. 137. The proof is straightforward. Let $\mathbf{D}_0^{(p)}(N)$ be the closure of $L_0(X)$ in $\mathbf{D}^{(p)}(N)$.

Lemma 1.240 *Let $A \subset X$ be a nonempty finite set. Then*

$$\mathbf{D}_0^{(p)}(N;\ A) = \mathbf{D}^p(N;\ A) \cap \mathbf{D}_0^{(p)}(N).$$

Cf. Lemma 2.2 in [Yam1], p. 137. The choice of the node $x_0 \in X$ [used in the definition of the norms $\|\cdot\|_p$, $1 < p \leq \infty$] is rather immaterial. Indeed, if another node $x_0' \in X$ is used in the definitions of said norms, and $\mathbf{D}_0^{(p)}(N,\ x_0')$ is the closure of $L_0(X)$ in $\mathbf{D}^{(p)}(N)$ endowed with said norms, then $\mathbf{D}_0^{(p)}(N,\ x_0') = \mathbf{D}_0^{(p)}$.

A function $T : \mathbb{R} \to \mathbb{R}$ is a *normal contraction* if $T(0) = 0$ and

$$\left|T(t_1) - T(t_2)\right| \leq \left|t_1 - t_2\right|, \quad t_1,\ t_2 \in \mathbb{R}.$$

The functions

$$T^{(+)}(t) := \max\{t,\ 0\}, \quad T^{(-)}(t) := \max\{-t,\ 0\}, \quad t \in \mathbb{R},$$

are normal contractions. Given a normal contraction $T : \mathbb{R} \to \mathbb{R}$ and a function $u \in L(X)$ we set $Tu := T \circ u \in L(X)$.

Lemma 1.241 *Let T be a normal contraction and let $u \in \mathbf{D}^{(p)}(N)$. Then*

$$Tu \in \mathbf{D}^{(p)}(N), \quad D_p(Tu) \leq D_p(u), \quad \|Tu\|_p \leq \|u\|_p \, .$$

Cf. Lemma 2.3 in [Yam1], p. 138. The proof is straightforward.

1.12.8.2 Classifying Infinite Networks

Let $N = \{X, \ Y, \ K, \ r\}$ be an infinite network, and let $A \subset X$ be a nonempty finite subset. Together with [Yam1], let us consider the following extremum problem relative to the ideal boundary ∞ of N

$$d_p(A, \infty) := \inf \Big\{ D_p(u) \ : \ u \in L_0(X), \ \ u = 1 \ \text{ on } \ A \Big\}.$$

An explicit definition of the ideal boundary isn't needed, yet may be provided as follows. Let

$$\pi_0 : X \times X \to \mathbb{R}, \quad \pi_0(x, \, x') = \begin{cases} 1 & x = x', \\ \\ 0 & x \neq x', \end{cases}$$

$$\pi_k : X \times X \to \mathbb{R}, \quad k \geq 1,$$

$$\pi_{k+1}(x, \, x') = \sum_{x \in X} \pi_k(x, \, x'') \, \pi_k(x'', \, x'), \quad k \geq 0.$$

Also, let

$$G(x, \, x') = \sum_{k=0}^{\infty} \pi_k(x, \, x')$$

be the Green function of N. Let $x_0 \in X$ be fixed, and let us set

$$\Phi = \left\{ \frac{G(\,\cdot\,, \, x')}{G(\,\cdot\,, \, x_0)} \ : \ x' \in X \right\}.$$

X is a metric space, with the distance $d_X(x, \, x') =$ the length of the shortest set of arcs connecting x and x', and hence a topological space. By the Costantinescu-Cornea Theorem (cf. the Appendix to this section) there is a compact space $\hat{X}$ such that X is a dense open subspace of $\hat{X}$ and every function $f \in \Phi$ extends to a continuous function $\hat{f}$ on $\hat{X}$, separating points on $\infty := \hat{X} \setminus X$, the ideal boundary of X relative to Φ. The space ∞ is the *ideal boundary* of N.

Theorem 1.242 *Let $A \subset X$ be a finite set, $A \neq \emptyset$. Then*

$$d_p(A, \infty) = 0 \Longleftrightarrow 1 \in \mathbf{D}_0^{(p)}(N).$$

Cf. Theorem 3.1 in [Yam1], p. 138. The proof is straightforward.

Corollary 1.243 *Let $A \subset X$ and $A' \subset X$ be two nonempty finite subsets. Then*

$$d_p(A, \infty) = 0 \Longleftrightarrow d_p(A', \infty) = 0.$$

Cf. [Yam1], p. 138. As the most valuable results in mathematics, Theorem 1.242 and Corollary 1.243 lead to an important definition, by means of which one actually classifies infinite networks.

Definition 1.244 N is said to be of *parabolic type* of *order* p ($1 < p \leq \infty$) if there is a nonempty finite set $A \subset X$ such that $d_p(A, \infty) = 0$. Also N is of *hyperbolic type* of *order* p if N is not of the parabolic type of order p, i.e., $d_p(A, \infty) > 0$ for any nonempty finite subset $A \subset X$. □

The main result in [Yam1] is

Theorem 1.245 *Let $1 < p < \infty$, and let N be an infinite network. The following statements are equivalent*

(i) *N is of parabolic type of order p.*
(ii) $1 \in \mathbf{D}_0^{(p)}(N)$.
(iii) $\mathbf{D}_0^{(p)}(N) = \mathbf{D}^{(p)}(N)$.

Cf. Theorem 3.2 in [Yam1], p. 139.

Corollary 1.246 *Let $1 < p < \infty$ and let $A \subset X$ be a nonempty finite subset. The following statements are equivalent*

(i) *N is of parabolic type of order p.*
(ii) $\epsilon_{X \setminus A} \in \mathbf{D}_0^{(p)}(N;\ A)$ *and* $\mathbf{D}_0^{(p)}(N;\ A) = \mathbf{D}^{(p)}(N;\ A)$.

Cf. [Yam1], pp. 139–140. Here $\epsilon_A = \chi_A$ is the characteristic function of the set $A \subset X$, i.e., $\epsilon(x) = 1$ for any $x \in A$ and $\epsilon_A(x) = 0$ for any $x \in X \setminus A$.

1.12.9 Parabolic Index and Rough Isometries

Let G be an infinite, locally finite, connected graph of bounded degree, whose vertex set $V(G)$ and edge set $E(G)$ are countable. Let $K : V(G) \times E(G) \to \{0, \pm 1\}$ be the vertex-edge incidence function, or matrix. We shall work with the infinite network

$N = \{V(G), E(G), K, r\}$, where $r : V(G) \to \mathbb{R}$ is the constant function $r \equiv 1$, that throughout we identify to the graph G. In particular, we say that G is of parabolic (respectively hyperbolic) type of order p if the network N with $r \equiv 1$ is of that type.

Definition 1.247 The *parabolic index* $\operatorname{ind} G$ of the graph G is

$$\operatorname{ind} G := \inf \{p > 1 \ : \ G \ \text{ is of parabolic type of order } \ p\}.$$

□

The main results in [SoYa] are

Theorem 1.248 *Let G be a graph of bounded degree. Then* $\operatorname{ind} G = \operatorname{ind} G^k$.

Cf. Theorem 2.1 in [SoYa], p. 335.

Theorem 1.249 *Let G_1 and G_2 be two roughly isometric infinite connected graphs of bounded degree. Then* $\operatorname{ind} G_1 = \operatorname{ind} G_2$.

Cf. Theorem 2.4 in [SoYa], p. 337.

1.12.10 Appendix: Martin Boundary in Potential Theory

Let Ω be a locally compact, non-compact, topological space, and let $\Phi \subset C(\Omega, \overline{\mathbb{R}})$ be a family of continuous functions $f : \Omega \to \overline{\mathbb{R}}$, where $\overline{\mathbb{R}} = [-\infty, +\infty]$. A classical result in Potential Theory, known as the *Costantinescu-Cornea Theorem*, is (cf. Costantinescu and Cornea [CoCo]).

Theorem 1.250 *There is a compact topological space $\hat{\Omega}$, unique up to a homeomorphism, such that*

(1) *Ω is a dense subspace of $\hat{\Omega}$,*
(2) *Every $f \in \Phi$ extends continuously to a function $\hat{f} : \hat{\Omega} \to \overline{\mathbb{R}}$, separating points on $\Delta := \hat{\Omega} \setminus \Omega$,*
(3) *Ω is an open subset of $\hat{\Omega}$.*

Definition 1.251 $\Delta = \hat{\Omega} \setminus \Omega$ is the *ideal boundary* of Ω relative to Φ. □

Let $\Omega \subset \mathbb{R}^n$, $n \geq 2$, be a bounded domain, and let $G(x, y)$ be the Green function on Ω with pole $y \in \Omega$, and let $y_0 \in \Omega$ be fixed. Next, let us set

$$K(x, y) := \frac{G(x, y)}{G(x, y_0)}, \quad x \in \Omega, \quad \Phi := \Big\{K\big(\cdot, y\big) \ : \ y \in \Omega\Big\}.$$

Let $\hat{\Omega}$ be the compact space associated to the pair (Ω, Φ) by the Costantinescu-Cornea Theorem.

Definition 1.252 $\hat{\Omega}$ is the *Martin space*, or *Martin compactification*, of Ω, and the ideal boundary $\hat{\Omega} \setminus \Omega$ of Ω relative to Φ is the *Martin boundary* of Ω. The *Martin topology* is the topology of the Martin space $\hat{\Omega}$. □

Cf. also Martin [Mar], and Brelot [Bre]. Two Martin spaces $\hat{\Omega}'$ and $\hat{\Omega}''$, corresponding to two different poles $y_0', y_0'' \in \Omega$, are homeomorphic. Let

$$\hat{K}(\cdot, y) : \Delta \times \Omega \to [0, +\infty]$$

be the extension of $K(\cdot, y)$. Then $\hat{K}(\xi, y)$ is harmonic in y and jointly continuous in the variables (ξ, y). The Martin space $\hat{\Omega}$ is metrizable. By a result in [Mar] (known as *Martin's Fundamental Theorem*).

Theorem 1.253 *The class of all non-negative harmonic functions $u(y) \geq 0$ on Ω may be represented as*

$$u(y) = \int_{\Delta} K(\xi, y)\, d\mu(\xi) \tag{1.160}$$

for some positive Radon measure on Δ.

Formula (1.160) is the *Martin representation* of u. The measure μ in (1.160) is not uniquely determined by u.

Definition 1.254 A harmonic function $v \geq 0$ is called *minimal* in Ω if every harmonic function w such that $0 \leq w \leq v$ in Ω, is proportional to v. □

Minimal harmonic functions $v \neq 0$ are proportional to $\hat{K}(\xi, \cdot)$; the corresponding points $\xi \in \Delta$ are referred to as *minimal*, and the set of minimal points $\Delta_1 \subset \Delta$ is called the *minimal Martin boundary*. Let us look at a few examples.

(i) Let $\Omega = B_R(0) \subset \mathbb{R}^n$, $n \geq 2$, Then

$$\hat{K}(\xi, y) = \frac{R^{n-2}\left(R^2 - |y|^2\right)}{|\xi - y|^n}$$

is the Poisson kernel, $\hat{\Omega} = \overline{\Omega}$ (the closure of Ω), the Martin boundary is the sphere $\Delta = S^{n-1}(R)$, all points of which are minimal. The Martin representation (1.160) becomes the *Poisson-Herglotz formula*.

(ii) The Martin boundary is $\Delta = \Gamma$, where $\Gamma := \overline{\Omega} \setminus \Omega$, whenever $\Gamma \subset \mathbb{R}^n$ is a smooth hypersurface.

References

[Bla] D.E. Blair, *Contact Manifolds in Riemannian Geometry*, vol. 509 (Springer, Berlin-Heidelberg-New York, 1976)

[KoNo] S. Kobayashi, K. Nomizu, *Foundations of Differential Geometry*, vol. I. (Interscience Publishers, New York, 1963) (vol. II, 1969)

[Mil] J. Milnor, A note on curvature and fundamental group. J. Diff. Geom. **2**, 1–7 (1968)

[Mye] S. Myers, Riemannian manifolds with positive mean curvature. Duke Math. J. **8**, 401–404 (1941)

[Pre] A. Preissmann, Quelques propriétés globales des espaces de Riemann, Comment. Math. Helv. **15**, 175–216 (1942–43)

[Bis] R.L. Bishop, A relation between volume, mean curvature, and diameter. Notices Amer. Math. Soc. **10**, 364 (1963)

[Gun] P. Günter, Einige Sätze über das Volumenelement eines Riemannschen Raumes. Publ. Math. Debrecen **7**, 78–93 (1960)

[Kan1] M. Kanai, Rough isometries, and combinatorial approximations of geometries of non-compact Riemannian manifolds. J. Math. Soc. Japan **37**(3), 391–413 (1985)

[Kan2] M. Kanai, Rough isometries and the parabolicity of Riemannian manifolds. J. Math. Soc. Japan **38**(2), 228–238 (1986)

[Cro] C.B. Croke, Some isoperimetric inequalities and eigenvalue estimates. Ann. Sci. École Norm. Sup. **13**, 419–435 (1980)

[Bus] P. Buser, A note on the isoperimetric constant. Ann. Sci. École Norm. Sup. **15**, 213–230 (1982)

[Oss] R. Osserman, The isoperimetric inequality. Bull. Amer. Math. Soc. **84**, 1182–1238 (1978)

[Yau] S.-T. Yau, Survey on Partial Differential Equations in Differential Geometry. Annals of Mathematics Studies 102 (Princeton University Press, Princeton, 1982), pp. 3–71

[Mos] J. Moser, On Harnack's theorem for elliptic differential equations. Comm. Pure Appl. Math. **14**, 577–591 (1961)

[GrWu] R.E. Greene, H. Wu, *Function Theory on Manifolds which Possess a Pole*. Lecture Notes in Mathematics, 699 (Springer, 1979)

[Ito1] S. Itô, Fundamental solutions of parabolic differential equations and boundary value problems. Japan J. Math. **27**, 55–102 (1957)

[Dod1] J. Dodziuk, Maximum principle for parabolic inequalities and the heat flow on open manifolds. Indiana Univ. Math. J. **32**, 703–716 (1983)

[Ito2] S. Itô, On existence of Green function and positive superharmonic functions for linear elliptic operators of second order. J. Math. Soc. Japan **16**, 299–306 (1964)

[LySu] T. Lyons, D. Sullivan, Function theory, random paths, and covering spaces. J. Diff. Geom. **19**, 299–323 (1984)

[DyYu] E.B. Dynkin, A.H. Yushkevic, *Markov Processes* (Plenum Press, New York, Theorems and Problems, 1969)

[Var] N.Th. Varopoulos, Brownian motion and random walks on manifolds. Ann. Inst. Fourier, Grenoble **34**, 243–269 (1984)

[FCSc] D. Fischer-Colbrie, R. Schoen, The structure of complete stable minimal surfaces in 3-manifolds of non-negative scalar curvature. Comm. Pure appl. Math. **33**, 199–211 (1980)

[GiTr] D. Gilbarg, N.S. Trudinger, *Elliptic Partial Differential Equations of Second Order* (Springer, New York, 1983)

[Dod2] J. Dodziuk, Difference equations, isoperimetric inequality, and transience of certain random walks. Trans. Amer. Math. Soc. **284**, 787–794 (1984)

[Yam1] M. Yamasaki, Parabolic and hyperbolic infinite networks. Hiroshima Math. J. **7**, 135–146 (1977)

[Yam3] M. Yamasaki, Ideal boundary limit of discrete Dirichlet functions. Hiroshima Math. J. **16**, 353–360 (1986)

[KaYa] T. Kayano, M. Yamasaki, Boundary limits of discrete Dirichlet potentials. Hiroshima J. Math. **14**, 401–406 (1984)

[MGT] S. Markvorsen, S. McGuinness, C. Thomassen, Transient random walks on graphs and metric spaces with applications to hyperbolic surfaces. Proc. London Math. Soc. 64(3), 1–20 (1992)

[Yam2] M. Yamasaki, Discrete potentials on an infinite network. Mem. Fac. Sci. Shimane Univ. **13**, 31–44 (1979)

[Soa] P.M. Soardi, Rough isometries and Dirichlet finite harmonic functions on graphs. Proc. Amer. Math. Soc. **119**(4), 1239–1248 (1993)

[Yam4] M. Yamasaki, Extremum problems on an infinite network. Hiroshima Math. J. **5**, 223–250 (1975)

[SoYa] P.M. Soardi, M. Yamasaki, Parabolic index and rough isometries. Hiroshima Math. J. **23**, 333–342 (1993)

[CoCo] C. Costantinescu, A. Cornea, Ideale Ränder Riemannscher Flächen (Springer, 1963)

[Mar] R.S. Martin, Minimal positive harmonic functions. Trans. Amer. Math. Soc. **49**, 137–172 (1941)

[Bre] M. Brelot, *On Topologies and Boundaries in Potential Theory* (Springer, 1971)

Chapter 2
Finslerian Geometry

2.1 Finslerian Geometry

2.1.1 Pullback Bundle

Let M be a n-dimensional C^∞ manifold and let $T(M)$ be the total space of the tangent bundle over M. Let $\Pi : T(M) \to M$ be the canonical projection. Given a local coordinate system $\left(U,\ \tilde{x}^i\right)$ on M let $\left(\Pi^{-1}(U),\ x^i,\ y^i\right)$ be the induced local coordinate system on $T(M)$, i.e.,

$$x^i = \tilde{x}^i \circ \Pi, \quad v = y^i(v) \left(\frac{\partial}{\partial x^i}\right)_{\Pi(v)}, \quad v \in \Pi^{-1}(U).$$

The very pedantic notational distinction between the point coordinates $\tilde{x}^i : U \to \mathbb{R}$ and $x^i : \Pi^{-1}(U) \to \mathbb{R}$ will be dropped in the sequel. Finslerian metrics (energy functions) are defined on $T(M)$ yet are commonly not sufficiently smooth along the zero section in $T(M)$. For this reason, we shall work with

$$V(M) = T(M) \setminus \sigma(M),$$

$$\sigma : M \to T(M), \quad \sigma(x) = 0_x \in T_x(M), \quad x \in M.$$

$V(M)$ is an open subset in $T(M)$ and $\left(\Pi^{-1}(U),\ x^i,\ y^i\right)$ induces the local coordinate system $\left(\pi^{-1}(U),\ x^i,\ y^i\right)$ on $V(M)$, where $\pi : V(M) \to M$ is the projection.

Definition 2.1 The *pullback bundle* $\pi^{-1}T(M) \to V(M)$ is the ordinary pullback of $T(M)$ by π. □

The fiber in $\pi^{-1}T(M)$ over $v \in V(M)$ is $\left(\pi^{-1}T(M)\right)_v = T_{\pi(v)}(M)$. The following diagram

E. Barletta et al., *Differential Geometry*, Infosys Science Foundation Series,
https://doi.org/10.1007/978-981-96-1631-2_2

$$\begin{array}{ccc} \pi^{-1}T(M) & \xrightarrow{p} & V(M) \\ \hat{\pi}\downarrow & & \downarrow\pi \\ T(M) & \xrightarrow{\Pi} & M \end{array}$$

is commutative. Here p and $\hat{\pi}$ are, respectively, the maps

$$p(X) = v, \quad \hat{\pi}(X) = X, \quad X \in \left(\pi^{-1}T(M)\right)_v,$$

for any $v \in V(M)$. Let $U \subset M$ be an open set.

Definition 2.2 The *natural lift* of the tangent vector field $X : U \to T(M)$ is the cross section

$$X^\pi \in C^\infty\left(\pi^{-1}(U),\ \pi^{-1}T(M)\right), \quad X^\pi = X \circ \pi\,.$$

□

Definition 2.3 The *Liouville vector* is the (globally defined) cross section

$$\mathcal{L} \in C^\infty\left(\Pi^{-1}T(M)\right), \quad \mathcal{L}(v) = v, \quad v \in V(M).$$

□

Given a local coordinate system (U, x^i) on M let

$$X_i = \left(\frac{\partial}{\partial x^i}\right)^\pi : \pi^{-1}(U) \to \pi^{-1}T(M)$$

be the natural lifts of the vector fields $\partial/\partial x^i : U \to T(M)$. Then $\{X_i \ : \ 1 \le i \le n\}$ is a local frame in the pullback bundle $\pi^{-1}T(M) \to V(M)$, defined on the open set $\pi^{-1}(U) \subset V(M)$. Therefore, for any $v \in \pi^{-1}(U)$

$$\mathcal{L}(v) = v = y^i(v)\left(\frac{\partial}{\partial x^i}\right)_{\pi(v)} = y^i(v)\, X_i(v)$$

hence the Liouville vector admits the local representation

$$\mathcal{L} = y^i\, X_i \quad \text{on } \pi^{-1}(U). \tag{2.1}$$

2.1.2 *The Vertical Lift*

We shall need the vector bundle morphism

$$L : T(V(M)) \to \pi^{-1}T(M), \quad L_v A = (d_v\pi)A, \quad A \in T_v(V(M)), \quad v \in V(M).$$

We shall build a vector bundle morphism

$$\gamma : \pi^{-1}T(M) \to T(V(M))$$

as follows. Let $v \in V(M)$ and $X \in \left(\pi^{-1}T(M)\right)_v = T_{\pi(v)}(M)$. If $x = \pi(v) \in M$, then $v \neq 0_x$ hence there is $\epsilon > 0$ such that $v + tX \in T_x(M) \setminus \{0_x\}$ for any $|t| < \epsilon$. Let $C : (-\epsilon, \epsilon) \to V(M)$ be the curve given by

$$C(t) = v + tX, \quad |t| < \epsilon.$$

Finally, we consider the $\mathbb{R}$-linear maps

$$\gamma_v : \left(\pi^{-1}T(M)\right)_v \to T_v(V(M)), \quad v \in V(M),$$

given by

$$\gamma_v(X) = \frac{dC}{dt}(0).$$

Definition 2.4 The *vertical lift* is the vector bundle morphism $\gamma : \pi^{-1}T(M) \to T(V(M))$. □

The definition of $\gamma_v(X)$ doesn't depend upon the various choices and

$$\gamma\left[\pi^{-1}T(M)\right] = \text{Ker}(d\pi).$$

In particular $\gamma : \pi^{-1}T(M) \to \text{Ker}(d\pi)$ is a vector bundle isomorphism and

$$0 \to \pi^{-1}T(M) \xrightarrow{\gamma} T(V(M)) \xrightarrow{L} \pi^{-1}T(M) \to 0 \tag{2.2}$$

is a short exact sequence of vector bundles and vector bundle morphisms. All these statements are an easy consequence of the local representation of γ, i.e.,

$$\gamma(X_i) = \frac{\partial}{\partial y^i}, \quad 1 \leq i \leq n, \quad \text{on } \pi^{-1}(U).$$

Definition 2.5 A section $X \in C^\infty(\pi^{-1}TM)$ is referred to as a *Finslerian vector field* on M. □

No Finslerian structure has been assigned to M so far, yet locally $X = f^i X_i$ on $\pi^{-1}(U)$ for some C^∞ functions $f^i : \pi^{-1}(U) \to \mathbb{R}$. The functions f^i [the *local components* of X with respect to the local coordinate system (U, x^i) on M] depend both on the position arguments x^i and the directional arguments y^i, thus motivating the chosen terminology.

2.1.3 Connections in the Pullback Bundle

The canonical connections (similar to the Levi-Civita connection of a Riemannian manifold) in Finslerian geometry occur as connections in the pullback bundle $\pi^{-1}T(M) \to V(M)$. Let $\mathcal{C}(\pi^{-1}TM)$ be the affine space of all connections in $\pi^{-1}T(M)$. To every connection $\nabla \in \mathcal{C}(\pi^{-1}TM)$ one may associate its curvature 2-form

$$R^\nabla : \Omega^0(\pi^{-1}TM) \to \Omega^2(\pi^{-1}TM).$$

Here we use the customary notation $\Omega^k(E) = C^\infty(\Lambda^k T^*(M) \otimes E)$ for every vector bundle $E \to M$, so that $\Omega^0(E) = C^\infty(E)$ is the space of all C^∞ sections in E. Then

$$R^\nabla X : \mathfrak{X}(V(M)) \times \mathfrak{X}(V(M)) \to C^\infty(V(M)),$$

$$(R^\nabla X)(A, B) = [\nabla_A, \nabla_B]X - \nabla_{[A,B]}X, \quad X \in \Omega^0(\pi^{-1}TM), \quad A, B \in \mathfrak{X}(V(M)).$$

As well known, the notion of torsion makes sense for a linear connection, that is for a connection in the tangent bundle over a differentiable manifold. However, a concept of "torsion" may be associated to a connection $\nabla \in \mathcal{C}(\pi^{-1}TM)$ in a natural manner, by using the bundle morphism L. Indeed, one sets

$$T_L(A, B) = \nabla_A LB - \nabla_B LA - L[A, B], \quad A, B \in \mathfrak{X}(V(M)).$$

Lemma 2.6 *For every connection,* $\nabla \in \mathcal{C}(\pi^{-1}TM)$

$$T_L(\gamma X, \gamma Y) = 0 \tag{2.3}$$

for any $X, Y \in C^\infty(\pi^{-1}TM)$.

Proof By the exactness of the sequence (2.2) $\text{Ker}(L) = \text{Ker}(d\pi)$, i.e., $\text{Ker}(L)$ is precisely the vertical distribution on $V(M)$. As well known the vertical distribution is involutive, hence

$$[\gamma X, \gamma Y] \in \text{Ker}(L)$$

for any pair of Finslerian vector fields X and Y on M. Then $L \circ \gamma = 0$ yields (2.3). Q.e.d.

2.1.4 Nonlinear Connections

Definition 2.7 A *nonlinear connection* on M is a C^∞ distribution

$$N : v \in V(M) \longmapsto N_v \subset T_v(V(M))$$

such that

$$T_v(V(M)) = N_v \oplus \mathrm{Ker}(d_v \pi), \quad v \in V(M). \tag{2.4}$$

□

Cf. N. Kawaguchi, [Kaw]. Given a nonlinear connection N on M and a tangent direction $v \in V(M)$ the restriction of L_v to N_v

$$L_v : N_v \to \left(\pi^{-1} T M\right)_v$$

is a $\mathbb{R}$-linear isomorphism. Hence, $L : N \to \pi^{-1} T(M)$ is a vector bundle isomorphism whose inverse is denoted by $\beta : \pi^{-1} T(M) \to N$, i.e.,

$$\beta_v = \left(L_v : N_v \to (\pi^{-1} T M)_v\right)^{-1}, \quad v \in V(M).$$

Definition 2.8 The vector bundle isomorphism $\beta : \pi^{-1} T M \simeq N \subset T(V(M))$ is referred to as the *horizontal lift* with respect to the nonlinear connection N. □

Let (U, x^i) be a local coordinate system on M and $\left(\pi^{-1}(U), x^i, y^i\right)$ the induced local coordinates on $V(M)$. Let us set for simplicity

$$\left\{\partial_i, \dot{\partial}_i \ : \ 1 \le i \le n\right\} \equiv \left\{\frac{\partial}{\partial x^i}, \frac{\partial}{\partial y^i} \ : \ 1 \le i \le n\right\}.$$

For any $v \in \pi^{-1}(U)$, the space N_v is the span of $\{\delta_i(v) \ : \ 1 \le i \le n\}$ where

$$\delta_i \equiv \frac{\delta}{\delta x^i} \equiv \beta X_i, \quad 1 \le i \le n.$$

As $\{\partial_i, \dot{\partial}_i \ : \ 1 \le i \le n\}$ is a local frame in $T(V(M))$

$$\delta_i = M_i^j \, \partial_j - N_i^j \, \dot{\partial}_j \tag{2.5}$$

for some C^∞ functions $M_j^i, \ N_j^i : \pi^{-1}(U) \to \mathbb{R}$. Let us apply L to both sides of (2.5) and take into account that

$$L \circ \beta = 1_{\pi^{-1} T M}, \quad L \, \partial_i = X_i \, .$$

It follows that $M^i_j = \delta^i_j$. Therefore,

$$\delta_i = \partial_i - N^j_i \dot{\partial}_j, \quad 1 \le i \le n. \tag{2.6}$$

Definition 2.9 The C^∞ functions $N^i_j : \pi^{-1}(U) \to \mathbb{R}$ are the *local coefficients* of the nonlinear connection N [with respect to (U, x^i)]. □

For further use, let us consider the local differential 1-forms

$$\delta y^i \equiv dy^i + N^i_j dx^j, \quad 1 \le i \le n.$$

Then $\{dx^i, \ \delta y^i \ : \ 1 \le i \le n\}$ is a local frame in $T^*(V(M))$, defined on the open set $\pi^{-1}(U)$, dual to $\{\delta_i, \ \dot{\partial}_i \ : \ 1 \le i \le n\}$, i.e.,

$$dx^i\big(\delta_j\big) = \delta^i_j, \quad dx^i\big(\dot{\partial}_j\big) = 0, \quad \delta y^i\big(\delta_j\big) = 0, \quad \delta y^i\big(\dot{\partial}_j\big) = \delta^i_j.$$

In particular, in the older language of Pfaffian systems (rather than C^∞ distributions on manifolds) a nonlinear connection N is locally the Pfaffian system

$$dy^i + N^i_j dx^j = 0, \quad 1 \le i \le n.$$

Definition 2.10 Given a nonlinear connection N on M the *Dombrowski map* is the vector bundle morphism

$$K : T(V(M)) \to \pi^{-1}T(M), \quad K = \gamma^{-1} \circ V,$$

where $V : T(V(M)) \to \operatorname{Ker}(d\pi)$ is the projection associated to the direct sum decomposition

$$T(V(M)) = N \oplus \operatorname{Ker}(d\pi).$$

□

Then locally $K\big(\delta_i\big) = 0$ and $K\big(\dot{\partial}_i\big) = X_i$. Also the short sequence of vector bundles and vector bundle morphisms

$$0 \to \pi^{-1}T(M) \xrightarrow{\beta} T(V(M)) \xrightarrow{K} \pi^{-1}T(M) \to 0$$

is exact.

Let $\nabla \in \mathcal{C}\big(\pi^{-1}TM\big)$ and let $N_{\nabla, v} \subset T_v(V(M))$ be the subspace consisting of all $w \in T_v(V(M))$ such that $\big(\nabla_A \mathcal{L}\big)_v = 0$ for some $A \in \mathfrak{X}(V(M))$ such that $A_v = w$. Following B.T.M. Hassan (cf. [Has]), we adopt the following:

Definition 2.11 A connection ∇ in $\pi^{-1}T(M) \to V(M)$ is *regular* if

$$\nabla_{\gamma X}\mathcal{L} = X \tag{2.7}$$

for any $X \in C^\infty\left(\pi^{-1}TM\right)$. □

Lemma 2.12 *For every regular connection, $\nabla \in \mathcal{C}\left(\pi^{-1}TM\right)$ the assignment*

$$N_\nabla : v \in V(M) \longmapsto N_{\nabla,v} \subset T_v(V(M))$$

is a nonlinear connection on M and for every local coordinate system (U, x^i) on M the local coefficients of N_∇ are given by $N^i_j = \Gamma^i_{jk} y^k$ where $\nabla_{\partial_j} X_k = \Gamma^i_{jk} X_i$.

Proof Let $v \in V(M)$ and let us set $x = \pi(v) \in M$. Let (U, x^i) be a local coordinate system on M such that $x \in U$ [and in particular $v \in \pi^{-1}(U)$]. Let $w \in T_v(V(M))$. Let $A \in \mathfrak{X}(V(M))$ be a C^∞ extension of w to a tangent vector field A defined on the whole of $V(M)$, i.e., $A_v = w$. If $w \in N_{\nabla,v}$, then locally

$$w = w^i \left(\partial_i\right)_v + w^{i+n}\left(\dot{\partial}_i\right)_v, \quad A = A^i\,\partial_i + A^{i+n}\,\dot{\partial}_i, \quad A^i(v) = w^i, \quad A^{i+n}(v) = w^{i+n}$$

and

$$0 = \left(\nabla_A \mathcal{L}\right)_v = \left\{\nabla_A\left(y^i X_i\right)\right\}_v = \left\{A(y^i)\,X_i + y^i\,\nabla_A X_i\right\}_v =$$

$$= \left\{A^{i+n} X_i + y^i\left[A^j \nabla_{\partial_j} X_i + A^{j+n}\nabla_{\dot{\partial}_j} X_i\right]\right\}_v = \left\{w^{k+n} + y^i(v)\left[w^j\,\Gamma^k_{ji}(v) + w^{j+n}\,C^k_{ji}(v)\right]\right\} X_k(v)$$

where we have set

$$\nabla_{\partial_i} X_j = \Gamma^k_{ij} X_k, \quad \nabla_{\dot{\partial}_i} X_j = C^k_{ij} X_k\,.$$

On the other hand, let us apply (2.7) for $X = X_j$, i.e., $\nabla_{\dot{\partial}_j}\mathcal{L} = 0$ yielding

$$C^i_{jk} y^k = 0. \tag{2.8}$$

Consequently

$$w^{k+n} + y^i(v)\,\Gamma^k_{ji}(v)\,w^j = 0$$

so that

$$w = w^j\left\{\partial_j - y^i\Gamma^k_{ji}\dot{\partial}_k\right\}_v,$$

i.e., N_∇ is locally the span of

$$\partial_j - y^i\,\Gamma^k_{ji}\,\dot{\partial}_k, \quad 1 \le j \le n.$$

This proves smoothness, i.e., N_∇ is a C^∞ distribution on $V(M)$, of real rank n, whose local coefficients are $N^i_j = \Gamma^i_{jk} y^k$. To end the proof let $w \in N_{\nabla,v} \cap \text{Ker}(d_v \pi)$, i.e.,

$$w = \lambda^j \left(\partial_j - y^i \, \Gamma^k_{ji} \, \dot{\partial}_k\right)_v, \quad w = \mu^i \left(\dot{\partial}_i\right)_v,$$

for some λ^i, $\mu^i \in \mathbb{R}$. As $\{\partial_i, \ \dot{\partial}_i\}$ is a local frame of $T(V(M))$ it follows that $\lambda^i = 0$ so that

$$N_{\nabla,v} \cap \text{Ker}(d_v \pi) = (0)$$

proving that the sum $N_{\nabla,v} + \text{Ker}(d_v \pi)$ is direct. Finally, the equality

$$T_v(V(M)) = N_{\nabla,v} \oplus \text{Ker}(d_v \pi)$$

follows by comparing dimensions. □

2.1.5 *Finslerian Connections*

Following the terminology of M. Matsumoto (cf. [Mat]), we adopt the following:

Definition 2.13 A *Finslerian connection* on M is a pair $(\nabla, \ N)$ consisting of a connection ∇ in the pullback bundle $\pi^{-1}T(M) \to V(M)$ and a nonlinear connection N on M. □

For any regular connection, $\nabla \in C^\infty\left(\pi^{-1}TM\right)$ the pair $\left(\nabla, \ N_\nabla\right)$ is a Finslerian connection on M.

Let (∇, N) be a Finslerian connection on M and let $\beta : \pi^{-1}T(M) \to N$ be the corresponding horizontal lift. Another analog to the torsion of a linear connection [on $V(M)$] is

$$T_K(A, B) = \nabla_A K B - \nabla_B K A - K[A, B], \quad A, B \in \mathfrak{X}(V(M)).$$

Unlike T_L the new "torsion" T_K depends on the given nonlinear connection N. Several fragments of interest (of the torsions T_L and T_K) are defined by

$$T(X, Y) = T_L(\beta X, \beta Y), \quad C(X, Y) = T_L(\gamma X, \beta Y),$$

$$R^1(X, Y) = T_K(\beta X, \beta Y), \quad P^1(X, Y) = T_K(\gamma X, \beta Y), \quad S^1(X, Y) = T_K(\gamma X, \gamma Y),$$

for any $X, Y \in C^\infty(\pi^{-1}TM)$.

Definition 2.14 T and C will be referred to as the *horizontal* and *mixed* components of T_L, while R^1, P^1 and S^1 are, respectively, the *horizontal*, *mixed*, and *vertical* components of T_K. □

As argued above $T_L(\gamma X, \gamma Y) = 0$ so there is no "vertical" component of T_L. A similar decomposition of the curvature form R^∇ is feasible and useful in the sequel. Precisely we set

$$R(X, Y)Z = R^\nabla(\beta X, \beta Y)Z, \quad P(X, Y)Z = R^\nabla(\gamma X, \beta Y)Z, \quad S(X, Y)Z = R^\nabla(\gamma X, \gamma Y)Z,$$

for any $X, Y, Z \in C^\infty(\pi^{-1}TM)$.

2.1.6 Finslerian Metrics

Definition 2.15 A C^1 function $E : T(M) \to \mathbb{R}$ is said to be a *Finslerian metric* if

i) $E \in C^\infty(V(M))$.

ii) $E(0_x) = 0$ for any $x \in M$.

iii) For any $\lambda > 0$ and any $w \in T(M)$,

$$E\big(\lambda w\big) = \lambda^2 E(w). \tag{2.9}$$

□

One often accepts that E be defined only on some open cone $\Omega \subset T(M)$ and refers $E : \Omega \to \mathbb{R}$ as a Finslerian metric, as well.

Let $E : T(M) \to \mathbb{R}$ be a Finslerian metric on M and let (U, x^i) be a local coordinate system on M. Let $\big(\pi^{-1}(U),\ x^i,\ y^i\big)$ be the induced local coordinates on $V(M)$ and let us consider the functions

$$g_{ij} : \pi^{-1}(U) \to \mathbb{R}, \quad g_{ij} = \frac{1}{2}\frac{\partial^2 E}{\partial y^i\, \partial y^j}\,.$$

Next, let g be the C^∞ section in the vector bundle

$$\big(\pi^{-1}TM\big)^* \otimes \big(\pi^{-1}TM\big)^* \to V(M) \tag{2.10}$$

defined as follows. Let $v \in V(M)$ and let us set $x = \pi(v)$. Let then (U, x^i) be a local coordinate system on M about x. Given $X, Y \in \big(\pi^{-1}TM\big)_v$ locally represented as

$$X = \lambda^i\, X_i(v), \quad Y = \mu^i\, X_i(v),$$

we set by definition

$$g_v(X, Y) = g_{ij}(v)\, \lambda^i\, \mu^j\,.$$

The definition of g_v doesn't depend upon the choice of local coordinates about x, so that

$$g\ :\ v \in V(M) \longmapsto g_v$$

is a globally defined section in the vector bundle (2.10). As $g_{ij} \in C^\infty\big(\pi^{-1}(U)\big)$ the section g is C^∞.

Note that [by axiom (iii) above] E is only required to be *positively* homogeneous (of degree 2). When that is supplemented by the request

$$E(-w) = E(w), \quad w \in T(M),$$

the Finslerian metric E is referred to as *symmetric*. Let us differentiate with respect to λ in

$$E(\lambda v) = \lambda^2 E(v), \quad v \in V(M), \quad \lambda > 0, \tag{2.11}$$

so that to obtain

$$\frac{\partial E}{\partial y^i}(\lambda v)v^i = 2\lambda E(w).$$

Differentiating once more with respect to λ gives

$$\frac{\partial^2 E}{\partial y^i\, \partial y^j}(\lambda\, v)v^i v^j = 2\, E(w)$$

hence (for $\lambda = 1$)

$$g_{ij} y^i y^j = E \tag{2.12}$$

on $\pi^{-1}(U)$. The identity (2.12) admits the following coordinate free reformulation $g(\mathcal{L}, \mathcal{L}) = E$. As another consequence of positive homogeneity (iii) the functions $g_{ij} : \pi^{-1}(U) \to \mathbb{R}$ are positively homogeneous of degree 0. Indeed, the identity (2.11) yields the following relationship between the local expressions of the functions involved:

$$E(x, \lambda y) = \lambda^2 E(x, y).$$

Differentiation with respect to y^i then yields

$$\frac{\partial E}{\partial y^j}(x, \lambda y)\lambda \delta_i^j = \lambda^2 \frac{\partial E}{\partial y^i}$$

or

$$\frac{\partial E}{\partial y^i}(x, \lambda y) = \lambda \frac{\partial E}{\partial y^i}(x, y).$$

Finally, one may differentiate once more with respect to y^j so that to obtain

$$\frac{\partial^2 E}{\partial y^i\, \partial y^j}(x, \lambda y) = \frac{\partial^2 E}{\partial y^i\, \partial y^j}(x, y)$$

or

$$g_{ij}(\lambda v) = g_{ij}(v) \tag{2.13}$$

for any $v \in \pi^{-1}(U)$ and any $\lambda > 0$. Q.e.d.

Definition 2.16 A Finslerian metric E on M is *positive definite* if g is positive definite, i.e.,

$$g_v(X, X) \geq 0, \quad g_v(X, X) = 0 \Longleftrightarrow X = 0,$$

for any $X \in \left(\pi^{-1}TM\right)_v$ and any $v \in V(M)$. □

If E is a positive definite Finslerian metric, then g is a Riemannian bundle metric on the pullback bundle $\pi^{-1}T(M) \to V(M)$. In particular, any such E is nonnegative, for one has $E = g(\mathcal{L}, \mathcal{L}) > 0$ and $E(w) = 0 \Longleftrightarrow w \in \sigma(M)$.

Lemma 2.17 *If $E : T(M) \to [0, +\infty)$ is a positive definite Finslerian metric such that $E \in C^2(T(M))$, then for any $X, Y \in \mathfrak{X}(M)$ the function $g\left(X^\pi, Y^\pi\right) : V(M) \to \mathbb{R}$ is invariant along the fibers of π and descends to a unique C^∞ function $G(X, Y) : M \to \mathbb{R}$ such that G is a Riemannian metric on M*

Proof Let (U, x^i) be a local coordinate system on M and let $\left(\Pi^{-1}(U), x^i, y^i\right)$ be the induced local coordinates on $T(M)$. If $E \in C^2(T(M))$, and in particular E is C^2 along the zero section, then

$$g_{ij} = \frac{1}{2} \frac{\partial^2 E}{\partial y^i \, \partial y^j}$$

are defined and continuous on $\Pi^{-1}(U)$. The same calculations as above, performed on $\Pi^{-1}(U)$ rather than on $\pi^{-1}(U)$ only, lead to the identity

$$g_{ij}(\lambda w) = g_{ij}(w) \tag{2.14}$$

for every $w \in \Pi^{-1}(U)$. Let $\{\lambda_\nu\}_{\nu \geq 1} \subset (0, +\infty)$ be a sequence of positive numbers such that $\lim_{\nu \to \infty} \lambda_\nu = 0$. Then, for each fixed $w \in \Pi^{-1}(U)$ with $x = \Pi(w)$, the sequence

$$\{\lambda_\nu w\}_{\nu \geq 1} \subset \Pi^{-1}(x)$$

converges to 0_x in the topology of $T(M)$. By (2.14) for $\lambda = \lambda_\nu$

$$g_{ij}(\lambda_\nu \, w) = g_{ij}(w), \quad \nu \geq 1,$$

and passing to the limit with $\nu \to \infty$ one has (by the continuity of g_{ij})

$$g_{ij}(0_x) = g_{ij}(w).$$

In other words g_{ij} depend on the position arguments x^i alone and determine a (globally defined) Riemannian metric G on M. Q.e.d.

2.1.7 Cartan Connection

Theorem 2.18 *Let* $E : T(M) \to [0, +\infty)$ *be a positive definite Finslerian metric and* g *the corresponding Riemannian bundle metric on* $\pi^{-1}T(M) \to V(M)$*. There is a unique regular connection* $\nabla \in \mathcal{C}(\pi^{-1}TM)$ *such that* i) $\nabla g = 0$, ii) $T = 0$, *and* iii) $S^1 = 0$.

Definition 2.19 The connection ∇ furnished by Theorem 2.18 is referred to as the *Cartan connection* of (M, E). □

Theorem 2.18 is most easily proved by a calculation in local coordinates. A few local formulas established in the sequel hold for an arbitrary Finslerian connection and that will be emphasized as we proceed. Let (∇, N) be a Finslerian connection on M. The following notations are adopted, as to the local connection coefficients of ∇:

$$\nabla_{\delta_i} X_j = F^k_{ij} X_k, \quad \nabla_{\dot{\partial}_i} X_j = C^k_{ij} X_k, \quad \nabla_{\partial_i} X_j = \Gamma^k_{ij} X_k .$$

Moreover, we set

$$T(X_i, X_j) = T^k_{ij} X_k, \quad S^1(X_i, X_j) = S^k_{ij} X_k .$$

As $K\dot{\partial}_i = X_i$ and $[\dot{\partial}_i, \dot{\partial}_j] = 0$

$$S^k_{ij} = C^k_{ij} - C^k_{ji} . \tag{2.15}$$

If N^i_j are the local coefficients of the given nonlinear connection, then $\delta_i = \partial_i - N^j_i \dot{\partial}_j$ yields

$$[\delta_i, \delta_j] = \left(\frac{\partial N^k_i}{\partial x^j} - \frac{\partial N^k_j}{\partial x^i} \right) \dot{\partial}_k \tag{2.16}$$

so that $L\left[\delta_i, \delta_j\right] = 0$ and then

$$T^k_{ij} = F^k_{ij} - F^k_{ji} . \tag{2.17}$$

We first attack uniqueness. Let ∇ be a regular connection in $\pi^{-1}T(M) \to V(M)$ such that $\nabla g = 0$ and $T = S^1 = 0$. The identity $\left(\nabla_{\dot{\partial}_i} g\right)(X_j, X_k) = 0$ implies

$$\frac{\partial g_{jk}}{\partial y^i} = C^\ell_{ij}\, g_{\ell k} + C^\ell_{ik}\, g_{j\ell} . \tag{2.18}$$

Cyclic permutation $i \mapsto j \mapsto k \mapsto i$ of the indices i, j and k in (2.18) leads to two more identities of the sort

$$\frac{\partial g_{ki}}{\partial y^j} = C^\ell_{jk}\, g_{\ell i} + C^\ell_{ji}\, g_{k\ell}, \tag{2.19}$$

$$\frac{\partial g_{ij}}{\partial y^k} = C^\ell_{ki}\, g_{\ell j} + C^\ell_{kj}\, g_{i\ell}\,. \tag{2.20}$$

The so-called "Christoffel process" consists in adding the first two equations and subtracting the third

$$(2.18) + (2.19) - (2.20).$$

Using the symmetry $C^k_{ij} = C^k_{ji}$ one obtains

$$2\, C^\ell_{ij}\, g_{k\ell} = \frac{\partial g_{jk}}{\partial y^i} + \frac{\partial g_{ki}}{\partial y^j} - \frac{\partial g_{ij}}{\partial y^k}$$

or [by $g_{ij} = (1/2)\, \partial^2 E/\partial y^i\, \partial y^j$]

$$C^\ell_{ij} = \frac{1}{4}\, g^{\ell k}\, \frac{\partial^3 E}{\partial y^i\, \partial y^j\, \partial y^k}\,. \tag{2.21}$$

To establish uniqueness, besides C^k_{ij} we ought to determine N^i_j and F^i_{jk} [equivalently N^i_j and Γ^k_{ij}]. Similar to the above, the identity $\big(\nabla_{\delta_i} g\big)(X_j,\, X_k) = 0$ together with the Christoffel process (relying on $F^k_{ij} = F^k_{ji}$) leads to

$$2\, F^\ell_{ij}\, g_{k\ell} = \frac{\delta g_{jk}}{\delta x^i} + \frac{\delta g_{ki}}{\delta x^j} - \frac{\delta g_{ij}}{\delta x^k}\,. \tag{2.22}$$

Formula (2.22) doesn't determine F^ℓ_{ij} yet, because the coefficients

$$N^\ell_i = y^k\, F^\ell_{ik} \tag{2.23}$$

of the nonlinear connection N_∇ are unknown so far. An ingredient we shall need is

$$y^j\, C^k_{ij} = 0 \tag{2.24}$$

[a consequence of the regularity condition $\nabla_{\gamma X}\mathcal{L} = X$, for any $X \in C^\infty\big(\pi^{-1}TM\big)$]. Next let us substitute from $\delta_i = \partial_i - N^j_i\, \dot\partial_j$ into (2.61) so that to obtain

$$2\, F^\ell_{ij}\, g_{\ell k} = \frac{\partial g_{jk}}{\partial x^i} + \frac{\partial g_{ki}}{\partial x^j} - \frac{\partial g_{ij}}{\partial x^k} - N^\ell_i\, \frac{\partial g_{jk}}{\partial y^\ell} - N^\ell_j\, \frac{\partial g_{ki}}{\partial y^\ell} + N^\ell_k\, \frac{\partial g_{ij}}{\partial y^\ell}$$

or

$$F^\ell_{ij}\, g_{\ell k} = \gamma_{ijk} - N^\ell_i\, C_{jk\ell} - N^\ell_j\, C_{ki\ell} + N^\ell_k\, C_{ij\ell} \tag{2.25}$$

where we have set

$$\gamma_{ijk} = \frac{1}{2}\left(\frac{\partial g_{jk}}{\partial x^i} + \frac{\partial g_{ki}}{\partial x^j} - \frac{\partial g_{ij}}{\partial x^k}\right), \quad C_{ijk} = \frac{1}{4}\frac{\partial^3 E}{\partial y^i\,\partial y^j\,\partial y^k}\,.$$

in particular [by (2.24)]

$$C_{ijk}\,y^i = 0, \quad C_{ijk}\,y^j = 0, \quad C_{ijk}\,y^k = 0.$$

Contraction by y^i in (2.25) yields [by (2.23)]

$$N_j^\ell\, g_{\ell k} = y^i\,\gamma_{ijk} - N_i^\ell\, y^i\, C_{jk\ell}\,. \tag{2.26}$$

Next, contraction with y^j in (2.26) gives [by $C_{jk\ell}y^j = 0$]

$$N_j^\ell\, y^j\, g_{\ell k} = \gamma_{ijk}\, y^i\, y^j\, y^k$$

or

$$N_j^\ell\, y^j = g^{\ell k}\,\gamma_{00k} \tag{2.27}$$

where we have set

$$\gamma_{00k} = \gamma_{ijk}\, y^i\, y^j\,.$$

Moreover, substitution from (2.27) into (2.26) leads to

$$N_j^\ell = g^{\ell k}\left(y^i\,\gamma_{ijk} - C_{jk}^m\,\gamma_{00\,m}\right) \tag{2.28}$$

so the coefficients of the nonlinear connection N_∇ are determined. Finally, we may substitute from (2.28) into (2.25) so that to determine F_{ij}^k. Q.e.d. Viceversa the connection ∇ defined by (2.21), (2.25), and (2.28) satisfies the requirements in Theorem 2.18. □

While unnecessary towards establishing the existence and uniqueness of Cartan's connection, one may of course determine Γ_{ij}^k as well

$$\Gamma_{ij}^k = F_{ij}^k + N_i^\ell C_{\ell j}^k\,.$$

Next, we state a couple of useful properties of the torsion of the Cartan connection.

Lemma 2.20 *Let ∇ be the Cartan connection of $(M,\ E)$.*
(i) *For any $A, B \in \mathfrak{X}(V(M))$,*

$$g\big(T_L(A, B),\ \mathcal{L}\big) = 0.$$

(ii) *For any $X, Y \in C^\infty(\pi^{-1}TM)$,*

$$P(X, Y)\mathcal{L} = X^i\, Y^j\, y^k\, \frac{\partial F^m_{jk}}{\partial y^i}\, X_m\,.$$

(iii) *For any* $X \in C^\infty\big(\pi^{-1}TM\big)$,

$$P(X,\ \mathcal{L})\mathcal{L} = 0.$$

Proof (i) In local coordinates

$$A = A^i\,\partial_i + A^{i+n}\,\dot{\partial}_i, \quad B = B^j\,\partial_j + B^{j+n}\,\dot{\partial}_j\,.$$

Note that [as $L\partial_i = X_i$ and $L\dot{\partial}_i = 0$]

$$T_L(\partial_i,\ \partial_j) = \nabla_{\partial_i} X_j - \nabla_{\partial_j} X_i = \left(\Gamma^k_{ij} - \Gamma^k_{ji}\right) X_k,$$

$$T_L(\partial_i,\ \dot{\partial}_j) = -\nabla_{\dot{\partial}_j} X_i = -C^k_{ji} X_k, \quad T_L(\dot{\partial}_i,\ \dot{\partial}_j) = 0.$$

Hence,

$$T_L(A,\, B) = A^i\, B^j\, T_L(\partial_i,\ \partial_j) + A^i\, B^{j+n}\, T_L(\partial_i,\ \dot{\partial}_j) + A^{i+n}\, B^j\, T_L(\dot{\partial}_i,\ \partial_j) =$$

$$= \left\{A^i\, B^j\left(\Gamma^k_{ij} - \Gamma^k_{ji}\right) - A^i\, B^{j+n} C^k_{ji} + A^{i+n}\, B^j\, C^k_{ji}\right\} X_k\,.$$

On the other hand [by $\mathcal{L} = y^\ell\, X_\ell$] $g(\mathcal{L},\ X_k) = y^\ell g_{\ell k}$ so that

$$g\big(T_L(A,\, B),\ \mathcal{L}\big) = \left\{A^i\, B^j\left(\Gamma^k_{ij} - \Gamma^k_{ji}\right) - A^i\, B^{j+n} C^k_{ji} + A^{i+n}\, B^j\, C^k_{ji}\right\} y^\ell\, g_{\ell k}\,.$$

Let us recall that $C^k_{ji} g_{\ell k} y^\ell = C_{ij\ell}\, y^\ell = 0$ so that

$$g\big(T_L(A,\, B),\ \mathcal{L}\big) = A^i\, B^j\left(\Gamma^k_{ij} - \Gamma^k_{ji}\right) y^\ell\, g_{\ell k}\,.$$

Moreover, we recall $\Gamma^i_{jk} = F^i_{jk} + N^\ell_j C^i_{\ell k}$ and contract with $y^m\, g_{mi}$ so that to obtain

$$\Gamma^i_{jk}\, y^m\, g_{mi} = F^i_{jk}\, y^m\, g_{mi}\,.$$

On the other hand [by (2.25)]

$$F^\ell_{ij}\, g_{\ell k}\, y^k = \Big(\gamma_{ijk} - N^\ell_i\, C_{jk\ell} - N^\ell_j\, C_{ki\ell} + N^\ell_k\, C_{ij\ell}\Big)\, y^k = \Big(\gamma_{ijk} + N^\ell_k\, C_{ij\ell}\Big)\, y^k =$$

[by substitution from $N^\ell_j\, y^j = g^{\ell k}\, \gamma_{00k}$ (cf. (2.27))]

$$= \gamma_{ij0} + g^{\ell m}\, \gamma_{00m}\, C_{ij\ell} = \gamma_{ij0} + C^k_{ij}\, \gamma_{00k}\,.$$

Finally,

$$\left(\Gamma^k_{ij} - \Gamma^k_{ji}\right) y^\ell g_{\ell k} = \left(F^k_{ij} - F^k_{ji}\right) y^\ell g_{\ell k} = 0.$$

Q.e.d.

(ii) Locally $X = X^i X_i$ and $Y = Y^j X_j$ hence

$$P(X, Y)\mathcal{L} = X^i Y^j y^k P(X_i, X_j)X_k .$$

Moreover,

$$P(X_i, X_j)X_k = R^\nabla(\gamma X_i, \beta X_j)X_k = R^\nabla(\dot{\partial}_i, \delta_j)X_k$$

where δ_j is short for $\delta/\delta x^j$. Next

$$\delta_j = \partial_j - N^m_j \dot{\partial}_m \Longrightarrow \left[\dot{\partial}_i, \delta_j\right] = -\frac{\partial N^m_j}{\partial y^i} \dot{\partial}_m,$$

so that

$$R^\nabla(\dot{\partial}_i, \delta_j)X_k = \nabla_{\dot{\partial}_i} \nabla_{\delta_j} X_k - \nabla_{\delta_j} \nabla_{\dot{\partial}_i} X_k - \nabla_{\left[\dot{\partial}_i, \delta_j\right]} X_k =$$

$$= \nabla_{\dot{\partial}_i}\left(F^\ell_{jk} X_\ell\right) - \nabla_{\delta_j}\left(C^\ell_{ik} X_\ell\right) + \frac{\partial N^m_j}{\partial y^i} C^\ell_{mk} X_\ell =$$

$$= \left\{\frac{\partial F^m_{jk}}{\partial y^i} + F^\ell_{jk} C^m_{i\ell} - \frac{\delta C^m_{ik}}{\delta x^j} - C^\ell_{ik} F^m_{j\ell} + \frac{\partial N^\ell_j}{\partial y^i} C^m_{\ell k}\right\} X_m .$$

Hence [by $C^m_{ik} y^k = 0$],

$$y^k R^\nabla(\dot{\partial}_i, \delta_j)X_k = y^k \left\{\frac{\partial F^m_{jk}}{\partial y^i} + F^\ell_{jk} C^m_{i\ell} - \frac{\delta C^m_{ik}}{\delta x^j}\right\} X_m .$$

The last term is

$$y^k \frac{\delta C^m_{ik}}{\delta x^j} = y^k \left(\frac{\partial C^m_{ik}}{\partial x^j} - N^\ell_j \frac{\partial C^m_{ik}}{\partial y^\ell}\right) = \frac{\partial}{\partial x^j}\left(y^k C^m_{ik}\right) - y^k N^\ell_j \frac{\partial}{\partial y^\ell}\left(g^{mp} C_{ikp}\right) =$$

[by $y^k C^m_{ik} = 0$]

$$= -y^k N^\ell_j g^{mp} \frac{\partial C_{ikp}}{\partial y^\ell} .$$

Also

$$y^k C_{ikp} = 0 \Longrightarrow \delta^k_\ell C_{ikp} + y^k \frac{\partial C_{ikp}}{\partial y^\ell} = 0 \Longrightarrow y^k \frac{\partial C_{ikp}}{\partial y^\ell} = -C_{i\ell p}$$

so that

$$y^k \frac{\delta C^m_{ik}}{\delta x^j} = N^\ell_j \, C^m_{i\ell} \, .$$

Finally [by $y^k \, F^\ell_{jk} = N^\ell_j$],

$$y^k \left\{ F^\ell_{jk} \, C^m_{i\ell} - \frac{\delta C^m_{ik}}{\delta x^j} \right\} = 0$$

so that

$$y^k \, R^\nabla(\dot{\partial}_i, \, \delta_j) X_k = y^k \, \frac{\partial F^m_{jk}}{\partial y^i} \, X_m \, .$$

Q.e.d.

(iii) The statement is a rather involved consequence of (ii)

$$P(X, Y)\mathcal{L} = X^i \, Y^j \, y^k \, \frac{\partial F^m_{jk}}{\partial y^i} \, X_m$$

for $Y = \mathcal{L}$, i.e.,

$$P(X, \, \mathcal{L})\mathcal{L} = X^i \, y^j \, y^k \, \frac{\partial F^m_{jk}}{\partial y^i} \, X_m \, .$$

For the proof, we start from $y^k \, F^m_{jk} = N^m_j$ and differentiate with respect to y^i so that to get

$$F^m_{ji} + y^k \, \frac{\partial F^m_{jk}}{\partial y^i} = \frac{\partial N^m_j}{\partial y^i} \, .$$

Next, let us contract with y^j so that [by $y^j \, F^m_{ji} = N^m_i$]

$$y^j \, y^k \, \frac{\partial F^m_{jk}}{\partial y^i} = y^j \, \frac{\partial N^m_j}{\partial y^i} - N^m_i \, . \tag{2.29}$$

Next [by (2.27)] $N^m_j \, y^j = g^{mk} \, \gamma_{00k}$ hence [differentiating with respect to y^i]

$$y^j \, \frac{\partial N^m_j}{\partial y^i} = \frac{\partial}{\partial y^i} \left\{ g^{mk} \, \gamma_{00k} \right\} - N^m_i \, . \tag{2.30}$$

Next, we start from

$$\gamma_{ijk} = \frac{1}{2} \left(\frac{\partial g_{jk}}{\partial x^i} + \frac{\partial g_{ki}}{\partial x^j} - \frac{\partial g_{ij}}{\partial x^k} \right)$$

and contract with $y^i \, y^j$ so that to obtain

$$\gamma_{00k} = \frac{1}{2} \left(\frac{\partial g_{jk}}{\partial x^i} + \frac{\partial g_{ki}}{\partial x^j} - \frac{\partial g_{ij}}{\partial x^k} \right) y^i \, y^j$$

or

$$\gamma_{00k} = y^i\, y^j\, \frac{\partial g_{jk}}{\partial x^i} - \frac{1}{2}\frac{\partial E}{\partial x^k}\,. \tag{2.31}$$

Also

$$g^{mk}\, g_{k\ell} = \delta^m_\ell \Longrightarrow \frac{\partial g^{mk}}{\partial y^i}\, g_{k\ell} + g^{mk}\, \frac{\partial g_{k\ell}}{\partial y^i} = 0 \Longrightarrow \frac{\partial g^{mk}}{\partial y^i} = -g^{\ell k}\, g^{ms}\, \frac{\partial g_{s\ell}}{\partial y^i} \Longrightarrow$$

$$\frac{\partial g^{mk}}{\partial y^i} = -2\, g^{\ell k}\, g^{ms}\, C_{i\ell s}\,. \tag{2.32}$$

Going back to (2.29)–(2.30)

$$y^j\, y^k\, \frac{\partial F^m_{jk}}{\partial y^i} = \frac{\partial}{\partial y^i}\left\{g^{mk}\, \gamma_{00k}\right\} - 2\, N^m_i = \frac{\partial g^{mk}}{\partial y^i}\, \gamma_{00k} + g^{mk}\, \frac{\partial \gamma_{00k}}{\partial y^i} - 2\, N^m_i =$$

[by $N^m_i = g^{mj}\left(\gamma_{0ij} - C^k_{ij}\, \gamma_{00k}\right)$, cf. (2.28)]

$$= \frac{\partial g^{mk}}{\partial y^i}\, \gamma_{00k} + g^{mk}\, \frac{\partial \gamma_{00k}}{\partial y^i} - 2\, g^{mj}\left(\gamma_{0ij} - C^k_{ij}\, \gamma_{00k}\right) =$$

[by (2.32)]

$$= -2\, g^{\ell k}\, g^{ms}\, C_{i\ell s}\, \gamma_{00k} + g^{mk}\, \frac{\partial \gamma_{00k}}{\partial y^i} - 2\, g^{mj}\left(\gamma_{0ij} - C^k_{ij}\, \gamma_{00k}\right) =$$

[by observing the cancellation of terms]

$$= g^{mj}\left(\frac{\partial \gamma_{00j}}{\partial y^i} - 2\, \gamma_{0ij}\right).$$

As customary, let us differentiate with respect to λ in $E(x, \lambda y) = \lambda^2\, E(x, y)$ and set $\lambda = 1$ in the resulting equation. We obtain

$$\frac{\partial E}{\partial y^i}\, y^i = 2\, E.$$

Taking the partial derivative with respect to y^j yields

$$2\, g_{ij}\, y^i = \frac{\partial E}{\partial y^j} \tag{2.33}$$

Going back to (2.31) [by (2.33)]

$$\gamma_{00k} = y^i y^j \frac{\partial g_{jk}}{\partial x^i} - \frac{1}{2}\frac{\partial E}{\partial x^k} = y^i \frac{\partial}{\partial x^i}\left(\frac{1}{2}\frac{\partial E}{\partial y^k}\right) - \frac{1}{2}\frac{\partial E}{\partial x^k}$$

or

$$2\,\gamma_{00k} = y^i \frac{\partial}{\partial x^i}\left(\frac{\partial E}{\partial y^k}\right) - \frac{\partial E}{\partial x^k}.$$

Let us take partials with respect to y^ℓ so that to obtain

$$2\,\frac{\partial \gamma_{00k}}{\partial y^\ell} = \frac{\partial^2 E}{\partial x^\ell\,\partial y^k} - \frac{\partial^2 E}{\partial x^k\,\partial y^\ell} + y^i \frac{\partial^3 E}{\partial x^i\,\partial y^k\,\partial y^\ell}. \tag{2.34}$$

Let us contract with y^i in $\gamma_{ijk} = \frac{1}{2}\left(\frac{\partial g_{ik}}{\partial x^j} + \frac{\partial g_{jk}}{\partial x^i} - \frac{\partial g_{ij}}{\partial x^k}\right)$ and use again (2.33). We obtain

$$\gamma_{0jk} = \frac{1}{2}\left[\frac{\partial}{\partial x^j}\left(\frac{1}{2}\frac{\partial E}{\partial y^k}\right) + y^i \frac{\partial g_{jk}}{\partial x^i} - \frac{\partial}{\partial x^k}\left(\frac{1}{2}\frac{\partial E}{\partial y^j}\right)\right]. \tag{2.35}$$

Finally [by (2.34)–(2.35)],

$$\frac{\partial \gamma_{00j}}{\partial y^i} - 2\,\gamma_{0ij} = \frac{1}{2}\,y^s \frac{\partial}{\partial x^s}\left(\frac{\partial^2 E}{\partial y^j\,\partial y^i}\right) - y^s \frac{\partial g_{ij}}{\partial x^s} = 0.$$

Q.e.d.

2.2 Geodesics in Finslerian Manifolds

2.2.1 Regular Curves

Let $I \subset \mathbb{R}$ be an interval.

Definition 2.21 A C^1 curve $\alpha : I \to M$ is *regular* if $\dot\alpha(t) \neq 0$ for any $t \in I$. □

If this is the case one may consider the continuous curve $\dot\alpha : I \to V(M)$. Let (U, x^i) be a local coordinate system on M such that $\gamma(t_0) \in U$ for some $t_0 \in I$. By continuity there is $\epsilon > 0$ such that $\gamma(t) \in U$ for any $|t - t_0| < \epsilon$. The local components of α are

$$\alpha^i = \tilde{x}^i \circ \alpha : (t_0 - \epsilon,\ t_0 + \epsilon) \to U, \quad 1 \le i \le n,$$

while the local components of $\dot\alpha$ are

$$\dot\alpha^i,\ \dot\alpha^{i+n} : (t_0 - \epsilon,\ t_0 + \epsilon) \to \pi^{-1}(U),$$

$$\dot{\alpha}^i = x^i \circ \dot{\alpha} = \tilde{x}^i \circ \pi \circ \dot{\alpha} = \alpha^i, \quad \dot{\alpha}^{i+n} = y^i \circ \dot{\alpha}^i = \frac{d\alpha^i}{dt}.$$

Let $\alpha : I \to M$ be a regular curve of class C^2. Let N be a nonlinear connection on M. The tangent vector field along $\dot{\alpha}$ is

$$\frac{d\dot{\alpha}}{dt}(t) = \frac{d\dot{\alpha}^i}{dt}(t)\ (\partial_i)_{\dot{\alpha}(t)} + \frac{d\dot{\alpha}^{i+n}}{dt}(t)\ \left(\dot{\partial}_i\right)_{\dot{\alpha}(t)} = \frac{d\alpha^i}{dt}(t)\ \left(\delta_i + N_i^j\, \dot{\partial}_j\right)_{\dot{\alpha}(t)} + \frac{d^2\alpha^i}{dt^2}(t)\ \left(\dot{\partial}_i\right)_{\dot{\alpha}(t)}$$

or

$$\frac{d\dot{\alpha}}{dt}(t) = \frac{d\alpha^i}{dt}(t)\ (\delta_i)_{\dot{\alpha}(t)} + \left[\frac{d^2\alpha^j}{dt^2}(t) + N_i^j\left(\dot{\alpha}(t)\right)\frac{d\alpha^i}{dt}(t)\right]\left(\dot{\partial}_j\right)_{\dot{\alpha}(t)} \tag{2.36}$$

which is the decomposition of $d\dot{\alpha}/dt$ adapted to the direct sum decomposition

$$T_{\dot{\alpha}(t)}\big(V(M)\big) = N_{\dot{\alpha}(t)} \oplus \mathrm{Ker}\big(d_{\dot{\alpha}(t)}\pi\big).$$

Definition 2.22 A regular C^2 curve $\alpha : I \to M$ is a *geodesic* in (M, N), or a geodesic of the nonlinear connection N, if

$$K_{\dot{\alpha}(t)}\,\frac{d\dot{\alpha}}{dt}(t) = 0 \tag{2.37}$$

for any $t \in I$, where $K : T(V(M)) \to \pi^{-1}T(M)$ is the Dombowski map associated to N. □

Lemma 2.23 *Let ∇ be a regular connection in $\pi^{-1}T(M) \to V(M)$ and let $\alpha : I \to M$ be a regular C^2 curve. The following statements are equivalent:*

i) *α is a geodesic of the nonlinear connection N_∇.*

ii) *For any local coordinate system (U, x^i) and for any $t_0 \in I$ such that $\alpha(t_0) \in U$, there is $\epsilon > 0$ such that $\alpha(t) \in U$ and*

$$\frac{d^2\alpha^i}{dt^2}(t) + N_j^i\left(\dot{\alpha}(t)\right)\frac{d\alpha^j}{dt}(t) = 0, \quad 1 \le i \le n, \tag{2.38}$$

for any $\left|t - t_0\right| < \epsilon$.

Proof (i) $\Longrightarrow$ (ii)

Let α be a geodesic of N_∇. If $K : T(V(M)) \to \pi^{-1}T(M)$ is the Dombrowski map associated to the nonlinear connection N_∇, then

$$0 = K_{\dot{\alpha}(t)}\,\frac{d\dot{\alpha}}{dt}(t) = \gamma_{\dot{\alpha}(t)}^{-1}\, V_{\dot{\alpha}(t)}\frac{d\dot{\alpha}}{dt}(t) =$$

[by (2.36)]

$$= \left[\frac{d^2\alpha^j}{dt^2}(t) + N_i^j\big(\dot{\alpha}(t)\big)\frac{d\alpha^i}{dt}(t)\right](X_j)_{\dot{\alpha}(t)}$$

which is (2.38). □

For a regular connection ∇, the local coefficients of the nonlinear connection N_∇ are given by $N_j^i = F_{jk}^i y^k$ and $y^k\big(\dot{\alpha}(t)\big) = (d\alpha^k/dt)(t)$ hence the geodesics equation (2.38) may also be written in the more familiar form

$$\frac{d^2\alpha^i}{dt^2}(t) + F_{jk}^i\big(\dot{\alpha}(t)\big)\frac{d\alpha^j}{dt}(t)\frac{d\alpha^k}{dt}(t) = 0. \tag{2.39}$$

For any regular connection ∇ and any $A \in \mathfrak{X}(V(M))$,

$$\nabla_A \mathcal{L} = KA. \tag{2.40}$$

Indeed if $H : T(V(M)) \to N_\nabla$ is the projection, then $A = HA + VA$ and $\nabla_{HA}\mathcal{L} = 0$ hence

$$\nabla_A \mathcal{L} = \nabla_{VA}\mathcal{L} = \nabla_{\gamma(\gamma^{-1}(VA))}\mathcal{L} = \nabla_{\gamma(KA)}\mathcal{L} = KA.$$

Lemma 2.24 *Let $E : T(M) \to [0, +\infty)$ be a positive definite Finslerian metric on M and ∇ the Cartan connection of (M, E). Let $\alpha : I \to M$ be a geodesic of N_∇. Then $E(\dot{\alpha}(t)) =$ constant.*

Proof One has

$$\frac{d}{dt}\big[E(\dot{\alpha}(t))\big] = \frac{d}{dt}\Big[g(\mathcal{L}, \mathcal{L})_{\dot{\alpha}(t)}\Big] =$$

[by $\nabla g = 0$]

$$= 2g\left(\nabla_{d\dot{\alpha}/dt}\mathcal{L}, \mathcal{L}\right)_{\dot{\alpha}(t)} =$$

[by (2.67)]

$$= 2g\left(K_{\dot{\alpha}(t)}\frac{d\dot{\alpha}}{dt}(t), \mathcal{L}_{\dot{\alpha}(t)}\right) = 0$$

hence

$$E(\dot{\alpha}(t)) = \epsilon(\alpha)$$

for some constant $\epsilon(\alpha) > 0$. □

Definition 2.25 The constant $\epsilon(\alpha)$ in Lemma 2.24 is the *energy* of the geodesic α. □

The definition of the notion of a geodesic in a Finslerian manifold, as formulated in Definition 2.22, doesn't resemble much its Riemannian counterpart. In the sequel we make a few remarks, still following the exposition in B.T.M. Hassan (cf. [Has]) and aiming to establish precise analogies between the theory of geodesics in Riemannian and Finslerian geometry, starting from the basic fact that, at least for the time beeing,[1] Cartan's connection replaces the Levi-Civita connection of a Riemannian manifold.

Let $\tilde{\alpha} : I \to V(M)$ be a C^1 curve, where $I \subset \mathbb{R}$ is an interval. Curves of the sort occur mostly in relation with given regular C^2 curves $\alpha : I \to M$, for we may set $\tilde{\alpha}(t) = \dot{\alpha}(t)$ for any $t \in I$. Let V be a section in $\pi^{-1}T(M) \to V(M)$ defined *along* $\tilde{\alpha}$, i.e., a C^1 map

$$V : I \to \pi^{-1}T(M), \quad V(t) \in \left(\pi^{-1}TM\right)_{\tilde{\alpha}(t)} = T_{\pi(\tilde{\alpha}(t))}(M), \quad t \in I.$$

Given a connection ∇ in the vector bundle $\pi^{-1}T(M) \to V(M)$ the *covariant derivative* of V is a section defined along $\tilde{\alpha}$

$$\frac{DV}{dt} : I \to \pi^{-1}T(M)$$

$$\frac{DV}{dt}(t) := \left(\nabla_{d\tilde{\alpha}/dt}\overline{V}\right)_{\tilde{\alpha}(t)}, \quad t \in I, \tag{2.41}$$

where $\overline{V} : V(M) \to \pi^{-1}T(M)$ is a C^1 section extending V to the whole of $V(M)$, i.e., $\overline{V}(\tilde{\alpha}(t)) = V(t)$ for any $t \in I$. Here we assume tacitly that $\tilde{\alpha}$ is injective (the curve $\tilde{\alpha}$ has no self intersections) so that each value of the parameter identifies precisely one point of the curve. Then V may be thought of as a map $V : A \to \pi^{-1}T(M)$ defined on the closed set $A = \tilde{\alpha}(I) \subset V(M)$ by $V_{\tilde{\alpha}(t)} := V(t)$. Of course a section which is C^1 on a closed set may be extended to a C^1 section defined on the whole manifold. To simplify exposition let us assume that $0 \in I$ and let (U, x^i) be a local coordinate system about the point $\alpha(0)$, where $\alpha(t) := \pi(\tilde{\alpha}(t))$. Let $(\pi^{-1}(U), x^i, y^i)$ be the induced local coordinate system on $V(M)$, so that $\pi^{-1}(U)$ is an open neighborhood of $\tilde{\alpha}(0)$. By continuity, there is $\epsilon > 0$ such that $\alpha(t) \in U$ for every $|t| < \epsilon$, and in particular $\tilde{\alpha}(t) \in \pi^{-1}(U)$ for every $|t| < \epsilon$. Let $X_i = \left(\partial/\partial\tilde{x}^i\right)^\pi$ be the natural lift of $\partial/\partial\tilde{x}^i$. Starting from the local representations

[1] Unlike the case of Riemannian geometry, an array of canonical connections [mostly in the vector bundle $\pi^{-1}T(M) \to V(M)$] are equally frequent substitutes of the Levi-Civita connection, useful for various specific purposes (e.g., the Chern connection, the Berwald connection, the Rund connection, etc., cf. M. Matsumoto, [Mat]). These will not be further discussed in the present chapter, which is but introductory to Finslerian geometry, yet are needed for applications of sorts [e.g., Chern's connection turns out to be useful in the construction of a theory of harmonic maps from a Finslerian manifold].

$$\alpha^i(t) = \tilde{x}^i\big(\alpha(t)\big),$$

$$\tilde{\alpha}^i(t) = x^i\big(\tilde{\alpha}(t)\big), \quad \tilde{\alpha}^{i+n}(t) = y^i)\tilde{\alpha}(t)\big),$$

$$V(t) = V^i(t)\,X_{i,\,\tilde{\alpha}(t)}, \quad \overline{V}\Big|_{\pi^{-1}(U)} = \overline{V}^i\,X_i, \quad |t| < \epsilon,$$

one first observes that

$$\alpha = \pi \circ \tilde{\alpha} \Longrightarrow \tilde{\alpha}^i(t) = \alpha^i(t),$$

$$\frac{d\tilde{\alpha}}{dt}(t) = \frac{d\alpha^i}{dt}(t)\left(\frac{\partial}{\partial x^i}\right)_{\tilde{\alpha}(t)} + \frac{d\tilde{\alpha}^{i+n}}{dt}(t)\left(\frac{\partial}{\partial y^i}\right)_{\tilde{\alpha}(t)},$$

$$\overline{V}_{\tilde{\alpha}(t)} = V(t) \Longrightarrow \overline{V}^i\big(\tilde{\alpha}(t)\big) = V^i(t),$$

and then [by plugging into (2.41)]

$$\left(\nabla_{d\tilde{\alpha}/dt}\overline{V}\right)_{\tilde{\alpha}(t)} = \frac{d\alpha^i}{dt}(t)\,\left(\nabla_{\partial/\partial x^i}\overline{V}\right)_{\tilde{\alpha}(t)} + \frac{d\tilde{\alpha}^{i+n}}{dt}(t)\,\left(\nabla_{\partial/\partial y^i}\overline{V}\right)_{\tilde{\alpha}(t)},$$

$$\nabla_{\partial/\partial x^i}\overline{V} = \Big[\frac{\partial \overline{V}^k}{\partial x^i} + \overline{V}^j\,\Gamma^k_{ij}\Big]X_k, \quad \nabla_{\partial/\partial y^i}\overline{V} = \Big[\frac{\partial \overline{V}^k}{\partial y^i} + \overline{V}^j\,C^k_{ij}\Big]X_k,$$

where we have set as customary

$$\nabla_{\partial_i}X_j = \Gamma^k_{ij}\,X_k, \quad \nabla_{\dot{\partial}_i}X_j = C^k_{ij}\,X_k,$$

and $\partial_i = \partial/\partial x^i$ and $\dot{\partial}_i = \partial/\partial y^i$ for the sake of simplicity. Then

$$\left(\nabla_{d\tilde{\alpha}/dt}\overline{V}\right)_{\tilde{\alpha}(t)} = \left\{\frac{d\alpha^i}{dt}(t)\Big[\frac{\partial \overline{V}^k}{\partial x^i} + \overline{V}^j\,\Gamma^k_{ij}\Big]_{\tilde{\alpha}(t)} + \frac{d\tilde{\alpha}^{i+n}}{dt}(t)\Big[\frac{\partial \overline{V}^k}{\partial y^i} + \overline{V}^j\,C^k_{ij}\Big]_{\tilde{\alpha}(t)}\right\} X_k(\tilde{\alpha}(t)) =$$

$$= \left\{\frac{dV^k}{dt} + \Big[\Gamma^k_{ij}\big(\tilde{\alpha}(t)\big)\,\frac{d\alpha^i}{dt}(t) + C^k_{ij}\big(\tilde{\alpha}(t)\big)\,\frac{d\tilde{\alpha}^{i+n}}{dt}(t)\Big]\,V^j(t)\right\} X_k\big(\tilde{\alpha}(t)\big)$$

thus showing that the definition of $\dfrac{DV}{dt}(t)$ doesn't depend upon the choice of extension $\overline{V}$ to V. It is also customary to start with the last expression as a (local) definition of $\dfrac{DV}{dt}(t)$, good for the portion of $\tilde{\alpha}(t)$ which is surely contained in $\pi^{-1}(U)$, and then cover the whole of $\tilde{\alpha}$ with (induced) local coordinate neighborhoods. Of course, one then needs to check coherence on overlaps, and that the constructions do not depend upon the choice of local coordinate systems.

Let $\tilde{\alpha} : I \to V(M)$ be a C^1 curve and let V, W be two sections in $\pi^{-1}T(M)$ defined along $\tilde{\alpha}$. Then

$$\frac{d}{dt}\Big[g_{\tilde{\alpha}(t)}\big(V(t),\ W(t)\big)\Big] = g_{\tilde{\alpha}(t)}\big(\frac{DV}{dt}(t),\ W(t)\big) + g_{\tilde{\alpha}(t)}\big(V(t),\ \frac{DW}{dt}(t)\big) \tag{2.42}$$

for any connection ∇ in $\pi^{-1}T(M)$ such that $\nabla g = 0$. The proof of (2.42) relies on an easy local coordinate calculation (and is left as an exercise to the reader).

By following the conventions in [Has], one may regard the very tangent vector field of a regular curve α as a section in $\pi^{-1}T(M)$, defined along α

$$\frac{\overline{d}\alpha}{dt} : I \to \pi^{-1}T(M),$$

$$\frac{\overline{d}\alpha}{dt}(t) := \dot{\alpha}(t) \in T_{\alpha(t)}(M) = T_{\pi(\dot{\alpha}(t))}(M) = \big(\pi^{-1}TM\big)_{\dot{\alpha}(t)}, \quad t \in I.$$

By recalling the Liouville vector field $\mathcal{L} : V(M) \to \pi^{-1}T(M)$, one may easily see that

$$\frac{\overline{d}\alpha}{dt}(t) = \mathcal{L}\big(\dot{\alpha}(t)\big), \quad t \in I.$$

Let $(M,\ E)$ be a Finslerian manifold.

Definition 2.26 A regular curve $\alpha : I \to M$ of class C^2 is a *geodesic* of $(M,\ E)$ if it is a geodesic of the nonlinear connection N_∇ associated to the Cartan connection of $(M,\ E)$. □

Lemma 2.27 *Let $\alpha : I \to M$ be a regular curve, of class C^2. The following statements are equivalent:*

(i) *α is a geodesic.*

(ii) $\dfrac{D}{dt}\left(\dfrac{\overline{d}\alpha}{dt}\right) = 0$, *where the covariant derivative of* $\dfrac{\overline{d}\alpha}{dt}$ *is meant with respect to the Cartan connection ∇.*

(iii) $\left(\nabla_{d\tilde{\alpha}/dt}\mathcal{L}\right)_{\tilde{\alpha}(t)} = 0$ *where $\tilde{\alpha}(t) = \dot{\alpha}(t)$.*

The reader will certainly note that the (equivalent) descriptions (ii) and (iii) are closer to the formal definition of the notion of geodesic in Riemannian geometry.

Proof of Lemma 2.27. (i) $\Longrightarrow$ (iii). Let us assume that α is a geodesic, i.e., $K_{\tilde{\alpha}(t)}\dfrac{d\tilde{\alpha}}{dt}(t) = 0$ where $K : T(V(M)) \to \pi^{-1}T(M)$ is the Dombrowski map associated to the nonlinear connection N_∇. Let us recall [by (2.67)] that for any regular connection ∇, and in particular for the Cartan connection, and for any $A \in \mathfrak{X}(V(M))$ one has $\nabla_A \mathcal{L} = KA$. Let A be an extension of $d\tilde{\alpha}/dt$ (thought of as a tangent vector field on $V(M)$, defined along $\tilde{\alpha}$) to the whole of $V(M)$. Then

$$\left(\nabla_{d\tilde{\alpha}/dt}\mathcal{L}\right)_{\tilde{\alpha}(t)} = (\nabla_A\mathcal{L})_{\tilde{\alpha}(t)} = \left(KA\right)_{\tilde{\alpha}(t)} = K_{\tilde{\alpha}(t)}\frac{d\tilde{\alpha}}{dt}(t) = 0.$$

Q.e.d. The converse (iii) $\Longrightarrow$ (i) follows from the same calculation. Finally, (ii) $\Longleftrightarrow$ (iii) follows from

$$(\nabla_A\mathcal{L})_{\tilde{\alpha}(t)} = \left(\nabla_{d\tilde{\alpha}/dt}\frac{\overline{d\alpha}}{dt}\right)_{\tilde{\alpha}(t)}.$$

Q.e.d.

Let $\alpha : I \to M$ and $\beta : J \to M$ be two geodesics in (M, E) such that $\beta(\phi(t)) = \alpha(t)$ for some C^1 diffeomorphism $\phi : I \to J$ and any $t \in I$. Let us assume that $\phi'(t) > 0$. So α and β are equivalent curves, and have the same orientation. Then (locally)

$$\frac{d\alpha^i}{dt}(t) = \phi'(t)\,\frac{d\beta^i}{ds}\big(\phi(t)\big), \quad \frac{d^2\alpha^i}{dt^2}(t) = \phi''(t)\,\frac{d\beta^i}{ds}\big(\phi(t)\big) + \phi'(t)^2\,\frac{d^2\beta^i}{ds^2}\big(\phi(t)\big),$$

so that [as α is a geodesic]

$$0 = \frac{d^2\alpha^i}{dt^2} + \left(N^i_j y^j\right)\left[\frac{d\alpha}{dt}(t)\right] =$$

[as $N^i_j y^j$ are positively homogeneous of degree 2]

$$= \phi''(t)\,\frac{d\beta^i}{ds}\big(\phi(t)\big) + \phi'(t)^2\left\{\frac{d^2\beta^i}{ds^2}\big(\phi(t)\big) + \left(N^i_j y^j\right)\left[\frac{d\beta}{ds}\big(\phi(s)\big)\right]\right\} =$$

[as β is a geodesic]

$$= \phi''(t)\,\frac{d\beta^i}{ds}\big(\phi(t)\big) \Longrightarrow$$

[as β is regular]

$$\Longrightarrow \phi''(t) = 0 \Longrightarrow \phi(t) = at + b$$

for some $a,\ b \in \mathbb{R}$. If one also postulates $L(-w) = L(w)$, then any two equivalent geodesics in (M, E) differ by an affine transformation of the parameter.

Definition 2.28 The *length* of a regular C^1 curve $\gamma : [a, b] \to M$ of class C^1 is

$$L(\alpha) = \int_a^b E\big(\dot{\alpha}(t)\big)^{1/2}\,dt\,.$$

□

Let $\phi : [a,\, b] \to [0,\, L(\alpha)]$ be the C^1 diffeomorphism given by

$$\phi(t) = \int_a^t E\big(\dot\alpha(\tau)\big)^{1/2}\, d\tau.$$

Let $\beta : [0,\, L(\alpha)] \to M$ be defined by $\beta(s) = \alpha\big(\phi^{-1}(s)\big)$.

Lemma 2.29 *β is parametrized by arc length, i.e., $L\big(\dot\beta(s)\big) = 1$ for any value of the parameter s.*

Proof One has

$$\dot\alpha(t) = \phi'(t)\,\dot\beta(\phi(t)), \quad \phi'(t) = E\big(\dot\alpha(t)\big)^{1/2} > 0,$$

so that

$$E\big(\dot\alpha(t)\big) = \phi'(t)^2\, E\big(\dot\beta\big(\phi(t)\big)\big) \Longrightarrow E\big(\dot\beta\big(\phi(t)\big)\big) = 1.$$

Q.e.d.

2.2.2 *Parametrized Surfaces in V(M)*

Following [Has], we introduce the following notion.

Definition 2.30 A *parametrized surface* in $V(M)$ is a C^1 map $s : D \to V(M)$, where $D \subset \mathbb{R}^2$ is a domain. □

This very general notion works for our purposes through the present section.

Definition 2.31 A C^1 map $V : D \to \pi^{-1}T(M)$ is a *Finslerian vector field* along s if

$$V(u,t) \in \big(\pi^{-1}TM\big)_{s(u,t)}, \quad (u,t) \in D.$$

□

If s is injective than V may also be thought of as a C^1 section in $\pi^{-1}T(M)$ defined on the closed subset $A = s(D) \subset V(M)$, mapping every point $s(u,t) \in A$ onto $V(u,t)$.

For every connection $\nabla \in \mathcal{C}\big(\pi^{-1}TM\big)$, the *covariant derivative* of V is the Finslerian vector field along s

$$\frac{DV}{\partial t} : D \to \pi^{-1}T(M)$$

is built as follows. Let $(u_0,\, t_0) \in D$. Let s_{u_0} be the curve in $V(M)$ defined by $s_{u_0}(t) = s(u_0,\, t)$. Then $\dfrac{DV}{\partial t}(u_0,\, t_0)$ is by definition the covariant derivative of the restriction

V_{u_0} of V to s_{u_0} [which is a Finslerian vector field along that curve, i.e., $V_{u_0}(t) = V(u_0, t)$]

$$\frac{DV}{\partial t}(u_0, t_0) = \frac{DV_{u_0}}{dt}(t_0) = \left(\nabla_{ds_{u_0}/dt} V_{u_0}\right)_{s(u_0, t_0)}.$$

The Finslerian vector field along s

$$\frac{DV}{\partial u} : D \to \pi^{-1}T(M)$$

in defined in the similar manner, i.e.,

$$\frac{DV}{\partial u}(u_0, t_0) = \frac{DV_{t_0}}{du}(u_0) = \left(\nabla_{ds_{t_0}/du} V_{t_0}\right)_{s(u_0, t_0)}.$$

Here

$$s_{t_0}(u) = s(u, t_0), \quad V_{t_0} = V(u, t_0).$$

Next let

$$\frac{\partial s}{\partial t}, \frac{\partial s}{\partial u} : D \to T(V(M)) \tag{2.43}$$

be the tangent vector fields on $V(M)$ defined along s, i.e.,

$$\frac{\partial s}{\partial t}(u, t) = (d_{(u,t)}s)\left(\frac{\partial}{\partial t}\right)_{(u,t)}, \quad \frac{\partial s}{\partial u}(u, t) = (d_{(u,t)}s)\left(\frac{\partial}{\partial u}\right)_{(u,t)}.$$

Then

$$T, U : D \to \pi^{-1}T(M), \quad T = L\frac{\partial s}{\partial t}, \quad U = L\frac{\partial s}{\partial u},$$

are Finslerian vector fields associated to (2.43).

Lemma 2.32 *For any parametrized surface, $s : D \to V(M)$*

(i) $\frac{D}{\partial u}\left(L\frac{\partial s}{\partial t}\right) - \frac{D}{\partial t}\left(L\frac{\partial s}{\partial u}\right) = T_L\left(\frac{\partial s}{\partial u}, \frac{\partial s}{\partial t}\right).$

(ii) *If $X : D \to \pi^{-1}T(M)$ is a Finslerian vector field defined along s, then*

$$\left[\frac{D}{\partial t}, \frac{D}{\partial s}\right] X = R^{\nabla}\left(\frac{\partial s}{\partial t}, \frac{\partial s}{\partial u}\right) X.$$

The proof is a straightforward calculation in local coordinates. Precisely

$$s^i = x^i \circ s, \quad s^{i+n} = y^i \circ s,$$

$$\frac{\partial s}{\partial u}(u,t) = \frac{\partial s^i}{\partial u}(u,t)\left(\frac{\partial}{\partial x^i}\right)_{s(u,t)} + \frac{\partial s^{i+n}}{\partial u}(u,t)\left(\frac{\partial}{\partial y^i}\right)_{s(u,t)},$$

$$\frac{\partial s}{\partial t}(u,t) = \frac{\partial s^i}{\partial t}(u,t)\left(\frac{\partial}{\partial x^i}\right)_{s(u,t)} + \frac{\partial s^{i+n}}{\partial t}(u,t)\left(\frac{\partial}{\partial y^i}\right)_{s(u,t)},$$

$$U(u,t) = L_{s(u,t)}\frac{\partial s}{\partial u}(u,t) = \frac{\partial s^i}{\partial u}(u,t)\, X_{i,\, s(u,t)},$$

$$T(u,t) = L_{s(u,t)}\frac{\partial s}{\partial t}(u,t) = \frac{\partial s^i}{\partial t}(u,t)\, X_{i,\, s(u,t)},$$

so that

$$\frac{DT}{\partial u}(u,t) - \frac{DU}{\partial t}(u,t) = \frac{DT_t}{du}(u) - \frac{DU_u}{dt}(t) = \left(\nabla_{ds_t/du} T_t\right)_{s(u,t)} - \left(\nabla_{ds_u/dt} U_u\right)_{s(u,t)} =$$

$$= \Big\{\Big[\Gamma^k_{ij}(s(u,t)) - \Gamma^k_{ji}(s(u,t))\Big]\frac{\partial s^i}{\partial u}(u,t)\,\frac{\partial s^j}{\partial t}(u,t) + C^k_{ij}(s(u,t))\frac{\partial s^{i+n}}{\partial u}(u,t)\frac{\partial s^j}{\partial t}(u,t) +$$

$$- C^k_{ji}(s(u,t))\frac{\partial s^{j+n}}{\partial t}(u,t)\frac{\partial s^i}{\partial u}(u,t)\Big\}\, X_{k,\, s(u,t)}\,.$$

Similarly, a local coordinate calculation of the torsion term $T_L\left(\frac{\partial s}{\partial u}, \frac{\partial s}{\partial t}\right)$ leads to (i). The proof of (ii) is left as an exercise to the reader.

Definition 2.33 Let $f : D \to M$ be a regular surface, of class C^∞. Then

$$s = \tilde{f} : D \to V(M), \quad s(u,t) = \frac{df_u}{dt}(t), \quad f_u(t) = f(u,t), \quad (u,t) \in D,$$

is called the *lift* of f to $V(M)$. □

If $\tilde{f}$ is the lift of f, then $\pi \circ \tilde{f} = f$. By the chain rule

$$\left(d_{f(u,t)}\pi\right) \circ \left(d_{(u,t)}\tilde{f}\right) = d_{(u,t)} f$$

which applied to $(\partial/\partial u)_{(u,t)}$, respectively, to $(\partial(\partial t)_{(u,t)}$, yields

$$\left(d_{f(u,t)}\pi\right)\frac{\partial \tilde{f}}{\partial u}(u,t) = \frac{\partial f}{\partial u}(u,t), \quad \left(d_{f(u,t)}\pi\right)\frac{\partial \tilde{f}}{\partial t}(u,t) = \frac{\partial f}{\partial t}(u,t). \qquad (2.44)$$

Let $X : D \to T(M)$ be a tangent vector field along f, i.e.,

$$X(u,t) \in T_{f(u,t)}(M) = \left(\pi^{-1}TM\right)_{\tilde{f}(u,t)}$$

so that X may be thought of as a Finslerian vector field $\overline{X}$ along $\tilde{f}$

$$\overline{X} : D \to \pi^{-1}T(M), \quad \overline{X}(u,t) = X(u,t).$$

In particular

$$\frac{\partial f}{\partial u}, \ \frac{\partial f}{\partial t} \ : \ D \to T(M)$$

give rise, in a natural manner, to Finslerian vector fields along $\tilde{f}$

$$\frac{\overline{\partial} f}{\partial u}, \ \frac{\overline{\partial} f}{\partial t} \ : \ D \to \pi^{-1}T(M).$$

Lemma 2.34 *Let $f : D \to M$ be a regular surface. Then*
(i) $L\,\dfrac{\partial \tilde{f}}{\partial u} = \dfrac{\overline{\partial} f}{\partial u}$,
(ii) $L\,\dfrac{\partial \tilde{f}}{\partial u} = \dfrac{\overline{\partial} f}{\partial u}$.

Proof Follows from the very definition of the morphism $L : T(V(M)) \to \pi^{-1}T(M)$ [by (2.44)].

2.2.3 Existence and Uniqueness of Geodesics in Finslerian Manifolds

Next we wish to apply the classical Cauchy-Lipschitz-Picard (existence and uniqueness) theorem to the Cauchy problem for the ODE system (2.38). This aims of course to establish

Theorem 2.35 *Let $E : T(M) \to [0, +\infty)$ be a positive definite Finslerian metric on M and let ∇ be the Cartan connection of (M, E). Then, for every $x_0 \in M$, there is a neighborhood $U \subset M$ of x_0 and a number $\epsilon > 0$ such that for every $x \in U$ and every $v \in T_x(M) \setminus \{0_x\}$ with $L(v) < \epsilon$ there is a unique geodesic of N_∇*

$$\alpha_v : \big(-2,\, 2\big) \to M$$

satisfying the initial conditions

$$\alpha_v(0) = x, \quad \frac{d\alpha_v}{dt}(0) = v.$$

Here $L(w) = \sqrt{E(w)}$ for any $w \in T(M)$. We give a proof to Theorem 2.35 that willingly includes the proof to Cauchy-Lipschitz-Picard theorem. In doing so one pinpoints the precise dependence of the domain of the solution on the various data involved. On the other hand the proof of Cauchy-Lipschitz-Picard theorem relies, as well known, on the contraction theorem, hence one may indicate an iterative process for computing the geodesic α_v as the unique fixed point of a certain contraction.

2.2.4 Geodesics with Given Initial Data

The indices through the present section range as follows:

$$i, j, k, \cdots \in \{1, 2, \cdots, n\}, \quad a, b, c, \cdots \{1, 2, \cdots, 2n\}.$$

Let $x_0 \in M$ and let $\chi = (x^1, \cdots, x^n) : U \to \mathbb{R}^n$ be a local coordinate system on M such that $x_0 \in U$. Let

$$\tilde{\chi} = (x^1, \cdots, x^n, y^1, \cdots, y^n) : \pi^{-1}(U) \to \mathbb{R}^{2n}$$

be the induced local coordinate system on $V(M)$ and let us set

$$A = \tilde{\chi}\big[\pi^{-1}(U)\big] = \chi(U) \times \big(\mathbb{R}^n \setminus \{0\}\big) \subset \mathbb{R}^{2n}.$$

Next let

$$\mathbf{f} = (f^1, \cdots, f^n) : A \to \mathbb{R}^n, \quad f^i(\mathbf{x}, \mathbf{u}) = -N^i_j\left(\tilde{\chi}^{-1}(\mathbf{x}, \mathbf{u})\right) u^j,$$

$$1 \le i \le n, \quad \mathbf{x} \in \chi(U), \quad \mathbf{u} = (u^1, \cdots, u^n) \in \mathbb{R}^n,$$

where $N^i_j \in C^\infty(\pi^{-1}(U))$ are the local coefficients of the nonlinear connection N_∇ [associated to the unique regular Cartan connection of (M, E)]. Of course

$$\tilde{\chi}^{-1}(\mathbf{x}, \mathbf{u}) = u^i \left(\frac{\partial}{\partial \tilde{x}^i}\right)_{\chi^{-1}(\mathbf{x})}.$$

Let $x \in U$ and let $w \in T_x(M) \setminus \{0_x\}$. We set

$$\mathbf{U}_0 = \left(\chi(x), w^1, \cdots, w^n\right) \in A.$$

With the notations in the previous section, the problem

$$\frac{d^2\alpha^i}{dt^2}(t) + N^i_j\big(\dot\alpha(t)\big)\,\frac{d\alpha^j}{dt}(t) = 0, \quad 1 \le i \le n,$$

$$\alpha(0) = x, \quad \frac{d\alpha}{dt}(0) = w,$$

is equivalent to

$$\frac{d\mathbf{x}}{dt} = \mathbf{u}(t), \quad \frac{d\mathbf{u}}{dt} = \mathbf{f}\big(\mathbf{x}(t),\, \mathbf{u}(t)\big), \tag{2.45}$$

$$\mathbf{x}(0) = \chi(x), \quad \mathbf{u}(0) = \big(w^1, \cdots, w^n\big), \tag{2.46}$$

where

$$\mathbf{x}(t) = \big(x^1(t), \cdots, x^n(t)\big), \quad \mathbf{u}(t) = \big(u^1(t), \cdots, u^n(t)\big),$$

$$x^i(t) = \alpha^i(t), \quad u^i(t) = \frac{d\alpha^i}{dt}(t), \quad 1 \le i \le n.$$

Next we set $\mathbf{U}(t) = \big(\mathbf{x}(t),\, \mathbf{u}(t)\big)$ so that the problem (2.45)–(2.46) may be written

$$\frac{d\mathbf{U}}{dt} = \mathbf{F}\big(\mathbf{U}(t)\big), \quad \mathbf{U}(0) = \mathbf{U}_0, \tag{2.47}$$

where

$$\mathbf{F}(\mathbf{x},\, \mathbf{u}) = \big(\mathbf{u},\, \mathbf{f}(\mathbf{x},\, \mathbf{u})\big), \quad (\mathbf{x},\, \mathbf{u}) \in A.$$

For every $r > 0$, we adopt the notation $I_r = (-r,\, r) \subset \mathbb{R}$. Let us set $\mathbf{U}_0 = \big(\mathbf{x}_0,\, \mathbf{u}_0\big) \in \mathbb{R}^n \times \mathbb{R}^n$. Moreover, let $R > 0$ such that $K = \overline{B_R\big(\mathbf{x}_0\big) \times B_R\big(\mathbf{u}_0\big)} \subset A$. Here $B_R\big(\mathbf{x}_0\big) = \big\{\mathbf{x} \in \mathbb{R}^n \,:\, \|\mathbf{x} - \mathbf{x}_0\| < R\big\}$ is the open ball of center $\mathbf{x}_0 \in \mathbb{R}^n$ and radius $R > 0$. Note that $\mathbf{F} \in C^\infty(A,\, \mathbb{R}^{2n})$ and in particular $\mathbf{F}$ is continuous on A. Let $\mathbf{V},\, \mathbf{W} \in K$ and let

$$\varphi(t) = \mathbf{F}\big[(1-t)\mathbf{V} + t\,\mathbf{W}\big], \quad 0 \le t \le 1.$$

Then

$$\varphi_a'(t) = \frac{\partial F_a}{\partial \xi^b}\big[(1-t)\mathbf{V} + t\,\mathbf{W}\big]\,\big(W^b - V^b\big)$$

hence [by the theorem of Lagrange] there is $0 < \theta < 1$ such that

$$\big\|\mathbf{F}(\mathbf{V}) - \mathbf{F}(\mathbf{W})\big\| \le \left\|\frac{\partial \mathbf{F}}{\partial x^b}\big[\mathbf{V} + \theta(\mathbf{W} - \mathbf{V})\big]\right\|\, \big|V^b - W^b\big| \le L\, \big\|\mathbf{V} - \mathbf{W}\big\|$$

where

$$L = \sqrt{2n} \max_{a=1}^{2n} \sup_{\xi \in K} \left\| \frac{\partial \mathbf{F}}{\partial \xi^i}(\xi) \right\| .$$

Also we set

$$M = \sup_{\xi \in K} \left\| \mathbf{F}(\xi) \right\|$$

and consider $R_0 \in \mathbb{R}$ such that

$$0 < R_0 < \min \left\{ \frac{R}{M}, \frac{1}{L} \right\} .$$

The solution to the problem (2.47), as furnished by the Cauchy-Lipschitz-Picard theorem, is defined in the interval $I_0 = I_{R_0} \subset \mathbb{R}$. Let $C\big(I_0, \mathbb{R}^{2n}\big)$ be the Banach space of all continuous and bounded functions $\mathbf{U} : I_0 \to \mathbb{R}^{2n}$ with the supremum norm

$$\left\| \mathbf{U} \right\|_\infty = \sup_{t \in I_0} \left\| \mathbf{U}(t) \right\|$$

and let $\mathfrak{X}$ be the closed ball of center $\mathbf{U}_0$ [thought of as a (constant) function on I_0] and radius R, i.e.,

$$\mathfrak{X} = \Big\{ \mathbf{U} \in C\big(I_0, \mathbb{R}^{2n}\big) \; : \; \left\| \mathbf{U} - \mathbf{U}_0 \right\|_\infty \le R \Big\}.$$

$\mathfrak{X}$ is a closed subset of the complete metric space $C\big(I_0, \mathbb{R}^{2n}\big)$, hence a complete metric space itself, with the induced metric $d_\infty(\mathbf{U}, \mathbf{V}) = \|\mathbf{U} - \mathbf{V}\|_\infty$. For every $\mathbf{V} \in \mathfrak{X}$, we consider the function

$$\Phi(\mathbf{V}) : I_0 \to \mathbb{R}^{2n}, \quad \Phi(\mathbf{V})(t) = \mathbf{U}_0 + \int_0^t \mathbf{F}\big[\mathbf{V}(s)\big] \, ds, \quad t \in I_0 \, .$$

One checks easily that $\Phi(\mathbf{V}) \in C\big(I_0, \mathbb{R}^{2n}\big)$. Moreover,

$$\left\| \Phi(\mathbf{V})(t) - \mathbf{U}_0 \right\| = \left\| \int_0^t \mathbf{F}\big[\mathbf{V}(s)\big] \, ds \right\| \le \int_0^t \left\| \mathbf{F}\big[\mathbf{V}(s)\big] \right\| \, ds \le M \left| t \right| < M \, R_0 < R$$

so that [by passing to the least upper bound with $t \in I_0$]

$$\left\| \Phi(\mathbf{V}) - \mathbf{U}_0 \right\|_\infty \le R,$$

i.e., $\Phi(\mathbf{V}) \in \mathfrak{X}$. Hence, we built a map $\Phi : \mathfrak{X} \to \mathfrak{X}$. Let us show that Φ is a contraction. Indeed if $\mathbf{V}, \mathbf{W} \in \mathfrak{X}$

$$\left\| \Phi(\mathbf{V})(t) - \Phi(\mathbf{W})(t) \right\| = \left\| \int_0^t \big\{ \mathbf{F}\big[\mathbf{V}(s)\big] - \mathbf{F}\big[\mathbf{W}(s)\big] \big\} \, ds \right\| \le$$

$$\leq L\,|t|\,\|\mathbf{V}-\mathbf{W}\|_{\infty} < L\,R_0\,\|\mathbf{V}-\mathbf{W}\|_{\infty}$$

and $LR_0 < 1$ so that Φ is a contraction. By the Banach fixed point theorem Φ has a unique fixed point $\mathbf{U} \in \mathfrak{X}$, i.e., $\Phi(\mathbf{U}) = \mathbf{U}$ or

$$\mathbf{U}_0 + \int_0^t \mathbf{F}\big[\mathbf{U}(s)\big]\,ds = \mathbf{U}(t) \tag{2.48}$$

for any $t \in I_0$. If we set $\mathbf{U}(t) = \big(\mathbf{x}(t),\ \mathbf{u}(t)\big)$ and $\mathbf{U}_0 = \big(\mathbf{x}_0,\ \mathbf{u}_0\big)$, then (2.48) may be written

$$\mathbf{x}_0 + \int_0^t \mathbf{u}(s)\,ds = \mathbf{x}(t), \quad \mathbf{u}_0 + \int_0^t \mathbf{f}\big[\mathbf{x}(s),\ \mathbf{u}(s)\big]\,ds = \mathbf{u}(t),$$

or

$$\frac{d\mathbf{x}}{dt}(t) = \mathbf{u}(t), \quad \frac{d\mathbf{u}}{dt} = \mathbf{f}\big[\mathbf{x}(t),\ \mathbf{u}(t)\big), \quad \mathbf{x}(0) = \mathbf{x}_0, \quad \mathbf{u}(0) = \mathbf{u}_0,$$

or

$$\frac{d^2\mathbf{x}}{dt^2}(t) = \mathbf{f}\left[\mathbf{x}(t),\ \frac{d\mathbf{x}}{dt}(t)\right], \quad \mathbf{x}(0) = \mathbf{x}_0, \quad \frac{d\mathbf{x}}{dt}(0) = \mathbf{u}_0,$$

or finally,

$$\frac{d^2\mathbf{x}}{dt^2}(t) + \left(N^i_j y^j\right)\left[\tilde{\chi}^{-1}\left(\mathbf{x}(t),\ \frac{d\mathbf{x}}{dt}(t)\right)\right] = 0, \quad \mathbf{x}(0) = \mathbf{x}_0, \quad \frac{d\mathbf{x}}{dt}(0) = \mathbf{u}_0\,.$$

Note that

$$\mathbf{U} \in \mathfrak{X} \Longrightarrow R \geq \|\mathbf{U}(t) - \mathbf{U}_0\| \geq \|\mathbf{x}(t) - \mathbf{x}_0\| \Longrightarrow \mathbf{x}(t) \in \overline{B_R(\mathbf{x}_0)}$$

for any $t \in I_0$. Similarly $\mathbf{u}(t) \in \overline{B_R(\mathbf{u}_0)}$ so that $\mathbf{U}(t) \in K$ yielding $\mathbf{x}(t) \in \chi(U)$ for any $t \in I_0$. Let $\gamma_w : I_0 \to U$ be given by

$$\gamma_w(t) = \chi^{-1}\big[\mathbf{x}(t)\big], \quad t \in I_0\,.$$

Next let $v \in T_x(M) \setminus \{0_x\}$ and let us consider the curve

$$\alpha_v : \big(-2,\ 2\big) \to U, \quad \alpha_v(s) = \gamma_{(2/R_0)\,v}\left(\frac{R_0}{2}\,s\right), \quad |s| < 2.$$

This is well defined because $|s| < 2 \Longrightarrow (sR_0)/2 \in I_0$. Also

$$\alpha_v(0) = x, \quad \dot{\alpha}_v(0) = v.$$

On the other hand

$$\frac{d\alpha_v^i}{ds}(s) = \frac{R_0}{2}\frac{d\gamma_w^i}{dt}(t), \quad \frac{d^2\alpha_v^i}{ds^2}(s) = \left(\frac{R_0}{2}\right)^2 \frac{d^2\gamma_w^i}{dt^2}(t),$$

$$t = \frac{R_0}{2}s, \quad w = \frac{2}{R_0}v.$$

The formulas

$$N_j^i y^j = g^{ik}\gamma_{00k}, \quad \gamma_{00k} = \gamma_{ijk}\, y^i\, y^j, \quad \gamma_{ijk} = \frac{1}{2}\left(\frac{\partial g_{jk}}{\partial x^i} + \frac{\partial g_{ki}}{\partial x^j} - \frac{\partial g_{ij}}{\partial x^k}\right),$$

together with the fact that both g^{ij} and g_{ij} are positively homogeneous of degree 0 [with respect to the directional arguments] show that the functions $N_j^i\, y^j$: $\pi^{-1}(U) \to \mathbb{R}$ are positively homogeneous of degree 2, so that

$$\frac{d^2\alpha_v^i}{ds^2}(s)\big(N_j^i y^j\big)\big[\dot{\alpha}_v(s)\big] = \left(\frac{R_0}{2}\right)^2 \left\{\frac{d^2\gamma_w^i}{dt}(t) + \big(N_j^i y^j\big)\big[\dot{\gamma}_w(t)\big]\right\} = 0.$$

2.2.5 *Applying Banach's Fixed Point Theorem*

Let $(\mathfrak{X},\, d_\infty)$ be the metric space considered in the previous section. The map Φ : $\mathfrak{X} \to \mathfrak{X}$ used there was shown to be a contraction, i.e.,

$$d_\infty\big(\Phi(\mathbf{V}),\, \Phi(\mathbf{W})\big) \le \alpha\, d_\infty(\mathbf{V},\, \mathbf{W}), \quad \mathbf{V},\, \mathbf{W} \in \mathfrak{X},$$

where $\alpha = L R_0$ (and $0 < \alpha < 1$). The remainder of the section is devoted to the proof of the existence and uniqueness of the fixed point $\mathbf{U} \in \mathfrak{X}$. The proof is a *verbatim* repetition of the proof of the Banach fixed point theorem, holding of course for an arbitrary contraction of any complete metric space. The proof is however known to contain an explicit approximation scheme for the calculation of $\mathbf{U}$, with any desired accuracy, and explicitly recalling the iterative construction of $\mathbf{U}$ will be useful for our later purposes. Let $\mathbf{U}_0 \in \mathfrak{X}$ be an arbitrary point. We set

$$\mathbf{U}_1 = \Phi(\mathbf{U}_0), \quad \mathbf{U}_2 = \Phi(\mathbf{U}_1), \quad \cdots \quad, \quad \mathbf{U}_\nu = \Phi(\mathbf{U}_{\nu-1}), \quad \cdots$$

for any $\nu \in \mathbb{N}$. As Φ is a contraction

$$d_\infty\big(\mathbf{U}_{\nu+1},\, \mathbf{U}_\nu\big) = d_\infty\big(\Phi(\mathbf{U}_\nu),\, \Phi(\mathbf{U}_{\nu-1})\big) \le \alpha\, d_\infty(\mathbf{U}_\nu,\, \mathbf{U}_{\nu-1})$$

for any $\nu \in \mathbb{N}$. Iteration gives

$$d_\infty\big(\mathbf{U}_{\nu+1}, \mathbf{U}_\nu\big) \le \alpha\, d_\infty(\mathbf{U}_\nu, \mathbf{U}_{\nu-1}) \le \alpha^2 d_\infty\big(\mathbf{U}_{\nu-1}, \mathbf{U}_{\nu-2}\big) \le \cdots \le \alpha^\nu d_\infty(\mathbf{U}_1, \mathbf{U}_0)$$

hence

$$d_\infty\big(\mathbf{U}_{\nu+1}, \mathbf{U}_\nu\big) \le \alpha^\nu d_\infty(\mathbf{U}_1, \mathbf{U}_0) \tag{2.49}$$

for any $\nu \in \mathbb{N} \cup \{0\}$. Then, for any $\nu \in \mathbb{N} \cup \{0\}$ and any $p \in \mathbb{N}$,

$$d_\infty\big(\mathbf{U}_{\nu+p}, \mathbf{U}_\nu\big) \le \quad \text{(by the triangle inequality)}$$

$$\le \sum_{k=\nu}^{\nu+p-1} d_\infty\big(\mathbf{U}_{k+1}, \mathbf{U}_k\big) \le \quad \text{[by (2.49)]}$$

$$\le \sum_{k=n}^{\nu+p-1} \alpha^k d_\infty(\mathbf{U}_1, \mathbf{U}_0) = \alpha^n d_\infty(\mathbf{U}_1, \mathbf{U}_0) \sum_{\ell=0}^{p-1} \alpha^\ell =$$

$$= \alpha^\nu d_\infty(\mathbf{U}_1, \mathbf{U}_0)\, \frac{1-\alpha^p}{1-\alpha} \le \alpha^\nu d_\infty(\mathbf{U}_1, \mathbf{U}_0)\, \frac{1}{1-\alpha}$$

or

$$d_\infty\big(\mathbf{U}_{n+p}, \mathbf{U}_n\big) \le d_\infty(\mathbf{U}_1, \mathbf{U}_0)\, \frac{\alpha^\nu}{1-\alpha}. \tag{2.50}$$

As $0 < \alpha < 1$ it follows that $\lim_{\nu\to\infty} \alpha^\nu = 0$ hence for any $\epsilon > 0$ there is $\nu_\epsilon \in \mathbb{N}$ such that

$$\alpha^n < \frac{\epsilon(1-\alpha)}{d_\infty(\mathbf{U}_1, \mathbf{U}_0)}$$

for any $\nu \ge \nu_\epsilon$. Then

$$d_\infty\big(\mathbf{U}_{\nu+p}, \mathbf{U}_\nu\big) < \epsilon$$

for any $\nu \ge \nu_\epsilon$ and any $p \in \mathbb{N}$, i.e., $\{\mathbf{U}_\nu\}_{\nu\ge 0}$ is a Cauchy sequence and hence as $(\mathfrak{X}, d_\infty)$ is a complete metric space $\{\mathbf{U}_\nu\}_{\nu\ge 0}$ is convergent, i.e., there is a unique point $\overline{\mathbf{U}} \in \mathfrak{X}$ such that $\lim_{\nu\to\infty} \mathbf{U}_\nu = \overline{\mathbf{U}}$. A contraction is clearly a continuous map hence

$$\Phi(\overline{\mathbf{U}}) = \Phi(\lim_{\nu\to\infty} \mathbf{U}_\nu) = \lim_{\nu\to\infty} \Phi(\mathbf{U}_\nu) = \lim_{\nu\to\infty} \mathbf{U}_{\nu+1} = \overline{\mathbf{U}},$$

i.e., $\overline{\mathbf{U}}$ is a fixed point of Φ, as seek for. It remains that one proves uniqueness of such $\overline{\mathbf{U}}$. Indeed if $\overline{\overline{\mathbf{U}}} \in \mathfrak{X}$ is another fixed point, i.e., $\Phi\big(\overline{\overline{\mathbf{U}}}\big) = \overline{\overline{\mathbf{U}}}$ allora

$$d_\infty\big(\overline{\mathbf{U}}, \overline{\overline{\mathbf{U}}}\big) = d_\infty\big(\Phi(\overline{\mathbf{U}}), \Phi(\overline{\overline{\mathbf{U}}})\big) \le \alpha\, d_\infty\big(\overline{\mathbf{U}}, \overline{\overline{\mathbf{U}}}\big) \tag{2.51}$$

and the proof may be completed by a contradiction argument. Indeed if one assumes that $\overline{\mathbf{U}} \neq \overline{\overline{\mathbf{U}}}$, then $d_\infty\big(\overline{\mathbf{U}}, \overline{\overline{\mathbf{U}}}\big) > 0$ hence [by (2.51)] $\alpha > 1$, a contradiction.

Definition 2.36 Let $\epsilon > 0$. Two points $\mathbf{U}, \mathbf{V} \in \mathfrak{X}$ are *ϵ-close* if $d_\infty(\mathbf{U}, \mathbf{V}) \leq \epsilon$. □

Definition 2.37 Let $\epsilon > 0$. A map $\Psi : \mathfrak{X} \to \mathfrak{X}$ is *ϵ-close* to Φ if $d_\infty(\Phi(\mathbf{U}), \Psi(\mathbf{U})) \leq \epsilon$ for any $\mathbf{U} \in \mathfrak{X}$. □

Theorem 2.38 *Let $\Psi : \mathfrak{X} \to \mathfrak{X}$ be a contraction, i.e.,*

$$d_\infty\big(\Psi(\mathbf{U}), \Psi(\mathbf{V})\big) \leq \beta\, d_\infty(\mathbf{U}, \mathbf{V})$$

with $0 < \beta < 1$. If Ψ is ϵ-close to Φ, then the fixed points of Φ and Ψ are δ-close with

$$\delta = \frac{\epsilon}{1-\gamma}, \quad \gamma = \max\big\{\alpha, \beta\big\}.$$

Proof Let $\mathbf{U}, \mathbf{V} \in \mathfrak{X}$ be the fixed points of Φ and Ψ, respectively. Let us set $\mathbf{V}_0 = \mathbf{U}$ and let $\{\mathbf{V}_\nu\}_{\nu \geq 0}$ be the sequence of iterates of Ψ starting with $\mathbf{V}_0$, i.e., $\mathbf{V}_{\nu+1} = \Psi\big(\mathbf{V}_\nu\big)$ for any $\nu \geq 0$. By the proof of the Cauchy-Lipschitz-Picard theorem $\lim_{\nu\to\infty} \mathbf{V}_\nu = \mathbf{V}$. Moreover, let us observe that

$$d_\infty(\mathbf{V}_1, \mathbf{V}_2) = d_\infty\big(\Psi(\mathbf{V}_0), \Psi(\mathbf{V}_1)\big) \leq \gamma\, d_\infty(\mathbf{U}, \mathbf{V}_1),$$

$$d_\infty\big(\mathbf{V}_2, \mathbf{V}_3\big) = d_\infty\big(\Psi(\mathbf{V}_1), \Psi(\mathbf{V}_2)\big) \leq \gamma\, d_\infty(\mathbf{V}_1, \mathbf{V}_2) \leq \gamma^2\, d_\infty(\mathbf{U}, \mathbf{V}_1)$$

hence (by iterating these inequalities)

$$d_\infty(\mathbf{V}_\nu, \mathbf{V}_{\nu+1}) \leq \gamma^\nu\, d_\infty(\mathbf{U}, \mathbf{V}_1), \quad \nu \geq 0, \tag{2.52}$$

Hence [by the triangle inequality together with (2.52)],

$$d_\infty(\mathbf{U}, \mathbf{V}_\nu) \leq \sum_{i=0}^{\nu-1} d_\infty(\mathbf{V}_i, \mathbf{V}_{i+1}) \leq \sum_{i=0}^{\nu-1} \gamma^i\, d_\infty(\mathbf{U}, \mathbf{V}_1) = \frac{1-\gamma^\nu}{1-\gamma}\, d_\infty(\mathbf{U}, \mathbf{V}_1)$$

so that [for $\nu \to \infty$]

$$d_\infty(\mathbf{U}, \mathbf{V}) \leq \frac{1}{1-\gamma}\, d_\infty(\mathbf{U}, \mathbf{V}_1) = \frac{1}{1-\gamma}\, d_\infty\big(\Phi(\mathbf{U}), \Psi(\mathbf{U})\big) \leq \frac{\epsilon}{1-\gamma} = \delta\,.$$

Q.e.d.

2.2.6 Convex Neighborhoods

Let $X \in T_p(M)$ and assume there is a geodesic $\alpha_X : [0, 1] \to M$ with the initial data

$$\alpha_X(0) = p, \quad \frac{d\alpha_X}{dt}(0) = X.$$

Definition 2.39 The point $\alpha_X(1)$ is the *exponential* of the tangent vector X and one adopts the notation $\exp_p X = \alpha_X(1)$. □

As a mere consequence of the differential equations governing geodesics

$$\alpha_X(t) = \exp_p tX$$

for every value of the parameter t, for which both sides are defined. By Theorem 2.35 for every point $p_0 \in M$ there is a neighborhood $U \subset M$ of p_0 and a positive number $\epsilon > 0$ such that for every $p \in U$ and every $X \in T_p(M)$ with $L(X) < \epsilon$ there is a unique geodesic $\alpha_{p,X} : (-2, 2) \to U$ with the initial data (p, X). Let us set

$$N_\epsilon(p) = \{X \in T_p(M) : L(X) < \epsilon\}.$$

The exponential map may then be thought of as a map

$$\exp : B_\epsilon(U) = \bigcup_{p \in U} N_\epsilon(p) \subset T(U) \to U, \quad \exp : X \in N_\epsilon(p) \longmapsto \exp_p X = \alpha_{p,X}(1).$$

As such exp is C^∞ on $B_\epsilon(U) \setminus \sigma(U)$ yet only of class C^1 on the whole of $B_\epsilon(U)$, i.e., exp is only of class C^1 along the zero section. Here $\sigma : M \to T(M)$ is $\sigma(p) = 0_p$ for every $p \in U$. Also $T(U)$ is the portion of $T(M)$ over U.

Lemma 2.40 *For each point $p \in M$, there is a neighborhood N_0 of the origin in $T_p(M)$ and a neighborhood N_p of p in M such that* $\exp_p$ *is a diffeomorphism of N_0 onto N_p.*

Let $p_0 \in M$ be an arbitrary point, and let us choose an open neighborhood $U \subset M$ of p_0 and a positive number $\epsilon > 0$ as furnished by Theorem 2.35. Without loss of generality, we may assume that U is also the domain of a local coordinate system (U, x^i) on M. Let $\left(\Pi^{-1}(U), x^i, y^i\right)$ be the induced local coordinates on $T(M)$. For every point $p \in U$ let $\eta_p : T_p(M) \to T_{0_p}\left(T_p(M)\right)$ be the natural map, whose construction, we now recall, as follows. Let $X \in T_p(M)$ and let us set $\alpha(t) = tX$ for any $t \in \mathbb{R}$. By definition $\eta_p(X)$ is the tangent vector $X_0 \in T_{0_p}\left(T_p(M)\right)$ to the curve $\alpha : \mathbb{R} \to T_p(M)$ at the point $\alpha(0) = 0_p$, i.e.,

$$\eta_p\left(X\right) = X_0, \quad X_0 = \dot{\alpha}(0).$$

Let $y^i_p : T_p(M) \to \mathbb{R}$ be the (global) coordinate system on $T_p(M)$ [a linear space, thought of as a manifold] associated to the linear basis $\left\{ \left(\frac{\partial}{\partial x^i}\right)_p : 1 \le i \le n \right\} \subset T_p(M)$, i.e.,

$$y^i_p(X) = X^i, \quad X = X^i \left(\frac{\partial}{\partial x^i}\right)_p .$$

That is y^i_p is the restriction of $y^i : \Pi^{-1}(U) \to \mathbb{R}$ to $T_p(M)$. The resulting components of the curve α are $\alpha^i(t) = y^i_p\big(\alpha(t)\big) = t\, X^i$. Hence,

$$\dot{\alpha}(t) = \frac{d\alpha^i}{dt}(t) \left(\frac{\partial}{\partial y^i_p}\right)_{\alpha(t)} = X^i \left(\frac{\partial}{\partial y^i_p}\right)_{\alpha(t)}$$

so that

$$\eta_p(X) = \dot{\alpha}(0) = X^i \left(\frac{\partial}{\partial y^i_p}\right)_{\alpha(0)} = y^i(X) \left(\frac{\partial}{\partial y^i_p}\right)_{0_p} .$$

We shall also need the inclusion $j_p : T_p(M) \to T(M)$. Its differential at 0_p

$$d_{0_p} j_p : T_{0_p}\big(T_p(M)\big) \to T_{0_p}\big(T(M)\big)$$

maps $\left(\frac{\partial}{\partial y^i_p}\right)_{0_p}$ onto $\left(\frac{\partial}{\partial y^i}\right)_{0_p}$ for every $1 \le i \le n$. The maps

$$\eta_p : T_p(M) \to T_{0_p}\big(T_p(M)\big), \quad \big(d_{0_p} j_p\big) \circ \eta_p : T_p(M) \to T_{0_p}\big(T(M)\big),$$

are often identified[2].

Proof of Lemma 2.40. The differential of the exponential map $\exp_p : N_\epsilon(p) \to U$ at 0_p is [as $\exp_p 0_p = p$]

$$d_{0_p} \exp_p : T_{0_p}\big(T_p(M)\big) \to T_p(M).$$

Then, for every $X \in T_p(M)$ [by the chain rule]

$$\big(d_{0_p} \exp_p\big) \circ \eta_p(X) = \big(d_{0_p} \exp_p\big)\, \dot{\alpha}(0) = \big(d_{0_p} \exp_p\big) \circ \big(d_0 \alpha\big) \left(\frac{d}{dt}\right)_0 =$$

$$= d_0(\exp_p \circ \alpha) \left(\frac{d}{dt}\right)_0 = \big(d_0 \alpha_{p,X}\big) \left(\frac{d}{dt}\right)_0 = \dot{\alpha}_{p,X}(0) = X,$$

---

[2] We shall not do so through the present section, mostly for didactical purposes.

i.e.,

$$\left(d_{0_p} \exp_p\right) \circ \eta_p = 1_{N_\epsilon(p)}$$

[where $1_A : A \to A$ is the identical map of the set A]. By the inverse function theorem there is a neighborhood $N_0 \subset N_\epsilon(p)$ of the origin 0_p and a neighborhood $U_p \subset U$ of $p = \exp_p 0_p$ such that $\exp_p : N_0 \to U_p$ is a diffeomorphism. Q.e.d.

Definition 2.41 An open neighborhood $N_p \subset T_p(M)$ of the origin 0_p is *normal* if
(i) The map $\exp_p$ is a defined on N_p and a diffeomrphism of N_p onto some open neighborhood $U_p \subset M$ of p.
(ii) N_0 is star shaped.[3]
Also U_p is called a *normal neighborhood* of p in M. □

Let $\Lambda : \mathbb{R}^n \to \mathbb{R}$ be a function such that
(i) $\Lambda\left(\lambda\, \xi\right) = \lambda^2\, \Lambda(\xi)$ for any $\lambda > 0$ and any $\xi \in \mathbb{R}^n$,
(ii) The matrix $\left[\frac{\partial^2 \Lambda}{\partial \xi^i\, \partial \xi^j}(\xi)\right]$ is positive definite for any $\xi \in \mathbb{R}^n$.
For instance, one may work with

$$\Lambda(\xi) = \sum_{i,j=1}^{n} a_{ij} \xi^i \xi^j, \quad \xi = \left(\xi^1, \cdots, \xi^n\right) \in \mathbb{R}^n, \tag{2.53}$$

where $[a_{ij}]_{1 \le i,j \le n} \in \mathbb{R}^{n^2}$ is a symmetric positive definite $n \times n$ matrix with real coefficients. Occasionally we shall need the more general choice we have started with. Let $(U,\ x^i)$ be a local coordinate system on M. We adopt the notation

$$B_c = \left\{x \in U\ :\ \Lambda\left[\chi(\mathbf{x})\right] < c\right\}, \quad c > 0.$$

Here $\chi = \left(x^1, \cdots, x^n\right) : U \to \mathbb{R}^n$. The family $\{B_c\ :\ c > 0\}$ is a basis for the topology of M.

Lemma 2.42 *There is $c > 0$ such that for any $0 < \delta < c$ the points of any geodesic tangent to $\Sigma(\delta) = \left\{x \in U\ :\ \Lambda\left[\chi(x)\right] = \delta\right\}$ lying near the contact point, lie outside $\Sigma(\delta)$.*

Definition 2.43 A curve $\gamma : I \to M$ is *tangent* to the hypersurface $\Sigma \subset M$ at the point $p_0 \in \Sigma$ if there $t_0 \in I$ such that $\gamma(t_0) = p_0$ [$p_0 = \gamma(t_0)$ is the *contact point*] and $\dot\gamma(t_0) \in T_{\gamma(t_0)}(\Sigma)$. □

Here $I \subset \mathbb{R}$ is an interval. With this definition, a more formal statement of Lemma 2.42 is

[3] That is $X \in N_0$ and $0 \le t \le 1$ imply $tX \in N_0$.

$$\exists\, c > 0, \quad \forall\, 0 < \delta < c,$$

$$\forall\, \gamma : I \to M, \quad \text{geodesic tangent to } \Sigma(\delta) \text{ at } \gamma(t_0)$$

$$\exists\, r > 0, \quad \text{such that } I_r(t_0) = (t_0 - r,\, t_0 + r) \subset I \text{ and}$$

$$\forall\, t \in I_r(t_0) \setminus \{t_0\} \;\; : \;\; \gamma(t) \in U \setminus \Sigma(\delta).$$

Proof of Lemma 2.42. Let $\delta > 0$ and let $\varphi = \Lambda \circ \chi - \delta$ be the defining function of $\Sigma(\delta)$. Let $p_0 \in \Sigma(\delta)$ and let $\gamma : I \to U$ be a geodesic tangent to $\Sigma(\delta)$ at the (contact) point $\gamma(t_0) = p_0$. Then

$$\frac{d\gamma}{dt}(t_0) \in T_{p_0}\big(\Sigma(\delta)\big) \Longrightarrow 0 = \frac{d\gamma}{dt}(t_0)(\varphi) = \left[(d_{t_0}\gamma) \left(\frac{d}{dt}\right)_{t_0} \right](\varphi) =$$

$$= \left(\frac{d}{dt}\right)_{t_0} \big(\varphi \circ \gamma\big) = \frac{\partial \Lambda}{\partial \xi^i}(\xi_0) \frac{d\gamma^i}{dt}(t_0)$$

or

$$\frac{\partial \Lambda}{\partial \xi^i}(\xi_0) \frac{d\gamma^i}{dt}(t_0) = 0 \tag{2.54}$$

where $\xi_0 = \chi(p_0) \in \mathbb{R}^n$. On the other hand

$$\big(\Lambda \circ \gamma \circ \gamma\big)'(t) = \frac{\partial \Lambda}{\partial \xi^i}\big(\gamma^1(t), \cdots, \gamma^n(t)\big)\frac{d\gamma^i}{dt}(t),$$

$$\big(\Lambda \circ \gamma \circ \gamma\big)''(t) = \frac{d}{dt}\left[\frac{\partial \Lambda}{\partial \xi^i}\big(\gamma^1(t), \cdots, \gamma^n(t)\big)\frac{d\gamma^i}{dt}(t)\right] =$$

$$= \frac{\partial^2 \Lambda}{\partial \xi^i\, \partial \xi^j}\big(\gamma^1(t), \cdots, \gamma^n(t)\big)\frac{d\gamma^j}{dt}(t)\,\frac{d\gamma^i}{dt}(t) + \frac{\partial \Lambda}{\partial \xi^i}\big(\gamma^1(t), \cdots, \gamma^n(t)\big)\frac{d^2\gamma^i}{dt^2}(t) =$$

[as γ is a geodesic, i.e., $d^2\gamma^i/dt^2 = -N^i_j\big(\dot\gamma(t)\big)\, d\gamma^j/dt$]

$$= \frac{\partial^2 \Lambda}{\partial \xi^i\, \partial \xi^j}\big(\gamma^1(t), \cdots, \gamma^n(t)\big)\frac{d\gamma^j}{dt}(t)\,\frac{d\gamma^i}{dt}(t) - \frac{\partial \Lambda}{\partial \xi^i}\big(\gamma^1(t), \cdots, \gamma^n(t)\big)\, N^i_k\big(\dot\gamma(t)\big)\,\frac{d\gamma^k}{dt}(t). \tag{2.55}$$

Let us differentiate with respect to λ in $\Lambda\big(\lambda\xi\big) = \lambda^2\, \Lambda(\xi)$ and set $\lambda = 1$ in the resulting equation. We obtain

$$\frac{\partial \Lambda}{\partial \xi^j}(\xi)\, \xi^j = 2\, \Lambda(\xi)\,.$$

Next, let us differentiate with respect to ξ^i

$$\frac{\partial^2 \Lambda}{\partial \xi^j \, \partial \xi^i}(\xi)\, \xi^j = \frac{\partial \Lambda}{\partial \xi^i}\,. \tag{2.56}$$

At this point, we substitute $\dfrac{\partial \Lambda}{\partial \xi^i}$ from (2.56) into (2.55) so that to obtain

$$\big(\Lambda \circ \gamma \circ \gamma\big)''(t) = \frac{\partial^2 \Lambda}{\partial \xi^i \, \partial \xi^j}\big(\gamma^1(t), \cdots, \gamma^n(t)\big)\Big[\frac{d\gamma^j}{dt}(t)\, \frac{d\gamma^i}{dt}(t) - \gamma^j(t)\, N_0^i\big(\dot{\gamma}(t)\big)\Big] \tag{2.57}$$

where we have set $N_0^i = N_j^i y^j$. For every $v \in \pi^{-1}\big(\Sigma(\delta)\big) \subset \pi^{-1}(U) \subset V(M)$, we set

$$\overline{\Lambda}(v) = \Lambda(v^1, \cdots, v^n), \quad v = v^i \left(\frac{\partial}{\partial x^i}\right)_{\pi(v)}.$$

Next, let us set

$$K = \Big\{ v \in \pi^{-1}\big(\Sigma(\delta)\big) \, : \, \overline{\Lambda}(v) = 1 \Big\}.$$

Then K is a compact subset of $\pi^{-1}(U)$ and the functions $N_0^i : \pi^{-1}(U) \to \mathbb{R}$ are continuous. Then the functions

$$\frac{\partial^2 \Lambda}{\partial \xi^i \, \partial \xi^j}\, N_0^i, \quad 1 \le j \le n,$$

are bounded on K so that

$$\left| \left(\frac{\partial^2 \Lambda}{\partial \xi^i \, \partial \xi^j} \right)^{\chi \circ \pi} N_0^i \right| \le A, \quad 1 \le j \le n, \tag{2.58}$$

on K, for some constant $A > 0$. Here $\left(\dfrac{\partial^2 \Lambda}{\partial \xi^i \, \partial \xi^j} \right)^{\chi \circ \pi}$ is short for $\dfrac{\partial^2 \Lambda}{\partial \xi^i \, \partial \xi^j} \circ \chi \circ \pi$. If $v \in \pi^{-1}\big(\Sigma(\delta)\big)$ is an arbitrary vector, then $v \neq 0_{\pi(v)}$ and hence $w = \big[1/\sqrt{\overline{\Lambda}(v)}\big]\, v \in K$. Consequently [as N_0^i are positively homogenous of degree 2 and $\pi(w) = \pi(v)$]

$$\left| \left(\frac{\partial^2 \Lambda}{\partial \xi^i \, \partial \xi^j} \right)^{\chi \circ \pi} (v)\, \big(x^j \circ \pi\big)(v)\, N_0^i(v) \right| = \overline{\Lambda}(v) \left| \left(\frac{\partial^2 \Lambda}{\partial \xi^i \, \partial \xi^j} \right)^{\chi \circ \pi} (w)\, \big(x^j \circ \pi\big)(w)\, N_0^i(w) \right| \le$$

[by (2.58)]

$$\le A\, \overline{\Lambda}(v) \sum_{j=1}^{n} \big| x^j\big(\pi(v)\big) \big|.$$

The function $\dfrac{1}{\Lambda(\xi)}\sum_{j=1}^{n}(\xi^j)^2$ is positively homogeneous of degree 0 and continuous on $\mathbb{R}^n \setminus \{0\}$. Therefore,

$$\sum_{j=1}^{n}\left(\xi^j\right)^2 \le C\,\Lambda(\xi) \tag{2.59}$$

for some $C > 0$ and every $\xi \in \mathbb{R}^n$. Also, if $\mathbf{1} = \big(1, \cdots, 1\big) \in \mathbb{R}^n$

$$\sum_{j=1}^{n}\left|x^j\right| = \left|\mathbf{1}\cdot\big(|x^1|, \cdots, |x^n|\big)\right| \le$$

[by the Cauchy-Schwartz inequality, and $\|\mathbf{1}\| = \sqrt{n}$, and (2.59)]

$$\le \|\mathbf{1}\| \sqrt{\sum_{j=1}^{n}\left|x_j\right|^2} = \sqrt{n\sum_{j=1}^{n}(x^j)^2} \le \sqrt{n\,C\,\Lambda(\mathbf{x})}$$

where $\mathbf{x} = (x^1, \cdots, x^n)$. Therefore,

$$\left|\frac{\partial^2\Lambda}{\partial\xi^i\,\partial\xi^j}\big(\gamma^1(t), \cdots, \gamma^n(t)\big)\,\gamma^j(t)\,N_0^i\big(\dot\gamma(t)\big)\right| \le \tag{2.60}$$

$$\le A\,\overline{\Lambda}\big(\dot\gamma(t)\big)\sum_{j=1}^{n}\left|\gamma^j(t)\right| \le A\,\overline{\Lambda}\big(\dot\gamma(t)\big)\sqrt{n\,C\,\Lambda\big(\gamma^1(t), \cdots, \gamma^n(t)\big)}\,.$$

Next, let us observe that

$$\frac{1}{\overline{\Lambda}(v)}\left(\frac{\partial^2\Lambda}{\partial\xi^i\,\partial\xi^j}\right)^{\chi\circ\pi}(v)\;v^i\;v^j$$

is positive definite and positively homogeneous of degree 0 hence

$$\frac{\partial^2\Lambda}{\partial\xi^i\,\partial\xi^j}\big(\gamma^1(t), \cdots, \gamma^n(t)\big)\,\frac{d\gamma^i}{dt}(t)\,\frac{d\gamma^j}{dt}(t) \ge B\,\overline{\Lambda}\big(\dot\gamma(t)\big) \tag{2.61}$$

for some constant $B > 0$. In particular from (2.60)

$$-\frac{\partial^2\Lambda}{\partial\xi^i\,\partial\xi^j}\big(\gamma^1(t), \cdots, \gamma^n(t)\big)\,\gamma^j(t)\,N_0^i\big(\dot\gamma(t)\big) \ge -A\,\overline{\Lambda}\big(\dot\gamma(t)\big)\sqrt{n\,C\,\Lambda\big(\gamma^1(t), \cdots, \gamma^n(t)\big)}\,. \tag{2.62}$$

Finally [by (2.57) and (2.61)–(2.62)],

$$\big(\Lambda\circ\chi\circ\gamma\big)''(t_0)\geq\overline{\Lambda}\big(\dot\gamma(t_0)\big)\Big[B-A\sqrt{n\,C\,\Lambda(\xi_0)}\Big]=$$

$$=\overline{\Lambda}\big(\dot\gamma(t_0)\big)\Big[B-A\sqrt{n\,C\,\delta}\Big]>0\Longleftrightarrow\delta<\frac{B^2}{A^2\,n\,C}=:c.$$

Summing up, there is $c>0$ [precisely $c=B^2\big/(A^2\,n\,C)$] such that for any $0<\delta<c$ one has

$$\big(\Lambda\circ\chi\circ\gamma\big)'(t_0)=0,\quad \big(\Lambda\circ\chi\circ\gamma\big)''(t_0)>0,$$

so that t_0 is a minimum point, at which $\Lambda\circ\chi\circ\gamma$ attains the minimum value $\Lambda(\chi(\gamma(t_0)))=\delta$. Hence, there is $r>0$ such that $\Lambda(\chi(\gamma(t)))>\delta$ for all $t\in I_r(t_0)\setminus\{t_0\}$. Q.e.d.

Lemma 2.44 *Let $c>0$ be chosen as in Lemma 2.42. There is $0<a<c$ such that*
(i) *Any two points in B_a can be joined by a geodesic lying entirely in B_c.*
(ii) *Each point of B_a has a normal neighborhood containing B_a.*

Proof Let $\sigma:M\to T(M)$ be the natural embedding, i.e., $\sigma(p)=0_p$ for any $p\in M$. Let us consider the map

$$\Phi:T(M)\to M\times M,\quad \Phi(X)=\big(\Pi(X),\,\exp_{\Pi(X)}X\big),\quad X\in T(M).$$

For each $p\in M$, the exponential map $\exp_p$ is only defined on some neighborhood of 0_p in $T_p(M)$ hence Φ is only defined on some open neighborhood of $\sigma(M)$ in $T(M)$. For every $p\in M$,

$$\Phi(0_p)=\big(\Pi(0_p),\,\exp_{\Pi(0_p)}0_p\big)=\big(p,\;p\big).$$

The differential of Φ at $p\in M$ [identified with $0_p\in T_p(M)$] is

$$d_{0_p}\Phi:T_{0_p}(T(M))\to T_{(p,\,p)}(M\times M).$$

We shall make use of the maps

$$\beta_p:M\to M\times M,\quad \alpha_q:M\to M\times M,\quad \beta_p(q)=(p,q),\quad \alpha_q(p)=(p,q),\quad p,q\in M,$$

so that the tangent space at (p,q) to the product manifold $M\times M$ is

$$T_{(p,q)}(M\times M)=\big[(d_p\alpha_q)T_p(M)\big]\oplus\big[(d_q\beta_p)T_q(M)\big].$$

Let $X\in T_p(M)$ and let us compute

$$(d_{0_p}\Phi)\circ(d_{0_p}j_p)\circ\eta_p(X)=d_{0_p}(\Phi\circ j_p)\dot\alpha(0)=d_{0_p}(\Phi\circ j_p)\,(d_0\alpha)\left(\frac{d}{dt}\right)_0=d_0(\Phi\circ j_p\circ\alpha)\left(\frac{d}{dt}\right)_0$$

and

$$\Phi \circ j_p \circ \alpha(t) = \Phi(j_p(tX)) = \big(p, \exp_p tX\big) = \big(p, \alpha_{p,X}(t)\big) =: \gamma(t)$$

hence

$$(d_{0_p}\Phi) \circ (d_{0_p} j_p) \circ \eta_p(X) = \dot{\gamma}(0) \in T_{(p,p)}(M \times M).$$

To compute the tangent vector along the curve

$$\gamma : \mathbb{R} \to M \times M, \quad \gamma(t) = (p, \alpha_{p,X}(t)), \quad t \in \mathbb{R},$$

one endows $M \times M$ with the C^∞ atlas $\{h \times k : U \times V \to \mathbb{R}^{2n}\}$ got as h and k run over an atlas of M. Let $\pi_a : M \times M \to M$, $a \in \{1, 2\}$, be the projections. We may use the local chart $\chi \times \chi : U \times U \to \mathbb{R}^{2n}$ about the point $(p, p) \in M \times M$. If the coordinate functions relative to χ are $\chi = (x^1, \cdots, x^n)$, then

$$\chi \times \chi = \big(x^1 \circ \pi_1, \cdots, x^n \circ \pi_1, x^1 \circ \pi_2, \cdots, x^n \circ \pi_2\big) =: \big(x^1, \cdots, x^{2n}\big).$$

Therefore, the components of the curve γ with respect to the chart $\chi \times \chi$ are

$$\gamma^i(t) = x^i(p), \quad \gamma^{i+n}(t) = x^i\big(\alpha_{p,X}(t)\big) = \alpha^i_{p,X}(t).$$

Moreover,

$$\dot{\gamma}(0) = \frac{d\gamma^i}{dt}(0)\left(\frac{\partial}{\partial x^i}\right)_{\beta(0)} + \frac{d\gamma^{i+n}}{dt}(0)\left(\frac{\partial}{\partial x^{i+n}}\right)_{\beta(0)} = \frac{d\alpha^i_{p,X}}{dt}(0)\left(\frac{\partial}{\partial x^{i+n}}\right)_{(p,p)}.$$

Note that the differential of β_p at p maps $(\partial/\partial x^i)_p$ onto $(\partial/\partial x^{i+n})_{(p,p)}$. Hence,

$$\dot{\gamma}(0) = \frac{d\alpha^i_{p,X}}{dt}(0)\,(d_p\beta_p)\left(\frac{\partial}{\partial x^i}\right)_p = (d_p\beta_p)\,\dot{\alpha}_{p,X}(0) = (d_p\beta_p)X.$$

Finally,

$$(d_{0_p}\Phi) \circ (d_{0_p} j_p) \circ \eta_p = d_p\beta_p$$

hence $d_{0_p}\Phi$ is nonsingular. Consequently there is a neighborhood N' of[4] 0_p in $T(M)$ and there is a positive number $0 < a' < c$ such that $\Phi : N' \to B_{a'} \times B_{a'}$ is a diffeomorphism. Also the differential of $\exp : T(M) \to M$ is nonsingular at 0_p hence there is a star shaped neighborhood N'' of 0_p in $T(M)$ such that $\exp\big(N''\big) \subset B_c$. We may choose N'' such that $N'' \subset N'$. Also let us choose a positive number $0 < a < a'$ such that $\Phi\big(N''\big) \supset B_a \times B_a$. Let us set

[4] Morally, of the point p (thinking of M as immersed into $T(M)$ as the zero section).

$$N = \Phi^{-1}\big(B_a \times B_a\big).$$

It follows that $\exp(tX) \in B_c$ for every $X \in N$ and every $0 \le t \le 1$. At this point, we may prove statement (i) in Lemma 2.44.

(i) Let $q_1,\ q_2 \in B_a$ be two arbitrary points. Then $X := \Phi^{-1}(q_1,\ q_2) \in N$. Let $\alpha = \alpha_{q_1,X}$ be the geodesic with the initial data $\big(q_1,\ X\big)$. Then α is a geodesic joining q_1 and q_2 in B_c. Q.e.d.

At this point, we may prove statement (ii) in Lemma 2.44.

(ii) For every $q \in B_a$, let us set

$$N_q := N'' \cap T_q(M).$$

Then N_q is star shaped and exp is a diffeomorphism of N_q onto some neighborhood of q in M containing B_a. Q.e.d.

Definition 2.45 A set B_δ is said to be *simple* if for any $q_1,\ q_2 \in B_\delta$ there is at most one geodesic joining q_1 and q_2, and lying entirely in B_δ. □

Definition 2.46 A set B_δ is said to be *convex* if for any $q_1,\ q_2 \in B_\delta$ there is a geodesic joining q_1 and q_2, and lying entirely in B_δ. □

Theorem 2.47 *For every point $p \in M$, there is a positive number $a > 0$ such that for any $0 < \delta < a$ the set B_δ is both simple and convex.*

Proof Let $p \in M$ and let $(U,\ x^i)$ be a local coordinate system on M such that $p \in U$ and $x^i(p) = 0$. Let us choose $c > 0$ and $a > 0$ as in Lemma 2.44. Let $0 < \delta < a$. For any $q_1,\ q_2 \in B_\delta$, let $\gamma(t) = \exp(tX)$ be a geodesic joining q_1 and q_2 and lying in B_c, as in Lemma 2.44. Let us suppose that $\gamma(t_1)$ lies outside B_δ, for some value t_1 of the parameter. As $\Lambda \circ \chi \circ \gamma$ is continuous and $[0,\ 1]$ is compact, there is $0 \le t_0 \le 1$ such that

$$\Lambda\big[\chi\big(\gamma(t_0)\big)\big] = \sup_{0\le t\le 1} \Lambda\big[\chi\big(\gamma(t)\big)\big] < +\infty\,.$$

As $\gamma(t_1) \in U \setminus B_\delta$ one has $\Lambda\big[\chi\big(\gamma(t_1)\big)\big] \ge \delta = \Lambda\big(\chi(q_1)\big) = \Lambda\big(\chi(q_2)\big)$ the maximum isn't reached at a boundary point, i.e., $0 < t_0 < 1$. Thus, t_0 is a critical point

$$0 = \big(\Lambda \circ \chi \circ \gamma\big)'(t_0) = \frac{\partial \Lambda}{\partial \xi^i}(\xi_0)\,\frac{d\gamma^i}{dt}(t_0) = \dot\gamma(t_0)(\Lambda \circ \chi - \delta_0) \Longrightarrow$$

$$\Longrightarrow \dot\gamma(t_0) \in T_{\gamma(t_0)}(\Sigma(\delta_0))$$

where $\delta_0 := \Lambda\big[\chi\big(\gamma(t_0)\big)\big]$, that is γ is tangent to $\Sigma(\delta_0)$ at $\gamma(t_0)$. Thus [by Lemma 2.42], $\big(\Lambda \circ \chi \circ \gamma\big)''(t_0) > 0$. Yet t_0 is a maximum point, so that $\big(\Lambda \circ \chi \circ \gamma\big)''(t_0) < 0$, a contradiction. The contradiction comes from our assumption that γ lies outside B_δ. So B_δ is indeed convex. Q.e.d. By Lemma 2.44 every point of B_δ has a normal neighborhood containing[5] B_δ. Thus, B_δ is simple. Q.e.d.

---

[5] Actually B_δ is itself a normal neighborhood of any of its points.

The set N in the proof of Lemma 2.44 may be realized as follows. Start with a neighborhood U of a point p in M and a number $\epsilon > 0$ such that Φ is a diffeomorphism on $\Omega := \Pi^{-1}(U) \cap E^{-1}\big([0, \epsilon)\big)$. Then choose N to be a subset of Ω. The set B_δ in Theorem 2.47 is such that $\Phi(N) \supset B_\delta \times B_\delta$. Let γ be the unique geodesic in B_δ joining two given points $q_1, q_2 \in B_\delta$ and let $X \in N$ such that $\exp_{q_1} X = q_2$. Then the length of γ is $L(\gamma) = \|X\| < \epsilon$. We have established the following:

Corollary 2.48 *For every point $p \in M$, there is a positive number $\epsilon > 0$ and a neighborhood B_δ such that any two points in B_δ can be joined by a unique geodesic in M of length $< \epsilon$.*

The existence of both simple and convex neighborhoods is now established. Our next task is to study the relationship between geodesics and arc length.

Theorem 2.49 *Let B_δ and $\epsilon > 0$ be as in Corollary 2.48. Let $\alpha : [0, 1] \to M$ be the geodesic of length $< \epsilon$, contained in B_δ, joining two points $q_1, q_2 \in B_\delta$, and let $\omega : [0, 1] \to M$ be any other piecewise differentiable path joining q_1 to q_2. Then $L(\alpha) \leq L(\omega)$ with equality if and only if the point sets $\omega([0, 1])$ and $\alpha([0, 1])$ coincide and the parameter t is a monotonic increasing function of arc length.*

To prove Theorem 2.49 we need to prepare two lemmas. To simplify the exposition, we adopt the following:

Definition 2.50 Let $v \in V(M)$ and $Y \in T_{\pi(v)}(M)$ be two tangent vectors. We say Y is *transversal* to v if $g_v\big(v, Y\big) = 0$. □

Lemma 2.51 *Let $q \in M$ and let $\epsilon > 0$ be a positive number such that the exponential map $\exp_q$ is well defined on $N_0 = \big\{X \in T_q(M) \,:\, E(X) < \epsilon^2\big\}$. Let $u \longmapsto X(u)$ be a curve in $T_q(M)$ such that $E\big(X(u)\big) = 1$. Then each curve*

$$u \longmapsto \exp_q \big(t_0\, X(u)\big), \quad 0 < t_0 < \epsilon,$$

is transversal to each geodesic

$$t \longmapsto \exp_q \big(t\, X(u_0)\big).$$

Proof Let $f : D \to M$ be given by

$$f(u, t) = \exp_q \big(t\, X(u)\big), \quad 0 \leq t < \epsilon.$$

f is a parametrized surface in M and $D \subset \mathbb{R}^2$ is the domain of the parametrization, i.e., $D = J \times (-\epsilon, \epsilon)$ for some interval $J \subset \mathbb{R}$ [the interval where the parameter u ranges]. The differential of f at $(u, t) \in D$ is

$$d_{(u,t)} f \,:\, T_{(u,t)}(\mathbb{R}^2) \to T_{f(u,t)}(M).$$

Let $\tilde{f}$ be the *lift* of f to $V(M)$, i.e.,

$$\tilde{f}: D \to V(M), \quad \tilde{f}(u,t) = \left(d_{(u,t)} f\right) \left(\frac{\partial}{\partial t}\right)_{(u,t)} .$$

Then $\pi \circ \tilde{f} = f$. The notion of lift of f to $V(M)$ is taken by us from B.T.M. Hassan, [Has]. Despite the fact that the variable t seems to be privileged, the notion is certainly natural. One thinks of the (parametrized) surface f as the family of curves $\{f_u\}_{u \in J}$, where $f_u(t) = f(u,t)$, and then one lifts independently each of the curves f_u. The lift of f_u is $\tilde{f}_u(t) = (df_u/dt)(t) \in V(M)$ and then the lift of f to $V(M)$ is declared to be the surface in $V(M)$ generated by the curve $\tilde{f}_u$ as u ranges over J. We set

$$\frac{\overline{\partial} f}{\partial u}(u,t) = (d_{(u,t)} f) \left(\frac{\partial}{\partial u}\right)_{(u,t)} \in T_{f(u,t)}(M) = \left(\pi^{-1} TM\right)_{\tilde{f}(u,v)}$$

so that $\dfrac{\overline{\partial} f}{\partial u} : D \to \pi^{-1} T(M)$ is a section in $\pi^{-1} T(M)$ defined *along* $\tilde{f}$. The section $\dfrac{\overline{\partial} f}{\partial t} : D \to \pi^{-1} T(M)$ in $\pi^{-1} T(M)$ along $\tilde{f}$ is defined in the same manner

$$\frac{\overline{\partial} f}{\partial t}(u,t) = (d_{(u,t)} f) \left(\frac{\partial}{\partial t}\right)_{(u,t)} \in T_{f(u,t)}(M) = \left(\pi^{-1} TM\right)_{\tilde{f}(u,v)} .$$

The statement in Lemma 2.51 is that

$$g_{\tilde{f}(u,t)} \left(\frac{\overline{\partial} f}{\partial u}(u,t), \ \frac{\overline{\partial} f}{\partial t}(u,t) \right) = 0$$

for any $(u,t) \in D$. For arbitrary fixed $u \in J$, the maps $t \mapsto \dfrac{\overline{\partial} f}{\partial u}(u,t)$ and $t \mapsto \dfrac{\overline{\partial} f}{\partial t}(u,t)$ are sections in $\pi^{-1} T(M)$ defined along the curve $t \mapsto \tilde{f}(u, t)$. Then [by (2.42)]

$$\frac{\partial}{\partial t} \left[g_{\tilde{f}(u,t)} \left(\frac{\overline{\partial} f}{\partial u}(u,t), \ \frac{\overline{\partial} f}{\partial t}(u,t) \right) \right] = \tag{2.63}$$

$$= g_{\tilde{f}(u,t)} \left(\frac{D}{\partial t} \left(\frac{\overline{\partial} f}{\partial u} \right)(u,t), \ \frac{\overline{\partial} f}{\partial t}(u,t) \right) + g_{\tilde{f}(u,t)} \left(\frac{\overline{\partial} f}{\partial u}(u,t), \ \frac{D}{\partial t} \left(\frac{\overline{\partial} f}{\partial t} \right)(u,t) \right).$$

We adopt the notation

$$f_u(t) = f(u,t), \quad f_t(u) = f(u,t), \quad (u,t) \in D,$$

so that

$$\frac{D}{\partial t}\left(\frac{\overline{\partial} f}{\partial t}\right)(u,t) = \frac{D}{dt}\left(\frac{\overline{d} f_u}{dt}\right)(t).$$

Then $f_u(t) = \exp_q\big(t\,X(u)\big)$ is a geodesic [the geodesic of initial data $(q,\ X(u))$] hence

$$0 = \frac{D}{dt}\left(\frac{\overline{d} f_u}{dt}\right)(t) = \left(\nabla_{d\tilde{f}_u/dt}\frac{\overline{d} f_u}{dt}\right)_{\tilde{f}_u(t)} = \left(\nabla_{d\tilde{f}_u/dt}\mathcal{L}\right)_{\tilde{f}_u(t)} = K_{\tilde{f}_u(t)}\frac{d\tilde{f}_u}{dt}(t)$$

where $\tilde{f}_u(t) = (df_u/dt)(t)$. The formula (2.63) becomes

$$\frac{\partial}{\partial t}\left[g_{\tilde{f}(u,t)}\left(\frac{\overline{\partial} f}{\partial u}(u,t),\ \frac{\overline{\partial} f}{\partial t}(u,t)\right)\right] = g_{\tilde{f}(u,t)}\left(\frac{D}{\partial t}\left(\frac{\overline{\partial} f}{\partial u}\right)(u,t),\ \frac{\overline{\partial} f}{\partial t}(u,t)\right). \tag{2.64}$$

If we set $f^i(u,t) = \tilde{x}^i(f(u,t))$, then

$$\tilde{f}_u(t) = \tilde{f}(u,t) = \frac{\partial f^i}{\partial t}(u,t)\left(\frac{\partial}{\partial \tilde{x}^i}\right)_{f_u(t)},$$

$$\tilde{f}_u^i(t) = x^i\big(\tilde{f}_u(t)\big) = \tilde{x}^i\big(f_u(t)\big) = f^i(u,t),\quad \tilde{f}_u^{i+n}(t) = y^i\big(\tilde{f}_u(t)\big) = \frac{\partial f^i}{\partial t}(u,t),$$

$$\begin{aligned}
\frac{d\tilde{f}_u}{dt}(t) &= \frac{d\tilde{f}_u^i}{dt}(t)\left(\frac{\partial}{\partial x^i}\right)_{\tilde{f}_u(t)} + \frac{d\tilde{f}_u^{i+n}}{dt}(t)\left(\frac{\partial}{\partial y^i}\right)_{\tilde{f}_u(t)} = \\
&= \frac{\partial f^i}{\partial t}(u,t)\left(\frac{\partial}{\partial x^i}\right)_{\tilde{f}_u(t)} + \frac{\partial^2 f^i}{\partial t^2}(u,t)\left(\frac{\partial}{\partial y^i}\right)_{\tilde{f}_u(t)} = \qquad (2.65)\\
&= \frac{\partial f^i}{\partial t}(u,t)\left(\frac{\delta}{\delta x^i} + N_i^j\frac{\partial}{\partial y^j}\right)_{\tilde{f}_u(t)} + \frac{\partial^2 f^i}{\partial t^2}(u,t)\left(\frac{\partial}{\partial y^i}\right)_{\tilde{f}_u(t)} = \\
&= \frac{\partial f^i}{\partial t}(u,t)\left(\frac{\delta}{\delta x^i}\right)_{\tilde{f}_u(t)} + \left[\frac{\partial^2 f^j}{\partial t^2}(u,t) + N_i^j\big(\tilde{f}_u(t)\big)\frac{\partial f^i}{\partial t}(u,t)\right]\left(\frac{\partial}{\partial y^j}\right)_{\tilde{f}_u(t)},
\end{aligned}$$

hence

$$0 = K_{\tilde{f}_u(t)}\frac{d\tilde{f}_u}{dt}(t) = \left[\frac{\partial^2 f^j}{\partial t^2}(u,t) + N_i^j\big(\tilde{f}(u,t)\big)\frac{\partial f^i}{\partial t}(u,t)\right]X_{j,\tilde{f}(u,t)}$$

yielding

$$\frac{\partial^2 f^j}{\partial t^2}(u,t) + N_i^j\big(\tilde{f}(u,t)\big)\frac{\partial f^i}{\partial t}(u,t) = 0.$$

In particular

$$\frac{d\tilde{f}_u}{dt}(t) = \frac{\partial f^i}{\partial t}(u,t)\left(\frac{\delta}{\delta x^i}\right)_{\tilde{f}(u,t)} \in N_{\tilde{f}(u,t)},$$

i.e., $\dfrac{d\tilde{f}_u}{dt}(t)$ is horizontal. We shall need the vector fields on $V(M)$

$$\frac{\partial \tilde{f}}{\partial u}, \frac{\partial \tilde{f}}{\partial t} : D \to T(V(M))$$

defined along $\tilde{f}$ by

$$\frac{\partial \tilde{f}}{\partial u}(u,t) = \big(d_{(u,t)}\tilde{f}\big)\left(\frac{\partial}{\partial u}\right)_{(u,v)}, \quad \frac{\partial \tilde{f}}{\partial t}(u,t) = \big(d_{(u,t)}\tilde{f}\big)\left(\frac{\partial}{\partial t}\right)_{(u,v)}.$$

Note that $\dfrac{\partial \tilde{f}}{\partial t}(u,t)$ is the same as $\dfrac{d\tilde{f}_u}{dt}(t)$. Indeed

$$\tilde{f}^i(u,t) = x^i\big(\tilde{f}(u,t)\big) = x^i\left[(d_{(u,t)}f)\left(\frac{\partial}{\partial t}\right)_{(u,t)}\right] = x^i\big[f(u,t)\big] = f^i(u,t),$$

$$\tilde{f}^{i+n}(u,t) = y^i\big(\tilde{f}(u,t)\big) = y^i\left[(d_{(u,t)}f)\left(\frac{\partial}{\partial t}\right)_{(u,t)}\right] = \frac{\partial f^i}{\partial t}(u,t),$$

hence

$$\frac{\partial \tilde{f}}{\partial t}(u,t) = (d_{(u,t)}\tilde{f})\left(\frac{\partial}{\partial t}\right)_{(u,t)} = \frac{\partial \tilde{f}^i}{\partial t}(u,t)\left(\frac{\partial}{\partial x^i}\right)_{\tilde{f}(u,t)} + \frac{\partial \tilde{f}^{i+n}}{\partial t}(u,t)\left(\frac{\partial}{\partial y^i}\right)_{\tilde{f}(u,t)} =$$

$$= \frac{\partial f^i}{\partial t}(u,t)\left(\frac{\partial}{\partial x^i}\right)_{\tilde{f}(u,t)} + \frac{\partial^2 f^i}{\partial t^2}(u,t)\left(\frac{\partial}{\partial y^i}\right)_{\tilde{f}(u,t)}$$

which compared to (2.65) yields the result, i.e.,

$$\frac{\partial \tilde{f}}{\partial t}(u,t) = \frac{d\tilde{f}_u}{dt}(t).$$

In particular

$$\frac{\partial \tilde{f}}{\partial t}(u,t) \in N_{\nabla,\,\tilde{f}(u,t)}.$$

Going back to (2.64) let us observe that [by Lemma 2.32]

$$\frac{D}{\partial u}\left(L\,\frac{\partial s}{\partial t}\right)-\frac{D}{\partial t}\left(L\,\frac{\partial s}{\partial u}\right)=T_L\left(\frac{\partial s}{\partial u},\,\frac{\partial s}{\partial t}\right)$$

[with $s=\tilde{f}$] or

$$\frac{D}{\partial u}\left(\frac{\overline{\partial} f}{\partial t}\right)-\frac{D}{\partial t}\left(\frac{\overline{\partial} f}{\partial u}\right)=T_L\left(\frac{\partial \tilde{f}}{\partial u},\,\frac{\partial \tilde{f}}{\partial t}\right).$$

so that (2.64) may be written

$$\frac{\partial}{\partial t}\left[g_{\tilde{f}(u,t)}\left(\frac{\overline{\partial} f}{\partial u}(u,t),\,\frac{\overline{\partial} f}{\partial t}(u,t)\right)\right]=$$

$$=g_{\tilde{f}(u,t)}\left(\frac{D}{\partial u}\left(\frac{\overline{\partial} f}{\partial t}\right)(u,t),\,\frac{\overline{\partial} f}{\partial t}(u,t)\right)-g_{\tilde{f}(u,t)}\left(T_L\left(\frac{\partial s}{\partial u},\,\frac{\partial s}{\partial t}\right)_{\tilde{f}(u,t)},\,\frac{\overline{\partial} f}{\partial t}(u,t)\right).$$

The last term vanishes [by Lemma 2.20] because of

$$\mathcal{L}[\tilde{f}(u,t)]=\frac{\overline{\partial} f}{\partial t}(u,t)$$

where $\mathcal{L}$ is the Liouville vector field. We are left with

$$\frac{\partial}{\partial t}\left[g_{\tilde{f}(u,t)}\left(\frac{\overline{\partial} f}{\partial u}(u,t),\,\frac{\overline{\partial} f}{\partial t}(u,t)\right)\right]=g_{\tilde{f}(u,t)}\left(\frac{D}{\partial u}\left(\frac{\overline{\partial} f}{\partial t}\right)(u,t),\,\frac{\overline{\partial} f}{\partial t}(u,t)\right)=$$

[by $\nabla g=0$]

$$=\frac{1}{2}\frac{\partial}{\partial u}\left[g_{\tilde{f}(u,t)}\left(\frac{\overline{\partial} f}{\partial t}(u,t),\,\frac{\overline{\partial} f}{\partial t}(u,t)\right)\right]. \tag{2.66}$$

Since $f_u(t)=\exp_q\left[t\,X(u)\right]$

$$\frac{\partial f}{\partial t}(u,t)=(d_{(u,t)}f)\left(\frac{\partial}{\partial t}\right)_{(u,t)}=(d_t f_u)\left(\frac{d}{dt}\right)_t=\frac{df_u}{dt}(t)=t\,X(u)$$

and

$$E(X(u))=1\Longrightarrow t^2=E(t\,X(u))=g\big(\mathcal{L},\,\mathcal{L}\big)_{t\,X(u)}=$$

$$=g\big(\mathcal{L},\,\mathcal{L}\big)_{\tilde{f}(u,t)}=g_{\tilde{f}(u,t)}\left(\frac{\overline{\partial} f}{\partial t}(u,t),\,\frac{\overline{\partial} f}{\partial t}(u,t)\right)$$

the term (2.66) vanishes, implying that

$$g_{\tilde{f}(u,t)}\left(\frac{\overline{\partial} f}{\partial u}(u,t),\ \frac{\overline{\partial} f}{\partial t}(u,t)\right) = C(u)$$

for some "constant" $C(u) \in \mathbb{R}$. Yet for $t = 0$

$$f(u,0) = \exp_q 0_q = q \Longrightarrow \frac{\partial f}{\partial u}(u,0) = \frac{df_0}{du}(u) = 0 \Longrightarrow \frac{\overline{\partial} f}{\partial u}(u,0) = 0 \Longrightarrow C(u) = 0$$

hence

$$g_{\tilde{f}(u,t)}\left(\frac{\overline{\partial} f}{\partial u}(u,t),\ \frac{\overline{\partial} f}{\partial t}(u,t)\right) = 0.$$

Q.e.d.

Lemma 2.52 *Let N_0 be as in the previous lemma and let N_q be the image of N_0 by* $\exp_q$. *Let*

$$\omega : [a,b] \to N_q \setminus \{q\}$$

be any piecewise differentiable curve. Each point $\omega(t)$ may be expressed uniquely as $\omega(t) = \exp_q\left[r(t)\, X(t)\right]$ for some

$$0 < r(t) < \epsilon, \quad X(t) \in T_q(M), \quad L(X(t)) = 1.$$

Also the length of ω obeys to

$$L(\omega) \geq r(b) - r(a), \quad r(b) > r(a),$$

with equality if and only if $r(t)$ is monotonically increasing, while $X(t)$ is constant.

Let $X \in V(M)$ and $Y \in T_{\pi(X)}(M) = \left(\pi^{-1}TM\right)_X$ be two tangent vectors having the same origin (*a fortiori* $X \neq 0$). We chose to call Y transversal to X if $g_X(X,\ Y) = 0$, a requirement equivalent to $g_X\left(\mathcal{L}(X),\ Y\right) = 0$. To prove Lemma 2.52 we shall first show that

$$E(X+Y) \geq E(X)$$

provided that Y is transversal to X, with equality if and only if $Y = 0$. To this end let us set $q = \pi(X) \in M$ and let $(U,\ x^i)$ a local coordinate system about q. Let us consider the function

$$F\left(v^1, \cdots, v^n\right) = E\left[v^i \left(\frac{\partial}{\partial x^i}\right)_q\right].$$

If

$$X = X^i \left(\frac{\partial}{\partial x^i}\right)_q, \quad Y = Y^j \left(\frac{\partial}{\partial x^j}\right)_q,$$

then

$$E(X+Y) = E\left[(X^i + Y^i)\left(\frac{\partial}{\partial x^i}\right)_q\right] = F\big(X^1 + Y^1, \cdots, X^n + Y^n\big) =: G(Y^1, \cdots, Y^n).$$

Then

$$\frac{\partial G}{\partial Y^i}(Y^1, \cdots, Y^n) = \frac{\partial F}{\partial v^i}(X^1 + Y^1, \cdots, X^n + Y^n) = \frac{\partial E}{\partial y^i}(X+Y),$$

$$\frac{\partial^2 G}{\partial y^i\, \partial y^j}(Y^1, \cdots, Y^n) = \frac{\partial^2 E}{\partial y^i\, \partial y^j}(X+Y) = 2\, g_{ij}(X+Y).$$

The truncated Taylor development of G about the origin is

$$G(Y^1, \cdots, Y^n) = G(0) + \sum_{i=1}^{n} \frac{\partial G}{\partial Y^i}(0)\, Y^i + \frac{1}{2} \sum_{i,j=1}^{n} \frac{\partial^2 G}{\partial Y^i\, \partial Y^j}(\theta Y)\, Y^i\, Y^j$$

[for some $0 < \theta < 1$] or

$$E(X+Y) = E(X) + \sum_{i=1}^{n} \frac{\partial E}{\partial y^i}(X)\, Y^i + \sum_{i,j=1}^{n} g_{ij}(X + \theta Y)\, Y^i\, Y^j. \tag{2.67}$$

As Y is transversal to X and $\mathcal{L} = y^i\, X_i$

$$0 = g_X(X, Y) = g_X\big(\mathcal{L}(X), Y\big) = y^i Y^j\, g_{ij}(X) = Y^j\, \frac{\partial E}{\partial y^j}(X),$$

$$g_{ij}(X + \theta Y)\, Y^i\, Y^j \geq 0,$$

it follows that $E(X+Y) \geq E(X)$. Finally, [by (2.67)]

$$E(X+Y) = E(X) \Longleftrightarrow g_{ij}(X + \theta Y)\, Y^i\, Y^j 0 \Longleftrightarrow Y = 0.$$

Q.e.d.

Proof of Lemma 2.52. Let us set $f(r,t) = \exp_q\big(r\, X(t)\big)$, so that $\omega(t) = f\big(r(t), t\big)$. Hence,

$$\frac{d\omega}{dt}(t) = \frac{d\omega^i}{dt}(t)\left(\frac{\partial}{\partial x^i}\right)_{\omega(t)} = \left[\frac{\partial f^i}{\partial r}(r(t),\, t)\, r'(t) + \frac{\partial f^i}{\delta t}(r(t),\, t)\right]\left(\frac{\partial}{\partial x^i}\right)_{\omega(t)} =$$

$$= \frac{\partial f}{\partial r}(r(t),\, t)\, r'(t) + \frac{\partial f}{\partial t}(r(t),\, t)$$

According to Lemma 2.51 the vector $\frac{\partial f}{\partial t}(r, t)$ is transversal to $\frac{\partial f}{\partial r}(r, t)$. Also $E\left(\frac{\partial f}{\partial r}(r, t)\right) = 1$ hence

$$E\left[\frac{d\omega}{dt}(t)\right] \geq r'(t)^2$$

with equality if and only if $\frac{\partial f}{\partial t}(r, t) = 0$ and $r'(t) \geq 0$. This is equivalent to $\frac{dX}{dt}(t) = 0$ and $r'(t) \geq 0$. Thus,

$$\int_a^b E\left[\frac{d\omega}{dt}(t)\right]^{1/2} dt \geq \int_a^b r'(t)\, t = r(b) - r(a)$$

with equality if and only if $r(t)$ is monotonically increasing and $X(t)$ is constant. Q.e.d.

Proof of theorem 2.49. Let ω be a piecewise differentiable curve joining q_1 to $q_2 = \exp_{q_1}(rX) \in B_\delta$ with $0 < r < \epsilon$ and $E(X) = 1$. Then, for any $\delta' > 0$, the curve ω must contain a segment joining the spherical shell $\Lambda[\chi(x)] = \delta'$ [of radius δ'] to the spherical shell $\Lambda[\chi(x)] = r$ [of radius r] and lying between the two shells. The length of that segment is $\geq r - \delta'$. In the limit as $\delta' \to 0^+$ the length of ω is $\geq r$. On the other hand $L(\alpha) = r$ so that $L(\alpha) \leq L(\omega)$. Next $L(\omega) = r$ if and only if $X(t)$ is constant and the parameter t is a monotonically increasing function of arc length. As $X(t)$ is contant, ω is a geodesic. We assumed that α is the unique geodesic joining q_1 to q_2, hence $L(\alpha) = L(\omega)$ if and only if the point sets $\omega([0, 1])$ and $\alpha([0, 1])$ coincide and the parameter t is a monotonically increasing function of the arc length. Q.e.d.

2.3 Finslerian Distance Functions

Definition 2.53 A geodesic $\omega : [a, b] \to M$ is said to be *minimal* if its length is $\leq$ the length of any other piecewise differentiable curve joining $\alpha(a)$ to $\alpha(b)$. □

Let (M, E) be a Finslerian manifold. We recall

Definition 2.54 The *length* of a C^1 curve $\gamma : [a, b] \to M$ is

$$L(\gamma) = \int_a^b \sqrt{E\left[\frac{d\gamma}{dt}(t)\right]}\, dt.$$

□

The definition of the notion of length extends to piecewise C^1 curves, in a natural manner.

Definition 2.55 Let $(p, q) \in M \times M$ be an ordered pair of points in M. The *distance* $d(p, q)$ from the point $p \in M$ to the point q is the greatest lower bound of all lengths $L(\gamma)$ as γ ranges over all piecewise C^1 functions joining p and q. □

One may easily show that $d : M \times M \to \mathbb{R}$ is a *generalized* distance function on M, which is to say that

(i) $d(p, q) \geq 0$ and $d(p, q) = 0 \Longleftrightarrow p = q$, and

(iii) $d(p, q) \leq d(p, r) + d(r, q)$, for any $p, q, r \in M$.

However, unless one assumes that $E(-w) = E(w)$, the symmetry property (ii) $d(p, q) = d(q, p)$ is not guaranteed, in general.

Let $B(p, \delta) = \{q \in M \, : \, d(p, q) < \delta\}$ be the *open ball* of center p and radius $0 < \delta < \infty$, associated to the generalized distance function d. The set of all open balls is a base for a topology on M, referred to as the *metric topology* of (M, E). The metric topology is known to coincide with the original manifold topology (cf. e.g., [Mat]). The proof requires a bit of preparation, as follows.

Proposition 2.56 *Let $\delta > 0$ such that $N_0 = \{X \in T_p(M) \, : \, E(X) < \delta^2\}$ is a normal neighborhood of 0_p in $T_p(M)$. Then*

$$B(p, \delta) = \exp_p N_0 .$$

Proof If $q \in \exp_p N : 0$, then $d(p, q) < \delta$, so that $\exp_p N_0 \subset B(p, \delta)$. Conversely let $q \in B(p, \delta)$ and let α be a curve joining p and q and such that $L(\alpha) < \delta$. Let us assume that $\alpha(t) \in \exp_p N_0$ for any $0 \leq t \leq u$. Let u_0 be the least upper bound of all such u. Each point $\alpha(t)$, $0 \leq t < u_0$ can be uniquely represented as

$$\alpha(t) = \exp_p \left[r(t)\, X(t)\right]$$

for some

$$0 \leq r(t) < u_0, \quad X(t) \in T_p(M), \quad F(X(t)) = 1.$$

Then [by Lemma 2.52]

$$L(\alpha(t)) \leq d(p, \alpha(t)) = r(t), \quad 0 \leq t < u_0 .$$

Thus, $L(\alpha(u_0)) \geq r(u_0) = \delta$, which is a contradiction. Hence, α lies entirely in $\exp_p N_0$. Q.e.d.

Corollary 2.57 *The metric topology on (M, E) and the original manifold topology on M coincide.*

Proof The set of balls $\{B(p, \delta) : p \in M, \ \delta > 0\}$ is a base for the metric topology. On the other hand $\{\exp_p N_0\}$ with $N_0 = \{X \in T_p(M) : E(X) < \delta^2\}$ a normal neighborhood of $0_p \in T_p(M)$ is a local base at p for the original topology of M.

Proposition 2.58 *Let $\{U_\alpha\}_{\alpha \in A}$ be an open cover of a compact subset $K \subset M$. Then there is $\delta > 0$ such that*

$$d(p, q) < \delta \Longrightarrow \exists\, \alpha \in A \ \textit{ such that } \ p, q \in U_\alpha\,.$$

Proof Let us assume there is none such $\delta > 0$. Then there are two sequences $\{p_\nu\}_{\nu \geq 1} \subset K$ and $\{q_\nu\}_{\nu \geq 1} \subset K$ such that $\lim_{\nu \to \infty} d(p_\nu, q_\nu) = 0$ and no set of the cover contains both p_ν and q_ν. Yet K is compact, hence we may extract convergent subsequences

$$\{p_{\nu_i}\}_{i \geq 1} \subset \{p_\nu\}_{\nu \geq 1}, \quad \{q_{\nu_i}\}_{i \geq 1} \subset \{q_\nu\} - \nu \geq 1,$$

[of course with $\nu_1 < \nu_2 < \cdots$] such that $\lim_{i \to \infty} p_{\nu_i} = p$ and $\lim_{i \to \infty} q_{\nu_i} = q$. Since d is a continuous function of both its arguments, $d(p, q) = 0$ yields $p = q$. Finally, as $p \in K$ there is $\alpha \in A$ such that $p \in U_\alpha$, and one may choose $i \geq 1$ such that $\{p_{\nu_i}, q_{\nu_i}\} \subset U_\alpha$, a contradiction. Q.e.d.

Corollary 2.59 *For every compact subset $K \subset M$, there is $\delta > 0$ such that any two points in K of distance $< \delta$ can be joined by a unique geodesic of length $< \delta$. This geodesic is minimal and depends smoothly on its endpoints.*

Proof By Theorem 2.47 every point $p \in M$ admits a neighborhood which is both simple and convex. The family of all such neighborhoods as p ranges over K is an open cover $\mathfrak{U}$ of K. Then [by Proposition 2.58] there is $\delta > 0$ such that $d(p, q) < \delta$ with $p, q \in K$ yields $p, q \in U$ for some $U \in \mathfrak{U}$. Q.e.d.

Theorem 2.60 *Let (M, E) be a Finslerian manifold, and let d be the corresponding distance function. Then*

$$\forall\, p \in M, \quad \exists\, a > 0, \quad \forall\, 0 < \delta < a \ :$$

(i) *$B(p, \delta)$ is a normal neighborhood of each of its points,*

(ii) *$\forall\, x, y \in B(p, \delta)$ the unique geodesic joining x and y in $B(p, \delta)$ is also the only curve of length $d(x, y)$ from x to y.*

Proof Let $p \in M$ and let (U, x^i) be a local coordinate system on M such that $p \in U$ and $x^i(p) = 0$. Let us consider the function Λ given by $\Lambda(\xi) = \sum_{i=1}^{n} (\xi^i)^2$ for any $\xi \in \mathbb{R}^n$. Let $a > 0$ be the positive number furnished by Theorem 2.47 and let $0 < \delta < a$. Once again by Theorem 2.47

$$B_\delta = \{q \in U \,:\, \Lambda(\chi(q)) < \delta\}$$

is a normal neighborhood of each of its points. Then $B(p, \delta) = B_\delta$ for sufficiently small δ, so that property (i) is proved.

Let $x, y \in B(p, \delta)$. As $B(p, \delta)$ is a normal neighborhood of x, α is the only shortest curve in $B(p, \delta)$ from x to y. Finally [by Proposition 2.56], a curve from x to y that doesn't lie in $B(p, \delta)$ has length $> \delta$.

Lemma 2.61 *Let $(M,\ E)$ be a Finslerian manifold. There is a continuous function $r : M \to (0,\ +\infty)$ such that any two points of the closed ball*

$$S(p,\ r(p)) = \{q \in M \,:\, d(p,\ q) \le r(p)\}$$

can be joined by a minimizing geodesic.

Proof Let $p \in M$ and let $r(p)$ be the least upper bound of all $r > 0$ such that any two points $q, q' \in M$ with $d(p, q) < r$ and $d(p, q') < r$ can be joined by a minimizing geodesic. Then [by Theorem 2.60] $r(p) > 0$. If $r(p) = +\infty$ for some $p \in M$, then $r(q) = +\infty$ for every $q \in M$, hence any continuous function on M will do. If $r(p) < \infty$ for every $p \in M$, then one may assume w.l.o.g. that $r(p) > r(q)$. If $d(p, q) \ge r(p)$, then

$$\left| r(p) - r(q) \right| < d(p, q) \le d_m(p, q),$$

$$d_m(p, q) := \max \left\{ d(p, q),\ d(q, p) \right\}.$$

If in turn $d(p, q) < r(p)$, then we set $r' := r(p) - d(p, q)$. Let $q' \in B(q, r')$. Then

$$d(p, q') \le d(p, q) + d(q, q') \le d(p, q) + r = r(p).$$

Thus, $q' \in S(p,\ r(p))$ yielding $S(q, r') \subset S(p, r(p))$ and then

$$r(q) \ge r' - d(p, q)$$

meaning that

$$\left| r(p) - r(q) \right| \le d(p, q) \le d_m(p, q).$$

The proof will be complete as soon as we ascertain that d_m is a distance function on M. Indeed d_m satisfies all axioms defining a distance function on M. It may be easily shown that the family of all opne balls $B_m(p, \delta) = \{q \in M \,:\, d_m(p, q) < \delta\}$ is a base for a topology on M (the topology *induced* by d_m on M).

2.4 Complete Finslerian Manifolds

Let (M, E) be a Finslerian manifold and let d be its generalized distance function. By following [Has], we present the Finslerian generalization of the well known Hopf-Rinow theorem for complete Riemannian manifolds. We shall make use of the generalized distance function, very much that way the (genuine) distance function (associated to a given Riemannian metric) is made use of.

Definition 2.62 A subset $B \subset M$ is said to be *bounded* if there are $p \in M$ and $\delta > 0$ such that $B \subset B(p, \delta)$. □

Definition 2.63 M is said to be *finitely compact* if every infinite bounded subset $B \subset M$ has an accumulation point (not necessarily belonging to B). □

Definition 2.64 A sequence $\{p_\nu\}_{\nu \geq 1} \subset M$ is a *Cauchy sequence* if

$$\forall\, \epsilon > 0, \quad \exists\, N = N_\epsilon \geq 1, \quad \forall\, \nu > \mu \geq N \,:\, d(p_\mu, p_\nu) < \epsilon .$$

□

A convergent sequence in a generalized metric space need not be Cauchy. If a Cauchy sequence $\{p_\nu\}_{\nu \geq 1}$ admits a convergent subsequence, then $\{p_\nu\}_{\nu \geq 1}$ is convergent.

Definition 2.65 A Finslerian manifold (M, E) is said to be *metrically complete*, with respect to the generalized distance function d associated to E, if every Cauchy sequence in (M, d) converges. □

We have seen that in general the exponential mapping $\exp_p$ is only defined on some open subset of $T_p(M)$.

Definition 2.66 (M, E) is said to be *geodesically complete* if for every $p \in M$ the exponential mapping $\exp_p$ is defined on the whole of $T_p(M)$. □

For further use, we adopt the following notations. For every point $p \in M$, let $\overline{B}(p, r)$ denote the closure of the ball $B(p, r)$. Also let $E(r)$ be the set of all $q \in \overline{B}(p, r)$ such that p can be joined to q by a minimizing geodesic.

Proposition 2.67 *If M is geodesically complete, then $E(r)$ is compact.*

Proof The tangent vector at the origin of a geodesic arc, having the same sense as the geodesic and length equal to that of the geodesic, is referred to as the *initial vector* of the geodesic. Let $\{q_\nu\}_{\nu \geq 1} \subset E(r)$ and let $\{\tau_\nu\}_{\nu \geq 1}$ be a sequence of minimizing geodesics from p to q_ν, for every $\nu \geq 1$. Let X_ν be the initial vector of τ_ν. As $E\big(X_\nu\big) \leq r^2$ one may assume, by passing to a subsequence if necessary, that $\{X_\nu\}_{\nu \geq 1}$

converges to some $X \in T_p(M)$. Also, as $d(p, q_\nu) \le r$ one may assume, once again by passing to a subsequence if necessary, that $\{d(p, q_\nu)\}_{\nu \ge 1}$ converges to some $r_0 \ge 0$. As every τ_ν is minimizing

$$q_\nu = \exp_p \big(d(p, q_\nu) X_\nu\big).$$

As M is geodesically complete the exponential map $\exp_p$ may be applied to any $X \in T_p(M)$, of whatever length, so that every $X \in T_p(M)$ is the initial vector of a unique geodesic arc issuing at p. In particular $q_0 := \exp_p \big(r_0 X\big)$ is well defined. As the exponential map is continuous

$$\lim_{\nu \to \infty} q_\nu = \lim_{\nu \to \infty} \exp_p \big[d(p, q_\nu)\, X_\nu\big] = \exp_p \Big[\Big(\lim_{\nu \to \infty} d(p, q_\nu)\Big) \lim_{\nu \to infty} X_\nu\Big] = \exp_p (r_0 X = q_0 .$$

Thus, $q_\nu \to q_0$ as $\nu \to \infty$ so that $d(p, q_0) = r_0$. It follows that $\exp_p(tX), 0 \le t \le t_0$, is a minimizing geodesic and $q_0 \in E(r)$. Hence, $E(r)$ is compact. Q.e.d.

Lemma 2.68 *If $E(r) = \overline{B}(p, r)$ for some $r > 0$, then, for any $q \in M$ with $d(p, q) > r$, there is a point $x \in M$ such that $d(p, x) = r$ and $d(p, q) = r + d(x, q)$.*

Proof The distance between every ordered pair of points $p, q \in M$ was defined to be the greatest lower bound of the lengths of curves from p to q. One may then join p and q by a curve of length as close to $d(p, q)$ as desired. For every positive integer $\nu \in \mathbb{N}$, let τ_ν be a curve from p to q such that

$$L(\tau_\nu) < d(p, q) + \frac{1}{\nu} .$$

Let $\tau_\nu(t) \in B(p, r)$ for $0 \le t \le u$. Let u_0 be the least upper bound of all such u and let us set $x_\nu = \tau_\nu(u_0)$. Then $d(p, x_\nu) = r$. As $E(r)$ is compact, the sequence $\{x_\nu\}_{\nu \ge 1}$ has some limit point x. By eventually passing to a subsequence one may assume that $\lim_{\nu \to \infty} x_\nu = x$. Then

$$d(p, q) \le d(p, x) + d(x, q) = r + d(x, q).$$

On the other hand

$$d(x_\nu, q) \le L\big(\tau_\nu(x_\nu, q)\big) = L(\tau_\nu) - L\big(\tau_\nu(p, x_\nu)\big) < d(p, q) + \frac{1}{\nu} - L\big(\tau_\nu(o, x_\nu)\big) < d(p, q) + \frac{1}{\nu} - r$$

where $\tau_\nu(y, z)$ denotes the segment of τ_ν joining y to z (an ordered pair of points on τ_ν). Passing to the limit with $\nu \to \infty$ gives

$$d(x, q) \le d(p, q) - r.$$

Q.e.d.

Corollary 2.69 *$E(r) = \overline{B}(p, r)$ for every $r \ge 0$.*

Proof The statement is clearly true for $r = 0$. Also, if $E(r_0) = \overline{B}(p, r_0)$ for some $r_0 > 0$, then $E(r) = \overline{B}(p, r)$ for all $r < r_0$. It remains to be proved that

Claim 1 $E(r) = \overline{B}(p, r)$ *for some* $r_0 > 0$ *and all* $r < r_0$ implies $E(r_0) = \overline{B}(p, r_0)$.

Claim 2 *If* $E(r) = \overline{B}(p, r)$ *for some* $r > 0$, *then there is* $c > 0$ *such that* $E(r + c) = \overline{B}(p, r + c)$.

Proof of Claim 1 It suffices to show that every point $q \in \overline{B}(p, r_0)$ such that $d(p, q) = r_0$ belongs to $E(r_0)$. For, if $q \in \overline{B}(p, r_0)$ and $d(p, q) = r$ for some $r < r_0$, then $q \in \overline{B}(p, r) = E(r)$ [by hypothesis] and hence $q \in E(r_0)$. Next, one may choose a sequence $\{q_\nu\}_{\nu \geq 1}$ for which q is a limit point, and such that $q_h \in \overline{B}(p, r_\nu)$ for some $r_\nu < r_0$. Then each q_ν belongs to $E(r_\nu) \subset E(r_0)$. Since $E(r_0)$ is compact the very point q belongs to $E(r_0)$. Q.e.d.

Proof of Claim 2 Let c be the minimum of the continuous function in Lemma 2.61, attained on the compact set $E(r)$. Then

$$\forall\, q \in \overline{B}(p,\, r + c) \setminus \overline{B}(p, r) \;:\; d(p, q) > r.$$

Therefore [by Lemma 2.68], there is a point x such that

$$d(p, x) = r, \quad f(p, q) = r + d(x, q).$$

Yet $d(p, q) \leq r + c$ so that $d(x, q) \leq c$. Hence, there is a geodesic γ from p to x with $L(\gamma) = d(x, q)$. Let σ be a geodesic from p to x with $L(\sigma) = d(p, x) = r$. Such a geodedesic may be chosen because $x \in \overline{B}(p, r) = E(r)$. The juxtaposition $\sigma \cdot \gamma$ is then a piecewise C^∞ curve (a "broken" C^∞ curve) from p to q such that

$$L(\sigma \cdot \gamma) = L(\sigma) + L(\gamma) = r + d(x, q) = d(p, q).$$

Thus, $\sigma \cdot \gamma$ is a "broken" C^∞ reparametrization of a geodesic. Therefore, $q \in E(r + c)$. Q.e.d.

Corollary 2.70 *If* (M, E) *is geodesically complete, then any two points can be joined by a minimal geodesic.*

Corollary 2.70 is a straightforward consequence of Proposition 2.67 and of Corollary 2.69.

Theorem 2.71 *Let* (M, E) *be a Finslerian manifold. The following statements are equivalent:*

(i) M *is complete as a generalized metric space, with respect to the generalized distance function* d.

(ii) *Every closed and bounded subset of* M *is compact.*

(iii) (M, E) *is geodesically complete.*

Proof (iii) $\Longrightarrow$ (ii). Let $S \subset M$ be a bounded subset. There are $p \in M$ and $r > 0$ such that $S \subset \overline{B}(p, r)$. As M Is geodesically complete $E(r) = \overline{B}(p, r)$ is compact. Therefore, S is relatively compact. Q.e.d.

(ii) $\Longrightarrow$ (i). Let $\{x_\nu\}_{\nu \geq 1}$ be a Cauchy sequence in (M, d). The closure S of the set $\{x_\nu : \nu \geq 1\}$ is bounded. So S is closed and bounded, and hence, by hypothesis, S is compact. Hence, we may extract a convergent subsequence $\{x_{\nu_k}\}_{k\geq 1}$. Yet $\{x_\nu\}_{\nu \geq 1}$ is Cauchy, so it must be convergent. Q.e.d.

(i) $\Longrightarrow$ (iii). To simplify the presentation, we adopt the following terminology. A *maximal geodesic* is a geodesic which is not the proper restriction of any other geodesic. The proof is by contradiction, i.e., we assume that M is complete as a generalized metric space, yet M is not geodesically complete. By the "contradiction assumption" there is $p_0 \in M$ such that $\exp_{p_0}$ is not defined on the whole of $T_{p_0}(M)$, and in particular there exists a maximal geodesic ray $\alpha(t)$, $0 \leq t < \ell$, parametrized by arc length, with $\ell < +\infty$. We shall show that α may be extended to a geodesic defined beyond ℓ, thus contradicting maximality. The arc length, measured from the origin of the initial vector $X = \dfrac{d\alpha}{dt}(0)$, equals t. Let

$$\{t_\nu\}_{\ nu\geq 1} \subset I = [0,\ \ell)$$

be an infinite increasing sequence such that $t_\nu \to \ell$ for $\nu \to \infty$. The sequence $\{t_\nu\}_{\nu\geq 1}$ is a Cauchy sequence of real numbers. Hence, $\{\alpha(x_\nu)\}_{\nu\geq 1}$ is a Cauchy sequence in M, for

$$d\big(\alpha(t_\nu),\ \alpha(t_\mu)\big) \leq t_\mu - t_\nu\,.$$

As the generalized metric space (M, d) is complete, $\{\alpha(t_\nu)\}_{\nu\geq 1}$ converges to some $p \in M$.

Claim1. *The limit point p doesn't depend upon the choice of the sequence* $\{t_\nu\}_{\nu\geq 1}$.
Proof of Claim 1. Let $\{t'_\nu\}_{\nu\geq 1}$ be another sequence such that

$$\{t'_\nu\}_{\nu\geq 1} \subset I, \quad t'_\nu < t'_{\nu+1}, \quad \lim_{\nu\to\infty} t'_\nu = \ell.$$

Let $\{t''_\nu\}_{\nu\geq 1}$ be the increasing sequence consisting of the terms of the two sequences $\{t_\nu\}_{\nu\geq 1}$ and $\{t'_\nu\}_{\nu\geq 1}$ set together. Then $\lim_{\nu\to\infty} t''_\nu = \ell$. Let p' be a limit point of the Cauchy sequence $\{\alpha(t''_\nu)\}_{\nu\geq 1}$. In any Hausdorff space the subsequences of a convergent sequence converge to the same point. Hence, $\lim_{\nu\to\infty} \alpha\big(t''_\nu\big) = p$, as well. Claim 1 is proved.

Let $B(p, \delta)$ be a normal neighborhood of p, which is a normal neighborhood of each of its points. Let $\big(U,\ x^i\big)$ be a local coordinate system about p, whose domain U contains $B(p, \delta)$. Let us set $\alpha^i(t) = x^i(\alpha(t))$ for all $\alpha(t) \in B(p, \delta)$. Then $\lim_{t\to\ell^-} \alpha^i(t) = x^i(p)$. So if we extend $\alpha : [0, \ell) \to B(p, \delta)$ by setting $\alpha(\ell) := p$, then α is continuous at $t = \ell$. By continuity, there is $0 < \rho < \ell$ such that $\alpha(t) \in B(p, \delta)$ for any $\rho \leq t \leq \ell$. We shall show that α is differentiable at $t = \ell$.

The functions $d\alpha^i/dt$ are bounded on $[\rho, \ell)$. Let

$$f : [\rho, \ell] \times \mathbb{R}^n \to [0, +\infty), \quad f(t, \xi) = E\left(\xi^i \left(\frac{\partial}{\partial x^i}\right)_{\alpha(t)}\right), \quad \delta \le t \le \ell, \quad \xi \in S^{n-1}.$$

As f is continuous and $[\rho, \ell] \times S^{n-1}$ is compact, there is $K > 0$ such that

$$0 < K \le f(t, \xi), \quad \rho \le t \le \ell, \quad \xi \in S^{n-1}.$$

Let $\rho \le t < \ell$ and let

$$\xi = \frac{1}{\sqrt{\sum_{i=1}^{n} \left[\frac{d\alpha^i}{dt}(t)\right]^2}} \left(\frac{d\alpha^1}{dt}(t), \cdots, \frac{d\alpha^n}{dt}(t)\right) \in S^{n-1}.$$

Then [as E is positively homogeneous of degree 2 and α is parametrized by arc length, hence $E(\dot{\alpha}(t)) = 1$]

$$K < E\left(\xi^i \left(\frac{\partial}{\partial x^i}\right)_{\alpha(t)}\right) = \frac{E(\dot{\alpha}(t))}{\sum_{i=1}^{n} \left[\frac{d\alpha^i}{dt}(t)\right]^2} = \frac{1}{\sum_{i=1}^{n} \left[\frac{d\alpha^i}{dt}(t)\right]^2}$$

so that

$$\left|\frac{d\alpha^i}{dt}(t)\right| \le \left\{\sum_{i=1}^{n} \left[\frac{d\alpha^i}{dt}(t)\right]^2\right\}^{1/2} < \frac{1}{\sqrt{K}}.$$

Q.e.d. Next, the functions $N_0^i = N_j^i y^j$ are positively homogeneous of degree 2 and continuous, hence

$$\left|N_0^i(\dot{\alpha}(t))\right| \le A$$

for some constant $A > 0$ and any $\rho \le t < \ell$. Then, by inspecting the geodesics differential system

$$\frac{d^2\alpha^i}{dt^2}(t) + N_j^i(\dot{\alpha}(t)) \frac{d\alpha^j}{dt}(t) = 0 \tag{2.68}$$

we may conclude that the functions $\frac{d^2\alpha^i}{dt^2}$ are bounded. The function $\frac{d\alpha^i}{dt}$ is uniformly continuous near ℓ, hence it has a limit as $t \to \ell$. By Lagrange's (mean-value) theorem

$$\frac{\alpha^i(t) - \alpha^i(\ell)}{t - \ell} = \frac{d\alpha^i}{dt}\big(\ell + \theta(t - \ell)\big)$$

for some $0 \leq \theta \leq 1$, implying that the left derivative $\dfrac{d\alpha^i}{dt}(\ell)$ satisfies

$$\frac{d\alpha^i}{dt}(\ell) = \lim_{t \to \ell} \frac{d\alpha^i}{dt}(t). \tag{2.69}$$

Therefore, again from the geodesics differential equations, $\dfrac{d^2\alpha^i}{dt^2}(t)$ has a limit as $t \to \ell$. Then, again by Lagrange's theorem, the left derivative $\dfrac{d^2\alpha^i}{dt}(\ell)$ satisfies

$$\frac{d^2\alpha^i}{dt^2}(\ell) = \lim_{t \to \ell} \frac{d^2\alpha^i}{dt^2}. \tag{2.70}$$

The tangent vector

$$v = \frac{d\alpha^i}{dt}(\ell)\left(\frac{\partial}{\partial x^i}\right)_{\alpha(\ell)}$$

satisfies $E(v) = 1$. Let $\sigma(t)$ be the geodesic of initial vector v defined for $\ell \leq t \leq \ell + \epsilon$. The mapping

$$t \longmapsto \Gamma(t) = \begin{cases} \alpha(t) & \text{if } 0 \leq t \leq \ell, \\ \\ \sigma(t) & \text{if } \ell \leq t \leq \ell + \epsilon, \end{cases}$$

satisfies the geodesics differential system for every $\rho \leq t \leq L + \epsilon$. By (2.69)–(2.70) equations (2.68) are satisfied by the left derivative at $t = \ell$. Then Γ is a geodesic, in contradiction with the maximality of α. Q.e.d.

Corollary 2.72 *Every compact Finslerian manifold is complete.*

2.5 First and Second Variations of the Energy Function

2.5.1 The Energy Function

Let (M, E) be a n-dimensional Finslerian manifold. Let $p, q \in M$. A *piecewise differentiable* curve is a map $\gamma : [0, 1] \to M$ such that there is a partition $0 = t_0 < t_1 < \cdots < t_N = 1$ such that $\gamma\big|_{[t_{i-1}, t_i]}$ is differentiable of class C^∞, $1 \leq i \leq N$, and $\gamma(0) = p$, $\gamma(1) = q$. Let $\Omega = \Omega(M;\ p, q)$ be the set of all piecewise differentiable curves from p to q.

Definition 2.73 The *tangent space* $T_\gamma(\Omega)$ at $\gamma \in \Omega$ consists of all piecewise differentiable vector fields W along γ such that $W(0) = 0$ and $W(1) = 0$. □

Definition 2.74 The *energy* of a path $\gamma \in \Omega$ from a to b, where $0 \leq a < b \leq 1$, is

$$E_a^b(\gamma) = \int_a^b E\left[\frac{d\gamma}{dt}(t)\right] dt.$$

We set merely $E = E_0^1$ when $a = 0$ and $b = 1$. □

We shall also make use of the arc length from a to b

$$L_a^b(\gamma) = \int_a^b L\left[\frac{d\gamma}{dt}(t)\right] dt\,.$$

By the Schwartz inequality

$$\left(\int_a^b fg\, dt\right)^2 \leq \left(\int_a^b f^2\, dt\right)\left(\int_a^b g^2\, dt\right)$$

with $f(t) = 1$ and $g(t) = E[(d\gamma/dt)(t)]$

$$L_a^b(\gamma)^2 \leq (b-a)\, E_a^b(\gamma)$$

with equality if and only if $L[(d\gamma/dt)(t)]$ is constant (equivalently the arc length is a linear function of the parameter t).

If there is a minimal geodesic α from p to q, then

$$E(\alpha) = L(\alpha)^2 \leq L(\gamma)^2 \leq E(\gamma).$$

The equality $L(\alpha)^2 = L(\gamma)^2$ holds if and only of γ is also a minimal geodesic, eventually reparametrized. Yet, as noticed above, $L^2(\gamma) = E(\gamma)$ if and only if the parameter is proportional to the arc length along γ. Therefore, $E(\alpha) < E(\gamma)$ unless γ is also a minimal geodesic.

2.5.2 *Critical Points of the Energy Function*

Let $\omega \in \Omega = \Omega(M;\ p, q)$.

Definition 2.75 An *n-parameter variation* of ω, keeping the end points fixed, is a map

$$\alpha^* : U \to \Omega(M;\ p, q)$$

where $U \subset \mathbb{R}^n$ is a neighborhood of the origin $0 \in \mathbb{R}^n$ such that

(i) $\alpha^*(0) = \omega$,

(ii) There is a partition of $0 = t_0 < t_1 < \cdots < t_N = 1$ of $[0, 1]$ such that the map

$$\alpha : U \times [0,1] \to M, \quad \alpha\big(u^1, \cdots, u^n, t\big) = \alpha^*\big(u^1, \cdots, u^n\big)(t),$$

is differentiable of class C^∞ on each strip $U \times [t_{i-1}, t_i]$, $1 \le i \le N$. □

Let $\epsilon > 0$ and let $\alpha^* : (-\epsilon, \epsilon) \to \Omega$ be a 1-parameter variation of ω (a *differentiable path* in Ω).

Definition 2.76 The *velocity vector* of α^* is the vector field $W = \dfrac{d\alpha^*}{du}(0) \in T_\omega(\Omega)$ along ω given by

$$W(t) = \frac{\partial \alpha}{\partial u}(0, t).$$

The vector field $W \in T_\omega(\Omega)$ is also called the *infinitesimal variation* induced by the variation α^* of ω. □

Viceversa, for any $W \in T_\omega(\Omega)$ there is a 1-parameter variation $\alpha^* : (-\epsilon, \epsilon) \to \Omega$ such that $\alpha^*(0) = \omega$ and $\dfrac{d\alpha^*}{du}(0) = W$.

The energy function E, thought of as a real-valued function on Ω, induces the map

$$E_* : T_\omega(\Omega) \to T_{E(\omega)}(\mathbb{R})$$

built as follows. For every $W \in T_\omega(\Omega)$, let α^* be a 1-parameter variation α^* such that $\alpha^*(0) = \omega$ and $\dfrac{d\alpha^*}{du}(0) = W$. Then we set

$$E_*(W) := \left\{ \frac{d\, E\big(\alpha^*(u)\big)}{du} \right\}_{u=0} \left(\frac{d}{dt}\right)_{E(\omega)}.$$

Definition 2.77 A path $\omega \in \Omega$ is a *critical point* of the energy function if

$$\left\{ \frac{d\, E\big(\alpha^*(u)\big)}{du} \right\}_{u=0} = 0$$

for every 1-parameter variation α^* of ω. □

A basic ingredient in the study of critical points of the energy function is the first variation formula. Let $\alpha^* : (-\epsilon, \epsilon) \to \Omega$ be a 1-parameter variation of the piecewise regular path $\omega \in \Omega(M;\ p,\ q)$ and let W be the induced infinitesimal variation. Let $\tilde{\alpha} : (-\epsilon, \epsilon) \times [0,1] \to V(M)$ be the lift of $\alpha : (-\epsilon, \epsilon) \times [0,1] \to M$, i.e.,

$$\tilde{\alpha}(u, t) = \frac{\partial \alpha}{\partial t}(u, t) = (d_{(u,t)}\alpha)\left(\frac{\partial}{\partial t}\right)_{(u,t)}.$$

Let

$$\overline{W} : [0, 1] \to \pi^{-1}T(M), \quad \overline{V} = \frac{\overline{d\omega}}{dt} : [0, 1] \to \pi^{-1}T(M),$$

be the Finslerian vector fields along ω got as the natural lifts of W and $d\omega/dt$, i.e.,

$$\overline{W}(t) = W(t) \in T_{\omega(t)}(M) = \left(\pi^{-1}TM\right)_{\tilde{\omega}(t)},$$

$$\overline{V}(t) = \frac{d\omega}{dt}(t) \in T_{\omega(t)}(M) = \left(\pi^{-1}TM\right)_{\tilde{\omega}(t)}.$$

Theorem 2.78 (*First variation formula*)

$$\frac{1}{2}\left\{\frac{d\,E\big(\alpha^*(u)\big)}{du}\right\}_{u=0} = -\sum_{i=1}^{N-1} g_{\tilde{\omega}(t)}\big(\overline{V}(t),\, \overline{V}(t)\big)\Big|_{t_i^-}^{t_i^+} - \int_0^1 g_{\tilde{\omega}(t)}\left(\overline{W}(t),\, \frac{\nabla\overline{V}}{dt}\right) dt.$$

Proof The energy of the curve $\alpha^*(u)(t) = \alpha(u, t)$ is

$$E\big(\alpha^*(u)\big) = \int_0^1 \left\|\frac{\overline{\partial\alpha}}{\partial t}(u, t)\right\|^2 dt.$$

Therefore,

$$\frac{d\,E(\alpha^*(u))}{du} = \frac{d}{du}\int_0^1 g_{\tilde{\alpha}(u,t)}\Big(\frac{\overline{\partial\alpha}}{\partial t}(u, t),\, \frac{\overline{\partial\alpha}}{\partial t}(u, t)\Big)\, dt.$$

We shall make use of the identities

$$\frac{\partial}{\partial u}\left[g_{\tilde{\alpha}(u,t)}\left(\frac{\overline{\partial\alpha}}{\partial t}(u, t),\, \frac{\overline{\partial\alpha}}{\partial t}(u, t)\right)\right] = 2\, g_{\tilde{\alpha}(u,t)}\left(\frac{\nabla}{\partial u}\left(\frac{\overline{\partial\alpha}}{\partial t}\right)(u, t),\, \frac{\overline{\partial\alpha}}{\partial t}(u, t)\right),$$

$$\frac{\nabla}{\partial u}\left(\frac{\overline{\partial\alpha}}{\partial t}\right) = \frac{\nabla}{\partial t}\left(\frac{\overline{\partial\alpha}}{\partial u}\right) + T_L\left(\frac{\partial\tilde{\alpha}}{\partial u},\, \frac{\partial\tilde{\alpha}}{\partial t}\right).$$

Let $0 = t_0 < t_1 < \cdots < t_N = 1$ be a partition of the interval $[0, 1]$ such that α is differentiable of class C^∞ on every strip $(-\epsilon, \epsilon) \times [t_{i-1}, t_i]$. We integrate by parts on each interval $[t_{i-1}, t_i]$, in the following manner. One starts from

$$\frac{\partial}{\partial t}\left[g\left(\frac{\overline{\partial\alpha}}{\partial u},\, \frac{\overline{\partial\alpha}}{\partial t}\right)\right] = g\left(\frac{\nabla}{\partial t}\left(\frac{\overline{\partial\alpha}}{\partial u}\right),\, \frac{\overline{\partial\alpha}}{\partial t}\right) + g\left(\frac{\overline{\partial\alpha}}{\partial u},\, \frac{\nabla}{\partial t}\left(\frac{\overline{\partial\alpha}}{\partial t}\right)\right)$$

so that

$$\int_{t_{i-1}}^{t_i} g\left(\frac{\nabla}{\partial t}\left(\frac{\overline{\partial}\alpha}{\partial u}\right), \frac{\overline{\partial}\alpha}{\partial t}\right) dt = g\left(\frac{\overline{\partial}\alpha}{\partial u}, \frac{\overline{\partial}\alpha}{\partial t}\right)\Big|_{t_{i-1}}^{t_i} - \int_{t_{i-1}}^{t_i} g\left(\frac{\overline{\partial}\alpha}{\partial u}, \frac{\nabla}{\partial t}\left(\frac{\overline{\partial}\alpha}{\partial t}\right)\right) dt.$$

Since $\frac{\overline{\partial}\alpha}{\partial u}(u, 0) = 0$ and $\frac{\overline{\partial}\alpha}{\partial u}(u, 1) = 0$ one has [by summing over i in the previous identity]

$$\int_0^1 g\left(\frac{\nabla}{\partial t}\left(\frac{\overline{\partial}\alpha}{\partial u}\right), \frac{\overline{\partial}\alpha}{\partial t}\right) dt = -\sum_{i=1}^{N-1} g\left(\frac{\overline{\partial}\alpha}{\partial u}, \frac{\overline{\partial}\alpha}{\partial t}\right)\Big|_{t_i^-}^{t_i^+} - \int_0^1 g\left(\frac{\overline{\partial}\alpha}{\partial u}, \frac{\nabla}{\partial t}\left(\frac{\overline{\partial}\alpha}{\partial t}\right)\right) dt.$$

Finally, let us set $u = 0$ and use

$$g_{\tilde{\alpha}(0,t)}\left(T_L\left(\frac{\partial\tilde{\alpha}}{\partial u}(0, t), \frac{\partial\tilde{\alpha}}{\partial t}(0, t)\right), \frac{\overline{\partial}\alpha}{\partial t}(0, t)\right) = 0$$

[holding because $(\overline{\partial}\alpha/\partial t)(0, t)$ is precisely the restriction of the Liouville vector field to $\tilde{\alpha}(0, t)$. Q.e.d.

Corollary 2.79 *A path $\omega \in \Omega$ is a critical point of the energy function if and only if ω is a geodesic.*

2.6 A Postface on Finslerian Geometry

The present chapter was devoted merely to a few introductory notions in Finslerian geometry, treated by mainly following the very inspiring Ph.D. thesis of B.T.M. Hassan, [Has], and the authors relegate a more involved presentation of Finslerian geometry, old and new, to a further volume. A very good history of Finslerian geometry, starting with the work by Bernhard Riemann (1826–1866) and with the romantic contribution by Paul Finsler (1894–1970), and reporting on the growth of the subject up to the year 1980, was written by M. Matsumoto, as a postface to his monograph [Mat]. According to [Mat], the main character, to whom the credit for setting the foundations of Finslerian geometry is actually due, is Ludwig Berwald (1993–1942). L. Berwald's main contribution[6] seems to be the geometric treatment of the variational problem $\delta \int L\big(x(s), \dot{x}(s)\big)\, ds = 0$ relying on a Finslerian connection built for the purpose (and nowadays referred to as the *Berwald connection*, cf. e.g., Definition 18.2 in [Mat], p. 117). To give correct historical credit, one should keep in mind that the work by L. Berwald quoted previous foundational contributions to

[6] Cf. L. Berwald, *Untersuchung der Krümmung allgemeiner metrischer Räume auf Grund des in ihnen herrschenden Parallelism*, Math. Z., 25(1926), 40–73.

Finslerian geometry by John Lighten Synge[7] and by James Henry Taylor.[8] The firm axiomatic presentation of what was to be from there on the *Theory of Finsler Spaces* was exhibited by Élie Cartan[9] (1869–1951). The history of the subject after 1950 as interpreted in [Mat] recognizes the relevance of a few Western contributors, e.g., A. Deicke,[10] R.S. Ingarden,[11] and F. Brickell.[12] The major role played by the contributions of the French school, to which we shall come in a moment, isn't fully grasped by [Mat] and only the scientific creation by Hassan Akbar-Zadeh[13] is discussed in [Mat] at a certain extent,[14]. Although quoted in [Mat], the works by A. Lichnerowicz[15] P. Dazord,[16] and J. Grifone,[17] whose review should be *sine-qua-non* to any tentative to understand the development of Finslerian geometry, are only marginally discussed. A sequel to the treatment of modern Finslerian geometry in the present chapter will start mainly from P. Dazord's thesis (cf. *op. cit.*) and further published work[18] and, faithful to the more PDE-oriented standpoint in the forthcoming volumes of the present book, will mainly aim at the study of natural generalizations

---

[7] Cf. J.L. Synge, *A generalization of the Riemannian line element*, Trans. Amer. Math. Soc., 27(1925), 61–67.

[8] Cf. J.H. Taylor, *A generalization of Levi-Civita's parallelism and the Frenet formulas*, Trans. Amer. Math. Soc., 27(1925), 246–264.

[9] Cf. É. Cartan, *Les espaces de Finsler*, Actualités 79, Paris, 1934.

[10] Cf. e.g., A. Deicke, *Finsler spaces as non-holonomic subspaces of Riemannian spaces*, J. London Math. Soc., 30(1955), 53–58.

[11] Cf. R.S. Ingarden, *On the geometrically absolute optical representation in the electron microscope*, Trav. Soc. Sci. Lett. Wroclaw, 45(1957), 60 pp.; *Differential geometry and physics*, Tensor, N.S., 39(1976), 201–209. The possibility of using Finslerian geometry as a geometric basis for the treatment of certain problems in mathematical physics was to lead (in the second half of the twentieth century) to a huge amount of work (on which we partially report in the sequel volumes to the present book).

[12] Cf. F. Brickell, *A new proof of Deicke's theorem on homogeneous functions*, Proc. Amer. Math. Soc., 16(1965), 190–191; *A theorem on homogeneous functions*, J. London Math. Soc., 42(1967), 325–329.

[13] Cf. H. Akbar-Zadeh, *Les espaces de Finsler et certaines de leurs généralisations*, Ann. Sci. École Norm. Sup., (3)80(1963), 1–79.

[14] The reason for that is perhaps to be looked for within M. Matsumoto's interest in the realization of Finslerian geometry as a chapter of the more modern (in comparison with local tensor calculus) theory of fiber bundles, and connections in fiber bundles, whether principal or vector bundles.

[15] Cf. e.g., A. Lichnerowicz, *Sue une extension de la formule d'Allendoerfer-Weil à certains variétés finslériennes*, C.R. Acad. Sci. Paris, 223(1946), 12–14.

[16] Cf. e.g., P. Dazord, *Propriétés globales des géodésiques des espaces de Finsler*, Thèse, Lyon, 1969, 193 pp.

[17] Cf. e.g., J. Grifone, *Sur les connexions d'une variété finslérienne et d'un système méchanique*, C.R. Acad. Sci. Paris, 272(1971), 1510–1513.

[18] Cf. P. Dazord, *Sur une généralisation de la notion de "spray"*, C.R. Acad. Sci. Paris, 263(1966), 543–546; *Connexion de direction symmétrique associée à un "spray" généralisé*, Ibidem, 263(1966), 576–578; *Variétés finslériennes à géodésiques fermées*, Ibidem, 266(1968), 348–350; *Variétés finslériennes de dimension paire δ-pincées*, Ibidem, 266(1968), 496–498; *Variétés finslériennes en forme de sphères*, Ibidem, 267(1968), 353–355; *Sue la formule de Gauss-Bonnet en géométrie finslérienne*, Ibidem, 270(1970), 1241–1243.

(from Riemannian to Finslerian geometry) of the Leplace-Beltrami operator, such as produced by Paul Centore,[19], Brad Lackey[20] David Bao & Brad Lackey,[21] and (last but not the least) Zhongmin Shen.[22] It is noteworthy that the canonical Finslerian connection (the *Chern connection*) devised by Shiing-Shen Chern hasn't been discussed in [Mat] although quoted there.[23] Despite S-S. Chern repeated attempts,[24] in the last years of his long and proficuous mathematical life, to render more popular his discovery of the Chern connection, it wasn't until the arrival of Zhongmin Shen[25] at Berkley that the relevance of the Chern connection was recognized.[26] Again staying at the cultural preferences of the Authors of the present book, the discussion of Finslerian Laplacians will be followed by a report on the recent works recovering[27] (from Riemannian to Finslerian geometry) elements of the theory of harmonic maps between Riemannian manifolds. Ample portions of the first volume of the present book are concerned with the theory of characteristic classes and, as a reflection of that, the sequel to this volume will regard the characteristic classes of Finslerian manifolds, and in particular the characteristic classes of the low-dimensional special Finsler spaces (cf. [Mat]) by starting from Z. Kovács attempt[28] to bring the

---

[19] Cf. P. Centore, *A mean-value Laplacian for Finsler spaces*, Ph.D. Thesis, University of Toronto, 1998.

[20] Cf. B. Lackey, *A Bochner vanishing theorem for elliptic complexes*, preprint.

[21] Cf. D. Bao & B. Lackey, *A geometric inequality and a Weitzenböck formula for Finsler surfaces*, preprint; *A Hodge decomposition theorem for Finsler spaces*, C.R. Acad. Sci. Paris, 323(1996), 51–56.

[22] Cf. Z. Shen, *The nonlinear Laplacian for Finsler manifolds*, preprint.

[23] Cf. S-S. Chern, *On the Euclidean connections in a Finsler space*, Proc. Nat. Acad. Sci. U.S.A., 29(1943), 33–37; *Local equivalence and Euclidean connections in Finsler spaces*, Sci. Rep. Nat. Tsing Hua Univ., 5(1948), 95–121.

[24] Cf. S-S. Chern *Riemannian geometry as a special case of Finsler geometry*, Cont. Math., 196(1996), 51–58; *Finsler geometry is just Riemannian geometry without the quadratic restriction*, Notices of Amer. Math. Soc., 43(1996), 959–963.

[25] At the time a young and highly gifted Ph.D. recipient from Stony Brook University, whose field of expertise was Riemannian geometry.

[26] Cf. also the monograph by D. Bao & S-S. Chern & Z. Shen, *An introduction to Riemann-Finsler geometry*, Lecture Notes in Math., Springer-Verlag, 2000.

[27] Cf. e.g., A. Tachikawa, *A partial regularity result for harmonic maps into a Finsler manifold*, preprint; X. Mo, *Harmonic maps from Finsler spaces*, preprint; J. Li, *Some properties of p-harmonic maps for Finsler manifolds*, preprint; Y. Shen & Y. Zhang, *Second variation of harmonic maps between Finsler manifolds*, Science in China, Ser A, Mathematics, (1)47(2004), 39–51; X. Mo & Y. Yang, *The existence of harmonic maps from Finsler manifolds to Riemannian manifolds*, Science in China, Ser. A, Mathematics, (1)48(2005), 115–130; Y. Shen & S.W. Wei, *The stability of harmonic maps on Finsler manifolds*, preprint; J. Li, *Stable $\mathscr{F}$-harmonic maps for Finsler manifolds*, preprint.

[28] Cf. Z. Kovács, *On the Chern-Weil homomorphism in Finsler spaces*, Acta Mathematica Academiae Paedagogicae Nyíregyháziensis, 17(2001), 131–135.

Chern-Weil homomorphism to the realm of Finslerian geometry. It should be mentioned that the fundamental work by S-S. Chern & J. Simons[29] on characteristic classes and their applications to Riemannian geometry, is still awaiting for an appropriate generalization to Finslerian geometry.

References

[Kaw] A. Kawaguchi, On the theory of nonlinear connections. Introduction to the theory of general nonlinear connections. Tensor, N.S., **2**, 123–142 (1952); On the theory of nonlinear connections. Theory of Minkowski spaces and of nonlinear connections in a Finsler space. Tensor, N.S. **6**, 165–199 (1956)

[Has] B.T.M. Hassan, *The Theory of Geodesics in Finsler Spaces* (Southampton, Thesis, 1967), p.108

[Mat] M. Matsumoto, Foundations of Finsler Geometry and Special Finsler Spaces (College of Liberal Arts, Kyoto University, Kyoto, Department of Mathematics, 1980)

29 Cf. S-S. Chern & J. Simons, *Characteristic forms and geometric invariants*, Ann. of Math., 99(1974), 48–69; *Some cohomology classes in principal fiber bundles and their application to Riemannian geometry*, Proc. Nat. Acad. Sci. U.S.A., (4)68(1971), 71–794.

Chapter 3
Isometric Immersions

3.1 Fundamental Forms of an Immersed Manifold

Let M be a n-dimensional differentiable manifold and let $f : M \to \overline{M}$ be an immersion of M in a semi-Riemannian manifold $(\overline{M}, \overline{g})$ of dimension $n + r$.

Definition 3.1 The manifold $\overline{M}$ is commonly referred to as the *ambient space*. The manifold M is said to be an *(immersed) submanifold* of $\overline{M}$. □

The semi-Riemannian metric $\overline{g}$ induces a tensor field of type (0, 2) on M given by

$$g_p(v, w) = \overline{g}_{f(p)}((d_p f)v, (d_p f)w), \quad v, w \in T_p(M), \quad p \in M.$$

For every point $p \in M$ we have an inner product space $(T_{f(p)}(\overline{M}), \overline{g}_{f(p)})$ and a subspace $(d_p f)T_p(M)$. Clearly

$$\dim_{\mathbb{R}}(d_p f)T_p(M) = n$$

as the map $d_p f : T_p(M) \to T_{f(p)}(\overline{M})$ is linear and injective.

From now on we assume that for any point $p \in M$ the subspace $(d_p f)T_p(M)$ is nondegenerate in $(T_{f(p)}(\overline{M}), \overline{g}_{f(p)})$, i.e., $\overline{g}_{f(p)}\big|_{(d_p f)T_p(M)}$ is an inner product on $(d_p f)T_p(M)$. Then the space

$$T_p(M)^\perp \equiv \left[(d_p f)T_p(M)\right]^\perp = \{v \in T_{f(p)}(\overline{M}) : v \perp (d_p f)T_p(M)\}$$

is a complement of $(d_p f)T_p(M)$, i.e.,

$$T_{f(p)}(\overline{M}) = \left[(d_p f)T_p(M)\right] \oplus T_p(M)^\perp, \quad p \in M. \tag{3.1}$$

E. Barletta et al., *Differential Geometry*, Infosys Science Foundation Series,
https://doi.org/10.1007/978-981-96-1631-2_3

Clearly

$$\dim_{\mathbb{R}} T_p(M)^{\perp} = r.$$

Moreover, by the preceding assumption [that $(d_p f)T_p(M)$ be nondegenerate for any $p \in M$] it follows that g_p is an inner product on $T_p(M)$ for any point $p \in M$.

Exercise 3.2 Show that the assignment

$$p \in M \longmapsto g_p \in T_p^*(M) \otimes T_p^*(M)$$

defines a differentiable map $g : M \to \mathcal{T}^{0,2}(M)$.

If M is connected, an assumption we shall always adopt, then

$$\text{ind}\left(T_p(M), g_p\right) = \text{ costant},$$

hence g is a semi-Riemannian metric on M.

Definition 3.3 The semi-Riemannian metric g on M is called the *first fundamental form* of the immersion $f : M \to \overline{M}$. □

Definition 3.4 Let (M, g) and $(\overline{M}, \overline{g})$ be two semi-Riemannian manifolds [with $\dim(M) = n$ and $\dim(\overline{M}) = n + r$]. A differentiable map $f : M \to \overline{M}$ is called an *isometric immersion* if

(i) f is an immersion and
(ii) $\overline{g}_{f(p)}((d_p f)v, (d_p f)w) = g_p(v, w)$ for any $v, w \in T_p(M)$ and any $p \in M$.

□

If $f : M \to \overline{M}$ is an immersion of a connected manifold M in the semi-Riemannian manifold $(\overline{M}, \overline{g})$ and one endows M with the first fundamental form g [provided that $(d_p f)T_P(M)$ be nondegenerate in $(T_{f(p)}(\overline{M}), \overline{g}_{f(p)})$] then $f : M \to \overline{M}$ is an isometric immersion of (M, g) into $(\overline{M}, \overline{g})$.

3.2 Normal Bundle

Definition 3.5 A tangent vector $v \in T_{f(p)}(\overline{M})$ is said to be a *normal vector* at p on M if $v \in T_p(M)^{\perp}$. A *normal field* on M is a map $\xi : M \to T(\overline{M})$ such that $\xi_p \in T_p(M)^{\perp}$ for any $p \in M$. □

A posteriori the vector fields $X \in \mathfrak{X}(M)$ are referred to as *tangential fields*. Let $p \in M$. Let us choose a local coordinate system (U, x^i) on M such that $p \in U$ and another local coordinate system $(\overline{U}, y^{\alpha})$ on $\overline{M}$ such that $f(U) \subset \overline{U}$. Moreover let

$$\underline{f} = \overline{\varphi} \circ f \circ \varphi^{-1} : \varphi(U) \to \overline{\varphi}(\overline{U})$$

be the local expression of f where

$$\varphi = (x^1, \ldots, x^n), \quad \overline{\varphi} = (y^1, \ldots, y^{n+r}).$$

Moreover let us set $\underline{f} = (\underline{f}^1, \ldots, \underline{f}^{n+r})$ and $f^\alpha = y^\alpha \circ f$ for any $1 \leq \alpha \leq n + r$ so that $\underline{f}^\alpha = f^\alpha \circ \varphi^{-1}$. In the older language of nineteenth-century local differential geometry, it was customary to say that

$$y^\alpha = \underline{f}^\alpha \circ (x^1, \ldots, x^n), \quad 1 \leq \alpha \leq n + r \tag{3.2}$$

are the *equations of the submanifold M* in $\overline{M}$. With a common abuse of notation Eq. (3.2) may also be written

$$y^\alpha = y^\alpha(x^1, \ldots, x^n).$$

A normal field $\xi : M \to T(\overline{M})$ may be locally written

$$\xi_p = \xi^\alpha(p) \left(\frac{\partial}{\partial y^\alpha} \right)_{f(p)}, \quad p \in U,$$

where the functions $\xi^\alpha : U \to \mathbb{R}$ are uniquely determined by this representation. It may be easily shown that the map $\xi : M \to T(\overline{M})$ is differentiable if and only if $\xi^\alpha \in \mathcal{F}(U)$ for any choice of local coordinate systems (U, x^i) and $(\overline{U}, y^\alpha)$ on M and $\overline{M}$ such that $f(U) \subset \overline{U}$.

Theorem 3.6 *Let us set* $T(M)^\perp \equiv \bigcup_{p \in M} T_p(M)^\perp$. *Then*

$$\tau_M^\perp(f) \equiv (T(M)^\perp, \ \pi^\perp, \ M, \ \mathbb{R}^r)$$

is a vector bundle or rank r, where $\pi^\perp : T(M)^\perp \to M$ *is the canonical projection. Any differentiable normal field* $\xi : M \to T(\overline{M})$ *is a differentiable section in* $\tau_M^\perp(f)$, *i.e.,* $\xi \in C^\infty(\tau_M^\perp(f))$.

Proof Exercise.

Definition 3.7 The vector bundle $\tau_M^\perp(f)$ is called the *normal bundle* of the immersion $f : M \to \overline{M}$. □

One often adopts the notation $\mathfrak{X}(M)^\perp = C^\infty(\tau_M^\perp(f))$.

From now on we assume that $f : M \to \overline{M}$ is an embedding, i.e., f is an injective immersion. Each vector field $X \in \mathfrak{X}(M)$ determines a differentiable section $\hat{X} \in C^\infty(f(M), \tau_{\overline{M}})$ defined on the closed set $f(M) \subset \overline{M}$

$$\hat{X}_{f(p)} = (d_p f) X_p \in T_{f(p)}(\overline{M}), \quad p \in M.$$

As f is injective the tangent vector $\hat{X}_{f(p)}$ is well defined. As $f(M)$ is a closed set there is a C^∞ extension of $\hat{X}$ to the whole of $\overline{M}$, i.e., there is a vector field $\overline{X} \in \mathfrak{X}(\overline{M})$ such that $\overline{X}\big|_{f(M)} = \hat{X}$. The vector field $\overline{X}$ is merely referred to as an *extension* of the field X to $\overline{M}$. Let $\overline{h} \in \mathcal{F}(\overline{M})$ and let $h \equiv \overline{h} \circ f \in \mathcal{F}(M)$. Then for any point $p \in M$

$$\overline{X}(\overline{h})(f(p)) = \overline{X}_{f(p)}(\overline{h}) = \hat{X}_{f(p)}(\overline{h}) =$$
$$= \left[(d_p f)X_p\right](\overline{h}) = X_p(\overline{h} \circ f) = X_p(h) = X(h)(p)$$

and then

$$\overline{X}(\overline{h}) \circ f = X(\overline{h} \circ f), \quad \overline{h} \in \mathcal{F}(\overline{M}). \tag{3.3}$$

Let $\overline{X}, \overline{Y} \in \mathfrak{X}(\overline{M})$ be extensions of the fields $X, Y \in \mathfrak{X}(M)$. Then

$$\left[\overline{X}, \overline{Y}\right]_{f(p)}(\overline{h}) = \overline{X}_{f(p)}\left(\overline{Y}(\overline{h})\right) - \overline{Y}_{f(p)}\left(\overline{X}(\overline{h})\right) =$$

[using the identity (3.3) several times]

$$= X_p\left(\overline{Y}(\overline{h}) \circ f\right) - Y_p\left(\overline{X}(\overline{h}) \circ f\right) =$$

$$= X_p\left(Y(\overline{h} \circ f)\right) - Y_p\left(X(\overline{h} \circ f)\right) = [X, Y]_p(\overline{h} \circ f) =$$

$$= \left\{(d_p f)[X, Y]_p\right\}(\overline{h}) = \widehat{[X, Y]}_{f(p)}(\overline{h})$$

or

$$\left[\overline{X}, \overline{Y}\right]\big|_{f(M)} = \widehat{[X, Y]d} \tag{3.4}$$

hence $[\overline{X}, \overline{Y}]$ is an extension of $[X, Y]$ to $\overline{M}$.

The next lemma plays a first-magnitude role in the construction of the theory of submanifolds of semi-Riemannian manifolds.

Lemma 3.8 *Let $\overline{\nabla} \in \mathcal{C}(\tau_{\overline{M}})$ be a linear connection on $\overline{M}$ and let $X, Y \in \mathfrak{X}(M)$. Then $\left(\overline{\nabla}_{\overline{X}}\overline{Y}\right)_{f(p)}$ doesn't depend on the choice of the extensions $\overline{X}, \overline{Y} \in \mathfrak{X}(\overline{M})$ of the fields X, Y, for any $p \in M$.*

Proof Let us consider the differentiable curve $\gamma : I \to M$ defined on the interval $I \subset \mathbb{R}$. Let us set $\overline{\gamma} = f \circ \gamma : I \to \overline{M}$. Then

$$\dot{\overline{\gamma}}(t_0) = (d_{t_0}\overline{\gamma})\left(\frac{\partial}{\partial t}\right)_{t_0} = \left\{d_{t_0}(f \circ \gamma)\right\}\left(\frac{\partial}{\partial t}\right)_{t_0} =$$

$$= \left\{(d_{\gamma(t_0)}f) \circ (d_{t_0}\gamma)\right\}\left(\frac{\partial}{\partial t}\right)_{t_0} = (d_{\gamma(t_0)}f)\dot{\gamma}(t_0)$$

or

$$\dot{\overline{\gamma}}(t) = (d_{\gamma(t)} f)\dot{\gamma}(t), \quad t \in I. \tag{3.5}$$

Formula (3.5) implies that for any integral curve $\gamma : I \to M$ of the field X, passing through the point p,

$$\overline{\gamma} = f \circ \gamma : I \to \overline{M}$$

is an integral curve of $\overline{X}$ passing through $f(p)$. Indeed

$$\overline{X}_{\overline{\gamma}(t)} = \overline{X}_{f(\gamma(t))} = \hat{X}_{f(\gamma(t))} = (d_{\gamma(t)} f) X_{\gamma(t)} =$$

$$= (d_{\gamma(t)} f)\dot{\gamma}(t) = \dot{\overline{\gamma}}(t)$$

for any $t \in I$. Let now $\overline{X}', \overline{Y}' \in \mathfrak{X}(\overline{M})$ be other extensions of X, Y. Then

$$\left(\overline{\nabla}_{\overline{X}'} \overline{Y}'\right)_{f(p)} = \quad \left[\text{as } \overline{X}'_{f(p)} = \overline{X}_{f(p)}\right]$$

$$= \left(\overline{\nabla}_{\overline{X}} \overline{Y}'\right)_{f(p)} = \left(\nabla_{\dot{\overline{\gamma}}} \overline{Y}'\right)_{f(p)}$$

where $\overline{\gamma} = f \circ \gamma$ and $\gamma : I \to M$ is an integral curve of X passing through p. Finally as

$$\overline{Y}'_{\overline{\gamma}(t)} = \overline{Y}'_{f(\gamma(t))} = \hat{Y}_{f(\gamma(t))} = \overline{Y}_{f(\gamma(t))} = \overline{Y}_{\gamma(t)}, \quad t \in I,$$

and the tangent vector $(\overline{\nabla}_{\dot{\overline{\gamma}}} \overline{Y}')_{\overline{\gamma}(0)}$ is determined by the values of $\overline{Y}'$ along $\overline{\gamma}$, we may conclude that

$$\left(\overline{\nabla}_{\dot{\overline{\gamma}}} \overline{Y}'\right)_{f(p)} = \left(\overline{\nabla}_{\dot{\overline{\gamma}}} \overline{Y}\right)_{f(p)} = \left(\overline{\nabla}_{\overline{X}} \overline{Y}\right)_{f(p)} .$$

Q.e.d.

By Lemma 3.8 the following notation is legitimate

$$\left(\overline{\nabla}_X Y\right)_p = \left(\overline{\nabla}_{\overline{X}} \overline{Y}\right)_{f(p)} . \tag{3.6}$$

3.3 Gauss and Weingarten Formulas

From now on let $\overline{\nabla} \in \mathcal{C}(\tau_{\overline{M}}, \overline{g})$ be the Levi-Civita connection of the semi-Riemannian manifold $(\overline{M}, \overline{g})$. Let

$$\tan_p : T_{f(p)}(\overline{M}) \to T_p(M), \quad \text{nor}_p : T_{f(p)}(\overline{M}) \to T_p(M)^{\perp}$$

be the canonical projections associated to the direct sum decomposition (3.1). If $v \in T_{f(p)}(\overline{M})$ then

$$v = (d_p f)\tan_p v + \mathrm{nor}_p v\,.$$

Definition 3.9 $\tan_p v$ and $\mathrm{nor}_p v$ are respectively the *tangential component* and the *normal component* of v. □

As $(\overline{\nabla}_X Y)_p \in T_{f(p)}(\overline{M})$ we may set

$$(\nabla_X Y)_p = \tan_p \left(\overline{\nabla}_X Y\right)_p, \quad B(X,Y)_p = \mathrm{nor}\left(\overline{\nabla}_X Y\right)_p,$$

so that

$$(\overline{\nabla}_X Y)_p = (d_p f)(\nabla_X Y)_p + B(X,Y)_p \tag{3.7}$$

for any $X, Y \in \mathfrak{X}(M)$ and any $p \in M$. This yields a tangential field and a normal field on M

$$\nabla_X Y : M \to T(M), \quad B(X,Y) : M \to T(\overline{M}).$$

The identity (3.7) may be rewritten as

$$\overline{\nabla}_X Y = f_* \nabla_X Y + B(X,Y). \tag{3.8}$$

Here, for any $X \in \mathfrak{X}(M)$, we have adopted the notation

$$f_* X \in C^\infty(f^{-1}T(\overline{M}))$$

to indicate the section in the pullback bundle $f^{-1}T(\overline{M}) \to M$ given by

$$f_* X : M \to f^{-1}T(\overline{M}),$$

$$(f_* X)(p) = (d_p f)X_p \in T_{f(p)}(\overline{M}) = (f^{-1}T(\overline{M}))_p, \quad p \in M.$$

Clearly [by (3.6)]

$$\overline{\nabla}_X Y,\ B(X,Y) \in C^\infty(f^{-1}T(\overline{M})).$$

Hence (3.8) is an identity in $C^\infty(f^{-1}T(\overline{M}))$.

Theorem 3.10 *Let $f : M \to \overline{M}$ be an embedding of the manifold M into the semi-Riemannian manifold $(\overline{M}, \overline{g})$. Let g be the first fundamental form of $f : M \to \overline{M}$. Then*

(i) *The map $\nabla : \mathfrak{X}(M) \times \mathfrak{X}(M) \to \mathfrak{X}(M)$ is the Levi-Civita connection of the semi-Riemannian manifold (M, g).*

(ii) *The map $B : \mathfrak{X}(M) \times \mathfrak{X}(M) \to \mathfrak{X}(M)^\perp$ is $\mathcal{F}(M)$-bilinear and symmetric.*

Definition 3.11 ∇ and B are respectively called the *induced connection* and the *second fundamental form* of the embedding $f : M \to \overline{M}$. Also (3.8) is the *Gauss formula* (of the embedding f). □

Proof of Theorem 3.10 We shall need the identities

$$\overline{\nabla}_{hX} Y = h\overline{\nabla}_X Y, \tag{3.9}$$

$$\overline{\nabla}_X (hY) = X(h) f_* Y + h\overline{\nabla}_X Y, \tag{3.10}$$

for any $X, Y \in \mathfrak{X}(M)$ and $h \in \mathcal{F}(M)$. To prove (3.9) let $\hat{h} : f(M) \to \mathbb{R}$ be the function defined by $\hat{h}(f(p)) = h(p)$ per ogni $p \in M$. As $\hat{h}$ is differentiable on the closed set $f(M)$ one may extends $\hat{h}$ smoothly to the whole of $\overline{M}$, i.e., there is $\overline{h} \in \mathcal{F}(\overline{M})$ such that $\overline{h}\big|_{f(M)} = \hat{h}$. In particular $\overline{h} \circ f = h$. Then, for an arbitrary extension $\overline{X} \in \mathfrak{X}(\overline{M})$ of X, the vector field $\overline{hX}$ is an extension of hX so that

$$(\overline{\nabla}_{hX} Y)_p = (\overline{\nabla}_{\overline{h}\,\overline{X}} \overline{Y})_{f(p)} =$$

$$= \left\{\overline{h}\,\overline{\nabla}_{\overline{X}} \overline{Y}\right\}_{f(p)} = h(p)(\overline{\nabla}_X Y)_p \,.$$

Analogously

$$(\overline{\nabla}_X (hY))_p = (\overline{\nabla}_{\overline{X}} (\overline{h}\,\overline{Y}))_{f(p)} = \left\{\overline{X}(\overline{h})\overline{Y} + \overline{h}\,\overline{\nabla}_{\overline{X}} \overline{Y}\right\}_{f(p)} =$$

$$= X_p(\overline{h} \circ f) Y_p + h(p)(\overline{\nabla}_X Y)_p \,.$$

Exploiting the identities (3.9)–(3.10) one may easily show that ∇ is a linear connection and that B is $\mathcal{F}(M)$-bilinear. For instance

$$f_* \nabla_{hX} Y + B(hX, Y) = \overline{\nabla}_{hX} Y = h\overline{\nabla}_X Y =$$

$$= h\{f_* \nabla_X Y + B(X, Y)\}$$

and then (by the uniqueness of the direct sum decomposition)

$$\nabla_{hX} Y = h\nabla_X Y, \quad B(hX, Y) = hB(X, Y).$$

Let $X, Y \in \mathfrak{X}(M)$ and let $\overline{X}, \overline{Y} \in \mathfrak{X}(\overline{M})$ be extensions of X and Y to $\overline{M}$. As $\overline{\nabla}$ is a symmetric connection, for any $p \in M$

$$0 = \overline{T}(\overline{X}, \overline{Y})_{f(p)} = \left\{\overline{\nabla}_{\overline{X}} \overline{Y} - \overline{\nabla}_{\overline{Y}} \overline{X} - [\overline{X}, \overline{Y}]\right\}_{f(p)} =$$

[by (3.4) together with (3.6)]

$$= (\overline{\nabla}_X Y)_p - (\overline{\nabla}_Y X)_p - \widehat{[X, Y]}_{f(p)} =$$

[by Gauss formula]

$$= (d_p f)(\nabla_X Y)_p + h(X, Y)_p - (d_p f)(\nabla_Y X)_p - h(Y, X)_p - (d_p f)[X, Y]_p =$$

$$= (d_p f)T(X, Y)_p + \{h(X, Y) - h(Y, X)\}_p$$

hence

$$T(X, Y) = 0, \quad h(X, Y) = h(Y, X), \quad X, Y \in \mathfrak{X}(M).$$

One may prove similarly that g is parallel with respect to ∇. Indeed let $X, Y, Z \in \mathfrak{X}(M)$ and let $\overline{X}, \overline{Y}, \overline{Z} \in \mathfrak{X}(M)$ be extensions of X, Y, Z to $\overline{M}$. One has

$$\overline{g}(\overline{Y}, \overline{Z})_{f(p)} = \overline{g}_{f(p)}(\hat{Y}_{f(p)}, \hat{Z}_{f(p)}) =$$

$$= \overline{g}_{f(p)}((d_p f)Y_p, (d_p f)Z_p) = g_p(X_p, Y_p)$$

or

$$\overline{g}(\overline{Y}, \overline{Z}) \circ f = g(Y, Z). \tag{3.11}$$

By the identity (3.11) it follows that $\overline{g}(\overline{Y}, \overline{Z}) \in \mathcal{F}(\overline{M})$ is a smooth extension to the whole of $\overline{M}$ of the function $\widehat{g(Y, Z)}$. Moreover [as $\overline{\nabla}\overline{g} = 0$] for any $p \in M$

$$0 = \left(\overline{\nabla}_{\overline{X}}\overline{g}\right)(\overline{Y}, \overline{Z})_{f(p)} =$$

$$= \left\{\overline{X}\left(\overline{g}(\overline{Y}, \overline{Z})\right) - \overline{g}\left(\overline{\nabla}_{\overline{X}}\overline{Y}, \overline{Z}\right) - \overline{g}\left(\overline{Y}, \overline{\nabla}_{\overline{X}}\overline{Z}\right)\right\}_{f(p)} =$$

[by (3.3)]

$$= X_p(\overline{g}(\overline{Y}, \overline{Z}) \circ f) - \overline{g}_{f(p)}((\overline{\nabla}_{\overline{X}}\overline{Y})_{f(p)}, \overline{Z}_{f(p)}) -$$

$$- \overline{g}_{f(p)}(\overline{Y}_{f(p)}, (\overline{\nabla}_{\overline{X}}\overline{Z})_{f(p)}) =$$

[by Lemma 3.8]

$$= X_p(g(Y, Z)) - \overline{g}_{f(p)}((\overline{\nabla}_X Y)_p, (d_p f)Z_p) -$$

$$- \overline{g}_{f(p)}((d_p f)Y_p, (\overline{\nabla}_X Z)_p) =$$

[by (3.7)]

$$= X_p(g(Y, X)) - \overline{g}_{f(p)}((d_p f)(\nabla_X Y)_p, (d_p f)Z_p) -$$

$$- \overline{g}_{f(p)}((d_p f)Y_p, (d_p f)(\nabla_X Z)_p) =$$

[by the very definition of the first fundamental form]

$$= X_p(g(Y, X)) - g_p((\nabla_X Y)_p, Z_p) - g_p(Y_p, (\nabla_X Z)_p) =$$

$$= (\nabla_X g)(Y, Z)_p$$

hence $\nabla g = 0$. By the uniqueness statement in the fundamental theorem of semi-Riemannian geometry ∇ must be the Levi-Civita connection of (M, g). Q.e.d.

Let $\xi \in \mathfrak{X}(M)^{\perp}$ and let us set

$$\hat{\xi}_{f(p)} = \xi_p \in T_p(M)^{\perp} \subset T_{f(p)}(\overline{M}), \quad p \in M.$$

As $f(M)$ is a closed subset of $\overline{M}$ and $\hat{\xi} \in C^{\infty}(f(M), \tau_{\overline{M}})$ there is a smooth extension $\overline{\xi} \in \mathfrak{X}(\overline{M})$ of $\hat{\xi}$ to the whole of $\overline{M}$, i.e., $\overline{\xi}\big|_{f(M)} = \hat{\xi}$. The vector field $\overline{\xi}$ is referred to merely as an *extension* of the normal field ξ.

Let $X \in \mathfrak{X}(M)$ and $\xi \in \mathfrak{X}(M)^{\perp}$ be a tangential field and a normal field on the submanifold M and let $\overline{X}, \overline{\xi} \in \mathfrak{X}(\overline{M})$ be smooth extensions of X and ξ respectively. Arguments similar to those in the proof of Lemma 3.8 lead to

Lemma 3.12 *For any linear connection $\overline{\nabla} \in \mathcal{C}(\tau_{\overline{M}})$ the tangent vector*

$$\left(\overline{\nabla}_{\overline{X}}\overline{\xi}\right)_{f(p)} \in T_{f(p)}(\overline{M})$$

doesn't depend upon the extensions $\overline{X}$ and $\overline{\xi}$ of X and ξ.

From now let $\overline{\nabla}$ be the Levi-Civita connection of the semi-Riemannian manifold $(\overline{M}, \overline{g})$. By Lemma 3.12 the following notation is legitimate.

$$\left(\overline{\nabla}_X \xi\right)_p = \left(\overline{\nabla}_{\overline{X}} \overline{Y}\right)_{f(p)} \cdot$$

Moreover let us set

$$(A_{\xi}X)_p = -\tan_p(\overline{\nabla}_X \xi)_p, \quad (\nabla^{\perp}_X \xi)_p = \mathrm{nor}_p(\overline{\nabla}_X \xi)_p,$$

so that

$$(\overline{\nabla}_X \xi)_p = -(d_p f)(A_{\xi}X)_p + (\nabla^{\perp}_X \xi)_p \,. \tag{3.12}$$

Let now $A_{\xi}X \in \mathfrak{X}(M)$ and $\nabla^{\perp}_X \xi \in \mathfrak{X}(M)^{\perp}$ be the fields given by

$$p \in M \longmapsto (A_{\xi}X)_p \in T_p(M), \quad p \in M \longmapsto (\nabla^{\perp}_X \xi)_p \in T_p(M)^{\perp},$$

so that formula (3.12) may also be written

$$\overline{\nabla}_X \xi = -f_* A_{\xi} X + \nabla^{\perp}_X \xi \tag{3.13}$$

for any $X \in \mathfrak{X}(M)$ and any $\xi \in \mathfrak{X}(M)^{\perp}$. Formula (3.13) is an identity in $C^{\infty}(f^{-1}T(\overline{M}))$, as well as Gauss formula.

Definition 3.13 Formula (3.13) is referred to as the *Weingarten formula* of the immersion $f : M \to \overline{M}$. □

Let $\xi, \eta \in \mathfrak{X}(M)^\perp$ be two normal fields. We adopt the notation

$$\overline{g}(\xi, \eta)_p = \overline{g}_{f(p)}(\xi_p, \eta_p), \quad p \in M.$$

Proposition 3.14 *The map* $A : \mathfrak{X}(M) \times \mathfrak{X}(M)^\perp \to \mathfrak{X}(M)$ *defined by*

$$(X, \xi) \in \mathfrak{X}(M) \times \mathfrak{X}(M)^\perp \longmapsto A_\xi X \in \mathfrak{X}(M)$$

is $\mathcal{F}(M)$*-bilinear and satisfies the identity*

Julius Weingarten (1836–1910), German mathematician of Polish origin. **W.** studied at the University of Berlin, and is known to have devoted particular attention to Dirichlet's lectures on potential theory. At the same time **W.** studied chemistry. As part of a poor family of Polish immigrants, **W.** could not continue the formal studies leading to a doctoral degree. Although **W.** had to earn his living by teaching in schools, **W.** conducted research on the differential geometry of surfaces, and in particular constant mean curvature hypersurfaces in $\mathbb{E}^3$. The work by **W.** was particularly appreciated by Darboux (the work on the infinitesimal deformation of surfaces, 1886) and by Bianchi (Cayley's equation, Weingarten systems, 1924). Both Darboux and Bianchi included in their books ample portions devoted to the discoveries by **W.** Despite the recognition of his scientific merits (**W.** was awarded the *Grand Prix de l'Académie des Sciences* in Paris, 1894) **W.** hasn't obtained an academic position on the measure of his great mathematical talent until 1902. That is only partially due to the social inequities related to **W.**'s condition as an immigrant, and is prevalently explained by the status of the German university system at the time: prestigious German universities were offering significant academic positions yet accompanied by stipends whose level requested revenue from a third source. With a declining health, **W.** continued to teach only up to 1908. Worth to remember within the scientific creation by **W.** is his research work on the classification of the surfaces isometric to a given surface (and in particular to a surface of revolution): said classification was reduced by **W.** to solving a Monge-Ampère type equation.

$$g(A_\xi X, Y) = \overline{g}(B(X, Y), \xi) \tag{3.14}$$

for any $X, Y \in \mathfrak{X}(M)$ *and* $\xi \in \mathfrak{X}(M)^\perp$.

Definition 3.15 A_ξ is called the *Weingarten operator*, or the *shape operator*, of the immersion $f : M \to \overline{M}$ associated to the normal field ξ. □

Proof of Proposition 3.14 Let $h \in \mathcal{F}(M)$, $X \in \mathfrak{X}(M)$ and $\xi \in \mathfrak{X}(M)^\perp$. Then

$$\overline{\nabla}_{hX}\xi = h\overline{\nabla}_X\xi, \tag{3.15}$$

$$\overline{\nabla}_X(h\xi) = X(h)\xi + h\overline{\nabla}_X\xi\,. \tag{3.16}$$

For instance

$$(\overline{\nabla}_{hX}\xi)_p = (\overline{\nabla}_{\overline{h}\,\overline{X}}\overline{\xi})_{f(p)} = \left\{\overline{h}\,\overline{\nabla}_{\overline{X}}\overline{\xi}\right\}_{f(p)} = h(p)(\overline{\nabla}_X\xi)_p\,.$$

The proof of identity (3.16) is similar to the above and therefore left as an exercise to the reader. By identity (3.15)

$$-(d_pf)(A_\xi hX)_p + (\nabla^\perp_{hX}\xi)_p = (\overline{\nabla}_{hX}\xi)_p = \left\{h\overline{\nabla}_X\xi\right\}_p =$$

$$= h(p)(\overline{\nabla}_X\xi)_p = -h(p)\left\{-(d_pf)(A_\xi X)_p + (\nabla^\perp_X\xi)_p\right\}$$

and then (by the uniqueness of the direct sum decomposition, together with the fact that d_pf is a monomorphism)

$$A_\xi(hX) = hA_\xi X, \quad \nabla^\perp_{hX}\xi = h\nabla^\perp_X\xi.$$

One may use (3.16) in a similar manner to show that

$$A_{h\xi}X = hA_\xi X, \quad \nabla^\perp_X(h\xi) = X(h)\xi + h\nabla^\perp_X\xi.$$

Finally

$$g(A_\xi X, Y)_p = g_p((A_\xi X)_p, Y_p) = \overline{g}_{f(p)}((d_pf)(A_\xi X)_p, (d_pf)Y_p) =$$

[as $(\nabla^\perp_X\xi)_p \perp (d_pf)Y_p$]

$$= -\overline{g}_{f(p)}((\overline{\nabla}_X\xi)_p, (d_pf)Y_p) = -\overline{g}_{f(p)}((\overline{\nabla}_{\overline{X}}\overline{\xi})_{f(p)}, \overline{Y}_{f(p)}) =$$

$$= -\left\{\overline{g}(\overline{\nabla}_{\overline{X}}\overline{\xi}, \overline{Y})\right\}_{f(p)} = -\left\{\overline{X}(\overline{g}(\overline{\xi}, \overline{Y})) - \overline{g}(\overline{\xi}, \overline{\nabla}_{\overline{X}}\overline{Y})\right\}_{f(p)} =$$

$$= -X_p(\overline{g}(\overline{\xi}, \overline{Y}) \circ f) + \overline{g}_{f(p)}(\xi_p, (\overline{\nabla}_X\xi)_p) =$$

[as $\overline{g}(\overline{\xi}, \overline{Y})_{f(p)} = \overline{g}_{f(p)}(\xi_p, (d_p f)Y_p) = 0$]

$$= \overline{g}_{f(p)}(\xi_p, B(X, Y)_p) = \overline{g}(\xi, B(X, Y))_p .$$

Q.e.d.

To understand the nature of the object $\nabla^\perp$ one needs to introduce the notion of a connection in a vector bundle.[1]

Definition 3.16 Let $E \to M$ be a real vector bundle and let $\Omega^0(E) = C^\infty(E)$ be the space of smooth sections in E. A $\mathbb{R}$-linear map

$$D : \mathfrak{X}(M) \times \Omega^0(E) \to \Omega^0(E)$$

is called a *connection* in the vector bundle $E \to M$ if

$$D_{hX}s = hD_X s, \quad D_X(hs) = X(h)s + hD_X s$$

for any $h \in \mathcal{F}(M)$, $X \in \mathfrak{X}(M)$ and $s \in \Omega^0(E)$. □

Let us denote by $\mathcal{C}(E)$ the set of all connections in the vector bundle $E \to M$. Note that

$$\mathfrak{X}(M)^\perp = \Omega^0(T(M)^\perp).$$

Proposition 3.14, or rather its proof, yields.

Theorem 3.17 $\nabla^\perp : \mathfrak{X}(M) \times \Omega^0(T(M)^\perp) \to \Omega^0(T(M)^\perp)$ *is a connection in the normal bundle* $T(M)^\perp \to M$.

Definition 3.18 $\nabla^\perp$ is called the *normal connection* of the immersion $f : M \to \overline{M}$.

□

3.4 Covariant Calculus in $\Omega^0(f^{-1}T(\overline{M}))$

Gauss and Weingarten formulas may be "immersed" in a covariant differential calculus with sections in the pullback bundle $f^{-1}T(\overline{M}) \to M$. Learning said formalism will prepare the reader for the construction of harmonic map theory. Let $V \in \mathfrak{X}(\overline{M})$ be a vector field on $\overline{M}$.

Definition 3.19 The *natural lift* of V is the section

$$V^f \in \Omega^0(f^{-1}T(\overline{M}))$$

[1] In such a manner that linear connections appear as connections in a particular vector bundle, i.e., the tangent bundle over the given manifold.

given by $V^f = V \circ f$ that is $V^f(p) = V_{f(p)} \in T_{f(p)}(\overline{M}) = (f^{-1}T(\overline{M}))_p$ for any $p \in M$. □

Let $(A, y^1, \dots, y^{n+r})$ be a local coordinate system on $\overline{M}$. Then $\{\partial/\partial y^\alpha : 1 \le \alpha \le n+r\}$ is a local frame in the tangent bundle $\tau_{\overline{M}}$, defined on the open set $A \subset \overline{M}$ so that

$$\{s_\alpha \equiv (\partial/\partial y^\alpha)^f : 1 \le \alpha \le n+r\}$$

is a local frame in the vector bundle $f^{-1}T(\overline{M}) \to M$, defined on the open set $f^{-1}(A)$. The Levi-Civita connection $\overline{\nabla}$ induces a connection in the vector bundle $f^{-1}T(\overline{M}) \to M$

$$f^{-1}\overline{\nabla} \equiv \nabla^f : \mathfrak{X}(M) \times \Omega^0(f^{-1}T(\overline{M})) \to \Omega^0(f^{-1}T(\overline{M}))$$

defined as follows. Let $X \in \mathfrak{X}(\overline{M})$ and $s \in \Omega^0(f^{-1}T(\overline{M}))$ and $p \in M$. Moreover let us consider a local coordinate system (A, y^α) on $\overline{M}$ such that $f(p) \in A$. Let us locally represent s as

$$s|_A = h^\alpha s_\alpha, \quad h^\alpha \in \mathcal{F}(A).$$

Now let us set by definition[2]

$$(\nabla^f_X s)_p = \left\{ X(h^\alpha) + X(f^\beta) h^\gamma \left(\left\{ {\alpha \atop \beta\gamma} \right\} \circ f \right) \right\}(p)\, s_\gamma(p). \tag{3.17}$$

One easily checks that the definition of $(\nabla^f_X s)_p$ doesn't depend upon the choice of local coordinates (A, y^α) about $f(p)$.

Definition 3.20 The connection $\nabla^f \in \mathcal{C}(f^{-1}T(\overline{M}))$ is called the *pullback* by f of the connection $\overline{\nabla}$. □

---

[2] The heuristics leading to definition (3.17) is actually quite simple. Let (U, x^i) and (A, y^α) be local coordinate systems on M and $\overline{M}$ such that $f(U) \subset A$. We start by prescribing $\nabla^f_{\partial_i} s_\alpha$ and then we derive the expression (3.17) by requesting that ∇^f possesses the properties in Definition 3.16. The vector field $\nabla^f_{\partial_i} s_\alpha$ should be naturally induced by $\overline{\nabla}$. As we only know

$$\overline{\nabla}_{\partial_\alpha} \partial_\beta = \left\{ {\gamma \atop \alpha\beta} \right\} \partial_\gamma, \quad \partial_\alpha \equiv \frac{\partial}{\partial y^\alpha},$$

we first consider the "push-forward" of ∂_i by f

$$(d_p f)\partial_{i,p} = \frac{\partial f^\alpha}{\partial x^i}(p) \left(\frac{\partial}{\partial y^\alpha} \right)_{f(p)}$$

and we set by definition

$$\nabla^f_{\partial_i} s_\alpha = \frac{\partial f^\beta}{\partial x^i} \left(\left\{ {\gamma \atop \beta\alpha} \right\} \circ f \right) \left(\frac{\partial}{\partial y^\gamma} \right)^f.$$

Definition 3.21 Let $E \to M$ be a real vector bundle of rank r. A map $H : p \in M \mapsto H_p \in E_p^* \otimes E_p^*$ is a *semi-Riemannian bundle metric* on E if H_p is an inner product on E_p such that $\text{ind}(E_p, H_p) = \text{constant}$ and $H_{\alpha\beta} \equiv H(s_\alpha, s_\beta) \in \mathcal{F}(U)$ for any local frame $\{s_\alpha : 1 \leq \alpha \leq r\}$ in E defined on the open set $U \subset M$. □

Note that the semi-Riemannian metric $\overline{g}$ induces a semi-Riemannian bundle metric $f^{-1}\overline{g} = g^f$ on $f^{-1}T(\overline{M}) \to M$. Indeed if $p \in M$ and (A, y^α) is a local coordinate system on $\overline{M}$ such that $f(p) \in A$ then we set

$$g^f(r, s)_p = \lambda^\alpha \mu^\alpha \overline{g}_{\alpha\beta}(f(p)),$$

$$\overline{g}_{\alpha\beta} = \overline{g}\left(\frac{\partial}{\partial y^\alpha}, \frac{\partial}{\partial y^\beta}\right),$$

$$r|_A = \lambda^\alpha s_\alpha, \quad s|_A = \mu^\alpha s_\alpha, \quad \lambda^\alpha, \mu^\alpha \in \mathcal{F}(A).$$

One may easily check that the definition of $g^f(r, s)_p$ doesn't depend upon the choice of local coordinate system (A, y^α) about $f(p)$. The covariant differential calculus alluded to at the beginning of this section is completely described by Proposition 3.22. Essentially, once one realizes that any normal field $\xi : M \to T(M)^\perp$ is a section in $f^{-1}T(\overline{M})$ and that $f_*X \in \Omega^0(f^{-1}T(\overline{M}))$ for any $X \in \mathfrak{X}(M)$, one uses the bundle metric g^f to measure the lengths of sections in $\Omega^0(f^{-1}T(\overline{M}))$ and the connection $\nabla^f \in \mathcal{C}(f^{-1}T(\overline{M}))$ to take derivatives of these sections.

Moreover any section in $f^{-1}T(\overline{M}) \to M$ admits a canonical decomposition, as the sum of a tangential component and a normal component, relative to the direct sum decomposition (3.1). This leads to the Gauss and Weingarten formulas (3.18)–(3.19) and then to a description, as we shall shortly see, of the geometry of the embedding $f : M \to \overline{M}$ in terms of the objects ∇, B, A and $\nabla^\perp$.

These methods of study belong to the theory of submanifolds in semi-Riemannian manifolds yet also present the advantage, as mentioned above, to prepare in an "innocuous" manner the passage from the theory of minimal isometric immersions to the theory of harmonic maps of semi-Riemannian manifolds.

Proposition 3.22 (i) g^f *is a semi-Riemannian bunde metric on the pullback bundle* $f^{-1}T(\overline{M}) \to M$. *Moreover*

$$g^f(f_*X, f_*Y) = g(X, Y), \quad g^f(f_*X, \xi) = 0,$$

for any $X, Y \in \mathfrak{X}(M)$ *and any* $\xi \in \mathfrak{X}(M)^\perp$.

(ii) ∇^f *is a metric connection, i.e.,* $\nabla^f g^f = 0$ *where*

$$(\nabla^f_X g^f)(r, s) = X(g^f(r, s)) - g^f(\nabla^f_X r, s) - g^f(r, \nabla^f_X s).$$

(iii) *Gauss and Weingarten formulas may be written*

$$\nabla^f_X f_* Y = f_* \nabla_X Y + B(X, Y), \tag{3.18}$$

$$\nabla^f_X \xi = -f_* A_\xi X + \nabla^\perp_X \xi \tag{3.19}$$

for any $X, Y \in \mathfrak{X}(M)$ *and* $\xi \in \mathfrak{X}(M)^\perp$.

(iv) *Formula* (3.14) *may be written*

$$g(A_\xi X, Y) = g^f(B(X, Y), \xi).$$

3.5 Gauss–Ricci–Codazzi Equations

One may believe, based on the intuition acquired in the study of the geometry of curves and surfaces in $\mathbb{E}^3$ (cf. e.g., M. Do Carmo[3]), that the curvature $R = R^\nabla$ of the Levi-Civita connection of a semi-Riemanian manifold describes, in an intrinsic manner, the shape of that semi-Riemannian manifold (M, g). Given the human habit, within scientific thought, to dissociate objects (such as differentiable manifolds) from their additional structures (such as semi-Riemannian metrics) to better contemplate the former, one may think that adding a semi-Riemannian metric to a differentiable manifold is a rather arbitrary act and that the association (M, g) between a manifold M and a metric g is too artificial (after all a manifold admits, as we have previously seen, infinitely many Riemannian metrics) to allow speaking without ambiguity about the *shape* of (M, g). There would be as many *shapes* as many semi-Riemannian metrics may be assigned to M: wouldn't then shape be an entirely relative concept? It should be noticed that the majority of examples of semi-Riemannian manifolds that we know of, are submanifolds of some semi-Euclidean space $\mathbb{E}^n_\nu$ and that the very differential geometry is born as a discipline aiming to the study of shapes of submanifolds in $\mathbb{E}^3$. Also, by a celebrated result by J. Nash[4] for any Riemannian manifold (M, g) there is a positive integer $\mathrm{N} \in \mathbb{N}$ and an isometric immersion $f : M \to \mathbb{E}^{\mathrm{N}}$ di (M, g) into $\mathbb{E}^{\mathrm{N}}$. As a submanifold of $\mathbb{E}^{\mathrm{N}}$, any manifold M admits a *natural* Riemannian metric, the first fundamental form of the given immersion, and it is a difficult action, we may say a purely metaphysical action, to dissociate M from said metric, as it is to conceive M as dissociated from $f(M)$ where $f : M \to \mathbb{E}^{\mathrm{N}}$ is the given immersion. The space surrounding us is, at least at a not too large scale, the Euclidean space $\mathbb{E}^3$ which comes equipped with the Euclidean metric and

---

[3] Cf. M. Do Carmo, *Differential Geometry of Curves and Surfaces* (Prentice-Hall, Inc., Englewood Cliffs, New Jersey, 1976).

[4] Cf. J. Nash, The imbedding problem for Riemannian manifolds. Ann. Math. (1) **63**, 20–63 (1956).

all submanifolds of $\mathbb{E}^3$ appear as inseparable[5] from their first fundamental forms. All geometric and physical measurements are intimately related to the presence of said metrics. It follows that after all the concept of shape of (M, g) is related to the given isometric immersion in a given ambient space $(\overline{M}, \overline{g})$. However, as in the spirit of Nash's theorem the metric g on M is the first to be assigned and the isometric immersion is considered, or built, successively, the geometric properties of M that spring from g are referred to as *intrinsic*, while geometric properties of M thought of as $f(M)$, that is those properties springing from the presence of f, are termed *extrinsic*. Once the given semi-Riemannian manifold (M, g) is immersed, the distinction between intrinsic and extrinsic properties becomes, as we shall shortly show, rather immaterial (intrinsic and extrinsic properties may be deduced one from the other, in an explicit quantitative manner).

The main object entering the description of extrinsic geometric properties of a semi-Riemannian manifold isometrically immersed in another semi-Riemannian manifold $(\overline{M}, \overline{g})$ is the second fundamental form B of the given immersion. The Weingarten, or shape, operator is, by formula (3.14), but a contravariant manifestation of the second fundamental form. Intrinsic geometry, based on g, and extrinsic geometry, relying on B, of the given submanifold, are not independent: they are related in a quantitative manner by the classical Gauss-Ricci-Codazzi equations, capturing the relationship between the curvature of the submanifold, the curvature of the ambient space, and the second fundamental form of the given immersion. As a purpose of this section we shall derive said equations starting from the Gauss and Weingarten formulas. This leads to a very elegant mathematical theory making use of geometric methods, alternative to methods in the theory of partial differential equations, to study the Gauss-Ricci-Codazzi equations. It shouldn't be forgotten that Gauss and Weingarten equations are locally systems of second order PDEs whose unknown functions are the local components of the immersion $f : M \to \overline{M}$. Indeed let (U, x^i) and (A, y^α) be local coordinate systems on M and $\overline{M}$, such that $f(U) \subset A$. Then

---

[5] At what extent a semi-Riemannian manifold is inseparable from its metric is one of the deepest questions in semi-Riemannian geometry (one has in mind the subtle relationship between curvature and the topology of the manifold). By a classical result known as *Synge's theorem* (cf. e.g., Kobayashi and Nomizu [KoNo]) any Riemannian manifold of positive sectional curvature is simply connected. The latter is a topological property of the given manifold (the vanishing of the first homotopy group) and morally should be a property of the given Riemannian manifold even if it is mentally dissociated from its Riemannian structure. We may conclude that not only is the metric of the given Riemannian manifold intimately attached to the underlying manifold: the metric determines the properties of the manifold and of the underlying topological space. It also follows that in our metaphysical approach ("first" a Hausdorff space with a countable base, "then" a differentiable atlas, and "in the end" a metric) we have operated with notions ("first", "then", "in the end") that are entirely questionable.

$$\nabla_{\partial_i}\partial_j = \left|\begin{matrix} k \\ ij \end{matrix}\right| \partial_k, \quad B(\partial_i, \partial_j) = B_{ij}^{\alpha} s_\alpha,$$

$$f_*\partial_i = \frac{\partial f^\alpha}{\partial x^i} s_\alpha, \quad f_*\nabla_{\partial_i}\partial_j = \left|\begin{matrix} k \\ ij \end{matrix}\right| \frac{\partial f^\alpha}{\partial x^k} s_\alpha,$$

$$\nabla^f_{\partial_i} f_*\partial_j = \nabla^f_{\partial_i}\left(f^\alpha{}_{|j} s_\alpha\right) = \frac{\partial^2 f^\alpha}{\partial x^i \partial x^j} s_\alpha + f^\beta{}_{|i} f^\gamma{}_{|j} \left(\left|\begin{matrix} \alpha \\ \beta\gamma \end{matrix}\right| \circ f \right) s_\alpha .$$

Gauss equation with $X = \partial_i$ and $Y = \partial_j$ may be written

$$\frac{\partial^2 f^\alpha}{\partial x^i \, \partial x^j} + \left(\left|\begin{matrix} \alpha \\ \beta\gamma \end{matrix}\right| \circ f \right) \frac{\partial f^\beta}{\partial x^i} \frac{\partial f^\gamma}{\partial x^j} = \left|\begin{matrix} k \\ ij \end{matrix}\right| \frac{\partial f^\alpha}{\partial x^k} + B^\alpha_{ij} . \tag{3.20}$$

So (3.20) is a second order quasi-linear PDEs system. In the spirit of submanifold theory we shall not apply specific methods of PDEs but geometric methods, relying on the Gauss-Codazzi-Ricci equations yet belonging to the theory of connections on vector bundles. The geometric methods we shall use will nevertheless lead to the understanding of qualitative properties of solutions f^α.

Definition 3.23 Let $D \in \mathcal{C}(E)$ be a connection in the vector bundle $E \to M$. The *curvature* of D is the map

$$R^D : \mathfrak{X}(M) \times \mathfrak{X}(M) \times \Omega^0(E) \to \Omega^0(M),$$

$$R^D(X, Y)s = D_X D_Y s - D_Y D_X s - D_{[X,Y]} s$$

for any $X, Y \in \mathfrak{X}(M)$ and any $s \in \Omega^0(E)$. □

Let $f : M \to \overline{M}$ be a regular embedding of a n-dimensional manifold M in a $(n+r)$-dimensional semi-Riemannian manifold $(\overline{M}, \overline{g})$. Let $R^f \equiv R^{\nabla^f}$, $R \equiv R^\nabla$ e $R^\perp \equiv R^{\nabla^\perp}$ be respectively the curvatures of the connections

$$\nabla^f \in \mathcal{C}(f^{-1}\tau_{\overline{M}}), \quad \nabla \in \mathcal{C}(\tau_M), \quad \nabla^\perp \in \mathcal{C}(T(M)^\perp).$$

Then (by Gauss and Weingarten equations)

$$R^f(X, Y) f_* Z = \nabla^f_X \nabla^f_Y f_* Z - \nabla^f_Y \nabla^f_X f_* Z - \nabla^f_{[X,Y]} f_* Z =$$

$$= \nabla^f_X \{f_* \nabla_Y Z + B(Y, Z)\} - \nabla^f_Y \{f_* \nabla_X Z + B(X, Z)\} -$$

$$- f_* \nabla_{[X,Y]} Z - B([X, Y], Z) =$$

$$= f_* \nabla_X \nabla_Y Z + B(X, \nabla_Y Z) - f_* A_{B(Y,Z)} X + \nabla^\perp_X B(Y, Z) -$$

$$- f_* \nabla_Y \nabla_X Z - B(Y, \nabla_X Z) + f_* A_{B(X,Z)} Y - \nabla_Y^{\perp} B(X, Z) -$$

$$- f_* \nabla_{[X,Y]} Z - B([X, Y], Z) =$$

$$= f_* \left\{ R(X, Y)Z + A_{B(X,Z)} Y - A_{B(Y,Z)} X \right\} +$$

$$+ \nabla_X^{\perp} B(Y, Z) - B(Y, \nabla_X Z) - \nabla_Y^{\perp} B(X, Z) + B(X, \nabla_Y Z) - B([X, Y], Z) =$$

(by substitution from $[X, Y] = \nabla_X Y - \nabla_Y X$)

$$= f_* \left\{ R(X, Y)Z + A_{B(X,Z)} Y - A_{B(Y,Z)} X \right\} +$$

$$+ (\nabla_X B)(Y, Z) - (\nabla_Y B)(X, Z)$$

where we have set by definition

$$(\nabla_X B)(Y, Z) = \nabla_X^{\perp} B(Y, Z) - B(\nabla_X Y, Z) - B(Y, \nabla_X Z)$$

for any $X, Y, Z \in \mathfrak{X}(M)$.

Definition 3.24 ∇B is called the a *covariant derivative* of the second fundamental form B. □

We have obtained the following *Gauss-Codazzi equation*

$$R^f(X, Y) f_* Z = f_* \left\{ R(X, Y)Z + A_{B(X,Z)} Y - A_{B(Y,Z)} X \right\} + \tag{3.21}$$

$$+ (\nabla_X B)(Y, Z) - (\nabla_Y B)(X, Z).$$

Using the canonical projections

$$\tan : f^{-1} \tau_{\overline{M}} \to \tau_M, \quad \text{nor} : f^{-1} \tau_{\overline{M}} \to \tau_M^{\perp}(f),$$

one obtains

$$\tan \left\{ R^f(X, Y) f_* Z \right\} = R(X, Y)Z + A_{B(X,Z)} Y - A_{B(Y,Z)} X, \tag{3.22}$$

$$\text{nor} \left\{ R^f(X, Y) f_* Z \right\} = (\nabla_X B)(Y, Z) - (\nabla_Y B)(X, Z). \tag{3.23}$$

Definition 3.25 The identities (3.22)–(3.23) are respectively referred to as the *Gauss equation* and the *Codazzi equation*. □

Gauss equation (3.22) may also be written in the following form, which will prove useful in the sequel. Let us multiply in both sides by $W \in \mathfrak{X}(M)$ (the inner product we use is g). We obtain

$$g(\tan\left\{R^f(X,Y)f_*Z\right\},\ W) = g(R(X,Y)Z,\ W) +$$

$$+\, g(A_{B(X,Z)}Y, W) - g(A_{B(Y,Z)}X, W)$$

or [by (3.14)]

$$g^f(R^f(X,Y)f_*Z, f_*W) = g(R(X,Y)Z, W) + \tag{3.24}$$

$$+\, g^f(B(Y,W), B(X,Z)) - g^f(B(X,W), B(Y,Z)).$$

Now one computes $R^f(X,Y)\xi$ for any $\xi \in \mathfrak{X}(M)^{\perp}$. By Gauss and Weingarten formulas

$$R^f(X,Y)\xi = \nabla^f_X \nabla^f_Y \xi - \nabla^f_Y \nabla^f_X \xi - \nabla^f_{[X,Y]}\xi =$$

$$= \nabla^f_X \left\{-f_* A_\xi Y + \nabla^{\perp}_Y \xi\right\} -$$

$$- \nabla^f_Y \left\{-f_* A_\xi X + \nabla^{\perp}_X \xi\right\} -$$

$$+\, f_* A_\xi [X,Y] - \nabla^{\perp}_{[X,Y]}\xi =$$

$$= -f_* \nabla_X A_\xi Y - B(X, A_\xi Y) - A_{\nabla^{\perp}_Y \xi} X + \nabla^{\perp}_X \nabla^{\perp}_Y \xi +$$

$$+\, f_* \nabla_Y A_\xi X + B(Y, A_\xi X) + f_* A_{\nabla^{\perp}_X \xi} Y - \nabla^{\perp}_Y \nabla^{\perp}_X \xi +$$

$$+\, f_* A_\xi [X,Y] - \nabla^{\perp}_{[X,Y]}\xi =$$

$$= R^{\perp}(X,Y)\xi + B(Y, A_\xi X) - B(X, A_\xi Y) -$$

$$- f_* \left\{\nabla_X(A_\xi Y) - \nabla_Y(A_\xi X) + A_{\nabla^{\perp}_Y \xi} X - A_{\nabla^{\perp}_X \xi} Y - A_\xi [X,Y]\right\}$$

or (by substitution from $[X,Y] = \nabla_X Y - \nabla_Y X$)

$$R^f(X,Y)\xi = R^{\perp}(X,Y)\xi + B(Y, A_\xi X) - B(X, A_\xi Y) - \tag{3.25}$$

$$- f_* \left\{(\nabla_X A)(\xi, Y) - (\nabla_Y A)(\xi, X)\right\}$$

where we have set by definition

$$(\nabla_X A)(\xi, Y) = \nabla_X(A_\xi Y) - A_{\nabla^{\perp}_X \xi} - A_\xi \nabla_X Y.$$

Applying the projection nor : $f^{-1}\tau_{\overline{M}} \to \tau_M^{\perp}(f)$ to the identity (3.25) we obtain

$$\text{nor}\left\{R^f(X,Y)\xi\right\} = R^{\perp}(X,Y)\xi + B(Y, A_\xi X) - B(X, A_\xi Y) \tag{3.26}$$

for any $X, Y \in \mathfrak{X}(M)$ and $\xi \in \mathfrak{X}(M)^{\perp}$.

Definition 3.26 The identity (3.26) is referred to as the *Ricci equation.* □

Ricci's equation may also be written in the following form, useful in the sequel. Let us take the inner product (by using the inner product g^f) of (3.26) with $\eta \in \mathfrak{X}(M)^{\perp}$. We obtain [by (3.14)]

$$g^f(R^f(X,Y)\xi,\ \eta) = g^f(R^{\perp}(X,Y)\xi,\ \eta) + g(A_\eta Y, A_\xi X) - g(A_\eta X, A_\xi Y)$$

or

$$g^f(R^f(X,Y)\xi, \eta) = g^f(R^{\perp}(X,Y)\xi, \eta) - g([A_\xi, A_\eta]X, Y). \tag{3.27}$$

Here

$$[A_\xi, A_\eta]X = A_\xi A_\eta X - A_\eta A_\xi X.$$

To derive Eq. (3.27) we made use of the identity

$$g(A_\xi X, Y) = g(X, A_\xi Y)$$

Gregorio Ricci Curbastro (1853–1925), Italian mathematician. **R.** has studied at Technische Hochschule in Munich (from 1877) and successively under Ulisse Dini. **R.** became a professor of mathematics at the University of Padova (from 1880) and at this location studied Riemannian geometry [Absolute differential calculus (*calcolo differenziale assoluto*), according to the terminology of the time, mainly attributed to Levi-Civita alone] together with Tullio Levi-Civita. The Ricci tensor R_{ij} also belongs to the scientific creation by **R.** and is *sine qua non* (in its Lorentzian version) to the formulation of the gravitational field equations within GRG theory.

itself a consequence of the symmetry of B and of (3.14) One may of course consider, starting from (3.25), the tangential component, as well (i.e., apply the projection tan to both members). Said procedure leads however to an equation equivalent to Codazzi's equation.

The curvature R^f of ∇^f is essentially the curvature $\overline{R}$ of $\overline{\nabla}$. Indeed [by $[\partial_i, \partial_j] = 0$]

$$R^f(\partial_i, \partial_j)s_\alpha =$$

$$= \nabla^f_{\partial_i}\left\{f^\beta{}_{|j}\left(\left\{ {\gamma \atop \beta\alpha} \right\}\circ f\right)s_\gamma\right\} - \nabla^f_{\partial_j}\left\{f^\beta{}_{|i}\left(\left\{ {\gamma \atop \beta\alpha} \right\}\circ f\right)s_\gamma\right\} =$$

$$= \left\{\frac{\partial^2 f^\beta}{\partial x^i \partial x^j}\left(\left\{ {\gamma \atop \beta\alpha} \right\}\circ f\right) + f^\beta{}_{|j}\frac{\partial}{\partial x^i}\left(\left\{ {\gamma \atop \beta\alpha} \right\}\circ f\right)\right\}s_\gamma +$$

$$+ f^\beta{}_{|j}\left(\left\{ {\gamma \atop \beta\alpha} \right\}\circ f\right) f^\mu{}_{|i}\left(\left\{ {\lambda \atop \mu\gamma} \right\}\circ f\right)s_\lambda -$$

$$- \left\{\frac{\partial^2 f^\beta}{\partial x^j \partial x^i}\left(\left\{ {\gamma \atop \beta\alpha} \right\}\circ f\right) + f^\beta{}_{|i}\frac{\partial}{\partial x^i}\left(\left\{ {\gamma \atop \beta\alpha} \right\}\circ f\right)\right\}s_\gamma -$$

$$- f^\beta{}_{|i}\left(\left\{ {\gamma \atop \beta\alpha} \right\}\circ f\right) f^\mu{}_{|j}\left(\left\{ {\lambda \atop \mu\gamma} \right\}\circ f\right)s_\lambda =$$

[by observing the cancellation of the second order partials]

$$= \left\{f^\beta{}_{|j}\left(\frac{\partial\left\{ {\gamma \atop \beta\alpha} \right\}}{\partial y^\nu}\circ f\right) f^\nu{}_{|i} - f^\beta{}_{|i}\left(\frac{\partial\left\{ {\gamma \atop \beta\alpha} \right\}}{\partial y^\nu}\circ f\right) f^\nu{}_{|j}\right\}s_\gamma +$$

$$+\left\{\left(\left\{ {\gamma \atop \beta\alpha} \right\}\circ f\right)\left(\left\{ {\gamma \atop \mu\alpha} \right\}\circ f\right)\left(\left\{ {\lambda \atop \beta\gamma} \right\}\circ f\right)\right\} f^\beta{}_{|j} f^\mu{}_{|i} s_\lambda =$$

$$= f^\beta{}_{|j} f^\nu{}_{|i}\left\{\left(\frac{\partial\left\{ {\lambda \atop \beta\alpha} \right\}}{\partial y^\nu} - \frac{\partial\left\{ {\lambda \atop \nu\alpha} \right\}}{\partial y^\beta} +\right.\right.$$

$$\left.\left.+ \left\{ {\gamma \atop \beta\alpha} \right\}\left\{ {\lambda \atop \nu\gamma} \right\} - \left\{ {\gamma \atop \nu\alpha} \right\}\left\{ {\lambda \atop \beta\gamma} \right\}\right)\circ f\right\}s_\lambda =$$

$$= f^\beta{}_{|j} f^\nu{}_{|i}\left(\overline{R}^\lambda_{\nu\beta\alpha}\circ f\right)s_\lambda$$

where

$$\overline{R}^{\lambda}_{\nu\beta\alpha}\partial_\lambda = \overline{R}(\partial_\nu, \partial_\beta)\partial_\alpha =$$

$$= \left\{ \frac{\partial \left\{ \begin{matrix} \lambda \\ \beta\alpha \end{matrix} \right\}}{\partial y^\nu} - \frac{\partial \left\{ \begin{matrix} \lambda \\ \nu\alpha \end{matrix} \right\}}{\partial y^\beta} + \left\{ \begin{matrix} \gamma \\ \beta\alpha \end{matrix} \right\} \left\{ \begin{matrix} \lambda \\ \nu\gamma \end{matrix} \right\} - \left\{ \begin{matrix} \gamma \\ \nu\alpha \end{matrix} \right\} \left\{ \begin{matrix} \lambda \\ \beta\gamma \end{matrix} \right\} \right\} \partial_\lambda$$

is the local expression of the curvature $\overline{R}$ di $(\overline{M}, \overline{g})$. We have shown that

$$R^f(\partial_i, \partial_j)s_\gamma = \left(\overline{R}^{\lambda}_{\alpha\beta\gamma} \circ f\right) f^\alpha{}_{|i} f^\beta{}_{|j} s_\lambda . \tag{3.28}$$

Q.e.d.

It should be kept in mind that all equations obtained in the present section hold for any regular embedding $f : M \to \overline{M}$ of a manifold M in a semi-Riemannian manifold $(\overline{M}, \overline{g})$ such that $(d_p f)T_p(M)$ is nondegenerate as a subspace of the inner product space $(T_{f(p)}(\overline{M}), \overline{g}_{f(p)})$ for any point $p \in M$.

Let $(\overline{M}, \overline{g})$ be a Riemannian manifold of constant sectional curvature $\overline{k}$, i.e.,

$$\overline{R}(U, V)W = \overline{k}\,\{\overline{g}(W, V)U - \overline{g}(W, U)V\}$$

for any $U, V, W \in \mathfrak{X}(\overline{M})$. If (A, y^α) is a local coordinate system on $\overline{M}$ then

$$\overline{R}^{\alpha}_{\beta\gamma\delta}\partial_\alpha = \overline{R}(\partial_\beta, \partial_\gamma)\partial_\delta = \overline{k}\left\{\overline{g}_{\delta\gamma}\partial_\beta - \overline{g}_{\delta\beta}\partial_\gamma\right\}$$

and hence

$$\overline{R}^{\alpha}_{\beta\gamma\delta} = \overline{k}\left\{\overline{g}_{\delta\gamma}\delta^\alpha_\beta - \overline{g}_{\delta\beta}\delta^\alpha_\gamma\right\} \tag{3.29}$$

Let us substitute from (3.29) into (3.28). We obtain

$$R^f(\partial_i, \partial_j)s_\gamma = \overline{k}\left\{\overline{g}_{\gamma\beta}\delta^\lambda_\alpha - \overline{g}_{\gamma\alpha}\delta^\lambda_\beta\right\} f^\alpha{}_{|i} f^\beta{}_{|j} s_\lambda =$$

$$= \overline{k}\left\{g^f(s_\gamma, f_*\partial_j) f^\lambda{}_{|i} - g^f(s_\gamma, f_*\partial_i) f^\lambda{}_{|j}\right\} s_\lambda$$

and hence

$$R^f(X, Y)s = \overline{k}\left\{g^f(s, f_*Y)X(f^\lambda) - g^f(s, f_*X)Y(f^\lambda)\right\} s_\lambda$$

for any $X, Y \in \mathfrak{X}(M)$ and any $s \in \Omega^0(f^{-1}\tau_{\overline{M}})$. Also

$$(f_*X)(p) = (d_p f)X = X^i(p)(d_p f)(\partial_i)_p =$$

$$= X^i(p) f^\alpha{}_{|i}(p)(\partial_\alpha)_{f(p)} = X^i(p) f^\alpha{}_{|i}(p) s_\alpha(p)$$

or

$$f_*X = X(f^\alpha)s_\alpha \,. \tag{3.30}$$

We may conclude that the curvature $R^f \equiv R^{\nabla^f}$ of the connection $\nabla^f \in \mathcal{C}(f^{-1}\tau_{\overline{M}}, g^f)$ for an embedding $f : M \to \overline{M}$ in a constant curvature space $M^{n+r}(\overline{k}) = (\overline{M}, \overline{g})$ is given by

$$R^f(X,Y)s = \overline{k}\left\{g^f(s, f_*Y)f_*X - g^f(s, f_*X)f_*Y\right\} \tag{3.31}$$

for any $X, Y \in \mathfrak{X}(M)$ and $s \in \Omega^0(f^{-1}\tau_{\overline{M}})$. In particular for $s = f_*Z$ with $Z \in \mathfrak{X}(M)$ one has

$$R^f(X,Y)f_*Z = \overline{k}\left\{g^f(f_*Z, f_*Y)f_*X - g^f(f_*Z, f_*X)f_*Y\right\} =$$

[by Proposition 3.22]

$$= \overline{k}\,\{g(Z,Y)f_*X - g(Z,X)f_*Y\}$$

for any $X, Y, Z \in \mathfrak{X}(M)$. Taking the inner product with f_*W leads to [once again by Proposition 3.22]

$$g^f(R^f(X,Y)f_*Z,\ f_*W) = \tag{3.32}$$

$$= \overline{k}\,\{g(Z,Y)g(X,W) - g(Z,X)g(Y,W)\}\,.$$

Starting from (3.32) one may write Gauss equation (3.24) in the form

$$g(R(X,Y)Z,W) + g^f(B(Y,W), B(Y,Z)) - g^f(B(X,W), B(Y,Z)) =$$

$$= \overline{k}\,\{g(Z,Y)g(X,W) - g(Z,X)g(Y,W)\}$$

or

$$R(X,Y)Z = \overline{k}\,\{g(Z,Y)X - g(Z,X)Y\} + A_{B(Y,Z)}X - A_{B(Y,Z)}X \tag{3.33}$$

for any $X, Y, Z \in \mathfrak{X}(M)$. Let us now define the tensor field $X \wedge Y \in \mathcal{T}^{1,1}(M)$ by setting

$$(X \wedge Y)Z = g(Y,Z)X - g(X,Z)Y, \quad Z \in \mathfrak{X}(M). \tag{3.34}$$

Exercise 3.27 Show that $\wedge : \mathfrak{X}(M) \times \mathfrak{X}(M) \to \mathcal{T}^{1,1}(M)$ is $\mathcal{F}(M)$-bilinear skew-symmetric.

Using the *wedge product* of vector fields defined by (3.34) one may write Gauss equation (3.33) as

$$R(X,Y)Z = \overline{k}(X \wedge Y)Z + A_{B(Y,Z)}X - A_{B(X,Z)}Y. \tag{3.35}$$

Let us observe that the identity

$$R^f(X,Y)f_*Y = \overline{k}\,\{g(Z,Y)f_*X - g(Z,X)f_*Y\}$$

also yields

$$\text{nor } R^f(X,Y)f_*Z = 0$$

so that Codazzi's equation (3.23) becomes

$$(\nabla_X B)(Y,Z) - (\nabla_Y B)(X,Z) = 0 \tag{3.36}$$

for any $X, Y, Z \in \mathfrak{X}(M)$. Finally let us derive the Ricci equation for the embedding $f : M \to M^{n+r}(\overline{k})$. Equation (3.31) for $s = \xi \in \mathfrak{X}(M)^\perp$ becomes

$$R^f(X,Y)\xi = 0$$

and hence Ricci's equation (3.27) becomes

$$g^f(R^\perp(X,Y)\xi,\eta) = g([A_\xi, A_\eta]X, Y). \tag{3.37}$$

Let $\{\xi_a : 1 \le a \le r\}$ be a local g^f-orthonormal frame in the normal bundle $T(M)^\perp \to M$, defined on the open set $U \subset M$, and let us set $A_a = A_{\xi_a}$ for $1 \le a \le r$. Moreover

$$B(X,Y) = B^a(X,Y)\xi_a$$

where the local bilinear forms $B^a \in \mathcal{T}^{0,2}(U)$ are uniquely determined by the above representation. Then

$$g^f(B(X,Y),\xi_a) = g(A_a X, Y)$$

implies

$$B^a(X,Y) = g(A_a X, Y), \quad X, Y \in \mathfrak{X}(U). \tag{3.38}$$

In particular

$$A_{B(Y,Z)}X = B^a(Y,Z)A_a X = \sum_{a=1}^r g(A_a Y, Z)A_a X$$

and hence

$$A_{B(Y,Z)}X - A_{B(X,Z)}Y = \sum_{a=1}^r \{g(A_a Y, Z)A_a X - g(A_a X, Z)A_a Y\} =$$

$$= \sum_{a=1}^r (A_a X \wedge A_a Y)\, Z$$

and Gauss equation (3.35) may be locally written as

$$R(X, Y) = \overline{k}\; X \wedge Y + \sum_{a=1}^{r} A_a X \wedge A_a Y. \tag{3.39}$$

3.6 Totally Geodesic and Minimal Submanifolds

Let $(\overline{M}, \overline{g})$ be a $(n + r)$-dimensional semi-Riemannian manifold and let $f : M \to \overline{M}$ be an immersion such that $(d_p f)T_p(M)$ be nondegenerate as a subspace of the inner product space $(T_{f(p)}(\overline{M}), \overline{g}_{f(p)})$ for any point $p \in M$. Then the first fundamental form g is a semi-Riemannian metric on M.

Definition 3.28 An immersion $f : M \to \overline{M}$ is *totally geodesic* if for any geodesic γ of (M, g) the differentiable curve $\overline{\gamma} = f \circ \gamma$ is a geodesic of the ambient space $(\overline{M}, \overline{g})$. If this is the case M is called a *totally geodesic submanifold* of $(\overline{M}, \overline{g})$. □

Proposition 3.29 *Let $f : M \to \overline{M}$ be an embedding. Then M is a totally geodesic submanifold of $(\overline{M}, \overline{g})$ if and only if the second fundamental form of f vanishes identically $(B = 0)$.*

Proof Let us assume that $B = 0$. Then [by Gauss formula (3.26)]

$$f^{\alpha}{}_{|ij} + \left(\left\{ {\alpha \atop \beta\lambda} \right\} \circ f \right) f^{\beta}{}_{|i} f^{\lambda}{}_{|j} = \left\{ {k \atop ij} \right\} f^{\alpha}{}_{|k}\,. \tag{3.40}$$

If $\gamma : I \to M$ is a geodesic of (M, g) e $\overline{\gamma} = f \circ \gamma$ then

$$\frac{d^2\gamma^i}{dt^2}(t) + \left\{ {i \atop jk} \right\} (\gamma(t)) \frac{d\gamma^j}{dt}(t) \frac{d\gamma^k}{dt}(t) = 0,$$

$$\frac{d\overline{\gamma}^{\alpha}}{dt}(t) = f^{\alpha}{}_{|i}(\gamma(t)) \frac{d\gamma^i}{dt}(t), \tag{3.41}$$

$$\frac{d^2\overline{\gamma}^{\alpha}}{dt^2}(t) = f^{\alpha}{}_{|ij}(\gamma(t)) \frac{d\gamma^i}{dt}(t) \frac{d\gamma^j}{dt}(t) + f^{\alpha}{}_{|k}(\gamma(t)) \frac{d^2\gamma^k}{dt^2}(t), \tag{3.42}$$

hence

$$\frac{d^2\overline{\gamma}^\alpha}{dt^2}(t)+\left\{\begin{matrix}\alpha\\ \beta\lambda\end{matrix}\right\}(\overline{\gamma}(t))\frac{d\overline{\gamma}^\beta}{dt}(t)\frac{d\overline{\gamma}^\lambda}{dt}(t)=$$

$$=f^\alpha{}_{|ij}(\gamma(t))\frac{d\gamma^i}{dt}(t)\frac{d\gamma^j}{dt}(t)+f^\alpha{}_{|k}(\gamma(t))\frac{d^2\gamma^k}{dt^2}(t)+$$

$$+\left\{\begin{matrix}\alpha\\ \beta\lambda\end{matrix}\right\}(f(\gamma(t)))f^\beta{}_{|i}(\gamma(t))\frac{d\gamma^i}{dt}(t)f^\lambda{}_{|j}(\gamma(t))\frac{d\gamma^j}{dt}(t)=$$

$$=\left[f^\alpha{}_{|ij}+\left(\left\{\begin{matrix}\alpha\\ \beta\lambda\end{matrix}\right\}\circ f\right)f^\beta{}_{|i}f^\lambda{}_{|j}\right](\gamma(t))\,\dot{\gamma}^i(t)\,\dot{\gamma}^j(t)+$$

$$+f^\alpha{}_{|k}(\gamma(t))\frac{d^2\gamma^k}{dt^2}(t)=$$

[by Eq. (3.40)]

$$=\left[\left\{\begin{matrix}k\\ ij\end{matrix}\right\}(\gamma(t))+\frac{d^2\gamma^k}{dt^2}(t)\right]f^\alpha{}_{|k}(\gamma(t))=0$$

so that $\overline{\gamma}$ is an auto-parallel curve of $\overline{\nabla}$. Q.e.d.

Viceversa, let us assume that any geodesic of the submanifold is also a geodesic of the ambient space. Let $p\in M$ and $v\in T_p(M)$ and let $\gamma:(-\epsilon,\epsilon)\to M$ be the geodesic of (M,g) of initial data $\gamma(0)=p$ and $\dot{\gamma}(0)=v$. Then

$$B_p(v,v)=v^iv^jB^\alpha_{ij}(p)(\partial_\alpha)_{f(p)}=$$

[by the Gauss formula (3.26)]

$$=v^iv^j\left[f^\alpha{}_{|ij}+\left(\left\{\begin{matrix}\alpha\\ \beta\lambda\end{matrix}\right\}\circ f\right)f^\beta{}_{|i}f^\lambda{}_{|j}-\left\{\begin{matrix}k\\ ij\end{matrix}\right\}f^\alpha{}_{|k}\right](p)\,s_\alpha(p)=$$

$$=\big[\dot{\gamma}^i(0)\dot{\gamma}^j(0)f^\alpha{}_{|ij}(\gamma(0))+$$

$$+\left\{\begin{matrix}\alpha\\ \beta\lambda\end{matrix}\right\}(\overline{\gamma}(0))\dot{\gamma}^i(0)f^\beta{}_{|i}(\gamma(0))\dot{\gamma}^j(0)f^\lambda{}_{|j}(\gamma(0))\Big]\,s_\alpha(p)-$$

$$-\dot{\gamma}^i(0)\dot{\gamma}^j(0)\left\{\begin{matrix}k\\ ij\end{matrix}\right\}(\gamma(0))f^\alpha{}_{|k}(\gamma(0))s_\alpha(p)=$$

[as γ is a geodesic of (M, g)]

$$= \left[\dot{\gamma}^i(0)\dot{\gamma}^j(0) f^\alpha{}_{|ij}(\gamma(0)) + \right.$$

$$\left. + \left\{ {\alpha \atop \beta\lambda} \right\} (\overline{\gamma}(0))\dot{\gamma}^i(0) f^\beta{}_{|i}(\gamma(0))\dot{\gamma}^j(0) f^\lambda{}_{|j}(\gamma(0)) \right] s_\alpha(p) -$$

$$+ \frac{d^2\gamma^k}{dt^2}(0) f^\alpha{}_{|k}(\gamma(0)) s_\alpha(p) =$$

[by (3.41)–(3.42)]

$$= \left[\frac{d^2\overline{\gamma}^\alpha}{dt^2}(0) + \left\{ {\alpha \atop \beta\lambda} \right\} (\overline{\gamma}(0)) \frac{d\overline{\gamma}^\beta}{dt}(0) \frac{d\overline{\gamma}^\lambda}{dt}(0) \right] s_\alpha(p) = 0$$

as $\overline{\gamma} = f \circ \gamma$ is a geodesic of $(\overline{M}, \overline{g})$. We have proved that $B_p(v, v) = 0$ and then (by the symmetry of B) it follows that $B = 0$. Q.e.d.

3.7 Cartan's Theorem

The purpose of this section is to give a classical characterization (due to E. Cartan[6]) of constant curvature spaces in terms of their totally geodesic submanifolds.

Definition 3.30 Let (M, g) be a n-dimensional Riemannian manifold with $n \geq 3$. Let $r \in \mathbb{N}$ such that $2 \leq r \leq n - 1$. We say M satisfies the *axiom of r-planes* if for any point $p \in M$ and any r-dimensional subspace $V \subset T_p(M)$ there is a totally geodesic submanifold $S \subset M$ such that $p \in S$ e $T_p(S) = V$. □

Definition 3.30 was proposed by E. Cartan[7] who also established the following.[8]

Theorem 3.31 *If a Riemannian manifold M of dimension $n \geq 3$ satisfies the axiom of r-planes for some $2 \leq r < n$ then M is a space of constant sectional curvature.*

To prove Theorem 3.31 we need

Lemma 3.32 *Let (M, g) be a Riemannian manifold of dimension $n \geq 3$ and $p \in M$. If for every choice of g_p-orthonormal vectord $\{u, v, w\} \subset T_p(M)$ one has $g_p(R_p(u, v)w, u) = 0$ then M is isotropic at p, i.e., all 2-planes $\sigma \subset T_p(M)$ have the same sectional curvature.*

---

[6] Cf. E. Cartan, *Leçons sur la geometrie des espaces de Riemann* (Gauthier-Villars, Paris, 1946).

[7] Cf. *op. cit.*

[8] A generalization to Finslerian geometry was provided in the work by S. Dragomir, The theorem of É. Cartan on Finsler manifolds, *Proceedings of the National Colloquium on Geometry and Topology* (Buşteni, 1981), pp. 103–112.

Élie Cartan (1869–1951), French mathematician. **C.** published 186 papers in the period 1893–1947. Most of these papers are looked at as fundamental for differential geometry and the theory of partial differential equations. **C.** introduced the notion of differential form and created exterior differential calculus. It is difficult to overestimate the relevance of **C.**'s works (devoted mainly to the study of Lie groups - according to **C.** - but actually embracing all geometric differential themes at the level of the first half of the twentieth century) or the great influence exerted by **C.** on scientists of his time (cf. e.g., the correspondence [cf. *Hommage à Élie Cartan*, 1869–1951, Editura Academiei Republicii Socialiste Romania, Bucarest, 1975] between **C.** and other mathematicians such as G. Tzitzeica, A. Pantazi and G. Vrânceanu).

Proof We set

$$v' = \frac{1}{\sqrt{2}}(v + w), \quad w' = \frac{1}{\sqrt{2}}(v - w),$$

so that $\{v', w'\}$ are g_p-orthonormal. On the other hand, by he very hypothesis of Lemma 3.32

$$0 = g_p(R_p(u, v')w', u) =$$

$$= \frac{1}{2}\Big\{g_p(R_p(u, v)v, u) + \underbrace{g_p(R_p(u, v)w, u)}_{=0} +$$

$$+ \underbrace{g_p(R_p(u, w)v, u)}_{=0} - g_p(R_p(u, w)w, u)\Big\} =$$

$$= k(u \wedge v) - k(u \wedge w)$$

where $u \wedge v$ is the 2-plane in $T_p(M)$ spanned by the vectors $\{u, v\}$. It follows that $k(u \wedge v) = k(u \wedge w)$, i.e., the 2-planes $u \wedge v$ and $u \wedge w$ have the same sectional curvature.

Now let us consider two arbitrary 2-planes $\sigma, \sigma' \subset T_p(M)$. We distinguish two cases, as

(I) $\sigma \cap \sigma' \neq (0)$

or

(II) $\sigma \cap \sigma' = (0)$.

In the first case, let $u \in \sigma \cap \sigma'$ be a unit vector. Let us complete u to an orthonormal basis $\{u, v\} \subset \sigma$, respectively to an orthonormal basis $\{u, w\} \subset \sigma'$. Then

$$k(\sigma) = k(u \wedge v) = k(u \wedge w) = k(\sigma').$$

In the second case let us consider an orthonormal basis $\{u, v\} \subset \sigma$ and complete it to an orthonormal basis $\{u, v, w, z\} \subset \sigma \oplus \sigma'$ such that $\{w, z\}$ be an orthonormal basis of σ'. Then

$$k(\sigma) = k(u \wedge v) = k(u \wedge w) = k(z \wedge w) = k(\sigma').$$

Q.e.d.

Proof of Theorem 3.31 Let $V \subset T_p(M)$ be a r-dimensional subspace. As $r \geq 2$ there exists an orthonormal system $\{u, v\} \subset V$. Moreover by $r \leq n - 1$ there is a unit vector $w \in T_p(M)$ orthogonal to V. Then $\{u, v, w\}$ is an orthonormal system in $T_p(M)$. As M satisfies the axiom of r-planes we may consider a totally geodesic submanifold $S \subset M$ passing through p and such that $T_p(S) = V$. Let $X, Y \in \mathfrak{X}(S)$ and $\xi \in \mathfrak{X}(S)^\perp$ be vector fields such that $X_p = u$, $Y_p = v$ e $\xi_p = w$. The identity (??)

$$R^f(X, Y)\xi = R^\perp(X, Y)\xi + B(Y, A_\xi X) - B(X, A_\xi Y) -$$

$$- f_*\{(\nabla_X A)(\xi, Y) - (\nabla_Y A)(\xi, X)\}$$

for $f = \iota : S \subset M$ and with $B = 0$ [as S is totally geodesic] may be written

$$R^\iota(X, Y)\xi = R^\perp(X, Y)\xi$$

and hence at the point p

$$(R^\iota)_p(u, v)w = (R^\perp)_p(u, v)w \in T_p^\perp(S),$$

i.e., $(R^\iota)_p(u, v)w$ is orthogonal to V and in particular on u

$$g_p((R^\iota)_p(u, v)w, u) = 0. \tag{3.43}$$

On the other hand [by (3.28) with $s_\alpha = (\partial/\partial y^\alpha)^\iota$]

$$(R^\iota)_p(u,v)w = (R^\iota)(u,v)(d_p\iota)w = u^i v^j w(\iota^\gamma)\left\{R^\iota(\partial_i,\partial_j)s_\gamma\right\}_p =$$

$$= u^i v^j w(\iota^\gamma) R^\lambda_{\alpha\beta\gamma}(\iota(p))\iota^\alpha{}_{|i}(p)\iota^\beta{}_{|j}(p)s_\lambda(p) =$$

$$= u(\iota^\alpha)v(\iota^\beta)w(\iota^\gamma)\left\{R(\partial_\alpha,\partial_\beta)\partial_\gamma\right\}_{\iota(p)} =$$

$$= R_{\iota(p)}((d_p\iota)u,(d_p\iota)v)(d_p\iota)w = R_p(u,v)w$$

and then [by (3.43)] $g_p(R_p(u,v)w,u) = 0$ and one may apply Lemma 3.32 to conclude that M is isotropic in p. As p is an arbitrary point and $n \geq 3$ we may conclude [by Schur's theorem] that M is a space form. Q.e.d.

3.8 Nomizu's Theorem

One of the most difficult problems in the theory of submanifolds is to determine the totally geodesic submanifolds of a given Riemannian manifold. K. Nomizu contributed with a result relating totally geodesic submanifolds to the structure of the fixed point set of an isometry. It is our purpose in this section to present K. Nomizu's contribution.[9] To start with we need some preparation regarding isometries of Riemannian manifolds. Let $F : M \to N$ be a differentiable map and let $h \in \mathcal{T}^{0,2}(N)$ be a tensor field of type $(0,2)$ on N.

Definition 3.33 The *pullback* of h by F is the tensor field $F^*h \in \mathcal{T}^{0,2}(M)$ defined by

$$(F^*h)_p(u,v) = h_{F(p)}((d_pF)u,(d_pF)v), \quad u,v \in T_p(M), \quad p \in M.$$

□

Locally

$$(F^*h)_{ij}(p) = (F^*h)_p((\partial_i)_p,(\partial_j)_p) = F^\alpha{}_{|i}(p)F^\beta{}_{|j}(p)h_{\alpha\beta}(F(p)),$$

$$h_{\alpha\beta} = h(\partial_\alpha,\partial_\beta),$$

hence

$$(F^*h)_{ij} = F^\alpha{}_{|i}\,F^\beta{}_{|j}\,\left(h_{\alpha\beta}\circ F\right). \tag{3.44}$$

---

[9] Cf. K. Nomizu, *Introduzione alla geometria differenziale*, Appunti Scuola Matematica Interuniversitaria (1975).

Definition 3.34 An *isometry* of the Riemannian manifold (M, g) is a C^{∞} diffeomorphism $F : M \to M$ of M such that

$$g_{F(p)}(d_p F)u, (d_p F)v) = g_p(u, v)$$

for any $u, v \in T_p(M)$ and any $p \in M$. □

Equivalently $F : M \to M$ is an isometry if $F^* g = g$ or locally [by (3.44)]

$$g_{rs} = F^i{}_{|r} F^j{}_{|s} \left(g_{ij} \circ F\right). \tag{3.45}$$

This has implications in terms of the Christoffel symbols, implications that will be investigated shortly. As F is a diffeomorphism, the Jacobi matrix $\left[F^i{}_{|j}\right]$ is invertible. Then let us set

$$\left[G^i_j\right] = \left[F^i{}_{|j}\right]^{-1}.$$

Let us contract (3.45) with G^r_k. One obtains

$$g_{rs} G^r_k = F^j{}_{|s} \left(g_{kj} \circ F\right).$$

Now we may contract with g^{si} to obtain

$$G^i_k = g^{si} F^j{}_{|s} \left(g_{kj} \circ F\right).$$

The contraction with $g^{km} \circ F$ leads to

$$G^i_k \left(g^{km} \circ F\right) = g^{si} F^m{}_{|s}$$

and finally the contraction with G^{ℓ}_m furnishes

$$g^{\ell i} = G^i_k G^{\ell}_m \left(g^{km} \circ F\right). \tag{3.46}$$

Then [by (3.45)]

$$\left\{rs, \ell\right\} = \frac{1}{2} \left(g_{r\ell|s} + g_{s\ell|r} - g_{rs|\ell}\right) =$$

$$= F^i{}_{|rs} F^j{}_{|\ell} \left(g_{ij} \circ F\right) + F^i{}_{|r} F^j{}_{|\ell} F^m{}_{|s} \left(\left\{im, j\right\} \circ F\right),$$

hence [by (3.46)]

$$\left\{ \begin{matrix} k \\ rs \end{matrix} \right\} = g^{k\ell}\{rs, \ell\} =$$

$$= G^k_a G^{\ell}_b \left(g^{ab} \circ F\right) \left\{F^i{}_{|rs} F^j{}_{|\ell} \left(g_{ij} \circ F\right) + F^i{}_{|r} F^j{}_{|\ell} F^m{}_{|s} (|im, j| \circ F)\right\} =$$

$$= G^k_a \left[F^a{}_{|rs} + F^i{}_{|r} F^m{}_{|s} \left(\left\{ \begin{matrix} a \\ im \end{matrix} \right\} \circ F\right)\right]$$

and the contraction with $F^\ell{}_{|k}$ leads to

$$F^\ell{}_{|k} \left\{ {k \atop rs} \right\} = F^\ell{}_{|rs} + F^i{}_{|r} F^m{}_{|s} \left(\left\{ {\ell \atop im} \right\} \circ F \right). \tag{3.47}$$

Theorem 3.35 *Let $F : M \to M$ be an isometria of the Riemannian manifold M and let $K = \{p \in M : F(p) = p\}$ be the set of fixed points of F. Then any connected component of K is a totally geodesic submanifold of M.*

The proof includes a few arguments of topology.[10]

Proof[11] ***of Theorem*** 3.35 For any fixed point $p \in K$ we set by definition

$$V_p = \{u \in T_p(M) : (d_p F)u = u\}.$$

By the construction of the exponential map, there is a star-shaped neighborhood $N_0 \subset T_p(M)$ such that $0_p \in N_0$ and a neighborhood $U \subset M$ of p such that $\exp_p : N_0 \to U$ is a diffeomorphism. We shall show that

Lemma 3.36 $\exp_p$ *is a diffeomorphism of $V_p \cap N_0$ onto $K \cap U$.*

To prove Lemma 3.36 one needs

Lemma 3.37 *If $u \in V_p \cap N_0$ then the geodesic*

$$\gamma(t) = \exp_p(tu), \quad t \in [0, 1]$$

consists only of fixed points of the transformation $F : M \to M$.

---

[10] For a better understanding of those arguments we briefly recall the needed elements of topology. Let $\mathfrak{X}$ be a topological space and $x \in \mathfrak{X}$ a point. Let L_x be the largest connected subset containing x, i.e., L_x is the union of all connected sets $A \subset \mathfrak{X}$ such that $x \in A$. The set L_x is closed. Indeed the closure of a connected set is connected, as well. Hence [as $x \in \overline{L}_x$ and $\overline{L}_x$ connected] $\overline{L}_x \subset L_x$ hence $L_x = \overline{L}_x$. Q.e.d. The closed set L_x is the *connected component* of the point x in the topological space $\mathfrak{X}$. Let L_x and L_y be the connected components of the points $x, y \in X$ with $x \neq y$. Then either (I) $L_x = L_y$, or (II) $L_x \cap L_y = \emptyset$. Indeed if $L_x \cap L_y \neq \emptyset$ then let $z \in L_x \cap L_y$. On one hand the set L_x [respectively L_y) is connected and contains the point z so that $L_x \subset L_z$ [respectively $L_y \subset L_z$]. In particular $x, y \in L_z$ so that [as L_z is a connected set] it must be that $L_z \subset L_x$ and $L_z \subset L_y$. We may conclude that $L_x = L_z = L_y$. In the second case the sets L_x and L_y are separated because of $L_x \cap \overline{L}_y = L_x \cap L_y = \emptyset$ and similarly $\overline{L}_x \cap L_y = \emptyset$. Of course we may write $\mathfrak{X} = \bigcup_{x \in X} L_x$ and successively write just once each distinct connected component. We proved that any topological space may be decomposed as the union of its connected components, which are closed and mutually disjoint. In general, the connected components of $\mathfrak{X}$ aren't open (this may be proved by looking at appropriate examples). Nevertheless if $\mathfrak{X}$ possesses a finite number of connected components then these are open sets. Let us denote by $\pi_0(X) = \{L_x : x \in X\}$ the set of all connected components of $\mathfrak{X}$. If $\mathfrak{X}$ e $\mathfrak{Y}$ are homeomorphic topological spaces then $|\pi_0(X)| = |\pi_0(Y)|$ (where $|A|$ is the cardinality of the set A).

[11] A generalization of Nomizu's theorem (Theorem 3.35) to the realm of Finslerian geometry was provided in the work by S. Dragomir, The theorem of K. Nomizu on Finsler manifolds. An. Univ. Timişoara, Ser. Ştiinţ. Mat. **19**(1), 41–30 (1981).

Proof Let us consider the curve $\beta(t) = F(\gamma(t))$, $0 \le t \le 1$. Then

$$\frac{d\beta^i}{dt}(t) = F^i{}_{|j}(\gamma(t))\frac{d\gamma^j}{dt}(t),$$

$$\frac{d^2\beta^i}{dt^2}(t) = F^i{}_{|jk}(\gamma(t))\frac{d\gamma^k}{dt}(t)\frac{d\gamma^j}{dt}(t) + F^i{}_{|j}(\gamma(t))\frac{d^2\gamma^j}{dt^2}(t),$$

and then

$$\frac{d^2\beta^i}{dt^2}(t) + \left\{ \begin{matrix} i \\ jk \end{matrix} \right\}(\beta(t))\frac{d\beta^j}{dt}(t)\frac{d\beta^k}{dt}(t) =$$

$$= F^i{}_{|jk}(\gamma(t))\frac{d\gamma^k}{dt}(t)\frac{d\gamma^j}{dt}(t) + F^i{}_{|j}(\gamma(t))\frac{d^2\gamma^j}{dt^2}(t) +$$

$$+ \left\{ \begin{matrix} i \\ jk \end{matrix} \right\}(\beta(t))F^j{}_{|\ell}(\gamma(t))\frac{d\gamma^\ell}{dt}(t)F^k{}_{|m}(\gamma(t))\frac{d\gamma^m}{dt}(t) =$$

[substituting from (3.47) the term $\left(\left\{ \begin{matrix} i \\ jk \end{matrix} \right\} \circ F\right) F^j{}_{|\ell} F^k{}_{|m}$ evaluated at $\gamma(t)$]

$$= F^i{}_{|jk}(\gamma(t))\dot{\gamma}^k(t)\dot{\gamma}^j(t) + F^i{}_{|j}(\gamma(t))\frac{d^2\gamma^j}{dt^2}(t) +$$

$$+ \dot{\gamma}^\ell(t)\dot{\gamma}^m(t)\left\{\left| \begin{matrix} s \\ \ell m \end{matrix} \right|(\gamma(t))F^i{}_{|s}(\gamma(t)) - F^i{}_{\ell m}(\gamma(t))\right\} =$$

$$= \left\{\frac{d^2\gamma^j}{dt^2}(t) + \left| \begin{matrix} j \\ \ell m \end{matrix} \right|(\gamma(t))\dot{\gamma}^\ell(t)\dot{\gamma}^m(t)\right\} F^i{}_{|j}(\gamma(t)) = 0$$

ad hence β is a geodesic of (M, g), too. On the other hand

$$\beta(0) = F(\gamma(0)) = F(p) = p,$$

$$\dot{\beta}(t) = (d_{\gamma(t)}F)\dot{\gamma}(t) \Longrightarrow \dot{\beta}(0) = (d_pF)\dot{\gamma}(0) = (d_pF)u = u,$$

as $u \in V_p$ to start with. The geodesics γ and β have the same initial data and hence coincide. Then

$$F(\gamma(t)) = \beta(t) = \gamma(t), \quad 0 \le t \le 1,$$

i.e., $\gamma(t) \in K$ for any $t \in [0, 1]$. Lemma 3.37 is proved.

Now let us get back to the proof of Theorem 3.35, and precisely to the proof of Lemma 3.36. It suffices to check that

$$\exp_p\left(V_p \cap N_0\right) = K \cap U.$$

That equality may be proved by double inclusion. Let us consider a point $q \in \exp_p\left(V_p \cap N_0\right)$. Then there is $u \in V_p \cap N_0$ such that $q = \exp_p u = \gamma(1)$ where $\gamma(1)$ belongs to U because of $\exp_p(N_0) = U$ and at the same time $\gamma(1)$ belongs to K by Lemma 3.37. It follows that $q \in K \cap U$ and then the inclusion

$$\exp_p(V_p \cap N_0) \subset K \cap U$$

is proved.

Viceversa let $q \in K \cap U$ and [as $q \in U$] let $u \in N_0$ such that $\exp_p(u) = q$. Let us consider the geodesic $\gamma(t) = \exp_p(tu)$ and let us set

$$\beta(t) = F(\gamma(t)), \quad \alpha(t) = \exp_p(t\,(d_pF)u), \quad 0 \le t \le 1.$$

By the proof of Lemma 3.37 it follows that β is a geodesic of (M, g). Hence both curves α and β are geodesics of (M, g) of initial data $(p,\ (d_pF)u)$. Then $\alpha = \beta$. As $q \in K$, i.e., q is a fixed point of F one has

$$\exp_p u = q = F(q) = F(\exp_p u) = F(\gamma(1)) =$$

$$= \beta(1) = \alpha(1) = \exp_p\left[(d_pF)u\right]$$

and hence [by the injectivity of $\exp_p$]

$$u = (d_pF)u,$$

i.e., $u \in V_p$. It follows that $u \in V_p \cap N_0$ and then $q \in \exp_p\left(V_p \cap N_0\right)$. We have proved the opposite of inclusion

$$K \cap U \subset \exp_p\left(V_p \cap N_0\right)$$

and therefore the desired equality. Lemma 3.36 is proved.

Let $L \in \pi_0(K)$ be a connected component of K thought of as a topological subspace of M. We must show that L is a totally geodesic submanifold of (M, g). The most delicate part of the proof consists in organizing L as a differentiable manifold. Let us consider a point $p \in L$ so that $L = L_p$. Let (U, x^i) be a local coordinate system on M such that $p \in U$. The tangent space to M at p is a topological space with the topology induced by the isomorphism

$$h_p : T_p(M) \to \mathbb{R}^n, \quad h_p(u) = (u^1, \dots, u^n), \quad u = u^i \left(\frac{\partial}{\partial x^i}\right)_p .$$

Said topology coincides with the topology associated to the norm

$$\|u\| = g_p(u, u)^{1/2}, \quad u \in T_p(M). \tag{3.48}$$

Let $\tilde{U} \subset M$ be a normal neighborhood of p in M, i.e., there is a star-shaped neighborhood $\tilde{N}_0 \subset T_p(M)$ of the zero vector 0_p such that $\exp_p : \tilde{N}_0 \to \tilde{U}$ is a diffeomorphism. With respect to the topology induced by $T_p(M)$ on V_p [the set of all fixed points of the linear isomorphism $d_p F$] the set $\tilde{N}_0 \cap V_p$ is open in V_p. Moreover the topology induced by $T_p(M)$ on V_p coincides with the norm topology, that is with the topology determined by the restriction to V_p of the norm (3.48). Let us consider $r > 0$ such that the ball

$$N_0 \equiv \{u \in T_p(M) : \|u\| < r\}$$

be contained in $\tilde{N}_0$ and let us set $U = \exp_p(N_0) \subset \tilde{U}$ so that $\exp_p : N_0 \to U$ is a diffeomorphism. On the other hand

$$N_0 \cap V_p = \{u \in V_p : \|u\| < r\}$$

so that $N_0 \cap V_p$ is a ball in V_p and then $N_0 \cap V_p$ is a connected subset of V_p. By Lemma 3.36 the map

$$\exp_p : N_0 \cap V_p \to U \cap K$$

is a diffeomorphism and hence $U \cap K$ is a connected subset of K containing the point p. By the maximality of L it follows that $U \cap K \subset L$. In particular

$$U \cap K = (U \cap K) \cap L$$

and then $U \cap K$ is an open subset of L. We shall build a local chart on L about the point $p \in L$, whose domain will be the set $U \cap K$.

The topology of V_p coincides with the topology induced by the restriction of $h_p : T_p(M) \to \mathbb{R}^n$ to V_p, i.e., by the bijection $h_p : V_p \to \mathbb{R}^r$ where we have set $r = \dim_{\mathbb{R}} V_p$. It follows that $h_p : N_0 \cap V_p \to \mathbb{R}^r$ is a homeomorphism on its image and hence the map

$$\psi_p : U \cap K \to h_p\left(N_0 \cap V_p\right) \subset \mathbb{R}^r,$$

$$\psi_p = h_p \circ \left(\exp_p\big|_{N_0 \cap V_p}\right)^{-1}$$

is a local chart on L about the point p.

Thus we have organized L as a differentiable manifold of dimension r. Next, we shall check that the canonical inclusion $\iota : L \to M$ is an immersion (with respect to the differentiable structure just set on L). The elements of the problem to be solved may be summed up in the following commutative diagram

$$\begin{array}{ccc} L \supset \quad U \cap K & \stackrel{\iota}{\Longrightarrow} & U \subset M \\ \psi_p \downarrow\uparrow \psi_p^{-1} & & \downarrow \quad (x^1, \ldots, x^n) \equiv \psi \\ \mathbb{R}^r \supset h_p(N_0 \cap V_p) & \stackrel{\underline{\iota}}{\Longrightarrow} & \mathbb{R}^n, \end{array}$$

where $\underline{\iota}$ denotes the local expression of ι. As well known, we may use just any local chart on the target manifold. We choose to use a normal coordinate system centered at p, i.e., we fix an arbitrary linear frame $u : \mathbb{R}^n \to T_p(M)$ (a $\mathbb{R}$-linear isomorphism) and we set

$$\psi = (\exp_p \circ u)^{-1} : U \to N_0 .$$

Then

$$\underline{\iota}(\xi^1, \ldots, \xi^r) = (\psi \circ \psi_p^{-1})(\xi^1, \ldots, \xi^r) = (\xi^1, \ldots, \xi^r, \underbrace{0, \ldots, 0}_{n-r \text{ terms}}),$$

$$\forall \ (\xi^1, \ldots, \xi^r) \in h_p(N_0 \cap V_p),$$

i.e., locally ι is the canonical immersion $\mathbb{R}^r \hookrightarrow \mathbb{R}^n$.

It remains to be checked that L is a totally geodesic submanifold of M. Let ∇ and B be respectively the induced connection and the second fundamental form of the given immersion $\iota : L \to M$. Let $\gamma : (-\epsilon, \epsilon) \to L$ be an auto-parallel curve of ∇ and let us set $p = \gamma(0) \in L$. Let $(U \cap K, \psi_p)$ be a local chart of L about p, built as above. We may assume that $\epsilon > 0$ is sufficiently small, such that $\gamma(t) \in U \cap K$ for any $|t| < \epsilon$. In particular $F(\gamma(t)) = \gamma(t)$ fro any $|t| < \epsilon$. It follows that $(d_p F)u = u$ where we have set $u = \dot{\gamma}(0)$, or $u \in V_p$. For $\epsilon > 0$ sufficiently small we may also assume that $tu \in N_0$ for any $|t| < \epsilon$. Let us set

$$\gamma_1 : (-\epsilon, \epsilon) \to M, \quad \gamma_1(t) = \exp_p(t\,u), \quad |t| < \epsilon,$$

so that γ_1 is the geodesic of (M, g) of initial conditions (p, u). Moreover one has $tu \in V_p$ [as V_p is a vector space] and hence $tu \in N_0 \cap V_p$. It follows that $\exp_p(tu) \in U \cap K$ that is $\gamma_1(t) \in U \cap K$ for any $|t| < \epsilon$, and then $\dot{\gamma}_1(t)$ is tangent to L. By Lemma 3.38 (stated below) it follows that γ_1 is a geodesic of L, too, having the same initial data as the geodesic γ, so that it must be $\gamma = \gamma_1$, i.e., γ is a geodesic of (M, g), too. Q.e.d.

Lemma 3.38 *Let $f : M \to \overline{M}$ be an embedding. Let γ be a geodesic of $(\overline{M}, \overline{g})$ such that there is a curve a in M with $f \circ a = \gamma$. Then a is a geodesic of (M, g).*

Proof By $\gamma = f \circ a$

$$\frac{d\gamma^\alpha}{dt}(t) = f^\alpha{}_{|i}(a(t))\frac{da^i}{dt}(t),$$

$$\frac{d^2\gamma^\alpha}{dt}(t) = f^\alpha{}_{|ij}(a(t))\frac{da^i}{dt}(t)\frac{da^j}{dt}(t) + f^\alpha{}_{|i}(a(t))\frac{d^2a^i}{dt^2}(t),$$

so that

$$0 = \frac{d^2\gamma^\alpha}{dt^2}(t) + \left\{ {\alpha \atop \beta\lambda} \right\}(a(t))\frac{d\gamma^\beta}{dt}(t)\frac{d\gamma^\lambda}{dt}(t) =$$

$$= \left[f^\alpha{}_{|ij} + \left\{ {\alpha \atop \beta\lambda} \right\} f^\beta{}_{|i} f^\lambda{}_{|j} \right](a(t))\frac{da^i}{dt}(t)\frac{da^j}{dt}(t) + f^\alpha{}_{|k}(a(t))\frac{d^2a^k}{dt^2}(t) =$$

[by Gauss equation $f^\alpha{}_{|ij} + \left(\left\{ {\alpha \atop \beta\lambda} \right\} \circ f \right) f^\beta{}_{|i} f^\lambda{}_{|j} = \left\{ {k \atop ij} \right\} f^\alpha{}_{|k} + B^\alpha_{ij}$]

$$= \left[\left\{ {k \atop ij} \right\}(a(t))\frac{da^i}{dt}(t)\frac{da^j}{dt}(t) + \right.$$

$$\left. + \frac{d^2a^k}{dt^2}(t) \right] f^\alpha{}_{|k}(a(t)) + B^\alpha_{ij}(a(t))\frac{da^i}{dt}(t)\frac{da^j}{dt}(t)$$

or [by contraction with $(\partial/\partial y^\alpha)_{f(a(t))}$]

$$\left[\left\{ {k \atop ij} \right\}(a(t))\frac{da^i}{dt}(t)\frac{da^j}{dt}(t) + \frac{d^2a^k}{dt^2}(t) \right] (f_*\partial_k)_{a(t)} +$$

$$+ B_{a(t)}(\dot{a}(t), \dot{a}(t)) = 0$$

so that [by the uniqueness of the direct sum decomposition]

$$\left[\left\{ {k \atop ij} \right\}(a(t))\frac{da^i}{dt}(t)\frac{da^j}{dt}(t) + \frac{d^2a^k}{dt^2}(t) \right] (f_*\partial_k)_{a(t)} = 0.$$

Finally, as $d_{a(t)} f$ is injective

$$\left\{ {k \atop ij} \right\}(a(t))\frac{da^i}{dt}(t)\frac{da^j}{dt}(t) + \frac{d^2a^k}{dt^2}(t) = 0,$$

i.e., a is a geodesic of (M, g). Q.e.d.

Katsumi Nomizu (1924–2008), Japanese mathematician. **N.** earned his Ph.D. at the University of Chicago in 1953. In 1963 **N.** was already a professor (tenured) at Brown University where he taught until his retirement (in 1995). **N.** supervised a large number of doctoral students (thirteen) and in general put great devotion into teaching mathematics. At the same time **N.** conducted an intense research activity within differential geometry, and in particular affine differential geometry. The influence of **N.**'s scientific creation on the differential geometric community is best illustrated by the universal use of one of his five monographs, the famous *Foundations of differential geometry*, written in collaboration with Kobayashi, cfr. [KoNo]. (**N.**'s photograph is reproduced with permission from the Archives of Mathematisches Forschungsinstitut Oberwolfach).

3.9 Submanifolds of Brieskorn Spheres

We now give a few applications of Theorem 3.35. To this end let us consider the positive integers $a_i \in \mathbb{Z}$, $a_i \geq 2$, $1 \leq i \leq n+1$. If $(z^1, \ldots, z^{n+1})$ are the Cartesian complex coordinates on $\mathbb{C}^{n+1}$ then let us consider the algebraic hypersurface $X_a \subset \mathbb{C}^{n+1}$ of equation

$$\left(z^1\right)^{a_1} + \cdots + \left(z^{n+1}\right)^{a_{n+1}} = 0.$$

V has a singularity at the origin. Let us set

$$B^{2n}(a) \equiv B^{2n}(a_1, \ldots, a_{n+1}) = X_a \setminus \{0\},$$

$$\Sigma^{2n-1}(a) \equiv \Sigma^{2n-1}(a_1, \ldots, a_{n+1}) = B^{2n}(a_1, \ldots, a_{n+1}) \cap S^{2n+1},$$

so that $B^{2n}(a)$ and $\Sigma^{2n-1}(a)$ are differentiable manifolds.

Definition 3.39 $B^{2n}(a)$ [respectively $\Sigma^{2n-1}(a)$] is called the *Brieskorn manifold* (respectively the *Brieskorn sphere*) determined by $a = (a_1, \ldots, a_{n+1})$. □

Cf. E. Brieskorn.[12] We shall use Nomizu's theorem (Theorem 3.35) to determine certain totally geodesic submanifolds of a Brieskorn sphere $\Sigma^{2n-1}(2p, a_2, \dots, a_{n+1})$. Let us consider the transformation $F : \mathbb{C}^{n+1} \to \mathbb{C}^{n+1}$ given by

$$F(z^1, \dots, z^{n+1}) = (-z^1, z^2, \dots, z^{n+1}), \quad (z^1, \dots, z^{n+1}) \in \mathbb{C}^{n+1}.$$

Lemma 3.40 *$F : \mathbb{C}^{n+1} \to \mathbb{C}^{n+1}$ is an isometry of $\mathbb{C}^{n+1} \approx \mathbb{R}^{2n+2}$ endowed with the Euclidean metric g_0. Moreover*

$$F\left(\Sigma^{2n-1}(2p, a')\right) \subset \Sigma^{2n-1}(2p, a')$$

where $a' = (a_2, \dots, a_{n+1})$.

The proof is left as an exercise for the reader. By Lemma 3.40 the transformation

$$F : \Sigma^{2n-1}(2p, a') \to \Sigma^{2n-1}(2p, a_{n+1})$$

is an isometry of $\Sigma^{2n-1}(2p, a')$. Here $\Sigma^{2n-1}(2p, a')$ is equipped with the induced Riemannian metric $\iota^* g_0$ [$\iota : \Sigma^{2n-1}(2p, a') \to \mathbb{C}^{n+1}$ is the inclusion]. Let

$$K = \{z \in \Sigma^{2n-1}(2p, a') : F(z) = z\}$$

be the set of fixed points of F. If $z \in K$ then

$$z = (0, z^2, \dots, z^{n+1}), \quad \|z\| = 1, \quad \left(z^2\right)^{a_2} + \cdots + \left(z^{n+1}\right)^{a_{n+1}} = 0.$$

It follows that $z \in \Sigma^{2n-3}(a')$ and hence that $K = \Sigma^{2n-3}(a')$. Finally, by Theorem 3.35 it follows that $\Sigma^{2n-3}(a')$ is a totally geodesic submanifold of $\Sigma^{2n-1}(2p, a')$.

3.10 Topology of Brieskorn Spheres

We review a few topological properties of Brieskorn spheres, by essentially following E. Brieskorn, *op. cit.*, and E. Brieskorn and A. Van de Ven.[13] The relationship between the presented topological properties and the geometry of the second fundamental form (of an isometric immersion) has not been studied so far. Said topological properties appear to be related to the existence of complex structures on even dimensional (products of) homotopy spheres, so the similar relationship is to be expected

[12] E. Brieskorn, *Ein Satz über Komplexe Quadriken*. Math. Ann. **155**, 184–193 (1964).

[13] E. Brieskorn, A. Van de Ven, Some complex structures on products of homotopy spheres. Topology **7**, 389–393 (1968).

in the odd dimensional case, as to the existence of CR structures.[14] Once again, the problem of the existence of CR structures is open.

By a well-known result of D. Mumford[15] no 2-dimensional normal complex space with singularities admits a topological manifold structure. No such statement holds for higher dimensions. The first example was provided by E. Brieskorn.[16] The singular hypersurface

$$X = \{(z_1, \dots, z_n) \in \mathbb{C}^n \ : \ z_1^2 + \cdots + z_{n-1}^2 + z_n^3 = 0\}$$

is a topological manifold for every n. Equivalent to that is the statement that the intersection of X with the sphere S^{2n-1}

$$\Sigma = \Sigma^{2n-3}(2, \dots, 2, 3) = X \cap S^{2n-1}$$

is homeomorphic to the $(2n-3)$-dimensional sphere. The manifold Σ also admits a natural differentiable structure. By a result of F. Hirzebruch[17] when $n \equiv 2$, mod 4, said natural differentiable structure is distinct from the standard differentiable manifold structure of S^{2n-3}.

Let $a = (a_1, \dots, a_n)$ with $a_i \in \mathbb{Z}$ and $a_i > 1$, and let $X_a = X(a_1, \dots, a_n)$ be the hypersurface of equation $z_1^{a_1} + \cdots + z_n^{a_n} = 0$, and let $\Sigma_a = \Sigma(a_1, \dots, a_n)$ be the compact orientable $(2n-3)$-dimensional manifold $\Sigma_a = X_a \cap S^{2n-1}$. The main result by E. Brieskorn, *op. cit.*, is.

Theorem 3.41 (i) *Σ_a is a topological sphere precisely when the characteristic polynomial $\Delta_a(t)$ satisfies $\Delta_a(1) = 1$.*

(ii) *$\Delta_a(1) = 1$ precisely when the graph G_a possesses one of the following two properties*

(a) *G_a has at least two isolated vertices.*

(b) *G_a has an isolated vertex and a component satisfying condition* (B).

The graph G_a in Theorem 3.41 is the weighted graph on n vertices

$$V(G_a) = \{v_1, \dots, v_n\}$$

[14] Whose study is undertaken by us in Book I-C of the present book series [i.e., Differential Geometry: *Foundations of Cauchy-Riemann and Pseudohermitian Geometry*].

[15] D. Mumford, The topology of normal singularities of an algebraic surface and a criterion for simplicity. Publ. Math. de l'Institute des hautes études scientifiques (9), Paris (1961).

[16] Cf. E. Brieskorn, Examples of singular normal complex spaces which are topological manifolds. Proc. Nat. Acad. Sci. **55**, 1395–1397 (1966).

[17] Cf. F. Hirzebruch, O(n)-*Mannigfaltigkeiten, exotische Sphären, kuriose Involutionen*, preprint.

each having *weight*

$$w(v_i) = a_i, \quad 1 \le i \le n.$$

Two vertices v_i and v_j in G_a are connected by an edge if the greatest common divisor of their weights is $(a_i, a_j) > 1$. Such G_a may then have connected components and isolated vertices. Given a component K of G_a the condition (B) in Theorem 3.41 is

Condition (B). *K consists of an odd number of vertices, and for any $a_i, a_j \in K$ with $i \neq j$ one has $(a_i, a_j) = 2$.*

Moreover, the polynomial $\Delta_a(t)$ in Theorem 3.41 is

$$\Delta_a(t) = \prod_{\mathbf{0}<(j_1, \dots, j_n)<a} \left(t - \xi_1^{j_1} \cdots \xi_n^{j_n}\right), \quad \xi_k := \exp \frac{2\pi i}{a_k}, \quad 1 \le k \le n.$$

Given two multi-indices $\alpha, \beta \in \mathbb{Z}_+^n$ we write $\alpha < \beta$ if $\alpha_k < \beta_k$ for every $1 \le k \le n$. Also $\mathbf{0}$ is the multi-index $\mathbf{0} = (0, \dots, 0) \in \mathbb{Z}_+^n$.

Theorem 3.41 admits[18] a corollary on rings with unique prime factorization (UPF-rings). It is known that the local ring of germs of holomorphic functions in an isolated singularity of a hypersurface of dimension > 3 is a UPF-ring. For 3-dimensional singularities said corollary is

Corollary 3.42 *If the graph $G(a_1, a_2, a_3, a_4)$ enjoys one of the properties* (a) *or* (b) *in Theorem 3.41, the local ring*

$$\mathbb{C}\{z_1, \dots, z_4\}/\left(z_1^{a_1} + z_2^{a_2} + z_3^{a_3} + z_4^{a_4}\right)$$

is a UPF-ring.

To state the second[19] and third[20] main results by E. Brieskorn, *op. cit.*, we briefly recall a few facts from M.A. Kervaire and J. Milnor.[21] Let $b\,P_{2m}$ be the group of $(2m-1)$-dimensional homotopy spheres that bound a $2m$-dimensional parallelizable manifold ($m > 2$). For even m, $b\,P_{2m}$ is a high-order cyclic group. For odd m, $b\,P_{2m} = 0$ or $b\,P_{2m} = \mathbb{Z}_2$. Specifically for $m \equiv 1$, mod 4, one has $b\,P_{2m} = \mathbb{Z}_2$ by a result of E.H. Brown and F.P. Peterson.[22] In order to decide for a homotopy sphere $\Sigma = \partial M$, with M parallelizable, which element in $b\,P_{2m}$ it represents, one has to calculate the *signature* $\sigma(M)$ of the intersection matrix of $H_m(M, \mathbb{Z})$ for even m, and for m odd a mod 2-invariant, the *Arf-invariant* $c(M)$.

---

[18] Cf. E. Brieskorn, *op. cit.*

[19] Making use of the Arf-invariant $c(M)$ of a parallelizable manifold M.

[20] Making use of the signature $\sigma(M)$ of a parallelizable manifold M.

[21] M.A. Kervaire, J. Milnor, Groups of homotopy spheres: I. Ann. Math. **77**, 504–537 (1963).

[22] E.H. Brown, F.P. Peterson, The Kervaire invariant of $(8k + 2)$-manifolds. Bull. Amer. Math. Soc. **70**, 670–675 (1964).

Theorem 3.43 *Let us assume that $\Sigma(a_1, \ldots, a_n)$ is a topological sphere for any even integer $n \geq 4$. Then the Arf-invariant $c(M_a)$ enjoys the following properties*

(i)

$$c(M_a) = 0 \ \textit{precisely when} \ \Delta_a(-1) \equiv \pm 1, \quad \text{mod } 8,$$

$$c(M_a) = 1 \ \textit{precisely when} \ \Delta_a(-1) \equiv \pm 3, \quad \text{mod } 8.$$

(ii) $c(M_a) = 1$ *if and only if G_a has precisely one isolated vertex a_k, and only one other component, and $a_k \equiv \pm 3$,* mod 8.

Theorem 3.44 *Let $\Sigma(a_1, \ldots, a_n)$ be a topological sphere, with $n \geq 5$ odd. Then the oriented diffeomorphism type of Σ_a is determined by the signature $\sigma(M_a)$. The signature $\sigma(M_a)$ is*

$$\sigma(M_a) = \sigma_a^+ - \sigma_a^-,$$

where σ_a^+ is the number of n-tuples of integers

$$j = (j_1 \ldots j_n), \quad 0 < j_k < a_k$$

for which

$$0 < \sum_{k=1}^{n} \frac{j_k}{a_k} < 1 \quad \text{mod } 2$$

and σ_a^- is the number of $(j_1, \ldots, j_n)$ for which

$$-1 < \sum_{k=1}^{n} \frac{j_k}{a_k} \quad \text{mod } 2.$$

The complex Hopf manifolds $\mathbb{C}H^n(\lambda) = \big(\mathbb{C}^n \setminus \{0\}\big)/\Delta_\lambda$ are[23] quotients of $\mathbb{C}^n \setminus \{0\}$ by the group $\Delta_\lambda \simeq \mathbb{Z}$ of complex analytic transformations generated by $z \longmapsto \lambda z$, $z \neq 0$, where $\lambda \in \mathbb{C} \setminus S^1$, $\lambda \neq 0$. As a differentiable manifold $\mathbb{C}H^n(\lambda) \simeq S^1 \times S^{2n-1}$ (a diffeomorphism) and as a complex manifold $\mathbb{C}H^n(\lambda)$ is a homogeneous manifold and a fiber bundle over the complex projective space $\mathbb{C}P^{n-1}$ with an elliptic curve as a standard fiber. For $n \geq 2$ the complex Hopf manifolds $\mathbb{C}H^n(\lambda)$ admit no globally defined Kähler metric[24] and consequently no projective algebraic structure.

---

[23] As introduced by H. Hopf, Zur Topologie der komplexen Mannigfaltigkeiten, in *Studies and Essays presented to R. Courant*. New York (1948).

[24] Because the first Betti number is odd, i.e., $b_1\big(\mathbb{C}H^n(\lambda)\big) = 1$ [cf. e.g., S. Dragomir, L. Ornea, *Locally Conformal Kähler Geometry*. Progress in Mathematics, vol. 155 (Birkhäuser, Boston-Basel-Berlin, 1998)] while the odd order Betti numbers of a compact Kählerian manifold are even [cf. Theorem 5.6.2 in S.I. Goldberg, *op. cit.*, p. 178].

Successively E. Calabi and B. Eckmann[25] built complex structures on products of two arbitrary odd dimensional spheres.[26]

It is therefore a natural question whether a product $S^1 \times \Sigma^{2n-1}$ of a circle with an exotic sphere[27] Σ^{2n-1} can be equipped with a complex structure. A positive answer to the question was provided[28] by E. Brieskorn and A. Van de Ven, *op. cit.*, when Σ^{2n-1} is a homotopy sphere bounding a parallelizable manifold and $n \neq 2$. The result by E. Brieskorn and A. Van de Ven, *op. cit.*, materializes in the construction of their *generalized Hopf manifolds*,[29] that are neither homogeneous, nor holomorphic fiber bundles, and instead are holomorphic Seifert fiber spaces over a projective variety V, whose fibers are elliptic curves. In a number of cases $V \simeq \mathbb{C}P^{n-1}$, yet in general V is singular. As a corollary of Brieskorn's Theorem 3.44 one has.

Corollary 3.45 *For $n \neq 2$ and every $(2n-1)$-dimensional homotopy sphere Σ bounding a parallelizable manifold, there are infinitely many $a = (a_1, \ldots, a_n)$ such that $\Sigma_a \simeq \Sigma$ (a diffeomorphism).*

Let us consider the holomorphic action of $\mathbb{C}$ on $X_a \setminus \{0\}$ given by

$$t \cdot \left(z_1, \ldots, z_n\right) = \left(e^{t/a_1} z_1, \ldots, e^{t/a_n} z_n\right), \quad t \in \mathbb{C}.$$

Then $\mathbb{Z}$ and $\mathbb{R}$ act on $X_a \setminus \{0\}$ as subgroups of $\mathbb{C}$. The action of $\mathbb{Z}$ on $X_a \setminus \{0\}$ is free, hence the quotient

$$H_a = \left(X_a \setminus \{0\}\right)/\mathbb{Z}$$

is a complex manifold (the *generalized Hopf manifold*), in a natural manner. Note that

$$H_{\mathbf{1}} \simeq \mathbb{C}H^n(\lambda), \quad \mathbf{1} = (1, \ldots, 1) \in \mathbb{Z}^n_+ .$$

Moreover the map

$$\Sigma_a \times \mathbb{R} \to X_a \setminus \{0\}, \quad (z, t) \mapsto t \cdot z$$

is a diffeomorphism $\Sigma_a \times \mathbb{R} \simeq X_a \setminus \{0\}$, that induces a diffeomorphism

$$H_a \simeq \Sigma_a \times S^1 .$$

---

[25] E. Calabi, B. Eckmann, *A class of compact, complex manifolds which are not algebraic*. Ann. Math. **58**, 494–500 (1953).

[26] That is $M^{p,q}_{\tau_1\tau_2} \simeq S^{2p+1} \times S^{2q+1}$, cf. e.g., Kobayashi and Nomizu [KoNo], vol. II, p. 136.

[27] A differentiable manifold that is homeomorphic to a sphere, yet it is not diffeomorphic to that sphere.

[28] By exploiting the description of exotic spheres as neighborhood boundaries of isolated singularities of quasi-homogeneous affine varieties.

[29] Not to be confused with I. Vaisman's generalized Hopf manifolds (locally conformal Kähler manifolds with a parallel Lee form) [cf. I. Vaisman, Generalized Hopf manifolds. Geometriae Dedicata **13**, 231–255 (1982)] alternatively called *Vaisman manifolds* by S. Dragomir and L. Ornea, *op. cit.*, p. 34.

This observation together with Corollary 3.45 allows for.

Theorem 3.46 *For $n \neq 2$ and for every $(2n-1)$-dimensional homotopy sphere Σ such that $\Sigma = \partial M$ for some parallelizable manifold M, there exist complex structures on the product $S^1 \times \Sigma$.*

This is the main result by E. Brieskorn and A. Van de Ven, *op. cit.*. Let

$$\left(X_a \setminus \{0\}\right)/\mathbb{Z} \to \left(X_a \setminus \{0\}\right)/\mathbb{C}$$

be the canonical projection. Let $\Gamma_a \subset \mathbb{C}$ be the discrete subgroup generated by $\{1,\ 2\pi i\,[a]\}$, where $[a] = [a_1, \ldots, a_n]$ is the least common multiple of $a_1, \ldots, a_n$. The action of $\mathbb{C}$ on $X_a \setminus \{0\}$ induces a proper holomorphic action of the complex 1-torus $T_a := \mathbb{C}/\Gamma_a$ on H_a. Then, by a result of H. Holman[30]

(i) the quotient $V_a := H_a/T_a$ is a normal complex space,
(ii) the projection $\pi_a : H_a \to H_a/T_a$ is holomorphic,
(iii) $T_a \to H_a \xrightarrow{\pi_a} V_a$ is a holomorphic Seifert principal fiber space (with fiber T_a).

However, the action of T_a on H_a is in general not free, so that in general V_a may have singularities.

The second main result by E. Brieskorn and A. Van de Ven, *op. cit.*, is to determine those H_a for which Σ_a is a homotopy sphere and V_a is a complex manifold.

Theorem 3.47 *Let Σ_a be a homotopy sphere with $a_j \neq 1$ for every $1 \le j \le n$. Then H_a is a Seifert fiber space $(H_a,\ \pi_a,\ V_a)$ with elliptic curves as fibers.*

(i) *V_a is nonsingular if and only if one of the following conditions is satisfied*

(a) *The greatest common divisor of a_i and a_j is 1, i.e., $(a_i,\ a_j) = 1$, for any $i \neq j$,*
(b) *n is odd and, eventually up to a permutation of indices,*

$$a = \left(a_1,\ 2b_2,\ \ldots,\ 2b_n\right),$$

$$(a_1,\ 2) = 1, \quad (a_1,\ b_j) = 1, \quad (b_i,\ b_j) = 1, \quad i \neq j.$$

(ii) *If V_a is nonsingular then V_a is biregularly equivalent to the complex projective space $\mathbb{C}P^{n-1}$.*

[30] H. Holman, *Quotientenräume kmplexer Manigfaltigkeiten nach komplexen Lieschen Automorphismengruppen*, Math. Annalen **139**, 383–402 (1960); Seifertsche Faserräume. Math. Annalen **157**, 138–166 (1964).

The complex Hopf manifold H_1 admits a natural Hermitian metric (the *Boothby metric*) discovered by W. Boothby[31] [and recognized by I. Vaisman[32] as a locally conformal Kähler (l.c.K.) metric]. Also $H_{21} \simeq Q_{n-1}$ where $Q_{n-1} = \pi\big(X(2, \ldots, 2)\big)$ [the *complex sphere* in $\mathbb{C}H^n(\lambda)$] and $\pi : \mathbb{C}^n \setminus \{0\} \to \mathbb{C}H^n(\lambda)$ is the projection. By Theorem 3.2 in S. Dragomir and L. Ornea, *op. cit.*, pp. 30–31, for every $\lambda \in \mathbb{R}$, $0 < \lambda < 1$, the complex sphere Q_{n-1} is a compact l.c.K. manifold (with the metric induced on Q_{n-1} by the Boothby metric) admitting no globally defined Kähler metrics. For general $a = (a_1, \ldots, a_n)$ it is not known whether the generalized Hopf manifolds admit Hermitian metrics [compatible with the complex structures (whose existence is) provided by Theorem 3.46].

Little is known about the geometry of submanifolds in H_a. Let $0 < \alpha < 1$ and $\Delta_\alpha = \{\alpha^p I : p \in \mathbb{Z}\}$ where I is the identical transformation of $\mathbb{R}^m \setminus \{0\}$, $m > 1$. The quotient space

$$\mathbb{R}H^m_\alpha = \big(\mathbb{R}^m \setminus \{0\}\big)/\Delta_\alpha$$

is the *real Hopf manifold*[33] thought of as equipped with the metric

$$|x|^{-2}\, \delta_{ij}\, dx^i \otimes dx^j$$

(the *untwisting metric* in C. Reischer and I. Vaisman, *op. cit.*). The canonical injection $\mathbb{R}^{2n-1} \to \mathbb{R}^{2n}$, $\big(x^1, \ldots, x^{2n-1}\big) \mapsto \big(x^1, \ldots, x^{2n-1}, 0\big)$ induces an isometric immersion $\mathbb{R}H^{2n-1}_\alpha \to \mathbb{C}H^n(\alpha) \simeq H_1$. By Theorem 12.11 in S. Dragomir and L. Ornea, *op. cit.*, p. 175, $\mathbb{R}H^{2n-1}_\alpha$ is a totally umbilical submanifold of H_1 (endowed with the Boothby metric).

3.11 Germs of Totally Geodesic Submanifolds

Let $(\overline{M}, \overline{g})$ be a Riemannian manifold. Given a point $x \in \overline{M}$, the set $\text{Submfd}\big(\overline{M}\big)$ of all submanifolds in $\overline{M}$ is equipped with the relation of equivalence $\sim_x$ defined as follows. Let M and M' be two submanifolds of $\overline{M}$. Then

$$M \sim_x M' \Longleftrightarrow \begin{cases} x \in M \cap M' \\ \exists\, U \subset \overline{M}, \;\; x \in U = \text{open} \; : \; M \cap U = M' \cap U. \end{cases}$$

If $M \sim_x M'$ then M and M' are said to define the same germ at x.

---

[31] W. Boothby, Some fundamental formulas for Hermitian manifolds with vanishing torsion. American J. Math. **76**, 509–534 (1954).

[32] Cf. I. Vaisman, Locally conformal Kähler manifolds with parallel Lee form. Rendiconti di Matematica, Roma **12**, 263–284 (1979).

[33] Cf. C. Reischer, I. Vaisman, *Local similarity manifolds*. Ann. Matem. Pura App. **35**, 279–292 (1983).

Definition 3.48 The equivalence class

$$[M]_x := \{M' \in \mathrm{Submfd}(\overline{M}) \;:\; M \sim_x M'\}$$

is the *germ* of M at x. □

If M and M' have the same germ at $x \in \overline{M}$ then $T_x(M) = T_x(M')$.

Definition 3.49 Let $G_x \in \mathrm{Submfd}\big(\overline{M}\big)/\sim_x$ be a germ at $x \in \overline{M}$. The space

$$T\big(G_x\big) := T_x(M), \quad M \in G_x$$

is called the *tangent space* at the germ G_x. □

Definition 3.50 A germ $G_x \in \mathrm{Submfd}\big(\overline{M}\big)/\sim_x$ is said to be *totally geodesic* if it admits a totally geodesic representative, i.e., there is a totally geodesic submanifold M of $\overline{M}$ such that $G_x = \big[M\big]_x$. □

Proposition 3.51 *Let $(\overline{M}, \overline{g})$ be a Riemannian manifold and $x \in \overline{M}$ a point. Let $V \subset T_x\big(\overline{M}\big)$ be a vector subspace. There is at most one totally geodesic submanifold M of $\overline{M}$ such that the tangent space at the germ $\big[M\big]_x$ is V.*

Proof Let M and M' be two totally geodesic submanifolds of $\overline{M}$ such that $x \in M \cap M'$. One may consider a connected neighborhood $U \subset T_x\big(\overline{M}\big)$ of the origin 0_x such that the exponential mapping $\exp_x$ is a diffeomorphism of $U \cap V$ onto some open subset of M, and also a diffeomorphism of $U \cap V$ onto some open subset of M'. As U is connected, the set of all connected components of x in $M \cap \exp_x U$ coincides with the set of all connected components of x in $M' \cap \exp_x U$. By eventually replacing $\exp_x U$ by a smaller open set, one may find an open neighborhood Ω of x in $\overline{M}$ such that $M \cap \Omega = M' \cap \Omega$. Q.e.d.

Remark 3.52 If $1 < \dim_{\mathbb{R}} V < \dim(M)$ then Proposition 3.51 is but a uniqueness result, and in general there might be no totally geodesic germ G_x such that $T(G_x) = V$. When

$$\big(\overline{M}, \overline{g}\big) = \big(S^n, g_0\big)$$

is the sphere, then for every point $x \in S^n$ and every subspace $V \subset T_x(S^n)$ there is a closed totally geodesic submanifold[34] of S^n, tangent to V at x. A Riemannian manifold $(\overline{M}, \overline{g})$ such that for any $x \in \overline{M}$ and any subspace $V \subset T_x(M)$ admits a totally geodesic submanifold tangent to V at x, satisfies the axiom of r-planes for every $2 \le r \le n-1$. By Cartan's theorem $\big(\overline{M}, \overline{g}\big)$ is a space of constant sectional curvature. □

[34] The great sphere in S^n of dimension $\dim_{\mathbb{R}} V$.

3.12 Minimal Submanifolds

3.12.1 Basic Notions and Results, Examples

We start the study of minimal submanifolds of a given semi-Riemannian manifold. For the present section, we adopt a non-variational point of view. Let $(\overline{M}, \overline{g})$ be a semi-Riemannian manifold and $f : M \to \overline{M}$ an embedding obeying to the fundamental assumption in the previous sections, i.e., $(d_p f)T_p(M)$ is a nondegenerate subspace of the inner product space $(T_{f(p)}(\overline{M}), \overline{g}_{f(p)})$ for any point $p \in M$. As we have seen, this assumption allows for the construction of the basic objects of the theory, such that the induced connection ∇, the second fundamental form B, the Weingarten operator A_ξ, and the normal connection $\nabla^\perp$. Moreover, under the same assumption, said objects are, together with the curvature tensor fields $R = R^\nabla$, $R^f = R^{\nabla^f}$ [*alias* $\overline{R} = R^{\overline{\nabla}}$ restricted to $f(M)$] and $R^\perp = R^{\nabla^\perp}$, are related by the Gauss-Weingarten and Gauss-Codazzi-Ricci equations. Now let us introduce the notion of mean curvature vector of the given immersion. Let $p \in M$ and $\{e_i : 1 \le i \le n\} \subset T_p(M)$ be a g_p-orthonormal basis, i.e.,

$$g_p(e_i, e_j) = \epsilon_i \delta_{ij}, \quad \epsilon_i \in \{\pm 1\}, \quad 1 \le i, j \le n,$$

and let us set by definition

$$H_p = \frac{1}{n} \sum_{i=1}^{n} \epsilon_i B_p(e_i, e_i).$$

The definition of H_p doesn't depend upon the choice of g_p-orthonormal basis.

Definition 3.53 The normal field

$$H : p \in M \longmapsto H_p \in T_p(M)^\perp$$

is called the *mean curvature* vector field of the immersion f. □

A local expression of the mean curvature vector field may be obtained as follows. Let $\{\xi_a : 1 \le a \le r\}$ be a local g^f-orthonormal frame

$$g^f(\xi_a, \xi_b) = \epsilon_a \delta_{ab}, \quad \epsilon_a \in \{\pm 1\}, \quad 1 \le a, b \le r$$

in the normal bundle $\tau_M^\perp$, defined on the open set $U \subset M$, and let us set

$$A_a = A_{\xi_a}, \quad 1 \le a \le r.$$

Moreover let us consider a local g-orthonormal frame $\{E_i : 1 \leq i \leq n\}$ in the tangent bundle τ_M, defined on the same open set U. Then

$$n\, g^f(H, \xi_a) = \sum_{i=1}^{n} \epsilon_i g^f(B(E_i, E_i), \xi_a) = \sum_{i=1}^{n} \epsilon_i g(A_a E_i, E_i) = \text{trace}(A_a)$$

everywhere in U. On the other hand

$$H = \sum_{a=1}^{r} g^f(H, \xi_a)\, \xi_a$$

or

$$H = \frac{1}{n} \sum_{a=1}^{r} \text{trace}(A_a)\, \xi_a\,. \tag{3.49}$$

Similarly if $\xi \in \mathfrak{X}(M)^\perp$ then

$$g^f(\xi, H) = \frac{1}{n}\, \text{trace}(A_\xi).$$

For the remainder of this section we confine ourselves to the Riemannian category, i.e., to the study of (immersed) submanifolds of a Riemannian manifold $(\overline{M}, \overline{g})$, so that g^f is now a Riemannian bundle metric on $f^{-1}\tau_{\overline{M}}$.

Definition 3.54 The norm $\|H\| = g^f(H, H)^{1/2}$ is called the *mean curvature* of the immersion f. □

A useful local formula for the mean curvature is

$$\|H\|^2 = g^f(H, H) = \sum_{a=1}^{r} g^f(H, \xi_a)^2 = \frac{1}{n^2} \sum_{a=1}^{r} (\text{trace } A_a)^2\,.$$

Definition 3.55 The immersion $f : M \to \overline{M}$ is said to be *minimal* if $H = 0$. If this is the case one also says that M is a *minimal submanifold* of $(\overline{M}, \overline{g})$. □

Definition 3.56 We say that M is a submanifold with a *parallel* mean curvature vector field if $\nabla^\perp_X H = 0$ per ogni $X \in \mathfrak{X}(M)$. If this is the case one also says that H is *parallel* in the normal bundle. □

Definition 3.57 A submanifold M is *totally umbilical* if $B = H \otimes g$ or $B(X, Y) = g(X, Y)H$ for any $X, Y \in \mathfrak{X}(M)$. □

Let us look at a few examples.

Example 3.58 (a) To start with let us show that spheres are totally umbilical as submanifolds of the Euclidean space.

Proposition 3.59 *The sphere $S^n(1) \subset \mathbb{R}^{n+1}$ is a totally umbilical submanifold of $\mathbb{E}^{n+1}$. Moreover $S^n(1)$ is a real hypersurface of constant mean curvature* 1.

Proof As previously shown the tangent bundle to $S^n(r)$ admits the local frame $\{X_i : 1 \le i \le n\}$ defined on the open set $U_{n+1} = \{p \in S^n(r) : x^{n+1}(p) \ne 0\}$ by

$$X_i = \frac{\partial}{\partial x^i} - \frac{x^i}{x^{n+1}} \frac{\partial}{\partial x^{n+1}} \in \mathfrak{X}(U_{n+1}), \quad 1 \le i \le n.$$

Let g_0 be the Euclidean metric on $\mathbb{R}^{n+1}$

$$g_0 = dx^1 \otimes dx^1 + \cdots + dx^{n+1} \otimes dx^{n+1}$$

so that $\mathbb{E}^{n+1} = (\mathbb{R}^{n+1}, g_0)$. We may now compute the first fundamental form of $S^n(r)$ in $\mathbb{E}^{n+1}$, i.e.,

$$g_{ij} \equiv g(X_i, X_j) = g_0(X_i, X_j) = \delta_{ij} + \frac{x^i x^j}{\left(x^{n+1}\right)^2}.$$

Lemma 3.60 *The inverse $[g^{ij}] = [g_{ij}]^{-1}$ is given by $g^{ij} = \delta^{ij} - x^i x^j$.*

Proof We determine a function a such that $g^{ij} = \delta^{ij} + a x^i x^j$ be the inverse seek for, i.e.,

$$\delta^i_k = g^{ij} g_{jk} = \left(\delta^{ij} + a x^i x^j\right)\left(\delta_{jk} + \frac{x_j x_k}{(x^{n+1})^2}\right) =$$

$$= \delta^i_k + \frac{x^i x_k}{(x^{n+1})^2} + a x^i x_k + a \frac{x^i x_k}{(x^{n+1})^2} \sum_{j=1}^n x_j^2 =$$

$$= \delta^i_k + \frac{x^i x_k}{(x^{n+1})^2} \left\{1 + a\left[\left(x^{n+1}\right)^2 + \sum_{j=1}^n (x^j)^2\right]\right\} =$$

$$= \delta^i_k + (1 + a) \frac{x^i x_k}{(x^{n+1})^2}$$

and then we may choose $a \equiv -1$. Q.e.d.

Let us now compute the second fundamental form of the immersion $S^n(1) \to \mathbb{E}^{n+1}$. By Gauss formula

$$\overline{\nabla}_{X_i} X_j = \nabla_{X_i} X_j + B(X_i, X_j).$$

Moreover

$$\overline{\nabla}_{X_i} X_j = \left\{ {A \atop ij} \right\} \partial_A - \frac{x^j}{x^{n+1}} \left\{ {A \atop i\ n+1} \right\} \partial_A - \frac{1}{x^{n+1}} \delta_i^j \partial_{n+1} -$$

$$- \frac{x^i}{x^{n+1}} \left[\left\{ {A \atop n+1\ j} \right\} \partial_A - \frac{x^j}{x^{n+1}} \left\{ {A \atop n+1\ n+1} \right\} \partial_A + \frac{x^j}{(x^{n+1})^2} \partial_{n+1} \right] =$$

(as $\left\{ {A \atop BC} \right\} = 0$)

$$= - \frac{1}{x^{n+1}} \left\{ \delta_i^j + \frac{x_i x^j}{x^{n+1}} \right\} \partial_{n+1}$$

where $\partial_A \equiv \partial / \partial x^A$ and the range of indices is

$$i, j, k, \ldots \in \{1, \ldots, n\}, \quad A, B, C, \ldots \in \{1, \ldots, n+1\}.$$

It follows that

$$B(X_i, X_j) = - \frac{1}{x^{n+1}} \left\{ \delta_{ij} + \frac{x_i x_j}{x^{n+1}} \right\} \text{ nor } (\partial_{n+1}) = - \frac{1}{x^{n+1}} g_{ij} \text{ nor } (\partial_{n+1}) .$$

We need to determine a local frame in the normal bundle. To this end let us consider $v = v^A (\partial_A)_p \in T_p(\mathbb{R}^{n+1})$ with $p \in S^n(1)$. Then v is orthogonal to $T_p(S^n(1))$ if and only if

$$0 = g_{0,p}(v, X_{j,p}) = v^i g_{0,p}((\partial_i)_p, X_{j,p}) + v^{n+1} g_{0,p}((\partial_{n+1})_p, X_{j,p}) =$$

$$= v^j - \frac{x^j}{x^{n+1}} v^{n+1}$$

or

$$v^j = \frac{x^j}{x^{n+1}} v^{n+1}, \quad 1 \leq j \leq n.$$

Hence if

$$\xi \equiv \frac{\partial}{\partial x^{n+1}} + \frac{x^j}{x^{n+1}} \frac{\partial}{\partial x^j},$$

then $\{\xi\}$ is a local orthonormal frame in $T_{S^n(1)}^{\perp}$ defined on the open set $U_{n+1} \subset S^n(1)$. One should compute $\text{nor}(\partial_{n+1})$, i.e., determine $\lambda \in \mathcal{F}(U_{n+1})$ such that

$$\text{nor } (\partial_{n+1}) = \lambda \, \xi. \tag{3.50}$$

One has

$$\|\xi\|^2 = g_0(\xi, \xi) = 1 + \frac{x^j x_j}{(x^{n+1})^2} = \frac{1}{(x^{n+1})^2}$$

by $\sum_{A=1}^{n+1} x_A^2 = 1$. Hence $\|\xi\| = 1/|x^{n+1}|$. Let us take the inner product of (3.50) with ξ such that

$$g_0(\mathrm{nor}(\partial_{n+1}),\ \xi) = \lambda\ \|\xi\|^2$$

or $\lambda = \|\xi\|^{-2} = \left(x^{n+1}\right)^2$. Then by (3.50) it follows that $\mathrm{nor}(\partial_{n+1}) = (x^{n+1})^2\xi$ and hence

$$B(X_i, X_j) = -x^{n+1}\ g_{ij}\ \xi. \tag{3.51}$$

We may now compute the mean curvature vector of a sphere.

Lemma 3.61 *The mean curvature vector $S^n(1) \to \mathbb{E}^{n+1}$ is*

$$H = \frac{1}{n} g^{ij} B(X_i, X_j) = -x^{n+1}\ \xi.$$

In particular the sphere $S^n(1)$ is not a minimal submanifold of the Euclidean space $\mathbb{E}^{n+1}$.

Proof Let $\{E_i : 1 \leq i \leq n\}$ be a local g-orthonormal frame of $\tau_{S^n(1)}$ defined on the open set U_{n+1}. Let

$$\lambda = [\lambda^i_j] : U_{n+1} \to \mathrm{GL}(n, \mathbb{R})$$

be the unique function such that

$$E_i = \lambda_i^j X_j, \quad 1 \leq i \leq n.$$

Then

$$\delta_{ij} = g(E_i, E_j) = \lambda_i^k \lambda_j^\ell g(X_k, X_\ell)$$

hence

$$\delta_{ij} = \lambda_i^k \lambda_j^\ell g_{k\ell}\,. \tag{3.52}$$

By (3.52)

$$\sum_{i=1}^{n} \lambda_i^j \lambda_i^k = g^{jk}\,.$$

Then

$$H = \frac{1}{n} \sum_{i=1}^{n} B(E_i,\ E_i) = \frac{1}{n} \sum_{i=1}^{n} \lambda_i^j \lambda_i^k B(X_j,\ X_k) =$$

$$= \frac{1}{n}\ g^{jk} B(X_j,\ X_k)$$

hence

$$H = -x^{n+1}\xi. \tag{3.53}$$

Finally $B(X_i, X_j) = g_{ij} H$ hence $S^n(1)$ is totally umbilical in $\mathbb{E}^{n+1}$. Again by (3.53) the mean curvature is $\|H\| = 1$. Q.e.d.

Let $r_1, r_2 \in (0, +\infty)$ be two positive numbers such that $r_1^2 + r_2^2 = 1$. Moreover let us consider the integers

$$n_1, n_2 \in \mathbb{Z}, \quad n_1 \geq 0, \quad n_2 \geq 0, \quad n_1 + n_2 = n \geq 2.$$

Moreover let $(x^1, \dots, x^{n_1+1})$ and $(y^1, \dots, y^{n_2+1})$ be respectively the Cartesian coordinates on $\mathbb{R}^{n_1+1}$ and $\mathbb{R}^{n_2+1}$ so that

$$(x^1, \dots, x^{n_1+1}, y^1, \dots, y^{n_2+1})$$

are the Cartesian coordinates on $\mathbb{R}^{n+2} = \mathbb{R}^{n_1+1} \times \mathbb{R}^{n_2+1}$. Let us denote by $S^m(r)$ the sphere of radius r in $\mathbb{R}^{m+1}$ with center at the origin. Then $S^{n_1}(r_1) \times S^{n_2}(r_2)$ is a submanifold (a real hypersurface) in $S^{n+1}(1)$. Let us consider two points $p \in S^{n_1}(r_1)$ and $q \in S^{n_2}(r_2)$ and let us denote by $\vec{p}, \vec{q} \in T_{(p,q)}(\mathbb{R}^{n+2})$ the position vectors of the two points, i.e.,

$$\vec{p} = \sum_{i=1}^{n_1+1} x^i(p) \left(\frac{\partial}{\partial x^i}\right)_{(p,q)}, \quad \vec{q} = \sum_{a=1}^{n_2+1} y^a(q) \left(\frac{\partial}{\partial y^a}\right)_{(p,q)}.$$

Lemma 3.62 *The vector field*

$$\xi : (p, q) \in S^{n_1}(r_1) \times S^{n_2}(r_2) \longmapsto \xi_{(p,q)} = \frac{r_2}{r_1}\vec{p} - \frac{r_1}{r_2}\vec{q} \in T_{(p,q)}(\mathbb{R}^{n+2})$$

is tangent to $S^{n+1}(1)$ *and it is orthogonal to* $S^{n_1}(r_1) \times S^{n_2}(r_2)$.

Proof A vector field

$$\sum_{i=1}^{n_1+1} \lambda^i \frac{\partial}{\partial x^i} + \sum_{a=1}^{n_2+1} \mu^a \frac{\partial}{\partial y^a} \in \mathfrak{X}(\mathbb{R}^{n+2})$$

is tangent to $S^{n+1}(1)$ at (p, q) if and only if

$$\sum_{i=1}^{n_1+1} \lambda^i(p, q) x_i(p) + \sum_{a=1}^{n_2+1} \mu^a(p, q) y_a(q) = 0. \tag{3.54}$$

For the field ξ in Lemma 3.62 one has

$$\lambda^i(p,q) = \frac{r_2}{r_1}\, x^i(p), \quad \mu^a(p,q) = -\frac{r_1}{r_2}\, y^a(q),$$

and hence the tangency condition (3.54) is satisfied. One indeed has

$$\xi_{(p,q)} \in T_{(p,q)}(S^{n+1}(1)), \quad p \in S^{n_1}(r_1), \quad q \in S^{n_2}(r_2).$$

For any point $(p,q) \in S^{n_1}(r_1) \times S^{n_2}(r_2)$ we consider the injections

$$\iota^q : \mathbb{R}^{n_1+1} \to \mathbb{R}^{n+2}, \quad \iota_p : \mathbb{R}^{n_2+1} \to \mathbb{R}^{n+2},$$

$$\iota^q(x_1, \ldots, x_{n_1+1}) = (x_1, \ldots, x_{n_1+1},\ q),$$

$$\iota_p(y_1, \ldots, y_{n_2+1}) = (p,\ y_1, \ldots, y_{n_2+1}),$$

$$(x_1, \ldots, x_{n_1+1}) \in \mathbb{R}^{n_1+1}, \quad (y_1, \ldots, y_{n_2+1}) \in \mathbb{R}^{n_2+1}.$$

Let g_0 be the Euclidean metric on $\mathbb{R}^{n+1}$, i.e.,

$$g_0\left(\frac{\partial}{\partial x^i}, \frac{\partial}{\partial x^j}\right) = \delta_{ij}, \quad g_0\left(\frac{\partial}{\partial x^i}, \frac{\partial}{\partial y^b}\right) = 0, \quad g_0\left(\frac{\partial}{\partial y^a}, \frac{\partial}{\partial y^b}\right) = \delta_{ab}\,.$$

If $v = v^i(\partial/\partial x^i)_p \in T_p(\mathbb{R}^{n_1+1})$ is a tangent vector at $p \in S^{n_1}(r_1)$ on the sphere $S^{n_1}(r_1)$, i.e., $v^i x_i(p) = 0$ then

$$g_{0,(p,q)}\left(\xi_{(p,q)},\ (d_p\iota^q)v\right) =$$

$$= g_{0,(p,q)}\left(\xi_{(p,q)},\ v^i\left(\frac{\partial}{\partial x^i}\right)_{(p,q)}\right) = \frac{r_2}{r_1}\, x_i(p)v^i = 0$$

and hence $\xi_{(p,q)}$ is perpendicular to $(d_p\iota^q)T_p(S^{n_1}(r_1))$. One may check similarly that $\xi_{(p,q)}$ is perpendicular to $(d_q\iota_p)T_q(S^{n_2}(r_2))$. Hence $\xi_{(p,q)}$ is orthogonal to $S^{n_1}(r_1) \times S^{n_2}(r_2)$. Q.e.d.

Now let us consider the Levi-Civita connection ∇^0 of the Euclidean space $\mathbb{E}^{n+2} = (\mathbb{R}^{n+2}, g_0)$ and let $\nabla^{S^{n+1}}$ be the induced connection on $S^{n+1}(1)$. By Gauss formula

$$\nabla^0_X \xi = \nabla^{S^{n+1}}_X \xi + B(X,\xi), \quad X \in \mathfrak{X}(S^{n+1}(1)),$$

where B is the second fundamental form of the immersion $S^{n+1}(1) \to \mathbb{E}^{n+2}$.

Lemma 3.63 *If $X \in \mathfrak{X}(S^{n_1}(r_1))$ and $Y \in \mathfrak{X}(S^{n_2}(r_2))$ then one has*

$$\nabla^0_X \xi = \frac{r_2}{r_1} X, \quad \nabla^0_Y \xi = -\frac{r_1}{r_2} Y. \tag{3.55}$$

In particular $\nabla^0_X \xi$ and $\nabla^0_Y \xi$ are tangent vector fields on $S^{n+1}(1)$ and then $B(X, \xi) = 0$ and $B(Y, \xi) = 0$.

Proof As ∇^0 is the connection associated to the canonical full parallelism structure $\mathbb{R}^{n+2}$

$$(\nabla^0_X \xi)_{(p,q)} = X(\xi^A) \left(\frac{\partial}{\partial X^A} \right)_{(p,q)} .$$

Here $(X^1, \dots, X^{n+2}) \equiv (x^1, \dots, x^{n_1+1}, y^1, \dots, y^{n_2+1})$. On the other hand, if

$$X = X^j \frac{\partial}{\partial x^j}, \quad Y = Y^b \frac{\partial}{\partial y^b},$$

then

$$X(\xi^i) = X\left(\frac{r_2}{r_1} x^i \right) = \frac{r_2}{r_1} X^j \delta^i_j = \frac{r_2}{r_1} X^i,$$

$$X(\xi^{n_1+1+a}) = X\left(-\frac{r_1}{r_2} y^a \right) = -\frac{r_1}{r_2} Y^b \delta^a_b = -\frac{r_1}{r_2} Y^a,$$

and formulas (3.55) are proved. Q.e.d.

Now we need the following general fact:

Lemma 3.64 *Let M be a real hypersurface of the Riemannian manifold $(\overline{M}, \overline{g})$. If $\xi \in \mathfrak{X}(M)^\perp$ is a normal field of nonzero constant norm $\|\xi\| = c$, $c \in \mathbb{R} \setminus \{0\}$, then ξ is parallel in the normal bundle, i.e., $\nabla^\perp \xi = 0$.*

Proof For any $X \in \mathfrak{X}(M)$

$$0 = X\left(\|\xi\|^2 \right) = X(\overline{g}(\xi, \xi)) = 2\overline{g}(\overline{\nabla}_X \xi, \xi).$$

On the other hand $\|\xi\|_p = c \neq 0$ hence $\xi_p \neq 0$ for any $p \in M$ and then (as $T^\perp_M$ has rank 1) $\{\xi\}$ is a (global) frame of the normal bundle. Then

$$\nabla^\perp_X \xi = \lambda \xi$$

for some $\lambda \in \mathcal{F}(M)$ (depending on X) and

$$0 = \overline{g}(\overline{\nabla}_X \xi,\ \xi) = \overline{g}(\nabla^\perp_X \xi, \xi) = \lambda\, \overline{g}(\xi, \xi) = c^2 \lambda,$$

i.e., $\lambda = 0$ e quindi $\nabla^\perp_X \xi = 0$. Q.e.d.

We shall apply Lemma 3.64 to the real hypersurface $S^{n_1}(r_1) \times S^{n_2}(r_2)$ in $S^{n+1}(1)$. Let A_ξ be the Weingarten operator associated to the normal field ξ (described by Lemma 3.62). Indeed

$$\|\xi\|^2 = g_0(\xi, \xi) = \frac{r_2^2}{r_1^2} x^i x_i + \frac{{r_1^2}^2}{r_2} y^a y_a = r_2^1 + r_1^2 = 1.$$

By Weingarten formula

$$\nabla_X^{S^{n+1}} \xi = -A_\xi X, \quad X \in \mathfrak{X}(S^{n_1}(r_1) \times S^{n_2}(r_2)),$$

hence (by Lemma 3.63)

$$A_\xi X = -\frac{r_2}{r_1} X, \quad X \in \mathfrak{X}(S^{n_1}(r_1)),$$

$$A_\xi Y = \frac{r_1}{r-2} Y, \quad Y \in \mathfrak{X}(S^{n_2}(r_2)).$$

Finally let $\{E_i : 1 \leq i \leq n_1\}$ and $\{F_a : 1 \leq a \leq n_2\}$ be respectively local orthonormal frames in $\tau_{S^{n_1}(r_1)}$ and $\tau_{S^{n_2}(r_2)}$. It follows that

$$\operatorname{trace}(A_\xi) = \sum_{i=1}^{n_1} g_0(A_\xi E_i, E_i) + \sum_{a=1}^{n_2} g_0(A_\xi F_a, F_a) =$$

$$= -\frac{r_2}{r_1} \sum_{i=1}^{n_1} g_0(E_i, E_i) + \frac{r_1}{r_2} \sum_{a=1}^{n_2} g_0(F_a, F_a) = -\frac{r_2}{r_1} n_1 + \frac{r_1}{r_2} n_2$$

leading to the following.

Theorem 3.65 *The mean curvature vector of the immersion*

$$S^{n_1}(r_1) \times S^{n_2}(r_2) \to S^{n+1}(1), \quad n_1 + n_2 = n, \quad r_1^2 + r_2^2 = 1,$$

is given by

$$H = \frac{r_1^2 n_2 - r_2^2 n_1}{n r_1 r_2} \xi.$$

Consequently, the following statements are equivalent

(i) $S^{n_1}(r_1) \times S^{n_2}(r_2)$ *is minimal in* $S^{n+1}(1)$.

(ii) $r_1^2 n_2 = r_2^2 n_1$.

As a corollary to Theorem 3.65 one gets the complete list of the minimal hypersurfaces in the sphere $S^{n+1}(1)$ which are products of spheres

$$S^p\left(\sqrt{\frac{p}{n}}\right) \times S^{n-p}\left(\sqrt{\frac{n-p}{n}}\right), \quad 1 \le p \le n-1.$$

In particular the *Clifford torus* $S^1\left(\sqrt{\frac{1}{2}}\right) \times S^1\left(\sqrt{\frac{1}{2}}\right)$ is a minimal surface in $S^3(1)$.

(b) Let $\left(x^i\right)_{1\le i\le 3}$ and $\left(y^\alpha\right)_{1\le\alpha\le 5}$ be Cartesian coordinates on $\mathbb{E}^3$ and $\mathbb{E}^5$. Let us consider the map

$$F : \mathbb{E}^3 \to \mathbb{E}^5,$$

$$F\left(x^1, x^2, x^3\right) = \left(y^1, y^2, y^3, y^4, y^5\right),$$

$$y^1 = \frac{1}{\sqrt{3}} x^2 x^3, \quad y^2 = \frac{1}{\sqrt{3}} x^1 x^3, \quad y^3 = \frac{1}{\sqrt{3}} x^1 x^2,$$

$$y^4 = \frac{1}{2\sqrt{3}}\left[\left(x^1\right)^2 - \left(x^2\right)^2\right], \quad y^5 = \frac{1}{6}\left[\left(x^1\right)^2 + \left(x^2\right)^2 - 2\left(x^3\right)^2\right].$$

F is an isometric immersion of $S^2(\sqrt{3})$ into $S^4(1)$. The points

$$(x^1, x^2, x^3), \ (-x^1, -x^2, -x^3) \in S^2(\sqrt{3})$$

are mapped by F onto the same point of $S^4(1)$, hence F descends to an isometric embedding of the real projective plane $\mathbb{R}P^2$ into $S^4(1)$. The image $F\left(\mathbb{R}P^2\right)$ is the *Veronese surface*. The Veronese surface is a minimal submanifold of $S^4(1)$.

(c) H.B. Lawson built[35] examples of orientable, closed, minimal surfaces of arbitrary genus in $S^3(1)$.

(d) Every totally geodesic submanifold is minimal.

(e) We close with an example coming from Kählerian geometry, which requests a small preparation summing up a few basic notions. Said notions will be presented within real tensor calculus, avoiding the use of complex local coordinate systems.

Let M be a C^∞ manifold of dimension $2n$. A tensor field $J \in \mathcal{T}^{1,1}(M)$ is an *almost complex structure* on M if $J^2 \equiv J \circ J = -I$ where I is the identical transformation. A pair (M, J) consisting of a differentiable manifold M and an almost complex structure J is said to be an *almost complex manifold*.

Let (M, J) be an almost complex manifold and g a Riemannian metric on M. We say g is an *almost Hermitian metric* on M if $g(JX, JY) = g(X, Y)$ for any $X, Y \in \mathfrak{X}(M)$, i.e., J_p is a linear isometry of the inner product space $(T_p(M), g_p)$ for any point $p \in M$. A triad (M, J, g) consisting of an almost complex manifold (M, J) and an almost Hermitian metric g is called an *almost Hermitian manifold*.

---

[35] Cf. H.B. Lawson, Complete minimal surfaces in S^3. Ann. Math. **92**, 335–374 (1970).

Let (M, J, g) be an almost Hermitian manifold and let ∇ be the Levi-Civita connection of (M, g). If J is parallel with respect to ∇, i.e., $\nabla J = 0$ then (M, J, g) is said to be a *Kählerian manifold* and g is a *Kählerian metric*.

Let $A \in \mathcal{T}^{1,1}(M)$ be an arbitrary tensor field of type $(1, 1)$ on M. The *Nijenhuis torsion* of A is

$$N_A \in \mathcal{T}^{1,2}(M),$$

$$N_A(X, Y) = [AX, AY] + A^2[X, Y] - A\{[AX, Y] + [X, AY]\}$$

$$\forall\ X, Y \in \mathfrak{X}(M).$$

An almost complex structure J is said to be *integrable* if $N_J = 0$, i.e.,

$$[JX, JY] - [X, Y] = J\{[JX, Y] + [X, JY]\},$$

$$\forall\ X, Y \in \mathfrak{X}(M).$$

It may be easily shown that the almost complex structure underlying a Kählerian manifold is integrable. An integrable almost complex structure is also referred to as a *complex structure*. The choice of terminology ("integrable" almost complex structures) is related to deep results in complex differential geometry, such as the Newlander-Nirenberg theorem that will be discussed later[36] on in this book series.

Let $(\overline{M}, J)$ be an almost complex manifold and let $f : M \to \overline{M}$ be an embedding. Let the dimensions of M and $\overline{M}$ be

$$\dim(M) = 2p, \quad \dim(\overline{M}) = 2n.$$

We say M is an *invariant* submanifold if

$$J_p \circ (d_p f) = (d_p f) \circ J_{f(p)}, \quad p \in M.$$

Theorem 3.66 *Any invariant submanifold of a Kählerian manifold is minimal.*

Proof By Gauss formula and the invariance of M

$$\nabla_X JY = J\nabla_X Y,$$

$$B(X, JY) = B(JX, Y) = JB(X, Y)$$

for any $X, Y \in \mathfrak{X}(M)$. As we may choose a local orthonormal frame of τ_M of the form $\{X_i, JX_i : 1 \leq i \leq p\}$ the mean curvature vector field is

[36] In Chapter 1 of Book I-C (Differential Geometry: Foundations of Cauchy–Riemann and Pseudohermitian Geometry).

$$H = \frac{1}{2p} \sum_{i=1}^{p} \{B(X_i, X_i) + B(JX_i, JX_i)\} = 0.$$

Q.e.d.

Theorem 3.67 *There are no compact minimal submanifolds of an Euclidean space.*

Proof Let $M \subset \mathbb{E}^n$ be a submanifold of the n-dimensional Euclidean space. Let $p \in M$ and let $\vec{p}$ be the "position vector" associated to p, i.e.,

$$\vec{p} = y^i(p) \left(\frac{\partial}{\partial y^i}\right)_p \in T_p(\mathbb{E}^n),$$

where $(y^1, \ldots, y^n)$ are the Cartesian coordinates on $\mathbb{R}^n$. Let us consider the function

$$F : M \to \mathbb{R}, \quad F(p) := \frac{1}{2} g_{0,p}(\vec{p}, \vec{p}), \quad p \in M.$$

As M is compact, F achieves its maximum at some point $p_0 \in M$. Consequently

$$X(F)_{p_0} = 0, \quad X\big(X(F)\big)_{p_0} \leq 0 \tag{3.56}$$

for every $X \in \mathfrak{X}(M)$. Indeed, let $\chi = (x^1, \ldots, x^m) : U \to \mathbb{R}^m$ be a local coordinate system on M about p_0, and let us set

$$\Omega := \chi(U) \subset \mathbb{R}^n, \quad \xi_0 := \chi(p_0) \in \Omega.$$

Then ξ_0 is a maximum point of $\underline{F} := F \circ \chi^{-1} : \Omega \to \mathbb{R}$, hence

$$\frac{\partial \underline{F}}{\partial \xi^\alpha}(\xi_0) = 0, \quad 1 \leq \alpha \leq m, \tag{3.57}$$

$$\frac{\partial^2 \underline{F}}{\partial \xi^\alpha \partial \xi^\beta}(\xi_0)\, V^\alpha V^\beta \leq 0 \tag{3.58}$$

for any $(V^1, \ldots, V^m) \in \mathbb{R}^m$. Let $X \in \mathfrak{X}(M)$ be locally represented as

$$X = X^\alpha\, \partial/\partial x^\alpha$$

on U. Then [by (3.57)]

$$0 = X^\alpha(p_0) \frac{\partial \underline{F}}{\partial \xi^\alpha}(\xi_0) = X^\alpha(p_0) \frac{\partial F}{\partial x^\alpha}(p_0) = X(F)_{p_0}.$$

Moreover

$$X\big(X(F)\big)_{p_0} = X_{p_0}\big(X(F)\big) = X^\alpha(p_0)\Big[\frac{\partial X^\beta}{\partial x^\alpha}\frac{\partial F}{\partial x^\beta} + X^\beta \frac{\partial^2 F}{\partial x^\alpha \partial x^\beta}\Big]_{p_0} =$$

[again by (3.57)]

$$= X^\alpha(p_0)\, X^\beta(p_0)\frac{\partial^2 F}{\partial x^\alpha \partial x^\beta}(p_0) = X^\alpha(p_0)\, X^\beta(p_0)\frac{\partial^2 \underline{F}}{\partial \xi^\alpha \partial \xi^\beta}(\xi_0) \le 0$$

by (3.58). Statement (3.56) is proved. Let ∇^0 be the Levi-Civita connection of $\mathbb{E}^n$. Let

$$\frac{\partial}{\partial r} \equiv X^i \frac{\partial}{\partial X^i} \in \mathfrak{X}(\mathbb{E}^n)$$

be the radial field on $\mathbb{E}^n$, so that

$$\Big(\frac{\partial}{\partial r}\Big)_p = \vec{p}, \quad p \in M.$$

Then F is the restriction to M of the function

$$\tilde{F} := g_0\Big(\frac{\partial}{\partial r}, \frac{\partial}{\partial r}\Big) \in C^\infty(\mathbb{E}^n),$$

i.e., $F = \tilde{F} \circ j$, where $j : M \to \mathbb{E}^n$ is the canonical inclusion (the given immersion). Note that

$$X(F)_p = X_p(\tilde{F} \circ j) = \big[(d_p j)X_p\big](\tilde{F}) = \tilde{X}_p\Big(g_0\Big(\frac{\partial}{\partial r}, \frac{\partial}{\partial r}\Big)\Big)$$

where $\tilde{X}$ is any C^∞ tangent vector field on $\mathbb{E}^n$ such that

$$\tilde{X}_{j(p)} = (d_p j)X_p, \quad p \in M.$$

Then [by $\nabla^0 g_0 = 0$]

$$\tilde{X}\Big(g_0\Big(\frac{\partial}{\partial r}, \frac{\partial}{\partial r}\Big)\Big)_p = 2\, g_0\Big(\nabla^0_{\tilde{X}} \frac{\partial}{\partial r}, \frac{\partial}{\partial r}\Big)_p$$

or[37]

$$X(F)_p = g_{0,\,p}\Big((d_p j)\, X_p,\ \vec{p}\Big). \tag{3.59}$$

In particular [by (3.56)]

$$g_{0,\,p_0}\Big((d_p j)\, X_{p_0},\ \vec{p}_0\Big) = 0, \tag{3.60}$$

that is $\vec{p}_0 \in T_{p_0}(M)^\perp$. Next, we start from (3.59)

$$X(F) = 2\, g_0^j\Big(j_* X,\ \frac{\partial}{\partial r}\Big)$$

and compute

$$0 \geq X_{p_0}(X(F)) = 2\left\{X\Big(g_0^j\Big(j_* X,\ \frac{\partial}{\partial r}\Big)\Big)\right\}_{p_0} =$$

[by $(\nabla^0)^j g_0^j = 0$, where $(\nabla^0)^j \equiv j^{-1}\nabla^0$ and $g^j = j^{-1} g_0$]

$$0 \geq \left\{g_0^j\Big((\nabla^0)^j_X j_* X,\ \frac{\partial}{\partial r}\Big) + g_0^j\Big(j_* X,\ (\nabla^0)^j_X \frac{\partial}{\partial r}\Big)\right\}_{p_0}$$

[by the Gauss formula

$$(\nabla^0)^j_X Y = j_* \nabla_X Y + B(x, Y)$$

and (according to the footnote calculation) by $(\nabla^0)^j_X(\partial/\partial r) = j_* X$]

$$0 \geq g_0^j\Big(j_* \nabla_X X + B(X, X),\ \frac{\partial}{\partial r}\Big)_{p_0} + g_0^j\Big(j_* X,\ j_* X\Big)_{p_0} =$$

[as $j^* g_0 = g$ (the first fundamental form of $j : M \hookrightarrow \mathbb{E}^n$)]

$$= g_{0,\,p_0}\Big((d_{p_0} j)(\nabla_X X)_{p_0} + B(X, X)_{p_0},\ \vec{p}_0\Big) +$$

$$+\, g_{p_0}\Big((d_{p_0} j) X_{p_0},\ (d_{p_0} j) X_{p_0}\Big) =$$

---

[37] As (by setting $j^i := y^i \circ j$)

$$\Big(\nabla^0_{\tilde{X}} \frac{\partial}{\partial r}\Big)_p = \Big(\tilde{X}(y^i)\, \frac{\partial}{\partial y^i}\Big)_p = \tilde{X}_p(y^i)\Big(\frac{\partial}{\partial y^i}\Big)_p = [(d_p j) X_p](y^i)\Big(\frac{\partial}{\partial y^i}\Big)_p =$$

$$= X_p(y^i \circ j)\Big(\frac{\partial}{\partial y^i}\Big)_p = X^\alpha(p)\, \frac{\partial j^i}{\partial x^\alpha}(p)\Big(\frac{\partial}{\partial y^i}\Big)_p =$$

$$= X^\alpha(p)\, (d_p j)\Big(\frac{\partial}{\partial x^\alpha}\Big)_p = (d_p j)\, X_p\,.$$

[as $\vec{p}_0$ is orthogonal to $(d_{p_0} j)T_{p_0}(M)$]

$$= g_{0,\, p_0}\left(B(X, X)_{p_0},\ \vec{p}_0\right) + \left\|(d_{p_0} j)X_{p_0}\right\|^2 >$$

[for every $X \in \mathfrak{X}(M)$ such that $X_{p_0} \neq 0$]

$$> g_{0,\, p_0}\left(B(X, X)_{p_0},\ \vec{p}_0\right).$$

In particular, let $\{E_1, \ldots, E_m\}$ be a local g-orthonormal frame of $T(M)$, defined on an open neighborhood of p_0, and let us use the above inequality for $X = E_\alpha$, i.e.,

$$g_{0,\, p_0}\left(B(E_\alpha, E_\alpha)_{p_0},\ \vec{p}_0\right) < 0$$

and then (summing over $1 \leq \alpha \leq m$)

$$g_{0,\, p_0}\left(H_{p_0},\ \vec{p}_0\right) < 0$$

implying $H_{p_0} \neq 0$. Q.e.d.

An alternative treatment of Theorem 3.67 establishes a bond among the geometry of minimal submanifolds and the theory of harmonic functions.

Theorem 3.68 *Let $\Psi : M \to \mathbb{R}^n$ be an isometric immersion of a m-dimensional Riemannian manifold M into the Euclidean space. Then*

$$\Delta\Psi = -m\, H \tag{3.61}$$

where $\Delta\Psi = \left(\Delta\Psi_1, \ldots, \Delta\Psi_m\right)$, $\Psi = \left(\Psi^1, \ldots, \Psi^n\right)$ and H denotes[38] *the mean curvature vector of the given immersion Ψ.*

Proof Let $(x^1, \ldots, x^n)$ be the Cartesian coordinates on $\mathbb{R}^n$, and let $h_p : T_{\Psi(p)}(\mathbb{R}^n) \to \mathbb{R}^n$ be the naturally associated $\mathbb{R}$-linear isomorphism, for any $p \in M$. Without the customary language abuse in (3.61), one should check that

$$\left(\Delta\Psi\right)_p = -h_p\left(H_p\right), \quad p \in M.$$

Let $p \in M$ and let $(U, u^1, \ldots, u^m)$ be local coordinate system on M such that $p \in U$. One has

$$H = H^i\, \partial_i, \quad \partial_i \equiv \frac{\partial}{\partial x^i}\,.$$

[38] Here H_p is identified to a vector in $\mathbb{R}^n$ through the canonical isomorphism $T_{\Psi(p)}(\mathbb{R}^n) \simeq \mathbb{R}^n$, for every $p \in M$.

The equations of M in $\mathbb{R}^n$ are

$$x^i = \Psi^i\big(u^1, \ldots, u^m\big), \quad \text{rang}\, B^i_\alpha(u) = m, \quad B^i_\alpha := \frac{\partial \Psi^i}{\partial u^\alpha}. \tag{3.62}$$

One ought to check that $\Delta \psi^i = -m\, H^i$ on U. Let $\overline{\nabla}$ be the Euclidean connection, i.e., the Levi-Civita connection of $\mathbb{R}^n$. Let $\{E_1, \ldots, E_m\}$ be a local orthonormal frame of $T(M)$, defined on the opens et U. Then (with the customary identifications)

$$\frac{\partial}{\partial u^\alpha} = b^i_\alpha \frac{\partial}{\partial x^i}, \quad 1 \le \alpha \le m.$$

If

$$E_\alpha = E^\sigma_\alpha\, \partial_\sigma, \quad \partial_\sigma \equiv \frac{\partial}{\partial u^\sigma},$$

then

$$\overline{\nabla}_{E_\alpha} E_\alpha = E^\rho_\alpha\, E^\sigma_\alpha\, B^{ij}_{\rho\sigma}\, \overline{\nabla}_{\partial_j} \partial_i + E^\sigma_\alpha\, \partial_\sigma\big(E^\rho_\alpha\, B^k_\rho\big)\, \partial_k, \tag{3.63}$$

where $B^{ij}_{\alpha\sigma} := B^i_\alpha\, B^j_\sigma$. For the Euclidean connection $\overline{\nabla}$

$$\overline{\nabla}_{\partial_i} \partial_j = \begin{Bmatrix} k \\ i\, j \end{Bmatrix} \partial_k, \quad \begin{Bmatrix} k \\ i\, j \end{Bmatrix} = 0.$$

Moreover, let us observe that

$$E^\rho_\alpha\, B^k_\rho = E^\rho_\alpha\, \partial_\rho\, \Psi^k = E_\alpha\big(\Psi^k\big).$$

Then (3.63) becomes

$$\overline{\nabla}_{E_\alpha} E_\alpha = E_\alpha\big(E_\alpha\, \Psi^k\big)\, \partial_k\,. \tag{3.64}$$

The proof of Theorem 3.68 may be completed by conducting the following calculation [where ∇ is the induced connection on M]

$$\Delta \Psi^i = -\sum_{\alpha=1}^m \Big\{E_\alpha(E_\alpha\, \Psi^i) - \big(\nabla_{E_\alpha} E_\alpha\big)\, \Psi^i\Big\} = \sum_{\alpha=1}^m \Big\{\big(\overline{\nabla}_{E_\alpha} E_\alpha\big)\, \Psi^i - \big(\nabla_{E_\alpha} E_\alpha\big)\, \Psi^i\Big\} =$$

$$= -\sum_{\alpha=1}^m B\big(E_\alpha,\, E_\alpha\big)\, \Psi^i = -m\, H\big(\Psi^i\big) = -m\, H^i\,.$$

Q.e.d.

Definition 3.69 A C^2 function $f : M \to \mathbb{R}$ is *harmonic* if $\Delta f = 0$. □

Corollary 3.70 *Let $\Psi = (\Psi^1, \ldots, \Psi^n) : M \to \mathbb{R}^n$ be a minimal immersion. Then Ψ^i, $1 \le i \le n$, are harmonic functions on M.*

As another corollary to Theorem 3.68 one re-obtains Theorem 3.67. Indeed, let M be compact and let us assume that there is a minimal isometric immersion $\Psi : M \to \mathbb{R}^n$. By Corollary 3.70 the components Ψ^i of Ψ are harmonic functions hence, by the classical Hopf lemma,[39] it must be $\Psi^i =$ constant, so that $\text{rang} B^i_\alpha = 0$, a contradiction. Q.e.d.

An array of results on minimal submanifolds extends to the more general case of the submanifolds with a parallel mean curvature vector. We quote.[40]

Theorem 3.71 *Let S be a surface in the space form $M^n(k)$ such that S has a parallel mean curvature vector, i.e., $\nabla^\perp H = 0$. Then S lies in one of the following categories*

(i) *S is a minimal surface in $M^n(k)$.*
(ii) *S is a minimal surface in a hypersphere of $M^n(k)$.*
(iii) *S is a constant mean curvature surface in S^3, where S^3 is a 3-sphere in $M^n(k)$.*

3.12.2 Minimal Surfaces in $\mathbb{R}^n$

Let M be a 2-dimensional Riemannian manifold, and $\Psi : M \to \mathbb{R}n$ a minimal isometric immersion. Let g be the Riemannian metric on M.

Definition 3.72 A local coordinate system (U, x, y) on M is a system of *isothermal coordinates* if g is expressed as

$$g = 2\,F\left(d\,x^2 + d\,y^2\right) \tag{3.65}$$

for some $F \in C\infty(U)$, $F(p) > 0$ for any $p \in U$. □

Theorem 3.73 *Let (M, g) be a 2-dimensional Riemannian manifold. For any $p \in M$ there is a local system of isothermal coordinates (U, x, y) such that $p \in U$.*

Theorem 3.73 is false if the components g_{ij} of the metric tensor g are only continuous functions, cf. Hartman and Winter [HaWi1]. Conditions weaker than $g_{ij} \in C^\infty(U)$ under which local systems of isothermal coordinates exist were found by Lichtenstein [Lic]. We recall that

Definition 3.74 Let $0 < \lambda \leq 1$. A function $f(x^1, x^2)$ defined in a domain $D \subset \mathbb{R}^2$ *satisfies a Hölder condition of order* λ if there is a constant $C > 0$ such that

$$\left|f(x^1, x^2) - f(x'^1, x'^2)\right| < C\,r^\lambda, \quad (x^1, x^2), (x'^1, x'^2) \in D, \tag{3.66}$$

where $r^2 = \left(x^1 - x'^1\right)^2 + \left(x^2 - x'^2\right)^2$. □

---

[39] Cf. e.g., [KoNo], vol. II, p. 338.

[40] Theorem 3.71 is due to B.-Y. Chen, On the surfaces with parallel mean curvature vector. Indiana Univ. Math. J. **22**, 655–666 (1973), and S.T. Yau, *Submanifolds with constant mean curvature.* Amer. J. Math. **96**, 346–366 (1974).

Theorem 3.73 may be reformulated as.

Theorem 3.75 *Let (M, g) be a 2-dimensional Riemannian manifold. Let us assume that for any $p \in M$ there is a local coordinate system (U, ψ), $p \in U$, $\psi = (x^1, x^2)$ such that the functions $g_{ij} \circ \psi^{-1}$, $i, j \in \{1, 2\}$, satisfy a Hölder condition of order $0 < \lambda < 1$. Then for every point of M there is a local system of isothermal coordinates.*

Proof Let (U, ψ), $\psi = (x, y)$, be a local coordinate system on M. Then

$$g = E(x, y)\, dx^2 + 2F(x, y)\, dx \odot dy + G(x, y)\, dy^2 \quad \text{on } U, \tag{3.67}$$

$$E > 0, \quad EG - F^2 > 0, \quad E, F, G : D = \psi(U) \to \mathbb{R}.$$

Here we have set

$$E = g_{11} \circ \psi^{-1}, \quad F = g_{12} \circ \psi^{-1}, \quad G = g_{22} \circ \psi^{-1},$$

$$g_{11} = g\Big(\frac{\partial}{\partial x}, \frac{\partial}{\partial x}\Big), \quad g_{12} = g\Big(\frac{\partial}{\partial x}, \frac{\partial}{\partial y}\Big), \quad g_{22} = g\Big(\frac{\partial}{\partial y}, \frac{\partial}{\partial y}\Big).$$

There is a matrix

$$A = \begin{pmatrix} a_1 & b_1 \\ a_2 & b_2 \end{pmatrix}, \quad \det(A) > 0,$$

such that (3.67) may be written as

$$g = \theta_1 \otimes \theta_1 + \theta_2 \otimes \theta_2, \tag{3.68}$$

$$\theta_1 := a_1\, dx + b_1\, dy, \quad \theta_2 := a_2\, dx + b_2\, dy.$$

Let us consider the (complex valued) form $\Phi := \theta_1 + i\,\theta_2$ where $i = \sqrt{-1}$. Hence (3.68) may be written as $g = \Phi \odot \overline{\Phi}$ where $\Phi = \theta_1 - i\,\theta_2$. The determination of isothermal parameters u, v [that is functions u, v such that $g = P(u, v)\{du^2 + dv^2\}$ with $P > 0$] is equivalent to finding a complex valued function $w = u + iv$ such that

$$dw = \frac{1}{\rho}\, \Phi\,. \tag{3.69}$$

Indeed (3.69) would imply

$$g = \Phi \odot \overline{\Phi} = |\rho|^2\, d\,w \odot d\,\overline{w} = |\rho|^2 \big(d\,u^2 + d\,v^2\big).$$

We seek for a 1-form Φ of the form

$$\Phi = \big[1 - a(z, \overline{z})\big]\, d\,z + b(z, \overline{z})\, d\,\overline{z}$$

in a neighborhood of $z = 0$. We may assume[41] that a and b vanish at $z = 0$. Equation (3.69) is equivalent to

$$\frac{\partial w}{\partial z} = \frac{1-a}{\rho}, \quad \frac{\partial w}{\partial \overline{z}} = \frac{b}{\rho},$$

or (by eliminating ρ)

$$(1-a)\,\frac{\partial w}{\partial \overline{z}} - b\,\frac{\partial w}{\partial z} = 0.$$

Finally

$$\frac{\partial w}{\partial \overline{z}} = Z(w) \tag{3.70}$$

where the differential operator Z is defined by

$$Z \equiv a\,\frac{\partial}{\partial \overline{z}} + b\,\frac{\partial}{\partial z}\,.$$

Let σ be a holomorphic function on D with $\sigma(0) = 0$, and let us consider the equation

$$2\,\pi\,i\;w(\zeta,\,\overline{\zeta}) + \int\!\!\int_D \frac{(Zw)(z,\,\overline{z})}{z-\zeta}\,d\,\overline{z}\wedge d\,z = \sigma(\zeta), \quad \zeta \in D. \tag{3.71}$$

We shall need the following result from [Cher], p. 778.

Theorem 3.76 *Let $D = \{z \in \mathbb{C} \,:\, |z| \le R\}$, $R > 0$, and let $a(z,\,\overline{z})$, $b(z,\,\overline{z})$ be two functions satisfying a Hölder condition of order $0 < \lambda < 1$ in D. Then Eq. (3.71) has a unique solution $w(z,\,\overline{z})$ for $R > 0$ sufficiently small.*

We seek for a solution to (3.70). Let w be the solution to (3.71), as furnished by Theorem 3.76 for $R > 0$ sufficiently small. We are left with checking that the function

$$F(\zeta,\,\overline{\zeta}) = \frac{1}{2\pi i}\int\!\!\int_D \frac{(Zw)(z,\,\overline{z})}{z-\zeta}\,d\,\overline{z}\wedge dz$$

admits $\overline{\zeta}$ derivatives $\partial F/\partial\overline{\zeta}$, and that

$$\frac{\partial F}{\partial \overline{\zeta}} = (Zw)(\zeta,\,\overline{\zeta}). \tag{3.72}$$

Under the assumption that $f(z,\,\overline{z})$ admits first order partials with respect to x and y, where $x + iy := z$

[41] Cf. Chern [Cher], p. 781.

$$\frac{\partial f}{\partial \overline{z}} = \lim_{h \to 0} \frac{1}{h} \left\{ f\left(z + \frac{h}{2}, \overline{z} + \frac{h}{2}\right) - f(z, \overline{z}) + i\, f\left(z + \frac{i\,h}{2}, \overline{z} - \frac{i\,h}{2}\right) - i\, f(z, \overline{z}) \right\} \tag{3.73}$$

where $h \in \mathbb{R}$. To check the existence of $\partial F / \partial \overline{\zeta}$ we make use of (3.73). Precisely

$$\lim_{h \to 0} \frac{1}{h} \left\{ F\left(\zeta + \frac{h}{2}, \overline{\zeta} + \frac{h}{2}\right) - F(\zeta, \overline{\zeta}) + i\, F\left(\zeta + \frac{i\,h}{2}, \overline{\zeta} - \frac{i\,h}{2}\right) - i\, F(\zeta, \overline{\zeta}) \right\} =$$

$$= -\frac{1}{2\pi i} \lim_{h \to 0} \frac{1}{h} \left\{ \int\!\!\int_D \frac{(zw)(z, \overline{z})}{z - \zeta - \dfrac{h}{2}}\, d\overline{z} \wedge dz - \int\!\!\int_D \frac{(zw)(z, \overline{z})}{z - \zeta}\, d\overline{z} \wedge dz + \right.$$

$$\left. + i \int\!\!\int_D \frac{(zw)(z, \overline{z})}{z - \zeta - \dfrac{i\,h}{2}}\, d\overline{z} \wedge dz - i \int\!\!\int_D \frac{(zw)(z, \overline{z})}{z - \zeta}\, d\overline{z} \wedge dz \right\} =$$

$$= -\frac{1}{4\pi i} \lim_{h \to 0} \left\{ \int\!\!\int_D \left[(Zw)(z, \overline{z}) - (Zw)(\zeta, \overline{\zeta}) \right] \times \right.$$

$$\left. \times \left[\frac{1}{(z - \zeta)\left(z - \zeta - \dfrac{h}{2}\right)} - \frac{1}{(z - \zeta)\left(z - \zeta - \dfrac{i\,h}{2}\right)} \right] d\overline{z} \wedge dz \right\} +$$

$$- \frac{1}{2\pi i} (Zw)(\zeta, \overline{\zeta}) \frac{\partial}{\partial \overline{\zeta}} \int\!\!\int_D \frac{d\overline{z} \wedge dz}{z - \zeta} \Bigg\}.$$

As the domain is the closed ball $D = \{z \in \mathbb{C} : |z| \leq R\}$

$$\int\!\!\int_D \frac{d\,\overline{z} \wedge dz}{z - \zeta} = -2\pi i\, \overline{\zeta}\,.$$

On the other hand, one may show[42] that $(Zw)(z, \overline{z})$ satisfies a Hölder condition in D, and hence

$$\int\!\!\int_D \left[(Zw)(z, \overline{z}) - (Zw)(\zeta, \overline{\zeta}) \right] \frac{1}{z - \zeta} \frac{\dfrac{h}{2}(1 - i)}{\left(z - \zeta - \dfrac{h}{2}\right)\left(z - \zeta - \dfrac{i\,h}{2}\right)}\, d\,\overline{z} \wedge d\,z$$

tends to zero as $h \to 0$, so that (3.72) holds good. Q.e.d.

For other generalizations of Theorem 3.75 (the *Korn-Lichtenstein theorem*) one may see Morrey [Mor], and Hartman and Winter [HaWi2]

Let M be a 2-dimensional oriented Riemannian manifold, and let (U, x, y) be an oriented local system of isothermal coordinates on M. Then $z = x + i\,y$ is a

[42] Cf. formula (44) in [Cher], p. 779.

local complex coordinate on M, hence M becomes a complex manifold, of complex dimension 1, i.e., a Riemann surface.

Lemma 3.77 *Let M be a real 2-dimensional Riemannian manifold, and let $z = x + i\,y$ be the local complex coordinate associated to the local isothermal coordinates $(x,\ y)$ on M. Then the Laplace-Beltrami operator of M is (on C^2 functions)*

$$\Delta \equiv -\frac{2}{F}\frac{\partial}{\partial z}\frac{\partial}{\partial \bar{z}}. \tag{3.74}$$

Proof Let us set

$$\frac{\partial}{\partial \bar{z}} = \frac{1}{2}\Big(\frac{\partial}{\partial x} - i\,\frac{\partial}{\partial y}\Big), \quad \frac{\partial}{\partial \bar{z}} = \frac{1}{2}\Big(\frac{\partial}{\partial x} + i\,\frac{\partial}{\partial y}\Big).$$

The metric g on M may be represented as in (3.65) or, with respect to the complex coordinate z

$$g = 2\,F\,dz \odot d\bar{z} \tag{3.75}$$

where $dz = dx + i\,dy$ and $d\bar{z} = dx - i\,dy$. By (3.65)

$$g_{11} = g_{22} = 2\,F, \quad g_{12} = g_{21} = 0, \quad \mathfrak{g} = 4\,F^2,$$

$$g^{11} = g^{22} = 2\,F, \quad g^{12} = g^{21} = 0,$$

(with $\mathfrak{g} = \det\big[g_{ij}\big]_{1\le i,\,j\le 2}$). Let $f \in C^\infty(M)$. Then

$$\Delta f = -\frac{1}{2F}\left\{\frac{\partial^2 f}{\partial x^2} + \frac{\partial^2 f}{\partial y^2}\right\} \tag{3.76}$$

and (3.76) $\Longleftrightarrow$ (3.74). Q.e.d.

Corollary 3.78 *Let $\Psi : M \to \mathbb{R}^n$ be an isometric immersion of a real 2-dimensional Riemannian manifold in $\mathbb{R}^n$. Then Ψ is minimal if and only if*

$$\frac{\partial^2\,\Psi}{\partial\,z\,\partial\,\bar{z}} = 0.$$

The proof follows from (3.74) and (3.61) with $H = 0$.

An important remark is in order. According to Corollary 3.78, the minimality of Ψ doesn't depend on the metric of M, but only on the complex manifold structure of M, prompting[43] the following definitions

[43] According to a piece of mathematical folklore, a valuable mathematical result will prompt a relevant definition.

Definition 3.79 Let M be a Riemann surface and let $\Psi : M \to \mathbb{R}^n$ be a differentiable map. We say M is *conformal* if for every local complex coordinate z on M

$$\left\| (d\Psi) \frac{\partial}{\partial x} \right\| = \left\| (d\Psi) \frac{\partial}{\partial y} \right\|, \quad \overline{g}\Big((d\Psi) \frac{\partial}{\partial x}, \frac{\partial}{\partial y} \Big) = 0. \tag{3.77}$$

□

Here $\overline{g}$ is the Euclidean metric on $\mathbb{R}^n$, while $\| \cdot \|$ is the corresponding norm.

Definition 3.80 Let M be a Riemann surface. An immersion $\Psi : M \to \mathbb{R}^n$ is said to be a *minimal surface* in $\mathbb{R}^n$ if Ψ is conformal[44] and obeys to

$$\frac{\partial^2 \Psi}{\partial z \, \partial \overline{z}} = 0. \tag{3.78}$$

□

Theorem 3.81 *Let $\Psi : M \to \mathbb{R}^n$ be a conformal immersion, and let us endow M with the induced metric $g = \Psi^* \overline{g}$. Then Ψ is an isometric immersion of (M, g) into $(\mathbb{R}^n, \overline{g})$ and the induced metric is given by (3.65).*

Proof One has

$$g_{12} = g\Big(\frac{\partial}{\partial x}, \frac{\partial}{\partial y}\Big) = \overline{g}\Big((d\,\Psi)\frac{\partial}{\partial x}, (d\,\Psi)\frac{\partial}{\partial y}\Big) = 0,$$

$$g_{11} = g\Big(\frac{\partial}{\partial x}, \frac{\partial}{\partial x}\Big) = \overline{g}\Big((d\,\Psi)\frac{\partial}{\partial x}, (d\,\Psi)\frac{\partial}{\partial x}\Big) = \overline{g}\Big((d\,\Psi)\frac{\partial}{\partial y}, (d\,\Psi)\frac{\partial}{\partial y}\Big) = g\Big(\frac{\partial}{\partial y}, \frac{\partial}{\partial y}\Big) = g_{22}\,.$$

Finally, we set $F := (1/2)\, g_{11}$ and obtain (3.65). Q.e.d.

3.12.3 *Weierstrass Representation of Minimal Surfaces*

Let $D \subset \mathbb{C}$ be an open subset, and let z be the canonical complex coordinate on $\mathbb{C}$. Then D is a Riemann surface. Let $\Psi : D \to \mathbb{R}^n$ be a minimal surface, and let us consider

$$\Phi : D \to \mathbb{C}^n, \quad \Phi = \frac{\partial \Psi}{\partial z} = \frac{1}{2}\Big(\frac{\partial \Psi}{\partial x} - i\,\frac{\partial \Psi}{\partial y}\Big).$$

Theorem 3.82 (i) *Φ is holomorphic.*

[44] In the sense of Definition 3.79.

If $\Phi = \left(\phi^1, \ldots, \phi^n\right)$ *then*

(ii)

$$\sum_{j=1}^{n} \phi_j^2 = 0. \tag{3.79}$$

(iii) *The following identity holds*

$$\sum_{j=1}^{n} \left|\phi^j\right|^2 = F \neq 0. \tag{3.80}$$

Viceversa, if $D \subset \mathbb{C}$ *is a connected and simply connected open set, and* $\Phi : D \to \mathbb{C}^n$ *is a map satisfying* (i)–(iii)*, then*

$$\Psi : D \to \mathbb{R}^n, \quad \Psi^j(z) = \text{Re}\left\{\int_o^z \phi^j(\zeta)\, d\zeta\right\} + c^j, \quad 1 \leq j \leq n \tag{3.81}$$

is a minimal surface.

Here $o \in D$ is a fixed base point, and $c^j \in \mathbb{R}$, $1 \leq j \leq n$, are constants. Also $\phi_j = \phi^j$.

Proof of Theorem 3.82 (i) By (3.78) [i.e., by the minimality of Ψ]

$$\frac{\partial \Phi}{\partial \overline{z}} = \frac{\partial}{\partial \overline{z}}\left(\frac{\partial \Psi}{\partial z}\right) = 0.$$

Q.e.d.

(ii) Let $\langle \cdot, \cdot \rangle_{\mathbb{R}^n}$ be the Euclidean inner product on $\mathbb{R}^n$, so that $\overline{g}$ is induced by $\langle \cdot, \cdot \rangle_{\mathbb{R}^n}$ and the collection of isomorphisms $h_\xi : T_\xi(\mathbb{R}^n) \simeq \mathbb{R}^n$, $\xi \in \mathbb{R}^n$, i.e.,

$$\overline{g}_\xi\left(v, w\right) = \langle h_\xi\, v, h_\xi\, w \rangle_{\mathbb{R}^n}, \quad v, w \in T_\xi(\mathbb{R}^n).$$

Then [by (3.77)]

$$\left\langle \frac{\partial \Psi}{\partial x}, \frac{\partial \Psi}{\partial x} \right\rangle_{\mathbb{R}^n} = \left\langle \frac{\partial \Psi}{\partial y}, \frac{\partial \Psi}{\partial y} \right\rangle_{\mathbb{R}^n}, \quad \left\langle \frac{\partial \Psi}{\partial x}, \frac{\partial \Psi}{\partial y} \right\rangle_{\mathbb{R}^n} = 0, \tag{3.82}$$

$$\frac{\partial \Psi}{\partial x} = \left(\frac{\partial \Psi^1}{\partial x}, \ldots, \frac{\partial \Psi^n}{\partial x}\right), \quad \frac{\partial \Psi}{\partial y} = \left(\frac{\partial \Psi^1}{\partial y}, \ldots, \frac{\partial \Psi^n}{\partial y}\right), \quad \Psi = \left(\Psi^1, \ldots, \Psi^n\right).$$

Finally

$$\sum_{j=1}^{n} \phi_j^2 = \frac{1}{4}\sum_{j=1}^{n}\Big[\frac{\partial \Psi^j}{\partial x} - i\,\frac{\partial \Psi^j}{\partial y}\Big]^2 = \frac{1}{4}\sum_{j=1}^{n}\Big[\Big(\frac{\partial \Psi^j}{\partial x}\Big)^2 - \Big(\frac{\partial \Psi^j}{\partial y}\Big)^2\Big] - \frac{i}{2}\sum_{j=1}^{n}\frac{\partial \Psi^j}{\partial x}\,\frac{\partial \Psi^j}{\partial y} =$$

$$= \frac{1}{4}\Big\|\frac{\partial \Psi}{\partial x}\Big\|^2 - \frac{1}{4}\Big\|\frac{\partial \Psi}{\partial y}\Big\|^2 - \frac{i}{2}\Big\langle\frac{\partial \Psi}{\partial x},\, \frac{\partial \Psi}{\partial y}\Big\rangle_{\mathbb{R}^n} = 0.$$

Q.e.d.

(iii) To prove (3.80) one first observes that

$$\sum_{j=1}^{n} |\phi_j|^2 = \Big\|\frac{\partial \Psi}{\partial x}\Big\|^2 + \Big\|\frac{\partial \Psi}{\partial y}\Big\|^2 \geq 0 \tag{3.83}$$

and the inequality (3.83) is strict. Indeed, if one has equality in (3.83) at some point $x_0 + i\,y_0 \in D$ then

$$\frac{\partial \Psi}{\partial x}(x_0,\, y_0) = 0, \qquad \frac{\partial \Psi}{\partial y}(x_0,\, y_0) = 0,$$

so that $\Psi : D \to \mathbb{R}^n$ wouldn't be an immersion, a contradiction.

Viceversa, let $D \subset \mathbb{C}$ be a connected and simply connected open subset, and let $\Phi : D \to \mathbb{C}^n$ be a map possessing the properties (i)–(iii) above. Let Ψ be defined by (3.81). By the classical Cauchy theorem[45] the integral

$$\int_o^z \phi^j(\zeta)\, d\,\zeta$$

is well defined, i.e., it doesn't depend upon the choice of smooth path joining o and z in D. Then (3.81) yields

$$\Phi = 2\,\frac{\partial \Psi}{\partial z}\,. \tag{3.84}$$

Finally, the minimality of Ψ follows from

$$\frac{\partial^2 \Psi}{\partial \overline{z}\, \partial x} = \frac{1}{2}\,\frac{\partial \phi}{\partial \overline{z}} = 0$$

by (3.84) and (i). The property (ii) implies that Ψ is conformal, while with (iii) one makes sure that $\|\partial\Psi/\partial x\| \neq 0$, $\|\partial\Psi/\partial y\| \neq 0$ at any point of D, i.e., that Ψ is an immersion. Q.e.d.

---

[45] Cf. e.g., Mackey [Mac], Theorem 5.12, p. 90.

By Theorem 3.82 to find examples of minimal surfaces in $\mathbb{R}^n$ it suffices to produce maps $\Phi : D \to \mathbb{C}^n$ satisfying the requirements (i)–(iii).

Theorem 3.83 *Let $D \subset \mathbb{C}$ be a connected open set. Let $f, g : D \to \mathbb{C}$ such that g is meromorphic on D and f is holomorphic in D. Let us assume that*

(a) *At any point of D where g has a pole of order m, the function f has a zero or order at least $2m$.*

Then the map

$$\Phi = (\phi^1, \phi^2, \phi^3) : D \to \mathbb{C}^3,$$

$$\phi^1 = \frac{1}{2} f\,(1 - g^2), \quad \phi^2 = \frac{i}{2} f\,(1 + g^2), \quad \phi^3 = f\,g \tag{3.85}$$

obeys to (i)–(ii) *with $n = 3$. Moreover, every map $\Phi : D \to \mathbb{C}^3$ satisfying* (i)–(ii) *may be represented in the form (3.85), except for the case $\phi^3 = 0$. Moreover the map (3.85) satisfies* (iii) *provided that*

(b) *The zeros of f coincide with the poles of g and their order is precisely twice the order of the poles of g.*

Finally, all the maps $\Phi : D \to \mathbb{C}^3$ with $\phi^3 = 0$ and satisfying (i)–(ii) *are given by*

$$\phi_2 = i\,\phi_1, \quad \phi_3 = 0,$$

where $\phi_1 : D \to \mathbb{C}$ is an arbitrary holomorphic function. The condition (iii) *is satisfied if and only if $|\phi_1| \neq 0$ at every point of D.*

Proof Let us check that the components of Φ [given by (3.85)] are holomorphic in D. To this end, let $z_0 \in D$. As g is meromorphic, we may distinguish to cases, as follows. Either (I) g is holomorphic in a neighborhood of z_0 [and then ϕ^j are holomorphic in z_0] or (II) z_0 is an isolated singularity of g, of pole type. Let $m \geq 1$ be the order of the pole z_0 of g. Then [by (a)] the holomorphic function f has a zero in z_0 of order $k \geq 2m$. Hence, on one hand there exist $\delta' > 0$ and a function g_1 holomorphic in $B_{\delta'}(z_0)$ such that

$$\frac{1}{g(z)} = (z - z_0)^m\, g_1(z), \quad |z - z_0| < \delta', \tag{3.86}$$

and on the other hand there exist $\delta'' > 0$ and a function h holomorphic in $B_{\delta''}(z_0)$ such that

$$f(z) = (z - z_0)^k\, h(z), \quad |z - z_0| < \delta''. \tag{3.87}$$

Hence, for any $|z - z_0| < \delta = \min\{\delta', \delta''\}$

$$\phi^1(z) = \frac{1}{2} f(z) \left(1 - g(z)\right)^2 = \frac{1}{2} (z - z_0)^k h(z) \left\{1 - \frac{g_1(z)^2}{(z - z_0)^{2m}}\right\} =$$

$$= \frac{1}{2} h(z) \left\{(z - z_0)^k - (z - z_0)^{k-2m} g_1(z)^2\right\}$$

and $k - 2m \geq 0$, that is the isolated singularity $z = z_0$ of ϕ^1 is removable, so that ϕ^1 may be thought of as holomorphic[46] in $B_\delta(z_0)$, and in particular in z_0. The singularities of ϕ^2 and ϕ^3 may be removed in a similar manner. To check (ii) one computes

$$\sum_{j=1}^{3} \phi_j^2 = \frac{1}{4} f^2 \left\{(1 - g^2)^2 - (1 + g^2)^2 + 4 g^2\right\} = 0.$$

Moreover, let us observe that

$$\sum_{j=1}^{3} |\phi^j|^2 = \frac{1}{2} |f|^2 \left(1 + |g|^2\right)^2. \tag{3.88}$$

Now, it may be easily seen that (iii) holds when (b) does. Clearly [by (3.88)]

$$\sum_{j=1}^{3} |\phi_j|^2 \neq 0$$

at the points of D that aren't zeros of f. Let now $z_0 \in D$ be a zero of f. Then (3.87) holds on a neighborhood $B_{\delta''}(z_0)$, where h is holomorphic in $B_{\delta''}(z_0)$ and $h(z_0) \neq 0$. It follows [by (b)] that k is even and g has at z_0 a pole of order $m = k/2 \geq 1$. Therefore, there is $\delta' > 0$ such that (3.86) holds on $B_{\delta'}(z_0)$, where g_1 is holomorphic in $B_{\delta'}(z_0)$ and $g_1(z_0) \neq 0$. Let $\delta := \min\{\delta', \delta''\}$. For any $|z - z_0| < \delta$

$$\frac{1}{2} |f(z)|^2 \left(1 + |g(z)|^2\right)^2 = \frac{1}{2} |z - z_0|^{2k} |h(z)|^2 \left\{1 + \frac{|g_1(z)|^2}{|z - z_0|^k}\right\}^2 =$$

$$= \frac{1}{2} |h(z)|^2 \left\{(z - z_0)^k + |g_1(z)|^2\right\}^2.$$

Hence the function (172) $|f|^2(1 + |g|^2)$ has a removable singularity at z_0 and its natural extension to $B_\delta(z_0)$ takes at z_0 the value

[46] Let $\tilde{\phi}_1$ be the extension of ϕ_1 which is zero at $z = z_0$ if $k > 2m$, and equals $-(1/2)\, h(z_0)\, g_1(z_0)^2$ at $z = z_0$ if $k = 2m$. It is actually $\tilde{\phi}_1$ that deserves the adjective "holomorphic".

$$\frac{1}{2}\,\left|h(z_0)\right|^2\,\left|g_1(z_0)\right|^4 \neq 0\,.$$

Q.e.d. Finally, let $\Phi : D \to \mathbb{C}^3$ be a map obeying to (i)–(ii) and $\phi^3 \neq 0$. Let us set

$$f = \phi_1 - i\,\phi_2, \quad g = \frac{\phi_3}{\phi_1 - i\,\phi_2}\,.$$

Then f is holomorphic and g is meromorphic (in D) and Φ may be represented in the form (3.86).

Theorem 3.84 *Let $D \subset \mathbb{C}$ be a connected and simply connected open set. Let $o \in D$ be an arbitrary base point. Then every minimal surface $\Psi : D \to \mathbb{R}^3$ can be represented as*

$$\Psi^j(z) = \text{Re}\left\{\int_o^z \phi^j(\zeta)\,d\,\zeta\right\} + c^j \tag{3.89}$$

where $c^j \in \mathbb{R}$ are constants, and the functions ϕ^j are given by (3.86) with f holomorphic in D, g meromorphic in D and subject to[47] (b), *or $\phi^3 = 0$, $\phi_1 : D \to \mathbb{C}$ is holomorphic, and $\phi_2 = i\,\phi_1$.*

Proof By (3.81) formulas (3.89) provide a minimal surface. Viceversa, to check that any minimal surface may be recovered in this manner, let $\Phi = \partial\Psi/\partial z$. Then Φ satisfies (i)–(iii), and by Theorem 3.83 it follows that Φ can be represented in the form (3.85). Q.e.d.

Let us look at a few examples.

(1) Let $D = \mathbb{C}$. The functions

$$g(z) = -e^z, \quad f(z) = -e^{-z}, \quad z \in \mathbb{C}$$

posses the requirements needed to apply Theorem 3.83. Let Φ be given by (3.85), i.e.,

$$\Phi(z) = \big(\sinh z, \;\; -i\,\cosh z, \;\; 1\big), \quad z \in \mathbb{C}. \tag{3.90}$$

Here $\sinh z = (1/2)\big(e^z - e^{-z}\big)$ and $\cosh z = (1/2)\big(e^z + e^{-z}\big)$. Substitution of Φ into (3.89) yields

$$\Psi(x, y) = \Big(\text{Re}\,\big(\cosh z\big), \;\; -\text{Re}\big(i\,\sinh z\big), \;\; \text{Re}(z)\Big) + \big(c_1, c_2, c_3\big) \tag{3.91}$$

or

$$\Psi(x, y) = \big(\cos y\,\cosh x, \;\; \sin y\,\cosh x, \;\; x\big) + c \tag{3.92}$$

[47] That is possessing the property (b) in Theorem 3.83 above.

where $c = (c_1, c_2, c_3)$ is a vector constant. The surface $\Psi : (x, y) \mapsto (X, Y, Z)$ defined by (3.92) is given by the parametric equations

$$\begin{cases} X = \cos y \cosh x, \\ Y = \sin y \cosh x, \\ Z = x. \end{cases} \tag{3.93}$$

Let us eliminate the parameters x, y between Eq. (3.93) so that to obtain

$$X^2 + Y^2 = (\cosh z)^2 \tag{3.94}$$

which is the *catenoid*, i.e., the surface of revolution obtained by a rotation of axis OZ, of the "chain" of equation $X = \cosh Z$.

(2) Let

$$D = \mathbb{C}, \quad g(z) = -i\, e^z, \quad f(z) = e^{-z}, \quad z = u + i\, v.$$

Then (3.86) and (3.90) lead to

$$\Psi(u, v) = \big(\cos v, \ \sinh u, \ \sin v \ \sinh u, \ v\big). \tag{3.95}$$

Setting $\sinh u = \rho$ Eq. (3.95) becomes

$$\begin{cases} x = \rho \cos v \\ y = \rho \sin v \\ z = v \end{cases} \tag{3.96}$$

Equations (3.96) exhibit said surface as ruled.

3.12.4 Gauss Map of a Minimal Surface

The purpose of the present section is to explain the geometric meaning of the functions f and g in formulas (3.86) and (3.90). As we shall shortly see g is precisely the Gauss map[48] $g : M \to G(n, 1)$.

[48] Cf. e.g., Kobayashi and Nomizu [KoNo], vol. II, p. 51.

Let $\Psi : M \to \mathbb{R}^n$ be a conformal immersion of a Riemann surface M in $\mathbb{R}^n$. Let z be a local complex coordinate on M. The induced metric $g = \Psi^* \overline{g}$ may be expressed as

$$g = 2\,F\,dz \otimes d\overline{z}, \quad F > 0.$$

Let us tentatively set $\phi = \partial\Psi/\partial z$. If M is an open set $D \subset \mathbb{C}$, that is connected and simply connected, then $\phi : D \to \mathbb{C}^n$ is well defined, and leads to the Weierstrass representation. If in turn M is an arbitrary Riemann surface, then ϕ is defined only on the domain U of the considered local chart $(U,\ z)$. If $(V,\ w)$ is another complex coordinate system on M with $U \cap V \neq \emptyset$, then $\phi' := \partial\Psi/\partial w$ is related to ϕ by

$$\phi' = \frac{\partial z}{\partial w}\,\phi \quad \text{on } U \cap V,$$

so that $\phi(z)$ is only defined up to a multiplicative factor. In turn, a map $\Phi : M \to \mathbb{C}P^{n-1}$ may be built as follows. Let $z_0 \in M$ and let $(U,\ z)$ a local complex coordinate system about z_0, i.e., $z_0 \in U$. We set

$$\Phi(z_0) = \left[\frac{\partial\Psi}{\partial z}(z_0)\right] \in \mathbb{C}P^{n-1}. \tag{3.97}$$

Here $[\zeta]$ denotes the point of the complex projective space of homogeneous coordinates $\zeta \in \mathbb{C}^n \setminus \{0\}$. The definition of $\Phi(z_0)$ doesn't depend upon the choice of local chart $(U,\ z)$ about z_0 and the map $\Phi : M \to \mathbb{C}P^{n-1}$ is well defined.

Definition 3.85 The map $\Phi : M \to \mathbb{C}P^{n-1}$ defined by (3.97) is called the *Gauss map*. □

Theorem 3.86 *Let M be a Riemann surface and let $\Psi : M \to \mathbb{R}^n$ be a conformal immersion of M in $\mathbb{R}^n$. Let us consider the map $\Phi : M \to \mathbb{C}P^{n-1}$ defined by*

$$\Phi(p) = \left[\frac{\partial\Psi}{\partial z}(p)\right], \quad p \in M$$

for some local chart $(U,\ z)$ on M with $p \in U$. Then Ψ is minimal $\Longleftrightarrow$ Φ is holomorphic.

Proof As Ψ is an immersion

$$\frac{\partial\Psi}{\partial z} = \frac{1}{2}\left(\frac{\partial\Psi}{\partial x} - i\,\frac{\partial\Psi}{\partial y}\right) \neq 0$$

at any point of U, and hence it makes sense to consider the points $[(\partial\Psi/\partial z)(p)]$ with $p \in U$.

Lemma 3.87 *The projection $\pi : \mathbb{C}^n \setminus \{0\} \to \mathbb{C}P^{n-1}$ is holomorphic.*

Indeed the local expression of π with respect to the local chart

$$\varphi_j : U_j \to \mathbb{C}^{n-1}, \quad \varphi_j([w]) = \Big(\frac{w_1}{w_j}, \ldots, \frac{w_{j-1}}{w_j}, \frac{w_{j+1}}{w_j}, \ldots, \frac{w_n}{w_j}\Big),$$

$$U_j = \big\{[w_1, \ldots, w_n] \in \mathbb{C}P^{n-1} : w_j \neq 0\big\},$$

is

$$(\varphi_j \circ \pi)(\zeta_1, \ldots, \zeta_n) = \Big(\frac{\zeta_1}{\zeta_j}, \ldots, \frac{\zeta_{j-1}}{\zeta_j}, \frac{\zeta_{j+1}}{\zeta_j}, \ldots, \frac{\zeta_n}{\zeta_j}\Big)$$

and its components are holomorphic functions.

Note that

$$\Phi = \pi \circ \frac{\partial \Psi}{\partial z} \quad \text{on} \quad U.$$

By a result of S.-S. Chern[49] the converse of the above holds, i.e., Φ holomorphic $\Longrightarrow \Psi$ is minimal. The proof is rather involved, and we only give a brief argument. Let us assume that $\Phi : U \to \mathbb{C}P^{n-1}$ is holomorphic, i.e., $\partial\Phi/\partial\overline{z} = 0$ on U. Let $\phi = \partial\Psi/\partial z$. If $p \in U$ then, up to an isomorphism $\mathbb{C}^n \simeq T_\zeta(\mathbb{C}^n)$, $\zeta \in \mathbb{C}^n$, one has

$$\frac{\partial \phi}{\partial \overline{z}} \simeq (d_p\,\phi)\Big(\frac{\partial}{\partial \overline{z}}\Big)_p \in T_{\phi(p)}(\mathbb{C}^n).$$

Let us consider

$$d_{\Phi(p)}\pi : T_{\phi(p)}(\mathbb{C}^n) \to T_{\Phi(p)}\big(\mathbb{C}P^{n-1}\big), \quad \Phi := \pi \circ \phi.$$

As ϕ is holomorphic

$$\big(d_{\Phi(p)}\pi\big)(d_p\,\phi)\Big(\frac{\partial}{\partial \overline{z}}\Big)_p = d_p(\pi \circ \phi)\Big(\frac{\partial}{\partial \overline{z}}\Big)_p = (d_p\Phi)\Big(\frac{\partial}{\partial \overline{z}}\Big)_p = \frac{\partial \Phi}{\partial \overline{z}}(p) = 0.$$

Consequently

$$(d_p\,\phi)\Big(\frac{\partial}{\partial \overline{z}}\Big)_p \in \operatorname{Ker}\big(d_{\phi(p)}\,\pi\big), \quad p \in U,$$

i.e., $(d_p\,\phi)\big(\partial/\partial\overline{z}\big)_p$ is vertical vector with respect to the principal bundle $\pi : \mathbb{C}^n \setminus \{0\} \to \mathbb{C}P^{n-1}$ (with the structure Lie group $\mathbb{C}^* = \mathbb{C} \setminus \{0\}$). The vertical distribution is completely integrable and its maximal integral manifold is the fibers of π hence

$$(d_p\,\phi)\Big(\frac{\partial}{\partial \overline{z}}\Big)_p \in T_{\phi(p)}\Big(\pi^{-1}\big(\Phi(p)\big)\Big)$$

---

[49] Cf. the reference in Eells and Lemaire [EeLe].

so that [as the vertical distribution has (complex) rank 1]

$$(d_p \phi)\left(\frac{\partial}{\partial \overline{z}}\right)_p = \lambda\,(d_p \Psi)\left(\frac{\partial}{\partial z}\right)_p \tag{3.98}$$

for some $\lambda \in \mathbb{C}$. Formula (3.98) may also be written as

$$\frac{\partial^2 \Psi}{\partial \overline{z}\, \partial z}(p) = \lambda \frac{\partial \Psi}{\partial z}(p)$$

so that (3.67) prompts

$$2\,\Delta\,\Psi = -F\,\lambda \frac{\partial \Psi}{\partial z} \quad \text{at} \quad p.$$

Next [by (3.61)]

$$4\,H_p = F(p)\,\lambda \frac{\partial \Psi}{\partial z}(p)$$

and $(d_p\Psi)(\partial/\partial z)_p$ is tangent to M, while H_p is normal. It follows that $H_p = 0$, $p \in M$, i.e., M is minimal in $\mathbb{R}^n$. Viceversa, if Ψ is minimal, i.e., $\partial^2\Psi/\partial\overline{z}\,\partial z = 0$, then $\phi(z) = \partial\Psi/\partial z$ is holomorphic, and Φ follows to be holomorphic, as the composition of two holomorphic maps (i.e., $\Phi = \pi \circ \phi$). Q.e.d.

Theorem 3.88 *Let $\Psi : M \to \mathbb{R}^n$ be a conformal immersion of a Riemann surface M into $\mathbb{R}^n$. Let*

$$\Phi : M \to \mathbb{C}P^{n-1}, \quad \Phi = \left[\frac{\partial \Psi}{\partial z}\right]$$

be the Gauss map associated to Ψ. Then the image of Φ lies on the quadric

$$Q_{n-2} = \Big\{[w_1 \cdots w_n] \in \mathbb{C}P^{n-1} \ : \ \sum_{j=1}^{n} w_j^2 = 0\Big\}.$$

Proof One checks that

$$\sum_{j=1}^{n} \left(\frac{\partial \Psi^j}{\partial z}\right)^2 = 0$$

as a consequence of Ψ being conformal. Q.e.d.

Remark 3.89 (i) The quadric Q_{n-2} is a complex submanifold of $\mathbb{C}P^{n-1}$. It follows that $\Phi : M \to Q_{n-2}$ is holomorphic if and only if Ψ is minimal.

(ii) The Gauss map Φ admits the following geometric interpretation. Let $\tilde{G}_2(\mathbb{R}^n)$ be the Grassmann manifold of all oriented real 2-dimensional subspaces of $\mathbb{R}^n$. Let $\sigma \in \tilde{G}_2(\mathbb{R}^n)$ and let $\mathscr{O}$ be an orientation of σ. Let $(u,\ v) \in \mathscr{O}$ such that $\langle u,\ v\rangle = 0$ and $|u| = |v|$. Let us set $w = u + i\,v \in \mathbb{C}^n \setminus \{0\}$. The point $[w] \in \mathbb{C}P^{n-1}$ doesn't depend upon the choice[50] of oriented basis $\{u, v\} \in \mathscr{O}$ with $\langle u, v\rangle = 0$ and $|u| = |v|$. Also $|u| = |v|$ implies

$$\sum_{j=1}^{n} w_j^2 = \sum_{j=1}^{n} \left(u_j + i\, v_j\right)^2 = 0,$$

i.e., $[w] \in Q_{n-2}$. This prompts a bijective map $q : \tilde{G}_2(\mathbb{R}^n) \to Q_{n-2}$ and

$$q^{-1} \circ \overline{\Phi} = q^{-1} \circ \left[\overline{\frac{\partial \Psi}{\partial z}}\right] = q^{-1} \circ \left[\frac{\partial \Psi}{\partial x} + i\, \frac{\partial \Psi}{\partial y}\right],$$

i.e., for every $p \in M$ we may look at $\left(q^{-1} \circ \overline{\Phi}\right)(p)$ as the oriented tangent plane of Ψ at p. This is the geometric interpretation of Φ seek after.

Let $\Psi : M \to \mathbb{R}3$ be a minimal conformal immersion of a Riemann surface M into $\mathbb{R}^3$. Theorems 3.83 and 3.84 prompt the Weierstrass representation of Ψ, i.e.,

$$\Psi(z) = \mathrm{Re}\left\{\int_o^z \phi(\zeta)\, d\,\zeta\right\} + \text{const.},$$

$$\phi(z) = 2\,\frac{\partial \Psi}{\partial z} = 2\, f(z)\left(\frac{1}{2}\left(1 - g(z)^2\right),\ \frac{i}{2}\left(1 + g(z)^2\right),\ g(z)\right).$$

The Gauss map $\Phi : U \to Q_1 \subset \mathbb{C}P^2$ is

$$\Phi(z) = \left[\frac{1}{2}\left(1 - g(z)^2\right),\ \frac{i}{2}\left(1 + g(z)^2\right),\ g(z)\right].$$

It should be observed that the Gauss map Φ is fully determined by g (and the function f plays no role). This prompts the following geometric interpretation of g.

Definition 3.90 The *normal Gauss map* of Ψ is the map

$$N : M \to S^2, \quad p \in M \longmapsto N_p \in T_{\Psi(p)}\left(\mathbb{R}^3\right),$$

where the unit normal vector N_p is chosen such that the frame $\left\{\Psi_x,\ \Psi_y,\ N\right\}$ is positively oriented. □

[50] For any other basis $\{u', v'\} \in \mathscr{O}$ with $\langle u', v'\rangle = 0$ and $|u'| = |v'|$ (by assuming that $\mathscr{O}$ is the positive orientation) $\{u, v\}$ is mapped onto $\{u', v'\}$ by a transformation of the form

$$\begin{pmatrix} a & b \\ -b & a \end{pmatrix}, \quad a^2 + b^2 \neq 0.$$

Consequently, if $w' = u' + i\, v'$ then $w' = \overline{\lambda}\, w$ with $\lambda = a + i\, b \in \mathbb{C}^*$, so that $[w] = [w']$.

Here $\Psi_x = \partial\Psi/\partial x$ and $\Psi_y = \partial\Psi/\partial y$. To compute $N = \|\Psi_x \times \Psi_y\|^{-1}\left(\Psi_x \times \Psi_y\right)$ one makes use of

$$\Psi_x = 2\,\mathrm{Re}\,\frac{\partial\Psi}{\partial z} = \mathrm{Re}\left(\phi_1,\ \phi_2,\ \phi_3\right)$$

and obtains

$$N = \left(\frac{2\,\mathrm{Re}(g)}{|g|^2+1},\ \frac{2\,\mathrm{Im}(g)}{|g|^2+1},\ \frac{|g|^2-1}{|g|^2+1}\right). \tag{3.99}$$

Let $\pi : S^2 \setminus \{(0,\ 0,\ 1)\} \to \mathbb{C} \simeq \mathbb{R}^2$ be the stereographic projection. Its inverse is

$$\pi^{-1}(z) = \left(\frac{2\,\mathrm{Re}(z)}{|z|^2+1},\ \frac{2\,\mathrm{Im}(z)}{|z|^2+1},\ \frac{|z|^2-1}{|z|^2+1}\right). \tag{3.100}$$

A comparison of (3.99) and (3.100) shows that g is the local representation of N with respect to the local chart $\left(S^2 \setminus \{(0,\ 0,\ 1)\},\ \pi\right)$, which is the geometric interpretation of g seek for.

That $\Phi : M \to \mathbb{C}P^2$ and $N :\to S^2$ coincide up to appropriate identifications is emphasized by the following commutative diagram

$$\begin{array}{cccccc}
 & \mathbb{C}\cup\{\infty\} & = & \mathbb{C}\cup\{\infty\} & & \\
 & g\uparrow & & \downarrow \pi^{-1} & & \\
M \supset & U & \xrightarrow{N} & S^2 & \simeq \tilde{G}_1(\mathbb{R}^3) = \mathbb{R}P^2 & \\
 & \Phi\downarrow & & & \downarrow \perp & \\
 & Q_1 & \xrightarrow{q^{-1}} & & \tilde{G}_2(\mathbb{R}^3) &
\end{array}$$

Vis-a-vis to the previous diagram we should describe the identifications $S^2 \simeq \tilde{G}_1\left(\mathbb{R}^3\right) \simeq \tilde{G}_2\left(\mathbb{R}^3\right)$. The map $S^2 \to \tilde{G}_1\left(\mathbb{R}^3\right)$ assigns to every point $p \in S^2$ the line passing through p and the origin. Also, the map $\perp : \tilde{G}_1(\mathbb{R}^3) \to \tilde{G}_2\left(\mathbb{R}^3\right)$ assigns to every oriented line ℓ through the origin, the oriented plane passing through the origin and perpendicular to ℓ. □

Exercise 3.91 Show that the map $\mathbb{C}\cup\{\infty\} \simeq S^2 \simeq \tilde{G}_1\left(\mathbb{R}^3\right) \simeq \tilde{G}_2\left(\mathbb{R}^3\right) \simeq Q_1$ is explicitly given by

$$z \in \mathbb{C}\cup\{\infty\} \mapsto \left(\frac{2\,\mathrm{Re}(g)}{|g|^2+1},\ \frac{2\,\mathrm{Im}(g)}{|g|^2+1},\ \frac{|g|^2-1}{|g|^2+1}\right) \in S^2 \mapsto \left[\frac{1}{2}\left(1-z^2\right),\ \frac{i}{2}\left(1+z^2\right),\ z\right] \in Q_1\,.$$

□

Example 3.92 Let $\psi : \mathbb{C} \setminus \{0\} \to \mathbb{R}^3$ be defined by (3.89), with the functions ϕ^k, $k \in \{1, 2, 3\}$, given by (3.85) together with $f(z) = -z^{-2}$ and $g(z) = z$. Then Ψ is the catenoid. The normal Gauss map is

$$\mathbb{C} \setminus \{0\} \to \mathbb{C} \cup \{\infty\}, \quad z \longmapsto g(z) = z.$$

It should be observed that the normal Gauss map omits the two points $\{0, \infty\}$ in $\mathbb{C} \setminus \{\infty\}$ [that is $(0, 0, \pm 1)$ in S^2]. The following general problem is raised by the example at hand: how many points in S^2 can omit the normal Gauss map $N : M \to S^2$ of a complete[51] minimal surface $\Psi : M \to \mathbb{R}^3$? □

The following general problem is raised by the example at hand: how many points in S^2 can omit the normal Gauss map $N : M \to S^2$ of a complete[52] minimal surface $\Psi : M \to \mathbb{R}^3$ not contained in a plane? By a result of Fujimoto [Fuj].

Theorem 3.93 *Let $\Psi : M \to \mathbb{R}^3$ be a non-flat complete minimal surface. Then its normal Gauss map $N : M \to S^2$ can omit at most four points on S^2.*

The normal Gauss map of the Scherk surface (cf. e.g., [BaCo]) omits four points on S^2, hence H. Fujimoto's result is optimal.

3.12.5 Minimal Submanifolds of Spheres

Theorem 3.94 *Let $i : \overline{M} \to \mathbb{R}^n$ be a submanifold, where i is the inclusion, and let $\overline{g}$ be the induced metric on $\overline{M}$. Let $\psi : M \to \overline{M}$ be an isometric immersion of a m-dimensional Riemannian manifold (M, g) into $(\overline{M}, \overline{g})$. Then the mean curvature vector of ψ equals the tangential component [in $T(\overline{M})$] of $-(1/m)\,\Delta(i \circ \psi)$.*

Proof We may assume w.l.o.g. that M is a submanifold of $\overline{M}$, i.e., $\psi : M \to \overline{M}$ is the canonical inclusion. Let $H(\psi)$ [respectively $H(i \circ \psi)$] be the mean curvature vector of M in $\overline{M}$ [respectively of M in $\mathbb{R}^n$]. The following direct sum decompositions hold

$$T_x\left(\mathbb{R}^n\right) = T_x\left(\overline{M}\right) \oplus T\left(\overline{M}\right)_x^{\perp}, \quad x \in \overline{M}, \tag{3.101}$$

$$T_p\left(\overline{M}\right) = T_p(M) \oplus T(M)_p^{\perp}, \quad p \in M, \tag{3.102}$$

where $E(\psi) = T(M)^{\perp} \to M$ and $E(i) = T\left(\overline{M}\right)^{\perp} \to \overline{M}$ are the normal bundles of the given immersions $\psi : M \to \overline{M}$ and $i : \overline{M} \to \mathbb{R}^n$. For every tangent vector field

[51] That is the induced Riemannian metric $g = \Psi^* \overline{g}$ is complete.

[52] That is the induced Riemannian metric $g = \Psi^* \overline{g}$ is complete.

$X \in \mathfrak{X}(\mathbb{R}^n)$ let X^{T} and X^{N} denote its tangential and normal components with respect to the decomposition (3.101). We ought to show that

$$H(\psi) = H(i \circ \psi)^{\mathrm{T}}. \tag{3.103}$$

Let $T(M)^{\perp_o} = E(i \circ \psi) \to M$ be the normal bundle of the immersion $i \circ \psi : M \to \mathbb{R}^n$. Given $X \in \mathfrak{X}(\mathbb{R}^n)$ let X^{N_o} denote its normal component with respect to

$$T_p(\mathbb{R}^n) = T_p(M) \oplus T(M)_p^{\perp_o}, \quad p \in M. \tag{3.104}$$

Let $\{E_k : 1 \leq k \leq m\}$ be a local frame of $T(M)$. Let $\nabla^{\mathbb{R}^n}$ and $\overline{\nabla}$ be the Euclidean connection on $\mathbb{R}^n$ and the induced connection on $\overline{M}$, associated to the immersion $i : \overline{M} \to \mathbb{R}^n$. By the Gauss formula

$$H(\psi) = \Big(\sum_{k=1}^{m} \overline{\nabla}_{E_k} E_k\Big)^{\mathrm{N}} = \Big(\sum_{k=1}^{m} \big(\nabla^{\mathbb{R}^n}_{E_k} E_k\big)^{\mathrm{T}}\Big)^{\mathrm{N}} =$$

$$= \Big(\sum_{k=1}^{m} \big(\nabla^{\mathbb{R}^n}_{E_k} E_k\big)^{\mathrm{N}_o}\Big)^{\mathrm{T}} = H(i \circ \psi)^{\mathrm{T}}.$$

Q.e.d. Finally [by (3.61)]

$$H(\psi) = -\frac{1}{m}\Big[\Delta(i \circ \psi)\Big]^{\mathrm{T}} \tag{3.105}$$

and Theorem 3.94 is completely proved.

Lemma 3.95 *Let (M, g) be a Riemannian manifold. For every $f \in C^\infty(M)$*

$$\frac{1}{2}\Delta(f^2) = f\,\Delta f - \|df\|^2 \tag{3.106}$$

The proof is straightforward.

Theorem 3.96 *Let M be a m-dimensional Riemannian manifold, and let $\psi : M \to S^n(r)$ be an isometric immersion of M in the n-dimensional sphere of radius $r > 0$. Let $\Psi = i \circ \psi$, where $i : S^n(r) \to \mathbb{R}^{n+1}$ is the inclusion. Then ψ is minimal if and only if*

$$\Delta\Psi = \frac{m}{r^2}\,\Psi \tag{3.107}$$

Proof Equation (3.105) may be written

$$H(\psi) = -\frac{1}{m}\,\mathrm{pr}_{T(\overline{M})}\,\Delta(i \circ \psi) \tag{3.108}$$

where $\mathrm{pr}_{T(\overline{M})}$ denotes the projection on the first term of the direct sum (3.101). For $\overline{M} = S^n(r)$ one has

$$H(\psi) = -\frac{1}{m}\,\mathrm{pr}_{T(S^n(r))}\,\Delta\Psi\,. \tag{3.109}$$

Then $\psi : M \to S^n(r)$ is minimal if and only if [by (3.109)] $\Delta\Psi$ is vector in $\mathbb{R}^{n+1}$ orthogonal to $S^n(r)$. Yet, at any point $p \in M$, the normal space $T\big(S^n(r)\big)^{\perp}_{\psi(p)}$ is spanned by $\Psi(p) \in \mathbb{R}^{n+1} \simeq T_{\Psi(p)}\big(\mathbb{R}^{n+1}\big)$. Consequently

$$\Delta\Psi = -\lambda\,\Psi \tag{3.110}$$

for some $\lambda \in C^\infty(M)$. Let us set $\Psi = \big(\Psi^1, \ldots, \Psi^{n+1}\big)$. Also $\Psi(p) \in S^n(r)$ so that $\|\Psi(p)\|^2_{\mathbb{R}^{n+1}} = r^2 = \text{constant}$. Hence [by (3.106)]

$$0 = \frac{1}{2}\,\Delta\|\Psi\|^2_{\mathbb{R}^{n+1}} = \frac{1}{2}\sum_{k=1}^{n+1}\Delta\big(\Psi^k\big)^2 = \sum_{k=1}^{n+1}\Psi^k\,\Delta\Psi^k - \sum_{k=1}^{n+1}\big\|d\,\psi^k\big\|^2 .$$

Then [by (3.110)]

$$0 = \langle\Psi,\,\Delta\Psi\rangle_{\mathbb{R}^{n+1}} - \sum_{k=1}^{n+1}\big\|d\,\psi^k\big\|^2 = -\Big(\lambda\,\|\Psi\|^2_{\mathbb{R}^{n+1}} + \sum_{k=1}^{n+1}\big\|d\,\psi^k\big\|^2\Big)$$

where from

$$\lambda = -\frac{1}{r^2}\sum_{k=1}^{n+1}\|d\Psi^k\|^2 . \tag{3.111}$$

As $\Psi : M \to \mathbb{R}^{n+1}$ is an isometric immersion, its differential

$$d_p\Psi : T_p(M) \to T_{\Psi(p)}\big(\mathbb{R}^{n+1}\big) \simeq \mathbb{R}^{q+1}$$

is a linear isometry, so that

$$g^{\mathrm{Eucl}}_{\Psi(p)}\Big((d_p\Psi)X,\ (d_p\Psi)Y\Big) = g_p(X,\,Y), \quad X,\,Y \in T_p(M), \quad p \in M. \tag{3.112}$$

Let $(u^1, \ldots, u^m)$ be a system of local coordinates on M about p, and let us represent X and Y as

$$X = X^\alpha\left(\frac{\partial}{\partial u^\alpha}\right)_p, \quad Y = Y^\alpha\left(\frac{\partial}{\partial u^\alpha}\right)_p, \quad X^\alpha,\ Y^\alpha \in \mathbb{R}.$$

Then (3.112) reads

$$g_{\alpha\sigma}(p) = \sum_{k=1}^{n+1} \frac{\partial \Psi^k}{\partial u^\alpha}(p) \frac{\partial \Psi^k}{\partial u^\sigma}(p) \tag{3.113}$$

where $g_{\alpha\sigma}$ are the local components of the induced metric on M. Contraction with $g^{\sigma\mu}(p)$ in (3.113) leads to

$$\delta^\mu_\alpha = \sum_{k=1}^{n+1} g^{\sigma\mu}(p) \frac{\partial \Psi^k}{\partial u^\alpha}(p) \frac{\partial \Psi^k}{\partial u^\sigma}(p)$$

hence [by contracting α and μ]

$$m = \sum_{k=1}^{n+1} \|d\Psi^k\|^2 . \tag{3.114}$$

Now (3.111) and (3.114) lead

$$\lambda = -\frac{m}{r^2} . \tag{3.115}$$

Finally [by (3.110) and (3.115)] ψ is minimal if and only if (3.107) holds. Q.e.d.

Definition 3.97 Let M be a Riemannian manifold, and let $\Delta : C^\infty(M) \to C^\infty(M)$ be its Laplace-Beltrami operator. A function $h \in C^\infty(M)$ is an *eigenfunction* of Δ corresponding to the *eigenvalue* $\lambda \in \mathbb{R}$ if $-\Delta h = \lambda h$ on M. □

By Theorem 3.96 an isometric immersion of the Riemannian manifold M into the sphere $S^n(r)$ is minimal if and only if the components of $\Psi = i \circ \psi$ are eigenfunctions (for the Laplace-Beltrami operator Δ on M) corresponding to the same eigenvalue $\lambda = -m/r^2$.

The following (related) result is due to Takahasi [Tak].

Theorem 3.98 *Let $\Psi : M \to \mathbb{R}^{n+1}$ be an isometric immersion of a m-dimensional Riemannian manifold M into $\mathbb{R}^{n+1}$ such that*

$$\Delta\Psi = \mu\,\Psi \tag{3.116}$$

for some $\mu \in \mathbb{R} \setminus \{0\}$. Then

(i) $\mu > 0$,

(ii) $\Psi(M) \subset S^n(r)$ *where* $r^2 = m/\mu$,

(iii) $\Psi : M \to S^n(r)$ *is a minimal isometric immersion.*

Proof We may assume w.l.o.g. that $\Psi : M \to \mathbb{R}^{n+1}$ is the canonical inclusion. Let $H(\Psi)$ be the mean curvature vector of M in $\mathbb{R}^{n+1}$. Then [by (3.61)]

$$-\Delta\Psi = m\, H(\Psi).$$

Consequently [by (3.116)]

$$\Psi = -\frac{m}{\mu}\, H(\Psi)$$

so that Ψ is orthogonal to M. Let us consider the function $F \in C^\infty(M)$

$$F(p) = \frac{1}{2}\,\langle \Psi(p),\, \Psi(p)\rangle_{\mathbb{R}^{n+1}}, \quad p \in M.$$

For every $X \in \mathfrak{X}(M)$, $X \simeq (d\Psi)X$, one has

$$X(F) = 2\,g^{\mathrm{Eucl}}\left(\nabla_X^{\mathrm{Eucl}}\Psi,\, \Psi\right) = 2\,g^{\mathrm{Eucl}}\left(X(\Psi^k)\,\frac{\partial}{\partial x^k},\, \Psi\right) = 2\,g^{\mathrm{Eucl}}\left((d\Psi)X,\, \Psi\right) = 0$$

where $\Psi(p) \simeq \Psi^k(p)\left(\partial/\partial x^k\right)_{\Psi(p)}$ [and $(x^1, \ldots, x^{n+1})$ are the Cartesian coordinates on $\mathbb{R}^{n+1}$]. It follows that $F =$ constant, i.e., $\|\Psi\|^2_{\mathbb{R}^{n+1}} = r^2$ for some $r \geq 0$ and $r \neq 0$ because Ψ is an immersion. Hence $\Psi(M) \subset S^n(r)$ and (ii) in Theorem 3.98 is proved. Moreover

$$0 = -\frac{1}{2}\,\Delta\|\Psi\|^2_{\mathbb{R}^{n+1}} = -\langle\Psi,\, \Delta\Psi\rangle_{\mathbb{R}^{n+1}} + \sum_{k=1}^{n+1}\|d\Psi^k\|^2 = -\mu\, r^2 + \sum_{k=1}^{n+1}\|d\Psi^k\|^2$$

yields $\mu = m/r^2 > 0$, proving (i) in Theorem 3.98. To prove the statement (iii) one applies Theorem 3.96) to the isometric immersion $\Psi : M \to S^n\left(\sqrt{m/\mu}\right)$ and then [by (3.116)] the co-restriction of Ψ to $S^n\left(\sqrt{m/\mu}\right)$ follows to be minimal. Q.e.d.

We look at a number of applications of Theorem 3.98 (the *Takahashi theorem*). Let $C^\infty(S^m)$ consist of real-valued C^∞ functions on S^m. Let us consider the L^2 inner product

$$(\phi,\, \psi)_{L^2} = \int_{S^m} \phi\,\psi\; d\,\mathrm{v}_g \tag{3.117}$$

where g is the standard Riemannian metric on the sphere. Let $\lambda \in \mathbb{R}$ be an eigenvalue of the Laplacian $\Delta : C^\infty(S^m) \to C^\infty(S^m)$, and let $E_\lambda \subset C^\infty(S^m)$ be the corresponding eigenspace. Then E_λ is finite dimensional.[53] Let $N := \dim_{\mathbb{R}} E_\lambda$ and let $\{\phi_1, \ldots, \phi_N\}$ be a orthonormal (i.e., $(\phi_i,\, \phi_j)_{L^2} = \delta_{ij}$) basis of E_λ.

Theorem 3.99 *There exist constants $\alpha > 0$ and $r > 0$ such that $\Psi = \left(\alpha\phi_1, \ldots, \alpha\,\phi_N\right)$ is a minimal isometric immersion of S^m into $S^{N-1}(r) \subset \mathbb{R}^N$.*

---

[53] Cf. e.g., Theorem A.I.4 in [BGM], p. 142.

We shall need.

Lemma 3.100 *For any $A \in \mathrm{O}(m+1)$ the system $\{\phi_i \circ A \,:\, 1 \le i \le N\}$ is an orthonormal basis of E_λ.*

Proof An element $A \in \mathrm{O}(m+1)$ is thought of as a map $A : S^m \to S^m$. We ought to show that

(i) $\phi_i \circ A \in E_\lambda$ for any $1 \le i \le N$,

(ii) $(\phi_i \circ A, \ \phi_j \circ A)_{L^2} = \delta_{ij}$ for any $1 \le i, \ j \le N$.

As A is an isometry[54]

$$\Delta(\phi_i \circ A)(p) = (\Delta\phi_i)(A(p)) = -\lambda\, \phi_i(A(p)), \quad p \in S^m,$$

i.e., $\phi_i \circ A$ is an eigenfunction of the Laplacian. Next [by a change of variable $q = A(p)$ under the integral sign]

$$\left(\phi_i \circ A, \ \phi_j \circ A\right)_{L^2} = \int_{S^m} \left(\phi_i \phi_j\right) \circ A \; d\,\mathrm{v}_g = \int_{S^m} \phi_i \phi_j \; d\,\mathrm{v}_g = (\phi_i, \ \phi_j)_{L^2} = \delta_{ij} \,.$$

Q.e.d. That is, any orthogonal matrix $A \in \mathrm{O}(m+1)$ induces an isometry of E_λ.

Let $B = [b^i_j] \in \mathrm{O}(N)$ such that $\phi_j \circ A = b^i_j\, \phi_i$ or

$$\phi \circ A = B\, \phi, \quad \phi := \left(\phi_1, \ \ldots, \ \phi_N\right)^T . \tag{3.118}$$

Lemma 3.101 *For any $p, \ q \in S^m$ and any $X \in T_p(S^m)$ and $Y \in T_q(S^m)$ there is $A \in \mathrm{O}(m)$ such that $(d_p A)X = Y$.*

The proof is straightforward.[55] As B is an isometry

$$\|(d_q\phi)Y\| = \|(d_q\phi)(d_pA)X\| = \|d_p(\phi \circ A)X\| = \|d_p(B\,\phi)X\| = \|(d_p\phi)X\|,$$

i.e., $\|(d_p\phi)X\|$ doesn't depend upon the choice of $X \in T_p(S^m)$ with $\|X\| = 1$. Hence, there is a constant $\alpha > 0$ such that

$$\left\|d_p(\alpha\phi)X\right\| = 1, \quad p \in S^m, \quad X \in T_p(S^m), \quad \|X\|. = 1.$$

That is $\alpha\phi = \left(\alpha\phi_1, \ \ldots \alpha\, \phi_N\right) : S^m \to \mathbb{R}^N$ is an isometry. At this point we may apply Theorem 3.98 to conclude that

---

[54] Cf. e.g., [Hel], p. 387.

[55] Cf. also [Ste], p. 105.

$$\alpha\phi : S^m \to S^{N-1}\Big(\sqrt{\frac{m}{\lambda}}\Big)$$

is a minimal isometric immersion. Q.e.d.

Definition 3.102 A map $\Phi : \mathbb{R}^{m+1} \to \mathbb{R}$ is a *homogeneous harmonic polynomial* (of degree k) if Φ is a homogeneous polynomial $\Phi \in \mathbb{R}\big[x_1, \ldots, x_{m+1}\big]$ of degree k such that $\Delta_{\mathbb{R}^{n+1}} \Phi = 0$. □

Here $\Delta_{\mathbb{R}^{n+1}} \equiv - \sum_{i=1}^{m+1} \partial^2/\partial x_i^2$. We shall need.

Theorem 3.103 (i) *Let $\Phi : \mathbb{R}^{m+1} \to \mathbb{R}$ be a homogeneous harmonic polynomial of degree k. Then the restriction $\Phi\big|_{S^m}$ is an eigenfunction of the Laplacian Δ_{S^m} corresponding to the eigenvalue $\lambda_k = (k + m - 1)k$.*

(ii) *All the eigenfunctions and eigenvalues of Δ_{S^m} are obtained in this manner.*

Cf. Proposition C.I.1 and Corollary C.I.3 in [BGM], pp. 160–162. As a consequence of Theorem 3.99 one has $N > n$. A more precise statement is

$$N = \dim_{\mathbb{R}} E_{\lambda_k} = \binom{m+k}{k} - \binom{m+k+1}{k-1}.$$

Cf. [BGM], p. 162. Theorem 3.99 may be used to build examples of minimal isometric immersions in spheres. The homogeneous polynomials of degree $k = 1$ are

$$H : \mathbb{R}^{m+1} \to \mathbb{R}, \quad H(x_1, \ldots, x_{m+1}) = \mu_1 x_1 + \cdots + \mu_{m+1} x_{m+1}, \quad \big(\mu_1, \ldots, \mu_{m+1}\big) \in \mathbb{R}^{m+1} \setminus \{0\}.$$

Clearly any such H is harmonic. Hence $\lambda_1 = m \in \mathrm{Spec}\big(S^m, \Delta_{S^m}\big)$. To determine an orthonormal basis of the eigenspace $E_{\lambda_1} \subset C^\infty(S^m)$ we proceed as follows. We start with the case $k = 1$, i.e., we look at the polynomials $x_1, \ldots, x_{m+1}$. Then for any $i \neq j$

$$\big(x_i, x_j\big)_{L^2} = \int_{S^m} x_i x_j \, d\,\mathrm{v}_g = \int_{\substack{x_i \geq 0 \\ x \in S^m}} x_i x_j \, d\,\mathrm{v}_g + \int_{\substack{x_i \leq 0 \\ x \in S^m}} x_i x_j \, d\,\mathrm{v}_g = 0$$

$$\|x_i\|^2 = \int_{S^m} x_i^2 \, d\,\mathrm{v}_g = \frac{1}{m+1} \mathrm{Vol}(S^m) =: a^2 .$$

So $\{(1/a)\, x_i \, : \, 1 \leq i \leq m+1\}$ is an orthonormal basis of E_{λ_1}. By Theorem 3.99 there exist constants $\alpha > 0$ and $r > 0$ such that

$$\frac{\alpha}{a}\big(x_1, \ldots, x_{m+1}\big) : S^m \to \mathbb{R}^{m+1}$$

is a minimal isometric immersion $S^m \to S^m(r)$.

Let us look at the case $k = 2$, and for simplicity let $m = 2$. Points in $S^2 \subset \mathbb{R}^3$ are denoted by (x, y, z). By Theorem 3.103 one has $\lambda_2 = 6$. The eigenspace E_{λ_2} of Δ_{S^2} consists of the restriction to S^2 of the homogeneous harmonic polynomials

$$\Phi : \mathbb{R}^3 \to \mathbb{R}, \quad \Phi(x, y, z) = c_1\, xy + c_2\, xz + c_3\, yz + c_4\, x^2 + c_5\, y^2 + c_6\, z^2, \quad c_i \in \mathbb{R}, \quad c_4 + c_5 + c_6 = 0.$$

Thus $\dim_{\mathbb{R}} E_{\lambda_2} = 5$ and

$$\left\{xy, \quad xz, \quad yz, \quad x^2 - y^2, \quad x^2 + y^2 - 2\,z^2\right\} \subset E_{\lambda_2}$$

is a linear basis of eigenfunctions, whose elements are mutually orthogonal. Moreover, there is $a > 0$ such that

$$\|xy\| = \|xz\| = \|yz\| = a, \quad \|x^2 - y^2\| = 2a, \quad \|x^2 + y^2 - 2\,z^2\| = 2\sqrt{3}\,a.$$

Then [by Theorem 3.99]

$$\Phi = \frac{\alpha}{a}\left(x\,y, \;\; x\,z, \;\; y\,z, \;\; \frac{1}{2}(x^2 + y^2), \;\; \frac{1}{2\sqrt{3}}\,(x^2 + y^2 - 2\,z^2)\right) \; : \; S^2 \to \mathbb{R}^5$$

furnishes a minimal isometric immersion of S^2 in $S^4(r)$, with an appropriate choice of the constants $\alpha > 0$ and $r > 0$. The constant α/a can be determined from the requirement that Φ be an isometry. One starts with choosing an orthonormal basis in the tangent space $T_{(0,0,1)}(S^2)$. For Φ to be an isometry, the tangent vector $(\partial\Phi/\partial x)_{(0,0,1)}$ must be unitary. Then

$$\left(\frac{\partial\Phi}{\partial x}\right)_{(0,0,1)} = \frac{\alpha}{a}\left(y, \; z, \; 0, \; x, \; \frac{1}{\sqrt{3}}\,x\right) = \frac{\alpha}{a}\,(0, \; 1, \; 0, \; 0, \; 0) \Longrightarrow \frac{\alpha}{a} = 1$$

so one chooses $\alpha = a$. Here a was already determined e.g., $a = \|x\,y\|$. With $\alpha = a$ one has

$$\|\Phi\|^2 = x^2\,y^2 + x^2\,z^2 + y^2\,z^2 + \frac{1}{4}\left(x^2 - y^2\right) + \frac{1}{12}\,(x^2 + y^2 - 2\,z^2)^2 = \frac{1}{3}(x^2 + y^2 + z^2)^2 = \frac{1}{3}$$

so the image of Φ is contained in $S^4\left(1/\sqrt{3}\right)$. We conclude that

$$\Phi : S^2 \to S^4\left(\frac{1}{\sqrt{3}}\right),$$

$$\Phi(x, y, z) = \left(x\,y, \;\; x\,z, \;\; y\,z, \;\; \frac{1}{2}(x^2 + y^2), \;\; \frac{1}{2\sqrt{3}}\,(x^2 + y^2 - 2\,z^2)\right)$$

is a minimal isometric immersion.

We conclude the section with a few other examples. Let $\pi : S^2 \to \mathbb{R}P^2$ be the double cover $\pi(\mathbf{x}) = [\mathbf{x}]$, $\mathbf{x} \in S^2$. Then π is an isometry. The map $\Phi : S^2 \to S^4(1/\sqrt{3})$ in the previous example induces a minimal isometric immersion

$$\mathbb{R}P^2 \to S^4, \quad [\mathbf{x}] \longmapsto \frac{1}{3}\,\Phi(\mathbf{x}),$$

known as the *Veronese map*. This map is also injective (i.e., an embedding). A similar use of the covering map $S^4 \to \mathbb{R}P^4$ produces a minimal isometric immersion $\mathbb{R}P^2 \to \mathbb{R}^4$ known as the *Veronese surface*. Cf. also Safarevici [Saf], p. 72.

3.13 Harmonic Maps

Let (M, g) and (N, h) be two Riemannian manifolds, $\dim(M) = n$ and $\dim(N) = m$. We assume M is closed (i.e., M is compact and $\partial M = \emptyset$) and oriented. For the sake of simplicity, both manifolds and their Riemannian metrics are assumed to be smooth i.e., C^∞. However, all considerations below hold if manifolds and metrics are of class C^5. Let (U, xi) and (V, y^α) be local coordinate systems on M and N, respectively. For every C^2 map $f : M \to N$ we adopt the notations

$$g_{ij} = g\Big(\frac{\partial}{\partial x^i}, \frac{\partial}{\partial x^j}\Big), \quad h_{\alpha\beta} = h\Big(\frac{\partial}{\partial y^\alpha}, \frac{\partial}{\partial y^\beta}\Big),$$

$$f^\alpha = y^\alpha \circ f, \quad f_i^\alpha = \frac{\partial f^\alpha}{\partial x^i}, \quad f_{ij}^\alpha = \frac{\partial f^\alpha}{\partial x^i\,\partial x^j} - \begin{Bmatrix} k \\ ij \end{Bmatrix} f_k^\alpha,$$

$$\begin{Bmatrix} k \\ ij \end{Bmatrix} = g^{k\ell}\,\{i\,j,\,\ell\}, \quad \{i\,j,\,k\} = \frac{1}{2}\Big(\frac{\partial g_{ik}}{\partial x^j} + \frac{\partial g_{jk}}{\partial x^i} - \frac{\partial g_{ij}}{\partial x^k}\Big).$$

Definition 3.104 The *energy density* is the function

$$e(f) : M \to \mathbb{R}, \quad e(f) = \frac{1}{2}\,\mathrm{Trace}_g\,(f^*h). \tag{3.119}$$

□

Locally [on $U \cap f^{-1}(V)$]

$$e(f) = \frac{1}{2}\,g^{ij}\,f_i^\alpha\,f_j^\beta\,(h_{\alpha\beta} \circ f) \tag{3.120}$$

where $[g^{ij}] = [g_{ij}]^{-1}$.

Definition 3.105 The *energy* of $f \in C^2(M, N)$ is

$$E(f) = \int_M e(f)\, d\,\mathrm{v}_g\,. \tag{3.121}$$

□

Here $d\,\mathrm{v}_g$ is the canonical volume form on M, i.e., locally

$$d\,\mathrm{v}_g = \sqrt{\mathfrak{g}}\, dx^1 \wedge \cdots \wedge dx^n, \quad \mathfrak{g} = \det\left[g_{ij}\right].$$

Let $f^{-1}T(N) \to M$ be the pullback of the vector bundle $T(N) \to N$ by $f : M \to N$. Let ∇^h be the Levi-Civita connection of (N, h), and let $f^{-1}\nabla^h = \nabla^f$ be the connection in the pullback bundle $f^{-1}T(N) \to M$ induced by ∇^h by f (the *pullback* of the connection ∇^h). The connection ∇^f is most easily described in local coordinates, as follows.

Definition 3.106 The *natural lift* of a tangent vector field $Y : V \to T(N)$, defined on the open subset $V \subset N$, is the section $Y^f \in C^\infty\big(f^{-1}(V),\ f^{-1}T\,N\big)$ given by

$$Y^f(x) = Y_{f(x)}, \quad x \in f^{-1}(V).$$

□

Let

$$Y_\alpha := \left(\frac{\partial}{\partial y^\alpha}\right)^f \in C^\infty\big(f^{-1}(V),\ f^{-1}TN\big), \quad 1 \le \alpha \le m$$

be the natural lifts of the tangent vector fields $\partial/\partial y^\alpha$. Then $\{Y_\alpha : 1 \le \alpha \le m\}$ is a local frame in the pullback bundle $f^{-1}T(N)$, defined on the open set $f^{-1}(V)$. Then ∇^f is locally given by

$$\nabla^f_{\partial/\partial x^i} Y_\alpha = \left(\left\{ \begin{matrix} \beta \\ \alpha\gamma \end{matrix} \right\} \circ f\right) f_i^\gamma\, Y_\beta\,.$$

For every vector bundle $E \to M$ we denote by $\mathscr{C}(E)$ the affine space of all connections in E. Also, if h is a bundle metric on E, let $\mathscr{C}(E, h)$ be the affine subspace consisting of all $D \in \mathscr{C}(E)$ such that $Dh = 0$. Going back to the C^2 map $f : M \to N$, let $f^{-1}h = h^f$ be the pullback of h by f, i.e., locally

$$h^f\big(Y_\alpha,\ Y_\beta\big) = h_{\alpha\beta} \circ f.$$

Then h^f is a Riemannian bundle metric on $f^{-1}T(N)$ and $\nabla^f \in \mathscr{C}\big(f^{-1}TN,\ h^f\big)$.

For every tangent vector field $X \in \mathfrak{X}(M)$ let $f_* X$ denote the section in the pullback bundle $f^{-1}T(N)$ defined by

$$(f_* X)(x) = (d_x f)\, X_x, \quad x \in M.$$

Let ∇ be the Levi-Civita connection of (M, g).

Definition 3.107 The *second fundamental form* of $f : M \to N$ is

$$\beta(f)(X, Y) = \nabla^f_X f_* Y - f_* \nabla_X Y, \quad X, Y \in \mathfrak{X}(M).$$

□

Remark 3.108 (1) If $f : M \to N$ is an isometric immersion then $\beta(f)$ is its ordinary second fundamental form.

(2) For every $X \in \mathfrak{X}(M)$, the quantity $\beta(f)(X, X)$ admits the following geometric interpretation. Let $p \in M$ and $v = X_p \in T_p(M)$. Let $\gamma : (-\epsilon, \epsilon) \to M$ be the geodesic such that $\gamma(0) = p$ and $(d\gamma/dt)(0) = v$. Then $(\nabla_X X)_p = 0$ and

$$\beta(f)_p(v, v) = \left(\nabla^f_X f_* X\right)_p = \left(\nabla^h_{\dot{\gamma}_f} \dot{\gamma}_f\right)_{t=0}$$

where $\gamma_f = f \circ \gamma$. That is $\beta(f)_p(v, v)$ is the acceleration of the curve $f \circ \gamma$ [the image by f of the geodesic in M of initial data (p, v)]. □

Definition 3.109 A map $f \in C^\infty(M, N)$ is *totally geodesic* if $\beta(f) = 0$. □

Theorem 3.110 *A map $f \in C^\infty(M, N)$ is totally geodesic if and only if f maps geodesics of M into geodesics of N.*

Example 3.111 (1) Let $f : \mathbb{R}^n \to \mathbb{R}^m$ be a C^∞ map. Then f is totally geodesic (with respect to the Euclidean structures of $\mathbb{R}^n$ and $\mathbb{R}^m$) if and only if f is an affine map, i.e., $f(x) = Ax + b$ for some $A \in \mathfrak{M}_{n,m}(\mathbb{R})$ and some $b \in \mathbb{R}^m$, and any $x \in \mathbb{R}^n$. The statement is proved by showing that f maps straight lines into straight lines.

(2) Let (M, g) and (N, h) be two Riemannian metrics, and let G and H be the Sasaki metrics on $T(M)$ and $T(N)$, respectively. Let $f : M \to N$ be a C^∞ map, and let

$$f_* : T(M) \to T(N), \quad f_*(v) = (d_{\pi(v)} f)v, \quad v \in T(M).$$

By a result of Sanini [San], p. 61.

Theorem 3.112 *Let (M, g) and (N, h) be two Riemannian manifolds, and let $f : M \to N$ be a C^∞ map. The following statements are equivalent*

(i) *f is totally geodesic as a map of (M, g) into (N, h).*
(ii) *$f_* : T(M) \to T(N)$ is totally geodesic as a map of $(T(M), G)$ into $(T(N), H)$.*

□

Definition 3.113 The *tension field* of f is

$$\tau(f) = \mathrm{Trace}_g\, \beta(f) \in C^\infty\big(f^{-1}T(N)\big)\,.$$

□

Definition 3.114 Let $\{f_t\}_{|t|<\epsilon} \subset C^\infty(M,\, N)$ be a smooth 1-parameter variation of $f_0 = f \in C^\infty(M,\, N)$. Let $F : M \times (-\epsilon,\, \epsilon) \to N$ be defined by

$$F(x,\, t) = f_t(x), \quad x \in M, \quad |t| < \epsilon\,.$$

The *infinitesimal variation* induced by $\{f_t\}_{|t|<\epsilon}$ is the section $V \in C^\infty\big(f^{-1}TN\big)$ defined by

$$V(x) = \big(d_{(x,0)}\, F\big)\left(\frac{\partial}{\partial t}\right)_{(x,0)}, \quad x \in M.$$

□

Theorem 3.115 *Let $\{f_t\}_{|t|<\epsilon} \subset C^\infty(M,\, N)$ be a smooth 1-parameter variation of $f_0 = f \in C^\infty(M,\, N)$. Then*

$$\frac{d}{dt}\Big\{E\big(f_t\big)\Big\}_{t=0} = -\int_M h^f\big(\tau(f),\, V\big)\, d\,\mathrm{v}_g\,. \tag{3.122}$$

Definition 3.116 Equation (3.122) is the *first variation formula.* □

Proof of Theorem 3.115 Let us set $\tilde{M} = M \times (-\epsilon,\, \epsilon)$ and let us consider the injections

$$\alpha_t : M \to \tilde{M}, \quad \alpha_t(x) = (x,\, t), \quad x \in M, \quad |t| < \epsilon\,.$$

For every $X \in \mathfrak{X}(M)$ let $\tilde{X} \in \mathfrak{X}\big(\tilde{M}\big)$ be defined by

$$\tilde{X}_{(x,t)} = (d_x\alpha_t)\, X_x, \quad (x,\, t) \in \tilde{M}.$$

Let $F^{-1}T(N) \to \tilde{M}$ be the pullback of $T(N) \to N$ by $F : \tilde{M} \to N$. Let $h^F = F^{-1}h$ and $\nabla^F = F^{-1}\nabla^h$ be respectively the pullback of h and ∇^h by F, so that $\nabla^F \in \mathscr{C}\big(F^{-1}\,TN,\, h^F\big)$. Let $x \in M$ and let $\{E_i\, :\, 1 \le i \le n\}$ be a local g-orthonormal [i.e., $g(E_i,\, E_j) = \delta_{ij}$] in $T(M)$, defined on the open set $U \subset M$ with $x \in U$. Then for every $|t| < \epsilon$

$$\Big(\mathrm{Trace}_g\big(f_t^*h\big)\Big)(x) = \sum_{i=1}^n (f_t^*h)(E_i,\, E_i)_x = \sum_{i=1}^n h_{f_t(x)}\big((d_x f_t)\, E_{i,x},\, (d_x f_t)\, E_{i,x}\big) =$$

[as $f_t = F \circ \alpha_t \Longrightarrow d_x f_t = \big(d_{(x,t)} F\big) \circ (d_x \alpha_t)$]

$$= \sum_{i=1}^{n} h_{F(x,t)}\big((d_{(x,t)} F)\, \tilde{E}_{(x,t)},\ (d_{(x,t)} F)\, \tilde{E}_{(x,t)}\big) = \sum_{i=1}^{n} h^F\big(F_* \tilde{E}_i,\ F_* \tilde{E}_i\big)_{(x,t)}$$

that is

$$\Big\{\mathrm{Trace}_g\big(f_t^* h\big)\Big\}_x = \sum_{i=1}^{n} h^F\big(F_* \tilde{E}_i,\ F_* \tilde{E}_i\big)_{(x,t)}$$

for any $(x,\ t) \in U$. Let $\tilde{U} = U \times (-\epsilon,\ \epsilon)$ and let $\varphi : \tilde{M} \to \mathbb{R}$ be the function given by

$$\varphi(x,\ t) = \frac{1}{2}\,\mathrm{Trace}_g\big(f_t^* h\big)_x, \quad x \in M, \quad |t| < \epsilon.$$

Then, at each point of $\tilde{U}$

$$\frac{\partial \varphi}{\partial t} = \frac{1}{2}\sum_{i=1}^{n} \frac{\partial}{\partial t}\Big\{h^F\big(F_* \tilde{E}_i,\ F_* \tilde{E}_i\big)\Big\} =$$

[as $\nabla^F h^F = 0$]

$$= \sum_{i=1}^{n} h^F\big(\nabla^F_{\partial/\partial t} F_* \tilde{E}_i,\ F_* \tilde{E}_i\big) =$$

$\Big[$as $\nabla^F_{\partial/\partial t} F_* \tilde{E}_i = \nabla^F_{\tilde{E}_i} F_*\Big(\dfrac{\partial}{\partial t}\Big) + F_*\Big[\dfrac{\partial}{\partial t},\ \tilde{E}_i\Big]$ (because ∇^h is torsion-free), and $\Big[\dfrac{\partial}{\partial t},\ \tilde{E}_i\Big] = 0$

$$= \sum_{i=1}^{n} h^F\Big(\nabla^F_{\tilde{E}_i} F_*\Big(\frac{\partial}{\partial t}\Big),\ F_* \tilde{E}_i\Big) =$$

[by setting $\mathbb{V} := F_*\,(\partial/\partial t) \in C^\infty\big(F^{-1} T N\big)$ and using again $\nabla^F h^F = 0$]

$$= \sum_{i=1}^{n} h^F\Big(\nabla^F_{\tilde{E}_i} \mathbb{V},\ F_* \tilde{E}_i\Big) = \sum_{i=1}^{n} \Big\{\tilde{E}_i\Big(h^F\Big(\mathbb{V},\ F_* \tilde{E}_i\Big)\Big) - h^F\Big(\mathbb{V},\ \nabla^F_{\tilde{E}_i} F_* \tilde{E}_i\Big)\Big\}.$$

So far, the following identities were used in the calculations (whose proofs are left as an exercise to the reader)

Exercise 3.117 Show that, for every $X \in \mathfrak{X}(M)$

(1) $\nabla^F h^F = 0$,

(2) $\nabla^F_{\partial/\partial t} F_* \tilde{X} = \nabla^F_{\tilde{X}} F_* \left(\frac{\partial}{\partial t}\right) + F_* \left[\frac{\partial}{\partial t}, \tilde{X}\right]$,

(3) $\left[\frac{\partial}{\partial t}, \tilde{X}\right] = 0$.

□

To proceed, we need the following:

Lemma 3.118 *Let $X_f \in \mathfrak{X}(M)$ be determined by*

$$g(X_f, X) = h^f(V, f_* X), \quad X \in \mathfrak{X}(M). \tag{3.123}$$

Then, for any $x \in U$

$$\sum_{i=1}^n \tilde{E}_i\left(h^F(\mathbb{V}, F_* \tilde{E}_i)\right)_{(x,0)} = \operatorname{div}(X_f) + \sum_{i=1}^n h^f(V, f_* \nabla_{E_i} E_i)_x . \tag{3.124}$$

Proof For any $(x, t) \in \tilde{U}$

$$\sum_{i=1}^n \tilde{E}_i\left(h^F(\mathbb{V}, F_* \tilde{E}_i)\right)_{(x,t)} = \sum_{i=1}^n \tilde{E}_{i,(x,t)}\left(h^F(\mathbb{V}, F_* \tilde{E}_i)\right) =$$

$$= \sum_{i=1}^n ((d_x \alpha_t) E_{i,x})\left(h^F(\mathbb{V}, F_* \tilde{E}_i)\right)_{(x,t)} = \sum_{i=1}^n E_{i,x}\left(h^F(\mathbb{V}, F_* \tilde{E}_i) \circ \alpha_t\right).$$

Exercise 3.119 Show that

$$h^F(\mathbb{V}, F_* \tilde{X}) \circ \alpha_0 = h^f(V, f_* X), \quad h^F\left(\mathbb{V}, \nabla^F_{\tilde{X}} F_* \tilde{Y}\right) \circ \alpha_0 = h^f(V, \nabla^f_X f_* Y),$$

for any $X, Y \in \mathfrak{X}(M)$. □

Next (for $t = 0$)

$$\sum_{i=1}^n \tilde{E}_i\left(h^F(\mathbb{V}, F_* \tilde{E}_i)\right)_{(x,0)} = \sum_{i=1}^n E_i\left(h^f(V, f_* E_i)\right)_x =$$

[by (3.123) and $\nabla g = 0$]

$$= \sum_{i=1}^{n} E_i\big(g(X_f,\, E_i)\big)_x = \sum_{i=1}^{n} \Big\{ g\big(\nabla_{E_i} X_f,\, E_i\big) + g\big(X_f,\, \nabla_{E_i} E_i\big) \Big\}_x =$$

$$= \operatorname{div}\big(X_f\big)_x + \sum_{i=1}^{n} h^f\big(V,\, f_* \nabla_{E_i} E_i\big)_x .$$

Q.e.d. Then

$$\frac{\partial \varphi}{\partial t}(x,\, 0) = \operatorname{div}\big(X_f\big)_x - \sum_{i=1}^{n} h^f\big(V,\, \nabla^f_{E_i} f_* E_i - f_* \nabla_{E_i} E_i\big)_x =$$

$$= \operatorname{div}\big(X_f\big)_x - \sum_{i=1}^{n} h^f\big(V,\, \beta(f)(E_i,\, E_i)\big)_x = \operatorname{div}\big(X_f\big)_x - h^f\big(V,\, \tau(f)\big)_x$$

for every $x \in U$. Finally [by Green's lemma]

$$\frac{d}{dt}\Big\{E\big(f_t\big)\Big\}_{t=0} = \int_M \frac{\partial \varphi}{\partial t}(x,\, 0)\, d\,\mathrm{v}_g = - \int_M h^f\big(V,\, \tau(f)\big)\, d\,\mathrm{v}_g .$$

Q.e.d.

Let $f : M \to N$ be a C^∞ map of Riemannian manifolds, and let $(U,\, x^i)$ and $(V,\, y^\alpha)$ local coordinate systems on M and N such that $f(U) \subset V$. Let us consider the functions $\tau(f)^\alpha \in C^\infty)U)$ given by

$$\tau(f)^\alpha \equiv \Delta f^\alpha + g^{ij} \left(\left\{ {\alpha \atop \sigma\lambda} \right\} \circ f \right) f_i^\sigma\, f_j^\lambda . \tag{3.125}$$

Exercise 3.120 Show that

$$\tau(f)_x = \tau(f)^\alpha(x)\, Y_\alpha(x), \quad x \in U.$$

□

Definition 3.121 A C^∞ map $f : M \to N$ is said to be *harmonic* if $\tau(f) = 0$. □

By the first variation formula.

Corollary 3.122 *Let $f : M \to N$ be a C^∞ map of Riemannian manifolds, with M closed and orientable. The following statements are equivalent*

(i) *f is harmonic.*

(ii) *For every smooth 1-parameter variation $\{f_t\}_{|t|<\epsilon} \subset C^\infty(M, N)$ of $f_0 = f$*

$$\frac{d}{dt}\Big\{E(f_t)\Big\}_{t=0} = 0.$$

(iii) *For any local coordinate systems (U, x^i) on M and (V, y^α) on N with $f(U) \subset V$ the functions $f^\alpha := y^\alpha \circ f$ satisfy*

$$\Delta f^\alpha + g^{ij}\left(\left\{ {\alpha \atop \sigma\lambda} \right\} \circ f\right) f_i^\sigma f_j^\lambda = 0$$

in U.

Example 3.123 (1) Let $I = (-\epsilon, \epsilon) \subset \mathbb{R}$, $\epsilon > 0$, and let $f : I \to N$ be a C^∞ map into a Riemannian manifold N. Then $\tau(f)$ is the acceleration of f, so that f is harmonic if and only if f is a geodesic in N.

(2) Let M and N be Riemannian manifolds, with N flat. Then N admits a C^∞ atlas $\mathscr{A}$ such that for any local chart $(V, y^\alpha) \in \mathscr{A}$ with $\left\{ {\alpha \atop \gamma\sigma} \right\} = 0$ and then

$$\tau(f)^\alpha = -\Delta f^\alpha \tag{3.126}$$

for any $f \in C^\infty(M, N)$. In particular, if M is compact then f is a harmonic map if and only if f is constant.

(3) For the next example we need to recall a few notions from the theory of Riemannian submersions, cf. e.g., O'Neill [One]. Any submersion $f : M \to N$ is an open map. In particular $f(M)$ is an open subset of N. A submersion between equidimensional [i.e., $n = \dim(M) = \dim(N) = m$] manifolds has discrete fibers. Let (M, g) and (N, h) be two Riemannian manifolds ($n \geq m$) and $f : M \to N$ a surjective C^∞ submersion. The *vertical space* at $p \in M$ is $V_p = \mathrm{Ker}(d_p f)$ and the *horizontal space* at p is the orthogonal complement $H_p = V_p^\perp$ in the inner product space $(T_p(M), g_p)$. Then $V : p \in M \mapsto V_p \subset T_p(M)$ and $H : p \in M \mapsto H_p \subset T_p(M)$ are C^∞ distributions.

Definition 3.124 $f : M \to N$ is a *Riemannian submersion* if

$$g_p(v, w) = h_{f(p)}\big((d_p f)v,\ (d_p f)w\big), \quad v, w \in H_p, \quad p \in M. \tag{3.127}$$

□

The example we aim to present is furnished by:

Theorem 3.125 *A surjective Riemannian submersion $f : M \to N$ is a harmonic map of (M, g) into $(N\ h)$ if and only if every fiber $f^{-1}(y)$, $y \in N$, is a minimal submanifold of (M, g).*

(4) Let G and H be two Lie groups, equipped with bi-invariant Riemannian metrics, and let $f : G \to H$ be a group homomorphism and a C^∞ map. Then $f : G \to H$ is a harmonic map (cf. e.g., [EeSa], p. 118).

(5) Let M and N be two Riemannian manifolds, and $f : M \to N$ a C^∞ map. Let $p \in M$ and let (U, x^i) and (V, y^α) be local systems of normal coordinates on M and N, of centers $p \in U$ and $f(p) \in f(U) \subset V$. In particular $\left\{ {i \atop jk} \right\}(p) = 0$ and $\left\{ {\alpha \atop \beta\gamma} \right\}\big(f(p)\big) = 0$. Then

$$\tau(f)^\alpha_p = \sum_{i=1}^n \frac{\partial^2 f^\alpha}{\partial (x^i)^2}(p). \tag{3.128}$$

Theorem 3.126 *Let $M \xrightarrow{f} N$ longrightarrow^g P be C^∞ maps of Riemannian manifolds. If g is totally geodesic and f harmonic then $g \circ f$ is a harmonic map.*

Let $f^*\beta(g)$ be the field of bilinear forms on $T(M)$ defined by

$$\big(f^*\beta(g)\big)_p(v, w) = \beta(g)_{f(p)}\big((d_p f)v, (d_p f)w\big), \quad v, w \in T_p(M), \quad p \in M.$$

Theorem 3.126 is an immediate consequence of

Lemma 3.127 *Let $M \xrightarrow{f} N \xrightarrow{g} P$ be C^∞ maps of Riemannian manifolds. Then*

$$\tau(g \circ f)_p = (d_{f(p)} g)\tau(f)_p + \text{Trace}_g \left\{ f^* \beta(g) \right\}_p, \quad p \in M. \tag{3.129}$$

Proof Let (U, x^i), (V, y^α), and (W, u^a), be local systems of normal coordinates on M, N and P respectively, of centers $p \in U$, $f(p) \in f(U) \subset V$, and $g(f(p)) \in g(V) \subset W$. The convention one adopts as to the range of indices is

$$i, j, k, \ldots \in \{1, \ldots, n\}, \quad \alpha, \beta, \gamma, \cdots \in \{1, \ldots, m\}, \quad a, b, c, \ldots \in \{1, \ldots, p\}.$$

Then

$$\frac{\partial (g \circ f)^a}{\partial x^i} = \Big(\frac{\partial g^a}{\partial y^\alpha} \circ f\Big) \frac{\partial f^\alpha}{\partial x^i}, \tag{3.130}$$

$$\frac{\partial^2 (g \circ f)^a}{\partial x^i \, \partial x^j} = \Big(\frac{\partial g^a}{\partial y^\alpha \, \partial y^\sigma} \circ f\Big) \frac{\partial f^\alpha}{\partial x^i} \frac{\partial f^\sigma}{\partial x^j} + \Big(\frac{\partial g^a}{\partial y^\alpha} \circ f\Big) \frac{\partial^2 f^\alpha}{\partial x^i \, \partial x^j}. \tag{3.131}$$

Let us contract i and j in (3.131) so that to obtain

$$\tau(g\circ f)^a_p=\sum_{i=1}^{n}\Big\{\frac{\partial g^a}{\partial y^\alpha\partial y^\sigma}\big(f(p)\big)\,\frac{\partial^2 f^\alpha}{\partial x^i}(p)\,\frac{\partial f^\sigma}{\partial x^i}(p)+\frac{\partial g^a}{\partial y^\alpha}\big(f(p)\big)\,\frac{\partial^2 f^\alpha}{\partial(x^i)^2}(p)\Big\}=$$

$$=\Big\{(d_{f(p)}g)\,\tau(f)_p+\sum_{i=1}^{n}\beta(g)_{f(p)}\Big((d_pf)\Big(\frac{\partial}{\partial x^i}\Big)_p,\ (d_pf)\Big(\frac{\partial}{\partial x^i}\Big)_p\Big)\Big\}^a.$$

Q.e.d.

3.13.1 Harmonic Maps of Spheres

Definition 3.128 A *polynomial harmonic map* is a map $\Phi:\mathbb{R}^n\to\mathbb{R}^m$ such that every component $\Phi^\alpha:\mathbb{R}^n\to\mathbb{R}$ is a harmonic homogeneous polynomial of some degree k. □

Let $\Phi:\mathbb{R}^n\to\mathbb{R}^m$ be a polynomial harmonic map. The components ϕ^α of $\phi=\Phi\big|_{S^{n-1}}$ are eigenfunctions of the Laplacian $\Delta_{S^{n-1}}$ corresponding to the eigenvalue $k(k+m-2)$.

Theorem 3.129 *Let $\Phi:\mathbb{R}^n\to\mathbb{R}^m$ be a polynomial harmonic map. Let us assume that Φ descends to a map $\phi:S^{n-1}\to S^{m-1}$, i.e., $\Phi\big(S^{n-1}\big)\subset S^{m-1}$. Then ϕ is harmonic.*

Proof Let $i:S^{m-1}\to\mathbb{R}^m$ be the inclusion, so that $\Phi=i\circ\phi$. The target space is Euclidean, so that $\left\{\begin{matrix}\alpha\\ \sigma\lambda\end{matrix}\right\}=0$. Hence

$$\tau(i\circ\phi)_p=\Delta_{S^{n-1}}\big(i\circ\phi\big)_p,\quad p\in S^{n-1}. \tag{3.132}$$

Also, for every orthonormal basis $\{e_i\,:\,1\le i\le n-1\}\subset T_p(S^{n-1})$

$$\tau(i\circ\phi)_p=\big(d_{\phi(p)}i\big)\,\tau(\phi)_p+\sum_{j=1}^{n-1}\beta(i)_{\phi(p)}\big((d_p\phi)e_j,\ (d_p\phi)e_j\big) \tag{3.133}$$

where $(d_{\phi(p)}i)\tau(\phi)_p$ is tangent to S^{m-1} at $\phi(p)$. Moreover, as $i:S^{m-1}\to\mathbb{R}^m$ is an isometric immersion, $\beta(i)$ is its second fundamental form. It follows that $\beta(i)_{\phi(p)}\big((d_p\phi)e_j,\ (d_p\phi)e_j\big)$ is orthogonal to S^{m-1} at $\phi(p)$. Formulas (3.132)–(3.133) imply

$$(d_{\phi(p)}i)\,\tau(\phi)_p=\Big\{\big(\Delta_{S^{n-1}}(i\circ\phi)_p\Big\}^{\mathrm{T}}. \tag{3.134}$$

Here u^{T} is the component of $u \in T_{\phi(p)}(\mathbb{R}^m)$ tangent to S^{m-1} with respect to

$$T_{\phi(p)}(\mathbb{R}^m) = T_{\phi(p)}(S^{m-1}) \oplus N_{\phi(p)}(S^{m-1})$$

[and $N_{\phi(p)}(S^{m-1}) = T_{\phi(p)}(S^{m-1})^{\perp}$ is the normal space]. There is $\lambda \in \mathbb{R}$ such that

$$\Delta_{S^{n-1}}(i \circ \phi) = \lambda\, (i \circ \phi). \tag{3.135}$$

Then [by (3.135)] $\Delta_{S^{n-1}}(i \circ \phi)$ is orthogonal to S^{m-1}. Consequently [by (3.134)] $(d_{\phi(p)} i)\, \tau(\phi)_p = 0$, i.e., $\tau(\phi)_p = 0$ for any $p \in S^{n-1}$. Q.e.d.

Theorem 3.129 prompts the natural question of looking for harmonic polynomial maps that descend to maps between spheres.

Example 3.130 Let $\Phi : \mathbb{C} \to \mathbb{C}, \mathbb{C} \simeq \mathbb{R}^2$, be the map given by $\Phi(z) = z^k$, for some fixed $k \in \mathbb{N}$. For instance, if $k = 2$ then Φ is the map

$$\Phi(x, y) = \big(x^2 - y^2, 2\, x\, y\big).$$

As Φ is holomorphic, its real and imaginary parts are harmonic functions. Hence Φ is a polynomial harmonic map of degree k. Note that Φ descends to the map $\phi : S^1 \to S^1$, $\phi(z) = z^k$, which is harmonic by Theorem 3.129. The map $\phi : S^1 \to S^1$ may also be thought of as the geodesic $[0, 2\pi) \to S^1$, $\theta \mapsto \exp(k\, i\, \theta)$, $i = \sqrt{-1}$. ϕ wraps S^1 around itself k times, i.e., its Brouwer degree is k. One may also look at the case $k \in \mathbb{Z}$, $k \geq 0$, and then one obtains a harmonic map $\phi : S^1 \to S^1$ which inverts the orientations (and has Brouwer degree k). Actually, all homotopy classes in the fundamental group $\pi_1(S^1)$ contain harmonic maps (since $\pi_1(S^1) \simeq \mathbb{Z}$ identifies every homotopy class with the Brouwer degree, cf. e.g., [Dol]). □

Definition 3.131 An *orthogonal multiplication* is a $\mathbb{R}$-bilinear map $f : \mathbb{R}^p \to \mathbb{R}^q \to \mathbb{R}^n$ such that

$$\|f(x, y)\| = \|x\|\, \|y\|, \quad x \in \mathbb{R}^p, \quad y \in \mathbb{R}^q. \tag{3.136}$$

□

Theorem 3.132 *Let $f : \mathbb{R}^m \to \mathbb{R}^m \to \mathbb{R}^n$ be an orthogonal multiplication, and let us consider the map $\Phi : \mathbb{R}^{2m} \to \mathbb{R}^{n+1}$*

$$\Phi(x, y) = \big(\|x\|^2 - \|y\|^2, 2\, f(x, y)\big).$$

Then Φ is a polynomial harmonic map of degree 2, that descends to a map $\phi : S^{2m-1} \to S^n$.

Proof The first component of Φ

$$\Phi^1(x, y) = x_1^2 + \cdots + x_m^2 - \big(y_1^2 + \cdots + y_m^2\big)$$

is a harmonic homogeneous polynomial of degree 2. Here $x = (x_1, \ldots, x_m)$. As f is $\mathbb{R}$-bilinear there exist $a_{ij}^{\alpha} \in \mathbb{R}$ such that

$$f^{\alpha}(x, y) = \sum_{1 \leq i, j \leq m} a_{ij}^{\alpha} x_i y_j, \quad 1 \leq \alpha \leq n.$$

Therefore the last n components of Φ are harmonic polynomials (homogeneous of degree 2). Thus Φ is a polynomial harmonic map of degree 2. Finally [by (3.136)]

$$\|\Phi(x, y)\|^2 = \|x\|^2 - \|y\|^2 + 4\,\|f(x, y)\|^2 = \left(\|x\|^2 + \|y\|^2\right)^2 = 1$$

for any $(x, y) \in S^{2m-1}$. Hence $\Phi(S^{2m-1}) \subset S^n$, i.e., Φ descends to a harmonic map $\phi : S^{2m-1} \to S^n$. Q.e.d.

Example 3.133 Let $f : \mathbb{R}^m \times \mathbb{R}^m \to \mathbb{R}^m$ be the multiplications of real numbers, of complex numbers, of quaternions, and of Cayley octonions. If $m = 2$, i.e.,

$$f : \mathbb{C} \times \mathbb{C} \to \mathbb{C}, \quad (z, w) \mapsto zw, \quad z, w \in \mathbb{C},$$

then $|zw| = |z|\,|w|$, i.e., f is an orthogonal multiplication. By Theorem 3.132 (i.e., the *Hopf construction*) one obtains the map

$$\phi : S^3 \to S^2, \quad \phi(z, w) = \left(|z|^2 - |w|^2, 2\,z\,w\right), \quad (z, w) \in S^3 \subset \mathbb{R}^4 \simeq \mathbb{C}^2.$$

Then [by Theorem 3.132] ϕ is harmonic (ϕ is the *Hopf fibration*). The homotopy class of the Hopf fibration $\phi : S^3 \to S^2$ is a generator of $\pi_3\left(S^2\right) \simeq \mathbb{Z}$. The following general problem springs from the example at hand: build harmonic maps that represent each $r \in \mathbb{Z} \simeq \pi_3\left(S^2\right)$. The problem is open.[56] By a result of A. Ratto (cf. [Rat]) each element of $\pi_3\left(S^2\right) \simeq \mathbb{Z}$ can be rendered harmonic, i.e., admits a harmonic representative, provided that the domain S^3 is endowed with a suitable Riemannian metric.

If $m = 4$ the quaternion multiplication $f : \mathbb{H} \times \mathbb{H} \to \mathbb{H}$ ($\mathbb{H} \simeq \mathbb{R}^4$) is an orthogonal multiplication. By the Hopf construction, one obtains the Hopf fibration $S^7 \to S^4$, which is a harmonic map.

If $m = 8$ the multiplication of Cayley octonions $f : \mathbb{O} \times \mathbb{O} \to \mathbb{O}$, $\mathbb{O} \simeq \mathbb{R}^8$, is an orthogonal multiplication hence it provides the Hopf fibration $S^{15} \to S^8$, which is a harmonic map. The Hopf fibrations $S^3 \to S^2$, $S^7 \to S4$ and $S^{15} \to S^8$ are also Riemannian submersions with totally geodesic fibers (cf. e.g., [EeLe]). □

[56] Solutions are known for $r = k^2$ and particular values of $k \in \mathbb{Z}$.

Definition 3.134 A harmonic polynomial $P : \mathbb{R}^m \to \mathbb{R}$ of degree k is said to be an *eikonal* if

$$\|\nabla P\|_x^2 = \|x\|^{2k-2}, \quad x \in \mathbb{R}^m . \tag{3.137}$$

□

Here $\nabla P = \left(P_{x_1}, \ldots, P_{x_m}\right)$.

Theorem 3.135 *If $P : \mathbb{R}^m \to \mathbb{R}$ is an eikonal then $\nabla P : \mathbb{R}^m \to \mathbb{R}^m$ descends to a harmonic map $S^{m-1} \to S^{m-1}$.*

Proof ∇P is a polynomial harmonic map of degree $k - 1$. It descends to a map $S^{m-1} \to S^{m-1}$ to which Theorem 3.129 applies. Q.e.d.

Theorem 3.136 *Let $\nu \in \{1, 2, 4, 8\}$. Let us consider the map $P : \mathbb{R}^{3\nu+2} \to \mathbb{R}$ given by*

$$P(u, v, X, Y, Z) = u^2 - 3uv^2 + \tag{3.138}$$

$$+ \frac{3}{2}\, u\left(X\overline{X} + Y\overline{Y} - 2Z\overline{Z}\right) + \frac{3\sqrt{3}}{2}\, v\left(X\overline{X} - Y\overline{Y}\right) + \frac{3\sqrt{3}}{2}\left(XYZ + \overline{Z}\,\overline{Y}\,\overline{X}\right),$$

$$u, v \in \mathbb{R}, \quad X, Y, Z \in \mathbb{R}^\nu,$$

i.e., X, Y, Z belong to one of the following fields $\{\mathbb{R}, \mathbb{C}, \mathbb{H}, \mathbb{O}\}$. Then P is an eikonal of degree 3 and hence $\Phi = \nabla P : \mathbb{R}^{3\nu+2} \to \mathbb{R}^{3\nu+2}$ descends to a harmonic map $\phi : S^{3\nu+1} \to S^{3\nu+1}$, $\nu \in \{1, 2, 4, 8\}$.

An overbar denotes conjugation in $\mathbb{C}$, or $\mathbb{H}$, or $\mathbb{O}$ (and the identical mapping of $\mathbb{R}$). The expression (3.138) comes from the theory of isoparametric functions, cf. § 3.13.5 below. Cf. also Baird [Bai].

3.13.2 Fundamental Problems in the Theory of Harmonic Maps

The following fundamental problem arises in the theory of harmonic maps of Riemannian manifolds (cf. e.g., [EeLe]).

Problem. *Let M and N be two Riemannian manifolds, with M connected. Given a continuous map $\phi_0 : M \to N$, is there a harmonic map $\phi : M \to N$ homotopic to ϕ_0?*

Any continuous map $M \to N$ is homotopic to some C^∞ map $M \to N$. Moreover, if $\phi_0, \phi_1 : M \to N$ are homotopic and smooth, there is a *smooth* homotopy deforming ϕ_0 into ϕ_1. Let $[M, N]$ denote the set of all homotopy classes of maps of M into N. The fundamental problem states at the beginning of the present section admits the following reformulation.

Problem. *Given two Riemannian manifolds M and N, with M connected, and given a homotopy class $H \in [M, N]$, is there a harmonic map $\phi \in H$?*

If such $\phi \in H$ exists, it is said to be a *harmonic representative* of H. A partial answer to the fundamental problem above is provided by Eells and Sampson [EeSa].

Theorem 3.137 *Let M and N be two compact and connected Riemannian manifolds. If N has sectional curvature* $\text{Riem}(N) \le 0$ *then every homotopy class $H \in [M, N]$ admits a harmonic representative $\phi \in H$. The harmonic representative $\phi \in H$ is uniquely provided that* $\text{Riem}(N) < 0$.

We give a hint to the proof.[57]

Lemma 3.138 *Let M and N be two Riemannian manifolds, with M compact and connected. Let ϕ be an absolute minimum of the energy functional $E : C^\infty(M, N) \to \mathbb{R}$ within its homotopy class, i.e., $E(\phi) \le E(\psi)$ for any $\psi \in C^\infty(M, N)$ homotopic to ϕ. Then ϕ is harmonic.*

Proof Let $\{\phi_t\}_{t_0<t<t_1}$ $(t_0 < 0 < t_1)$ be a smooth 1-parameter variation of $\phi_0 = \phi$. By hypothesis $E(\phi) \le E(\phi_t)$ for each value of the parameter t. Hence the one real variable function $(t_0, t_1) \to \mathbb{R}$, $t \mapsto E(\phi_t)$, has a minimum point at $t = 0$. The function $t \mapsto E(\phi_t)$ is also differentiable, and $t = 0$ is an interior point of its domain, hence $\left\{d\, E(\phi_t)/d\, t\right\}_{t=0} = 0$, i.e., ϕ is harmonic. Q.e.d.

Let $\{\phi_t\}_{t\ge 0}$ be a smooth 1-parameter family of C^∞ maps $\phi_t : M \to N$. By the first variation formula

$$\frac{d}{dt}\left\{E(\phi_t)\right\} = -\int_M h^{\phi_t}\left(\tau(\phi_t),\ \mathbb{V} \circ \alpha_t\right) d\,v_g, \tag{3.139}$$

$$\mathbb{V}_{(x,t)} = \left(d_{(x,t)}\,\Phi\right)\left(\frac{\partial}{\partial t}\right)_{(x,t)}, \qquad \Phi(x, t) = \phi_t(x).$$

One often adopts the (a bit ambiguous) notation $\mathbb{V} =: \partial\phi_t/\partial t$. Should a choice of ϕ_t be feasible such that

$$\frac{\partial \phi_t}{\partial t} - \tau(\phi_t) = 0, \quad t > 0, \tag{3.140}$$

then (3.139) would imply $\frac{d}{dt}\left\{E(\phi_t)\right\} \le 0$, that is to say the energy decreases [and $d\,E(\phi_t)/dt = 0$ if and only if $\tau(\phi_t) = 0$, i.e., a harmonic map has been reached already]. Therefore, the proof of Theorem 3.137 amounts to the determination of a solution $\Phi : M \times [0, +\infty) \to N$, $\Phi(x, t) = \phi_t(x)$, $x \in M$, $t \ge 0$, to the *heat equation for harmonic maps*, i.e., Eq. (3.140) with initial condition $\phi_0 = \phi$.

Let $\text{Riem}(N)$ denote the sectional curvature of N. Under the assumptions of Theorem 3.137 [i.e., if M and N are compact and connected and $\text{Riem}(N) \le 0$] for

[57] Which is rather involved. For details one may see [EeSa].

any continuous map $\phi : M \to N$ there is (by a result of Eells and Sampson [EeSa]) a solution Φ to (3.140) with $\Phi(x, 0) = \phi(x)$, $x \in M$, and ϕ_t tends for $t \to \infty$ to a harmonic map that is an absolute minimum [for the energy E] within its homotopy class. The proof relies on the theory of parabolic systems of PDEs (cf. e.g., Friedman [Fri]). For the last statement in Theorem 3.137 one may also see Hartman [Har].

We end the section with the following remark. Let us recall that $[S^1, S^1] \simeq \mathbb{Z}$ (to every map one associates its degree). Indeed two maps $\phi, \psi : S^1 \to S^1$ are homotopic if and only if ϕ and ψ have the same degree. Let $H_k \in [S^1, S^1]$ be the homotopy class of maps of degree $k \in \mathbb{Z}$. By Theorem 3.137, as $\text{Riem}(S^1) = 0$ the class H_k must contain a harmonic representative. Indeed, as previously shown the map $S^1 \to S^1, \theta \mapsto k\,\theta$ is a harmonic representative of H_k. It is a natural question whether one may find a harmonic representative in every homotopy class $H \in [S^2, S^2]$.

3.13.3 Lichnerowicz Theorem

Let (M, J, g) and (N, J^N, h) be two Kählerian manifolds, and let $f : M \to N$ be a C^∞ map.

Definition 3.139 $f : M \to N$ is *holomorphic* if

$$(d_x f) \circ J_x = J^N_{f(x)} \circ (d_x f), \quad x \in M. \tag{3.141}$$

□

Let $p \in M$ and let (U, z^j) and (V, w^α) be local systems of complex coordinates on M and N such that $p \in U$ and $f(U) \subset V$. Also, let (V, w^α) be chosen such that $f^\alpha(p) = 0$ [where $f^\alpha = w^\alpha \circ f$] and $\Gamma^\alpha_{\sigma\lambda}(f(p)) = 0$ where

$$\nabla^N_{\partial/\partial w^\sigma} \frac{\partial}{\partial w^\lambda} = \Gamma^\alpha_{\sigma\lambda} \frac{\partial}{\partial w^\alpha}$$

[$\Gamma_{\sigma\lambda}$ are the only surviving complex components of the Levi-Civita connection ∇^N of (N, h)].

Theorem 3.140 *Let (M, J, g) and (N, J^N, h) be two Kählerian manifolds, and let $f \in C^\infty(M, N)$. If $f : M \to N$ is holomorphic then f is harmonic.*

Proof Let $z^j = x^j + i\, y^j$ be the real and imaginary parts of z^j. Then

$$\tau(f)^\alpha_p = \sum_{j=1}^{n} \left\{ \frac{\partial^2 f^\alpha}{\partial (x^j)^2} + \frac{\partial^2 f^\alpha}{\partial (y^j)^2} \right\} = \frac{1}{4} \sum_{j=1}^{n} \frac{\partial^2 f^\alpha}{\partial z^j\, \partial \overline{z}^j} = 0 \tag{3.142}$$

because of f = holomorphic $\Longrightarrow \partial f^\alpha / \partial \overline{z} = 0$. A similar argument yields $\tau(f)^{\overline{\alpha}} = 0$. Q.e.d.

Let (M, J) be a complex m-dimensional manifold, and let $T(M) \otimes \mathbb{C}$ be the complexification of its tangent bundle. Let us extend $J : T(M) \to T(M)$ by $\mathbb{C}$-linearity to an endomorphism of $\mathbb{J}^{\mathbb{C}}$ of $T(M) \otimes \mathbb{C}$. Then $\left(J^{\mathbb{C}}\right)^2 = -I$ hence $\text{Spec}\left(J^{\mathbb{C}}\right) = \{\pm i\}$, $i = \sqrt{-1}$. Let $T^{1,0}(M) = \text{Eigen}\left(J^{\mathbb{C}}, +i\right)$ be the holomorphic tangent bundle, and $T^{0,1}(M) \equiv \overline{T^{1,0}(M)}$ (an overbar denotes complex conjugation). Let $g^{\mathbb{C}}$ be the $\mathbb{C}$-linear extension of the metric g to $T(M) \otimes \mathbb{C}$, and let us consider the Hermitian form

$$\langle A, B \rangle = g^{\mathbb{C}}\left(A, \overline{B}\right), \quad A, B \in C^\infty\left(T(M) \otimes \mathbb{C}\right). \tag{3.143}$$

Let $\Omega(X, Y) = g(X, JY)$, $X, Y \in \mathfrak{X}(M)$, be the Kähler 2-form of (M, J, g), and let $\Omega^{\mathbb{C}}$ be the $\mathbb{C}$-linear extension of Ω. It is customary to drop the upper index $\mathbb{C}$ (that is we shall not distinguish notationally between tensor field of sorts, e.g., g and Ω, and their $\mathbb{C}$-linear extensions). Let (U, z^j) be a local system of complex coordinates on M and let us set

$$Z_j = \frac{\partial}{\partial z^j}, \quad Z_{\overline{j}} = \overline{Z_j}, \quad 1 \le i \le m$$

for the sake of simplicity. Let us set as customary

$$g_{j\overline{k}} = g\left(Z_j, Z_{\overline{k}}\right) = \langle Z_j, Z_k \rangle, \quad \Omega_{j\overline{k}} = \Omega\left(Z_j, Z_{\overline{k}}\right),$$

so that

$$\Omega_{j\overline{k}} = -i\, g_{j\overline{k}}. \tag{3.144}$$

Let (N, J^N, h) be another Kählerian manifold, and let $\phi : M \to N$ be a C^∞ map. Let $d_x^{\mathbb{C}} f : T_x(M) \otimes_{\mathbb{C}} \to T_{f(x)}(N) \otimes_{\mathbb{C}}$ be the $\mathbb{C}$-linear extension of $d_x f : T_x(M) \to T_{f(x)}(N)$. By taking into account the decompositions

$$T(M) \otimes \mathbb{C} = T^{1,0}(M) \oplus T^{0,1}(M), \quad T(N) \otimes \mathbb{C} = T^{1,0}(N) \oplus T^{0,1}(N),$$

one may introduce natural maps

$$\partial f : T^{1,0}(M) \to f^{-1}T^{1,0}(N), \quad \partial \overline{f} : T^{1,0}(M) \to f^{-1}T^{0,1}(N),$$

$$\overline{\partial} f : T^{0,1}(M) \to f^{-1}T^{1,0}(N), \quad \overline{\partial f} : T^{0,1}(M) \to f^{-1}T^{0,1}(N).$$

For instance ∂f is defined as the commutative closure of the diagram

$$\begin{array}{ccccc} T^{1,0}(M) & \overset{i}{\hookrightarrow} & T^{1,0}(M) \oplus T^{0,1}(M) & = & T(M) \otimes \mathbb{C} \\ \partial f \downarrow & & & & \downarrow d^{\mathbb{C}} f \\ f^{-1}\, T^{1,0}(N) & \overset{\left(\pi^{1,0}\right)^f}{\longleftarrow} & f^{-1}\left\{T^{1,0}(N) \oplus T^{0,1}(N)\right\} & = & f^{-1}\left\{T(N) \otimes \mathbb{C}\right\} \end{array}$$

Here $i : T^{1,0}(M) \to T(M) \otimes \mathbb{C}$ is the inclusion, $\pi^{1,0} : T(N) \otimes \mathbb{C} \to T_{1,0}(N)$ is the projection, and $\left(\pi^{1,0}\right)^f_x = \pi^{1,0}_{f(x)}$ for any $x \in M$. Then f is holomorphic $\Longleftrightarrow \overline{\partial} f = 0$.

Definition 3.141 $f : M \to N$ is said to be *anti-holomorphic* if $\partial f = 0$. □

Let $\|\partial f\|, \|\overline{\partial} f\| : M \to [0, +\infty)$ be defined as follows. Let $p \in M$ and let (U, z^j) be a local systems of complex coordinates with $p \in U$. We set by definition

$$\|\partial f\|^2_p = g^{j\overline{k}}(p)\, h_{\alpha\overline{\sigma}}(f(p))\, f^\alpha_j(p)\, \overline{f^\sigma_k(p)}, \quad \|\overline{\partial} f\|^2_p = g^{j\overline{k}}(p)\, h_{\alpha\overline{\sigma}}(f(p))\, f^\alpha_{\overline{k}}(p)\, \overline{f^\sigma_{\overline{j}}(p)}.$$

The definition doesn't depend upon the choice of local complex coordinates about p. The functions f^α_j and $f^\alpha_{\overline{j}}$ are determined by

$$(\partial f)\, Z_j = \phi^\alpha_j\, W_\alpha, \quad (\overline{\partial} f)\, Z_{\overline{j}} = \phi^\alpha_{\overline{j}}\, W_\alpha, \quad W_\alpha \equiv \frac{\partial}{\partial w^\alpha}.$$

We adopt the customary notations

$$e(f) = \frac{1}{2}\, \|df\|^2, \quad e'(f) = \|\partial f\|^2, \quad e''(f) = \|\overline{\partial} f\|^2,$$

so that

$$e(f) = e'(f) + e''(f). \tag{3.145}$$

Let us also set

$$E'(f) = \int_M e'(f)\, d\,\mathrm{v}_g, \quad E''(f) = \int_M e''(f)\, d\,\mathrm{v}_g,$$

so that $E(f) = E'(f) + E''(f)$.

Theorem 3.142 *Let M and N be Hermitian manifolds, with M compact, and let $f : M \to N$ be a C^∞ map. Then f is holomorphic (respectively anti-holomorphic) if and only if $E''(f) = 0$ (respectively $E'(f) = 0$).*

The proof is straightforward. The remainder of the section is devoted to a celebrated result by Lichnerowicz [Lic]. Let $\phi : M \to N$ be a C^∞ map of Kählerian manifolds, and let us set

$$k(\phi) = e'(\phi) - e''(\phi), \quad K(\phi) = \int_M k(\phi)\, d\,\mathrm{v}_g,$$

so that $K(\phi) = E'(\phi) - E''(\phi)$.

Theorem 3.143 *Let M and N be two Kählerian manifolds, and let $\phi_t : M \to N$ be a smooth 1-parameter family of C^∞ maps. Then the function $t \mapsto K(\phi_t)$ is constant.*

We shall need

Lemma 3.144 *Let $f : M \to N$ be a C^∞ map of Kählerian manifolds. Then*

$$k(\phi) = \langle \Omega^M,\, \phi^* \Omega^N \rangle. \tag{3.146}$$

Here Ω^M and Ω^N are the Kähler 2-forms of M and N. The proof is straightforward.

Lemma 3.145 *Let $\phi_t : M \to N$ be a smooth 1-parameter family of C^∞ maps. Let Ω be a closed 2-form on N. Then*

$$\frac{\partial}{\partial t}(\phi_t^* \,\Omega) = d\left(\phi_t^* \left(\frac{\partial \phi_t}{\partial t} \rfloor \Omega\right)\right). \tag{3.147}$$

For a proof one may see [Lic], p. 49.

Proof of Theorem 3.143 By (3.146) together with (2.7.11) in [Gol], p. 70

$$k(\phi)\, d\,\mathrm{v}_g = \langle \Omega^M,\, \phi^* \Omega^N \rangle\, d\,\mathrm{v}_g = \Omega^M \wedge *(\phi^* \Omega^N).$$

Next (as the Hodge operator $*$ possesses the property (iii) in [Gol], p. 70)

$$k(\phi)\, d\,\mathrm{v}_g = (\phi^*\, \Omega^N) \wedge *\Omega^M\,. \tag{3.148}$$

Let us integrate in (3.148) so that to obtain

$$K(\phi_1) - K(\phi_0) = \int_M \left(\phi_1^*\, \Omega^N - \phi_0^*\, \Omega^N\right) \wedge *\, \Omega^M\,. \tag{3.149}$$

Let us set by definition

$$\alpha = \int_0^1 \phi_t^* \left\{ \frac{\partial \phi_t}{\partial t} \rfloor \Omega^N \right\} dt\,.$$

Then [by (3.147)]

$$\phi_1^*\, \Omega^N - \phi_0^*\, \Omega^N = \in_0^1 \frac{\partial}{\partial t}\left\{\phi_t^*\, \Omega^N\right\} dt = d\alpha$$

hence (3.149) becomes [by the theorem of Stokes[58] together with

$$d(*\, \Omega^M) = d\left\{ \frac{1}{(m-1)!} \left(\Omega^M\right)^{m-1} \right\} = 0$$

(because of $d\, \Omega^M = 0$)]

[58] Cf. e.g., Boothby [Boo], p. 257.

$$K(\phi_1) - K(\phi_0) = \int_M (d\alpha) \wedge * \Omega^M = \int_M d\big(\alpha \wedge * \Omega^M\big) = 0.$$

We have shown that $K(\phi_0) = K(\phi_1)$. The same arguments, starting with the interval $[0, t]$ rather than $[0, 1]$, lead to $K(\phi_0) = K(\phi_t)$ hence the function $t \mapsto K(\phi_t)$ is constant. Q.e.d.

Corollary 3.146 *Let M and N be two Kählerian manifolds, with M compact, and let $\phi : M \to N$ be a holomorphic (or anti-holomorphic) map. Then ϕ is a harmonic map and an absolute minimum within its homotopy class.*

Proof Let $\psi : M \to N$ be a smooth map homotopic to ϕ. Then

$$E(\psi) = E'(\psi) + E''(\psi) \geq \big|E'(\psi) - E''(\psi)\big| = \big|K(\psi)\big| = \big|K(\phi)\big|.$$

For the last equality we made use of Theorem 3.143 and of the possibility to choose a smooth homotopy between ϕ and ψ. Then (by Theorem 3.142)

$$E(\psi) \geq \big|E'(\phi) - E''(\phi)\big| = \begin{cases} E'(\phi) & \text{if } \phi \text{ is holomorphic} \\ \\ E''(\phi) & \text{if } \phi \text{ is anti-holomorphic} \end{cases} = E'(\phi) + E''(\phi) = E(\phi).$$

Q.e.d.

3.13.4 Harmonic Maps of Surfaces

Let $M = N = S^2$ be equipped with the standard Kählerian structure. The Riemann sphere S^2 is identified (by the stereographic projections) to the extended Gauss plane, i.e., $S^2 \simeq \mathbb{C} \cup \{\infty\}$, and we consider the holomorphic map $\phi : S^2 \to S^2$, $\phi(z) = z^k$, $k \in \mathbb{Z}_+$. Then ϕ is (by Corollary 3.146) a harmonic map and an absolute minimum within its homotopy class.

We need to recall a few notions about degrees of maps $f : S^n \to S^n$. Any group endomorphism $\phi : \mathbb{Z} \to \mathbb{Z}$ is determined by an integer number $d \in \mathbb{Z}$, i.e., $\phi(x) = d\, x$, for any $x \in \mathbb{Z}$. Let us apply this observation to the homomorphism $f_* : \tilde{H}_n(S^n) \to \tilde{H}_n(S^n)$ induced by f on (reduced) homology. As $\tilde{H}_n(S^n) \simeq \mathbb{Z}$ (cf. e.g., [Dol], p. 56) there is a unique integer number $\deg(f) \in \mathbb{Z}$ (the *degree* of $f : S^2 \to S^2$) such that $f_*(x) = \deg(f)\, x$.

The harmonic map

$$\phi : S^2 \to S^2, \quad z \mapsto z^k, \quad z \in \mathbb{C} \cup \{\infty\} \simeq S^2, \quad k \in \mathbb{Z}_+ \tag{3.150}$$

has degree $\deg(\phi) = k$. We recall that given $\phi, \psi \in [S^2, S^2]$, if $\phi \simeq \psi$ then $\deg(\phi) = \deg(\psi)$, cf. e.g., [Dol], p. 63. Therefore $\deg : [S^2, S^2] \to \mathbb{Z}$ is a bijection. Note also that the map

$$\psi : S^2 \to S^2, \quad z \mapsto \bar{z}^k, \quad z \in \mathbb{C} \cup \{\infty\} \simeq S^2, \quad k \in \mathbb{Z}_+ \tag{3.151}$$

is anti-holomorphic and has degree $-k$. We may conclude that all classes of homotopy $H \in [S^2, S^2]$ admit harmonic representatives. It should be observed that, although $\text{Riem}(S^2) \geq 0$ (and Theorem 3.137 doesn't apply) the fundamental problem in Section § 3.13.2 is at this point solved (for $M = N = S^2$).

Let M and N be two compact, connected, oriented differentiable manifolds such that $\dim(M) = \dim(N)$. Let $\phi : M \to N$ be a smooth map, and let $y \in N$ be a regular value of ϕ. Then $\phi^{-1}(y)$ is a finite set $\phi^{-1}(y) = \{x_1, \ldots, x_k\}$. For any $x_i \in \phi^{-1}(y)$ the differential $d_{x_i}\phi$ either preserves or reverses the orientations. We set by definition $\text{sign}(x_i) = +1$ if $d_{x_i}\phi$ is orientation-preserving, and $\text{sign}(x_i) = -1$ if $d_{x_i}\phi$ is orientation-reversing. Also, we define

$$\deg_y \phi = \sum_{x_i \in \phi^{-1}(y)} \text{sign}\,(x_i). \tag{3.152}$$

We need to recall the following facts (cf. e.g., [Spi], vol. I, p. 372). The integer number $\deg_y \phi$ doesn't depend on the choice of regular value $y \in N$, so the following notation is legitimate

$$\deg\,(\phi) = \deg_y \phi \tag{3.153}$$

where $y \in N$ is an arbitrary regular value. Also, the number $\deg(\phi)$ depends only on the homotopy class $[\phi] \in [M, N]$, hence it determines a well-defined map $\deg : [M, N] \to \mathbb{Z}$.

The proof of the fact that the map (3.150) $\deg(\phi) = k$ relies on the following:

Lemma 3.147 *The only critical points of $\phi : S^2 \to S^2$, $z \mapsto z^k$, may be 0 and ∞. If $k = 1$ then ϕ has no critical points.*

Hence the critical values of the map may be 0 and ∞. Let $w \in \mathbb{C} \cup \{\infty\} \simeq S^2$ such that $w \neq 0$ and $w \neq \infty$. The equation $z^k = w$ has k solutions. As ϕ is holomorphic, $d_z\phi$ is orientation-preserving for any non critical point z. It follows that $\deg(\phi) = \deg_w \phi = k$ (because w has k counter images, and for any counter image $z \in \phi^{-1}(w)$ one has $\text{sign}\,(z) = +1$.

Similar arguments show that the map (3.151) has degree $\deg(\psi) = -k$. If $k = 0$ then ψ is constant. Hence $\deg(\psi) = 0$, because each $w \neq 1$ is a regular value having 0 as a counter image.

Let M and N be two compact, connected Riemann surfaces of genus 1. Any compact, connected, oriented Riemann surface M is homeomorphic to a sphere with p handles, $p \in \mathbb{Z}_+$ (p is the *genus* of M). In the sequel we look at harmonic maps $\phi : M \to N$. Note that

$$M \simeq \mathbb{C}/\Gamma \;\; \text{(a biholomorphism)}, \quad \Gamma = \{\lambda_1 e_1 + \lambda_2 e_2 \,:\, \lambda_1, \lambda_2 \in \mathbb{Z}\},$$

where $\{e_1, e_2\}$ is a $\mathbb{R}$-linear basis of $\mathbb{C}$. A special case is got for $e_1 = (1, 0)$ and $e_2 = (0, 1)$: if $\mathbb{C}/\Gamma$ is equipped with the standard metric then $M \simeq S^1 \times S^1$ (an isometry). Let now $M = N = S^1 \times S^1$ with the product metric. Then $\deg : [M, N] \to \mathbb{Z}$ is only surjective (it is not injective), cf. e.g., [EeLe]. In turn

$$\left[S^1 \times S^1, S^1 \times S^1\right] \simeq \mathbb{Z} \oplus \mathbb{Z} \oplus \mathbb{Z} \oplus \mathbb{Z} \;\; \text{(a bijection)}. \tag{3.154}$$

Indeed, if $\phi : S^1 \times S^1 \to S^1 \times S^1$ is a continuous map, let

$$\phi_* : \pi_1\left(S^1 \times S^1\right) \to \pi_1\left(S^1 \times S^1\right), \quad \pi_1\left(S^1 \times S^1\right) \simeq \mathbb{Z} \oplus \mathbb{Z},$$

$$\phi_*(r, s) = \left(a\, r + b\, s,\; c\, r + d\, s\right), \quad (r, s) \in \mathbb{Z} \oplus \mathbb{Z}.$$

The integer numbers $a, b, c, d \in \mathbb{Z}$ depend on ϕ and on the isomorphism $\pi_1\left(S^1 \times S^1\right) \simeq \mathbb{Z} \oplus \mathbb{Z}$. The bijection (3.154) is then $[\phi] \mapsto (a, b, c, d)$.

Exercise 3.148 Compute $\deg(\phi)$ in terms of the integer numbers a, b, c, d. □

Example 3.149 (1) The map

$$\phi : S^1 \times S^1 \to S^1 \times S^1, \quad (\theta, \psi) \longmapsto \left(a\, \theta + b\, \psi,\; c\, \theta + d\, \psi\right)$$

is harmonic and belongs to the homotopy class determined by $(a, b, c, d) \in \mathbb{Z}^4$. Here $\theta, \psi \in \mathbb{R}/(2\pi\mathbb{Z}) \simeq S^1$. Actually ϕ is totally geodesic (i.e., $\beta(\phi) = 0$). Indeed the map

$$\Phi : \mathbb{R}^2 \to \mathbb{R}^2, \quad (x, y) \longmapsto (a\, x + b\, y,\; c\, x + d\, y),$$

descends to ϕ, i.e., the following diagram

$$\begin{array}{ccc} \mathbb{R}^2 & \xrightarrow{\Phi} & \mathbb{R}^2 \\ \pi \downarrow & & \downarrow \pi \\ S^1 \times S^1 & \xrightarrow{\phi} & S^1 \times S^1 \end{array}$$

is commutative, where $\pi : \mathbb{R}^2 \to S^1 \times S^1$ is the direct product of two copies of the canonical projection $\mathbb{R} \to \mathbb{R}/(2\pi\mathbb{Z}) \simeq S^1$. Finally, Φ is linear and hence totally geodesic. Note that ϕ isn't holomorphic, in general.

(2) The map

$$\phi : S^1 \times S^1 \to S^1 \times S^1, \quad (\theta, \psi) \longmapsto (2\theta, \psi)$$

is harmonic yet it is neither holomorphic nor anti-holomorphic. Indeed, ϕ is not conformal, since

$$\left\| (d\phi) \frac{\partial}{\partial\theta} \right\| = 2 \left\| (d\phi) \frac{\partial}{\partial\psi} \right\|.$$

(3) The map $(\theta, \psi) \mapsto (2\theta, 2\psi)$ is holomorphic, conformal

$$\left\| (d\phi) \frac{\partial}{\partial\theta} \right\| = \left\| (d\phi) \frac{\partial}{\partial\psi} \right\|, \quad \left\langle (d\phi) \frac{\partial}{\partial\theta}, (d\phi) \frac{\partial}{\partial\psi} \right\rangle = 0$$

and orientation-preserving. Note that the map in question descends from the holomorphic mapping $\mathbb{C} \to \mathbb{C}$, $z \mapsto 2z$. □

Through the remainder of the present section, we quote a few results on harmonic maps $\phi : M_p \to N_p$, where M_p and N_p are two compact Riemann surfaces of genus $p \in \mathbb{Z}_+$, equipped with Hermitian metrics.

Theorem 3.150 *Every harmonic map $\phi : M_0 \to N_0$ is either holomorphic or anti-holomorphic. There are no harmonic maps $\phi : M_1 \to M_0$ such that* $\deg(\phi) \in \{\pm 1\}$.

Cf. Eells and Wood [EeWo]. It should be observed (cf. e.g., [EeLe]) that the map $\deg : [M_p, M_0] \to \mathbb{Z}$ is a bijection; the elements of[59] $[M_p, M_0]$ are classified because of $\pi_1(S^2) = 0$.

Theorem 3.151 *Every harmonic map $\phi : M_p \to M_0$ with $|\deg(\phi)| \geq p$ is holomorphic, or anti-holomorphic.*

A complete knowledge of the holomorphic maps $\phi : M_p \to M_0$ comes from the theory of Riemann surfaces. Little is known on the case $|\deg(\phi)| \leq p$, cf. § 11 in [EeLe1].

Using the last quoted result, we may give a proof of the non-existence of harmonic maps $\phi : M_1 \to M_0$ with $\deg(\phi) = 1$. We argue by contradiction. If $\phi : M_1 \to M_0$ is a harmonic map with $p = 1 = \deg(\phi)$ then ϕ is holomorphic.[60] Yet there is no holomorphic map $\phi : M_1 \to M_0$ with $\deg(\phi = 1$, for such a map would have no ramification point.[61] In a sufficiently small neighborhood of a ramification point, every point has at least $k \geq 2$ counter-images, for ϕ is locally of the form $z \mapsto z^k$, $k \in \{2, 3, \cdots\}$. It then follows that ϕ is locally biholomorphic, a contradiction.[62]

---

[59] One has $M_0 \simeq S^2$ (an isometry, where S^2 is equipped with a non standard metric).

[60] As $\deg(\phi) > 0$ the map ϕ cannot be anti-holomorphic.

[61] That is no critical point (here the two notions coincide).

[62] Indeed, it follows that M_1 is a covering space of M_0, yet every covering space of $M_0 \simeq S^2$ is S^2.

3.13.5 Isoparametric Functions

Let $(M, \overline{g})$ be a real space form, i.e., a Riemannian manifold of constant sectional curvature.

Definition 3.152 A C^∞ map is a *isoparametric function* if there exist two smooth functions $\phi, \psi : \mathbb{R} \to \mathbb{R}$ such that

$$\|df\|^2 = \phi \circ f, \tag{3.155}$$

$$\Delta f = \psi \circ f. \tag{3.156}$$

□

Theorem 3.153 *Let $(M, \overline{g})$ be a n-dimensional Riemannian manifold, and let $f : M \to \mathbb{R}$ be a smooth map obeying to (3.155)–(3.156). Let $c \in f(M)$ be a regular value of f, and let $M_c = f^{-1}(c)$. Then M_c has constant mean curvature $\|H\|$.*

Proof One has

$$T_x(M_c) = \big\{c \in T_x(M) \ : \ v(f) = 0\big\}, \quad x \in M_c\,.$$

Let $\xi = \|df\|^{-1}\,\overline{\nabla} f$. If $X \in \mathfrak{X}(M_c)$ then

$$\overline{g}(X, \xi) = \|df\|^{-1}\,\overline{g}\big((df)^\sharp, X\big) = \|df\|^{-1}\,X(f) = 0,$$

i.e., ξ is a unit normal field on M_c. By Weingarten's formula

$$\overline{\nabla}_X \xi = -A_\xi X,$$

where $\overline{\nabla}$ is the Levi-Civita connection of $(M, \overline{g})$. Let $\{E_1, \dots, E_{n-1}\}$ be a local orthonormal frame of $T(M_c)$, defined on some open subset $U \subset M_c$. Then

$$\|H\| = \frac{1}{n-1}\,\text{Trace}\,(A_\xi) = \frac{1}{n-1}\sum_{i=1}^{n-1} g(A_\xi E_i, E_i) = -\frac{1}{n-1}\sum_{i=1}^{n-1}\overline{g}(\overline{\nabla}_{E_i}\xi, E_i) =$$

$$= -\frac{1}{n-1}\sum_{i=1}^{n-1}\Big\{\|df\|^{-1}\,\overline{g}\big((df)^\sharp, E_i\big) + E_i\big(\|df\|^{-1}\big)\,\overline{g}\big((df)^\sharp, E_i\big)\Big\} =$$

$$= \frac{1}{n-1}\,\|df\|^{-1}\sum_{i=1}^{n-1}\big(\overline{\nabla}_{E_i} df\big)E_i$$

as $E_i(f) = 0$, $1 \le i \le n-1$. Finally

$$\|H\| = \frac{1}{n-1}\|df\|^{-1}\left\{\operatorname{div}(\nabla f) - (\overline{\nabla}_\xi df)\xi\right\}$$

hence

$$\|H\| = -\frac{\Delta f}{(n-1)\,\|df\|} - \frac{1}{n-1}\|df\|^{-1}\left(\overline{\nabla}_\xi df\right)\xi. \tag{3.157}$$

Cf. Nomizu [Nom1,Nom2].

Lemma 3.154 *The integral curves of ξ are geodesics in $(M, \overline{g})$ i.e.*

$$\overline{\nabla}_\xi \xi = 0. \tag{3.158}$$

Proof Let $X \in \mathfrak{X}(M_c)$, so that $X(f) = 0$. Then [by (3.155)]

$$\xi(f) = \overline{g}\left(\nabla f, \xi\right) = \|df\|^{-1}\,\overline{g}\left(\nabla f, \nabla f\right) = \|df\| = (\phi \circ f)^{1/2},$$

i.e., $\xi(f)$ is a function of f. Consequently $X(f) = 0$ implies $X(\xi(f)) = 0$ and then

$$0 = [X, \xi](f) = \left(\overline{\nabla}_X \xi - \overline{\nabla}_\xi X\right)(f)$$

showing that $\overline{\nabla}_X \xi - \overline{\nabla}_\xi X$ is tangent to M_c. Then

$$\overline{g}\left(\overline{\nabla}_X \xi - \overline{\nabla}_\xi X, \xi\right) = 0. \tag{3.159}$$

Yet $\|\xi\| = 1$ implies $\overline{g}\left(\overline{\nabla}_X \xi, \xi\right) = 0$ hence (3.159) may be written

$$\overline{g}\left(\overline{\nabla}_\xi X, \xi\right) = 0. \tag{3.160}$$

Finally $\overline{g}(X, \xi) = 0$ and (3.160) imply $\overline{g}\left(X, \overline{\nabla}_\xi \xi\right) = 0$ for any $X \in \mathfrak{X}(M_c)$, i.e., $\overline{\nabla}_\xi \xi$ is orthogonal to M_c. Consequently $\overline{\nabla}_\xi \xi = \lambda \xi$ for some $\lambda \in C^\infty(M_c)$, hence

$$0 = \overline{g}\left(\overline{\nabla}_\xi \xi, \xi\right) = \lambda\, \overline{g}(\xi, \xi) = \lambda.$$

Q.e.d.

At this point we may prove Theorem 3.153. By (3.158)

$$\left(\overline{\nabla}_\xi\, df\right)\xi = \xi\left(\xi(f)\right)$$

so that (3.157) becomes

$$\|H\| = \frac{\left(\nabla f\right)\left(\|df\|\right) - \Delta f}{(n-1)\,\|df\|}. \tag{3.161}$$

By (3.155)

$$\big(\nabla f\big)\big(\|df\|\big) = \frac{1}{2}\,\|df\|\,\phi', \quad \phi' \equiv \frac{d\phi}{dt}\,.$$

Finally [by (3.156) and (3.161)]

$$\|H\|_x = \frac{1}{2(n-1)}\left\{\phi'(c) - \frac{\psi(c)}{\phi(c)^{1/2}}\right\}, \quad x \in M_c\,.$$

Q.e.d.

Example 3.155 (1) Let $M = \mathbb{R}^m$ and $p \in \mathbb{N}$, $p \leq m$, and let us consider

$$f : \mathbb{R}^m \to \mathbb{R}, \quad f\big(x_1, \ldots, x_m\big) = x_1^2 + \cdots + x_p^2\,.$$

Then $\|df\|^2 = 4f$ and $\Delta f = -2p$, hence $f : \mathbb{R}^m \to \mathbb{R}$ is an isoparametric function. Let us look at the case $m = 3$ and $p = 2$. Let $c = a^2$, $a \in \mathbb{R} \setminus \{0\}$. Then $M_c = f^{-1}(c)$ is the cylinder $M_c = \{(x, y, z) \in \mathbb{R}^3 : x^2 + y^2 = a^2\}$ and $M_c \simeq (s\,S^1) \times \mathbb{R}$ (an isometry).

(2) Let $M = S^{m-1}$ and $F : \mathbb{R}^m \to \mathbb{R}$, $m = p + q$,

$$F(x, y) = \|x\|^2 - \|y\|^2, \quad x \in \mathbb{R}^p, \quad y \in S^q,$$

and let $f := F\big|_{S^{m-1}}$. Let

$$z \in S^{m-1}, \quad z = \big((\cos s)\,x, \ (\sin s)\,y\big), \quad 0 \leq s \leq \frac{\pi}{2}, \quad x \in S^{p-1}, \quad y \in S^{q-1},$$

so that $f(z) = \cos(2s)$. Consequently $df = -2\,\sin(2s)\,ds$ yielding

$$\|df\|^2 = 4\,\sin^2(2s) = 4(1 - f^2),$$

and one may choose $\phi(t) = 4(1 - t^2)$. One may also check that

$$\Delta f = -2\Big\{(p-2)(1+f) - (q-2)(1-f)\Big\}.$$

Thus f is an isoparametric function. For each fixed $0 < s_0 < \pi/2$ let $M_{c_0} := f^{-1}(c_0)$ with $c_0 = \cos(2s_0)$.

Exercise 3.156 Show that

$$M_{\cos(2s_0)} = \Big\{\big((\cos s_0)\,x, \ (\sin s_0)\,y\big) \ : \ x \in S^{p-1}, \ y \in S^{q-1}\Big\},$$

$$M_{\cos(2s_0)} \simeq \big((\cos s_0)\,S^{p-1}\big) \times \big((\sin s_0)\,S^{q-1}\big) \quad \text{(an isometry)}.$$

□

Let $M^n(c)$ a real space form, of (constant) sectional curvature c. Let $f : M^n(c) \to \mathbb{R}$ be an isoparametric function, and let $a \in \mathbb{R}$ be a regular value of f. Let

$$M_a = f^{-1}(a), \quad \xi = \|df\|^{-1}\overline{\nabla} f, \quad A = A_\xi, \quad B(X, Y) = h(X, Y)\,\xi,$$

[A and B are respectively the Weingarten operator and the second fundamental form of $M_a \hookrightarrow M^n(c)$]. For every $x \in M_a$ the Weingarten operator $A_x : T_x(M_a) \to T_x(M_a)$ is self-adjoint, hence there is an orthonormal basis $\{e_\alpha : 1 \le \alpha \le n-1\} \subset T_x(M_a)$ such that

$$e_\alpha \in S_\alpha(x) := \text{Eigen}\big(A_x, \lambda_\alpha(x)\big) \subset T_x(M_a), \quad 1 \le \alpha \le n-1$$

for some $\lambda_\alpha(x) \in \mathbb{R}$.

Definition 3.157 The numbers $\lambda_\alpha(x)$ are the *principal curvatures* at x, and the eigenvectors e_α are the corresponding *principal directions*. □

Let $\lambda_\alpha(x)$ be the distinct principal curvatures (we assume $1 \le \alpha \le p$ for some $1 \le p \le n-1$, eventually up to relabeling indices). We shall need the Codazzi equation[63]

$$\overline{g}\big(\overline{R}(X, Y)Z, \xi\big) = \big(\nabla_X h\big)(Y, Z) - \big(\nabla_Y h\big)(X, Z), \quad X, Y, Z \in \mathfrak{X}(M_a), \tag{3.162}$$

where $\overline{R}$ is the curvature of $\big(M^n(c), \overline{g}\big)$ i.e.

$$\overline{R}(U, V)W = c\big\{\overline{g}(V, W)U - \overline{g}(U, W)V\big\}, \quad U, V, W \in \mathfrak{X}(M^n(c))). \tag{3.163}$$

By (3.163) the Codazzi equation is merely

$$\big(\nabla_X h\big)(Y, Z) - \big(\nabla_Y h\big)(X, Z). \tag{3.164}$$

We recall

Theorem 3.158 *M_a has constant principal curvatures.*

For a proof of Theorem 3.158 one may see Cartan [Car]. We adopt the following

Definition 3.159 A real hypersurface M^{n-1} of the space form $M^n(c)$ is said to be an *isoparametric hypersurface* if M^{n-1} has constant principal curvatures. □

Let $\alpha, \beta \in \{1, \dots, p\}$, $\alpha \ne \beta$, and let $X \in S_\alpha$ and $Y, Z \in S_\beta$. Then

$$\big(\nabla_X h\big)(Y, Z) = \lambda_\alpha \, (\nabla_X g)(Y, Z) = 0. \tag{3.165}$$

[63] Cf. e.g., Chen [Che0], p. 46.

Next [by (3.164)]

$$0 = (\nabla_Y h)(X, Z) = \big(\lambda_\beta - \lambda_\alpha\big)\, g\big(X,\, \nabla_Y Z\big), \quad \lambda_\alpha \neq \lambda_\beta\,.$$

It follows that $\nabla_Y Z$ is orthogonal to S_α for any $Y,\ Z \in S_\beta$. We may state:

Theorem 3.160 *The distributions S_α, $1 \le \alpha \le p$, are completely integrable and the maximal integral manifolds of S_α are totally geodesic in M_a. Moreover the leaves of S_α are totally umbilical in $M^n(c)$ and their mean curvature vector is parallel. In particular, if $M^n(c) = S^{m-1}$, $n = m-1$, then the leaves of S_α are small spheres of S^{m-1}.*

Proof One has

$$T_x(M_a) = S_1(x) \oplus \cdots \oplus S_p(x), \quad x \in M_a\,. \tag{3.166}$$

Let $Y,\ Z \in S_\alpha$. By the remark preceding Theorem 3.160, for every $\beta \neq \alpha$ the vector field $\nabla_Y Z$ is orthogonal to S_β, and then on $\bigoplus_{\beta \neq \alpha} S_\beta$. Consequently $\nabla_Y Z \in S_\alpha$ and hence (as ∇ is symmetric) S_α is involutive. By the classical Frobenius theorem S_α is completely integrable. Let L be a leaf of S_α and $Y,\ Z \in S_\alpha$. In particular $Y,\ Z$ are tangent to L. By the Gauss formulas associated to the immersions $L \hookrightarrow M_a \hookrightarrow M^n(c)$

$$\overline{\nabla}_Y Z = \nabla_Y Z + g(AY,\ Z)\,\xi = \nabla_Y Z + \lambda_j\, g(Y,\ Z)\,\xi,$$

$$\nabla_Y Z = \nabla^L_Y Z + \sigma(Y,\ Z),$$

where ∇^L and σ are respectively the Levi-Civita connection of L and the second fundamental form of the immersion $L \hookrightarrow M_a$. Next

$$\nabla_Y Z \in S_\alpha, \quad \nabla^L_Y Z \in S_\alpha \Longrightarrow \sigma(Y,\ Z) = 0.$$

Hence the second fundamental form of $L \hookrightarrow M^n(c)$ is $(X,\ Y) \longmapsto \lambda_\alpha\, g(Y,\ Z)\,\xi$, so that L is totally umbilical in $M^n(c)$. Q.e.d.

Let $f : S^{m-1} \to \mathbb{R}$ be an isoparametric function and let $M_a = f^{-1}(a)$. Let us consider

$$\gamma_t : M_a \to S^{m-1}, \quad Y_t(x) = \big(\cos t\big)\,\vec{x} + \big(\sin t\big)\,\xi_x, \quad t \ge 0, \ \ x \in M_a\,.$$

Here $\vec{x} \in T_x(\mathbb{R}^m)$ is the position vector of the point x. Let $\alpha \in \{1,\ \ldots,\ p\}$ be a fixed index and let $v \in S_\alpha(x)$ and $x \in M_a$. Let $\beta(s) \in M_a$, $|s| < \epsilon$, be the curve in M_a such that

$$\beta(0) = x, \quad \frac{d\beta}{ds}(0) = v.$$

Then

$$(d_x\gamma_t)v = \frac{d}{ds}\Big\{Y_t \circ \beta\Big\}_{s=0} = (\cos t)\, v + (\sin t)\,\frac{d}{ds}\Big\{\xi_{\beta(s)}\Big\}_{s=0}.$$

As $s \mapsto \xi_{\beta(s)}$ is a curve in S^{m-1} it follows that $\frac{d}{ds}\left\{\xi_{\beta(s)}\right\}_{s=0}$ is tangent to S^{m-1}. Then (by he Gauss formula of $S^{m-1} \hookrightarrow \mathbb{R}^m$)

$$(d_x\gamma_t)v = (\cos t)\, v + (\sin t)\big(\nabla_X^{\text{Eucl}}\xi\big)_x = (\cos t)\, v + (\sin t)\, \big(\overline{\nabla}_X\xi\big)_x,$$

where $X \in \mathfrak{X}(M_a)$ is an arbitrary smooth extension of v, i.e., $X_x = v$. Finally [by the Weingarten formula of $M_a \hookrightarrow S^{m-1}$]

$$(d_x\gamma_t)v = (\cos t)\, v - (\sin t)\, A_x v = (\cos t)\, v - (\sin t)\, \lambda_\alpha\, v$$

so that $d_x Y_t = 0$ if and only if $(\cos t)\, v - (\sin t)\, \lambda_\alpha\, v = 0$ for any $v \in S_\alpha(x)$, or

$$d_x\gamma_t = 0 \Longleftrightarrow t = \arctan \lambda_\alpha \in (0,\ \pi).$$

Definition 3.161 Let $\theta_\alpha := \arctan \lambda_\alpha$, $1 \le \alpha \le p$. Then $\gamma_{\theta_\alpha}\big(M_a\big)$ is called a *focal manifold* of f. □

By a result of Nomizu [Nom1].

Theorem 3.162 *Any focal manifold of an isoparametric function $f : S^{m-1} \to \mathbb{R}$ is a minimal submanifold of S^{m-1}.*

Example 3.163 Let $m = 3$ and let $F : \mathbb{R}^3 \to \mathbb{R}$ be given by

$$F(z,\ x) = |z|^2 - |x|^2, \quad (z,\ x) \in \mathbb{C} \oplus \mathbb{R} \simeq \mathbb{R}^2 \oplus \mathbb{R} = \mathbb{R}^3 .$$

Let $f := F\big|_{S^2}$. Then $f : S^2 \to \mathbb{R}$ is an isoparametric function. If $p \in S^2$ then

$$p = \big((\cos s)\, z,\ \ (\sin s)\, x\big), \quad 0 \le s \le \frac{\pi}{2}, \quad z \in S^1, \quad x \in S^0 = \{\pm 1\} \subset \mathbb{R}.$$

Then

$$M_{c_0} = f^{-1}\big(\cos(2s_0)\big) \simeq (\cos\, s_0)\, S^1 \times \big\{\pm \sin s_0\big\},$$

i.e., M_{c_0} consists of two (parallel) circles of radius $\cos s_0$ on S^2. Let $M^+_{\cos(2s_0)} = (\cos\, s_0)\, S^1 \times \big\{\sin s_0\big\}$ be the upper circle, and let

$$x \in M^+_{c_0}, \quad x = (x^1,\ x^2,\ x^3), \quad x^3 = \sin s_0 .$$

Moreover

$$x^1 = (\cos s_0)\, \cos\psi, \quad x^2 = (\cos s_0)\, \sin\psi, \quad 0 \le \psi \le 2\pi.$$

We wish to compute ξ_x, i.e., the unit normal vector on M_{c_0}, $c_0 = \cos(2s_0)$, at the point x. The maximal integral curves of ξ_x are geodesics in S^2, i.e., great circles on S^2. Then

$$\begin{cases} \dot{x}^1 = -(\sin s)\,\cos\psi \\ \\ \dot{x}^2 = -(\sin s)\,\sin\psi \\ \\ \dot{x}^3 = \cos s \end{cases} \tag{3.167}$$

Integration of (3.167) furnishes

$$\xi_x = \Big(-\big(\sin s_0\big)\,\cos\psi,\quad -\big(\sin s_0\big)\,\sin\psi,\quad \cos s_0\Big).$$

Then $\gamma_t(x) = (\cos t)\,\vec{x} + (\sin t)\,\xi_x$ [that is the geodesic in S^2 of initial data $(x,\ \xi_x)$] may be written

$$\gamma_t(x) = \Big(\cos\big(t + s_0\big)\,\cos\psi,\quad \cos\big(t + s_0\big)\,\sin\psi,\quad \sin\big(t + s_0\big)\Big)$$

and hence

$$\frac{\partial}{\partial\psi}\,\gamma_t(x) = \Big(-\cos\big(t + s_0\big)\,\sin\psi,\quad \cos\big(t + s_0\big)\,\cos\psi,\quad 0\Big).$$

It follows that

$$d_x\gamma_t = 0 \Longleftrightarrow \cos(t + s_0) = 0 \Longleftrightarrow t \in \Big\{-s_0 + k\pi + \frac{\pi}{2}\ :\ k \in \mathbb{Z}\Big\}.$$

Let $t_k := -s_0 + k\,\pi + \pi/2$. Then $1/\tan t_k = \tan s_0$. Therefore $\tan s_0$ is the principal curvature of $M_{\cos(2s_0)}$ with respect to ξ_x, $x \in M_{\cos(2s_0)}$. We obtain

$$\gamma_{t_0}(x) = \Big(\big(\cos\frac{\pi}{2}\big)\,\cos\psi,\ \big(\cos\frac{\pi}{2}\big)\,\sin\psi,\ \sin\frac{\pi}{2}\Big) = (0,\ 0,\ 1).$$

We may conclude that the focal manifolds of f are the North and South poles on S^2. □

Let $P : \mathbb{R}^m \to \mathbb{R}$ be an eikonal i.e.

$$\|\nabla P(x)\|^2 = \|x\|^{2k-2}, \quad x \in \mathbb{R}^m.$$

Then (by Theorem 3.135) $\nabla P : \mathbb{R}^m \to \mathbb{R}^m$ descends to a harmonic map $\nabla P : S^{m-1} \to S^{m-1}$. Moreover the restriction of P to S^{m-1} is an isoparametric function. The last statement follows from the more general result.

Theorem 3.164 *Let $F : \mathbb{R}^m \to \mathbb{R}$ be a homogeneous polynomial such that*

$$\|\nabla F(x)\|^2 = k^2 \, \|x\|^{2k-2}, \tag{3.168}$$

$$\Delta F(x) = a \, \|x\|^{k-2} \tag{3.169}$$

for some $a \in \mathbb{R}$, $k \in \mathbb{N}$, and any $x \in \mathbb{R}^m$. Then the restriction of F to S^{m-1} is an isoparametric function.

Cf. Baird [Bai], p. 33. We end the section by quoting a deep result by Munzner [Mun].

Theorem 3.165 *Let $\Psi : M \to S^{m-1}$ be an isoparametric hypersurface, i.e., M has constant principal curvatures. Let us assume M has k distinct principal curvatures $\lambda_1, \ldots, \lambda_k$, respectively of multiplicities $m_1, \ldots, m_k$, i.e., $\dim_{\mathbb{R}} S_\alpha = m_\alpha$, $1 \le \alpha \le k$. Then*

(i) *$m_{\alpha+2} = m_\alpha$, i.e., there are at most two distinct multiplicities.*
(ii) *$\Psi : M \to S^{m-1}$ is the level hypersurface of an isoparametric function $f : S^{m-1} \to \mathbb{R}$ obtained as the restriction to $S^{m-1}$of a homogeneous polynomial $F : \mathbb{R}^m \to \mathbb{R}$ obeying to (3.168)–(3.169) with*

$$\begin{cases} a = \dfrac{1}{2}\left(m_2 - m_1\right) & \text{if } k \text{ is even}, \\ 0 & \text{if } k \text{ is odd}. \end{cases}$$

(iii) *Viceversa, every such F determines a family of isoparametric hypersurfaces in S^{m-1}.*
(iv) *The integer number k can only take the values* $\{1, 2, 3, 4, 6\}$.

3.14 Submanifolds of Space Forms

Let $\left(\overline{M}, \overline{g}\right)$ be a Riemannian manifold of constant sectional curvature $\overline{k}$, and let M be a submanifold of $\overline{M}$. By

$$\overline{R}(X, Y)Z = \overline{k}\left\{g(Z, Y)X - g(Z, X)Y\right\} \tag{3.170}$$

for any $X, Y, Z \in \mathfrak{X}(M)$ the vector field $\overline{R}(X, Y)Z$ is tangent to M. Hence the equation

$$\tan\left\{\overline{R}(X, Y)Z\right\} = R(X, Y)Z + \tag{3.171}$$

$$+ \sum_{a=1}^{r} \left\{h^a(X, Z)\, A_a Y - h^a(Y, Z)\, A_a X\right\}$$

may be written [by taking into account $g(A_a Y, Z) = h^a(Y, Z)$]

$$R(X, Y)Z = \overline{R}(X, Y)Z + \tag{3.172}$$

$$+ \sum_{a=1}^{r} g(A_a Y, Z)\, A_a X - g(A_a X, Z)\, A_a Y.$$

Let $S = \mathrm{Ric}_\nabla$ and $\overline{S} = \mathrm{Ric}_{\overline{\nabla}}$ be the Ricci tensor fields of (M, g) and $(\overline{M}, \overline{g})$, respectively. If $\{E_i : 1 \leq i \leq n\}$ is a local g-orthonormal frame of $T(M)$ then

$$S(Y, Z) = \overline{S}(Y, Z) + \tag{3.173}$$

$$+ \sum_{i=1}^{n} \sum_{a=1}^{r} \{g(A_a Y, Z)\, g(A_a E_i, E_i) - g(A_a E_i, Z)\, g(A_a Y, E_i)\} =$$

$$= \overline{S}(Y, Z) + \sum_{a=1}^{r} \Big\{ g(A_a Y, Z) \operatorname{trace}(A_a) - g\Big(A_a Y, \sum_{i=1}^{n} g(A_a E_i, Z)\, E_i\Big)\Big\}.$$

Note that

$$\sum_{i=1}^{n} g(A_a E_i, Z)\, E_i = A_a Z\,.$$

Hence

$$S(Y, Z) = \overline{S}(Y, Z) + \sum_{a=1}^{r} \{g(A_a Y, Z) \operatorname{trace}(A_a) - g\big(A_a^2 Y, Z\big)\}.$$

Next [by (3.170)]

$$\overline{S}(Y, Z) = (n-1)\overline{k}\, g(Y, Z)$$

and then

$$S(Y, Z) = (n-1)\, \overline{k}\, g(Y, Z) + \tag{3.174}$$

$$+ \sum_{a=1}^{r} \{g(A_a Y, Z) \operatorname{trace}(A_a) - g(A_a Y, A_a Z)\}.$$

3.15 Chern's Theorem

Let ρ be the scalar curvature of (M, g). Then [by taking traces in (3.174)]

$$\rho = n(n-1)\overline{k} + \sum_{a=1}^{r} \left\{ \left(\text{trace}(A_a)\right)^2 - \text{trace}(A_a^2) \right\}. \tag{3.175}$$

By using (3.174)–(3.175) one obtains the following result by S.-S. Chern[64]

Theorem 3.166 *Let M be a minimal submanifold of a space form $M^n(\overline{k})$. Then the first fundamental form g of $M \hookrightarrow M^n(\overline{k})$, the Ricci curvature S, and the scalar curvature ρ of (M, g), satisfy*

(i) $S(Y, Z) \leq (n-1)\overline{k}\, g(Y, Z)$ *for any* $Y, Z \in \mathfrak{X}(M)$,
(ii) $\rho \leq n(n-1)\overline{k}$.

Moreover, equality holds in (i) [*respectively in* (ii)] *if and only if M is totally geodesic.*

Corollary 3.167 *Let $(\overline{M}, \overline{g})$ be a Riemannian manifold of constant sectional curvature $\overline{k}$. Let M be a submanifold of $\overline{M}$, with the induced metric g. If M is a minimal submanifold and (M, g) has constant sectional curvature $k \geq \overline{k}$ then*

(a) $k = \overline{k}$,
(b) *M is a totally geodesic submanifold of $\overline{M}$.*

Proof The scalar curvature of (M, g) is $\rho = n(n-1)k$. Then [by part (ii) in Theorem 3.166] $k = \overline{k}$ and M is totally geodesic in $\overline{M}$. Q.e.d.

Corollary 3.168 *The standard sphere doesn't admit a minimal isometric immersion in a Euclidean space.*

Corollary 3.169 *An Euclidean space doesn't admit a minimal isometric immersion in a hyperbolic space.*

3.16 Allendoerfer–Erbacher's Theorem

Let M be a submanifold of $(\overline{M}, \overline{g})$. Let A be the shape operator of the given immersion, and let

$$A_p : T_p(M)^{\perp} \times T_p(M) \to \mathbb{R}, \quad p \in M$$

---

[64] Cf. S.-S. Chern, *Minimal Surfaces in an Euclidean Space of n Dimensions*. Symposium Differential and Combinatorial Topology in honor of M. Morse (Princeton University Press, Princeton, New Jersey, 1965), pp. 187–198.

be the corresponding field i.e.

$$A_p(N, v) = \left(A_\xi X\right)_p, \quad N \in T_p(M)^\perp, \quad v \in T_p(M),$$

where $\xi \in C^\infty\left(T(M)^\perp\right)$ and $X \in \mathfrak{X}(M)$ are arbitrary extensions of N and v, i.e., $\xi_p = N$ and $X_p = v$. Next, let us set

$$N_p^0 = \left\{N \in T_p(M)^\perp \ : \ A_p(N, v) = 0, \ \ \forall\, v \in T_p(M)\right\}, \quad p \in M.$$

Definition 3.170 The orthogonal complement

$$N_p^1 := \left(N_p^0\right)^\perp \subset T_p(M)^\perp$$

of N_p^0 in $T_p(M)^\perp$ is called the *principal normal space* at p, of the given immersion. □

Note that

$$N_p^1 = \left\{B_p(u, v) \ : \ u, \ v \in T_p(M)\right\}.$$

Let $\overline{M}^m(\overline{k}) \in \left\{\mathbb{E}^m, \ S^m, \ H^m\right\}$.

Theorem 3.171 *Let M be a connected n-dimensional submanifold of $\overline{M}^{n+r}\left(\overline{k}\right)$. Let us assume that the principal normal spaces of $M \hookrightarrow \overline{M}^{n+r}(\overline{k})$ are parallel with respect to the normal connection $\nabla^\perp$. Then the principal normal spaces have constant dimension*

$$\ell := \dim_{\mathbb{R}} N_p^1, \quad p \in M.$$

If $\ell < r$ there is a $(n + \ell)$-dimensional totally geodesic submanifold $M^* \hookrightarrow \overline{M}^{n+r}(\overline{k})$ such that $M \subset M^*$.

Theorem 3.171 is referred to as a "codimension reduction" result (for a given immersion) and is due to C.B. Allendoerfer[65] and J.A. Erbacher.[66]

---

65 Cf. C. Allendoerfer, Rigidity for spaces of class greater than one. Amer. J. Math. **61**, 633–644 (1939).

66 Cf. J.A. Erbacher, Reduction of the codimension of an isometric immersion. J. Diff. Geom. **5**, 333–340 (1971).

3.17 Totally Umbilical Submanifolds

We recall that a submanifold M of the Riemannian manifold $(\overline{M}, \overline{g})$ is *totally umbilical* if $B = g \otimes H$.

Proposition 3.172 *A totally umbilical submanifold of a space form has a flat normal connection.*

Proof Follows from the Ricci equations.

Definition 3.173 A submanifold M of a Riemannian manifold $(\overline{M}, \overline{g})$ is said to be an *extrinsic sphere* if

(i) M is totally umbilical, but not totally geodesic.
(ii) M has a parallel mean curvature vector (i.e., $\nabla^{\perp} H = 0$). □

Extrinsic spheres were introduced and investigated by K. Nomizu.[67]

Example 3.174 Let $S^3(1) \times S^3(1)$ be equipped with the product metric $\overline{g} = g_1 \times g_1$, where g_1 is the standard Riemannian metric on $S^3(1)$. Let $\tau : S^3(\sqrt{2}) \to S^3(1) \times S^3(1)$ be the isometric immersion $\tau(p) = (p, \ p)$. Let $S^2(r)$ with $r > 0, r \neq \sqrt{2}$ be a sphere in $S^3(\sqrt{2})$ whose center is not the origin. Then $S^2(r)$ is an extrinsic sphere in $S^3(1) \times S^3(1)$. □

Definition 3.175 Let $n, \ r \in \mathbb{N}$ such that $n \geq 3$ and $2 \leq r < n$. A n-dimensional Riemannian manifold (M, g) is said to satisfy the *axiom of r-spheres* if for every point $p \in M$ and every r-dimensional subspace $V \subset T_p(M)$ there is a r-dimensional extrinsic sphere $S \subset M$ passing through p (i.e., $p \in S$) such that $T_p(S) = V$. □

Theorem 3.176 *If a n-dimensional ($n \geq 3$) Riemannian manifold M satisfies the axiom of r-spheres for some $2 \leq r < n$ then M has constant sectional curvature.*

Theorem 3.176 is due to D.S. Leung and K. Nomizu.[68]

Proof of Theorem 3.176 Let $\{u, \ v, \ w\} \subset T_p(M)$ be an orthonormal system, and let $V \subset T_p(M)$ be a r-dimensional subspace such that $\{u, \ v\} \subset V$ and V is orthogonal to w. By the Axiom of r-spheres there is a totally umbilical submanifold $S \subset M$ such that

$$p \in S, \quad \nabla^{\perp} H = 0, \quad T_p(S) = V,$$

where H is the mean curvature vector of $S \hookrightarrow M$. Let $U \subset M$ be a normal[69] neighborhood of p. Every point $q \in U$ may be joined to p by a geodesic γ. Let $\xi_q \in T_q(M)$ be the parallel translate of w along γ. Both vector fields ξ and H are parallel, so $g(\xi, \ H)$ is a constant $\lambda \in \mathbb{R}$

---

[67] Cf. K. Nomizu, *Generalized central spheres and the notion of spheres in Riemannian geometry*. Tohoku Math. J. **25**, 103–109 (1973).

[68] *The axiom of r-spheres in Riemannian geometry*. J. Diff. Geom. **5**, 487–489 (1971).

[69] With respect to the Levi-Civita connection of (M, g).

$$\lambda = g(\xi, H) = g(\xi, H)_p = g_p(w, H_p).$$

As M is totally umbilical

$$A_\xi = \lambda I$$

everywhere on U, implying that

$$\left(\nabla_X A_\xi\right)_p = \left(\nabla_Y A_\xi\right)_p = 0 \tag{3.176}$$

where $X, Y \in \mathfrak{X}(U)$ are such that $X_p = u$ and $Y_p = v$. By hypothesis, one also has

$$\left(\nabla^\perp_X \xi\right)_p = \left(\nabla^\perp_Y \xi\right)_p = 0. \tag{3.177}$$

Recall that

$$\begin{aligned} \tan R(X, Y)\xi = \nabla_Y(A_\xi X) - \nabla_X(A_\xi Y) + \\ - A_\xi([X, Y]) - A_{\nabla^\perp_X \xi} Y + A_{\nabla^\perp_Y \xi} X. \end{aligned} \tag{3.178}$$

Then [by (3.176) and the symmetry of ∇] at the point p

$$\tan R(X, Y)\xi = \lambda \nabla_Y X - \lambda \nabla_X Y + \lambda[X, Y] = 0.$$

Hence $R_p(u, v)w$ is orthogonal to S at p, and consequently

$$g_p\big(R_p(u, v)w, u\big) = 0. \tag{3.179}$$

Hence the sectional curvature of M is a point function, so that (by Schur's theorem) (M, g) has constant sectional curvature. Q.e.d.

Theorem 3.177 *Any totally umbilical submanifold of a Riemannian manifold of constant sectional curvature has constant sectional curvature.*

Proof The Gauss equation reads

$$\begin{aligned} K(X, Y, Z, W) = \\ = \left(\overline{k} + \|H\|^2\right)\left\{g(X, W)\, g(Y, Z) - g(X, Z)\, g(Y, W)\right\} \end{aligned} \tag{3.180}$$

where $\overline{k}$ is the (constant) sectional curvature of the ambient space, and H is the mean curvature vector of the submanifold. By Eq. (3.180) for $n \geq 3$ the Riemannian manifold M is isotropic, and then (again by Schur's theorem) M has constant sectional curvature[70] $\overline{k} + \|H\|^2$. Q.e.d.

Totally umbilical submanifolds of Euclidean spaces may be classified, as follows:

---

[70] When $n = 2$ one observes that $\|H\|$ is a constant, and the conclusion is the same.

Theorem 3.178 *Let M be a n-dimensional totally umbilical submanifold of the Euclidean space $\mathbb{E}^{n+r}$. Then either*

(i) *M is totally geodesic,*

or

(ii) *M is contained in an n-dimensional sphere, i.e., $M \subset S^n$.*

The proof of Theorem 3.178 is well known (cf. e.g., B.-Y. Chen[71]).

3.18 Characterizations of the Sphere

One of the interesting questions in the geometry of submanifolds in an Euclidean space is to obtain conditions under which the submanifold is isometric to a sphere.

C. Tompkins considered[72] the problem whether a n-dimensional Riemannian manifold may, or may not, be isometrically immersed into $\mathbb{E}^{2n-1}$. C. Tompkins' result (cf. *op. cit.*) is that a flat n-dimensional torus cannot be isometrically embedded into $\mathbb{E}^{2n-1}$. S.-S. Chern and N. Kuiper[73] and T. Otsuki[74] have come to the same conclusion for an arbitrary compact n-dimensional Riemannian manifold of nonpositive sectional curvature (it cannot be embedded in $\mathbb{E}^{2n-1}$). A successive result by H. Jacobowitz[75] shows that a compact n-dimensional Riemannian manifold of sectional curvature $\leq \lambda^{-2}$ ($\lambda > 0$) cannot be isometrically immersed into a ball of radius λ in $\mathbb{E}^{2n-1}$. S. Deshmukh et al.[76] considered the problem of non embeddability with bounds on the Ricci curvature.

As well known the scalar curvature τ of the sphere $S^n(c)$, of constant curvature c, in the Euclidean space $\mathbb{E}^{n+1}$ satisfies

$$\tau = \lambda_1(n-1),$$

where λ_1 is the first nonzero eigenvalue of the Laplacian operator Δ acting on $C^\infty\big(S^n(c)\big)$ and the Ricci curvature of $S^n(c)$ is positive. Using that as a main ingre-

---

[71] Cf. *op. cit.*

[72] C. Tompkins, Isometric embedding of flat manifolds in Euclidean space. Duke Math. J. **5**, 58–61 (1930).

[73] Cf. S.-S. Chern, N.H. Kuiper, Some theorems on the isometric imbedding of compact Riemannian manifolds in Euclidean space. Ann. Math. **56**, 442–430 (1952).

[74] Cf. T. Otsuki, On the existence of solutions of a system of quadric equations and its geometrical applications. Proc. Japan Acad. **29**, 99–100 (1953).

[75] Cf. H. Jacobowitz, Isometric embedding of a compact Riemannian manifold into Euclidean space. Proc. Amer. Math. Soc. **40**, 245–246 (1973).

[76] Cf. S. Deshmukh, M.A. Al-Gwaiz, Compact hypersurface in even dimensional Euclidean space and in the sphere. Quart. J. Math. **45**, 151–157 (1994).

dient, S. Deshmukh showed[77] that a compact and connected immersed hypersurface $M \hookrightarrow \mathbb{E}^{n+1}$ is isometric to a sphere $S^n(c)$, provided M has positive Ricci curvature and its scalar curvature satisfies $\tau = \lambda_1(n-1)$. Moreover, Theorem 3.183 shows that if a compact and connected immersed hypersurface in $M \hookrightarrow \mathbb{E}^{n+1}$ has vanishing Minkowski integrands, then M is isometric to a sphere in $\mathbb{R}^{n+1}$.

The study of real hypersurfaces in complex space forms $\overline{M}(c)$ [i.e., Kählerian manifolds of constant holomorphic sectional curvature c] is a very interesting and active research area. A survey paper by R. Niebergall and P.J. Ryan[78] describes the state-of-the-art up to 1996.

For $c = 0$ complex space forms are of course Euclidean complex space forms. Characterizations of the sphere $S^{2n+1} \subset \mathbb{C}^{n+1}$ were the purpose of several works, cf. e.g., Chen [Che2], Chen and Garay [Che3], Chen, Sun and Tang [Che4], Montiel and Ros [MoRo], Ripoll [Rip], Ros [Ros2]. K. Nomizu and Symth proved[79] that compact hypersurfaces with a non-negative sectional curvature and of constant mean curvature are spheres. Cf. also Ros [Ros3], for further characterizations of spheres. It should be noticed that Theorem 3.184 doesn't require sectional curvatures to be ≥ 0.

Several necessary and sufficient conditions for a curve in $\mathbb{R}^3$ to lie on the unit sphere S^2 are available (cf. Breuer and Gottlieb [BrGo], Takahashi [Tak]). As a natural generalization of the problem, given a compact n-dimensional Riemannian manifold M one should look for immersions $\Psi : M^n \to \mathbb{R}^{n+2}$ such that $\Psi(M) \subset S^{n+1}(c)$.

3.18.1 Hypersurfaces in Euclidean Spaces

Let M be a real hypersurface in Euclidean space $\mathbb{R}^{n+1}$ with the almost complex structure J. Then $n+1$ must be even, e.g., $n+1 = 2m$ [hence M is $(2m-1)$-dimensional]. Note that we start with an arbitrary almost complex structure J [not necessarily the complex structure of $\mathbb{C}^m$] and only require that J be compatible to the Euclidean metric $\overline{g} = \langle , \rangle$ on $\mathbb{R}^{n+1}$ i.e.

$$\overline{g}\big(JV,\ JW\big) = \overline{g}(V,W), \quad \forall\ V, W \in \mathfrak{X}(\mathbb{R}^{n+1}).$$

Let $\Psi : M \to R^{n+1}$ be the given C^∞ immersion. Let $g = \Psi^* \overline{g}$ be the induced metric on M. Let $\overline{\nabla}$ and ∇ be respectively the Levi-Civita connections on $\big(\mathbb{R}^{n+1},\ \langle , \rangle\big)$ and $\big(M,\ g\big)$. Let us assume that M is oriented and choose a unit normal vector field N on M. Let then $(\phi,\ \xi,\ \eta,\ g)$ be the induced almost contact metric structure on M. It is a classical fact that M inherits via Ψ such a structure, cf. e.g., Blair [Bla]. We give

[77] Cf. S. Deshmukh, Compact hypersurfaces in a Euclidean space. Quart. J. Math. Oxford **49**, 35–41 (1998), and Theorem 3.181 below.

[78] Cf. R. Niebergall, P.J. Ryan, *Real Hypersurfaces in Complex Space Forms*. Tight and Taut Submanifolds. MSRI Publications 32 (Cambridge University Press, Cambridge, 1997), pp. 233–205.

[79] Cf. K. Nomizu, B. Smyth, On the Gauss mapping for hypersurfaces of constant mean curvature in the sphere. Comm. Math. Helv. **44**, 484–490 (1969).

a brief description of (ϕ, ξ, η, g), so that to fix notations and conventions. First one sets $\xi = -JN$ and observes that ξ is tangent to M, i.e., $\xi \in \mathfrak{X}(M)$. Next let η be the differential 1-form on M given by

$$\eta(X) = g(\xi, X), \quad \forall \ X \in \mathfrak{X}(M).$$

Let $\tau_M^{\perp}(\Psi)$ be the normal bundle of the given immersion and let

$$\tan : \Psi^{-1}\tau_{\mathbb{R}^{n+1}} \to \tau_M, \quad \text{nor} : \Psi^{-1}\tau_{\mathbb{R}^{n+1}} \to \tau_M^{\perp}(\Psi)$$

be the projections associated to the direct sum decomposition

$$\Psi^{-1}\tau_{\mathbb{R}^{n+1}} = \tau_M \oplus \tau_M^{\perp}(\Psi).$$

Next let us set

$$\phi X = \tan\big(JX\big), \quad FX = \text{nor}(JX), \quad X \in \mathfrak{X}(M),$$

so that

$$JX = \phi X + FX, \quad FX = \eta(X)\, N.$$

Then ϕ is a (1, 1)-tensor field on M and the following identities hold good (and may be easily checked)

$$\begin{cases} \phi^2 = -I + \eta \otimes \xi, \\ \phi \circ \xi = 0, \quad \eta(\xi) = 1, \quad \eta(X) = g(X, \xi), \\ g(\phi X, Y) = -g(X, \phi Y), \\ g(\phi X, \phi Y) = g(X, Y) - \eta(X)\eta(Y), \end{cases} \tag{3.181}$$

for any $X, Y \in \mathfrak{X}(M)$. Also

$$\nabla_X \xi = \phi A X, \quad (\nabla_X \phi)(Y) = \eta(Y)\, AX - g(AX, Y)\xi \tag{3.182}$$

where $A = A_N$ is the shape operator (associated to the normal section N). We shall make use of the Gauss and Weingarten formulas

$$\overline{\nabla}_X Y = \nabla_X Y + B(X, Y)N, \tag{3.183}$$

$$\overline{\nabla}_X N = -A X \tag{3.184}$$

for any $X, Y \in \mathfrak{X}(M)$. Here B is the second fundamental form of the given immersion. As previously shown B and A are related by

$$g(AX, Y) = B(X, Y). \tag{3.185}$$

Let $R = R^{\nabla}$ be the curvature tensor field of ∇. We shall make use of the Gauss equation

$$R(X, Y)Z = g(AY, Z)\, AX - g(AX, Z)\, AY \tag{3.186}$$

for any $X, Y \in \mathfrak{X}(M)$. The shape operator A obeys to the Codazzi equation

$$(\nabla A)(X, Y) = (\nabla A)(Y, X) \tag{3.187}$$

for any $X, Y \in \mathfrak{X}(M)$. As customary $(\nabla A)(X, Y) = \nabla_X AY - A(\nabla_X Y)$.

Let $\{E_1, \ldots, E_n\}$ be a local orthonormal frame of the tangent bundle τ_M, defined on the open set $U \subset M$. Let $\text{Ric} = \text{Ric}_g$ and $\tau = \tau_g$ be respectively the Ricci and scalar curvature of (M, g). By passing to traces in Gauss equation (3.186) one obtains

$$Ric(X, Y) = n\,\alpha\, g(AX, Y) - g(AX, AY) \tag{3.188}$$

for any $X, Y \in \mathfrak{X}(M)$ and

$$\tau = n^2\,\alpha^2 - \|A\|^2, \tag{3.189}$$

where α is the mean curvature

$$\alpha = \frac{1}{n}\,\text{trace}(A).$$

Let $\Psi : M \to \mathbb{C}^n$ be an isometric immersion such that $\Psi(M)$ is contained in a ball of radius r. The complex structure J on $\mathbb{C}^n$ is parallel with respect to the Euclidean connection on $\mathbb{C}^n$. Let us think of Ψ as the position vector field of points in $\Psi(M)$. Let us consider the smooth function

$$\varrho^* : M \to \mathbb{R}, \quad \varrho^* =< J\Psi, N >,$$

where $< \cdot, \cdot >$ is the Euclidean inner product on $\mathbb{C}^n$ and N is the unit normal vector field to M. Next let

$$J\Psi = v + \varrho^* N \tag{3.190}$$

where v is the tangential component of $J\Psi$ to M. Let g and ∇ be respectively the Riemannian metric and the Levi-Civita connection on M. Let $A = A_N$ be the Weingarten operator associated of the given immersion Ψ. We may assume that

$\|\Psi\| \leq r$ [otherwise one may use some isometry of $\mathbb{C}^n$ to achieve that]. Here $\|\cdot\|$ is Euclidean the norm on $\mathbb{C}^n$.

Covariant differentiation of (3.190) together with $J\xi = N$, $JX = \phi X + \varrho^* N$, and the Gauss-Weingarten formulas, leads to

$$\nabla_X \upsilon = \phi X + \varrho^* AX. \tag{3.191}$$

Then [by (3.181)]

$$g(\phi X, X) = 0, \quad g(\phi X, \xi) = 0, \quad \forall\ X \in \mathfrak{X}(M).$$

We may then choose a local orthonormal frame of τ_M of the form

$$\{E_1, \ldots, E_{n-1}, \phi E_1, \ldots, \phi E_{n-1}, \xi\}.$$

Next [by the skew-symmetry of ϕ and the symmetry of A]

$$\sum_{i=1}^{2n-1} g(\phi e_i, Ae_i) = 0 \tag{3.192}$$

for any local orthonormal frame $\{E_1, \ldots, E_{2n-1}\}$ of τ_M. Then [by (3.191)]

$$\delta\beta = (2n-1)\varrho^*\alpha, \tag{3.193}$$

where β is the differential 1-form on M defined by

$$\beta(X) = g(X, \upsilon), \quad \forall\ X \in \mathfrak{X}(M),$$

and α is the mean curvature of M.

S. Deshmukh et al.[80] considered the question of non immergibility using upper and lower bounds for the Ricci curvature of M.

Theorem 3.179 *Let M be a compact $(2n-1)$-dimensional Riemannian manifold whose Ricci curvature Ric and scalar curvature τ satisfy*

$$\mathrm{Ric}(X, X) + \tau \geq 0, \quad Ric(X, X) < 2(n-1)r^{-2}$$

for every unit vector field X on M and some positive constant r. Then no isometric immersion of M into $\mathbb{C}^n$ is contained in a ball of radius r.

We start by establishing:

---

[80] Cf. *op. cit.*

Lemma 3.180 *Let $\Psi : M \to \mathbb{C}^n$ be an isometric immersion of a compact $(2n-1)$-dimensional Riemannian manifold M. Then*

(a) $\|\nabla v\|^2 = -\mathrm{tr}(\phi^2) + \varrho^{*2}\,\mathrm{tr}(A^2)$,
(b) $\|\mathcal{L}_v g\|^2 = 4\varrho^{*2}\,\mathrm{tr}(A^2)$,

where $\mathcal{L}_v$ is the Lie derivative at the direction v.

Proof Let $\{E_1, \ldots, E_{2n-1}\}$ be a local orthonormal frame of τ_M. Then

$$\|\nabla v\|^2 = \sum g(\nabla_{E_i} v, \nabla_{E_i} v) = \tag{3.194}$$

$$= -\mathrm{tr}(\phi^2) + 2\varrho^* \sum g(Ae_i, \phi e_i) + \varrho^{*2}\,\mathrm{tr}(A^2) = -\mathrm{tr}(\phi^2) + \varrho^{*2}\,\mathrm{tr}(A^2),$$

where we have used Eq. (3.192). Furthermore

$$\|\mathcal{L}_v g\|^2 = \sum_{i,j} [(\mathcal{L}_v f)(E_i, E_j)]^2 = 4\varrho^{*2} \sum_{i,j} g(AE_i, E_j)^2 = \tag{3.195}$$

$$= 4\varrho^{*2} \sum_i g\left(AE_i, \sum_j g(AE_i, E_j)\, E_j\right) = 4\varrho^{*2} \sum_i g(AE_i, E_i)^2 = 4\varrho^{*2}\,\mathrm{tr}(A^2).$$

Q.e.d.

Proof of Theorem 3.179 The following integral formula holds on any compact orientable Riemannian manifold M

$$\int_M [Ric(v, v) + \frac{1}{2}||\mathcal{L}_v g||^2 - ||\nabla v||^2 - (\delta\beta)^2] d\,\mathrm{v}_g = 0.$$

Using (3.193)–(3.195) together with $\mathrm{tr}(\phi^2) = -2(n-1)$ [a consequence of (3.181)] we get

$$\int_M [Ric(v, v) + \frac{1}{2}||\mathcal{L}_v g||^2 - ||\nabla v||^2 - (\delta\beta)^2] d\,\mathrm{v}_g = 0, \tag{3.196}$$

where [by the Gauss equation] $\tau = (2n-1)^2\alpha^2 - \mathrm{tr}(A^2)$.

Note that v is not identically zero. For if $v = 0$ then $J\Psi = \varrho^* N$ yields $\Psi = \varrho^* \xi$ hence $< \Psi, N >= 0$. Then [by Minkowski formula for a compact hypersurface in $\mathbb{C}^n$ (cf. [Hsi])]

$$\int_M [1+ < \Psi, N >] d\,\mathrm{v}_g = 0$$

leads to a contradiction. Let $\hat{\upsilon}$ be the unit vector field defined on the open subset of M where υ is nonzero. Substitution from $||\Psi||^2 = ||\upsilon||^2 + \varrho^{*2}$ into the integral formula (3.196) gives

$$\int_M \Big\{\varrho^{*2}\big[\mathrm{Ric}(\hat{\upsilon}, \hat{\upsilon}) + \tau\big] + \big[2(n-1) - \|\Psi\|^2 \,\mathrm{Ric}(\hat{\upsilon}, \hat{\upsilon})\big]\Big\}\, d\,\mathrm{v}_g = 0. \qquad (3.197)$$

The (by hypothesis) $\mathrm{Ric}(\hat{\upsilon}, \hat{\upsilon}) + \tau \geq 0$ and

$$||\Psi||^2 \,\mathrm{Ric}(\hat{\upsilon}, \hat{\upsilon}) \leq r^2 \,\mathrm{Ric}(\hat{\upsilon}, \hat{\upsilon}) < 2(n-1)$$

contradicting (3.197) and proving the theorem. □

The mean curvature $\alpha = (1/n)\,\mathrm{tr}(A)$ satisfies

$$\begin{cases} \mathrm{grad}(\alpha) = (1/n) \sum_{i=1}^{n} (\nabla A)(E_i,\ E_i), \\ X(\alpha) = (1/n) \sum_{i=1}^{n} g((\nabla A)(E_i,\ E_i), X) \end{cases} \qquad (3.198)$$

for any $X \in \mathfrak{X}(M)$. If we consider $f : M \to \mathbb{R}$, $f = \frac{1}{2}\,\|\Psi\|^2$, and think of Ψ as the position vector field of M in $\mathbb{R}^{n+1}$ then

$$\Psi = \mathrm{grad} f + \varrho N, \quad \varrho : M \to \mathbb{R}, \quad \varrho = \langle \Psi, N\rangle. \qquad (3.199)$$

Then [by Gauss and Weingarten formulas]

$$\nabla_X \mathrm{grad} f = X + \varrho A X, \quad X(\varrho) = -g\big(A(X),\ \mathrm{grad} f\big), \quad X \in \mathfrak{X}(M). \; (3.200)$$

By the first equation in (3.231)

$$\Delta f = n(1 + \varrho\alpha),$$

and integration yields the following *Minkowski formula*

$$\int_M (1 + \varrho\alpha)\, d\,\mathrm{v}_g = 0. \qquad (3.201)$$

Next we report on a result by S. Deshmukh[81] showing that a compact and connected immersed hypersurface of Euclidean space $\mathbb{R}^{n+1}$, with upper bounds on the scalar curvature, is isometric to a sphere $S^n(c)$. Precisely

Theorem 3.181 *Let M be a compact and connected immersed hypersurface of $\mathbb{R}^{n+1}$ with positive Ricci curvature and scalar curvature τ satisfying $\tau \leq \lambda_1(n-1)$ where*

[81] Cf. *op. cit.*

λ_1 is the first nonzero eigenvalue of the Laplacian Δ on M. Then M is isometric to a sphere $S^n(c)$.

Proof We may assume w.l.o.g. that the center of mass of M is at the origin of $\mathbb{R}^{n+1}$ (otherwise one may choose an isometry $\phi : \mathbb{R}^{n+1} \to \mathbb{R}^{n+1}$ mapping the center of mass of M to the origin of $\mathbb{R}^{n+1}$ and then $\Psi' = \phi \circ \Psi$ would be an immersion with the desired property). Thus the immersion $\psi : M \to \mathbb{R}^{n+1}$ satisfies $\int_M \psi \, d\,\mathrm{v}_g = 0$. Hence [by the minimum principle]

$$\lambda_1 \leq n\,\mathrm{vol}(M) \int_M ||\Psi||^2 \, d\,\mathrm{v}_g$$

where λ_1 is the first nonzero eigenvalue of the Laplacian Δ on M. Consequently

$$\int_M \|\Psi\|^2 \, d\,\mathrm{v}_g \leq \frac{n\,\mathrm{vol}(M)}{\lambda_1}. \tag{3.202}$$

Next [by (3.188)]

$$\begin{aligned}\int_M \mathrm{Ric}(\mathrm{grad} f, \mathrm{grad} f)\, d\,\mathrm{v}_g =& n \int_M \alpha\, g(A \mathrm{grad} f, \mathrm{grad} f)\, d\,\mathrm{v}_g \\ & - \int_M ||A(\mathrm{grad} f)||^2 \, d\,\mathrm{v}_g.\end{aligned} \tag{3.203}$$

The second equation in (3.231) yields

$$\mathrm{grad}\varrho = -A(\mathrm{grad} f)$$

and we obtain

$$\begin{aligned} g(A(\mathrm{grad} f), \mathrm{grad} f) &= -g(\mathrm{grad}\varrho, \mathrm{grad} f) = \mathrm{grad} f(\varrho) \\ &= -\mathrm{div}(\varrho \mathrm{grad} f) + \varrho \Delta f \\ &= -\mathrm{div}(\varrho \mathrm{grad} f) + n\, \varrho(1 + \varrho\alpha). \end{aligned}$$

Thus

$$\alpha\, g\big(A(\mathrm{grad} f), \mathrm{grad} f\big) = -\alpha\, \mathrm{div}(\varrho \mathrm{grad} f) + n\, \alpha\, \varrho(1 + \varrho\alpha). \tag{3.204}$$

Given a local orthonormal frame $\{E_1, \ldots, E_n\}$ of τ_M

$$\begin{aligned} \mathrm{div}(A(\mathrm{grad} f)) &= \sum g(\nabla_{E_i} A(\mathrm{grad} f), E_i) \\ &= \sum [g((\nabla A)(E_i, \mathrm{grad} f), E_i) + g(A(\nabla_{E_i} \mathrm{grad} f), E_i)] \end{aligned}$$

which together with $\alpha = (1/n)\,\mathrm{tr}(A)$ gives

$$\mathrm{grad}\alpha = \frac{1}{n}\sum_{i=1}^{n}(\nabla A)(E_i,\ E_i)$$

and then [by (3.231)]

$$\mathrm{div}(A(\mathrm{grad} f)) = n\,(\mathrm{grad} f)\alpha + n\,\alpha + \varrho\,||A||^2. \tag{3.205}$$

By $\mathrm{div}(fX) = X(f) + f\mathrm{div}X$ for any C^∞ function $f : M \to \mathbb{R}$

$$\begin{aligned}\varrho\,\mathrm{div}(A(\mathrm{grad} f)) &= \mathrm{div}(\varrho A(\mathrm{grad} f)) - A(\mathrm{grad} f)\varrho \\ &= \mathrm{div}(\varrho A(\mathrm{grad} f)) + ||A(\mathrm{grad} f)||^2\end{aligned}$$

and together with (3.204)–(3.205) one arrives at

$$n\,\varrho(\mathrm{grad} f)\alpha + n\,\alpha\varrho + \varrho^2||A||^2 = \mathrm{div}(\varrho A(\mathrm{grad} f)) + ||A(\mathrm{grad} f)||^2. \tag{3.206}$$

By $\mathrm{div}(\alpha\varrho\mathrm{grad} f) = \varrho(\mathrm{grad} f)\alpha + \alpha\mathrm{div}(\varrho\mathrm{grad} f)$ and (3.206)

$$\begin{aligned}&-n\,\alpha\,\mathrm{div}(\varrho\mathrm{grad} f) + \mathrm{div}(n\,\alpha\,\varrho\mathrm{grad} f) + n\,\alpha\,\varrho + \varrho^2\,||A||^2 \\ &\qquad = \mathrm{div}(\varrho A(\mathrm{grad} f)) + ||A(\mathrm{grad} f)||^2.\end{aligned} \tag{3.207}$$

Substitution of $-n\,\alpha\,\mathrm{div}(\varrho\mathrm{grad} f)$ from (3.207) into (3.204), yields [by Green's lemma]

$$\int_M n\,\alpha\,g(A(\mathrm{grad} f),\ \mathrm{grad} f)\,d\,\mathrm{v}_g = \tag{3.208}$$

$$= \int_M [||A(\mathrm{grad} f)||^2 - n\,\alpha\,\varrho - \varrho^2||A||^2 + n^2\,\varrho\alpha(1 + \alpha\,\varrho)]\,d\,\mathrm{v}_g =$$

$$= \int_M [n(n-1)\varrho\alpha + \varrho^2\tau + ||A(grad\, f)||^2]\,d\,\mathrm{v}_g =$$

$$= -n(n-1)\,\mathrm{vol}(M) + \int_M \varrho^2\tau\,d\,\mathrm{v}_g + \int_M ||A(\mathrm{grad} f)||^2\,d\,\mathrm{v}_g\,.$$

[Use was made of Minkowski formula (3.201), too].

By

$$||\Psi||^2 = ||\mathrm{grad} f||^2 + \varrho^2, \quad \tau \le \lambda_1(n-1),$$

together with (3.202)

$$\int_M \varrho^2 \tau \, d\,\mathrm{v}_g \leq \int_M ||\Psi||^2 \tau \, d\,\mathrm{v}_g \leq$$

$$\leq \lambda_1 \, (n-1) \int_M ||\Psi||^2 \, d\,\mathrm{v}_g \leq n(n-1) \, \mathrm{vol}(M).$$

Thus [by (3.208)]

$$\int_M n\,\alpha\,(A(\mathrm{grad} f),\ \mathrm{grad} f)\, d\,\mathrm{v}_g \leq \int_M ||\mathrm{grad} f||^2 \, d\,\mathrm{v}_g,$$

hence [by (3.203)]

$$\int_M \mathrm{Ric}(\mathrm{grad} f,\ \mathrm{grad} f)\, d\,\mathrm{v}_g \leq 0.$$

As (by hypothesis) Ricci curvature is positive, it must be $\mathrm{grad} f = 0$. Thus (M is tacitly connected)

$$f = \text{constant} = \frac{r^2}{2}$$

hence M is a sphere of radius r. Q.e.d.

Next we report on a result by S. Deshmukh[82] on compact immersed hypersurfaces in $\mathbb{R}^{n+1}$ with the first and second Hsiung integrands zero. Precisely

Theorem 3.182 *Let M be a compact immersed hypersurface of $\mathbb{R}^{n+1}$. If the support function ϱ and the mean curvature α of M satisfy $1 + \alpha\,\varrho \leq 0$ then M is isometric to a sphere.*

Proof Given the immersion $\Psi : M \to \mathbb{R}^{n+1}$ the function $f = (1/2)\,||\psi||^2$ satisfies (3.231). Calculation of $\mathrm{div}(f \,\mathrm{grad} f)$ leads to

$$\mathrm{div}(f \,\mathrm{grad} f) = ||\mathrm{grad} f||^2 + \frac{n}{2}\,||\Psi||^2 (1 + \alpha\,\varrho). \tag{3.209}$$

Then [integrating over M]

$$\int_M ||\mathrm{grad} f||^2 \, d\,\mathrm{v}_g + \frac{n}{2} \int_M ||\Psi||^2 (1 + \alpha\varrho)\, d\,\mathrm{v}_g = 0. \tag{3.210}$$

As $1 + \varrho\,\alpha \leq 0$, Minkowski formula (3.201) gives $1 + \varrho\alpha = 0$ and then [by (3.210)] $\mathrm{grad} f = 0$ proving that M is isometric to a sphere. Q.e.d.

---

[82] Cf. *op. cit.*

By a result of Hsiung, one has the following integral formula [for a compact immersed hypersurface M in $\mathbb{R}^{n+1}$]

$$\int_M H_{s-1}\, d\,\mathrm{v}_g + \int_M H_s \varrho\, d\,\mathrm{v}_g = 0, \qquad s = 1, \ldots, n, \tag{3.211}$$

which includes Minkowski's formula (3.201).

S. Deshmukh established[83] the following:

Theorem 3.183 *Let M be a compact immersed hypersurface of $\mathbb{R}^{n+1}$. If the second fundamental form of M is negative definite and the support function ϱ, the mean curvature α, and the scalar curvature τ of M satisfy $n(n-1)\,\alpha + \varrho\,\tau \leq 0$ then M is isometric to a sphere.*

Proof Let $\Psi : M \to \mathbb{R}^{n+1}$ be the given isometric immersion and $f = \frac{1}{2}||\Psi||^2$. By (3.199)

$$A(\mathrm{grad}\, f) f = g(A(\mathrm{grad}\, f),\ \mathrm{grad}\, f)$$

and then [by (3.205)]

$$\begin{aligned} \mathrm{div}(f A(\mathrm{grad}\, f)) &= A(\mathrm{grad}\, f) f + f \mathrm{div}(A(\mathrm{grad}\, f)) \\ &= B(\mathrm{grad}\, f,\ \mathrm{grad}\, f) + nf(\mathrm{grad}\, f)\alpha + n\alpha f + f\varrho||A||^2. \end{aligned} \tag{3.212}$$

Now [by (3.209)]

$$\mathrm{div}(n\ f\ \alpha\, \mathrm{grad}\, f) = n\ f\ (\mathrm{grad}\, f)\alpha + n\, \alpha\, \mathrm{div}(f \mathrm{grad}\, f) = \tag{3.213}$$

$$= n\ f\ (\mathrm{grad}\, f)\alpha + n\, \alpha\, ||\mathrm{grad}\, f||^2 + \frac{n^2\alpha}{2}||\Psi||^2(1 + \alpha\varrho).$$

Substitution from (3.212) into (3.213) gives

$$\mathrm{div}(f\ A(\mathrm{grad}\, f) - n\ f\ \alpha\, \mathrm{grad}\, f) =$$

$$= B(\mathrm{grad}\, f, \mathrm{grad}\, f) + n\, \alpha\ f + \varrho\ f\ ||A||^2 +$$

$$- n\alpha||\mathrm{grad}\, f||^2 - n^2\, \alpha^2\, \varrho\, f - n^2\, \alpha\ f =$$

$$= B(\mathrm{grad}\, f,\ \mathrm{grad}\, f) - \varrho\ f\ \tau + n\, \alpha\ f - n\alpha||\mathrm{grad}\, f||^2 - n^2\, \alpha\ f =$$

$$= B(\mathrm{grad}\, f,\ \mathrm{grad}\, f) - f(n(n-1)\, \alpha + \varrho\,\tau) +$$

$$+ n(n-1)\ f\, \alpha + n\, \alpha\ f - n^2\, \alpha\ f - n\, \alpha\, ||\mathrm{grad}\, f||^2.$$

[83] Cf. *op. cit.*

Next [by integration over M]

$$\int_M [B(\operatorname{grad} f,\ \operatorname{grad} f) - n\,\alpha\,||\operatorname{grad} f||^2]\,d\,\mathrm{v}_g =$$

$$= \int_M f\big[n(n-1)\,\alpha + \varrho\,\tau\big]\,d\,\mathrm{v}_g.$$

If E is the unit vector field $E = ||\operatorname{grad} f||^{-1}\operatorname{grad} f$ [defined on the open set U where $\operatorname{grad} f$ is nonzero]

$$\int_M ||\operatorname{grad} f||^2[B(E,E) - n\,\alpha]\,d\,\mathrm{v}_g = \tag{3.214}$$

$$= \int_M f\big[n(n-1)\,\alpha + \varrho\tau\big]\,d\,\mathrm{v}_g.$$

Setting $s = 2$ in (3.211)

$$\int_M \big[n(n-1)\alpha + \varrho\tau\big]\,d\,\mathrm{v}_g = 0.$$

As $n(n-1)\,\alpha + \varrho\,\tau \leq 0$ one has $n(n-1)\,\alpha + \varrho\,\tau = 0$ and then (3.214) becomes

$$\int_M \|\operatorname{grad} f\|^2[B(E,E) - n\,\alpha]\,d\,\mathrm{v}_g = 0. \tag{3.215}$$

Let us complete E to a local orthonormal frame $\{E,\ E_2,\ \dots,\ E_n\}$ on U. Then

$$n\,\alpha = B(E,E) + B(E_2,\ E_2) + \cdots + B(E_n,\ E_n).$$

As the second fundamental form is negative definite $n\,\alpha < B(E,E)$ and then (3.215) yields $\|\operatorname{grad} f\| = 0$ proving that M is isometric to a sphere. Q.e.d.

3.18.2 Real Hypersurfaces in Euclidean Complex Space Forms

Let $\overline{M}(c)$ be a complex space form, of constant holomorphic sectional curvature c, and let J be its complex structure. Let M be an orientable real hypersurface in $\overline{M}(c)$ and let N be a unit normal vector field on M. The unit vector field $\xi \in \mathfrak{X}(M)$ determined by $J\xi = N$ is the *characteristic vector* field, or *Reeb vector* field, of M. The C^∞ function $f : M \to \mathbb{R}$, $f = g(A\xi,\ \xi)$, is the *characteristic function* of M. As customary g is the induced Riemannian metric on M and $A = A_N$ is the shape operator of the given immersion. For the sphere $S^n(c) \subset \mathbb{C}^{(n+1)/2}$ the characteristic function is a constant, i.e., $f = -\sqrt{c}$. The mean curvature α and the characteristic function of the real hypersurface $S^n(c)$ in $\mathbb{C}^{(n+1)/2}$ are related by

$$n^2 \alpha^2 = (n^2 - 1)c + f^2. \tag{3.216}$$

If k_0 is the infimum of the sectional curvature of $S^n(c) \subset C^{(n+1)/2}$ then Eq. (3.216) becomes

$$n^2\alpha^2 = (n^2 - 1)k_0 + f^2. \tag{3.217}$$

Our next task is to report on a result by Deshmuk [Des3], characterizing the sphere by means of Eq. (3.217).

Theorem 3.184 *Let M be a compact orientable real hypersurface in $\mathbb{C}^{(n+1)/2}$ and let $n(n-1)\delta$ be the infimum of the Ricci curvature of M. If the mean curvature α and the characteristic function f of M satisfy $n^2\alpha^2 = (n^2 - 1)\delta + f^2$, then α is a constant and M is the sphere $S^n(\alpha^2)$.*

The following lemmas are needed.

Lemma 3.185 *Let M be an orientable real hypersurface in $\mathbb{C}^{(n+1)/2}$ with characteristic function f. Then*

$$\nabla f = (\nabla A)(\xi, \xi) - 2Au.$$

Proof The definition of u and (3.311) imply $v = \phi A\xi$. Applying ϕ to both sides and using $\phi(v) = v$ gives

$$A\xi = -v + f\xi. \tag{3.218}$$

Next

$$(\nabla A)(X, \xi) + A\phi AX = -\nabla_X v + X(f)\xi + f\phi AX.$$

Let us take the inner product with ξ. As v is orthonormal to ξ

$$g((\nabla A)(\xi, \xi), X) - g(Au, X) = g(Au, X) + X(f).$$

Q.e.d. [Here one used $g((\nabla A)(X, Y), Z) = g(Y, (\nabla A)(X, Z))$ and $A\xi = -v + f\xi$ together with the definitions of u and v].

Lemma 3.186 *Let M be an orientable real hypersurface in $\mathbb{C}^{(n+1)/2}$, with characteristic function f. The vector fields u and v satisfy*

$$\nabla_X v = fAX - g(AX, A\xi)\xi + \phi(\nabla A)(\xi, X) + (\phi A)^2 X, \tag{3.219}$$

and

$$\nabla_X v = X(f)\xi + f\phi AX - (\nabla A)(\xi, X) - A\phi AX. \tag{3.220}$$

Proof Note that

$$\nabla_X v = R(X, \xi)\xi + \nabla_\xi \phi A X + \phi A[X, \xi].$$

Then (3.219) follows from (3.187) and (3.311). Moreover

$$\nabla_X v = (\nabla_X \phi)(v) + \phi \nabla_X u$$

and (3.219) yield (3.220). Q.e.d.

Lemma 3.187 *Let M be a compact orientable hypersurface in $\mathbb{C}^{(n+1)/2}$. Then*

$$\int_M \left[Ric(\xi, \xi) + \operatorname{tr}(\phi A)^2 \right] d\, \mathrm{v}_g = 0.$$

Proof Let us consider a local orthonormal frame of the form

$$\{E_1, \ldots, E_n\} \equiv \{E_1, \ldots, E_k, \phi E_k, \ldots, \phi E_k, \xi\}, \quad 2k = n - 1,$$

(an *adapted frame* of τ_M). Then

$$\begin{aligned} \sum_{i=1}^n g(\phi(\nabla A)(\xi, E_i), E_i) &= -\sum_{i=1}^n g((\nabla A)(\xi, E_i), \phi E_i) \\ &= -\sum_{i=1}^k g((\nabla A)(\xi, e_i), \phi e_i) + \sum_{i=1}^k g((\nabla A)(\xi, \phi e_i), e_i) \\ &= 0, \end{aligned}$$

where we have used $g((\nabla A)(X, Y), Z) = g(Y, (\nabla A)(X, Z))$ and the Codazzi equation. Now, we use (3.219) to compute the divergence of vector field u to arrive at [by (3.188)]

$$\operatorname{div} v = \operatorname{Ric}(\xi, \xi) + \operatorname{tr}(\phi A)^2.$$

Integrating over M leads to the result. Q.e.d.

Proof of Theorem 3.184 Let M be a compact orientable real hypersurface in $\mathbb{C}^{(n+1)/2}$. As the operator $A\phi - \phi A$ is symmetric

$$||A\phi - \phi A||^2 = tr(A\phi - \phi A)^2.$$

Also

$$\operatorname{tr}(\phi A)^2 = \operatorname{tr}(A\phi)^2.$$

Thus [by (3.181)]

$$||A\phi - \phi A||^2 = 2tr(\phi A)^2 + 2||A||^2 - 2||A\xi||^2. \tag{3.221}$$

Next we use (3.188), (3.309) and (3.221) in Lemma 3.215 to conclude that

$$\int_M \left[\mathrm{Ric}(\xi, \xi) + \tau + \frac{1}{2}||A\phi - \phi A||^2 + ||A\xi||^2 - n^2\alpha^2\right] d\,\mathrm{v}_g = 0. \tag{3.222}$$

As $(n-1)\delta$ is the infimum of the Ricci curvature

$$\tau \geq n(n-1)\delta, \quad ||A\xi||^2 \geq f^2.$$

Equation (3.222) yields

$$\int_M \left[n(n-1)\delta + f^2 + \frac{1}{2}||A\phi - \phi A||^2 - n^2\alpha^2\right] d\,\mathrm{v}_g \leq 0.$$

Thus [by $A\phi = \phi A$]

$$(n^2 - 1)\,\delta + f^2 = n^2\alpha^2. \tag{3.223}$$

Also [by (3.222)]

$$\mathrm{Ric}(\xi, \xi) = (n-1)\delta, \qquad \tau = n(n-1)\delta. \tag{3.224}$$

As ϕ and A commute it must be $\phi A\xi = 0$, yielding $A\xi = f\xi$ and consequently $v = v = 0$. Then

$$A^2 - fA = 0.$$

Consequently there are only two principal curvatures, 0 and f, at each point. However, as M is compact, there is a point $x_0 \in M$ where all principal curvatures have the same sign (say $+1$). Then all principal curvatures have the same value f_0 at that point. In a neighborhood of x_0 shape operator A must have rank n, so $A = f_0\, I$ holds near x_0. The set of points where $A = f_0 I$ is open and closed and therefore it must equal the whole of M. This completes the proof of the theorem.

3.18.3 Compact Submanifolds in Euclidean Spaces

Let M be a n-dimensional Riemannian submanifold in $\mathbb{R}^{n+p}$ and let $\Psi : M \to \mathbb{R}^{n+p}$ be the given immersion. Let $\overline{g} = \langle \cdot, \cdot \rangle$ and $\overline{\nabla}$ be the Euclidean metric and connection on $\mathbb{R}^{n+p}$. Also g and ∇ be the induced metric and connection on M. Then

$$\overline{\nabla}_X Y = \nabla_X Y + B(X, Y), \overline{\nabla}_X N = -A_N(X) + \nabla^{\perp}_X N \tag{3.225}$$

for any $X, Y \in \mathfrak{X}(M)$ and $N \in C^{\infty}(\mu)$. We shall also make use of the Gauss and Codazzi equations

$$\begin{cases} R(X, Y, Z, W) = g(B(Y, Z),\ B(X, W)) - g(B(X, Z),\ B(Y, W)), \\ \\ (DB)(X, Y, Z) = (DB)(Y, Z, X) = (DB)(Z, X, Y). \end{cases} \tag{3.226}$$

The Ricci tensor Ric of the submanifold is given by

$$\mathrm{Ric}(X, Y) = n\, g(H,\ B(X, Y)) - \sum_{i=1}^{n} g(B(X, E_i),\ B(Y, E_i)), \tag{3.227}$$

where $\{E_1, \ldots, E_n\}$ is a local orthonormal frame and

$$H = \frac{1}{n} \sum_{i=1}^{n} B(E_i,\ E_i).$$

A constant vector field $\nu \in \mathfrak{X}(R^{n+p})$ can be written as $\nu = \upsilon + v$, where $\upsilon \in \mathfrak{X}(M)$ is the tangential component of ν and $v \in \mathbb{C}^{\infty}(\mu)$ is the normal component of ν. Then [by (3.225)]

$$\nabla_X \upsilon = A_v(X), \qquad \nabla^{\perp}_X v = -B(X, \upsilon) \tag{3.228}$$

for any $X \in \mathfrak{X}(M)$. Here A_v denotes the Weingarten operator with respect to the normal vector field v. Let $\varphi : M \to \mathbb{R}$ be a C^{∞} function. The Hessian operator is

$$O : \mathfrak{X}(M) \to \mathfrak{X}(M), \quad g(OX, Y) = H_{\varphi}(X, Y), \quad \forall\ X, Y \in \mathfrak{X}(M).$$

Here H_{φ} is the Hessian of φ. Next we set

$$\zeta(\nu, M) = \int_M [(n-1)\, \| O + c\, A\|^2 - 2\, n\, c\, (\mathrm{tr}(AO) + \|\nabla\varphi\|^2)]\, d\, \mathrm{v}_g \tag{3.229}$$

where $c = 1/(n-1) \inf\ \mathrm{Ric}$ [and $\nabla\varphi = \mathrm{grad}(\varphi)$].

Next we report on a result by Deshmukh [Des4], characterizing spheres as n-dimensional compact submanifolds of positive curvature in $\mathbb{R}^{n+p}$ satisfying $\zeta(\xi, M) = 0$ for any constant vector field $\xi \in \mathfrak{X}(R^{n+p})$ and some non constant C^{∞} function φ.

Theorem 3.188 *Let M be a n-dimensional compact submanifold in $\mathbb{R}^{n+p}$, with the mean curvature vector H. Let* $\inf\left\{[1/(n-1)]\,\mathrm{Ric}\right\} = c > 0$ *and let $\xi \in \mathfrak{X}(\mathbb{R}^{n+1})$ be a constant vector field with $\varphi = < H, \xi >$. If φ is not a constant and*

$$\zeta(\xi, M) = \int_M [(n-1)||\mathcal{B} + cA||^2 - 2nc(tr(A\mathcal{B}) + ||\nabla\varphi||^2)]\, d\, \mathrm{v}_g = 0,$$

then M is isometric to $S^n(c)$.

We shall need the following lemmas.

Lemma 3.189 *Let M^n be a compact submanifold of $\mathbb{R}^{n+p}$ and let $\nu \in \mathbb{R}^{n+p}$ be a constant vector field. Then $\int_M \varphi \, d\,\mathrm{v}_g = 0$.*

Proof Let us consider the function $\varphi =< H, \nu >$ where $\nu \in \mathfrak{X}(\mathbb{R}^{n+p})$. The fact that ν is parallel yields the result. Q.e.d.

Lemma 3.190 *Let M^n be a compact submanifold of $\mathbb{R}^{n+p}$ and let $\nu \in R^{n+p}$ be a constant vector field. Then*

(a) $\operatorname{tr}(A) = n\,\varphi$,

(b) $(\nabla A)(X, Y) - (\nabla A)(Y, X) = R(X, Y)u$,

(c) $\sum_i (\nabla A)(E_i, E_i) = n\nabla\varphi + Q(\upsilon)$.

Here $\{E_1, \ldots, E_n\}$ is a local orthonormal frame of τ_M. Also

$$Ric(X, Y) = g(Q(X), Y)), \quad (\nabla A)(X, Y) = \nabla_X AY - A(\nabla_X Y),$$

$$\forall \; X, Y \in \mathfrak{X}(M).$$

[Q is the *Ricci operator*].

Lemma 3.191 *Let M^n be a compact submanifold of $\mathbb{R}^{n+p}$ and let $\nu \in \mathbb{R}^{n+p}$ be a constant vector field. Then*

$$\int_M [\mathrm{Ric}(\upsilon, \upsilon) + ||A||^2 - n^2 \varphi^2] \, d\,\mathrm{v}_g = 0.$$

Proof Let $\{E_1, \ldots, E_n\}$ be a pointwise constant local orthonormal frame. By Lemma 3.190 the divergence of the vector field $A(\upsilon)$ is

$$\operatorname{div} A\upsilon = \sum_i E_i g(\upsilon, AE_i) = ||A||^2 + n\, g(\upsilon, \nabla\varphi) + \mathrm{Ric}(\upsilon, \upsilon). \tag{3.230}$$

On another hand [by (3.228)]

$$g(\upsilon, \nabla\varphi) = \operatorname{div}(\varphi\upsilon) - n\varphi^2. \tag{3.231}$$

Now (3.230) and (3.231) imply the result. Q.e.d.

Lemma 3.192 *Let M^n be a compact submanifold of $\mathbb{R}^{n+p}$ and let $\nu \in \mathbb{R}^{n+p}$ be a constant vector field. Then*

$$\int_M [Ric(\nabla\varphi, v) + \sum_i g(\nabla_{e_i}\nabla\varphi, Ae_i) + n||\nabla\varphi||^2]\, d\,\mathrm{v}_g = 0,$$

where $\{E_1, \ldots, E_n\}$ *is a local orthonormal frame of* τ_M.

Proof By (3.227) and (3.228)

$$\begin{aligned}\mathrm{Ric}(\nabla\varphi, v) =& n\, g(H,\ B(\nabla\varphi, v)) - \sum_{i=1}^n g(B(\nabla\varphi,\ E_i),\ B(v, E_i))\\ =& - n||\nabla\varphi||^2 + n\, g(\nabla^\perp_{\nabla\varphi} H, v) + \sum_i E_i\big(g(A(\nabla\varphi), E_i)\big)\\ &- \sum_i g((DB)(e_i, \nabla\varphi, E_i), v) - \sum_i g(B(\nabla_{E_i}\nabla\varphi, E_i), v).\end{aligned}$$

Using $\sum_i (Dh)(\nabla\varphi, E_i,\ E_i) = n\ \nabla^\perp_{\nabla\varphi} H$

$$\mathrm{Ric}(\nabla\varphi, v) = -n\ ||\nabla\varphi||^2 + \operatorname{div} A(\nabla\varphi) - \sum_i g(\nabla_{E_i}\nabla\varphi, AE_i).$$

Let $\Delta\varphi$ be the Laplacian of φ, i.e., $\Delta\varphi = -\operatorname{div}(\nabla\varphi)$.

Lemma 3.193 *The operator* O *enjoys the properties*

(a) $\mathrm{tr}(O) = \nabla\varphi$,

(b) $(\nabla O)(X, Y) - (\nabla O)(Y, X) = R(X, Y)\nabla\varphi$,

(c) $\sum_i (\nabla O)(E_i,\ E_i) = \nabla(\Delta\varphi) + Q(\nabla\varphi)$,

where $\{E_1, \ldots, E_n\}$ *is a local orthonormal frame of* τ_M.

Lemma 3.194 *Let* (M, g) *a* n*-dimensional compact Riemannian manifold. Then*

$$\int_M [Ric(\nabla\varphi, \nabla\varphi) + ||O||^2 - (\Delta\varphi)^2]\, d\,\mathrm{v}_g = 0.$$

Proof of Theorem 3.188. By Lemmas 3.191, 3.192 and 3.194 together with

$$||O||^2 \geq \frac{(\Delta\varphi)^2}{n}, \quad ||A||^2 \geq n\,\varphi^2,$$

one has

$$\int_M [\mathrm{Ric}(\nabla\varphi + cu, \nabla\varphi + cv)]\, d\,\mathrm{v}_g =$$

$$= \int_M [\mathrm{Ric}(\nabla\varphi) + c^2\,\mathrm{Ric}(v, v) + 2c\,\mathrm{Ric}(\nabla\varphi, v)]\, d\,\mathrm{v}_g =$$

$$= \int_M [-||O||^2 + (\Delta\varphi)^2 - c^2||A||^2 + n^2\,c^2\varphi^2 +$$

$$-\,2c\sum_i g(\nabla_{E_i}(\nabla\varphi), AE_i) - 2nc||\nabla\varphi||^2]\, d\,\mathrm{v}_g \le$$

$$\le \int_M [(n-1)\,(||O||^2 + c^2\,||A||^2) - 2\,n\,c\,||\nabla\varphi||^2 +$$

$$-\,2c\sum_i g(\nabla_{E_i}(\nabla\varphi), AE_i)]\, d\,\mathrm{v}_g\,.$$

Next, by

$$\sum_i g(\nabla_{E_i}(\nabla E_i),\ AE_i) = \mathrm{tr}(AO)$$

and

$$||O + cA||^2 = ||O||^2 + c^2A^2 + 2c\,\mathrm{tr}(AO)$$

the last inequality becomes

$$\int_M [\mathrm{Ric}(\nabla\varphi + cu\,\nabla\varphi + c\,v)]\, d\,\mathrm{v}_g \le \zeta(\nu, M).$$

Consequently $\nabla\varphi = -c\,u$. Moreover [by Lemma 3.192 with $\zeta(\nu, M) = 0$]

$$\int_M [\mathrm{Ric}(\nabla\varphi, \nabla\varphi) - c(n-1)\,||\nabla\varphi||^2]\, d\,\mathrm{v}_g = 0.$$

As

$$\mathrm{Ric}(\nabla\varphi, \nabla\varphi) \ge (n-1)c||\nabla\varphi||^2$$

one has

$$\mathrm{Ric}(\nabla\varphi, \nabla\varphi) = (n-1)c||\nabla\varphi||^2\,.$$

so that [by $\nabla\varphi = -c\,u$ and Lemma 3.191]

$$\int_M [(n-1)||\nabla\varphi||^2 - n(n-1)c\varphi^2 + c(||A||^2 - n\varphi^2)]\, d\,\mathrm{v}_g = 0. \tag{3.232}$$

Thus (3.232) becomes

$$(n-1)(\lambda_1 - nc)\int_M \varphi^2 d\,\mathrm{v}_g + c\int_M (||A||^2 - n\varphi^2)\, d\,\mathrm{v}_g \leq 0. \tag{3.233}$$

Here we also used

$$\int_M ||\nabla\varphi||^2 d\,\mathrm{v}_g \geq \lambda_1 \int_M \varphi^2 d\,\mathrm{v}_g$$

where λ_1 is the first nonzero eigenvalue of the Laplacian. On the other hand [by Lichnerowicz's Theorem]

$$\lambda_1 \geq n\,c$$

with equality if and only if M is isometric to $S^n(c)$. Thus (3.233) yields the theorem. □

Corollary 3.195 *Let M be a n-dimensional compact submanifold of $\mathbb{R}^{n+p}$, with the mean curvature vector H. Let us assume* $\inf\{[1/(n-1)\,\mathrm{Ric}\} = c > 0$ *and let $\xi \in \mathfrak{X}(\mathbb{R}^{n+1})$ be a constant vector field with $\varphi =< H, \xi >$. If φ is not a constant and*

$$\zeta(\xi, M) = \int_M [(n-1)\,||\mathcal{B} + c\,A||^2 - 2\,n\,c\,\mathrm{tr}(A\mathcal{B})]\, d\,\mathrm{v}_g = 0,$$

then M is a submanifold of the hyperplane $\mathbb{R}^{n+p-1}$.

Let $\Psi : M^n \to \mathbb{R}^{n+p}$ be the given immersion. Then $\Psi = \Psi^T + \Psi^\perp$, where $\Psi^T \in \mathfrak{X}(M^n)$ and $\Psi^\perp \in C^\infty(\tau_M^\perp)$. Then [by (3.226)]

$$\nabla_X \Psi^T = X + A_{\Psi^\perp} X, \qquad \nabla_X^\perp \Psi^\perp = -B(X, \Psi^T) \tag{3.234}$$

for any $X \in \mathfrak{X}(M^n)$. On the other hand let $F \in C^\infty(M^n)$ be the function

$$F : M^n \to \mathbb{R}, \quad F =< H, \Psi^\perp > .$$

Then [by (3.234)] $\mathrm{div}\Psi^T = n\,(1 + F)$. Consequently

Lemma 3.196 *Let M^n be a compact orientable submanifold of $\mathbb{R}^{n+p}$. Then*

$$\int_{M^n} (1 + F)\, d\,\mathrm{v}_g = 0.$$

Together with Haila et al. [OdDe], we consider the symmetric (1, 1) tenor field $\mathcal{B}$ given by

$$\mathcal{B}(X) = A_{\Psi^\perp}(X), \quad X \in \mathfrak{X}(M^n).$$

Lemma 3.197 *$\mathcal{B}$ enjoys the following properties*

(a) $\operatorname{tr}(\mathcal{B}) = n\, F$,

(b) $(\nabla\mathcal{B})(X, Y) - (\nabla\mathcal{B})(Y, X) = R(X, Y)\Psi^T$,

(c) $\sum_i (\nabla\mathcal{B})(E_i,\, E_i) = n \operatorname{grad} F + Q(\Psi^T)$,

where $\{E_1, \ldots, E_n\}$ *is a local orthonormal frame of* τ_{M^n} *and*

$$Q(X) = n\, A_H X - \sum_i A_{B(E_i, X)} E_i, \quad (\nabla\mathcal{B})(X, Y) = \nabla_X \mathcal{B}(Y) - \mathcal{B}(\nabla_X Y)$$

for any $X, Y \in \mathfrak{X}(M^n)$.

Proof Statement (a) follows from the very definition of $\mathcal{B}$. As $\mathcal{B}$ is symmetric

$$g((\nabla\mathcal{B})(X, Y), Z) = g(Y, (\nabla\mathcal{B})(X, Z)). \tag{3.235}$$

Then [by (3.234) and (3.235)]

$$g((\nabla\mathcal{B})(X, Y) - (\nabla\mathcal{B})(Y, X), Z) = g(R(X, Y)\Psi^T, Z)$$

proving (b). To prove (c) note first that [by (a)]

$$\sum_i g(\mathcal{B}E_i,\, E_i) = n\, F.$$

Thus for any $X \in \mathfrak{X}(M^n)$

$$\sum_i g((\nabla\mathcal{B})(X,\, E_i),\, E_i) = n\, X(F) = n\, g(X < grad F).$$

By (b) and the definition of the Ricci tensor one obtains (c). Q.e.d.

We shall need the following integral formula (by Haila et al. [OdDe]).

Theorem 3.198 *Let (M^n, g) be a compact, orientable, positively curved Riemannian manifold. If there exists a positive constant $r > 0$ such that the scalar curvature τ of M^n satisfies $\tau < n(n-1)r^{-2}$ then there is no isometric immersion $M^n \to \mathbb{R}^{n+2}$ with a parallel mean curvature vector field, and contained in a ball of radius r.*

Proposition 3.199 *Let M^n be a compact orientable Riemannian manifold and $\Psi : M^n \to \mathbb{R}^{n+p}$ an isometric immersion. Then*

$$\int_{M^n} [\mathrm{Ric}(\Psi^T, \Psi^T) + ||\mathcal{B}||^2 - n^2 F^2 + n(n-1)]\, d\, \mathrm{v}_g = 0. \tag{3.236}$$

Proof Let us consider a local orthonormal frame $\{E_1, \ldots, E_n\}$ such that $(\nabla E_i)_{x_0} = 0$ for some $x_0 \in M$. Then [by Lemma 3.204] we may compute $\mathrm{div}(\mathcal{B}\Psi^T)$ and obtain

$$\mathrm{div}(\mathcal{B}\Psi^T) = n\, F + ||\mathcal{B}||^2 + n\, g(\Psi^T,\ \mathrm{grad} F) + \mathrm{Ric}(\Psi^T, \Psi^T). \tag{3.237}$$

We have

$$\mathrm{div}(F\Psi^T) = g(\psi^T,\ \mathrm{grad} F) + F\, \mathrm{div}\psi^T\, .$$

By the first equation in (3.234) $\mathrm{div}\psi^T = n\,(1+F)$. Thus

$$\mathrm{div}(F\psi^T) = g(\psi^T,\ \mathrm{grad} F) + n\, F + n\, F^2\, .$$

Then [by replacing $g(\psi^T,\ \mathrm{grad} F)$ in (3.237)]

$$\mathrm{div}(\mathcal{B}\psi^T - n\, F\, \psi^T) = -n(n-1)\, F - n^2\, F^2 + ||\mathcal{B}||^2 + \mathrm{Ric}(\psi^T, \psi^T).$$

Integration over M yields [by Lemma 3.196] the result. Q.e.d.

Let Ψ be an isometric immersion of a compact manifold with parallel mean curvature vector field, i.e., $\nabla^\perp H = 0$. Then $||H|| = \alpha$ for some $\alpha \in (0, +\infty)$. Let $\{N_1, N_2\} \subset C^\infty(U,\ \tau_M^\perp)$ be a local orthonormal frame[84] such that $H = \alpha N_1$. Then

$$\Psi^T = \frac{F}{\alpha} N_1 + f\, N_2, \qquad f = < \Psi^T, N_2 > . \tag{3.238}$$

As $\nabla_X^\perp H = 0$ it follows that $\nabla_X^\perp N_1 = 0$ and then [by $N_1 \perp N_2$] one has $\nabla_X^\perp N_2 = 0$. Let us set

$$\mathcal{C}(X) = A_{N_2}(X), \quad \forall\ X \in \mathfrak{X}(M^n).$$

It may be easily shown that $\mathcal{C}$ satisfies

$$\mathrm{tr}\,(\mathcal{C}) = 0, \quad (\nabla\mathcal{C})(X, Y) = (\nabla\mathcal{C})(Y, X), \quad \sum_i (\nabla\mathcal{C})(E_i,\ E_i) = 0 \tag{3.239}$$

for any $X, Y \in \mathfrak{X}(M^n)$.

Proof of Theorem 3.198. Let $\Psi : M^n \to \mathbb{R}^{n+2}$ be an isometric immersion with parallel mean curvature vector and $||\psi|| \le r$. Let us consider the C^∞ function

[84] One tacitly assumes that M has codimension 2.

$$\varphi : M^n \to \mathbb{R}^{n+2}, \quad \varphi = \frac{1}{2}||\mathcal{C}||^2.$$

The Laplacian of φ is given by

$$\Delta\varphi = ||\nabla\mathcal{C}||^2 + \frac{1}{2}\sum_{i<j}(a_i - a_j)^2 K_{ij}$$

where a_i are the eigenvalues of $\mathcal{C}$. Integrating the above equation and using $K_{ij} > 0$, we get $\nabla\mathcal{C} = 0$. This proves that $\mathcal{C} = aI$ which, together with $\mathrm{tr}\,(\mathcal{C}) = 0$, gives $\mathcal{C} = A_{N_2} = 0$. Now, from (3.238), for any $X \in \mathfrak{X}(M^n)$, we get

$$X(f) = X < \Psi^\perp, N_2 >= - < B(X, \Psi^T), N_2 >= 0,$$

consequently, f is constant, call it β. Thus

$$||\Psi||^2 = ||\Psi^T||^2 + \frac{F^2}{\alpha^2} + \beta^2, \quad \beta = A_{\Psi^\perp} = \frac{F}{\alpha}A_{N_1} \tag{3.240}$$

so that

$$||\mathcal{B}||^2 = \frac{F^2}{\alpha^2}||A_{N_1}||^2.$$

Using this in Proposition 3.199 we arrive at

$$\int_{M^n}[Ric(\Psi^T, \Psi^T) + \frac{F^2}{\alpha^2}(||A_{N_1}||^2 - n^2\alpha^2) + n(n-1)]\,d\,\mathrm{v}_g = 0.$$

The length $||h||^2$ of the second fundamental form is given by

$$||h||^2 = \sum_{\gamma=1}^{2}||A_{N_\gamma}||^2$$

and $A_{N_2} = 0$. Thus $||A_{N_1}||^2 = ||h||^2$ and consequently the above integral becomes

$$\int_{M^n}[Ric(\Psi^T, \Psi^T) + (n(n-1) - \frac{F^2}{\alpha^2}\tau)]\,d\,\mathrm{v}_g = 0,$$

where

$$\tau = n^2||H||^2 - ||h||^2$$

is the scalar curvature of M^n. Using (3.240) and $||\Psi|| \leq r$ in the above integral we get

$$\int_{M^n} [(Ric(\Psi^T, \Psi^T) + ||\Psi^T||^2\tau + \beta^2\tau) + r^2(n(n-1)r^{-2} - \tau)]\, d\,\mathrm{v}_g \leq 0. \tag{3.241}$$

The first term is non-negative and $\tau < n(n-1)r^{-2}$ hence the integrand is strictly positive, a contradiction.

As a consequence of Theorem 3.198.

Corollary 3.200 *Let (M^n, g) be a positively curved, compact orientable submanifold of $\mathbb{R}^{n+2}$, with a parallel mean curvature vector. If M^n is contained in a ball of radius r in $\mathbb{R}^{n+2}$ and the scalar curvature τ of M^n satisfies $\tau \leq n(n-1)r^{-2}$, then M^n*

(a) *lies on the hypersphere $S^{n+1}(r^{-2})$,*

(b) *is isometric to $S^n(r^{-2})$.*

Corollary 3.201 *Let (M^n, g) be a positively curved, compact orientable submanifold of $\mathbb{R}^{n+2}$, with a parallel mean curvature vector field H. If the scalar curvature τ of M^n is constant, then M^n is isometric to $S^n(\alpha^2)$, where $\alpha = ||H||$.*

Corollary 3.202 *Let (M^n, g) be a positively curved, compact orientable submanifold of $\mathbb{R}^{n+2}$, with a parallel mean curvature vector field. If $Ric(\Psi^T, \Psi^T) \geq n^2(1 + F^2)$, then M^n is isometric to some sphere $S^n(c)$.*

3.19 Submanifolds of Spheres

An ample portion of modern theory of isometric immersions is devoted to submanifolds of arbitrary codimension in spheres. Tompkins (cf. [Tom]), Chern and Kuiper (cf. [ChKu]), Otsuki (cf. [Ots23]), and H. Jacobowitz, cf. *op. cit.*, proved non existence theorems for isometric immersions of Riemannian manifolds into certain (balls in) Euclidean spaces, in the presence of assumptions on the sectional curvature. These works turn out to be motivational for Beltagy [Bel]. In 1983, he proved that there is no isometric immersion of a compact n-dimensional Riemannian manifold, whose sectional curvatures are everywhere less than some constant, into S^{n+p} and contained in some geodesic ball of S^{n+p} of radius r. However, so far there were no such results in the presence of requirements on the Ricci curvature. In 1994, Deshmukh and Al-Gwaiz [DeGw], considered the question of non embeddability in the presence of bounds on the Ricci curvature. Theorem 3.203 is proved for a compact Riemannian manifold of codimension 1 in the odd dimensional sphere.

De Giorgi [Gio], and Simons [Sim24], showed that the Gauss image of a minimal hypersurface other than a great hypersphere cannot lie in an open hemisphere. In 1969, Nomizu and Smyth [NoSm], generalized the above result to the hypersurfaces of constant mean curvature. For this, they obtained a characterization of the hyperspheres (great or small) of spheres by using their Gauss images. Later on, Markvorsen (cf. [Mar]) also obtained another characterization of small hyperspheres of spheres. A characterization of small spheres in a sphere was obtained by Coghlan and Itokawa [CoIt], in terms of a pinching relation on the sectional curvature. Drawing inspiration from the above works, Deshmukh and Al-Gwaiz [DsGw], used a pinching on the Ricci curvature and the position of a hypersurface in a sphere to characterize a small sphere (see Theorem 3.206 below).

In the geometry of submanifolds, an interesting question is to obtain conditions under which minimal submanifolds of the unit sphere S^{n+p} are totally geodesic. Generally, these conditions are in terms of pinching relations on the scalar, Ricci, or sectional curvatures. U. Simon proved a pinching theorem for compact minimal submanifolds for arbitrary codimension in the unit sphere. This result was improved by Sakaki [Sak25], in 1989. He showed that if the scalar curvature of a minimal submanifold M^n in S^{n+p} is greater than some constant, then either M^n is totally geodesic or for $n = 2$, the scalar curvature is $\tau = \frac{2}{3}$ and M^2 is a Veronese surface in a totally geodesic 4-sphere.

In 1979, for $n \geq 4$, N. Ejiri proved [Eji], that a n-dimensional minimal submanifold, whose Ricci curvature satisfies $\mathrm{Ric} > n - 2$, in S^{n+p} is totally geodesic. Later on, in 1992, Y-B. Shen improved (cf. [She]) the result of M. Sakaki for 3-dimensional compact minimal submanifolds of S^{3+p} with a pinching relation on the scalar curvature, i.e., $\tau > 4$.

S. Deshmukh proved (cf. [Des11]) Theorem 3.209, which can be considered as a partial generalization of the result by Y-B. Shen. It was an interesting question whether Shen's result could be extended beyond dimension 3. In [Des12], he also showed that for $p \leq 2$, an n-dimensional minimal submanifold in S^{n+p}, with a pinching relation on the scalar curvature, i.e., $\tau > n(n-2)$, is totally geodesic (see Theorem 3.216 below). For $p \geq 2$, he used some additional restrictions on the curvature tensor with respect to normal connection to achieve the result (see Theorem 3.212 below). Furthermore, in [Des12], he obtained an inequality for the scalar curvature of a non-totally geodesic, non-negatively curved minimal submanifold in S^{n+p} with a flat normal connection (see Theorem 3.219 below). From Corollary 3.223, one may conclude that a non-negatively curved minimal hypersurface of S^{n+1}, which is diffeomorphic to S^n, is a totally geodesic sphere.

According to Chen [Che5], a submanifold M of the Euclidean space $\mathbb{R}^n$ s *biharmonic* if its mean curvature vector is harmonic i.e. $\Delta H = (\Delta H_1, \ldots, \Delta H_n) = 0$ where $H = (H_1, \ldots, H_n)$ is the mean curvature vector and Δ is the Laplace-Beltrami operator on M.

The study of biharmonic immersions into a non-flat Riemannian manifold may be regarded as a generalization of B.-Y. Chen's study of biharmonic submanifolds. Jiang showed (cf. [Jia]) that a compact, orientable, biharmonic submanifold into a manifold M such that $R^M \leq 0$ is minimal. C. Oniciuc proved[85] that CMC biharmonic isometric immersions into a manifold M with $R^M \leq 0$ are minimal.

---

[85] Cf. C. Oniciuc, Biharmonic maps between Riemannian manifolds. An. Ştiinţ. Univ. *Al.I. Cuza*, Iaşi, Mat (N.S.) **48**, 237–248 (2002).

Chen's conjecture is that biharmonic submanifolds into a non positive constant sectional curvature manifold are minimal. Up to 2001 Chen's conjecture was expected to hold for biharmonic submanifolds into a Riemannian manifold of non positive sectional curvature, yet Ou and Tang produced (cf. [OuTa]) a counterexample, prompting the investigation of biharmonic submanifolds of spheres.

3.19.1 Hypersurfaces in Spheres

Let $S^{2n-1} \subset \mathbb{C}^n$ be the standard sphere. Let N be a unit normal vector field on S^{2n-1}. A rotation of N by $\pi/2$ gives rise to the Reeb vector field $\xi \in \mathfrak{X}(S^{2n-1})$ [i.e., $J\xi = N$ where J is the complex structure of $\mathbb{C}^n$]. Let ϕ and η be the natural f-structure and contact form on the sphere i.e. $JX = \phi X + \eta(X)\, N$. Moreover

$$\begin{cases} \phi^2 = -I + \eta \otimes \xi, \quad \phi(\xi) = 0, \quad \eta \circ \xi = 0, \\ \\ g(\phi X, \phi Y) = g(X, Y) - \eta(X)\eta(Y), \quad g(\phi X, Y) = -g(X, \phi Y), \end{cases} \tag{3.242}$$

$$\overline{\nabla}_X \xi = \phi X, \quad (\overline{\nabla}_X \phi)(Y) = \eta(Y) X - g(X, Y)\xi \tag{3.243}$$

for any $X, Y \in \mathfrak{X}(S^{2n-1})$. That is S^{2n-1} is a Sasakian manifold with the contact metric structure (ϕ, ξ, η, g), where g and $\overline{\nabla}$ are respectively the standard Riemannian metric and Levi-Civita connection on S^{2n-1}. Let (M, g) be a $(2n-2)$-dimensional Riemannian manifold admitting an isometric immersion in S^{2n-1}. For the unit normal vector field to M in S^{2n-1} we have

$$\xi|_M = u + f\overline{N}, \quad \phi X = \Psi X + \alpha(X)\overline{N}, \tag{3.244}$$

$$f = g(\xi, \overline{N}), \quad \alpha(X) = g(X, t), \quad \phi\overline{N} = -t.$$

Then [by (3.243) and (3.244)]

$$\nabla_X u = \Psi X + f\, A\, X \tag{3.245}$$

for any $X \in \mathfrak{X}(M)$, where $A = A_N$ is the Weingarten operator of the given isometric immersion. On the other hand

$$\nabla_X t = -f\, X + \Psi\, A\, X. \tag{3.246}$$

Then [by (3.244)] we get $\Psi(u) = f\, t$ hence $g(u, t) = 0$. Finally [by (3.245) and (3.246)] we arrive at $\Psi(t) = -f\, u$.

Next we report on a result by Deshmukh and Al-Gwaiz (cf. [DeGw]) on the non-existence of an isometric immersion $M \to S^{2n-1}$ with JN tangent to the image of M. Precisely

Theorem 3.203 *Let M be a $(2n-2)$-dimensional compact Riemannian manifold. If M satisfies $\mathrm{Ric} < 2(n-2)$ then there is no isometric immersion of M into S^{2n-1} such that JN is tangent to the image of M.*

To prove the theorem we need the following:

Lemma 3.204 *Let M be a $(2n-2)$-dimensional submanifold of S^{2n-1}. Then*

(1) $\mathrm{trace}(\Psi^2) = -2f^2 - 2n + 4.$

(2) $||\nabla u||^2 = f^2\,\mathrm{trace}(A^2) + 2f^2 + 2n - 4.$

(3) $||\mathcal{L}g||^2 = 4f^2\,\mathrm{trace}(A^2).$

(4) $\mathrm{div}(u) = 2(n-1)\,f\,\alpha,$

where α is the mean curvature vector $M \to S^{2n-1}$.

Proof On an open subset of M, we consider the unit vector fields $\mathcal{E}_1$ and $\mathcal{E}_2$ collinear to u and t, respectively. Then for any unit vector field $\mathcal{E}$ orthogonal to $\mathcal{E}_1$ and $\mathcal{E}_2$

$$\begin{cases} g(\Psi\mathcal{E},\mathcal{E}_1) = -g(\mathcal{E},\Psi\mathcal{E}_1) = \dfrac{-1}{||u||}g(\mathcal{E},\Psi u) = \dfrac{-f}{||u||}||t||g(\mathcal{E},\mathcal{E}_2) = 0, \\[2ex] g(\Psi\mathcal{E},\mathcal{E}_2) = -g(\mathcal{E},\Psi\mathcal{E}_2) = \dfrac{-1}{||u||}g(\mathcal{E},\Psi t) = \dfrac{-f}{||t||}||u||g(\mathcal{E},\mathcal{E}_1) = 0. \end{cases} \tag{3.247}$$

Also $\Psi(u) = ft \Longrightarrow t = 0$ and $\Psi(t) = -fu \Longrightarrow u = 0$. Then $\xi = \overline{N}$. Thus, on an open set where u and t are nonzero, we may choose as a local orthonormal frame either

$$\{\mathcal{E}_1, \ldots, \mathcal{E}_{n+1}, \Psi\mathcal{E}_3, \ldots, \Psi\mathcal{E}_{n+1}\}$$

or

$$\{E_1, \ldots, E_{n-1}, \Psi E_1, \ldots, \Psi E_{n-1}\}.$$

For the first case we have

$$\sum_{i=1}^{2n-2} g(Ae_i, \Psi e_i) = g(A\mathcal{E}_1, \Psi\mathcal{E}_1) + g(A\mathcal{E}_2, \Psi\mathcal{E}_2) + \sum_{i=3}^{n+1}[g(Ae_i, \Psi e_i) + g(A\Psi e_i, \Psi^2 e_i)].$$

For $i > 2$, E_i is perpendicular to u and t, which further implies that $\alpha(E_i) = 0$ and $\eta(E_i) = g(E_i, t) = 0$. Hence $\phi E_i = \Psi(E_i)$ and $\Psi^2(E_i) = -E_i$. Thus

$$\sum_{i=1}^{2n-2} g(AE_i, \Psi E_i) = 0. \tag{3.248}$$

The remaining case may be treated in a similar manner. On the open subset where u and t are nonzero, we also have

$$\mathrm{trace}(\Psi^2) = \sum_{i=1}^{2n-2} g(\Psi E_i, \Psi E_i) = -2f^2 - 2n + 4. \tag{3.249}$$

By (3.245)

$$\mathrm{div}(u) = 2(n-1)f\alpha. \tag{3.250}$$

Next [by (3.248) and (3.249)]

$$\begin{aligned} ||\nabla u||^2 &= -\mathrm{trace}(\Psi^2) + f^2\,\mathrm{trace}(A^2) \\ &= f^2\,\mathrm{trace}(A^2) + 2f^2 + 2n - 4 \end{aligned} \tag{3.251}$$

and

$$\begin{aligned} ||\mathcal{L}g||^2 &= \sum_{i,j}[g(\Psi E_i + fAE_i, E_j) + g(\Psi E_j + fAE_j, E_i)]^2 \\ &= 4f^2\,\mathrm{trace}(A^2). \end{aligned} \tag{3.252}$$

Q.e.d.

Proof of Theorem 3.203 By Lemma 3.204

$$\mathrm{Ric}(u, u) + \frac{1}{2}||\mathcal{L}g||^2 - ||\nabla u||^2 - (\delta u)^2 =$$

$$= \mathrm{Ric}(u, u) + f^2((2n-2)(2n-3) - \tau) - 2f^2 - 2n + 4, \tag{3.253}$$

where the scalar curvature τ of M is given by

$$\tau = (2n-2)(2n-3) + 4(n-1)^2\alpha^2 - \mathrm{trace}(A^2).$$

Next [by $||u||^2 = 1 - f^2$]

$$\int_M \left[||u||^2(\mathrm{Ric}(E_1, E_1) - (2n-2)) + f^2((4n-4)(n-2) - \tau)\right] d\,\mathrm{v}_g = 0. \tag{3.254}$$

By $JN = \xi \in \mathfrak{X}(M)$ and (3.244) we get $f = 0$ and $||u|| = 1$. Hence

$$\int_M [\mathrm{Ric}(u, u) - (2n-2)]\, d\,\mathrm{v}_g = 0. \tag{3.255}$$

This [by our assumption on Ricci curvature] is in conflict with (3.255). Q.e.d.

3.19.2 Small Spheres in Spheres

Let M be an orientable hypersurface in the unit sphere $S^{n+1} \subset \mathbb{R}^{n+2}$. Let N be a global unit normal vector field of M in S^{n+1}. We shall use the Gauss and Weingarten formulas

$$\overline{\nabla}_X Y = \nabla_X Y + g(AX, Y)N, \tag{3.256}$$

$$\overline{\nabla}_X N = -AX \tag{3.257}$$

for any $X, Y \in \mathfrak{X}(M)$. Let $R = R^{\nabla}$ be the curvature tensor field of M. The Gauss equation is

$$R(X, Y)Z = g(AY, Z)AX - g(AX, Z)AY \tag{3.258}$$

for any $X, Y \in \mathfrak{X}(M)$. Let $\{E_1, \ldots, E_n\}$ be a local orthonormal frame in τ_M. Then [by (3.258)]

$$\mathrm{Ric}(X, Y) = (n-1)g(X, Y) + n\,\alpha\, g(AX, Y) - g(AX, AY) \tag{3.259}$$

for any $X, Y \in \mathfrak{X}(M)$. Moreover the scalar curvature of M is

$$\tau = n(n-1) + n^2\alpha^2 - \mathrm{trace}(A^2), \tag{3.260}$$

where α is the mean curvature of M

$$\alpha = \frac{1}{n}\sum_{i=1}^{n} g(AE_i,\ AE_i).$$

By the symmetry of the shape operator A and by the Codazzi equation

$$g((\nabla_X A)Y, Z) = g((\nabla_X A)Z, Y) \tag{3.261}$$

for any $X, Y, Z \in \mathfrak{X}(M)$. For any parallel unit vector field Z on $\mathbb{R}^{n+2}$

$$Z = Z^T + Z^N,$$

where $Z^T \in \mathfrak{X}(S^{n+1})$ and $Z^N \in C^\infty\left(\tau^\perp_{S^{n+1}}\right)$ are the tangential and normal components of Z. Let $\overline{N}$ be a unit normal vector field on S^{n+1} in $\mathbb{R}^{n+2}$. If we set

$$f =< Z^N, \overline{N} >$$

then

$$\overline{\nabla}_X Z^T = -fX, \qquad Xf = g(X, Z^T) \tag{3.262}$$

for any $X \in \mathfrak{X}(M)$. Furthermore, let us consider the function

$$\rho : M \to \mathbb{R}, \quad Z^T = t + \rho N, \quad t \in \mathfrak{X}(M).$$

We have

$$\begin{cases} \nabla_X t = -fX + \rho AX, \quad X\rho = g(AX, t), \quad Xf = g(X, t), \\ \\ \operatorname{grad}\rho = -At, \quad \operatorname{grad} f = t \end{cases} \tag{3.263}$$

for any $X \in \mathfrak{X}(M)$.

Next we report on a result by Deshmukh and Al-Gwaiz (cf. [DsGw]) on hypersurfaces of constant mean curvature in S^{n+1}.

Theorem 3.205 *Let M be a compact orientable hypersurface of constant mean curvature in S^{n+1}. If some relative support function of M with respect to a parallel unit vector field Z in R^{n+2} is nowhere zero on M, then M is a hypersphere in S^{n+1}.*

Proof The Hessian $\mathcal{H}ess_\rho$ is [by (3.263)]

$$\mathcal{H}ess_\rho(X, Y) = -g((\nabla_X A)Y, t) + f\, g(AX, Y) - \rho\, g(AX, AY). \tag{3.264}$$

By (3.264) and (3.261)

$$\Delta\rho = -n\,t\,\alpha + n\,f\,\alpha - \rho\,\mathrm{trace}(A^2). \tag{3.265}$$

By (3.263) $\mathrm{div}(t\,\alpha) = t\,\alpha - n\,\alpha\,(\rho\alpha - f)$. By the last equality together with (3.265) and Lemma 3.207

$$\Delta\big[\rho - (n-1)\alpha f\big] = \rho\big[n\alpha^2 - \mathrm{trace}(A^2)\big] - \mathrm{div}(n\,t\,\alpha).$$

Integrating over M

$$\int_M [\rho(n\alpha^2 - trace(A^2))]\,d\,\mathrm{v}_g = 0. \tag{3.266}$$

Then (3.266) yields either $\rho = 0$ or $\alpha^2 = (1/n)\,\mathrm{trace}(A^2)$. In the first case [by Schwarz inequality]

$$\alpha^2 - \frac{1}{n}\,\mathrm{trace}(A^2) \geq 0$$

with equality if and only if M is totally umbilical. In the second case M is again totally umbilical and as M is compact and connected, it must be a hypersphere in S^{n+1}. Q.e.d.

Deshmukh and Al-Gwaiz (cf. [DsGw]) gave the following characterization of a small sphere in a sphere, in terms of a pinching relation on the Ricci curvature.

Theorem 3.206 *Let M be a compact orientable hypersurface of S^{n+1} with a relative support function ρ with respect to a parallel unit vector field Z in $\mathbb{R}^{n+2}$. If the Ricci curvature of M satisfies $\rho^2(n-1)(n+2) < \rho^2\,\mathrm{Ric} \leq n-1$, then*

(1) *ρ is constant,*

(2) $M = S^n(\rho^{-2})$.

We need the following lemmas.

Lemma 3.207 (1) $\Delta f = n(-f + \rho\alpha)$,

(2) $\frac{1}{2}\Delta f^2 = n(-f^2 + \alpha f \rho) + ||t||^2$.

Proof The first statement follows from (3.263). As to the second statement, one uses the first statement together with (3.263) and with

$$\Delta f^2 = 2\,f\,\Delta f + 2\,||\mathrm{grad}\,f||^2\,.$$

Q.e.d.

Lemma 3.208 *Let M be a hypersurface of S^{n+1} with a relative support function ρ with respect to a parallel unit vector field Z in $\mathbb{R}^{n+2}$. Then*

$$\begin{aligned}\frac{1}{2}\Delta\rho^2 =&\rho^2(\tau - n(n-1)) - f\,\rho\,\alpha\, n(n-1)\\ &- \left[\mathrm{Ric}(t,t) - (n-1)||t||^2\right] - \mathrm{div}(n\,\alpha\,\rho\, t).\end{aligned} \tag{3.267}$$

Proof Substitution from

$$\mathrm{div}(t\,\alpha) = t\alpha - n\alpha(\rho\alpha - f)$$

and from (3.260) into (3.265) yields

$$\rho\Delta\rho = \rho\,(\tau - n(n-1)) - n(n-1)\,f\,\rho\alpha - \rho\,\mathrm{div}\,(\mathrm{n\ t}). \tag{3.268}$$

By

$$\mathrm{div}(n\alpha\rho t) = n\alpha\rho t + \rho\,\mathrm{div}(n\alpha t)$$

and by (3.263) we obtain

$$-\rho\,\mathrm{div}(n\alpha t) = n\,\alpha\, g(At, t) - \mathrm{div}(n\,\alpha\,\rho\, t). \tag{3.269}$$

Substitution from (3.269) into (3.268) yields

$$\rho\,\Delta\rho = \rho\left[\tau - n(n-1)\right] - n(n-1)\,f\,\rho\alpha - n\,\alpha\, g(At,t) - \mathrm{div}(n\,\alpha\,\rho\, t).$$

Finally by (3.263) and

$$\Delta\rho^2 = 2\left[\rho\,\Delta\rho + ||\mathrm{grad}\,\rho||^2\right],$$

we obtain (3.267). Q.e.d.

Proof of Theorem 3.206 By Lemma 3.207 and (3.267)

$$\begin{aligned}\frac{1}{2}\,\Delta\left[\rho^2 + (n-1)f^2\right] + \mathrm{div}(n\,\alpha\,\rho\, t) =&\rho^2\tau - n(n-1)(\rho^2 + f^2 + ||t||^2)\\ &+ (n-1)(n+2)||t||^2 - \mathrm{Ric}(t,t).\end{aligned} \tag{3.270}$$

For a unit vector field $Z = t + \rho N + f\overline{N}$ we have $||t||^2 + \rho^2 + f^2 = 1$. Thus, by integrating (3.270)

$$\int_M \left\{||t||^2\left[\mathrm{Ric}(s,s) - (n-1)(n+2)\right] + n(n-1) - \rho^2\tau\right\}\, d\,\mathrm{v}_g = 0,$$

where $s = ||t||^{-1} t$ (a unit vector field defined on the open subset of M where t is nonzero). Then it must be $t = 0$ and

$$\rho^2 \tau = n(n-1).$$

Moreover ρ is a nonzero constant, $f|_M$ is a constant, and $A = \frac{f}{\rho} I$. Hence [by Gauss equation] M is isometric to $S^n(1/\rho^2)$. Q.e.d.

3.19.3 Minimal Submanifolds in Spheres

Let M^n be a n-dimensional minimal compact submanifold in S^{n+p}, with the normal bundle $\tau_M^\perp$. The Riemannian metric on S^{n+p} and the induced Riemannian metric on M are denoted by the same symbol g. The Levi-Civita connections on S^{n+p} and M are denoted by $\overline{\nabla}$ and ∇. Gauss-Weingarten formulae are

$$\overline{\nabla}_X Y = \nabla_X Y + B(X, Y), \qquad \overline{\nabla}_X N = -A_N X + \nabla_X^\perp Y \tag{3.271}$$

for any $X, Y \in \mathfrak{X}(M)$ and $N \in C^\infty(\tau_M^\perp)$. We shall also use

$$g(A_N X, Y) = g(B(X, Y), N). \tag{3.272}$$

Let $R = R^\nabla$ and $R^\perp = R^{\nabla^\perp}$ be the curvature tensor fields of ∇ and $\nabla^\perp$. Gauss-Codazzi-Ricci equations are

$$\begin{cases} R(X, Y)Z = g(Y, Z)X - g(X, Z)Y + A_{B(Y,Z)}X - A_{B(X,Z)}Y, \\ (\nabla B)(X, Y, Z) = (\nabla B)(Y, Z, X), \\ R^\perp(X, Y, N_1, N_2) = g([A_{N_1}, A_{N_2}](X), Y) \end{cases} \tag{3.273}$$

for any $X, Y \in \mathfrak{X}(M)$ and $N_1, N_2 \in C^\infty(\tau_M^\perp)$. Let $\{E_1, \ldots, E_n\}$ be a local orthonormal frame of τ_M. The Ricci operator R^* is

$$\mathrm{Ric}(X, Y) = g(R^*(X), Y), \quad \forall\ X, Y \in \mathfrak{X}(M).$$

Then [by Gauss equation]

$$R^*(X) = (n-1)X - \sum_{i=1}^n A_{B(E_i, X)} E_i \tag{3.274}$$

for any $X \in \mathfrak{X}(M)$. Next the scalar curvature is

$$\tau = \sum_{i}^{n} \mathrm{Ric}(E_i, E_i) = \tag{3.275}$$

$$= n(n-1) - \sum_{i,j} ||B(E_i, E_j)||^2 = n(n-1) - ||B||^2.$$

As M is minimal

$$\begin{cases} \sum_{i=1}^{n} B(E_i, B_i) = 0, \qquad \sum_{i=1}^{n} (\nabla B)(X, E_i, E_i) = 0, \\ \sum_{i=1}^{n} (\nabla^2 B)(X, Y, E_i, E_i) = 0. \end{cases} \tag{3.276}$$

Next we define

$$\begin{cases} \sigma = \sum_{i,j} ||B(E_i, E_j)||^2 = ||B||^2, \\ ||A_B||^2 = \sum_{i,j,k} ||A_{B(E_i, E_j)} E_k||^2, \\ ||\nabla B||^2 = \sum_{i,j,k} ||(\nabla B)(E_i, E_j, E_k)||^2. \end{cases} \tag{3.277}$$

Next we report on a result by Deshmukh (cf. [Des11]) concerning normal curvature of minimal submanifolds in spheres.

Theorem 3.209 *Let M^n be a minimal compact submanifold of S^{n+p}. If the scalar curvature τ and the normal curvature $K^\perp$ of M satisfy $\tau > (n-1)^2$, $\quad K^\perp \le \sigma$, then M is totally geodesic.*

We need the following lemmas.

Lemma 3.210 *Let M^n be a minimal submanifold of S^{n+p}. Then*

$$\sum_{i,j,k} R(E_k, E_i; E_j, A_{B(E_i, E_j)} E_k) =$$

$$= ||A_B||^2 + \frac{1}{2} K^\perp - \sum_{i,j,a,b} g(A_a E_i, A_b E_j)^2 - \sigma,$$

$$A_a \equiv A_{n_a}, \quad 1 \le a \le p$$

for any local orthonormal frames

$$\{E_i : 1 \le i \le n\} \subset C^\infty(U, \tau_M), \quad \{n_a : 1 \le a \le p\} \subset C^\infty(U, \tau_M^\perp).$$

Proof By (3.273)

$$K^\perp = \sum_{i,j,a,b} [R^\perp(E_i, E_j; n_a, n_b)]^2 = \tag{3.278}$$

$$= 2 \sum_{i,j,a,b} [g(A_a E_i, A_b E_j)^2 - g(A_a A_b E_i, E_j) g(A_b A_a E_i, E_j)].$$

On the other hand [by Gauss equation]

$$\sum_{i,j,k} R(E_k, E_i; E_j, A_{B(E_i,E_j)} E_k) =$$

$$= \delta_{ij} g(B(E_k, E_k), B(E_i, E_j)) - \delta_{kj} g(B(E_i, E_j), B(E_i, E_k)) +$$

$$+ g(B(E_i, E_j), B(E_k, A_{B(E_i,E_j)} E_k)) - g(B(E_k, E_j), B(E_i, A_{B(E_i,E_j)} E_k)) =$$

$$= \delta_{ij} g(B(E_k, E_k), B(E_i, E_j)) - \delta_{kj} g(B(E_i, E_j), B(E_i, E_k)) +$$

$$+ ||A_B||^2 - \sum_{i,j,a,b} g(A_a A_b E_i, E_j)\, g(A_b A_a E_i, E_j).$$

Finally [by (3.278) and minimality of M]

$$\sum_{i,j,k} R(E_k, E_i; E_j, A_{B(E_i,E_j)} E_k) =$$

$$= -\sigma + ||A_B||^2 - \sum_{i,j,a,b} g(A_a E_i, A_b E_j)^2 + \frac{1}{2} K^\perp.$$

Q.e.d.

Lemma 3.211 *Let M^n be a minimal submanifold of S^{n+p}. Then*

$$\Delta F = (n-1)\sigma - ||A_h||^2 + (\sigma - K^\perp) + ||\nabla h||^2. \tag{3.279}$$

Proof We consider the function $F : M \to \mathbb{R}$, $2F = \sigma$. Then

$$\Delta F = \sum_{i,j,k} g((\nabla^2 B)(E_k, E_k, E_i, E_j), B(E_i, E_j)) + \sum_{i,j,k} ||(\nabla h)(e_i, e_j, e_k)||^2.$$

By the Ricci identity

$$\Delta F = \sum_{i,j,k}[R^{\perp}(E_k, E_i; B(E_k, E_j), B(E_i, E_j)) - R(E_k, E_i; E_k, A_{B(E_j,E_i)}E_j)$$
$$- R(E_k, E_i; E_j, A_{B(E_i,E_j)}E_k)] + ||\nabla B||^2. \quad (3.280)$$

On the other hand [by (3.273)]

$$\sum_{i,j,k} R^{\perp}(E_k, E_i; B(E_k, E_j), B(E_i, E_j)) = \quad (3.281)$$

$$= \sum_{i,j,k}[g(A_{B(E_i,E_j)}E_k, A_{B(E_i,E_j)}E_k) +$$

$$+ g(A_{B(E_i,E_j)E_k}, R(E_i, E_k)E_j - \delta_{kj}E_i + \delta_{ij}E_k) - g(A_{B(E_k,E_j)}E_k, A_{B(E_i,E_j)}E_i)] =$$

$$= ||A_B||^2 - \sigma + \sum_{i,j,k}[R(E_i, E_k; E_j, A_{B(E_i,E_j)}E_k) - g(A_{B(E_k,E_j)}E_k, A_{B(E_i,E_j)}E_i)].$$

Next we evaluate [by (3.274)] the last term (in the equation above)

$$\sum_{i,j,k} g(A_{B(E_k,E_j)}E_k, A_{B(E_i,E_j)}E_i) = \quad (3.282)$$

$$= \sum_{i,j} g((n-1)e_j - R^*(E_j), A_{h(e_i,e_j)}e_i) = (n-1)\sigma + \sum_{i,j,k} R(E_k, E_i; E_k, A_{B(E_i,E_j)}E_j).$$

By (3.281) and (3.282)

$$\sum_{i,j,k} R^{\perp}(e_k, e_i; h(e_k, e_j), h(e_i, e_j)) = \quad (3.283)$$

$$= -n\sigma + ||A_h||^2 + \sum_{i,j,k}[R(e_i, e_k; e_j, A_{h(e_i,e_j)}e_k) - R(e_k, e_i; e_k, A_{h(e_i,e_j)}e_j)].$$

Moreover

$$\sum_{i,j,k} R(e_k, e_i; e_k, A_{h(e_i,e_j)}e_j) = -\sum_{i,j} Ric(e_i, A_{h(e_i,e_j)}e_j) = \quad (3.284)$$

$$= -\sum_{i,j} g(R^*(e_i), A_{h(e_i,e_j)}e_j) = -\sum_{j,a} g(R^*(A_a e_j), A_a e_j) =$$

$$= -\sum_{j,a} Ric(A_a e_j, A_a e_j) = -(n-1)\sigma + \sum_{i,j,a,b} g(A_a e_i, A_b e_j)^2 .$$

Substitution from Lemma 3.210 and from (3.283)–(3.284) into (3.280) yields (3.279). Q.e.d.

Proof of Theorem 3.209 Integrating (3.279) over M

$$\int_M \left[(n-1)\sigma - ||A_h||^2 + (\sigma - K^{\perp}) + ||\nabla h||^2\right] d\,\mathrm{v}_g = 0. \tag{3.285}$$

By

$$\sigma = \sum_a ||A_a||^2, \quad ||A_h||^2 = \sum_{i,j,k} ||A_{h(e_i,e_j)}e_k||^2 = \sum_a ||A_a||^2.$$

Equation (3.285) becomes

$$\int_M \left[\sum_a ((n-1) - ||A_a||^2)||A_a||^2 + (\sigma - K^{\perp}) + ||\nabla h||^2\right] d\,\mathrm{v}_g = 0.$$

Then $\sigma - K^{\perp} \geq 0$ and $||A_a||^2 < n-1$. Thus $||A_a|| = 0$, i.e., M is totally geodesic. Q.e.d.

Theorem 3.212 *Let M be a n-dimensional compact minimal submanifold in S^{n+p}. If $\tau > n(n-2)$ and $K^{\perp} \leq \varphi$ then M is totally geodesic.*

Cf. Deshmukh [Des12]. To prove the theorem we need the following lemmas.

Lemma 3.213 *Let M be a n-dimensional minimal compact submanifold in S^{n+p}. Then*

$$\begin{aligned} 0 = \int_M [||\nabla h||^2 + \sum_{i,j,k} (R^{\perp}(E_k, E_i; B(E_j, E_k), \ B(E_i, E_j)) + \\ - R(E_k, E_i; E_j, A_{B(E_i,E_j)}E_k)) + \sum_{i,j} \mathrm{Ric}(A_{B(E_i,E_j)}E_j, E_i)]\, d\,\mathrm{v}_g. \end{aligned} \tag{3.286}$$

Lemma 3.214 *Let M be a n-dimensional compact minimal submanifold in S^{n+p}. Then*

$$\begin{aligned} ||R||^2 = &\sum_{i,j,k} ||R(E_i, E_j)E_k||^2 = \\ =&2n(n-1) - 4\sigma + 2||A_B||^2 + \\ &- 2\sum_{i,j,k} g(A_{B(E_j,E_k)}E_i, A_{B(E_i,E_k)}E_j). \end{aligned} \tag{3.287}$$

Lemma 3.215 *Let M be a n-dimensional compact minimal submanifold M in S^{n+p}. Then*

$$2\sum_{i,j,k} R(E_k, E_i; E_j, A_{B(E_i,E_j)}E_k) = ||R||^2 - 2\tau. \tag{3.288}$$

Let us consider the function

$$\varphi : M \to \mathbb{R}, \quad \varphi = 2\sum_{a<b} ||A_a||^2||A_b||^2.$$

Proof of Theorem 3.212 Let us use (3.273) and (3.274) [by taking inner product with $A_{B(E_i,E_j)}E_k$]

$$\begin{cases} \sum_{i,j,k} g(A_{B(E_j,E_k)}E_i,\ A_{B(E_i,E_j)}E_k) = \sum_{i,j,k} g(R(E_i,\ E_k;\ E_j,\ A_{B(E_i,E_j)}E_k) - \sigma + ||A_B||^2. \\ \sum_{i,j,k} g(A_{B(E_j,E_k)}E_k,\ A_{B(E_i,E_j)}E_i) = (n-1)\sigma - \sum_{i,j} \mathrm{Ric}(E_j,\ A_{B(E_i,E_j)}E_i). \end{cases} \tag{3.289}$$

Substitution from (3.289) into the third equation of (3.273) yields

$$R^{\perp}(E_k, E_i; B(E_i, E_j),\ B(E_j,\ E_k)) = \sum_{i,j,k} R(E_i,\ E_k;\ E_j,\ A_{B(E_i,E_j)}E_k) + \tag{3.290}$$

$$+\sum_{i,j} \mathrm{Ric}(E_j,\ A_{B(E_i,E_j)}E_i) + ||A_B||^2 - n\sigma.$$

Moreover

$$\sum_{i,j} Ric(E_j,\ A_{B(E_i,E_j)}E_i) = \tag{3.291}$$

$$= \sum_{a,i,j} g(A_aE_i,\ E_j)\mathrm{Ric}(E_j,\ A_aE_i) = \sum_{a,i} \mathrm{Ric}(A_aE_i,\ A_aE_i).$$

Substitution from (3.274) into (3.291) yields

$$\sum_{i,j} \mathrm{Ric}(E_j,\ A_{B(E_i,E_j)}E_i) = (n-1)\sigma - \sum_{a,b} ||A_aA_b||^2. \tag{3.292}$$

On the other hand [by (3.287)]

$$||R||^2 = 2n(n-1) - 4\sigma + \tag{3.293}$$

$$+ 2||A_B||^2 + \sum_{a,b}[||A_aA_b - A_bA_a||^2 - 2||A_aA_b||^2].$$

Here one also made use of

$$\begin{cases} \sum_{i,j,k} g(A_{B(E_j,E_k)}E_i,\, A_{B(E_i,E_k)}E_j) = \\ = \sum_{i,j,a,b} g(A_a E_i,\, A_b E_j) g(A_a E_j,\, A_b E_i), \\ \sum_{a,b} ||A_a A_b - A_b A_a||^2 = \\ = -2\sum_{i,j,a,b} g(A_a E_i,\, A_b E_j) g(A_a E_j,\, A_b E_i) + 2\sum_{a,b} ||A_a A_b||^2. \end{cases}$$

Then [by Lemma 3.215 and (3.293)]

$$2\sum_{i,j,k} R(E_k,\, E_i;\, E_j,\, A_{B(E_i,E_j)}E_k) = \tag{3.294}$$

$$= 2n(n-1) - 4\sigma + 2||A_B||^2 = -2\tau + \sum_{a,b}[||A_a A_b - A_b A_a||^2 - 2||A_a A_b||^2].$$

By Lemma 3.213 and (3.290)

$$0 = \int_M [||\nabla h||^2 + ||A_B||^2 - n\sigma - 2\sum_{i,j,k} R(E_k,\, E_i;\, E_j,\, A_{B(E_i,E_j)}E_k) +$$

$$+ 2\,\mathrm{Ric}(E_j,\, A_{B(E_i,E_j)}E_i)]\, d\,\mathrm{v}_g$$

or [by (3.292), (3.294) and (3.275)]

$$0 = \int_M [||\nabla B||^2 + ||A_B||^2 - n\sigma - \sum_{a,b} ||A_a A_b - A_b A_a||^2]\, d\,\mathrm{v}_g\,. \tag{3.295}$$

To simplify the above integral formula, we use the identities

$$\begin{aligned} K^\perp &= \sum_{i,j,a,b} [R^\perp(E_i,\, E_j;\, n_a,\, n_b)]^2 \\ &= \sum_{i,j,a,b} g([A_a,\, A_b](E_i),\, E_j)^2 \\ &= \sum_{a,b}[\sum_{i,j} g((A_a A_b - A_b A_a)(e_i),\, e_j)^2] \\ &= \sum_{a,b} ||A_a A_b - A_b A_a||^2 \end{aligned}$$

and

$$\begin{aligned} ||A_B||^2 &= \sum_{i,j,k} ||A_{B(E_i,E_j)}e_k||^2 \\ &= \sum_{i,j,k,a} g(A_a E_i, E_j)^2 ||A_a E_k||^2 \\ &= \sum_{i,j,a} ||A_a||^2 g(A_a E_i, E_j)^2 \\ &= \sum_a ||A_a||^2 \\ &= ||h||^4 - 2\sum_{a<b} ||A_a||^2 ||A_b||^2. \end{aligned} \tag{3.296}$$

Thus

$$\int_M [||\nabla B||^2 + (n - \sigma)\sigma + (\varphi - K^{\perp})]\, d\,\mathrm{v}_g = 0. \tag{3.297}$$

By $\sigma < n$ together with (3.297) it follows that M is totally geodesic. Q.e.d.

Theorem 3.216 *Let M be a n-dimensional compact minimal submanifold in S^{n+2}. Suppose that $\tau > n(n-2)$. Then M is totally geodesic.*

Cf. Deshmukh [Des12].

Proof Let $\{n_1, n_2\}$ be a local orthonormal frame in $\tau_M^{\perp}$. Then (3.295) may be written [by (3.296)]

$$\int_M [||\nabla h||^2 + (n - \sigma)\sigma + \tag{3.298}$$

$$+ 2||A_1||^2||A_2||^2 - ||A_1A_2 - A_2A_1||^2]\, d\,\mathrm{v}_g = 0.$$

Using (3.298) we may conclude that $\sigma = 0$, that is M is totally geodesic. Q.e.d.

The previous theorem admits the following consequence.

Corollary 3.217 *Let M be a n-dimensional compact minimal hypersurface in S^{n+1}. If $\tau \geq n(n-2)$ then either*

(1) *M is totally geodesic, or*

(2) *M is isometric to $S^m\left(\sqrt{\frac{m}{n}}\right) \times S^{n-m}\left(\sqrt{\frac{n-m}{n}}\right)$.*

Proof Note that $K^{\perp} = \varphi = 0$. Consequently (3.297) becomes

$$\int_M [||\nabla A||^2 + (n - ||A||^2)||A||^2] \, d\, v_g = 0. \tag{3.299}$$

If M is not totally geodesic then (3.299) implies that A is parallel and $||A||^2 = n$. Consequently M is isometric to

$$S^m\left(\sqrt{\frac{m}{n}}\right) \times S^{n-m}\left(\sqrt{\frac{n-m}{n}}\right).$$

Q.e.d.

Theorem 3.218 *Let M be a n-dimensional compact minimal submanifold in S^{n+p}, with a flat normal connection. If M has constant sectional curvature $c \geq 0$ then*

(1) *M is flat and $p \geq n - 1$, or*

(2) *M is totally geodesic.*

Cf. Deshmukh [Des12]. To prove the theorem we need to use.

Theorem 3.219 *Let M a n-dimensional compact non-negatively curved minimal submanifold in S^{n+p} with a flat normal connection. Then either*

(1) *M is totally geodesic, or*

(2) *the scalar curvature τ of M satisfies*

$$n(n - p - 1) \leq \tau \leq n(n - 2).$$

To prove Theorem 3.219 we need the following lemma.

Lemma 3.220 *Let M be a n-dimensional non-negatively curved minimal submanifold in S^{n+p} with a flat normal connection. Then*

(i) $||A_h||^2 = n\sigma$.

(ii) $\sigma(\sigma - n) \geq 0$.

Proof We choose local orthonormal frames

$$\{E_1, \ldots, E_n\} \subset C^{\infty}(U, \tau_M), \quad \{n_1, \ldots, n_p\} \subset C^{\infty}(U, \tau_M^{\perp}),$$

such that $A_a(E_i) = \lambda_i^a E_i$ [with $A_a = A_{n_a}$] for some

$$\lambda_i^a \in C^\infty(U), \quad i \in \{1, \ldots, n\}, \quad a \in \{1, \ldots, p\}.$$

$$\sum_{i,j} \text{Ric}(E_j, A_{B(E_i,E_j)}E_j) - \sum_{i,j,k} R(E_k, E_i; E_j, A_{B(E_i,E_j)}E_k) = \tag{3.300}$$

$$= \sum_a [\sum_{i,j} g(A_a E_i, E_j) \text{Ric}(E_i, A_a E_j) - \sum_{i,j,k} g(A_a E_i, E_j) R(E_k, E_i; E_j, A_a E_k)] =$$

$$= \sum_{a,j,k} [(\lambda_j^a)^2 R(E_k, E_j; E_j, E_k) - \lambda_j^a \lambda_k^a R(E_k, E_j; E_j, E_k)] =$$

$$= \frac{1}{2} \sum_{a,j,k} [2\lambda_j^a)^2 R(E_k, E_j; E_j, E_k) - 2\lambda_j^a \lambda_k^a R(E_k, E_j; E_j, E_k)] =$$

$$= \frac{1}{2} [\sum_{a,j,k} (\lambda_j^a - \lambda_k^a)^2 R(E_k, E_j; E_j, E_k)] \geq 0,$$

where $R(E_k, E_j; E_j, E_k)$ is the sectional curvature of the 2-plane spanned by $\{E_k, E_j\}$. Equation (3.300) yields [by Lemma 3.213 and $R^\perp = 0$] $||\nabla h|| = 0$ and

$$\sum_{i,j} \text{Ric}(E_j, A_{B(E_i,E_j)}E_j) = \sum_{i,j,k} R(E_k, E_i; E_j, A_{B(E_i,E_j)}E_k).$$

This last equation together with (3.290) and $R^\perp = 0$ yields

$$||A_h||^2 = n\sigma. \tag{3.301}$$

Also

$$\begin{aligned} ||A_h||^2 &= \sum_a ||A_a||^4 \\ &= (\sum_a ||A_a||^2)^2 - 2\sum_{a<b} ||A_a||^2 ||A_b||^2 \\ &= ||h||^4 - 2\sum_{a<b} ||A_a||^2 ||A_b||^2. \end{aligned} \tag{3.302}$$

Equations (3.301) and (3.302) imply

$$\sigma(\sigma - n) \geq 0. \tag{3.303}$$

Q.e.d.

Proof of Theorem 3.219 By (3.301)

$$\sum_a (||A_a||^2 - \frac{n}{2})^2 = \frac{n^2 p}{4},$$

and then [by Schwarz inequality]

$$\sum_a \Big[||A_a||^2 - \frac{n}{2}\Big]^2 \geq \frac{1}{p}\Big(\sigma - \frac{np}{2}\Big)^2.$$

The last two relations imply

$$\sigma(\sigma - np) \leq 0. \tag{3.304}$$

Hence, either $\sigma = ||h||^2 = 0$ or $\sigma \leq np$. In the first case M is totally geodesic. If M is not totally geodesic, then by (3.303) $n \leq \sigma \leq np$ or equivalently [by (3.275)] $n(n-p-1) \leq \tau \leq n(n-2)$. Q.e.d.

Theorem 3.221 *Let M be a n-dimensional compact minimal submanifold in S^{n+p}, with a flat normal connection. If $\tau \geq n(n-2)$ then either*

(1) *M is totally geodesic, or*

(2) *M is isometric to $S^m\Big(\sqrt{\frac{m}{n}}\Big) \times S^{n-m}\Big(\sqrt{\frac{n-m}{n}}\Big)$.*

Cf. Deshmukh [Des12]. ***Proof*** By $R^\perp = 0$ and (3.290)

$$\sum_{i,j,k} R(E_i, E_k; E_j, A_{B(E_i,E_j)}E_k) - \sum_{i,j} \mathrm{Ric}(E_j, A_{B(E_i,E_j)}E_i) = ||A_B||^2 - n\sigma. \tag{3.305}$$

By Lemma 3.213 and (3.305)

$$\int_M [||\nabla B||^2 + n\sigma - ||A_B||^2]\, d\,\mathrm{v}_g = 0.$$

By (3.302) and the above integral formula

$$\int_M [||\nabla B||^2 + (n-\sigma)\sigma + 2\sum_{a<b} ||A_a||^2 ||A_b||^2]\, d\,\mathrm{v}_g = 0.$$

As $\sigma \leq n$ the integral formula yields

$$||\nabla B|| = 0, \quad (n-\sigma)\sigma = 0, \quad A_b = 0, \quad b \in \{2, \ldots, p\}.$$

When $\sigma = n$ we may conclude that M is isometric to

$$S^m\left(\sqrt{\frac{m}{n}}\right) \times S^{n-m}\left(\sqrt{\frac{n-m}{n}}\right).$$

When $\sigma = 0$ we may conclude that M is totally geodesic. Q.e.d.

A few immediate consequences may be stated as follows.

Corollary 3.222 *Let M be a n-dimensional compact minimal submanifold in S^{n+p}, with a flat normal connection. If $\tau > n(n-2)$ then M is totally geodesic.*

Corollary 3.223 *Let M be a n-dimensional compact non-negatively curved minimal hypersurface in S^{n+1}. Then either*

(1) *M is totally geodesic, or*

(2) *M is isometric to* $S^m\left(\sqrt{\frac{m}{n}}\right) \times S^{n-m}\left(\sqrt{\frac{n-m}{n}}\right)$.

Proof By Theorem 3.219 either M is totally geodesic or $\tau = n(n-2)$. If M is not totally geodesic then $\sigma = n$ and then M is isometric to

$$S^m\left(\sqrt{\frac{m}{n}}\right) \times S^{n-m}\left(\sqrt{\frac{n-m}{n}}\right).$$

Q.e.d.

Corollary 3.224 *Let M be a n-dimensional compact non-negatively curved minimal hypersurface in S^{n+1}. If M is diffeomorphic to S^n then M is totally geodesic.*

Lemma 3.225 *Let M be a n-dimensional minimal submanifold in S^{n+p} with a flat normal connection. Then*

$$||A_h||^2 = n(1-c)\sigma.$$

Proof The first equation in (3.273) (Gauss equation) is

$$\sum_{i,j,k}(c-1)\, g(\delta_{jk}E_i - \delta_{ik}E_j, A_{B(E_j,E_k)}E_i) =$$

$$= \sum_{i,j,k}[g(A_{B(E_j,E_k)}E_i, A_{B(E_j,E_k)}E_i) - g(A_{B(E_i,E_k)}E_j, A_{B(E_j,E_k)}E_i)].$$

This gives

$$(1-c)\,\sigma = ||A_h||^2 - \sum_{i,j,k} g(A_{B(E_i,E_k)}E_j,\ A_{B(E_j,E_k)}E_i). \tag{3.306}$$

By the third equation in (3.273)

$$\sum_{i,j,k} g(A_{B(E_i,E_k)}E_j,\ A_{B(E_j,E_k)}E_i) = \tag{3.307}$$

$$= \sum_{i,k} g((n-1)e_k - R^*(e_k),\ A_{B(E_i,E_k)}E_i) =$$

$$= (n-1)\sigma - \sum_{i,k} \mathrm{Ric}(E_k,\ A_{B(E_i,E_k)}E_i).$$

The Ricci curvature for a minimal submanifold is

$$Ric(X,Y) = (n-1)cg(X,Y), \quad X, Y \in \mathfrak{X}(M).$$

Then the above equation reduces to

$$\sum_{i,j,k} g(A_{B(E_i,E_k)}E_j,\ A_{B(E_j,E_k)}E_i) = (n-1)(1-c)\sigma. \tag{3.308}$$

Substitution from (3.308) into (3.306) leads to $||A_h||^2 = n(1-c)\,\sigma$. Q.e.d.

Proof of Theorem 3.218. By (3.301) and Lemma 3.225 we get $c\sigma = 0$ yielding either $c = 0$ or $\sigma = 0$. In the first case $\tau = 0$ implying $p \geq n-1$. In the second case M is totally geodesic. Q.e.d.

Theorem 3.218 admits the following immediate consequence.

Corollary 3.226 *There exist no compact flat minimal hypersurfaces in the sphere* S^4.

3.19.4 Biharmonic Immersions in Spheres

The notion of a biharmonic map generalizes that of a harmonic map. Biharmonic maps are critical points of the bienergy functional

$$E_2(\varphi) = \frac{1}{2}\int_M \|\tau(\varphi)\|^2 \, d\,\mathrm{v}_{\overline{g}}.$$

The Euler-Lagrange equations of the variational principle $\delta E_2(\varphi) = 0$ are

$$\tau_2(\varphi) \equiv -J^{\varphi}(\tau(\varphi)) \equiv -\Delta\tau(\varphi) - \text{trace}(R^N(d\varphi, \tau(\varphi))\, d\varphi) = 0,$$

where J^{φ} formally denotes the Jacobi operator of φ, Δ is the rough Laplacian on sections of $\varphi^{-1}T(N)$ and R^N is the curvature tensor field of (N, g). Any harmonic map is biharmonic. Non harmonic biharmonic maps are *proper* biharmonic.

Let (M, g) be an immersed submanifold into a Riemannian manifold (N, g), with the induced metric g. If the immersion is a biharmonic map, then that immersion is called a *biharmonic submanifold.* On the other hand, submanifolds with non harmonic (non minimal) biharmonic inclusion map are called *proper* biharmonic submanifolds.

We report on results by A. Balmus et al.[86] on proper biharmonic submanifolds, with a parallel mean curvature vector field, in spheres.

Proposition 3.227 *Let M be a proper biharmonic submanifold with a parallel mean curvature vector field in S^n. Then*

(1) $||A_H||^2 = m||H||^2 = constant.$
(2) $\text{trace}(\nabla A_H) = 0.$
(3) $\langle \text{trace}((\nabla^{\perp} B)(X, \cdot, A_H(\cdot))), H\rangle = \langle \text{trace}((\nabla^{\perp} B)(\cdot, X, A_H(\cdot))), H\rangle = 0$

for any $X \in \mathfrak{X}(M)$.

Theorem 3.228 *Let M^m, $m \geq 4$, be a compact proper biharmonic constant mean curvature hypersurface in S^{m+1}. Then*

(1) $||H|| \in \left(0, \dfrac{m-2}{m}\right] \cup \{1\}.$

(2) $||H|| = 1$ *if and only if M is isometric to* $S^m\left(\dfrac{1}{\sqrt{2}}\right).$

(3) $||H|| = \dfrac{m-2}{m}$ *if and only if M is isometric to* $S^1\left(\dfrac{1}{\sqrt{2}}\right) \times S^{m-1}\left(\dfrac{1}{\sqrt{2}}\right).$

Cf. Balmus et al.[87] To prove the result we need some preparation.

Theorem 3.229 (1) *The canonical inclusion $\varphi : M^m \to S^n$ is biharmonic if and only if*

$$\Delta^{\perp} H + \text{trace}(B(\cdot, A_H(\cdot))) - m\, H = 0, \quad 4\,\text{trace}(A_{\nabla^{\perp}_{(\cdot)} H}(\cdot)) + m\,\text{grad}(||H||^2) = 0.$$

---

[86] Cf. *op. cit.*

[87] Cf. A. Balmus, S. Montaldo, C. Oniciuc, Properties of biharmonic submanifolds in spheres. JGSP **17**, 87–102 (2010).

(2) *If M is a submanifold with parallel mean curvature vector field in S^n, then M is biharmonic if and only if*

$$\mathrm{trace}(B(\cdot, A_H(\cdot))) = mH.$$

(3) *A hypersurface M with nowhere zero mean curvature vector field in S^{m+1} is biharmonic if and only if*

$$\Delta^{\perp}H - (m - ||A||^2)H = 0, \quad 2A(\mathrm{grad}(||H||)) + m||H||\,\mathrm{grad}(||H||) = 0.$$

(4) *If M is a nonzero constant mean curvature hypersurface in S^{m+1}, then M is proper biharmonic if and only if*

$$||A||^2 = m.$$

Here $\nabla^{\perp}$ and $\Delta^{\perp}$ are the connection and the Laplacian in the normal bundle $\tau_M^{\perp}$ and grad is the gradient on M.

Proposition 3.230 *Let M be a proper biharmonic hypersurface with constant mean curvature in S^{m+1} and $m \geq 2$. Then M has positive constant scalar curvature*

$$\tau = m^2(1 + ||H||^2) - 2m.$$

The *normalized* scalar curvature ρ of a hypersurface M^m in S^{m+1} is

$$\rho = \frac{\tau}{m(m-1)}.$$

Theorem 3.231 *Let M be a compact hypersurface with constant normalized scalar curvature ρ in S^{m+1}. Let us assume that $\rho \geq 1$ and*

$$m(\rho - 1) \leq ||h||^2 \leq (m-1)\frac{m(\rho-1)+2}{m-2} + \frac{m-2}{m(\rho-1)+2}. \tag{3.309}$$

Then either

(1) $||h||^2 = m(\rho - 1)$ *and M is a totally umbilical hypersurface, or*

(2) $||h||^2 = (m-1)\dfrac{m(\rho-1)+2}{m-2} + \dfrac{m-2}{m(\rho-1)+2}$ *and*

$$M = S^1\left(\sqrt{1-c^2}\right) \times S^{m-1}(c), \quad c^2 = \frac{m-2}{m\rho}.$$

Proof of Theorem 3.228 By Theorem 3.229

$$||h||^2 = ||A||^2 = m. \tag{3.310}$$

Let us set $t = m||H||^2 - 1$ for simplicity. We assume that $||H|| \in \left(\frac{m-2}{m}, 1\right)$ or equivalently

$$t \in \left(\frac{(m-4)(m-1)}{m},\ m-1\right).$$

By Proposition 3.230

$$\rho = 1 + \frac{t}{m-1}. \tag{3.311}$$

We observe that the first condition of Theorem 3.231 is equivalent to $t \geq 0$, which is satisfied. Also, the first inequality of (3.309) is satisfied [by (3.310) and $t < m - 1$]. As

$$t > \frac{(m-4)(m-1)}{m}$$

the second inequality of (3.309) becomes

$$0 \leq mt^2 - (m^2 - 6m + 4)t - (m-4)(m-1),$$

which is satisfied. By Theorem 3.231 we get either $\rho = 2$, i.e., $||H|| = 1$, or $\rho = \frac{2(m-2)}{m}$, i.e., $||H|| = \frac{m-2}{m}$, a contradiction. Thus (1) is proved. The statement (2) follows by Theorem 3.231. As previously shown, if

$$M \approx S^1\left(\frac{1}{\sqrt{2}}\right) \times S^{m-1}\left(\frac{1}{\sqrt{2}}\right)$$

(an isometry) then $||H|| = \frac{m-2}{m}$.

Conversely, if $||H|| = \frac{m-2}{m}$ then $\rho = \frac{2(m-2)}{m}$. Thus (3) follows by Theorem 3.231. Q.e.d.

S. Montaldo et al.[88] listed the main examples of proper biharmonic immersions in S^m.

---

[88] Cf. S. Montaldo, C. Oniciuc, A. Ratto, Biharmonic immersions into spheres and ellipsoids. Rev. Roumaine Math. Pures Appl. **59**, 425–442 (2014).

(1) The canonical inclusion of the small hypersphere

$$S^{m-1}\left(\frac{1}{\sqrt{2}}\right) = \left\{(x, \frac{1}{\sqrt{2}}) \in R^{m+1} : x \in R^m, |x|^2 = \frac{1}{2}\right\} \subset S^m .$$

(2) The canonical inclusion of standard (extrinsic) products of spheres

$$S^{m_1}\left(\frac{1}{\sqrt{2}}\right) \times S^{m_2}\left(\frac{1}{\sqrt{2}}\right) = \left\{(x, y) \in R^{m_1+1} \times R^{m_2+1}, |x|^2 = |y|^2 = \frac{1}{2}\right\} \subset S^m,$$

$$m_1 + m_2 - 1, \quad m_1 \neq m_2 .$$

(3) The maps $\varphi = \iota \circ \phi : M \to S^m$, where $\phi : M \to S^{m-1}\left(\frac{1}{\sqrt{2}}\right)$ is a minimal immersion, and $\iota : S^{m-1}\left(\frac{1}{\sqrt{2}}\right) \subset S^m$ is the canonical inclusion.

(4) Maps $\varphi = \iota \circ (\phi_1 \times \phi_2) : M_1 \times M_2 \to S^m$, where $\phi_i : M_i^{n_i} \to S^{m_i}\left(\frac{1}{\sqrt{2}}\right)$, $0 < n_i \leq m_i$, $i = 1, 2$, are minimal immersions, $n_1 \neq n_2$, $m_1 + m_2 - 1$, and $\iota : S^{m_1}\left(\frac{1}{\sqrt{2}}\right) \times S^{m_2}\left(\frac{1}{\sqrt{2}}\right) \to S^m$ is the canonical inclusion.

The following result (by R. Caddeo, et al.[89] reduces the search for biharmonic submanifolds in S^n to that of minimal, and thus harmonic, immersions in $S^{n-1}(1/\sqrt{2})$.

Theorem 3.232 *Let M be a pseudo-umbilical submanifold of S^n with a parallel mean vector field, of norm* 1. *Then*

(1) *M is biharmonic in S^n.*
(2) *M is a minimal submanifold in a hypersphere $S^{n-1}(\frac{1}{\sqrt{2}}) \subset S^n$.*

Proof Let us denote the mean curvature vector field of M in $\mathbb{R}^{n+1}$ by $\overline{H}$. Then for every $p \in M$ we have $\overline{H}(p) = H(p) - p$, and for any $X\mathfrak{X}(M)$

$$\nabla_X^{\mathbb{R}^{n+1}} \overline{H} = \overline{\nabla}_X^{\perp} \overline{H} - \overline{A}_{\overline{H}}(X) = \nabla_X^{\perp} H - A_H(X) - X.$$

Thus

$$\overline{\nabla}^{\perp} \overline{H} = \nabla^{\perp} H = 0, \quad \overline{A}_{\overline{H}} = 2I.$$

[89] Cf. R. Caddeo, S. Montaldo, C. Oniciuc, Biharmonic Submanifolds in Spheres. Israel J. Math. **130**, 109–123 (2002).

On the other hand, let us consider the map

$$\psi : M \to \mathbb{R}^{n+1}, \quad \psi(p) = p + \frac{1}{2}\overline{H}(p).$$

Then

$$X(\psi) = \nabla_X^{\mathbb{R}^{n+1}} p + \frac{1}{2}\nabla_X^{\mathbb{R}^{n+1}} \overline{H} = 0.$$

Thus ψ is a constant vector. Consequently $|p - \psi|^2 = \frac{1}{2}$. Then

$$M \subset S\left(\psi; \frac{1}{\sqrt{2}}\right).$$

Since $|\psi| = 1/\sqrt{2}$ we may assume w.l.o.g. that $\psi = (0, \ldots, 0, \frac{1}{\sqrt{2}}) \in \mathbb{R}^{n+1}$. So

$$M \subset S^n\left(\psi; \frac{1}{\sqrt{2}}\right) \cap S^n = S^{n-1}\left(\frac{1}{\sqrt{2}}\right) \times \left\{\frac{1}{\sqrt{2}}\right\}.$$

Also, for every $p \in M$ the vector $(p - \psi)$ is collinear to $\overline{H}(p)$, hence M is minimal in $S^n\left(\psi; \frac{1}{\sqrt{2}}\right)$ and then in $S^{n-1}\left(\frac{1}{\sqrt{2}}\right) \times \left\{\frac{1}{\sqrt{2}}\right\}$. Q.e.d.

3.20 Calabi's Theorem

The purpose of the present section is to report on a classical result by E. Calabi[90] on weak elliptic inequalities in Riemannian geometry.

Theorem 3.233 *Let M be a complete Riemannian manifold, whose metric is of class C^3 and has non-negative Ricci curvature. If $f(t)$ is a positive, monotone increasing, differentiable function, defined for all $t \in \mathbb{R}$ and satisfying*

$$\int_0^\infty \left(\int_0^s f(t)\,dt\right)^{-1/2} ds < \infty \tag{3.312}$$

then there is no global solution in M to the inequality

$$\Delta u \ \text{(weakly)} \ \geq f(u). \tag{3.313}$$

---

[90] Cf. E. Calabi, An extension of E. Hopf's maximum principle with an application to Riemannian geometry. Duke Math. J. **25**, 45–56 (1958).

Theorem 3.233 generalizes (to complete Riemannian manifolds of non-negative Ricci curvature) a result by R. Osserman.[91] The proof of Theorem 3.233 relies on a version of Hopf's maximum principle, also established by E. Calabi[92] and presented by us in Chapter 2 of Book II-B[93]. To prove Theorem 3.233 we need to recall a few elements of Riemannian geometry.

Let (M, g) be a n-dimensional Riemannian manifold with metric tensor $g \in C^3$. Let $d = d_g : M \times M \to [0, +\infty)$ be the distance function associated to g. The resulting metric space (M, d) isn't necessarily complete. Let $\hat{M}$ be the Cauchy completion of (M, d). The completion is a metric space with the natural distance[94] function $\hat{d}$. Then M admits a natural injection $\mathbf{j}$ into $\hat{M}$ such that

$$d(x, y) = \hat{d}\big(\mathbf{j}(x), \mathbf{j}(y)\big), \quad x, y \in M.$$

When (M, d) is not complete, for every point $x \in M$ one sets

$$\gamma(x) := \hat{d}\big(x, \hat{M} \setminus \mathbf{j}(M)\big). \tag{3.314}$$

When (M, d) is complete we set

$$\gamma(x) = +\infty\,.$$

For every $x \in M$ and every $r > 0$ let $B_r(x)$ denote the open metric ball of center x and radius r in (M, d). Also $\overline{B}_r(x)$ denotes the closure of $B_r(x)$ in M. The classical Hopf-Rinow theorem best describes the relationship between the distance function d and geodesics in (M, g).

Theorem 3.234 *Let M be a Riemannian manifold with metric tensor $g \in C^3$. Then for every $x \in M$:*

(i) *The closed metric ball $\overline{B}_r(x)$ is compact $\Longleftrightarrow r < \gamma(x)$.*

(ii) *Every geodesic segment issuing at x can be continued to any length $< \gamma(x)$. However, there is a geodesic segment of length $\gamma(x)$, issuing at x, which has no end-point in M.*

(iii) *If $\hat{d}(x, \mathbf{y}) \le \gamma(x)$ for some $\mathbf{y} \in \hat{M}$ then the greatest lower bound of the arc length of all rectifiable curves joining x and $\mathbf{y}$ can be achieved by at least one curve, necessarily a geodesic segment.*

(iv) *For any $y \in M$*

---

[91] Cf. R. Osserman, On a differential inequality and conformal mapping of surfaces. Technical Report no. 3, December 16. 1956, Stanford University.

[92] Cf. E. Calabi, *op. cit.*

[93] Functional Analysis and Partial Differential Equations: Basics of Partial Differential Equations Theory (Book II-B).

[94] If $\mathbf{x}, \mathbf{y} \in \hat{M}$ are two classes of equivalent Cauchy sequences in (M, d) and $\{x_\nu\}_{\nu \ge 1} \in \mathbf{x}$ and $\{y_\nu\}_{\nu \ge 1} \in \mathbf{y}$ are representatives of $\mathbf{x}$ and $\mathbf{y}$, then $\hat{d}(\mathbf{x}, \mathbf{y}) := \lim_{\nu \to \infty} d(x_\nu, y_\nu)$.

$$\gamma(y) - d(x, y) \le \gamma(x) \le \gamma(y) + d(x, y). \tag{3.315}$$

Theorem 3.233 was first proved by H. Hopf and W. Rinow[95] in the case M is a complete (i.e., $\gamma(x) = +\infty$) surface. For higher dimensions Theorem 3.233 was proved by G. de Rham.[96] Of course the result is classical and an excellent account is provided in [KoNo], vol. I, pp. 172–179. Yet Theorem 3.234 is a slightly generalized version of Hopf-Rinow theorem, and reported on here for that reason[97].

Eugenio Calabi (Milan, May 11, 1923) is an Italian mathematician established in United States. **C.** is the brother of Tullia Calabi-Zevi [(Milan, February 2, 1919 - Rome, January 22, 2011), an Italian journalist and writer, and a distinguished member of the Italian Jewish community]. The family of **C.** moved to United States after the promulgation of the racial laws in Italy (in 1938). **C.** graduated from MIT in 1946, and earned his Ph.D. at the University of Princeton in 1950, under the supervision of Salomon Bochner. **C.** was first a professor at the University Minnesota, and worked for the Institute for Advanced Study of Princeton in the years 1958–1959. Since 1967, **C.** was a professor at the University of Pennsylvania, where **C.** became an emerite professor in 1994. **C.** is best known for his findings on the geometry of 10 dimensional manifolds endowed with a nonzero spinor, exploited in nowadays string theory and known as *Calabi-Yau spaces*. **C.** was granted the Steele Prize of the American Mathematical Society in 1991, for his results within differential geometry. **C.** became *Commendatore dell'Ordine al Merito della Repubblica Italiana*, the 27th of December 2021.

[95] Cf. H. Hopf, W. Rinow, Über den Begriff der vollständigen differentialgeometrischen Fläche. Commentarii Mathematici Helvetici **3**, 209–225 (1931).

[96] Cf. G. de Rham, Sur la réductibilité d'un espace de Riemann. Commentarii Mathematici Helvetici **26**, 328–344 (1952).

[97] Cf. Theorem 2 by E. Calabi, *op. cit.*, pp. 49–50. In the opinion of the Authors of the present book, Calabi's deep understanding of Riemannian geometry, together with Calabi's improvements of classical results, aren't sufficiently accounted for in the present day mathematical literature devoted to differential geometry.

Let $p_0 \in M$ be a fixed point and let $N = N(p_0)$ consist of all points $p \in M$ that can be joined to p_0 by a curve[98] of length $d(p_0, p)$. The set N is open and

$$\overline{B}_{\gamma(p_0)}(p_0) \subset N.$$

Let N_1 consist of all points $p \in N$ whose shortest connection with p_0 [*per force* a geodesic $\gamma(p_0, p)$, of length $d(p_0, p)$] is unique and has no conjugate points.[99] Let N_2 consist of all points $p \in N$ that can be joined to p_0 by more than one geodesic of length $d(p_0, p)$, or by at least one such geodesic on which p_0 and p are conjugate points. Then

$$N = N_1 \cup N_2, \quad N_1 \cap N_2 = \emptyset.$$

For each geodesic issuing at p_0 and continued in N there is at most one point of that geodesic that lies in N_2 before the geodesic ceases to be a shortest connection.[100] Note that $N_1 \subset N$. Thus N_1 is an open, dense set in N whose complement N_2 has n-dimensional measure zero. Let

$$r = r_{p_0} N \to \mathbb{R}, \quad r(p) = d(p_0, p), \quad p \in N$$

be the restriction of $d(p_0, \cdot)$ to N (the *geodesic distance* function[101] from p_0). If $g \in C^3$ then the function $r = r_{p_0}$ is of class C^2 in $N_1 \setminus \{p_0\}$

$$r \in C^2\big(N_1 \setminus \{p_0\}\big),$$

while in N it is but Lipschitz

$$\big|r(p_1) - r(p_2)\big| \le d\big(p_1, p_2\big). \tag{3.316}$$

Next, we recall a few differential geometric properties of the geodesic distance function

$$r(p) = d(p_0, p), \quad p \in N(p_0) \tag{3.317}$$

from a point $p_0 \in M$. We have already recalled that r is of class C^2 in an open dense subset $N_1 \setminus \{p_0\}$ of N. Also $N_1 \setminus \{p_0\}$ is a neighborhood of p_0 and is starlike with respect to p_0. For every constant $r_1 > 0$ we set

$$S_{r_1} = S(p_0, r_1) := \big\{p \in N \ : \ r(p) = r_1\big\} \tag{3.318}$$

---

[98] Not necessarily a geodesic.

[99] Since $\gamma(p_0, p)$ is a shortest connection, the only possible conjugate points are its end points p_0 and p.

[100] This follows from the triangle inequality for distances and the unique continuation property for geodesics.

[101] The distinction between the geodesic distance r_{p_0} and the Riemannian distance function $d(p_0, \cdot)$ consists in their distinct—if M is not complete—domains of definition.

(the *geodesic sphere* of radius r_1 and center p_0). Then S_{r_1} is the boundary of $B_{r_1}(p_0) \cap N$ relative to N. The set

$$S_{r_1,1} = S_{r_1} \cap N_1$$

is the *regular part* of S_{r_1}. The regular part is[102] a C^2 hypersurface, as r is of class C^2 in $N_1 \setminus \{p_0\}$ and $\nabla r \neq 0$ there.

The unit tangent vectors at p_0 on M form a Euclidean unit $(n-1)$-dimensional sphere $S_{p_0}(M)$. Let $(y^1, \ldots, y^{n-1})$ be a local coordinate system on $S_{p_0}(M)$. Let γ_1 be a geodesic segment contained in N_1, of initial data (p_0, y_1), i.e.,

$$\gamma(0) = p_0, \quad \dot{\gamma}(0) = y_1, \quad \gamma(s) \in N_1$$

with the initial tangent vector y_1 lying in the domain of the chosen local chart, and having the local coordinates $\left(y_1^1, \ldots, y_1^{n-1}\right)$. Let p_1 be a point on the geodesic γ_1, at an arc distance r_1 from p_0 along γ_1. Then

$$r\left(p_0, p_1\right) = r_1$$

and the numbers $\left(y_1^1, \ldots, y_1^{n-1}, r_1\right)$ may be regarded as local coordinates of p_1. More precisely, there is $\epsilon > 0$ such that

$$C\left(p_1, \epsilon\right) \subset N_1 .$$

Here $C\left(p_1, \epsilon\right)$ consists of points p similarly determined by the numbers $\left(y^1, \ldots, y^{n-1}, r\right)$ satisfying

$$\sum_{\alpha=1}^{n-1} \left(y^\alpha - y_1^\alpha\right)^2 < \epsilon^2, \quad 0 < r < r_1 + \epsilon . \tag{3.319}$$

The quantities $\left(y^1, \ldots, y^{n-1}, r\right)$, thought of as functions of $p \in C\left(p_1, \epsilon\right)$, form a local coordinate system on $C\left(p_1, \epsilon\right)$, of class C^2.

Definition 3.235 The region $C\left(p_1, \epsilon\right)$ is called an *open, conical neighborhood* of p_1 with vertex at p_0. □

Without loss of generality, one may choose the local coordinate system $(x^1, \ldots, x^n)$ in $C(p_1, \epsilon)$ to be precisely $(y^1, \ldots, y^{n-1}, r)$. The part of the geodesic sphere S_r contained in $C(p_1, \epsilon)$ is described by the equation

$$x^n = r \quad (0 < r < r_1 + \epsilon) \tag{3.320}$$

and then the local parametric equations of S_r will be

$$x^\alpha = y^\alpha, \quad x^n = r, \quad 1 \leq \alpha \leq n-1. \tag{3.321}$$

---

[102] But locally, for $S_{r_1,1}$ isn't necessarily connected.

Let

$$a_{\alpha\beta}(y,r)\,dy^\alpha \otimes dy^\beta, \quad b_{\alpha\beta}(y,r)\,dy^\alpha \otimes dy^\beta$$

be respectively the first and second fundamental forms of $S_r \cap C(p_1,\ \epsilon)$. Then

$$r_{,i} = \mathbf{N}_i$$

where $\mathbf{N}_i$ are the components of the unit normal vector field on S_r. Also

$$r_{,ij} = -b_{\alpha\beta}\,a^{\alpha\rho}\,a^{\beta\sigma}\,\frac{\partial x^k}{\partial y^\rho}\,\frac{\partial x^\ell}{\partial y^\sigma}\,g_{ik}\,g_{j\ell}\,.$$

Let $\partial/\partial r$ be the operation of differentiation along the geodesic from p_0, i.e., along the normal vector $\mathbf{N}^i$, or

$$\frac{\partial}{\partial r} = \mathbf{N}^i\,\frac{\partial}{\partial x^i}\,.$$

Then

$$\frac{\partial}{\partial r}\,b^\beta_\alpha = b^\gamma_\alpha\,b^\beta_\gamma + a^{\beta\gamma}\,R_{ijkh}\,\frac{\partial x^i}{\partial y^\gamma}\,\mathbf{N}^j\,\frac{\partial x^k}{\partial y^\alpha}\,\mathbf{N}^h$$

where $b^\beta_\alpha = b_{\alpha\gamma}\,a^{\gamma\beta}$. Consequently

$$\Delta r = -b^\alpha_\alpha = -\sum_{\nu=1}^{n-1} \kappa_\nu = -(n-1)H \tag{3.322}$$

where κ_ν and H are respectively the principal normal curvatures and the mean curvature of S_r. Also

$$\frac{\partial H}{\partial r} = \frac{1}{n-1}\Big(b^\gamma_\alpha\,b^\alpha_\gamma + R^i_{jih}\,\mathbf{N}^j\,\mathbf{N}^h\Big) = \tag{3.323}$$

$$= \frac{1}{n-1}\Big[\sum_{\nu=1}^{n-1}\kappa_\nu^2 + R_{jh}\,\mathbf{N}^j\,\mathbf{N}^h\Big]$$

where $R_{jh} = R^i_{jih}$ is the Ricci curvature of M. As $r \to 0$ the geodesic spheres S_r are asymptotic to Euclidean spheres of radius r with the normal vector $\mathbf{N}^i$ pointing outward. Consequently

$$\lim_{r\to 0} H = -\infty\,. \tag{3.324}$$

Also (3.323) yields the differential inequality

$$\frac{\partial H}{\partial r} \geq H^2 + \frac{1}{n-1}\,R_{jh}\,\mathbf{N}^j\,\mathbf{N}^h\,. \tag{3.325}$$

Theorem 3.236 *Let M be a n-dimensional Riemannian manifold, with metric tensor $g \in C^3$. Let $p_0 \in M$ be a point and let $N = N(p_0)$ be the domain of definition of the geodesic distance function $r(p) = d(p_0, p)$ from p_0. If the Ricci tensor R_{ij} is positive semi-definite and Δ is the Laplace-Beltrami operator then the inequality*

$$\Delta r \ \text{(weakly)} \ \leq \frac{n-1}{r} \tag{3.326}$$

holds everywhere in N. In particular, if M is complete then (3.326) *holds everywhere in M.*

Proof We decompose N into two complementary subsets N_1 (open and dense) and N_2 (of measure zero) as we did previously, and prove that (3.326) holds in N_1 in the strict sense. Indeed [by (3.324)–(3.325) and $R_{ij}\xi^i\xi^j \geq 0$]

$$H \geq -\frac{1}{r}\,. \tag{3.327}$$

Then (3.327) together with (3.322) yields the seek for result in N_1. Let $p \in N_2$. The geodesic distance function r is continuous at p. Let us choose a geodesic γ_1 of shortest length $r(p)$ joining p_0 to p. Let p_ϵ be a point of γ_1 at a distance $\epsilon > 0$, arbitrarily small, from p_0. By replacing p_0 by p_ϵ one observes that $p \in N_1(p - \epsilon)$, as the portion of γ_1 between p_ϵ and p is the unique geodesic of length $r(p) - \epsilon$ joining p_ϵ to p, and the end points aren't conjugate. Let $V_{p,\epsilon}$ be a neighborhood of p such that

$$V_{p,\epsilon} \subset N_1(p_\epsilon) \cap N(p_0).$$

Let

$$r_{p,\epsilon}(q) := d(p_\epsilon, q), \quad q \in V_{p,\epsilon}\,.$$

Then

$$\Delta r_{p,\epsilon} \leq \frac{n-1}{r_{p,\epsilon}}\,.$$

Next (by the triangle inequality)

$$r - r_{p,\epsilon} \leq \epsilon$$

with equality holding at p. Thus

$$\Delta r_{p,\epsilon} \leq \frac{n-1}{r-\epsilon}$$

everywhere in $V_{p,\epsilon}$, and $r - r_{p,\epsilon}$ achieves its maximum value ϵ at p. Yet ϵ is arbitrary, so (3.326) holds in N.

Corollary 3.237 *Let $\phi(t)$ be a function of class C^2, monotone increasing (respectively, decreasing), defined for $0 < a \le t \le b$. Under the assumptions of Theorem 3.236, let $v(p)$ be the function*

$$v(p) = \phi\big(r(p)\big), \quad p \in N(p_0) \cap \Big[B_b(p_0) \setminus \overline{B}_a(p_0)\Big].$$

Then

$$\Delta v \ \text{(weakly)} \ \le \phi''(r) + \frac{n-1}{r}\,\phi'(r) \tag{3.328}$$

$\Big[$*respectively*

$$\Delta v \ \text{(weakly)} \ \ge \phi''(r) + \frac{n-1}{r}\,\phi'(r)\Big].$$

Proof We may assume w.l.o.g. that $\phi'(t) \ge 0$. A calculation shows that in $N_1 \cap \Big[B_b(p_0) \setminus \overline{B}_a(p_0)\Big]$

$$\Delta v = \phi''(r)\, g^{ij}\, \frac{\partial r}{\partial x^i}\, \frac{\partial r}{\partial x^j} + \phi'(r)\, \Delta r \le$$

$$\le \phi''(r) + \frac{n-1}{r}\,\phi'(r).$$

On the other hand, if $p \in N_2 \cap \Big[B_b(p_0) \setminus \overline{B}_a(p_0)\Big]$ then a repetition of the arguments in the proof of Theorem 3.236 yields the seek for inequality in the weak sense.

Let us apply Corollary 3.237 to the function

$$\phi(r) = \begin{cases} \dfrac{1}{2\pi}\, \log r & \text{for } n = 2, \\[2ex] -\dfrac{1}{4}\, \pi^{-n/2}\, \Gamma\Big(\dfrac{n-2}{2}\Big)\, r^{2-n} & \text{for } n > 2, \end{cases}$$

(Green's function in Euclidean space). If M is a complete Riemannian manifold of non-negative Ricci curvature then Corollary 3.237 yields an approximation of the local fundamental singularity of the Laplace-Beltrami equation, namely

$$G(p_0, p) = \phi\big(d(p_0, p)\big)$$

which is everywhere superharmonic in each variable, i.e.,

$$\Delta_p\, G(p_0, p) \ \text{(weakly)} \ \le 0 \tag{3.329}$$

for all points $p_0, p \in M$.

As another application of Theorem 3.236 one may prove Theorem 3.233. To this end, let us consider the singular ODE

$$\frac{d^2\phi}{dt^2} + \frac{n-1}{t}\frac{d\phi}{dt} = f\big(\phi(t)\big) \tag{3.330}$$

with the initial conditions

$$\phi(0) = a, \quad \phi'(0) = 0, \tag{3.331}$$

where a is to specified later on. The problem (3.330)–(3.331) has a unique differentiable solution $\phi(t)$ in a neighborhood of $t = 0$. We start by showing that, under the assumption (3.312), for every $a \in \mathbb{R}$ there is $R_a > 0$ such that $\phi(t) \to \infty$ as $t \to R_a$.

The solution $\phi(t)$ to (3.330)–(3.331) is monotone increasing and strictly convex for $t > 0$. Indeed, by multiplying (3.330) by t^{n-1}

$$\frac{d}{dt}\Big(t^{n-1}\,\phi'(t)\Big) = t^{n-1}\,f\big(\phi(t)\big) > 0$$

hence (by the mean value theorem) $\phi'(t) > 0$. Next [by multiplying (3.330) b y t and differentiating the resulting identity]

$$t\,\frac{d^3\phi}{dt^3} + n\,\frac{d^2\phi}{dt^2} = f(\phi(t)) + t\,f'(\phi(t))\,\phi'(t)$$

or

$$\frac{d}{dt}\left(t^n\,\frac{d^2\phi}{dt^2}\right) = t^{n-1}\,f(\phi(t)) + t^n\,f'(\phi(t))\,\phi'(t) > 0$$

hence (again by the mean value theorem)

$$\phi''(t) \geq 0, \quad ; \forall\, t > 0.$$

Thus $\phi'(t)$ is positive and monotone increasing for $t > 0$. Let us multiply (3.330) by $\phi'(t)$ and integrate with respect to t in the resulting identity

$$\int_{\phi(t_1)}^{\phi(t)} f(s)\,ds = (n-1)\int_{t_1}^{t}\frac{\phi'(s)^2}{s}\,d\,s + \frac{1}{2}\,\phi'(t)^2 - \frac{1}{2}\,\phi'(t_1)^2 \geq$$

$$\geq \frac{1}{2}\,\phi'(t)^2 + \phi'(t-1)^2\left[(n-1)\log\frac{t}{t_1} - \frac{1}{2}\right]$$

for $0 < t_1 < t$. In particular for $t \geq 2t_1$ (and $n \geq 2$)

$$\frac{\phi'(t)}{\left(2\int_{\phi(t_1)}^{\phi(t)} f(s)\,ds\right)^{1/2}} \geq 1, \quad 0 < 2t_1 \leq t.$$

Let us integrate with respect to t in the last inequality. We obtain

$$t - 2t_1 \leq \int_{\phi(2t_1)}^{\phi(t)} \left[2\int_{\phi(t_1)}^{y} f(s)\,ds\right]^{-1/2} dy, \quad 0 < 2t_1 \leq t.$$

By (3.312) the right-hand side of the last inequality is bounded for all values of $\varphi(t)$ (with t_1 fixed), thus providing an upper bound for the domain of regularity of $\phi(t)$. Thus there is $R_a > 0$ such that either $\phi(t) \to \infty$, or $\phi'(t) \to \infty$, as $t \to R_a$. Yet [by the convexity of $\phi(t)$ together with (3.330)] $\phi'(t) \to \infty \Longrightarrow \phi(t) \to \infty$. Q.e.d.

Let us assume there is a function u on M satisfying (3.313). Let $p_0 \in M$ and let $a \in \mathbb{R}$ such that

$$a < u(p_0). \tag{3.332}$$

Let $r(p)$ be the geodesic distance from p_0, i.e., $r(p) = d(p_0, p)$, and let us consider the solution $\phi(t)$ to the problem (3.330)–(3.331) with initial value a. By the previous argument, there is $R_a > 0$ such that $\phi(t) \to \infty$ as $t \to R_a$. Let

$$v(p) := \phi(r(p)), \quad p \in B_{R_a}(p_0).$$

As M is complete, the closed ball $\overline{B}_{R_a}(p_0)$ is compact, hence $v(p) \to \infty$ as $p \to \partial B_{R_a}(p_0)$. By Corollary 3.237

$$\Delta v \ \text{(weakly)} \ \leq f(v) \tag{3.333}$$

everywhere[103] in $B_{R_a}(p_0)$. Let us consider the function

$$w = u - v \quad \text{in} \ B_{R_a}(p_0).$$

As $w \to -\infty$ at the boundary, it must achieve its maximum value at an interior point p_1. By (3.332)

$$w(p_0) > 0,$$

so the maximum value $w(p_1)$ is positive. Yet [by (3.313) and (3.333)]

$$\Delta w \ \text{(weakly)} \ \geq f(u) - f(v)$$

[103] This includes the point p_0 because $\phi'(0) = 0$ so that v is differentiable at p_0.

so that in a neighborhood of p_1 [by the monotonicity of f]

$$\Delta w \geq 0$$

a contradiction. This accounts for the impossibility of solving (3.313) globally in M, as stated. Q.e.d.

References

[KoNo] S. Kobayashi, K. Nomizu, *Foundations of Differential Geometry*, vol. I (Interscience Publishers, New York, 1963) (vol. II, 1969)

[HaWi1] P. Hartman, A. Winter, On the existence of Riemannian manifolds which cannot carry non-constant analytic or harmonic functions in the small. Amer. J. Math. **75**, 260–276 (1953)

[Lic] A. Lichnerowicz, *Applications harmoniques et variétés kähkeriennes*, Symposia Mathematica, vol. III. Istituto Nazionale di Alta Matematica. (Academic Press, London-New York, 1970), pp. 341–402

[Cher] S.-S. Chern, An elementary proof of the existence of isothermal parameters on a surface. Proc. Amer. Math. Soc. **6**, 771–782 (1955)

[Mor] C.B. Morrey, On the solutions of quasi-linear elliptic partial differential equations. Trans. Amer. Math. Soc. **43**, 126–166 (1938)

[HaWi2] P. Hartman, A. Winter, On uniform Dini conditions in the theory of linear partial differential equations of elliptic type. Amer. J. Math. **77**, 329–354 (1955)

[Mac] G.W. Mackey, *Lectures on the Theory of Functions of a Complex Variable* (D. Van Nostrand Co., Inc, Princeton, New Jersey, 1967)

[EeLe] J. Eells, L. Lemaire, *Selected topics in harmonic maps, CBMS Regional Conference Series in Mathematics 50* (American Mathematical Society, Providence, RI, 1983)

[Fuj] H. Fujimoto, On the number of exceptional values of the Gauss map of minimal surfaces. J. Math. Soc. Japan **40**, 235–247 (1988)

[BaCo] J.L.M. Barbosa, A.G. Colares, Minimal Surfaces in $\mathbb{R}^3$. Lecture Notes in Mathematics, No. 1195 (Springer, Berlin-New York, 1986)

[Tak] T. Takahashi, Minimal immersions of Riemannian manifolds. J. Math. Soc. Japan **18**, 380–385 (1966)

[BGM] M. Berger, P. Gauduchon, E. Mazet, *Le spectre d'une variété Riemanniene*, Lecture Notes in Mathematics, vol. 194. (Springer-Verlag, Berlin-New York, 1971)

[Hel] S. Helgason, *Differential Geometry and Symmetric Spaces* (Academic, New York-London, 1962)

[Ste] N. Steenrod, *The Topology of Fibre Bundles* (Princeton University Press, Princeton-New Jersey, 1974)

[Saf] R. Safarevici, *Bazele Geometriei Algebrice*, Editura Ştiinţifică şi Enciclopedică, Bucureşti (1976) (translated from Russian into Romanian by Andrei Iacob)

[San] A. Sanini, Applicazioni armoniche fra i fibrati tangenti di varietà Riemanniane. Boll. U.M.I., (6)2-**A**, 55–63 (1983)

[One] B. O'Neill, The fundamental equations of a submersion. Michigan Math. J. **13**, 459–469 (1966)

[EeSa] J. Eells, J.H. Sampson, Harmonic mappings of Riemannian manifolds. Amer. J. Math. **86**, 109–160 (1964)

[Dol] A. Dold, *Lectures on Algebraic Topology* (Springer, Berlin-Heidelberg-New York, 1980)

[Rat] A. Ratto, On harmonic maps between S^3 and S^2 of prescribed Hopf invariant. Mathem. Proc. Cambridge Phil. Soc. (2) **104**, 273–276 (1988)

[Bai] P. Baird, *Harmonic Maps with Symmetry, Harmonic Morphisms and Deformations of Metrics* (Pitman Books Ltd., London, 1983)

[Fri] A. Friedman, Interior estimates for parabolic systems of partial differential equations. J. Math. & Mech. **7**, 393–418 (1958)

[Har] P. Hartman, On homotopic harmonic maps. Canad. J. Math. **19**, 673–687 (1967)

[Gol] S.I. Goldberg, *Curvature and Homology*,

[Boo] W.M. Boothby, *An Introduction to Differentiable Manifolds and Riemannian Geometry* (Academic, New York-London, 1975)

[Spi] M. Spivak, *Calculus on Manifolds* (W.A. Benjamin Inc, Reading-Mass, 1965)

[EeWo] J. Eells, J.C. Wood, Restrictions of harmonic maps of surfaces. Topology **15**, 263–266 (1976)

[EeLe1] J. Eells, L. Lemaire, A report on harmonic maps. Bull. Lond. Math. Soc. **10**, 1–68 (1978)

[Nom1] K. Nomizu, Some results in Élie Cartan's theory of isoparametric families of hypersurfaces. Bull. Amer. Math. Soc. (6) **79**, 1184–1188 (1973)

[Nom2] K. Nomizu, Élie Cartan's work on isoparametric families of hypersurfaces. Proc. Symp. Pure Math. **27**, 191–200 (1975)

[Che0] B-Y. Chen, *Geometry of Submanifolds* (Dekker, N.Y., 1973)

[Car] E. Cartan, Familles de surfaces isomparamétriques dans l'espace à courbure constante. Annali di Matematica **17**, 177–191 (1938)

[Mun] H.F. Munzner, Isoparametrische Hyperflächen in Sphären, I-II. Math. Ann. **251**, 57–71 (1980); **256**, 215–232 (1981)

[Che2] B.-Y. Chen, Constant ratio hypersurfaces. Soochow J. Math. **27**, 353–362 (2001)

[Che3] B.-Y. Chen, O.J. Garay, Constant mean curvature hypersurfaces with constant $\delta-$ invariant. Int. J. Math. Math. Sci. **67**, 4205–4216 (2003)

[Che4] C. Chen, H. Sun, L. Tang, On translation hypersurfaces with constan mean curvature in $(n+1)-$ dimensional spaces. J. Bejing Inst. Technol. **12**, 322–325 (2003)

[MoRo] S. Montiel, A. Ros, Compact hypersurfaces: the Alexandrov theorem for higher order mean curvatures. Diff. Geom. (Pitman monographs and surveys in pure and applied mathematics, Longman Scientific & Technical, Harlow 1991), pp. 279–296

[Rip] J.B. Ripoll, A characterization of spheres in Euclidean space. Arch. Math. **51**, 474–475 (1988)

[Ros2] A. Ros, Compact hypersurfaces with constant higher order mean curvatures. Rev. Mat. Iberoamericana **3**, 447–453 (1987)

[Ros3] A. Ros, Compact hypersurfaces with constant scalar curvature and a congruence theorem. J. Diff. Geom. **27**, 215–223 (1988)

[BrGo] S. Breuer, D. Gottlieb, Explicit characterization of spherical curves. Proc. Amer. Math. Soc. **27**, 126–127 (1971)

[Bla] D.E. Blair, *Contact Manifolds in Riemannian Geometry*, vol. 509 (Springer, Berlin-Heidelberg-New York, 1976)

[Hsi] C.C. Hsiung, Some integral formulas for closed hypersurfaces. Math. Scand. **2**, 286–294 (1954)

[Des3] S. Deshmukh, Real hypersurfaces in a Euclidean complex space form. Quart. J. Math. **58**, 313–317 (2007)

[Des4] S. Deshmukh, Submanifolds of positive Ricci curvature in a Euclidean space. Ann. di Math. **187**, 59–65 (2008)

[OdDe] H. Al-Odan, S. Deshmukh, Spherical submanifolds of a Euclidean space. Quart. J. Math. **53**, 249–256 (2002)

[Tom] T. Tompkins, Isometric embedding of flat manifolds in Euclidean space. Duke Math. J. **5**, 58–61 (1930)

[ChKu] S.-S. Chern, N.H. Kuiper, Some theorems on the isometric imbedding of compact Riemannian manifolds in Euclidean space. Ann. Math. **56**, 422–430 (1952)

[Ots23] T. Otsuki, On the existence of solutions of a system of quadric equations and its geometrical applications. Proc. Japan Acad. **29**, 99–100 (1953)

[Bel] M.A. Beltagy, Isometric embedding of a compact Riemannian manifold into a sphere. Indian J. Pure Appl. Math. **14**, 1406–1411 (1983)

[DeGw] S. Deshmukh, M.A. Al-Gwaiz, Compact hypersurface in even dimensional Euclidean space and in the sphere. Quart. J. Math. **45**, 151–157 (1994)

[Gio] E. De Giorgi, Una estensione del teorema di Bernstein. Ann. Della Scuola, Normale Superiore di Pisa, Scienze Fis. Mat.III, XIX, I (1965), pp. 79–85

[Sim24] J. Simons, Minimal varieties in Riemannian manifolds. Ann. Math. **88**, 62–105 (1968)

[NoSm] K. Nomizu, B. Smyth, On the Gauss mapping for hypersurfaces of constant mean curvature in the sphere. Comm. Math. Helv. **44**, 484–490 (1969)

[Mar] S. Markvorsen, A sufficient condition for a compact immersion to be spherical. Math. Z. **183**, 407–411 (1983)

[CoIt] L. Coghlan, Y. Itokawa, On the sectional curvature of compact hypersurfaces. Proc. Amer. Math. Soc. **109**, 215–221 (1990)

[DsGw] S. Deshmukh, M.A. Al-Gwaiz, A sufficient condition for a compact hypersurfaces in a sphere to be a sphere. Yokohama Math. **40**, 105–109 (1993)

[Sak25] M. Sakaki, Remarks on the rigidity and stability of minimal submanifolds. Proc. Amer. Math. Soc. **106**, 793–795 (1989)

[Eji] N. Ejiri, Compact minimal submanifolds of a sphere with positive Ricci curvature. J. Math. Soc. Japan **31**, 251–256 (1979)

[She] Y.-B. Shen, Curvature pinching for the three-dimensional minimal submanifolds in a sphere. Proc. Amer. Math. Soc. **115**, 791–795 (1992)

[Des11] S. Deshmukh, Normal curvature of mimimal submanifolds in a sphere. Glas. Math. J. **39**, 29–33 (1997)

[Des12] S. Deshmukh, Rigidity of compact minimal submanifolds in a sphere. Pacific J. Math. **193**, 31–44 (2000)

[Che5] B.-Y. Chen, Some open problems and conjectures on submanifolds of finite type. Soochow J. Math. **17**, 169–188 (1991)

[Jia] G.Y. Jiang, 2- Harmonic maps and their first and second variation formulas. Chinese Ann. Math. Ser. A **77**, 130–144 (1986)

[OuTa] Y.-L. Ou, L. Tang, On the generalized Chen's conjecture on biharmonic submanifolds. Michigan Math. J. **61**, 531–542 (2012)

Chapter 4
Riemannian Foliations

Let M be a n-dimensional C^∞ manifold, and let $\mathscr{F}$ be a codimension q foliation of M. Let $B_T^1 = B_T^1(\mathscr{F}) = B_T^1(M, \mathscr{F})$ be the total space of the transverse frame bundle $\mathrm{GL}(q, \mathbb{R}) \to B_T^1 \to M$, and let $p_T^1 : B_T^1 \to M$ be the projection. Let $\theta_T^1 \in C^\infty\big(T^*\big(B_T^1\big) \otimes \mathbb{R}^q\big)$ be the canonical 1-form, i.e.,

$$\theta^1_{T,z} : T_z\big(B_T^1\big) \to \mathbb{R}^q , \quad z \in B_T^1 ,$$

$$\theta^1_{T,z} := z^{-1} \circ \pi_x \circ (d_z p_T^1), \quad x = p_T^1(z) \in M.$$

Here $\pi_x : T_x(M) \to Q_x$ is the projection on the transverse vector space at x

$$Q_x = \nu(\mathscr{F})_x = T_x(M)\big/T(\mathscr{F})_x .$$

Let $\mathscr{F}_T^1$ be the lifted foliation, i.e., the foliation of B_T^1 whose tangent bundle $P_T^1 = T(\mathscr{F}_T^1)$ is the p-dimensional distribution on B_T^1 ($p := n - q$) defined by

$$P^1_{T,z} = \big\{X \in T_z(B_T^1) \;:\; i_X\, \theta^1_{T,z} = 0, \quad i_X\, (d\theta_T^1)_z = 0\big\}, \quad z \in B_T^1 .$$

4.1 Transverse Riemannian Structures, Transverse Metrics, Bundle-Like Metrics

Let E_T^1 be a transverse $O(q)$-structure on $(M, \mathscr{F})$, i.e., E_T^1 is the total space of a principal subbundle $O(q) \to E_T^1 \to M$ of $\mathrm{GL}(q, \mathbb{R}) \to B_T^1 \to M$ such that

$$P^1_{T,z} \subset T_z\big(E_T^1\big), \quad z \in E_T^1 .$$

E. Barletta et al., *Differential Geometry*, Infosys Science Foundation Series,
https://doi.org/10.1007/978-981-96-1631-2_4

Definition 4.1 A pair $(\mathscr{F}, E_T^1)$ consisting of a codimension q foliation $\mathscr{F}$ and a transverse $\mathrm{O}(q)$-structure is called a Riemannian foliation. □

For every $x \in M$ let us consider the $\mathbb{R}$-bilinear form $\langle \cdot , \cdot \rangle_x$ on Q_x defined as follows. Let $z \in (E_T^1)_x$ and let $r, s \in Q_x$, so that

$$r = \lambda^j z(e_j), \quad s = \mu^j z(e_j), \quad \lambda^j, \mu^j \in \mathbb{R}, \quad 1 \le j \le q,$$

where $\{e_j : 1 \le j \le q\} \subset \mathbb{R}^q$ is the canonical linear basis. Finally, let us set

$$\langle r, s \rangle_x := \lambda \cdot \mu,$$

$$\lambda = (\lambda^1, \cdots, \lambda^q), \quad \mu = (\mu^1, \cdots \mu^q), \quad \lambda, \mu \in \mathbb{R}^q,$$

where a dot denotes the ordinary Euclidean inner product on $\mathbb{R}^q$. The definition of $\langle r, s\rangle_x$ doesn't depend upon the choice of $z \in (E_T^1)_x$. Composing $\langle \cdot , \cdot \rangle_x$ with the projection $\pi_x : T_x(M) \to Q_x$ yields the symmetric $\mathbb{R}$-bilinear form

$$g_{T,x} : T_x(M) \times T_x(M) \to \mathbb{R},$$

$$g_{T,x}(v, w) = \langle \pi_x v, \pi_x w \rangle_x, \quad v, w \in T_x(M).$$

Moreover, for any $X, Y \in \mathfrak{X}(M)$ let $g_T(X, Y) : M \to \mathbb{R}$ be defined by

$$g_T(X, Y)(x) := g_{T,x}(X_x, Y_x), \quad x \in M.$$

Exercise 4.2 Show that $g_T(X, Y) \in C^\infty(M, \mathbb{R})$. □

Let us also consider the positive semidefinite $C^\infty(M, \mathbb{R})$-bilinear form

$$g_T : \mathfrak{X}(M) \times \mathfrak{X}(M) \to C^\infty(M, \mathbb{R}), \quad (X, Y) \longmapsto g_T(X, Y).$$

Definition 4.3 The $(0, 2)$-tensor field g_T is called the *transverse metric* associated to the transverse $O(q)$-structure E_T^1. □

Let $\mathrm{Ker}(g_T) \to M$ be the vector bundle defined by

$$\mathrm{Ker}(g_T)_x := \{v \in T_x(M) : g_{T,x}(v, w) = 0, \quad \forall\ w \in T_x(M)\}, \quad x \in M.$$

Then $\mathrm{Ker}(g_T) = T(\mathscr{F})$.

Let E_T^1 be a transverse $\mathrm{O}(q)$-structure on $(M, \mathscr{F})$, and let g_T be the corresponding transverse metric. Given a tangent vector field $X \in \mathfrak{X}(M)$, the *Lie derivative* $\mathscr{L}_X g_T$ is

$$(\mathcal{L}_X g_T)(Y, Z) := X(g_T(Y, Z)) - g_T([X, Y], Z) \\ - g_T(Y, [X, Z]), \quad Y, Z \in \mathfrak{X}(M).$$

Exercise 4.4 Show that $\mathcal{L}_X g_T = 0$ for any $X \in \mathfrak{X}_{\mathcal{F}}$. □

Here $\mathfrak{X}_{\mathcal{F}} := C^\infty(T(\mathcal{F}))$. The property in Exercise 4.4 characterizes transverse metrics.

Proposition 4.5 *Let M be a C^∞ manifold, and let $g_T : \mathfrak{X}(M) \times \mathfrak{X}(M) \to C^\infty(M)$ be a $C^\infty(M)$-bilinear form such that*

i) $g_T(Y, Y) \geq 0$ *everywhere on M, for any $Y \in \mathfrak{X}(M)$,*

ii) $\dim_{\mathbb{R}} \operatorname{Ker}(g_T)_x =$ *constant, $x \in M$,*

iii) $\mathcal{L}_X g_T = 0$ *for any $X \in C^\infty(\operatorname{Ker}(g_T))$. Then* $\operatorname{Ker}(g_T)$ *is integrable and g_T is a transverse metric for the foliation tangent to* $\operatorname{Ker}(g_T)$.

The proof is straightforward. A vector field $V \in \mathfrak{X}(M)$ is *foliate* if $X \in \mathfrak{X}_{\mathcal{F}} \Longrightarrow [V, X] \in \mathfrak{X}_{\mathcal{F}}$. The set $L(M, \mathcal{F})$ is a Lie subalgebra of $\mathfrak{X}_{\mathcal{F}}$.

Definition 4.6 A Riemannian metric g on M is said to be *bundle-like* for the foliation $\mathcal{F}$ if, for any open set $U \subset M$ and any $Y, Z \in \mathfrak{X}(U)$, if Y, Z are foliate and orthogonal to $T(\mathcal{F})$, then $g(Y, Z)$ is a basic function on U. □

The notion of a bundle-like metric is due to Reinhart, [Rei]. Let g be a bundle-like Riemannian metric on $(M, \mathcal{F})$, and let $P^\perp$ be the orthogonal complement of $P = T(\mathcal{F})$. Let $\pi^\perp : T(M) \to P^\perp$ be the projection associated to the direct sum decomposition $T(M) = P \oplus P^\perp$. Let us set

$$g_T(Y, Z) := g(\pi^\perp Y, \pi^\perp Z), \quad Y, Z \in \mathfrak{X}(M). \tag{4.1}$$

Exercise 4.7 Show that the $(0, 2)$-tensor field g_T defined by (4.1) is a transverse metric for $\mathcal{F}$. □

Therefore, for every bundle-like metric there is a naturally associated transverse metric. Conversely

Proposition 4.8 *Let $(\mathcal{F}, E_T^1)$ be a Riemannian foliation on M, and let g_T be the corresponding transverse metric. There exist Riemannian metrics g on M such that*

i) *g is bundle-like for $\mathcal{F}$,*

ii) *g_T is the transverse metric associated to g.*

Proof Let g' be an arbitrary Riemannian metric on M, and let P' be the orthogonal complement of $P = T(\mathcal{F})$ with respect to g'. Let $\pi' : T(M) \to P'$ be the orthogonal projection, and let us set

$$g(Y, Z) := g'(Y - \pi' Y, Z - \pi' Z) + g_T(Y, Z), \quad Y, Z \in \mathfrak{X}(M).$$

Q.e.d.

4.2 Bott Connection, Canonical Adapted Connection

Let $(M, \mathscr{F})$ be a foliated manifold, and let $Q = \nu(\mathscr{F}) = T(M)/T(\mathscr{F})$ be its transverse bundle. Let $\pi : T(M) \to Q$ be the projection. Let us consider the partial connection $\mathring{\nabla}$ given by

$$\mathring{\nabla}_X s = \pi \left[X, Y_s\right], \quad X \in \mathfrak{X}_{\mathscr{F}}, \quad s \in C^\infty(Q), \tag{4.2}$$

where $Y_s \in \mathfrak{X}(M)$ is a tangent vector field such that $\pi Y_s = s$.

Definition 4.9 $\mathring{\nabla}$ is called the *Bott connection* of $(M, \mathscr{F})$. □

The definition of $\mathring{\nabla}_X s$ doesn't depend upon the choice of $Y_s \in \mathfrak{X}(M)$ such that $\pi Y_s = s$. Indeed, if $Y'_s \in \mathfrak{X}(M)$ is another tangent vector field such that $\pi Y'_s = s$. Then $Y'_s - Y_s \in \mathrm{Ker}(\pi) = T(\mathscr{F})$, i.e., $Y'_s = Y_s + X'$ for some $X' \in \mathfrak{X}_{\mathscr{F}}$. Finally,

$$\pi \left[X, Y'_s\right] = \pi \left[X, Y_s\right] + \pi \left[X, X'\right]$$

and $\pi [X, X'] = 0$ because $P = T(\mathscr{F})$ is involutive. Q.e.d.

Let g be a Riemannian metric on M, and let $P^\perp$ be the g-orthogonal complement of P. Let $\pi^\perp : T(M) \to P^\perp$ be the projection. Let $\sigma : Q \to P^\perp$ be the bundle isomorphism given by

$$\sigma(s) = \pi^\perp Y_s, \quad s \in C^\infty(Q),$$

where $Y_s \in \mathfrak{X}(M)$ is chosen such that $\pi Y_s = s$. Let ∇^g be the Levi-Civita connection of the Riemannian manifold (M, g), and let us set

$$\nabla_X s = \begin{cases} \pi \left[X, \sigma(s)\right] & \text{if } X \in C^\infty(P), \\ \\ \pi \nabla^g_X \sigma(s) & \text{if } X \in C^\infty(P^\perp), \end{cases} \tag{4.3}$$

for any $s \in C^\infty(Q)$.

Lemma 4.10 1) *The mapping* $\nabla : \mathfrak{X}(M) \times C^\infty(Q) \to C^\infty(Q)$ *given by* (4.3) *is a connection in the vector bundle* $Q \to M$.

2) *For every* $s \in C^\infty(Q)$ *let* $\nabla s : \mathfrak{X}(M) \to C^\infty(Q)$ *be* Q*-valued differential* 1*-form on* M *given by* $(\nabla s)(X) = \nabla_X s$ *for any* $X \in \mathfrak{X}(M)$. *Then* $\left(\nabla s\right)\big|_{\mathfrak{X}_{\mathscr{F}}} = \mathring{\nabla} s$.

3) *If* R_∇ *is the curvature tensor field of* ∇, *then*

$$R_\nabla(X, X') = 0, \quad X, X' \in \mathfrak{X}_{\mathscr{F}} .$$

As customary, we denote by $\Omega^0(Q) = C^\infty(Q)$ the space of all (globally defined) C^∞ sections in Q. Also

$$\Omega^r(Q) = C^\infty\left(\Lambda^r T^*(M) \otimes Q\right), \quad r \geq 1,$$

[the $C^\infty(M)$-module of all Q-valued differential r-forms on M]. With these notations, a connection ∇ in Q is a map $\nabla : \Omega^0(Q) \to \Omega^1(Q)$.

Definition 4.11 A connection in the transverse bundle $Q = \nu(\mathscr{F})$ possessing the property 2) in Lemma 4.10 is said to be *adapted* to the foliation $\mathscr{F}$. □

An analog of the torsion tensor field for a linear connection may be associated with any connection ∇ in the transverse bundle $Q \to M$ by setting

$$T_\nabla(Y,\, Y') := \nabla_Y \pi\, Y' - \nabla_{Y'} \pi\, Y - \pi\left[Y,\, Y'\right], \quad Y,\, Y' \in \mathfrak{X}(M).$$

Lemma 4.12 *Let* ∇ *be the connection in* $Q \to M$ *given by* (4.3). *Then* $T_\nabla = 0$.

The proof is straightforward.

Let g be a Riemannian metric on the foliated manifold $(M,\, \mathscr{F})$, and let us set

$$g_Q(s,\, r) := g\big(\sigma(s),\, \sigma(r)\big), \quad s,\, r \in Q,$$

so that g_Q is a Riemannian bundle metric on Q.

Lemma 4.13 *g is bundle-like for $\mathscr{F}$ if and only if g_Q is holonomy invariant, i.e., $\mathscr{L}_X g_Q = 0$ for any $X \in \mathfrak{X}_{\mathscr{F}}$.*

The Lie derivative $\mathscr{L}_X g_Q$ is defined in terms of the Bott connection, i.e.,

$$\big(\mathscr{L}_X g_Q\big)(s,\, s') := X\big(g_Q(s,\, s')\big) - g_Q\big(\mathring{\nabla}_X s,\, s'\big) - g_Q\big(s,\, \mathring{\nabla}_X s'\big).$$

Example 4.14 (*Killing vector fields*) Let $(M,\, g)$ be a Riemannian manifold, and let $V \in \mathfrak{i}(M, g)$ be a nonsingular[1] Killing vector field, i.e., $\mathscr{L}_V g = 0$. Let $P_x := \big\{\lambda V_x \,:\, \lambda \in \mathbb{R}\big\}$ for any $x \in P$. Then P is an involutive rank 1 distribution on M, and hence completely integrable. Let $\mathscr{F}$ be the foliation of M tangent to P. *Then g is bundle-like for $\mathscr{F}$*. Indeed, let $X \in \mathfrak{X}_{\mathscr{F}}$, i.e., $X = f\, V$ for some C^∞ function $f : M \to \mathbb{R} \setminus \{0\}$, and for any $s,\, s' \in C^\infty(Q)$

$$\mathscr{L}_X g_Q = f\, \mathscr{L}_V g_Q\,,$$

$$\big(\mathscr{L}_V g_Q\big)(s,\, s') = V\big(g_Q(s,\, s')\big) - g_Q\big(\mathring{\nabla}_V s,\, s'\big) - g_Q\big(s,\, \mathring{\nabla}_V s'\big) =$$

$$= V\big(g_Q(s,\, s')\big) - g_Q\big(\pi\left[V,\, Z_s\right],\, s'\big) - g_Q\big(s,\, \pi\left[V,\, Z_{s'}\right]\big),$$

where we have set $Z_s := \sigma(s)$. Note that

[1] That is $V_x \neq 0$ for any $x \in M$.

$$\sigma \circ \pi = \pi^{\perp};$$

hence

$$(\mathscr{L}_V g_Q)(s, s') = V\big(g(Z_s, Z_{s'})\big) - g\big(\pi^{\perp}[V, Z_s], Z_{s'}\big) - g\big(Z_s, \pi^{\perp}[V, Z_{s'}]\big)$$

and $Z_s \in P^{\perp}$ so that

$$(\mathscr{L}_V g_Q)(s, s') = (\mathscr{L}_V g)(Z_s, Z_{s'}) = 0.$$

Q.e.d.

□

4.3 Second Fundamental Form and Mean Curvature

Let $\mathscr{F}$ be a codimension q foliation of the Riemannian manifold (M, g).

Definition 4.15 The *second fundamental form* of $\mathscr{F}$ in (M, g) is

$$\alpha(X, X') = \pi\, \nabla^g_X X', \quad X, X' \in C^{\infty}(P). \tag{4.4}$$

□

Exercise 4.16 Show that $\alpha : \mathfrak{X}_{\mathscr{F}} \times \mathfrak{X}_{\mathscr{F}} \to C^{\infty}(Q)$ is $C^{\infty}(M)$-bilinear symmetric. □

Definition 4.17 A foliation $\mathscr{F}$ of a Riemannian manifold (M, g) is *totally geodesic* if every leaf of $\mathscr{F}$ is a totally geodesic submanifold of (M, g). □

The pointwise restriction of α to a leaf $L \in M/\mathscr{F}$ is the second fundamental form of the immersion $L \hookrightarrow M$. Therefore, $\mathscr{F}$ is totally geodesic $\Longleftrightarrow \alpha = 0$.

For every $Z \in C^{\infty}(P^{\perp})$ we consider the bundle morphism $W_Z : P \to P$ determined by

$$g_Q\big(\alpha(X, X'), \pi Z\big) = g\big(W_Z X, X'\big), \quad X, X' \in \mathfrak{X}_{\mathscr{F}}.$$

Definition 4.18 The *mean curvature* of $\mathscr{F}$ in (M, g) is the differential 1-form $\kappa \in \Omega^1(M)$ given by

$$\kappa(Z) := \operatorname{Trace} W_Z, \quad \kappa(X) = 0, \quad Z \in P^{\perp}, \quad X \in P.$$

The *mean curvature vector* $\tau \in C^{\infty}(Q)$ is determined by

$$g_Q(\tau, \pi Z) = \kappa(Z), \quad Z \in P^{\perp}. \tag{4.5}$$

□

The pointwise restriction of τ to a leaf $L \in M/\mathscr{F}$ is the mean curvature vector[2] of $L \hookrightarrow M$ in (M, g). In particular for $Z = \sigma(\tau)$ [in (4.5), as $\pi \circ \sigma = \mathrm{id}_Q$]

$$|\tau|^2 = \kappa(\sigma(\tau)). \tag{4.6}$$

Definition 4.19 $\mathscr{F}$ is said to be *tangentially oriented* if $P = T(\mathscr{F})$ is oriented, i.e., a principal $\mathrm{GL}^+(p)$-subbundle $B \to M$ of $\mathrm{GL}(p) \to L(P) \to M$ is specified. □

Here $\mathrm{GL}^+(p) = \{a \in \mathrm{GL}(p, \mathbb{R} \; : \; \det(a) > 0\}$ and $p = \dim(M) - \mathrm{codim}(\mathscr{F})$. Also $L(P)_x$ consists of all $\mathbb{R}$-linear isomorphisms $u : \mathbb{R}^p \to P_x$, for any $x \in M$. To every tangentially oriented foliation $\mathscr{F}$ on (M, g) one associates a differential p-form $\chi_{\mathscr{F}} \in \Omega^p(M)$ defined as follows. Let $x \in M$ and let $\{E_i \; : \; 1 \le i \le p\}$ be a local g-orthonormal frame of P, adapted[3] to the given orientation $\mathrm{GL}^+(q) \to B \to M$ of P, defined on an open neighborhood U of the point x. We set by definition

$$\chi_{\mathscr{F},x}(v_1, \cdots, v_p) := \det\left[g(v_i, E_{j,x})\right]_{1 \le i,j \le p}, \quad v_i \in T_x(M), \quad 1 \le i \le p.$$

Exercise 4.20 Show that

1) The definition of $\chi_{\mathscr{F},x}(v_1, \cdots, v_p)$ doesn't depend upon the choice of orthonormal adapted frame at x.

2) $\chi_{\mathscr{F}}(E_1, \cdots, E_p) = 1$.

3) $i_Z \, \chi_{\mathscr{F}} = 0$ for any $Z \in P^\perp$. □

Definition 4.21 $\chi_{\mathscr{F}}$ is the *characteristic form* of the tangentially oriented foliation $\mathscr{F}$ of the Riemannian manifold (M, g). □

Theorem 4.22 *For every $Z \in C^\infty(P^\perp)$*

$$\left(\mathscr{L}_Z \, \chi_{\mathscr{F}}\right)(X_1, \cdots, X_p) = -\kappa(Z) \, \chi_{\mathscr{F}}(X_1, \cdots, X_p), \quad X_1, \cdots, X_p \in C^\infty(P). \tag{4.7}$$

In particular the p-form $i_Z \, d \, \chi_{\mathscr{F}} + \kappa(Z) \, \chi_{\mathscr{F}}$ evaluates to 0 along $T(\mathscr{F})$.

Theorem 4.22 is due to Rummler, [Rum].

Definition 4.23 Formula (4.7) is referred to as *Rummler's formula*. □

Proof of Theorem 4.22. For any $X_1, \cdots, X_p \in C^\infty(P)$

---

[2] Except for the $1/\dim(L)$ factor - customary in the theory of submanifolds of Riemannian manifolds - that has been suppressed through the present chapter.

[3] That is, the cross section $\sigma : U \to L(P)$, given by $\sigma(x) : \mathbb{R}^p \to P_x$, $\sigma(x) \, e_i := E_{i,x}$, has values in B (the total space of the given orientation of P).

$$(\mathscr{L}_Z \chi_{\mathscr{F}})(X_1, \cdots, X_p) = Z(\chi_{\mathscr{F}}(X_1, \cdots, X_p)) - \sum_{i=1}^{p} \chi_{\mathscr{F}}(X_1, \cdots, \mathscr{L}_Z X_i, \cdots, X_p) =$$

(by $P^\perp \rfloor \chi_{\mathscr{F}} = 0$)

$$= Z(\chi_{\mathscr{F}}(X_1, \cdots, X_p) - \sum_{i=1}^{p} \chi_{\mathscr{F}}(X_1, \cdots, \pi_P[Z, X_i], \cdots, X_p),$$

where $\pi_P : T(M) \to P$ is the projection. Note that

$$\pi_P[Z, E_i] = \sum_{j=1}^{p} g([Z, E_i], E_j) E_j$$

yielding [as $\chi_{\mathscr{F}}(E_1, \cdots, E_p) = 1$]

$$(\mathscr{L}_Z \chi_{\mathscr{F}})(E_1, \cdots, E_p) =$$

$$= -\sum_{i=1}^{p} \sum_{j=1}^{p} g([Z, E_i], E_j) \chi_{\mathscr{F}}(E_1, \cdots, E_{i-1}, E_j, E_{i+1}, \cdots, E_p) =$$

[by computing the sum over $1 \le j \le p$, as $\chi_{\mathscr{F}}$ is skew-symmetric[4]]

$$= -\sum_{i=1}^{p} g([Z, E_i], E_i) \chi_{\mathscr{F}}(E_1, \cdots, E_p). \tag{4.8}$$

On the other hand,

$$\kappa(Z) = \text{Trace } W_Z = \sum_{i=1}^{p} g(W_Z E_i, E_i) = \sum_{i=1}^{p} g_Q(\alpha(E_i, E_i), \pi Z) =$$

$$= \sum_{i=1}^{p} g_Q(\pi \nabla^g_{E_i} E_i, \pi Z) = \sum_{i=1}^{p} g(\nabla^g_{E_i} E_i, Z) =$$

[as $\nabla^g g = 0$]

$$= \sum_{i=1}^{p} \left\{ E_i(g(E_i, Z)) - g(E_i, \nabla^g_{E_i} Z) \right\} =$$

[4] So the only nonzero term is got for $j = i$.

[as $g(E_i\,,\,Z) = 0$ and ∇^g is torsion-free]

$$= -\sum_{i=1}^{p} g\big(\nabla^g_Z E_i + [E_i\,,\,Z]\,,\,E_i\big) =$$

[as $g(E_i\,,\,E_i) = 1 \Longrightarrow 0 = Z\big(g(E_i\,,\,E_i)\big) = 2\,g(\nabla^g_Z E_i\,,\,E_i)$]

$$= \sum_{i=1}^{p} g\big([Z,\,E_i]\,,\,E_i\big)$$

and substitution into (4.8) leads to

$$\big(\mathscr{L}_Z\,\chi_{\mathscr{F}}\big)\big(E_1\,,\,\cdots\,,\,E_p\big) = -\kappa(Z). \tag{4.9}$$

Finally [as $X_j = f_j^i E_i$ for some $f_j^i \in C^\infty(U)$],

$$\big(\mathscr{L}_Z\chi_{\mathscr{F}}\big)\big(X_1\,,\,\cdots\,,\,X_p\big) = f_1^{i_1}\,\cdots\,f_p^{i_p}\,\big(\mathscr{L}_Z\chi_{\mathscr{F}}\big)\big(E_{i_1}\,,\,\cdots\,,\,E_{i_p}\big) =$$

$$= \sum_{i_1\neq\cdots\neq i_p\neq i_1} f_1^{i_1}\,\cdots\,f_p^{i_p}\,\big(\mathscr{L}_Z\chi_{\mathscr{F}}\big)\big(E_{i_1}\,,\,\cdots\,,\,E_{i_p}\big) =$$

$$= \sum_{\sigma\in\Sigma_p} f_1^{\sigma(1)}\,\cdots\,f_p^{\sigma(p)}\,\big(\mathscr{L}_Z\chi_{\mathscr{F}}\big)\big(E_{\sigma(1)}\,,\,\cdots\,,\,E_{\sigma(p)}\big),$$

$$= \sum_{\sigma\in\Sigma_p} \epsilon(\sigma) f_1^{\sigma(1)}\,\cdots\,f_p^{\sigma(p)}\,\big(\mathscr{L}_Z\chi_{\mathscr{F}}\big)\big(E_1\,,\,\cdots\,,\,E_p\big)$$

[by (4.9)]

$$= -\kappa(Z)\,\det\big[f_j^i\big]_{1\le i,\,j\le p} = -\kappa(Z)\,\chi_{\mathscr{F}}\big(X_1\,,\,\cdots\,,\,X_p\big).$$

Q.e.d.

Let us set $Z = \sigma(\tau)$ in Rummler's formula [and use (4.6)]. We obtain

$$\mathscr{L}_{\sigma(\tau)}\,\chi_{\mathscr{F}} = -\big|\tau\big|^2\,\chi_{\mathscr{F}} \quad \text{on} \quad P\otimes\cdots\otimes P \;\; (p \text{ terms}), \tag{4.10}$$

i.e., $\tau = 0$ is roughly the invariance of $\chi_{\mathscr{F}}$ under transversal flows.

Definition 4.24 A foliation $\mathscr{F}$ of the Riemannian manifold $(M,\,g)$ with $\tau = 0$ is said to be *harmonic*. □

Everly leaf of a harmonic foliation of $(M,\,g)$ is a minimal submanifold of $(M,\,g)$. The term "minimal foliation" is already in use in foliation theory, with a different meaning.[5] The term "harmonic foliation" has been introduced in mathematical practice by F. Kamber[6] and Ph. Tondeur[7] in their

[5] A foliation $\mathscr{F}$ of M is *minimal* if every leaf of $\mathscr{F}$ is dense in M.

[6] Franz Kamber (1936 - 2017) was a Swiss-American mathematician.

Franz Kamber

K. was born in Olten, Switzerland on June 19, 1936, a son of Max and Lydia Winterhalter Kamber. **K.** served in the Swiss army and studied at the University of Zürich where he majored in Mathematics and Theoretical Physics, earning his Ph.D. in 1965. Successively **K.** moved to the United States and began his professional career at the University of California, Berkeley and later, Princeton. In 1968 he settled in Champaign/Urbana at the University of Illinois as a Professor of Mathematics. **K.** was proud to have obtained U.S. Citizenship in 1974. **K.** retired in 1994 and was a Professor Emeritus of Mathematics since 1995 at the University of Illinois. **K.** married Karen Hansmann in Denmark in 1989 and together they have enjoyed traveling the world. **K.** had eclectic cultural interests, including painting, music, and poetry. Among the Ph.D. students of **K.** we ought to mention Steven Edmond Hurder, himself a prominent protagonist of foliation theory, with the thesis *Dual Homotopy Invariants of G-Foliations*, 1980.

[7] Philippe Tondeur (born 1932) was a Swiss-American mathematician and educator, and a consultant on mathematics, science and technology policy.

Philippe Maurice Tondeur

T.'s interests included mathematics research and education, the role of mathematics in science and society, innovation and science policy, institutional governance, and leadership development. **T.** served as Director of the Division of Mathematical Sciences at the National Science Foundation (NSF). Previously, **T.** was Chair of the Department of Mathematics at the University of Illinois in Urbana/ Champaign. **T.** earned an Engineering degree in Zürich, and a Ph.D. degree in Mathematics from the University of Zürich in 1961. Successively **T.** was a Research Fellow and Lecturer at the University of Paris, at Harvard University, the University of California at Berkeley, an Associate Professor at Wesleyan University, before joining the University of Illinois at Urbana/Champaign faculty in 1968, where he became a Full Professor in 1970. **T.** published approximately 100 articles and monographs, on several differential geometry and topology themes, in particular the geometry of foliations and geometric applications of partial differential equations. **T.** authored nine books. **T.** was supported by grants from the NSF from 1967 to 1990. **T.** served as Editor and then as Managing Editor of the Illinois Journal of Mathematics, and edited the collected works of K.T. Chen. **T.** served on numerous committees at universities and professional societies. The following mathematicians had the privilege of earning their Ph.D.s under the guidance of **T.** (i) Pyng Wang, *Decomposition theorems of Riemannian manifolds*, 1973, (ii) Thomas Duchamp, *Characteristic invariants of G-foliations*, 1976, (iii) Michael Hvidsten, *Volume and energy stability for immersions*, 1985, (iv) Paul Scofield, *Symplectic and complex foliations*, 1990, (v) Hobum Kim, *Geometric and dynamical properties of Riemannian foliations*, 1990, (vi) Louis Kerofsky, *Harmonic forms under metric and topological perturbations*, 1995, (vii) Alan Mason, *An application of stochastic flows to*

famous[8] paper [KaTo6]. The choice of terminology may be motivated in several ways. The projection $\pi : T(M) \to Q$ may be viewed as a Q-valued differential 1-form on M, i.e.,

$$\pi \in \Omega^1(Q) = C^\infty\big(T^*(M) \otimes Q\big).$$

Let $d_\nabla : \Omega^r(Q) \to \Omega^{r+1}(Q)$ be defined by

$$(d_\nabla \alpha)\big(Y_0, \cdots, Y_r\big) := \frac{1}{r+1}\Big\{\sum_{j=0}^{r} (-1)^j \nabla_{Y_j} \alpha(Y_0, \cdots, \hat{Y}_j, \cdots, Y_r) +$$

$$+ \sum_{0 \le i < j \le r} (-1)^{i+j} \alpha\big([Y_i, Y_j], Y_0, \cdots \hat{Y}_i, \cdots \hat{Y}_j, \cdots, Y_r\big)\Big\},$$

$$Y_0, \cdots, Y_r \in \mathfrak{X}(M),$$

where a hat indicates as customary the suppression of a term. Here ∇ is the canonical adapted foliation of $\mathcal{F}$. Then

$$d_\nabla \pi = 0, \quad \delta_\nabla \pi = \tau,$$

where δ_∇ is the formal adjoint of d_∇. So $\mathcal{F}$ is a harmonic foliation $\Longleftrightarrow \pi$ is a harmonic 1-form (cf. [KaTo6]). Another motivation comes from the theory of harmonic maps (cf. e.g., [BaWo]) and the result there, that the foliation defined by a submersion from a Riemannian manifold is harmonic if and only is said submersion is a harmonic map.

Let $\mathcal{F}$ be a foliation of the Riemannian manifold (M, g), and let us assume that $\mathcal{F}$ is *transversally oriented*, i.e., the transverse bundle $Q \to M$ is equipped with an orientation

$$\begin{array}{ccccccc} \mathrm{GL}^+(q) & \to & B & \hookrightarrow & B^1_T(\mathcal{F}) & \leftarrow & \mathrm{GL}(q, \mathbb{R}) \\ & & \downarrow & & \downarrow & & \\ & & M & = & M & & \end{array}.$$

Let us define a differential q-form ν on M, as follows. Let $x \in M$ and let $\{s_j : 1 \le j \le q\}$ be a local g_Q orthonormal frame of Q, adapted to the transverse $\mathrm{GL}^+(q)$-structure B, defined on an open neighborhood $U \subset M$ of the point x. We set

---

Riemannian foliations, 1997, (viii) Amine Fawaz, *Transversely holomorphic flows on 3-manifolds and geodesible vector fields*, 1998.

[8] One may hardly overestimate the importance of the scientific creation of Franz Kamber and Philippe Maurice Tondeur and their role in the development of foliation theory, a field of scientific inquiry they have vigorously dominated for over thirty years. The reference list of the present book contains the magnificent series of papers [KaTo1]-[KaTo16] coauthored by F. Kamber and Ph. Tondeur between 1971 and 1987.

$$\nu_x\big(v_1\,,\,\cdots\,,\,v_q\big) := \det\big[g_x\big(v_i\,,\,\sigma(s_j)_x\big)\big]_{1\le i,\,j\le q}\,,\qquad v_1\,,\,\cdots\,,\,v_q\in T_x(M).$$

Exercise 4.25 Show that

1) The definition of ν_x doesn't depend upon the choice of adapted frame at x,
2) $\nu\big(\sigma(s_1)\,,\,\cdots\,,\,\sigma(s_q)\big)=1$,
3) $P\,\rfloor\,\nu=0$. □

Definition 4.26 The differential q-form $\nu\in\Omega^q(M)$ is the *transversal volume form* of the transversally oriented foliation $\mathscr{F}$ of the Riemannian manifold $(M,\,g)$. □

Moreover, let us consider the $C^\infty(M)$-multi-linear map

$$\mathscr{L}_X\nu : C^\infty\big(P^\perp\big)\times\cdots\times C^\infty\big(P^\perp\big)\to C^\infty(M),\quad X\in\mathfrak{X}_{\mathscr{F}}\,,$$

defined by

$$\begin{gathered}(\mathscr{L}_X\nu)(Z_1\,,\,\cdots\,,\,Z_q) := X(\nu(Z_1\,,\,\cdots\,,\,Z_q)) - \sum_{i=1}^{q}\nu(Z_1\,,\,\cdots\,\sigma\,\nabla_X\pi\,Z_i\,,\,\cdots\,,\,Z_q),\\ Z_1\,,\,\cdots\,,\,Z_q\in C^\infty(P^\perp).\end{gathered}\tag{4.11}$$

It should be observed that the right-hand side of (4.11) may actually be rewritten in terms of the Bott connection, for

$$X\in\mathfrak{X}_{\mathscr{F}}\Longrightarrow \sigma\,\nabla_X\pi\,Z_i=\sigma\,\mathring{\nabla}_X\pi\,Z_i=\sigma\,\pi\,\big[X,\,Z_j\big]=\pi^\perp\big[X,\,Z_j\big],$$

so that (4.11) is [by statement 3) in Exercise 4.25] merely the restriction of the Lie derivative $\mathscr{L}_X\nu$ to $P^\perp\otimes\cdots\otimes P^\perp$ (q terms). On the other hand, if one of the arguments $Y_1\,,\,\cdots\,,\,Y_q\in\mathfrak{X}(M)$ is tangent to the leaves, say $Y_1\in C^\infty(P)$, then [again by 3) in Exercise 4.25]

$$\big(\mathscr{L}_X\,\nu\big)\big(Y_1\,,\,\cdots\,,\,Y_q\big)=-\nu\big(\big[X,\,Y_1\big],\,Y_2\,,\,\cdots\,,\,Y_q\big)=0$$

because P is involutive, so that $\big[X,\,Y_1\big]\in C^\infty(P)$.

Proposition 4.27 *If ν is holonomy invariant, i.e., $(\mathscr{L}_X\nu)(Z_1\,,\,\cdots\,,\,Z_q)=0$ for any $X\in\mathfrak{X}_{\mathscr{F}}$ and any $Z_1\,,\,\cdots\,,\,Z_q\in C^\infty(P^\perp)$, then ν is a basic q-form on $(M,\,\mathscr{F})$, i.e., $\nu\in\Omega^q_B(\mathscr{F})$.*

Proof Our calculations preceding Proposition 4.27 together with the holonomy invariance assumption show that $\mathscr{L}_X\nu=0$ at any direction $X\in\mathfrak{X}_{\mathscr{F}}$. Then [by Cartan's formula $\mathscr{L}_X\nu=i_X\circ d+d\circ i_X$] $i_X\,d\,\nu=0$, implying [again by 3) in Exercise 4.25] that $\nu\in\Omega^q_B(\mathscr{F})$. Q.e.d.

A useful relationship between the transversal volume form ν and the characteristic form $\chi_{\mathscr{F}}$ of a foliation $\mathscr{F}$ of a n-dimensional Riemannian manifold $(M,\,g)$ may be exhibited in terms of the Hodge star operator $* : \Omega^r(M) \to \Omega^{n-r}(M)$, as follows. Let us assume that M is oriented, so that the Hodge star is fully determined by

$$(*\alpha)(Y_1, \cdots, Y_{n-r})\, d\,\mathrm{v}_g = \alpha \wedge Y_1^\flat \wedge \cdots \wedge Y_{n-r}^\flat ,$$

$$\alpha \in \Omega^r(M), \quad Y_1, \cdots, Y_{n-r} \in \mathfrak{X}(M).$$

Here $\flat : \mathfrak{X}(M) \to \Omega^1(M)$ is lowering of indices by g, i.e., locally $Y^\flat = g_{ij}\, Y^j\, dx^i$ whenever $Y = Y^j\, (\partial/\partial x^j)$, with respect to a local coordinate system (U, x^i) on M. Also $d\,\mathrm{v}_g$ is the canonical volume form of (M, g), i.e., locally

$$d\,\mathrm{v}_g = \sqrt{\mathfrak{g}}\, dx^1 \wedge \cdots \wedge dx^n ,$$

$$\mathfrak{g} = \det\left[g_{ij}\right]_{1 \le i, j \le n}, \quad g_{ij} = g\left(\partial_i , \partial_j\right), \quad \partial_i \equiv \frac{\partial}{\partial x^i} .$$

In particular $*1 = d\,\mathrm{v}_g$ and $*\, d\,\mathrm{v}_g = 1$. Also, if for any positively oriented[9] local g-orthonormal frame $\{E_1, \cdots, E_n\}$, defined on the open set $U \subset M$,

$$(*\alpha)\left(E_{i_1}, \cdots E_{i_{n-r}}\right) d\,\mathrm{v}_g = \alpha \wedge \omega^{i_1} \wedge \cdots \wedge \omega^{i_{n-r}}$$

for any $\alpha \in \Omega^r(M)$. Here $\{\omega^1, \cdots, \omega^n\}$ is the dual local frame, i.e., $\omega^i \in \Omega^1(U)$ and $\omega^i(E_j) = \delta^i_j$. As $\{E_i \,:\, 1 \le i \le n\}$ is orthonormal, one also has $\omega^i = E_i^\flat$, and of course $\left(\omega^i\right)^\sharp = E_i$ where $\sharp : \Omega^1(M) \to \mathfrak{X}(M)$ is raising the indices with the metric. Now let us consider a tangentially and transversally oriented foliation $\mathcal{F}$ on (M, g). Let $\{E_1, \cdots, E_n\}$ be a local g-orthonormal frame of $T(M)$, defined on the open set $U \subset M$ and adapted to the decomposition $T(M) = P \oplus P^\perp$, i.e.,

$$\{E_\alpha \,:\, 1 \le \alpha \le p\} \subset C^\infty\left(U, P\right), \quad \{E_{p+j} \,:\, 1 \le j \le q\} \subset C^\infty\left(U, P^\perp\right).$$

Lemma 4.28 *The transverse volume form* ν *satisfies* i)

$$(*\,\nu)(E_1, \cdots, E_p)\, d\,\mathrm{v}_g = \left(\omega_{p+1} \wedge \cdots \wedge \omega_n\right) \wedge \left(\omega_1 \wedge \cdots \wedge \omega_p\right),$$

ii) $E_{p+j} \rfloor (*\,\nu) = 0$, $1 \le j \le q$, *so that* $*\,\nu$ *is, up to a sign, the characteristic form* $\chi_{\mathcal{F}}$.

iii) *If orientations of* P *and* $P^\perp$ *are chosen such that* $*\,\nu = \chi_{\mathcal{F}}$, *then*

$$\nu \wedge \chi_{\mathcal{F}} = \nu \wedge *\nu = d\,\mathrm{v}_g .$$

The proof is straightforward.

---

[9] That is $(d\,\mathrm{v}_g)(E_1, \cdots, E_n) > 0$ everywhere on U.

4.4 $d_{\mathscr{F}}$-Cohomology

The present section is devoted to I. Vaisman's $d_{\mathscr{F}}$-cohomology[10] (cf. [Vai1]). Cf. also Molino, [Mol], pp. 58–59.

Let $(M, \mathscr{F})$ be a foliated manifold, with $\dim(M) = n$ and $\operatorname{cod}(\mathscr{F}) = q$. Let P' be any smooth q-dimensional distribution on M such that

$$T(M) = T(\mathscr{F}) \oplus P' . \tag{4.12}$$

Let us identify the dual $T(\mathscr{F})^*$ (respectively, P'^*) with the bundle $\{\alpha \in T^*(M) : P' \rfloor \alpha = 0\}$ (respectively, with the bundle $\{\beta \in T^*(M) : T(\mathscr{F}) \rfloor \beta = 0\}$). Then (4.12) yields the decomposition

$$T^*(M) = T(\mathscr{F})^* \oplus P'^* . \tag{4.13}$$

The decomposition (4.13) induces a decomposition at the level of differential forms on M. To describe it, let $\{\theta^\alpha : 1 \le \alpha \le n - q\}$ be a local frame of $T(\mathscr{F})^*$ defined on an open set U, i.e., each θ^α is a 1-form on U such that $P' \rfloor \theta^\alpha = 0$. Similarly, we consider a local frame $\{\eta^i : 1 \le i \le q\}$ of P'^* defined on the same open set U, i.e., η^i is a 1-form on U such that $T(\mathscr{F}) \rfloor \eta^i = 0$. Then

$$\{\theta^\alpha , \ \eta^i \ : \ 1 \le \alpha \le n - q, \quad 1 \le i \le q\}$$

is a local frame on $T^*(M)$, defined on the open set U, referred to as *adapted* to the decomposition (4.13). Every differential k-form on $\beta \in \Omega^k(M)$ may be locally represented as a sum of monomials of the form

$$\beta_{\alpha_1 \cdots \alpha_r i_1 \cdots i_r} \, \theta^{\alpha_1} \wedge \cdots \wedge \theta^{\alpha_r} \wedge \eta^{i_1} \wedge \cdots \wedge \eta^{i_s} , \quad \beta_{\alpha_1 \cdots \alpha_r i_1 \cdots i_r} \in C^\infty(U), \tag{4.14}$$

$$\alpha_1 , \cdots , \alpha_r \in \{1, \cdots , n - q\}, \quad i_1 , \cdots , i_s \in \{1, \cdots , q\}, \quad r + s = k,$$

i.e.,

$$\beta = \sum_{r+s=k} \sum_{\substack{1 \le \alpha_1 < \cdots < \alpha_r \le n-q \\ 1 \le i_1 < \cdots < i_s \le q}} \beta_{\alpha_1 \cdots \alpha_r i_1 \cdots i_r} \, \theta^{\alpha_1 \cdots \alpha_r} \wedge \eta^{i_1 \cdots i_s} , \tag{4.15}$$

$$\theta^{\alpha_1 \cdots \alpha_r} := \theta^{\alpha_1} \wedge \cdots \wedge \theta^{\alpha_r} , \quad \eta^{i_1 \cdots i_s} := \eta^{i_1} \wedge \cdots \wedge \eta^{i_s} .$$

---

[10] An ingredient needed in the discussion of the existence of projectable connections on foliated bundles.

Definition 4.29 A differential form $\beta \in \Omega^k(M)$ is said to be of *type* (r, s) if for every point $x_0 \in M$ there is an open neighborhood $U \subset M$ of x_0 and an adapted frame $\{\theta^\alpha, \eta^i : 1 \le \alpha \le n-q, \ 1 \le i \le q\}$ of $T^*(M)$ on U, such that β is represented as a sum of monomials of the form (4.14). □

Let $\beta \in \Omega^k(M)$ be a differential k-form of type (r, s) (with $r + s = k$), i.e.,

$$\beta = \sum_{\substack{1 \le \alpha_1 < \cdots < \alpha_r \le n-q \\ 1 \le i_1 < \cdots < i_s \le q}} \beta_{\alpha_1 \cdots \alpha_r \, i_1 \cdots i_r} \, \theta^{\alpha_1 \cdots \alpha_r} \wedge \eta^{i_1 \cdots i_s}$$

on an open neighborhood U of an arbitrarily fixed point $x_0 \in M$. Let $\{\theta'^\alpha, \eta'^i : 1 \le \alpha \le n-q, \ 1 \le i \le q\}$ be another adapted frame, defined on the open subset $x_0 \in U' \subset M$. Then

$$\theta'^\alpha = U^\alpha_\beta \, \theta^\beta \quad \eta'^i = f^i_j \, \eta^j$$
$$\left[U^\alpha_\beta\right] : U \cap U' \to \mathrm{GL}(n-q, \mathbb{R}), \quad \left[f^i_j\right] : U \cap U' \to \mathrm{GL}(q, \mathbb{R}).$$

Let us set $\left[V^\alpha_\beta\right] := \left[U^\alpha_\beta\right]^{-1}$ and $\left[g^i_j\right] := \left[f^i_j\right]^{-1}$. Then

$$\beta = \sum_{\substack{1 \le \gamma_1 < \cdots < \gamma_r \le n-q \\ 1 \le j_1 < \cdots < j_s \le q}} \beta'_{\gamma_1 \cdots \gamma_r \, j_1 \cdots j_r} \, \theta'^{\gamma_1 \cdots \gamma_r} \wedge \eta'^{j_1 \cdots j_s},$$

$$\beta'_{\gamma_1 \cdots \gamma_r \, j_1 \cdots j_r} := \beta_{\alpha_1 \cdots \alpha_r \, i_1 \cdots i_r} \, V^{\alpha_1}_{\gamma_1} \cdots V^{\alpha_r}_{\gamma_r} \, g^{i_1}_{j_1} \cdots g^{i_s}_{j_s}.$$

Let $\beta \in \Omega^k(M)$ and let (r, s) be a pair of nonnegative integers such that $r + s = k$. Let $x \in M$ and let $\{\theta^\alpha, \eta^i : 1 \le \alpha \le n-q, \ 1 \le i \le q\}$ be an adapted frame of $T^*(M)$, defined on an open neighborhood $U \subset M$ of x. Let us set

$$\beta_{r,s}(x) := \sum_{\substack{1 \le \alpha_1 < \cdots < \alpha_r \le n-q \\ 1 \le i_1 < \cdots < i_s \le q}} \beta_{\alpha_1 \cdots \alpha_r \, i_1 \cdots i_r}(x) \left(\theta^{\alpha_1 \cdots \alpha_r} \wedge \eta^{i_1 \cdots i_s}\right)_x.$$

We claim that $\beta_{r,s}$ is a globally defined k-form of type (r, s). To prove the claim we ought to compare the locally defined k-forms

$$\beta^U_{r,s} = \sum_{\substack{1 \le \alpha_1 < \cdots < \alpha_r \le n-q \\ 1 \le i_1 < \cdots < i_s \le q}} \beta_{\alpha_1 \cdots \alpha_r \, i_1 \cdots i_r} \, \theta^{\alpha_1 \cdots \alpha_r} \wedge \eta^{i_1 \cdots i_s},$$

$$\beta^{U'}_{r,s} = \sum_{\substack{1 \leq \alpha_1 < \cdots < \alpha_r \leq n-q \\ 1 \leq i_1 < \cdots < i_s \leq q}} \beta'_{\alpha_1 \cdots \alpha_r\, i_1 \cdots i_r}\, \theta'^{\alpha_1 \cdots \alpha_r} \wedge \eta'^{i_1 \cdots i_s},$$

starting from

$$\sum_{r+s=k} \beta^U_{r,s} = \sum_{r+s=k} \beta^{U'}_{r,s} \quad \text{on } U \cap U'. \tag{4.16}$$

Note that

$$\beta^{U'}_{r,s} = \lambda_{\gamma_1 \cdots \gamma_r\, j_1 \cdots j_s}\, \theta^{\gamma_1, \cdots, \gamma_r} \wedge \eta^{j_1 \cdots j_s},$$

$$\lambda_{\gamma_1 \cdots \gamma_r\, j_1 \cdots j_s} := \sum_{\substack{1 \leq \alpha_1 < \cdots < \alpha_r \leq n-q \\ 1 \leq i_1 < \cdots < i_s \leq q}} \beta'_{\alpha_1 \cdots \alpha_r\, i_1 \cdots i_s}\, U^{\alpha_1}_{\gamma_1} \cdots U^{\alpha_r}_{\gamma_r}\, f^{i_1}_{j_1} \cdots f^{i_s}_{j_s}$$

so that [by (4.16)]

$$\beta_{\gamma_1 \cdots \gamma_r\, j_1 \cdots j_s} = r!\, s!\, \lambda_{\gamma_1 \cdots \gamma_r\, j_1 \cdots j_s} \quad \text{on } U \cap U'.$$

We may conclude that the local k-forms $\beta^U_{r,s}$ glue up to a (globally defined) k-form $\beta_{r,s} \in \Omega^k(M)$, that is clearly of type (r, s).

Definition 4.30 The form $\beta_{r,s} \in \Omega^k(M)$ is called the (r, s)-*component* of $\beta \in \Omega^k(M)$. □

From now on let g be a Riemannian metric on M and let $P' := T(\mathscr{F})^\perp$ be the g-orthogonal complement of $T(\mathscr{F})$ in $T(M)$. Given a foliated chart $(U, x^\alpha, y^i) \in \mathscr{F}$ we choose a local frame of $T^*(M)$ adapted to the decomposition

$$T^*(M) = T(\mathscr{F})^* \oplus \left[T(\mathscr{F})^\perp\right]^*$$

as follows. Let us set

$$g_{\alpha\beta} = g\left(\frac{\partial}{\partial x^\alpha}, \frac{\partial}{\partial x^\beta}\right), \quad g_{\alpha,\, p+j} = g\left(\frac{\partial}{\partial x^\alpha}, \frac{\partial}{\partial y^j}\right), \quad g_{p+i,\, p+j} = g\left(\frac{\partial}{\partial y^i}, \frac{\partial}{\partial y^j}\right),$$

$$1 \leq \alpha, \beta \leq p, \quad 1 \leq i, j \leq q, \quad p = n - q,$$

,i.e.,

$$\begin{pmatrix} g_{\alpha\beta} & g_{\alpha\,,\, p+j} \\ g_{p+i\,,\,\alpha} & g_{p+i\,,\, p+j} \end{pmatrix}$$

are the local components of g with respect to the local coordinate system (U, x^α, y^i). Let $\sigma = \sigma_g : Q \to T(\mathscr{F})^\perp$ be the natural vector bundle isomorphism $Q \simeq T(\mathscr{F})^\perp$.

Let g_P and g_Q be the Riemannian metrics induced by g on $P = T(\mathcal{F})$ and $Q = \nu(\mathcal{F})$, i.e.,

$$g_P(X, X') = g(X, X'), \quad g_Q(r, s) = g(\sigma(r), \sigma(s)), \quad X, X' \in \mathfrak{X}_{\mathcal{F}}, \quad r, s \in C^\infty(Q).$$

Yet P is locally the span of $\{\partial/\partial x^\alpha : 1 \le \alpha \le p\}$, hence $g_{\alpha\beta} \in C^\infty(U)$ are also the components of the Riemannian bundle metric g_P with respect to that local frame. More is true, i.e., the pointwise restriction of g_P to a leaf $L \in M/\mathcal{F}$ is a Riemannian metric on L [and each $\partial/\partial x^\alpha$ is tangent to the leaf] hence the restrictions of $g_{\alpha\beta}$ to $U \cap L$ are the local coefficients of that Riemannian metric on L [with respect to the local chart $(U \cap L, x^\alpha)$]. In particular $\left[g_{\alpha\beta}\right]$ is invertible and we set by definition

$$\left[g^{\alpha\beta}\right]_{1\le\alpha,\,\beta\le p} := \left(\left[g_{\alpha\beta}\right]_{1\le\alpha,\,\beta\le p}\right)^{-1} \quad \text{on } U$$

so that $g^{\alpha\gamma}\, g_{\gamma\beta} = \delta^\alpha_\beta$. Next, let us set

$$\left(g_Q\right)_{ij} = g_Q\left(\pi\,\frac{\partial}{\partial y^i},\, \pi\,\frac{\partial}{\partial y^j}\right), \quad 1 \le i,\, j \le q.$$

Here $\pi : T(M) \to Q$ is the projection. As $\{\pi\,(\partial/\partial y^i) : 1 \le i \le q\}$ is a local frame of Q, the functions $(g_Q)_{ij} \in C^\infty(U)$ are the local components of the Riemannian bundle metric g_Q with respect to that local frame. In particular the matrix $\left[(g_Q)_{ij}\right]$ is invertible and we set

$$\left[\left(g_Q\right)^{ij}\right]_{1\le i,\,j\le q} := \left(\left[\left(g_Q\right)_{ij}\right]_{1\le i,\,j\le q}\right)^{-1} \quad \text{on } U,$$

so that $(g_Q)^{ik}\,(g_Q)_{kj} = \delta^i_j$. The $T(\mathcal{F})^\perp$-component pf $\partial/\partial y^i$ may be represented as

$$\left(\frac{\partial}{\partial y^i}\right)^\perp = \lambda^\alpha_i\,\frac{\partial}{\partial^\alpha} + \mu^j_i\,\frac{\partial}{\partial y^j}, \quad \lambda^\alpha_i,\ \mu^j_i \in C^\infty(U). \tag{4.17}$$

Let us take the inner product of (4.17) with $\partial/\partial x^\beta$ so that to obtain

$$0 = \lambda^\alpha_i\, g_{\alpha\beta} + \mu^j_i\, g_{p+j\,,\beta}\,.$$

Next, let us contract with $g^{\beta\gamma}$ so that

$$\lambda^\gamma_i = -\mu^j_i\, g_{p+j\,,\beta}\, g^{\beta\gamma}\,. \tag{4.18}$$

Let us set $s_i = \pi\left(\partial/\partial y^i\right)$ for the sake of simplicity. By applying π to (4.17) we get [as $\mathrm{Ker}(\pi) = T(\mathcal{F})$]

$$s_i = \mu^j_i\, s_j$$

yielding $\mu_i^j = \delta_i^j$. Going back to (4.18) we obtain $\lambda_i^\gamma = -g_{p+i\,,\beta}\, g^{\beta\gamma}$, so that (4.17) becomes

$$\left(\frac{\partial}{\partial y^i}\right)^\perp = \frac{\partial}{\partial y^i} - g_{p+i\,,\alpha}\, g^{\alpha\beta}\, \frac{\partial}{\partial x^\beta}\,. \tag{4.19}$$

The decomposition (4.19) will prove useful in local calculations on $(M,\,\mathcal{F})$ with $\mathcal{F}$ (transversally) Riemannian. The assumption that $\mathcal{F}$ is Riemannian wasn't however required so far. In particular one may relate the local components of g_Q to those of g, i.e.,

$$\left(g_Q\right)_{ij} = g_Q\left(s_i\,,\, s_j\right) = g\left(\sigma(s_i)\,,\,\sigma(s_j)\right) = g\left(\left(\frac{\partial}{\partial y^i}\right)^\perp,\,\left(\frac{\partial}{\partial y^j}\right)^\perp\right) =$$

[by (4.19)]

$$= g_{p+i\,,\,p+j} - g_{p+j\,,\lambda}\, g^{\lambda\mu}\, g_{p+i\,,\mu} - g_{p+i\,,\alpha}\, g^{\alpha\beta}\, g_{\beta,\,p+j}$$
$$+ g_{p+i\,,\alpha}\, g^{\alpha\beta}\, g_{p+j\,,\lambda}\, g^{\lambda\mu}\, g_{\beta\mu}$$

or

$$\left(g_Q\right)_{ij} = g_{p+i\,,\,p+j} - g^{\alpha\beta}\, g_{\alpha,\,p+i}\, g_{\beta,\,p+j}\,. \tag{4.20}$$

The identity (4.20) is useful in the calculation of the inverse

$$\left[G^{AB}\right] = \left[g_{AB}\right]^{-1} = \begin{pmatrix} G^{\alpha\beta} & G^{\alpha,\,p+j} \\ G^{p+i\,,\beta} & G^{p+i\,,\,p+j} \end{pmatrix}.$$

The range of the capital indices is $A,\ B,\ \cdots \in \{1,\ \cdots,\ n\}$. Indeed, we may use

$$g_{AB}\, G^{BC} = \delta_A^C \tag{4.21}$$

to show that

$$G^{p+i\,,\,p+j} = \left(g_Q\right)^{ij}\,. \tag{4.22}$$

Let us set $A = p + i$ and $B = p + k$ in (4.21) so that [by (4.20)]

$$\delta_i^k = g_{p+i\,,B}\, G^{B,\,p+k} = g_{p+i\,,\beta}\, G^{\beta,\,p+k} + g_{p+i\,,\,p+j}\, G^{p+j\,,\,p+k} =$$

$$= g_{p+i\,,\beta}\, G^{\beta,\,p+k} + \left[\left(gQ\right)_{ij} + g^{\alpha\beta}\, g_{\alpha,\,p+i}\, g_{\beta,\,p+j}\right] G^{p+j,\,p+k} =$$

$$= g_{p+i,\,\beta}\, G^{\beta,\,p+k} + \left(g_Q\right)_{ij}\, G^{p+j\,,\,p+k} - g^{\alpha\beta}\, g_{\alpha,\,p+i}\, g_{\beta\gamma}\, G^{\gamma,\,p+k}$$

as [again by (4.21)] $g_{\beta B}\, G^{B,\,p+k} = 0$. Finally, $g^{\alpha\beta}\, g_{\beta\gamma} = \delta_\gamma^\alpha$ yields

$$\delta_i^k = \left(g_Q\right)_{ij} G^{p+j\,,\,p+k}$$

hence (4.22) is proved. To determine the remaining entries of $\left[G^{AB}\right]$ we use (4.21) with $A = \alpha$ and $C = p + k$ so that

$$0 = g_{AB}\, G^{B,\,p+k} = g_{\alpha\beta}\, G^{\beta,\,p+k} + g_{\alpha,\,p+j}\, G^{p+j\,,\,p+k}$$

and then [by (4.22)]

$$G^{\alpha,\,p+j} = -g^{\alpha\beta}\, g_{\beta,\,p+i} \left(g_Q\right)^{ij} . \tag{4.23}$$

Similarly, (4.21) with $A = \alpha$ and $C = \gamma$ gives

$$\delta_\alpha^\gamma = g_{\alpha B}\, G^{B\gamma} = g_{\alpha\beta}\, G^{\beta\gamma} + g_{\alpha\;p+j}\, G^{p+j\,,\,\gamma}$$

so that [by contracting with $g^{\alpha\mu}$ and using (4.23)]

$$G^{\alpha\beta} = g^{\alpha\beta} + g^{\alpha\lambda}\, g^{\beta\mu}\, g_{\lambda\,,\,p+i}\, g_{\mu\,,\,p+j} \left(g_Q\right)^{ij} . \tag{4.24}$$

Summing up [by (4.22)–(4.24)],

$$\left[G^{AB}\right]_{1\le A,\,B\le n} = \begin{pmatrix} g^{\alpha\beta} + g^{\alpha\lambda}\, g^{\beta\mu}\, g_{\lambda,\,p+i}\, g_{\mu,\,p+j} \left(g_Q\right)^{ij} & -g^{\alpha\beta}\, g_{\beta,\,p+i} \left(g_Q\right)^{ij} \\ -g^{\alpha\beta}\, g_{\alpha,\,p+j} \left(g_Q\right)^{ij} & \left(g_Q\right)^{ij} \end{pmatrix} . \tag{4.25}$$

Let us go back to the decomposition of forms with respect to (4.13) with $P' = T(\mathscr{F})^\perp$. As

$$\left\{\frac{\partial}{\partial x^\alpha} \;:\; 1 \le \alpha \le p\right\} \subset C^\infty\big(U,\, T(\mathscr{F})\big),$$
$$\left\{\left(\frac{\partial}{\partial y^i}\right)^\perp \;:\; 1 \le i \le q\right\} \subset C^\infty\big(U,\, T(\mathscr{F})^\perp\big),$$

are local frames, it follows that

$$\left\{d\, y^i \;:\; 1 \le i \le q\right\} \subset C^\infty\big(U,\, (T(\mathscr{F})^\perp)^*\big), \quad \left\{\delta\, x^\alpha \;:\; 1 \le \alpha \le \right\} \subset C^\infty\big(U,\, T(\mathscr{F})^*\big),$$

$$\delta\, x^\alpha := d\, x^\alpha + g^{\alpha\beta}\, g_{\beta,\,p+j}\, d\, y^j\,, \quad 1 \le \alpha \le p, \tag{4.26}$$

are local coframes [as a consequence of (4.19)]. Of course, the local frame

$$\left\{\delta\, x^\alpha\,,\; d\, y^i \;:\; 1 \le \alpha \le p,\; 1 \le i \le q\right\} \subset C^\infty\big(U,\, T^*(M)\big)$$

is dual to

$$\left\{ \frac{\partial}{\partial x^{\alpha}}, \frac{\partial}{\partial y^{i}} \; : \; 1 \le \alpha \le p, \;\; 1 \le i \le q \right\} \subset C^{\infty}\big(U, \, T(M)\big),$$

and adapted to the decomposition (4.13). Let $\beta \in \Omega^k(M)$ and let $\beta_{r,\,s}$ be its $(r,\,s)$-component. Then $\beta_{r,\,s}$ may be locally represented as

$$\sum_{\substack{1 \le \alpha_1 < \cdots < \alpha_r \le p \\ 1 \le i_1 < \cdots < i_s \le q}} \beta_{\alpha_1 \cdots \alpha_r \, i_1 \cdots i_s} \, \delta x^{\alpha_1} \wedge \cdots \wedge \delta x^{\alpha_r} \wedge d\, y^{i_1} \wedge \cdots \wedge d\, y^{i_s} \, .$$

For every function $u \in C^{\infty}(U)$, one has [by (4.26)]

$$du = \frac{\partial u}{\partial x^{\alpha}} \, d\, x^{\alpha} + \frac{\partial u}{\partial y^{i}} \, d\, y^{i} = \frac{\partial u}{\partial x^{\alpha}} \left(\delta\, x^{\alpha} - N_i^{\alpha} \, d\, y^{i} \right) + \frac{\partial u}{\partial y^{i}} \, d\, y^{i} \, ,$$

$$N_i^{\alpha} := g^{\alpha\beta} \, g_{\beta,\, p+i} \in C^{\infty}(U),$$

or

$$d\, u = \frac{\partial u}{\partial x^{\alpha}} \, \delta\, x^{\alpha} + \frac{\delta\, u}{\delta\, y^{i}} \, d\, y^{i} \, , \tag{4.27}$$

$$\frac{\delta}{\delta\, y^{i}} \equiv \frac{\partial}{\partial\, y^{i}} - N_i^{\alpha} \, \frac{\partial}{\partial x^{\alpha}} \, .$$

Let us observe that [by (4.26)],

$$d\big(\delta\, x^{\alpha}\big) = d\, N_i^{\alpha} \wedge d\, y^{i} = \left(\frac{\partial\, N_i^{\alpha}}{\partial\, x^{\beta}} \, \delta\, x^{\beta} + \frac{\delta\, N_i^{\alpha}}{\delta\, u^{j}} \right) \wedge d\, y^{i}$$

hence

$$d\, \big(\delta\, x^{\alpha}\big) = \frac{\partial\, N_i^{\alpha}}{\partial\, x^{\beta}} \, \delta\, x^{\beta} \wedge d\, y^{i} + \frac{\delta\, N_i^{\alpha}}{\delta\, y^{i}} \, d\, y^{i} \wedge d\, y^{j} \, . \tag{4.28}$$

Moreover [by taking into account (4.27)–(4.28)], the $(k+1)$-form $d\, \beta_{r,\,s}$ may be locally written as

$$\sum_{\substack{1 \le \alpha_1 < \cdots < \alpha_r \le p \\ 1 \le i_1 < \cdots < i_s \le q}} \Big\{ \big(d\, \beta_{\alpha_1 \cdots \alpha_r \, i_1 \cdots i_s} \big) \wedge \delta\, x^{\alpha_1} \wedge \cdots \wedge \delta\, x^{\alpha_r} \wedge d\, y^{i_1} \wedge \cdots \wedge d\, y^{i_s} +$$

$$+ \sum_{\ell=1}^{s} (-1)^{\ell+1} \, \beta_{\alpha_1 \cdots \alpha_r \, i_1 \cdots i_s} \, \delta\, x^{\alpha_1} \wedge \cdots \wedge \big(d\, \delta\, x^{\alpha_\ell} \big) \wedge \cdots \wedge d\, y^{i_1} \wedge \cdots \wedge d\, y^{i_s} \Big\}$$

hence

$$d\,\beta_{r,s} = \left(d\,\beta_{r,s}\right)_{r-1,\,s+2} + \left(d\,\beta_{r,s}\right)_{r,\,s+1} + \left(d\,\beta_{r,s}\right)_{r+1,\,s} \tag{4.29}$$

where the differential forms on the right-hand side of (4.29) are (locally) given by

$$\left(d\,\beta_{r,s}\right)_{r-1,\,s+2} = \sum_{\substack{1\le\alpha_1<\cdots<\alpha_r\le p\\ 1\le i_1<\cdots<i_s\le q}} \beta_{\alpha_1\cdots\alpha_r\,i_1\cdots i_s}\,\times \tag{4.30}$$

$$\times\sum_{\ell=1}^{r}(-1)^{\ell+1}\frac{\delta\,N_k^{\alpha_\ell}}{\delta\,y^j}\,\delta\,x^{\alpha_1}\wedge\cdots\wedge\widehat{\delta\,x^{\alpha_\ell}}\wedge\cdots\wedge$$
$$\delta\,x^{\alpha_r}\wedge d\,y^j\wedge d\,y^k\wedge d\,y^{i_1}\wedge\cdots\wedge d\,y^{i_s}\,,$$

$$\left(d\,\beta_{r,s}\right)_{r,\,s+1} = \sum_{\substack{1\le\alpha_1<\cdots<\alpha_r\le p\\ 1\le i_1<\cdots<i_s\le q}} \left\{\frac{\delta\,\beta_{\alpha_1\cdots\alpha_r\,i_1\cdots i_s}}{\delta\,y^j}\,d\,y^j\wedge\right. \tag{4.31}$$

$$\wedge\,\delta\,x^{\alpha_1}\wedge\cdots\wedge\delta\,x^{\alpha_r}\wedge d\,y^{i_1}\wedge\cdots\wedge d\,y^{i_s}+$$

$$+\,\beta_{\alpha_1\cdots\alpha_r\,i_1\cdots i_s}\sum_{\ell=1}^{r}(-1)^{\ell+1}\,\delta\,x^{\alpha_1}\wedge\cdots\wedge$$
$$\left.\left(\frac{\partial\,N^{\alpha_\ell}}{\partial x^\gamma}\,\delta\,x^\gamma\wedge d\,y^j\right)\wedge\cdots\wedge\delta\,x^{\alpha_r}\wedge d\,y^{i_1}\wedge\cdots\wedge d\,y^{i_s}\right\} =$$

$$=(-1)^r\sum_{\substack{1\le\alpha_1<\cdots<\alpha_r\le p\\ 1\le i_1<\cdots<i_s\le q}}$$
$$\left\{\frac{\delta\,\beta_{\alpha_1\cdots\alpha_r\,i_1\cdots i_s}}{\delta\,y^{i_0}}\,\delta\,x^{\alpha_1}\wedge\cdots\wedge\delta\,x^{\alpha_r}\wedge d\,y^{i_0}\wedge d\,y^{i_1}\wedge\cdots\wedge d\,y^{i_s}+\right.$$

$$\left.-\,\beta_{\alpha_1\cdots\alpha_r\,i_1\cdots i_s}\sum_{\ell=1}^{r}\frac{\partial\,N_{i_0}^{\alpha_\ell}}{\partial\,x^{\alpha_\ell}}\,\delta\,x^{\alpha_1}\wedge\cdots\wedge\delta\,x^{\alpha_r}\wedge d\,y^{i_0}\wedge d\,y^{i_1}\wedge\cdots\wedge d\,y^{i_s}\right\},$$

$$\left(d\,\beta_{r,s}\right)_{r+1,\,s} = \sum_{\substack{1\le\alpha_1<\cdots<\alpha_r\le p\\ 1\le i_1<\cdots<i_s\le q}} \frac{\partial\,\beta_{\alpha_1\cdots\alpha_r\,i_1\cdots i_s}}{\partial\,x^{\alpha_0}}\,\times \tag{4.32}$$

$$\times\ \delta x^{\alpha_0} \wedge \delta x^{\alpha_1} \wedge \cdots \wedge \delta x^{\alpha_r} \wedge d y^{i_1} \wedge \cdots \wedge d y^{i_s} .$$

As customary, a hat denotes suppression of a term. Let $\Omega^{r,s}(\mathscr{F}) = \Omega^{r,s}(M, \mathscr{F})$ denote the $C^\infty(M)$-module of all $(r+s)$-forms of type (r, s) on $(M, \mathscr{F})$. We define the first-order differential operator

$$d^r_{\mathscr{F}} = d^{r,s}_{\mathscr{F}} : \Omega^{r,s}(\mathscr{F}) \to \Omega^{r+1,s}(\mathscr{F}) \tag{4.33}$$

by declaring $d^r_{\mathscr{F}}\, \beta_{r,s}$ to be the $(r+1,\ s)$-component $\left(d\,\beta_{r,s}\right)_{r+1,s}$ of $d\,\beta_{r,s}$.

Definition 4.31 The differential operator (4.33) is referred to as the *differentiation along the leaves* of $\mathscr{F}$. □

Lemma 4.32 *Let us consider the sequence*

$$\Omega^{r-1,s}(\mathscr{F}) \xrightarrow{d^{r-1}_{\mathscr{F}}} \Omega^{r,s}(\mathscr{F}) \xrightarrow{d^{r}_{\mathscr{F}}} \Omega^{r+1,s}(\mathscr{F}).$$

Then $d^r_{\mathscr{F}} \circ d^{r-1}_{\mathscr{F}} = 0$.

Proof Let $\beta_{r-1,s} \in \Omega^{r-1,s}$ be a form of type $(r-1,\ s)$ on $(M, \mathscr{F})$, thought of as the $(r-1,\ s)$-component of some form $\beta \in \Omega^{r+s-1}(M)$. Then

$$d\,\beta_{r-1,s} = \left(d\,\beta_{r-1,s}\right)_{r-2,s+2} + \left(d\,\beta_{r-1,s}\right)_{r-1,s+1} + d^{r-1}_{\mathscr{F}}\,\beta_{r-1,s} . \tag{4.34}$$

Let us set

$$\gamma_{r,s} = d^{r-1}_{\mathscr{F}}\,\beta_{r-1,s} \in \Omega^{r,s}(\mathscr{F})$$

for the sake of simplicity. Then

$$d\,\gamma_{r,s} = \left(d\,\gamma_{r,s}\right)_{r-1,s+2} + \left(d\,\gamma_{r,s}\right)_{r,s+1} + d^{r}_{\mathscr{F}}\,\gamma_{r,s} . \tag{4.35}$$

Let us differentiate (4.34) so that to get

$$0 = d\left\{\left(d\,\beta_{r-1,s}\right)_{r-2,s+2}\right\} + d\left\{\left(d\,\beta_{r-1,s}\right)_{r-1,s+1}\right\} + d\,\gamma_{r,s} .$$

By (4.29) there exist differential $(r+s+1)$-forms

$$A_{r-3,s+4},\quad A_{r-2,s+3},\quad A_{r-1,s+2},\quad B_{r-2,s+3},\quad B_{r-1,s+2},\quad B_{r,s+1} \in \Omega^{r+s+1}(M)$$

(whose type is indicated by the subscripts of each form) such that

$$0 = A_{r-3,s+4} + A_{r-2,s+3} + A_{r-1,s+2} + B_{r-2,s+3} + B_{r-1,s+2} + B_{r,s+1} + d\,\gamma_{r,s} .$$

Finally [by (4.35) and by comparing types], it follows that $\left(d\,\gamma_{r,s}\right)_{r+1,s} = 0$, which is $d^r_{\mathscr{F}}\,\gamma_{r,s} = 0$. Q.e.d.

As a consequence of Lemma 4.32

$$\Omega^{0,s}(\mathscr{F}) \xrightarrow{d^0_{\mathscr{F}}} \Omega^{1,s}(\mathscr{F}) \xrightarrow{d^1_{\mathscr{F}}} \cdots \xrightarrow{d^{p-1}_{\mathscr{F}}} \Omega^{p,s}(\mathscr{F}) \xrightarrow{d^p_{\mathscr{F}}} 0 \tag{4.36}$$

is a cochain complex.

Definition 4.33 The complex (4.36) is called the $d_{\mathscr{F}}$*-complex*, and its cohomology groups

$$H^{r,s}_{\mathscr{F}}(M) = H^r\big(\Omega^{\bullet,s}(\mathscr{F}),\ d^{\bullet}_{\mathscr{F}}\big) = \frac{\operatorname{Ker}\{d^r_{\mathscr{F}}\ :\ \Omega^{r,s}(\mathscr{F}) \to \bullet\}}{d^{r-1}_{\mathscr{F}}\ \Omega^{r-1,s}(\mathscr{F})}$$

are the $d_{\mathscr{F}}$*-cohomology groups*. □

It should be observed that

$$\Omega^0_B(\mathscr{F}) = H^{0,0}_{\mathscr{F}}(M). \tag{4.37}$$

Indeed

$$H^{0,0}_{\mathscr{F}}(M) = \operatorname{Ker}\left\{d^{0,0}_{\mathscr{F}}\ :\ \Omega^{0,0}(\mathscr{F}) \to \Omega^{1,0}(\mathscr{F})\right\},\quad \Omega^{0,0}(\mathscr{F}) = \Omega^0(M),$$

hence for any $u \in H^{0,0}_{\mathscr{F}}(M)$, one has [by (4.27)]

$$0 = d^0_{\mathscr{F}}\, u = \big(d\,u\big)_{1,0} = \Big(\frac{\partial u}{\partial x^\alpha}\,\delta\,x^\alpha + \frac{\delta\,u}{\delta\,yi}\,d\,y^i\Big)_{1,0} = \frac{\partial u}{\partial x^\alpha}\,\delta\,x^\alpha$$

that is $u \in \Omega^0_B(\mathscr{F})$ for any $(U,\ x^\alpha\ ,\ y^i) \in \mathscr{F}$. Q.e.d.

Of course Definition 4.33 makes sense for vector-valued forms, as well. Next, we generalize the notions above to tensorial forms of type $(\rho,\ V)$. Let $\big(E,\ M,\ p,\ G,\ \mathscr{F}_E\big)$ be a foliated principal bundle over $(M,\ \mathscr{F})$. Let $\rho : G \to \operatorname{End}_{\mathbb{R}}(V)$ be a representation of G on the finite-dimensional linear space V. Let $\varphi \in C^\infty\big(\Lambda^k\, T^*(E) \otimes V\big)$ be a tensorial k-form of type $(\rho,\ V)$.

Definition 4.34 Let $r,\ s \in \mathbb{Z}_+$ such that $r + s = k$. We say φ is a $(r,\ s)$*-form* if $s^*\,\varphi \in \Omega^{r,s}(\mathscr{F})$, i.e., $s^*\,\varphi$ is a k-form of type $(r,\ s)$ on U, for any local section $s : U \to E$. □

Let $\{s_i\}_{i\in I}$ be a family of foliate local sections $s_i\ :\ U_i \to E$ such that $\{U_i\}_{i\in I}$ is an open cover of M. Let $\Phi_i\ :\ p^{-1}(U_i) \to U_i \times G$ be the local trivialization associated to s_i. Also $s_j = s_i \cdot a_{ji}$ for some G-valued basic functions a_{ji} on $U_j \cap U_i$. Let $\varphi \in C^\infty\big(\Lambda^k\, T^*(M) \otimes V\big)$ be a tensorial form of type $(\rho,\ V)$ on E, and let us set

$$\varphi_i = s_i^*\,\varphi\,,\quad i \in I.$$

Then

$$\varphi_{j,x} = \rho\big(a_{ji}(x)^{-1}\big) \cdot \varphi_{i,x}\,, \quad x \in U_i \cap U_j\,. \tag{4.38}$$

The dot product in (4.38) is given by

$$(T \cdot \beta)_x(X_1\,, \cdots, X_k) = T\big[\beta_x(X_1\,, \cdots, X_k)\big],$$

$$T \in \mathrm{End}_{\mathbb{R}}(V), \quad \beta \in C^\infty\big(\Lambda^k T^*(M) \otimes V\big), \quad X_1\,, \cdots, X_k \in T_x(M), \quad x \in M.$$

Conversely, a family of forms $\varphi_i \in C^\infty\big(\Lambda^k T^*(U_i) \otimes V\big)$, $i \in I$, satisfying (4.38) determines a tensorial k-form φ of type (ρ, V) on E, in a natural manner. Indeed, if $z \in E$ and $x = p(z) \in M$, let $i \in I$ such that $x \in U_i$ and let us set

$$\varphi_z\big(v_1\,, \cdots, v_k\big) = \varphi_{i,x}\big((d_z\,p)\,v_1\,, \cdots, (d_z\,p)\,v_k\big), \quad v_1\,, \cdots, v_k \in T_z(E).$$

The definition of $\varphi_z(v_1\,, \cdots, v_k)$ doesn't depend upon the choice of U_i such that $x \in U_i$, precisely due to the condition (4.38).

Lemma 4.35 *Let $\varphi \in C^\infty\big(\Lambda^k T^*(E) \otimes V\big)$ be a tensorial k-form of type (ρ, V) on E. Let us assume that φ is also a (r, s)-form for some $r, s \in \mathbb{Z}_+$ with $r + s = k$. Then*

$$d^r_{\mathscr{F}}\,\varphi_j = \rho\big(a_{ji}^{-1}\big) \cdot d^r_{\mathscr{F}}\,\varphi_i \quad \text{on } U_i \cap U_j\,. \tag{4.39}$$

In particular the $(k+1)$-forms $\big\{d_{\mathscr{F}}\,\varphi_i\big\}_{i \in I}$ determine a tensorial $(k+1)$-form of type (ρ, V) on E.

The proof of Lemma 4.35 is easily got by taking the exterior derivative of $\varphi_j = \rho\big(a_{ji}^{-1}\big) \cdot \varphi_i$ [with a_{ji} basic, cf. (4.38)] and exploiting the fact that $d^0_{\mathscr{F}}$ annihilates the basic functions.

The $(r + s + 1)$-form furnished by Lemma 4.35 is denoted by

$$d_{\mathscr{F}}\,\varphi = d^r_{\mathscr{F}}\,\varphi \in C^\infty\big(\Lambda^{r+s+1}\,T^*(E) \otimes V\big).$$

Clearly $d_{\mathscr{F}}\,\varphi$ is a $(r + 1, s)$-form.

Let $\Omega^{r,s}_\rho(E, V)$ be the space of (r, s)-forms, i.e., tensorial $(r + s)$-forms $\varphi \in C^\infty\big(\Lambda^{r+s}\,T^*(M) \otimes V\big)$ of type (ρ, V) on E, which are also of type (r, s). Lemma 4.35 furnishes the first-order differential operator

$$d_{\mathscr{F}} \;:\; \Omega^{r,s}_\rho(E, V) \to \Omega^{r+1,s}_\rho(E, V) \tag{4.40}$$

Already $d_{\mathscr{F}}$ at the level of forms on M squares to zero (by Lemma 4.32); hence [by (4.39)], $d_{\mathscr{F}}$ on tensorial forms on E squares to zero, as well. Therefore, the cohomology groups

$$H^{r,s}_{\rho,\mathscr{F}}(E, V) = H^r\big(\Omega^{\bullet,s}_\rho(E, V), d^\bullet_{\mathscr{F}}\big) = \frac{\mathrm{Ker}\big\{d_{\mathscr{F}} \,:\, \Omega^{r,s}_\rho(E, V) \to \bullet\big\}}{d_{\mathscr{F}}\,\Omega^{r-1,s}_\rho(E, V)}$$

are well defined.

4.5 Atiyah Class

Let $\left(E, M, p, G, \mathcal{F}_E\right)$ be a foliated principal bundle over the foliated manifold $(M, \mathcal{F})$, and let $\omega \in C^\infty\left(T^*(E) \otimes \mathfrak{g}\right)$ be an adapted connection 1-form on E, i.e., $T\left(\mathcal{F}_E\right) \subset \Gamma = \mathrm{Ker}(\omega)$. Let $s : U \to E$ be a local foliate section in E. Then

$$T(\mathcal{F}) \rfloor s^* \omega = \omega^s\left(s_* T(\mathcal{F})\right) \subset \omega^s\left(s^{-1} T(\mathcal{F}_E)\right) \subset \omega^s\left(s^{-1} \Gamma\right) = (0),$$

where $\omega^s := \omega \circ s$. Also

$$(s_* X)_x = (d_x s) X_x, \quad X \in \mathfrak{X}_{\mathcal{F}}, \quad x \in U.$$

As customary

$$s^{-1} T(\mathcal{F}_E) \to U, \quad s^{-1} \Gamma \to U,$$

are pullback bundles. Therefore, $s^* \omega \in C^\infty\left(T^*(U) \otimes \mathfrak{g}\right)$ is a form of type (r, s) on U. Let $\Omega = D\omega$ be the curvature 2-form of ω. Then (by the second structure equation)

$$s^* \Omega = s^*\left(d\,\omega + \frac{1}{2}[\omega, \omega]\right) = d\left(s^* \omega\right) + \frac{1}{2}\left[s^* \omega, s^* \omega\right]. \tag{4.41}$$

We seek to decompose the form $s^* \Omega \in C^\infty\left(\Lambda^2 T^*(U) \otimes \mathfrak{g}\right)$ in a sum of components of type $(r, s) \in \left\{(0, 2), (1, 1), (2, 0)\right\}$. We recall that $\left[s^* \omega, s^* \omega\right]$ is the $\mathfrak{g}$-valued 2-form on U given by

$$\left[s^* \omega, s^* \omega\right](X, Y) = \left[(s^*\omega)(X), (s^*\omega)(Y)\right], \quad X, Y \in \mathfrak{X}(U),$$

and $\left[\cdot, \cdot\right]$ is the Lie algebra product on $\mathfrak{g}$. Let us assume that U is the domain of a foliated local chart $(U, x^\alpha, y^i) \in \mathcal{F}$ so that

$$s^* \omega = g_i \otimes d\,y^i \quad \text{on } U \tag{4.42}$$

for some $g_i \in C^\infty(U, \mathfrak{g})$. No loos of generality results from our working assumption[11] on U. The 2-form $\left[s^* \omega, s^* \omega\right]$ is (locally) a combination of

$$\left\{\delta x^\alpha \wedge \delta x^\beta, \ \delta x^\alpha \wedge d\,y^j, \ d\,y^i \wedge d\,y^j \ : \ 1 \le \alpha, \beta \le p, \ 1 \le i, j \le q\right\}$$

---

[11] Strictly speaking one chooses a point $x \in U$ and performs calculations on an open neighborhood of x contained in U and endowed with foliated local coordinates.

with coefficients in $C^\infty(U, \mathfrak{g})$. Yet [by (4.42)],

$$[s^*\omega,\, s^*\omega]\Big(\frac{\partial}{\partial x^\alpha},\, \frac{\partial}{\partial x^\beta}\Big) = 0, \quad [s^*\omega,\, s^*\omega]\Big(\frac{\partial}{\partial x^\alpha},\, \frac{\delta}{\delta y^j}\Big) = 0,$$

hence

$$[s^*\omega,\, s^*\omega] = [g_i\,,\, g_j] \otimes (d\,y^i \wedge d\,y^j)$$

so that $[s^*\omega,\, s^*\omega]$ is a $\mathfrak{g}$-valued form of type $(2, 0)$ on U. Here

$$[g_i\,,\, g_j] : U \to \mathfrak{g}, \quad [g_i\,,\, g_j](x) = [g_i(x)\,,\, g_j(x)]\,, \quad x \in U.$$

Moreover [by (4.29) for vector-valued forms],

$$d(s^*\omega) = \big(d(s^*\omega)\big)_{0,2} + \big(d(s^*\omega)\big)_{1,1}\,. \tag{4.43}$$

Let $\{X_a \,:\, 1 \le a \le \dim(G)\}$ be a linear basis of the Lie algebra $\mathfrak{g}$. Then $g_i = u_i^a\, X_a$ for some scalar functions $u_i^a \in C^\infty(U)$. Let us take the exterior derivative of (4.42) so that to get

$$\begin{aligned} d(s^*\omega) =& X_a \otimes d(u_\alpha^i\, d\,y^i) = X_a \otimes (d\,u_i^a \wedge d\,y^i) \\ =& X_a \otimes \Big(\frac{\partial\, u_i^a}{\partial\, x^\alpha}\,\delta\,x^\alpha + \frac{\delta\, u_i^a}{\delta\, y^j}\,d\,y^j\Big) = \end{aligned}$$

$$= \frac{\partial\, g_i}{\partial\, x^\alpha} \otimes (\delta\,x^\alpha \wedge d\,y^i) + \frac{\delta\, g_i}{\delta\, y^j} \otimes (d\,y^j \wedge d\,y^i)$$

so that [by (4.41) and (4.43)]

$$\Omega = \Omega_{1,\,1} + \Omega_{0,\,2} \tag{4.44}$$

for some 2-forms $\Omega_{1,1}\,,\ \Omega_{0,2} \in C^\infty\big(\Lambda^2\, T^*(E) \otimes \mathfrak{g}\big)$ such that

$$s^*\,\Omega_{1,1} = \frac{\partial g_i}{\partial x^\alpha} \otimes (\delta\,x^\alpha \wedge d\,y^i), \quad s^*\,\Omega_{0,2} = \Big(\frac{\delta\, g_j}{\delta y^i} + \frac{1}{2}\,[g_i\,,\, g_j]\Big) \otimes (d\,y^i \wedge d\,y^j),$$

on U. In particular

$$\Omega_{1,\,1} \in \Omega^{1,\,1}_{\mathrm{ad}}(E,\, \mathfrak{g}),$$

i.e., $\Omega_{1,1}$ is both a tensorial 2-form of type $(\mathrm{ad},\ \mathfrak{g})$ on E, and a $(1,\ 1)$-form.

Theorem 4.36 $d_{\mathscr{F}}\, \Omega_{1,\,1} = 0$ *and the cohomology class*

$$A(E,\, \mathscr{F}_E) = [\Omega_{1,\,1}] \in H^{1,\,1}_{\mathrm{ad}\,,\,\mathscr{F}}(E,\, \mathfrak{g})$$

doesn't depend upon the choice of adapted connection. Moreover, $A(E, \mathscr{F}_E) = 0$ *if and only if the foliated principal bundle* $(E, M, p, G, \mathscr{F}_E)$ *admits projectable connections.*

Cf. Molino, [Mol1].

Proof of Theorem 4.36. Let $\{s_i : U_i \to E\}_{i \in I}$ be a family of foliated local sections such that $\{U_i\}_{i \in I}$ is an open cover of M. Let us consider the $\mathfrak{g}$-valued 2-form of type $(1, 1)$ on U_i given by

$$(\Omega_{1,1})_i = s_i^* \Omega_{1,1}, \quad i \in I.$$

One has

$$(\Omega_{1,1})_j = \mathrm{ad}(a_{ji}^{-1}) \cdot (\Omega_{1,1})_i \quad \text{on } U_i \cap U_j$$

hence (by Lemma 4.35)

$$d_{\mathscr{F}} (\Omega_{1,1})_j = \mathrm{ad}(a_{ji}^{-1}) \cdot d_{\mathscr{F}} (\Omega_{1,1})_i \quad \text{on } U_i \cap U_j .$$

Consequently, the family of local forms $\{(\Omega_{1,1})_i\}_{i \in I}$ determines a unique globally defined tensorial form denoted by

$$d_{\mathscr{F}} \Omega_{1,1} \in \Omega_{\mathrm{ad}}^{2,1}(E, \mathfrak{g})$$

and such that

$$s_i^* (d_{\mathscr{F}} \Omega_{1,1}) = d_{\mathscr{F}} (\Omega_{1,1})_i, \quad i \in I.$$

Definition 4.37 The cohomology class

$$A(E, \mathscr{F}_E) \in H_{\mathrm{ad}, \mathscr{F}}^{1,1}(E, \mathfrak{g})$$

is called the *Atiyah class* of $(E, M, p, G, \mathscr{F}_E)$. The *Atiyah class* of $\mathscr{F}$ is

$$A(\mathscr{F}) = A(M, \mathscr{F}) = A\big(B_T^1(\mathscr{F}), \mathscr{F}_T^1\big) \in H_{\mathrm{ad}, \mathscr{F}}^{1,1}\big(B_T^1(\mathscr{F}), \mathfrak{gl}(q, \mathbb{R})\big).$$

□

The meaning of Theorem 4.36 is that the Atiyah class of a foliated principal bundle $(E, M, p, G, \mathscr{F}_E)$ may be looked at as an obstruction to the existence of projectable connections in that bundle.

4.6 A Spectral Sequence

4.6.1 Filtrations of the De Rham Complex and Rummler's Formula Revisited

In a series of works F. Kamber and Ph. Tondeur made use (cf. [KaTo1]-[KaTo3, KaTo11]) of a natural spectral sequence[12] associated to a foliated manifold $(M, \mathcal{F})$. For a better understanding of the constructions in the present section we add an Appendix on decreasing filtrations and spectral sequences of cohomological type.[13] Said spectral sequence is associated to the multiplicative filtration of the de Rham complex $\Omega^\bullet(M)$ given by

$$F^0\,\Omega^r(M) = \Omega^r(M),$$

$$F^k\,\Omega^r(M) = \Big\{\omega \in \Omega^r(M) \, : \, i_{X_1} \cdots i_{X_{r-k+1}}\,\omega = 0, \tag{4.45}$$

$$X_i \in \mathfrak{X}_{\mathcal{F}}, \quad 1 \le i \le r-k+1\Big\}, \quad k \ge 1.$$

Besides from being important in its own right, many calculus (with basic forms) assertions are, as we shall shortly show, easily stated in terms of the Kamber-Tondeur spectral sequence. Let us start by investigating a few basic properties of the filtration (4.45). A more advanced study of the spectral sequence of a Riemannian foliation was performed by Alvarez, [Alv1]-[Alv2]. Locally, with respect to a foliated local chart $(U, x^\alpha, y^i) \in \mathcal{F}$, the space $F^k\,\Omega^r(M)$ consists of the r-forms whose local representation exterior monomials contain at most $r-k$ differentials $d\,x^\alpha$. Note that

$$F^k\,\Omega^r(M) = (0), \quad k \ge r+1. \tag{4.46}$$

The filtration (4.45) is (according to the terminology in the next section) a decreasing filtration of the graded vector space $\Omega^\bullet(M)$ and looks like

$$(0) \subset F^r\,\Omega^r(M) \subset \cdots \subset F^{k+1}\,\Omega^r(M) \subset F^k\,\Omega^r(M) \subset \cdots$$
$$\subset F^1\,\Omega^r(M) \subset F^0\,\Omega^r(M) = \Omega^r(M).$$

In particular (4.45) is a bounded filtration [cf. Definition 4.52 in the Appendix to this section]. Let us also observe that (4.45) is stable under exterior differentiation, i.e.,

$$d\,F^k\,\Omega^r(M) \subset F^k\,\Omega^{r+1}(M).$$

---

[12] Book I-A in the present book series studies a few geometric properties of said spectral sequence that could be presented without using elements of Riemannian geometry (i.e., the geometry of transversally Riemanian foliations of Riemannian manifolds).

[13] While Book I-A in the present book series is provided with an algebraic preparation regarding mainly spectral sequences of homological type.

Consequently, Theorem 4.53 in the Appendix can be applied [the description of the spectral sequence furnished by (the proof) of Theorem 4.53 is given later in this section]. Moreover, if $\dim(\mathscr{F}) = p = n - q$, then

$$\begin{aligned} F^k\,\Omega^{p+k}(M) =& \Big\{\omega \in \Omega^{p+k}(M)\ :\ i_{X_1} \cdots i_{X_{p+1}}\,\omega = 0,\quad X_i \in \mathfrak{X}_{\mathscr{F}},\quad 1 \le i \le p+1\Big\} \\ =& \Omega^{p+k}(M) \end{aligned}$$

because $p+1$ vectors lying in (a fiber of) $T(\mathscr{F})$ cannot be linearly independent. Next

$$F^1\,\Omega^p(M) = \Big\{\omega \in \Omega^p(M)\ :\ i_{X_1} \cdots i_{X_p}\,\omega = 0,\quad X_\alpha \in \mathfrak{X}_{\mathscr{F}}\,,\quad 1 \le \alpha \le p\Big\}$$

so that $F^1\,\Omega^p(M)$ consists of the p-forms vanishing on $T(\mathscr{F}) \otimes \cdots \otimes T(\mathscr{F})$ (p terms). This observation leads to a useful reformulation of Rummler's formula [for a foliation $\mathscr{F}$ of a Riemannian manifold $(M,\ g)$]. Indeed [by (4.7)], for every $Z \in C^\infty\big(T(\mathscr{F})^\perp\big)$ the p-form $\mathscr{L}_Z\,\chi_{\mathscr{F}} + \kappa(Z)\,\chi_{\mathscr{F}}$ evaluates to zero on $T(\mathscr{F}) \otimes \cdots \otimes T(\mathscr{F})$ (p terms) hence the announced reformulation of Rummler's formula is

$$\mathscr{L}_Z\,\chi_{\mathscr{F}} + \kappa(Z)\,\chi_{\mathscr{F}} \in F^1\,\Omega^p(M), \quad Z \in C^\infty\big(T(\mathscr{F})^\perp\big). \tag{4.47}$$

In general, $F^{k+1}\,\Omega^{p+k}(M)$ consists of all $(p+k)$-forms evaluating to zero when p arguments lie in $\mathfrak{X}_{\mathscr{F}}$. We may also state

Proposition 4.38 *Let $\varphi_0 := d\,\chi_{\mathscr{F}} + (p+1)\,\kappa \wedge \chi_{\mathscr{F}} \in \Omega^{p+1}(M)$. Rummler's formula is equivalent to $\varphi_0 \in F^2\,\Omega^{p+1}(M)$.*

Proof If all $p+1$ arguments $X_0\,, \cdots, X_p$ lie in $\mathfrak{X}_{\mathscr{F}}$, then of course $\varphi_0(X_0\,, \cdots, X_p) = 0$. To prove Proposition 4.38, it suffices then to consider the case $X_0 = Z \in T(\mathscr{F})^\perp$, i.e., check that $i_Z\,\varphi_0$ vanishes on $T(\mathscr{F}) \otimes \cdots \otimes T(\mathscr{F})$ (p terms). Then[14] (by $i_Z\,\chi_{\mathscr{F}} = 0$)

$$(\kappa \wedge \chi_{\mathscr{F}})(Z,\ X_1\,, \cdots, X_p) = \frac{1}{(p+1)!} \sum_{\sigma \in \sigma_{p+1}} \epsilon(\sigma)\,(\kappa \otimes \chi_{\mathscr{F}})(X_{\sigma(0)}\,, \cdots, X_{\sigma(p)}) =$$

$$= \frac{1}{(p+1)!} \sum_{\sigma \in A} \epsilon(\sigma)\,\kappa\big(X_{\sigma(0)}\big)\,\chi_{\mathscr{F}}\big(X_{\sigma(1)}\,, \cdots, X_{\sigma(p)}\big)$$

where $A = \big\{\sigma \in \sigma_{p+1}\ :\ \sigma(0) = 0\big\}$. Also σ_{p+1} denotes the permutation group of order $(p+1)!$ [thought of as consisting of the permutations of $\{0,\ 1, \cdots, p\}$]. If $\pi \in \sigma_p$ we define $\sigma_\pi \in \sigma_{p+1}$ by setting

$$\sigma_\pi(0) = 0, \quad \sigma_\pi(\alpha) = \pi(\alpha), \quad 1 \le \alpha \le p,$$

---

[14] Our conventions as to multi-linear algebra calculations are those in [KoNo]. There is an extra $p+1$ factor in our definition of φ_0 with respect to the treatment in [Ton], p. 121.

so that $A = \{\sigma_\pi \,:\, \pi \in \sigma_p\}$. Then A has cardinality $|A| = p!$ and the sign of every $\sigma \in A$ is $\epsilon(\sigma) = (-1)^p$ so that

$$\chi_{\mathscr{F}}\big(X_{\sigma(1)}, \cdots, X_{\sigma(p)}\big) = (-1)^p \chi_{\mathscr{F}}\big(X_1, \cdots, X_p\big), \quad \sigma \in A.$$

We conclude that

$$\big(\kappa \wedge \chi_{\mathscr{F}}\big)(Z, X_1, \cdots, X_p) = \frac{1}{p+1}\kappa(Z)\,\chi_{\mathscr{F}}\big(X_1, \cdots, X_p\big).$$

Finally (by Cartan's formula),

$$(i_Z \varphi_0)\big(X_1, \cdots, X_p\big) = \big(\mathscr{L}_Z \chi_{\mathscr{F}} + \kappa(Z)\,\chi_{\mathscr{F}}\big)\big(X_1, \cdots, X_p\big) = 0.$$

Q.e.d.

A few remarks pertaining to the local description of $F^k\,\Omega^r(M)$ are in order. Let $(U, x^\alpha, y^i) \in \mathscr{F}$ be a foliated local chart. Every r-form $\omega \in \Omega^r(M)$ is then locally represented as a sum of exterior monomials of the form

$$d x^{\alpha_1} \wedge \cdots \wedge d x^{\alpha_u} \wedge d y^{i_1} \wedge \cdots \wedge d y^{i_v}, \tag{4.48}$$

$$1 \le \alpha_1 < \cdots < \alpha_u \le p, \quad 1 \le i_1 < \cdots < i_v \le q, \quad u + v = r,$$

with coefficients in $\Omega^0(U)$. Them the condition $\omega \in F^k\,\Omega^r(M)$ is locally that $o \le u \le r - k$. With these tools one may easily show that

Proposition 4.39 *If* $q = \operatorname{cod}(\mathscr{F})$, *then*

$$F^{p+1}\,\Omega^r(M) = (0). \tag{4.49}$$

In particular, if $q = 1$, *then* $F^2\,\Omega^{p+1}(M) = (0)$ *so that Rummler's formula becomes* $d\,\chi_{\mathscr{F}} + (p+1)\,\kappa \wedge \chi_{\mathscr{F}} = 0$.

Proof Let $\omega \in F^{q+1}\,\Omega^r(M)$. Then the exterior monomials (4.48) in the local representation of ω on U contain at most $r - (q+1)$ differentials $d x^\alpha$, that is $u \le r - (q+1)$. Equivalently $v \ge q + 1$ hence

$$d y^{i_1} \wedge \cdots \wedge d y^{i_v} = 0$$

and then $\omega = 0$ thus yielding (4.49). The particular form of Rummler's formula for codimension one foliations plays a major role in their study (cf. e.g., [Ton], p. 75).

By Theorem 4.53 (in the Appendix following the present section), given the graded vector space $\Omega^\bullet(M)$ and its filtration $\{F^k\,\Omega^\bullet(M) \,:\, k \in \mathbb{Z}\}$ [defined by (4.45)] there is a cohomological spectral sequence

$$E(\mathscr{F}) = E(M,\ \mathscr{F}) = \{E_s(M,\ \mathscr{F})\ :\ s \in \mathbb{N}\},$$

$$E_s(\mathscr{F}) = E_s(M,\ \mathscr{F}) = \left\{\left(E_s^{k,\ell}(M,\ \mathscr{F}),\ d_s^{k,\ell}\right)\ :\ k,\ \ell \in \mathbb{Z}\right\},$$

whose first terms is [cf. (4.63)]

$$E_1^{k,\ell}(\mathscr{F}) = E_1^{k,\ell}(M,\ \mathscr{F}) \simeq H^{k+\ell}\big(F^k\,\Omega(M)/F^{k+1}\,\Omega(M)\big). \tag{4.50}$$

Here $F^k\,\Omega(M) = \oplus_{r=0}^{n} F^k\,\Omega^r(M)$. If $\Omega(M) = \oplus_{r=0}^{n}\Omega^r(M)$, then

$$F^k\,\Omega^r(M) = F^k\,\Omega(M) \cap \Omega^r(M).$$

By the proof of Theorem 4.53

$$E_s^{k,\ell}(\mathscr{F}) = E_s^{k,\ell}(M,\ \mathscr{F}) = \frac{Z_s^{k,\ell}(\mathscr{F})}{Z_{s-1}^{k+1,\,\ell-1}(\mathscr{F}) + B_{s-1}^{k,\ell}(\mathscr{F})} \tag{4.51}$$

where

$$Z_s^{k,\ell}(\mathscr{F}) = Z_s^{k,\ell}(M,\ \mathscr{F}) = \{F^k\,\Omega^{k+\ell}(M)\} \cap d^{-1}\big[F^{k+s}\,\Omega^{k+\ell+1}(M)\big],$$

$$B_s^{k,\ell}(\mathscr{F}) = B_s^{k,\ell}(M,\ \mathscr{F}) = \{F^k\,\Omega^{k+\ell}(M)\} \cap d\,\big[F^{k-s}\,\Omega^{k+\ell+1}(M)\big].$$

That is $Z_s^{k,\ell}(\mathscr{F})$ consists of all $(k+\ell)$-forms in $F^k\,\Omega^{k+\ell}(M)$ which are boundaries in $F^{k+s}\,\Omega^{k+\ell+1}(M)$. Similarly, we may observe that $B_s^{k,\ell}(\mathscr{F})$ consists of all $(k+\ell)$-forms in $F^k\,\Omega^{k+\ell}(M)$ which are images of $(k+\ell-1)$-forms from $F^{k-s}\,\Omega^{k+\ell-1}(M)$ by d. In particular [by (4.51)],

$$E_1^{k,\ell}(\mathscr{F}) = \frac{Z_1^{k,\ell}(\mathscr{F})}{Z_0^{k+1,\,\ell-1}(\mathscr{F}) + B_0^{k,\ell}(\mathscr{F})}$$

and

$$Z_1^{k,\ell}(\mathscr{F}) = \{F^k\,\Omega^{k+\ell}(M)\} \cap d^{-1}\big[F^{k+1}\,\Omega^{k+\ell+1}(M)\big],$$

$$Z_0^{k+1,\,\ell-1}(\mathscr{F}) = \{F^{k+1}\,\Omega^{k+\ell}(M)\} \cap d^{-1}\big[F^{k+1}\,\Omega^{k+\ell+1}(M)\big] = F^{k+1}\,\Omega^{k+\ell}(M),$$

$$B_0^{k,\ell}(\mathscr{F}) = \{F^k\,\Omega^{k+\ell}(M)\} \cap d\,\big[F^{k+1}\,\Omega^{k+\ell-1}(M)\big] = d\,F^k\,\Omega^{k+\ell-1}(M),$$

so that

$$E_1^{k\,\ell}(\mathscr{F}) = \frac{\{F^k\,\Omega^{k+\ell}(M)\} \cap d^{-1}\big[F^{k+1}\,\Omega^{k+\ell+1}(M)\big]}{F^{k+1}\,\Omega^{k+\ell}(M) + d\,\big[F^k\,\Omega^{k+\ell-1}(M)\big]}. \tag{4.52}$$

In particular for $\ell = 0$ one has

$$F^k\,\Omega^k(M) = \{\omega \in \Omega^k(M)\ :\ i_X\,\omega = 0,\quad X \in \mathfrak{X}_{\mathscr{F}}\},$$

$$d^{-1}\left\{F^{k+1}\,\Omega^{k+1}(M)\right\} = \{\omega \in \Omega^k(M)\ :\ d\,\omega \in F^{k+1}\,\Omega^{k+1}(M)\} =$$

$$= \{\omega \in \Omega^k(M)\ :\ i_X\,d\,\omega = 0,\quad X \in \mathfrak{X}_{\mathscr{F}}\},$$

hence

$$F^k\,\Omega^k(M) \cap d^{-1}\left\{F^{k+1}\,\Omega^{k+1}(M)\right\} = \begin{cases} \Omega^k_B(\mathscr{F}) & \text{if } 0 \le k \le q, \\ 0 & \text{otherwise.} \end{cases}$$

Also $F^{k+1}\,\Omega^k(M) = (0)$ and in particular

$$F^{k+1}\,\Omega^k(M) + d\left\{F^k\,\Omega^{k-1}(M)\right\} = (0).$$

Finally [by (4.53) for $\ell = 0$],

$$E_1^{k,\,0}(\mathscr{F}) = \Omega^k_B(\mathscr{F}),\quad 0 \le k \le q. \tag{4.53}$$

We refer again to the Appendix (following the present section) for the construction of the differentials

$$d_s^{k,\,\ell} : E_s^{k,\,\ell}(\mathscr{F}) \to E_s^{k+s,\,\ell-s+1}(\mathscr{F}).$$

If $\pi_s^{k,\,\ell} : Z_s^{k,\,\ell}(\mathscr{F}) \to E_s^{k,\,\ell}(\mathscr{F})$ is the projection, then the diagram

$$\begin{array}{ccc} Z_s^{k,\,\ell}(\mathscr{F}) & \xrightarrow{d} & Z_s^{k+s,\,\ell-s+1}(\mathscr{F}) \\ \pi_s^{k,\,\ell}\downarrow & & \downarrow \pi_s^{k+s,\,\ell-s+1} \\ E_s^{k,\,\ell}(\mathscr{F}) & \xrightarrow{d_s^{k,\,\ell}} & E_s^{k+s,\,\ell-s+1}(\mathscr{F}) \end{array}$$

is commutative. In particular for $s = 1$

$$\begin{array}{ccc} Z_1^{k,\,\ell}(\mathscr{F}) & \xrightarrow{d} & Z_1^{k+1,\,\ell}(\mathscr{F}) \\ \pi_1^{k,\,\ell}\downarrow & & \downarrow \pi_1^{k+1,\,\ell} \\ E_1^{k,\,\ell}(\mathscr{F}) & \xrightarrow{d_1^{k,\,\ell}} & E_s^{k+1,\,\ell}(\mathscr{F}) \end{array}$$

is commutative. Next, for $\ell = 0$ one has (as shown above)

$$E_1^{k,0}(\mathcal{F}) = Z_1^{k,0}(\mathcal{F}) = \Omega_B^k(\mathcal{F})$$

hence $\pi_1^{k,0} = 1_{\Omega_B^k(\mathcal{F})}$ and $d_1^{k,0} = d_B : \Omega_B^k(\mathcal{F}) \to \Omega_B^{k+1}(\mathcal{F})$. Yet $E(\mathcal{F})$ is a spectral sequence hence (by Definition 4.42 in the Appendix) $E_2^{k,\ell}(\mathcal{F})$ must be the cohomology of the term $E_1(\mathcal{F})$, i.e.,

$$E_2^{k,\ell}(\mathcal{F}) \simeq H^{k,\ell}\big(E_1^{\bullet,\bullet}(\mathcal{F}),\, d_1^{k,\ell}\big) = \frac{\operatorname{Ker}\big\{d_1^{k,\ell} : E_1^{k,\ell}(\mathcal{F}) \to E_1^{k+1,\ell}(\mathcal{F})}{d_1^{k-1,\ell}\big\{E_1^{k-1,\ell}(\mathcal{F})\big\}}.$$

In particular for $\ell = 0$

$$E_2^{k,0}(\mathcal{F}) \simeq \frac{\operatorname{Ker}\big\{d_B : \Omega_B^k(\mathcal{F}) \to \Omega_B^{k+1}(\mathcal{F})\big\}}{d_B\, \Omega_B^{k-1}(\mathcal{F})} = H_B^k(\mathcal{F}). \tag{4.54}$$

4.6.2 *Appendix: Spectral Sequences*

The elements of the theory of spectral sequences presented in this Appendix rely on classical work by Leray, [Ler]. There are however a few monographs containing the material below, cf. e.g., Mac Lane, [MaLa], or McCleary, [McCl]. As spectral sequences are recalled solely for the sake of applications, our point of view is indeed close to that in [McCl].

4.6.2.1 Basic Definitions and Properties

Let R be a commutative ring with unit.

Definition 4.40 Let $a, b \in \mathbb{Z}$ and let $\{E^{p,q}\}_{p,q\in\mathbb{Z}}$ be a family of R-modules. Let $\{d^{p,q} : E^{p,q} \to E^{p+a,q+b}\}_{p,q\in\mathbb{Z}}$ be a family of R-linear maps such that

$$d^{p,q} \circ d^{p-a,q-b} = 0, \quad p, q \in \mathbb{Z}.$$

The synthetic object $\{E^{p,q}, d^{p,q}\}_{p,q\in\mathbb{Z}}$ is called a *differential bigraded module* over R. The map $d^{p,q} : E^{p,q} \to E^{p+a,q+b}$ is referred to as a *differential* and the pair of integers (a, b) is its *bidegree*. □

Let $\{E^{p,q}, d^{p,q}\}_{p,q\in\mathbb{Z}}$ be a differential bigraded module over R.

Definition 4.41 The *homology* $\{H^{p,q}(E^{\bullet,\bullet}, d^{\bullet,\bullet})\}_{p,q\in\mathbb{Z}}$ of $\{E^{p,q}, d^{p,q}\}_{p,q\in\mathbb{Z}}$ is the family of R-modules given by

$$H^{p,q}\big(E^{\bullet,\bullet}, d^{\bullet,\bullet}\big) = \frac{\operatorname{Ker}\big\{d^{p,q} : E^{p,q} \to E^{p+a,q+b}\big\}}{d^{p-a,q-b}\, E^{p-a,q-b}}, \quad p, q \in \mathbb{Z}.$$

□

Definition 4.42 Let

$$E = \left\{ \left(E_r^{p,q} \, , \, d_r^{p,q} \right)_{p,\, q \in \mathbb{Z}} \; : \; r \in \mathbb{N} \right\}$$

be a family of differential bigraded modules over R, where each differential $d_r^{p,q} : E_r^{p,q} \to \bullet$ has bidegree $\big(a(r),\, b(r)\big)$, for some functions $a,\, b : \mathbb{N} \to \mathbb{Z}$, and

$$E_{r+1}^{p,q} \simeq H^{p,q}\big(E_r^{\bullet,\bullet} \, , \, d_r^{\bullet,\bullet}\big) \tag{4.55}$$

(an isomorphism of R-modules). We call E a *spectral sequence of homological type* (respectively, of *cohomological type*) if $a(r) = -r$ and $b(r) = r - 1$ (respectively, $a(r) = r$ and $b(r) = 1 - r$) for any $r \in \mathbb{N}$. □

We adopt the notation $E_r = \left\{ E_r^{p,q} \, , \, d_r^{p,q} \right\}_{p,\, q \in \mathbb{Z}}$ for the sake of brevity. It is worth observing that, given a spectral sequence $E = \left\{ E_r \; : \; r \in \mathbb{N} \right\}$, the differential bigraded module E_r determines [according to (4.55)] the bigraded module $\left\{ E_{r+1}^{p,q} \right\}_{p,\, q \in \mathbb{Z}}$, yet it doesn't determine the differential $d_{r+1}^{p,q} : E_{r+1}^{p,q} \to \bullet$.

Definition 4.43 The differential bigraded modules E_r are the *terms*[15] of the spectral sequence. □

Let $E = \left\{ E_r \; : \; r \in \mathbb{N}, \; r \geq 2 \right\}$ be a spectral sequence, and let us set

$$Z_2^{p,q} = \operatorname{Ker}\big(d_2^{p,q}\big), \quad B_2^{p,q} = d_2^{p-a(2),\, q-b(2)} \, E^{p-a(2),\, q-b(2)} \, .$$

Definition 4.44 The elements of $Z_2^{p,q}$ (respectively, of $B_2^{p,q}$) are referred to as *cycles* (respectively, as *boundaries*). □

The differentials $d_2^{\bullet,\bullet}$ have bidegree $(-2,\, 1)$ if E is a spectral sequence of homological type [respectively, $(2,\, -1)$ if E is a spectral sequence of cohomological type].

As $d_2^{p,q} \circ d_2^{p-a(2),\, q-b(2)} = 0$, it follows that

$$B_2^{p,q} \subset Z_2^{p,q} \subset E_2^{p,q} \, .$$

By Definition 4.42

$$E_3^{p,q} \simeq H^{p,q}\big(E_2^{\bullet,\bullet} \, , \, d_2^{\bullet,\bullet}\big) = \frac{\operatorname{Ker}\big(d_2^{p,q}\big)}{d_2^{p-a(2),\, q-b(2)} \, E_2^{p-a(2),\, q-b(2)}} = \frac{Z_2^{p,q}}{B_2^{p,q}} \, .$$

[15] In mathematical practice, the terms of a spectral sequence are often labeled starting with the integer $r = 2$.

Next, we set

$$\overline{Z}_3^{p,q} := \operatorname{Ker}\big(d_3^{p,q}\big), \quad \overline{B}_3^{p,q} := d_3^{p-a(3),\,q-b(3)}\, E_3^{p-a(3),\,q-b(3)},$$

$$Z_3^{p,q} := \Big\{x \in Z_2^{p,q} \ :\ x - B_2^{p,q} \in \overline{Z}_3^{p,q}\Big\},$$
$$B_3^{p,q} := \Big\{x \in Z_2^{p,q} \ :\ x + B_2^{p,q} \in \overline{B}_3^{p,q}\Big\},$$

so that

$$\overline{B}_3^{p,q} \subset \overline{Z}_3^{p,q} \subset E_3^{p,q}, \quad \overline{Z}_3^{p,q} = \frac{Z_3^{p,q}}{B_2^{p,q}}, \quad \overline{B}_3^{p,q} = \frac{B_3^{p,q}}{B_2^{p,q}}.$$

Moreover [again by (4.55)],

$$\begin{aligned} E_4^{p,q} \simeq H^{p,q}\big(E_3^{\bullet,\bullet},\, d_3^{\bullet,\bullet}\big) &= \frac{\operatorname{Ker}\big(d_3^{p,q}\big)}{d_3^{p-a(3),\,q-b(3)}\, E_3^{p-a(3),\,q-b(3)}} \\ &= \frac{\overline{Z}_3^{p,q}}{\overline{B}_3^{p,q}} = \Big(\frac{Z_3^{p,q}}{B_2^{p,q}}\Big)\Big/\Big(\frac{B_3^{p,q}}{B_2^{p,q}}\Big) \simeq \frac{Z_3^{p,q}}{B_3^{p,q}}. \end{aligned}$$

Up to this point

$$B_2^{p,q} \subset B_3^{p,q} \subset Z_2^{p,q} \subset E_2^{p,q}.$$

Quite obviously, we may iterate these constructions to produce a tower of submodules of $E_2^{p,q}$ for each pair of integers $p,\, q \in \mathbb{Z}$, i.e.,

$$B_2^{p,q} \subset B_3^{p,q} \subset \cdots \subset B_n^{p,q} \subset \cdots \subset Z_n^{p,q} \subset \cdots \subset Z_3^{p,q} \subset Z_2^{p,q} \subset E_2^{p,q} \tag{4.56}$$

so that

$$E_{n+1}^{p,q} \simeq \frac{Z_n^{p,q}}{B_n^{p,q}}, \quad n \in \{2,\, 3,\, \cdots\}. \tag{4.57}$$

Hence, we may think of the differential

$$d_{n+1}^{p,q} \ :\ E_{n+1}^{p,q} \to E_{n+1}^{p+a(n+1),\,q+b(n+1)}$$

as a map

$$\frac{Z_n^{p,q}}{B_n^{p,q}} \to \frac{Z_n^{p+a(n+1),\,q+b(n+1)}}{B_n^{p+a(n+1),\,q+b(n+1)}}$$

with kernel $Z_{n+1}^{p,q}\big/B_n^{p,q}$ and image $B_{n+1}^{p+a(n+1),\,q+b(n+1)}\big/B_n^{p+a(n+1),\,q+b(n+1)}$. For instance, if E is a spectral sequence of homological type, then $d_{n+1}^{p,q}$ is a map $Z_n^{p,q}\big/B_n^{p,q} \to Z_n^{p-n,\,q-n+1}\big/B_n^{p-n,\,q-n+1}$ with kernel $Z_{n+1}^{p,q}\big/B_n^{p,q}$ and image

$B_{n+1}^{p-n,\, q-n+1} / B_n^{p-n,\, q-n+1}$. Going back to spectral sequences of arbitrary type, there is a short exact sequence of R-modules and R-module homomorphisms

$$0 \to \frac{Z_{n+1}^{p,q}}{B_n^{p,q}} \to \frac{Z_n^{p,q}}{B_n^{p,q}} \xrightarrow{d_{n+1}^{p,q}} \frac{B_{n+1}^{p+a(n+1),\, q+b(n+1)}}{B_n^{p+a(n+1),\, q+b(n+1)}} \to 0 \tag{4.58}$$

and then

$$\begin{aligned} \frac{B_{n+1}^{p+a(n+1),\, q+b(n+1)}}{B_n^{p+a(n+1),\, q+b(n+1)}} &= d_{n+1}^{p,q}\left(\frac{Z_n^{p,q}}{B_n^{p,q}}\right) \simeq \left(\frac{Z_n^{p,q}}{B_n^{p,q}}\right) \Big/ \operatorname{Ker}\left(d_{n+1}^{p,q}\right) \\ &= \left(\frac{Z_n^{p,q}}{B_n^{p,q}}\right) \Big/ \left(\frac{Z_{n+1}^{p,q}}{B_n^{p,q}}\right) \simeq \frac{Z_n^{p,q}}{Z_{n+1}^{p,q}} . \end{aligned}$$

To simplify notation one often writes exact sequences such as (4.58) or the isomorphisms following (4.58) without specifying the upper indices, i.e.,

$$0 \to \frac{Z_{n+1}^{\bullet,\bullet}}{B_n^{\bullet,\bullet}} \to \frac{Z_n^{\bullet,\bullet}}{B_n^{\bullet,\bullet}} \xrightarrow{d_{n+1}^{\bullet,\bullet}} \frac{B_{n+1}^{\bullet,\bullet}}{B_n^{\bullet,\bullet}} \to 0, \quad \frac{B_{n+1}^{\bullet,\bullet}}{B_n^{\bullet,\bullet}} \simeq \frac{Z_n^{\bullet,\bullet}}{Z_{n+1}^{\bullet,\bullet}} ,$$

yet the reader should keep in mind that the empty upper bullets may not stand for identical indices. The data contained in the tower (4.56) (a tower of submodules of $E_2^{\bullet,\bullet}$) together with the information furnished by the isomorphisms (4.57) may be used to reconstruct the spectral sequence E.

Definition 4.45 An element $x \in E_2^{\bullet,\bullet}$ such that $x \in Z_n^{\bullet,\bullet}$ is said to *survive to the* n^{th} *stage*. The submodule

$$Z_\infty^{\bullet,\bullet} := \bigcap_{n=2}^{\infty} Z_n^{\bullet,\bullet} \subset E_2^{\bullet,\bullet}$$

is referred to as the submodule of elements that *survive forever.* □

The reason one adopts the terminology in Definition 4.45 for a given $x \in Z_n^{\bullet,\bullet}$ is merely that x lies in the kernels of the previous $n-2$ differentials (i.e., $x \in Z_{n-1}^{\bullet,\bullet} \subset \cdots \subset Z_2^{\bullet,\bullet}$). The elements of $E_2^{\bullet,\bullet}$ that survive for ever are cycles at every stage.

Definition 4.46 An element $x \in E_2^{\bullet,\bullet}$ such that $x \in B_n^{\bullet,\bullet}$ is said to be a *boundary at the* n^{th} *stage*. The submodule

$$B_\infty^{\bullet,\bullet} = \bigcup_{n=2}^{\infty} B_n^{\bullet,\bullet} \subset E_2^{\bullet,\bullet}$$

is referred to as the submodule of elements which *eventually bound.* □

An inspection of the tower (4.56) shows that $B_\infty^{\bullet,\bullet} \subset Z_\infty^{\bullet,\bullet}$, so that we may consider the R-module

$$E_\infty^{\bullet,\bullet} = \frac{Z_\infty^{\bullet,\bullet}}{B_\infty^{\bullet,\bullet}} \tag{4.59}$$

Although the concept of convergence of a spectral sequence E hasn't been introduced as yet, it is quite clear that computing $E_\infty^{\bullet,\bullet}$ is the general task, and that the bigraded module (4.59) is what one is left with after the calculation of the infinite sequence of homologies pertaining to E. In mathematical practice, the calculation often ends at some finite stage. We adopt the following:

Definition 4.47 Let E be a spectral sequence. If $d_r^{\bullet,\bullet} = 0$ for any $r \geq N$, then we say that E *collapses at the* N^{th} *term.* □

If E is a spectral sequence collapsing at the Nth term, then the exactness of

$$0 \to \frac{Z_r^{\bullet,\bullet}}{B_{r-1}^{\bullet,\bullet}} \to \frac{Z_{r-1}^{\bullet,\bullet}}{B_{r-1}^{\bullet,\bullet}} \xrightarrow{d_r^{\bullet,\bullet}} \frac{B_r^{\bullet,\bullet}}{B_{r-1}^{\bullet,\bullet}} \to 0$$

yields $B_r^{\bullet,\bullet} = B_{r-1}^{\bullet,\bullet}$ (as the image of $d_r^{\bullet,\bullet}$ is zero) and $Z_r^{\bullet,\bullet} = Z_{r-1}^{\bullet,\bullet}$ [because the sequence

$$0 \to Z_r^{\bullet,\bullet}/B_{r-1}^{\bullet,\bullet} \hookrightarrow Z_{r-1}^{\bullet,\bullet}/B_{r-1}^{\bullet,\bullet} \to 0$$

is exact, i.e., $Z_r^{\bullet,\bullet}/B_{r-1}^{\bullet,\bullet} \simeq Z_{r-1}^{\bullet,\bullet}/B_{r-1}^{\bullet,\bullet}$] for any $r \geq N$. Consequently, the tower (4.56) may be written

$$B_2^{\bullet,\bullet} \subset B_3^{\bullet,\bullet} \subset \cdots \subset B_{N-1}^{\bullet,\bullet} = B_N^{\bullet,\bullet} = \cdots = B_\infty^{\bullet,\bullet} \subset$$

$$\subset Z_\infty^{\bullet,\bullet} = \cdots = Z_N^{\bullet,\bullet} = Z_{N-1}^{\bullet,\bullet} \subset \cdots \subset Z_3^{\bullet,\bullet} \subset Z_2^{\bullet,\bullet} \subset E_2^{\bullet,\bullet}$$

and then $E_\infty^{\bullet,\bullet} = E_N^{\bullet,\bullet}$.

4.6.2.2 Filtered Differential Modules

Let A be a R-module and let $F = \{F^pA : p \in \mathbb{Z}\}$ be a family of submodules $F^pA \subset A$.

Definition 4.48 We call $F = \{F^pA : p \in \mathbb{Z}\}$ a *decreasing* (respectively, *increasing*) *filtration* on A if $F^{p+1}A \subset F^pA$ (respectively, if $F^{p+1}A \supset F^pA$) for every $p \in \mathbb{Z}$. □

Definition 4.49 Let $F = \{F^pA : p \in \mathbb{Z}\}$ be a filtration on A. The *associated graded module* is the R-module

$$E_0^p(A) = \begin{cases} \dfrac{F^pA}{F^{p+1}A} & \text{if F is decreasing,} \\[2ex] \dfrac{F^{p+1}A}{F^pA} & \text{if F is increasing,} \end{cases} \quad , \quad p \in \mathbb{Z}. \tag{4.60}$$

□

Let $H^\bullet$ be a graded module. Let $\{F^p H^\bullet \,:\, p \in \mathbb{Z}\}$ be a filtration on $H^\bullet$. That is, we are given a filtration $F^n = \{F^p H^n \,:\, p \in \mathbb{Z}\}$ on H^n for very $n \in \mathbb{Z}$. If, for instance, F^n is an increasing filtration, then

$$\cdots \subset F^{p-1}H^n \subset F^p H^n \subset F^{p+1}H^n \subset \cdots \subset H^n, \quad n \in \mathbb{Z}.$$

The associated graded module is then bigraded. Precisely

$$E_0^p(H^n) = \begin{cases} \dfrac{F^p H^n}{F^{p+1}H^n} & \text{if } F^n \text{ is decreasing,} \\[2ex] \dfrac{F^{p+1}H^n}{F^p H^n} & \text{if } F^n \text{ is increasing.} \end{cases}$$

It is customary to adopt the notation

$$E_0^{p,q}(H^\bullet, F^\bullet) := E_0^p(H^{p+q}), \quad p, q \in \mathbb{Z}.$$

Convergence of a spectral sequence $E = \{E_r \,:\, r \geq 2\}$ is then defined as follows.

Definition 4.50 Let $E_r = \{(E_r^{p,q}, d_r^{p,q}) \,:\, p, q \in \mathbb{Z}\}, r \geq 2$, be a spectral sequence and let $H^\bullet$ be a graded module. We say that $E = \{E_r \,:\, r \geq 2\}$ *converges* to $H^\bullet$ if there is a filtration $F^\bullet$ on $H^\bullet$ such that

$$E_\infty^{p,q} \simeq E_0^{p,q}(H^\bullet, F^\bullet), \quad p, q \in \mathbb{Z}, \tag{4.61}$$

(a R-module isomorphism). □

If the spectral sequence E converges to the graded module $H^\bullet$ in the sense of Definition B.9 we write $E \Longrightarrow H^\bullet$. In practical problems one aims to the calculation of a certain graded module $H^\bullet$. If a spectral sequence E converging to $H^\bullet$ exists, one actually computes the associated graded module of each H^n (associated to some filtration on H^n) rather than H^n itself. Can one recover H^n from the knowledge of its associated graded module? Unfortunately, the problem of reconstructing a filtered module from the associated graded module is in general far from being trivial. Let us illustrate our possibilities by looking at a particular case. Let A be a filtered R-module and $F = \{F^p A \,:\, p \in \mathbb{Z}\}$ its decreasing filtration. The particular case we have in mind is that F ceases at some stage, i.e., there is $n \in \mathbb{Z}$ such that $n \geq 0$ and $F^p A = (0)$ for any $p > n$. Let us also assume that $F^p A = A$ for all $p < 0$. The given filtration on A then looks like

$$(0) \subset F^n A \subset F^{n-1}A \subset \cdots \subset F^1 A \subset F^0 A \subset F^{-1}A = A.$$

If $E_0^\bullet(A)$ is the associated graded module, then

$$E_0^p(A) = (0), \quad p \in \mathbb{Z} \setminus \{-1, 0, 1, \cdots, n\},$$

i.e., only the R-modules $E_0^p(A)$ with $-1 \le p \le n$ may be nonzero. With the eventually nontrivial associated graded modules one may build the following short exact sequences:

$$\begin{array}{ccccccccc}
 & & & & F^n A & & \\
 & & & & \| & & \\
0 & \to & F^n A & \to & E_0^n(A) & \to & 0, & & \\
 & & & & & & & & \\
0 & \to & F^n A & \to & F^{n-1} A & \to & E_0^{n-1}(A) & \to & 0, \\
 & & \vdots & & \vdots & & \vdots & & \\
0 & \to & F^p A & \to & F^{p-1} A & \to & E_0^{p-1}(A) & \to & 0, \\
 & & \vdots & & \vdots & & \vdots & & \\
0 & \to & F^1 A & \to & F^0 A & \to & E_0^0(A) & \to & 0, \\
 & & & & & & & & \\
0 & \to & F^0 A & \to & F^{-1} A & \to & E_0^{-1}(A) & \to & 0. \\
 & & & & \| & & & & \\
 & & & & A & & & &
\end{array}$$

Now, let us suppose that the graded module $E_0^\bullet(A)$ is known. Then, of course, $F^n A$ is known. However, determining $F^{n-1} A$ from the second short exact sequence above depends on whether it is split or not. Equivalently, one should know whether $F^{n-1} A \to E_0^{n-1}(A)$ possesses a right inverse $E_0^{n-1}(A) \to F^{n-1} A$ or not. If this were the case, then (cf. e.g., Proposition 4.3 in [MaLa], p. 16)

$$F^{n-1} A \simeq F^n A \oplus E_0^{n-1}(A).$$

The sequence in discussion is surely split if its modules are finite-dimensional vector spaces. However, if R is an arbitrary commutative ring, then ambiguities may arise. To put it in the algebraic language of [MaLa], pp. 63–64, the determination of $F^{n-1} A$ depends on the way $E_0^{n-1}(A)$ extends $F^n A$. One should look at the set $\mathrm{Ext}_R(C, F^n A)$ of all congruence[16] classes of extensions of $F^n A$ by a R-module C. An extension C of $F^n A$ (i.e., a short exact sequence $0 \to F^n A \to B \to C \to 0$) is commonly referred to as split if it is congruent to the direct sum extension $0 \to F^n A \to \big(FnA\big) \oplus C \to C \to 0$. How many elements does $\mathrm{Ext}_R\big(C, F^n A\big)$ possess? If C is a finite-dimensional vector space (or more generally a projective module), then any extension by C is split and $\mathrm{Ext}_R\big(C, F^n A\big)$ has but one element.

[16] Cf. [MaLa], p. 64, for *congruences* of extensions. The present monograph doesn't aim to an exhaustive presentation of the amount of homological algebra eventually required by problems in mathematical practice, and adopts the conventions and results in [MaLa], p. 63–65, instead.

Nontrivial examples are provided in [MaLa], p. 64. If all goes well (for instance when all modules involved are finite-dimensional vector spaces) one chases the short exact sequences above and eventually obtains

$$A = \bigoplus_{p=-1}^{n} E_0^p(A).$$

In general however, there will be some ambiguity about A unless further information becomes available.

To describe the setting in which spectral sequences arise, and which suffices for our purposes in the previous section, we need the notion of a filtered differential graded module.

Definition 4.51 Let A be a R-module. We say A is a *filtered differential graded module* (or a *filtered* DG-*module*) if

i) There is a family $\{A^n \,:\, n \in \mathbb{Z}_+\}$ of submodules of A such that $A = \oplus_{n=0}^{\infty} A^n$.

ii) There is a R-linear map $d : A \to A$ of *degree* $\epsilon \in \{\pm 1\}$ [i.e., $d\left(A^n\right) \subset A^{n+\epsilon}$ for any $n \geq 0$] such that $d \circ d = 0$.

iii) A possesses a filtration $F = \{F^p A \,:\, p \in \mathbb{Z}\}$ which is preserved by the differential, i.e., $d\big(F^p A\big) \subset F^p A$ for any $p \in \mathbb{Z}$. □

Let A be a filtered DG-module, and let $d : A \to A$ be its differential. Let then

$$H(A,\, d) = \frac{\operatorname{Ker}(d)}{d(A)}, \quad H\big(F^p A,\, d^p\big) = \frac{\operatorname{Ker}(d^p)}{d(F^p A)}, \quad d^p = d\big|_{F^p A},$$

be the homologies of $(A,\, d)$ and $\big(F^p A,\, d^p\big)$, respectively. The inclusion $\iota^p : F^p A \to A$ induces a map $\iota_*^p : H\big(F^p A,\, d^p\big) \to H(A,\, d)$. We set

$$F^p H(A,\, d) = \iota_*^p\Big\{H\big(F^p A,\, d^p\big)\Big\} \subset H(A,\, d), \quad p \in \mathbb{Z}, \tag{4.62}$$

so that $\big\{F^p\, H(A,\, d) \,:\, p \in \mathbb{Z}\big\}$ is a filtration on $H(A,\, d)$. This is used in Theorem 4.53 below.

Definition 4.52 Let $F = \{F^p A \,:\, p \in \mathbb{Z}\}$ be a filtration on the R-module A. We say F is *bounded* if there exist $s,\, t \in \mathbb{Z}$ such that

$$F^p A = (0), \quad F^q A = A, \quad p \geq s + 1, \quad q \leq t.$$

□

For instance, a decreasing bounded filtration on A looks like

$$(0) \subset F^s A \subset F^{s-1} A \subset \cdots \subset F^{t+1} A \subset F^t A = A.$$

Given a filtered DG-module as in Definition 4.51 we may set

$$F^pA^n := \big(F^pA\big) \cap A^n\,, \quad p,\, n \in \mathbb{Z}, \quad n \geq 0.$$

As $d\big(F^pA\big) \subset F^pA$ and $d(A^n) \subset A^{n+1}$, it follows that $D(F^pA^n) \subset F^pA^{n+1}$. Hence, we obtain a cochain complex

$$F^pA^0 \xrightarrow{d^{p,0}} F^pA^1 \xrightarrow{d^{p,1}} \cdots \xrightarrow{d^{p,n-1}} F^pA^n \xrightarrow{d^{p,n}} F^pA^{n+1} \xrightarrow{d^{p,n+1}} \cdots$$

where $d^{p,n} = d\big|_{F^pA^n}$ (hence $d^{p,n+1} \circ d^{p,n} = 0$). Then $d^{p,n}$ descends to a natural map

$$D^{p,n} : \frac{F^pA^n}{F^{p+1}A^n} \to \frac{F^pA^{n+1}}{F^{p+1}A^{n+1}}\,, \quad D^{p,n}\big(x + F^{p+1}A^n\big) = dx + F^{p+1}A^{n+1}\,, \quad x \in F^pA^n\,,$$

such that $D^{p,n+1} \circ D^{p,n} = 0$. That is, we obtain another cochain complex

$$\frac{F^pA^0}{F^{p+1}A^0} \to \frac{F^pA^1}{F^{p+1}A^1} \to \cdots \to \frac{F^pA^n}{F^{p+1}A^n} \to \frac{F^pA^{n+1}}{F^{p+1}A^{n+1}} \to \cdots$$

Let

$$H^n\big(F^pA/F^{p+1}A\big) = H^n\big(F^pA^\bullet/F^{p+1}A^\bullet,\, D^{p,\bullet}\big) = \frac{\operatorname{Ker}(D^{p,n})}{D^{p,n-1}\big(F^pA^{n-1}/F^{p+1}A^{n-1}\big)}$$

be its cohomology. A task of the present Appendix is to recall

Theorem 4.53 *Let A be a filtered* DG-*module whose differential $d : A \to A$ has degree $\epsilon = 1$, and whose filtration $F = \{F^pA \,:\, p \in \mathbb{Z}\}$ is decreasing. There is a spectral sequence $E = \{E_r \,:\, r \in \mathbb{N}\}$ whose differential $d_r^{\bullet,\bullet}$ has bidegree $(r,\, 1-r)$, and such that*

$$E_1^{p,q} \simeq H^{p+q}\big(F^pA/F^{p+1}A\big). \tag{4.63}$$

Moreover, if the filtration $\{F^pA^n \,:\, p \in \mathbb{Z}\}$ (induced by F on A^n) is bounded for each $n \geq 0$, then the spectral sequence E converges to $H(A,\, d)$.

A comment on the last statement in Theorem 4.53 is in order. To apply Definition 4.50 we must see $H(A,\, d)$ as a graded module. Indeed we may consider the cohomology

$$H^n(A,\, d) = H^n(A^\bullet\; d^\bullet) = \frac{\operatorname{Ker}(d^n)}{d^{n-1}\,(A^{n-1})}$$

of the cochain complex

$$A^0 \xrightarrow{d^0} A^1 \xrightarrow{d^1} \cdots \xrightarrow{d^{n-1}} A^n \xrightarrow{d^n} A^{n+1} \xrightarrow{d^{n+1}} \cdots$$

where $d^n = d\big|_{A^n}$ for any $n \geq 0$. Each $H^n(A,\, d)$ may be thought of as a submodule of $H(A,\ ,d)$, for the map

$$H^n(A, d) \to H(A, d), \quad x + d^{n-1} A^{n-1} \mapsto x + d\,A, \quad x \in \operatorname{Ker}(d^n) = \operatorname{Ker})(d) \cap A^n,$$

is injective. Next $A = \oplus_{n=0}^{\infty} A^n$ yields

$$H(A, d) = \bigoplus_{n=0}^{\infty} H^n(A, d)$$

hence $H(A, d)$ is a graded module in a natural manner. Let $\{F^p H(A, d) : p \in \mathbb{Z}\}$ be the natural filtration on $H(A, d)$ [induced by the filtration $\{F^p A : p \in \mathbb{Z}\}$ on A, according to (4.62)]. We set

$$F^p H^n(A, d) := \{F^p H(A, d)\} \cap H^n(A, d), \quad p, n \in \mathbb{Z}, \quad n \geq 0,$$

so that $\{F^p H^n(A, d) : p \in \mathbb{Z}\}$ is a filtration on $H^n(A, d)$ for each $n \geq 0$. We may then consider the associated graded module [of $H^n(A, d)$ as a filtered module]

$$E_0^p\big[H^n(A, d)\big] = \frac{F^p H^n(A, d)}{F^{p+1} H^n(A, d)}$$

and set (as earlier in this Appendix)

$$E_0^{p,q}\big[H(A, d), F^\bullet H(A, d)\big] = E_0^p\big[H^{p+q}(A, d)\big].$$

Let E be the spectral sequence converging to $H(A, d)$ as furnished by Theorem 4.53. Then $E \Longrightarrow H(A, d)$ that is (by Definition 4.50)

$$E_\infty^{p,q} \simeq E_0^{p,q}\big[H(A, d), F^\bullet H(A, d)\big] = \frac{F^p H^{p+q}(A, d)}{F^{p+1} H^{p+q}(A, d)}. \tag{4.64}$$

Theorem 4.53 is due to Leray, [Ler]. Cf. also Theorem 2.1 in [McCl], p. 32. We are mainly interested in its practical use, that may be roughly described as follows. Given a differential module (A, d) we wish to compute its homology $H(A, d)$. If A may be filtered (compatibly with d) and the associated spectral sequence E (provided by Theorem 4.53) possesses some computable term, then we may determine $H(A, d)$ by the calculation of the successive homologies (pertaining to E) and by reconstruction from the associated graded module.

For our needs in this book we only stated Theorem 4.53 in the cohomological setting, i.e., for decreasing filtrations and spectral sequences of cohomological type. A natural version of Theorem 4.53 predicting the existence of a spectral sequence of homological type is of course available (cf. [McCl], p. 32). The proof of Theorem 4.53 is rather involved (cf. [McCl], pp. 33–36). Through the present Appendix we confine ourselves to the construction of the spectral sequence E associated to the filtered DG-module A. To this end we need the following ingredients. First, we need the filtration

$$\cdots \subset F^{p+1} A^{p+q} \subset F^p A^{p+q} \subset F^{p-1} A^{p+q} \subset \cdots$$

with $d\big(F^p A^{p+qq}\big) \subset F^p A^{p+q+1}$. Moreover, let us set

$$Z_r^{p,q} = \big(F^p A^{p+q}\big) \cap d^{-1}\left(F^{p+r} A^{p+q+1}\right), \tag{4.65}$$

$$B_r^{p,q} = \big(F^p A^{p+q}\big) \cap d\left(F^{p-r} A^{p+q-1}\right), \tag{4.66}$$

$$Z_\infty^{p,q} = \mathrm{Ker}(d) \cap F^p A^{p+q}, \tag{4.67}$$

$$B_\infty^{p,q} = (d\,A) \cap F^p A^{p+q}. \tag{4.68}$$

It should be observed that $Z_r^{p,q}$ consists of all elements of $F^p A^{p+q}$ which are boundaries in $F^{p+r} A^{p+q+1}$. Also $B_r^{p,q}$ consists of all elements of $F^p A^{p+q}$ that are images via d of elements from $F^{p-r} A^{p+q+1}$. As d preserves the filtration on A

$$B_0^{p,q} \subset B_1^{p,q} \subset \cdots \subset B_\infty^{p,q} \subset Z_\infty^{p,q} \subset \cdots \subset Z_1^{p,q} \subset Z_0^{p,q}. \tag{4.69}$$

Also

$$\begin{aligned} d\big(Z_r^{p-r,\,q+r-1}\big) =& d\Big[\big(F^{p-r} A^{p+q-1}\big) \cap d^{-1}\big(F^p A^{p+q}\big)\Big] \\ =& \big(F^p A^{p+q}\big) \cap d\left(F^{p-r} A^{p+q-1}\right) \end{aligned}$$

that is

$$d\big(Z_r^{p-r,\,q+r-1}\big) = B_r^{p,\,q}. \tag{4.70}$$

Let us assume that the filtration $\{F^p A^n \,:\, p \in \mathbb{Z}\}$ on A^n is bounded for every $n \geq 0$. In other words, there exist functions $s,\, t : \mathbb{Z}_+ \to \mathbb{Z}$ such that

$$F^p A^n = (0), \quad F^q A^n = A^n, \quad p \geq s(n) + 1, \quad q \leq t(n).$$

The surviving terms of the filtration on A^n may then be written

$$(0) \subset F^{s(n)} A^n \subset F^{s(n)-1} A^n \subset \cdots \subset F^{t(n)+1} A^n \subset F^{t(n)} A^n = A^n.$$

Consequently [by the definitions (4.65)–(4.68)],

$$r > s(p+q+1) - p \Longrightarrow F^{p+r} A^{p+q+1} = (0) \Longrightarrow$$

$$Z_r^{p,q} = \big(F^p A^{p+q}\big) \cap d^{-1}\big(F^{p+r} A^{p+q+1}\big) = \big(F^p A^{p+q}\big) \cap d^{-1}(0) = Z_\infty^{p,q}$$

and

$$r \geq p - t(p+q-1) \Longrightarrow F^{p-r} A^{p+q-1} = A^{p+q-1} \Longrightarrow$$

$$B_r^{p,q} = \left(F^p A^{p+q}\right) \cap d\left(F^{p-r} A^{p+q-1}\right) = \left(F^p A^{p+q}\right) \cap d\left(A^{p+q-1}\right)$$
$$= \left(F^p A^{p+q}\right) \cap d(A) = B_\infty^{p,q}.$$

These are known to be the key facts ensuring the convergence of the spectral sequence we shall shortly build. The last calculation is a bit less obvious and follows from

$$\left(F^p A^n\right) \cap d\left(A^{n-1}\right) = \left(F^p A^n\right) \cap d(A).$$

To check the last identity one observes that

$$\left(F^p A^n\right) \cap d(A^{n-1}) = \left(F^p A^n\right) \cap A^n \cap d(A) = \left(F^p A^n\right) \cap d(A)$$

because $d\,A^{n-1} = A^n \cap d\,A$. Of course, the inclusion $d\,A^{n-1} \subset A^n \cap d\,A$ follows from $dA^{n-1} \subset A^n$ and $A^{n-1} \subset A$. Vice versa, if $y \in A^n \cap d\,A$, then $y = d\,x$ for some

$$x = \sum_{m=0}^{\infty} x_m, \quad x_m \in A^m,$$

(with $x_m \neq 0$ for finitely many $m \geq 0$ only). As the sum $\oplus_{m=0}^{\infty} A^m$ is direct

$$A^m \ni y = \sum_{m=0}^{\infty} d(x_m), \quad d\,x_m \in A^{m+1}, \quad m \geq 0,$$

yields $d\,x_m = 0$ whenever $m + 1 \neq n$, that is $y = d\,x_{n-1} \in d\,A^{n-1}$. Q.e.d.

We come to the construction of the spectral sequence $\left(E_r^{p,q}, d_r^{p,q}\right)$. Precisely, we set

$$E_r^{p,q} = \frac{Z_r^{p,q}}{Z_{r-1}^{p+1,q-1} + B_{r-1}^{p,q}}. \tag{4.71}$$

Moreover, let $\pi_r^{p,q} : Z_r^{p,q} \to E_r^{p,q}$ be the canonical projection [so that $\mathrm{Ker}\left(\pi_r^{p,q}\right) = Z_{r-1}^{p+1,q-1} + B_{r-1}^{p,q}$]. Let us substitute $p' \mapsto p - r$ and $q' \mapsto q + r - 1$ into (4.70) and use (4.69) so that to yield

$$d\,Z_r^{p,q} = B_r^{p+r,q-r+1} \subset Z_r^{p+r,q-r+1}.$$

Therefore, d induces homomorphisms

$$d : Z_r^{p,q} \to Z_r^{p+r,q-r+1}. \tag{4.72}$$

On the other hand [again by (4.70)],

$$
\begin{aligned}
d\big(Z_{r-1}^{p+1,\,q-1} + B_{r-1}^{p,q}\big) =&d\, Z_{r-1}^{p+1,\,q-1} + d\, B_{r-1}^{p,q} \subset B_{r-1}^{p+r,\,q-r+1} \\
&+(0) \subset B_{r-1}^{p+r,\,q-r+1} + Z_{r-1}^{p+r+1,\,q-r}
\end{aligned}
$$

so that (4.72) descends to a homomorphism among quotients

$$
d_r^{p,q} : E_r^{p,q} \to E_r^{p+r,\,q-r+1} \tag{4.73}
$$

such that the diagram

$$
\begin{array}{ccc}
Z_r^{p,q} & \xrightarrow{d} & Z_r^{p+r,\,q-r+1} \\
\pi_r^{p,q} \downarrow & & \downarrow \pi_r^{p+r,\,q-r+1} \\
E_r^{p,q} & \xrightarrow{d_r^{p,q}} & E_r^{p+r,\,q-r+1}
\end{array}
$$

is commutative. Following (4.73), let us consider the sequence of homomorphisms

$$
E_r^{p-r,\,q+r-1} \xrightarrow{d_r^{p-r,\,q+r-1}} E_r^{p,q} \xrightarrow{d_r^{p,q}} E_r^{p+r,\,q-r+1} .
$$

Then $d^2 = 0$ implies $d_r^{p,q} \circ d_r^{p-r,\,q+r-1} = 0$.

4.7 Foliations by Level Hypersurfaces

Let (V^{n+1}, g) be a $(n+1)$-dimensional Riemannian manifold, and let $f : V \to \mathbb{R}$ be a C^∞ function. Let $\text{Crit}(f) = \{x \in V : d_x f = 0\}$, and let us set $M = V^{n+1} \setminus \text{Crit}(f)$ (an open subset of V^{n+1}). Let $\mathscr{F}$ be the foliation of M by the level hypersurfaces of f, i.e., the leaf space $M/\mathscr{F}$ consists of all hypersurfaces $f^{-1}(c)$, $c \in f(M)$. Then

$$
Z := \frac{1}{|\nabla f|} \nabla f \in C^\infty\big(P^\perp\big)
$$

and

$$
\nu := \frac{1}{|\nabla f|} \nabla f \in \Omega^1(M)
$$

is dual to Z, i.e., $\nu(Z) = 1$ and $P \rfloor \nu = 0$. Let $\text{Hess}(f)$ be the Hessian of f, i.e.,

$$
\begin{aligned}
\text{Hess}(f)(X,\, Y) =&(\nabla^g\, df)(X,\, Y) = (\nabla^g_X df)Y \\
=&X(Y(f)) - (\nabla^g_X Y)(f), \quad X,\, Y \in \mathfrak{X}(M).
\end{aligned}
$$

As ∇^g is torsion-free, the Hessian Hess_g is symmetric. Also, note that

$$\nabla^g_X df = \left(\nabla^g_X \nabla f\right)^\flat.$$

Lemma 4.54 *For any* $X,\ X' \in C^\infty(P)$ i)

$$\mathrm{Hess}_f(X,\ X') = -|\nabla f|\, g(W_Z X\ ,\ X') \tag{4.74}$$

and consequently ii), *the second fundamental form of* $\mathscr{F}$ *in* $(M,\ g)$ *is*

$$\alpha(X,\ X') = -\frac{1}{|\nabla f|}\mathrm{Hess}(f)(X,\ X')\,\pi\, Z. \tag{4.75}$$

iii) *The mean curvature is*

$$\kappa(Z) = \frac{1}{|\nabla f|}\left\{\Delta f + \mathrm{Hess}(f)(Z,\ Z)\right\}. \tag{4.76}$$

Proof i) Let us set $\lambda = |\nabla f|$ for the sake of brevity. Then (by $\nabla f = \lambda\, Z$)

$$\mathrm{Hess}(f)(X,\ X') = \left(\nabla^g_X df\right)X' = g\left(\nabla^g_X \nabla f,\ X'\right) = g\left(X(\lambda)\, Z + \lambda\, \nabla^g_X Z,\ X'\right) =$$

[as $Z \in P^\perp \perp P \ni X' \Longrightarrow g(Z,\ X') = 0$]

$$= \lambda\, g\left(\nabla^g_X Z,\ X'\right) =$$

[as $\nabla^g g = 0$]

$$= \lambda \left\{X\left(g(Z,\ X')\right) - g\left(Z,\ \nabla^g_X X'\right)\right\} = -\lambda\, g\left(Z,\ \nabla^g_X X'\right).$$

On the other hand,

$$g_Q\left(\alpha(X,\ X'),\ \pi\, Z\right) = g_Q\left(\pi\ \nabla^g_X X'\ ,\ \pi\ Z\right) = g\left(\sigma\, \pi\ \nabla^g_X X'\ ,\ \sigma\, \pi\ Z\right) =$$

[as $\sigma \circ \pi = \pi^\perp$]

$$= g\left(\pi^\perp\, \nabla^g_X X'\ ,\ \pi^\perp\, Z\right) = g\left(\nabla^g_X X'\ ,\ Z\right).$$

Hence,

$$\mathrm{Hess}(f)(X,\ X') = -\lambda\, g_Q\left(\alpha(X,\ X'),\ \pi\ Z\right) = -\lambda\, g\left(W_Z X,\ X'\right).$$

Q.e.d.

ii) By the very definition of α

$$\alpha(X,\ X') \in Q \simeq P^\perp = \mathrm{Span}(Z)$$

hence

$$\sigma\,\alpha(X,\, X') = \mu\, Z$$

for some $\mu \in C^\infty(M)$. Taking the inner product with Z determines μ as

$$\mu = g\big(\sigma\,\alpha(X,\, X'),\; Z\big) = -\frac{1}{\lambda}\,\mathrm{Hess}(f)(X,\, X')$$

and finally,

$$\sigma\,\alpha(X,\, X') = -\frac{1}{\lambda}\,\mathrm{Hess}(f)(X,\, X')\, Z.$$

Q.e.d.

iii) Next, we compute the mean curvature κ of $\mathscr{F}$ in $(M,\, g)$. For every local orthonormal frame $\{E_i \,:\, 1 \le i \le n\} \subset C^\infty(U,\, P)$

$$\kappa(Z) = \mathrm{Trace}\, W_Z = \sum_{i=1}^{n} g\big(W_Z E_i\,,\; E_i\big)$$

hence

$$\kappa(Z) = -\frac{1}{|\nabla f|}\sum_{i=1}^{n}\mathrm{Hess}(f)(E_i\,,\; E_i) \tag{4.77}$$

on U. Let us set $E_{n+1} := Z$ so that $\{E_A \,:\, 1 \le A \le n+1\}$ is a local g-orthonormal frame on $T(M)$. The Laplace-Beltrami operator Δ of $(M,\, g)$ is

$$\Delta u = -\sum_{A=1}^{n+1}\Big\{E_A^2(u) - \big(\nabla_{E_A} E_A\big)(u)\Big\} = -\mathrm{Trace}_g\,\mathrm{Hess}(u)$$

for every $u \in C^2(M)$. Consequently,

$$\Delta f = -\sum_{i=1}^{n}\mathrm{Hess}(f)(E_i\,,\; E_i) - \mathrm{Hess}(f)(Z,\, Z) = \qquad \text{[by (4.77)]}$$

$$= |\nabla f|\,\kappa(Z) - \mathrm{Hess}(f)(Z,\, Z)$$

that is

$$\kappa(Z) = \frac{1}{|\nabla f|}\Big\{\Delta f + \mathrm{Hess}(f)(Z,\, Z)\Big\}.$$

Q.e.d.

An useful expression of the mean curvature κ [alternative to (4.76)] is got by computing

$$\mathrm{Hess}(f)(Z, Z) = (\nabla^g_Z df)Z = g\big(\nabla^g_Z \nabla f\,,\, Z\big) = \quad [\text{as } Z = (1/\lambda)\,\nabla f]$$

$$= \frac{1}{\lambda^2}\, g\big(\nabla^g_{\nabla f} \nabla f\,,\, \nabla f\big) = \frac{1}{2\,\lambda^2}\,(\nabla f)\big[|\nabla f|^2\big] = \frac{1}{2\,\lambda^2}\,(\nabla f)\big(\lambda^2\big) = \frac{1}{\lambda}\,(\nabla f)(\lambda).$$

Then (4.76) reads

$$\kappa(Z) = \frac{1}{|\nabla f|}\Big\{\Delta f + (\nabla f)\big(\log|\nabla f|\big)\Big\} = \frac{1}{|\nabla f|^2}\, g\big(\nabla f,\ \nabla(|\nabla f|)\big).$$

Also

$$\nabla^g_Z Z = \nabla^g_Z\Big(\frac{1}{\lambda}\,\nabla f\Big) = -\frac{1}{\lambda^2}\, Z(\lambda)\,\nabla f + \frac{1}{\lambda}\,\nabla^g_Z \nabla f,$$

$$\begin{aligned}\big(\nabla^g_Z Z\big)(f) &= -\frac{1}{\lambda^2}\, Z(\lambda)\,(\nabla f)(f) + \frac{1}{\lambda}\,\big(\nabla^g_Z \nabla f\big)(f)\\ &= -Z(\lambda) + \frac{1}{\lambda}\, g\big(\nabla^g_Z \nabla f\,,\, \nabla f\big) =\end{aligned}$$

$$= -Z(\lambda) + \frac{1}{2\,\lambda}\, Z\big(\nabla f,\ \nabla f\big) = -Z(\lambda) + \frac{1}{2\,\lambda}\, Z\big(\lambda^2\big) = 0,$$

meaning that $\nabla^g_Z Z \in C^\infty(P)$. Consequently,

$$\mathrm{Hess}(f)(Z, Z) = Z(Zf)$$

and another form of (4.76) is

$$\kappa(Z) = \frac{1}{|\nabla f|}\,\Big\{\Delta f + Z\big(Z(f)\big)\Big\}.$$

If f is a harmonic function (i.e., $\Delta f = 0$), then

$$\kappa(Z) = \frac{1}{|\nabla f|}\, Z\big(Z(f)\big)$$

so, in general, $\mathscr{F}$ isn't a harmonic foliation.

Example 4.55 Let $M = \mathbb{R}^{n+1} \setminus \{0\}$ with the Euclidean metric $g = g_0$, and let $\mathscr{F}$ be the foliation of M by level hypersurfaces of $f = r = \Big(\sum_{i=1}^{n+1} x_i^2\Big)^{1/2}$. Then

$$\nabla r = \frac{1}{r}\sum_{i=1}^{n+1} x_i\,\frac{\partial}{\partial x_i}, \qquad \Delta r = -\mathrm{div}\,\nabla r = \frac{n}{r}\,,$$

$$\text{Hess}(r)(Z,\ Z) = ZZr, \quad \nabla^g_Z Z = 0, \quad Zr = 1,$$

yield $\kappa(Z) = \Delta r = -(n/r)$. □

Theorem 4.56 *Let (V^{n+1}, g) be a $(n+1)$-dimensional Riemannian manifold, and let $f : V^{n+1} \to \mathbb{R}$ be a C^∞ function. Let $\mathcal{F}$ be the foliation of $M = V^{n+1} \setminus \text{Crit}(f)$ by level hypersurfaces of f. The following statements are equivalent:*

i) *$\mathcal{F}$ is Riemannian and g is bundle-like.*
ii) *$X(|\nabla f|^2) = 0$ for any $X \in \mathfrak{X}_{\mathcal{F}}$.*
iii) *$\text{Hess}(f)(X,\ \nabla f) = 0$ for any $X \in \mathfrak{X}_{\mathcal{F}}$.*
iv) *$[X,\ Z]f = 0$ for any $X \in \mathfrak{X}_{\mathcal{F}}$.*

Cf. Theorem 8.9 in [Ton], p. 107. The proof of Theorem 4.56 is straightforward, and left as an exercise to the reader.

Exercise 4.57 If $\nu := (1/|\nabla f|)\, df$ show that

$$(d\nu)(Z,\ X) = \frac{1}{2\,|\nabla f|^2}\, X\big(|\nabla f|^2\big)$$

for any $X \in \mathfrak{X}_{\mathcal{F}}$. □

Corollary 4.58 *$\mathcal{F}$ is Riemannian and g is bundle-like $\Longleftrightarrow d\nu = 0$.*

Harmonic foliations by level hypersurfaces are described by

Theorem 4.59 *Let $\mathcal{F}$ be the foliation on the Riemannian manifold $M = V^{n+1} \setminus \text{Crit}(f)$ by level hypersurfaces of the C^∞ function $f : V^{n+1} \to \mathbb{R}$. The following statements are equivalent:*

i) *$\mathcal{F}$ is harmonic, i.e., every leaf of $\mathcal{F}$ is a minimal submanifold in M.*
ii) $\Delta f + \dfrac{1}{|\nabla f|^2}\,\text{Hess}(f)(\nabla f,\ \nabla f) = 0.$
iii) $\text{div}\left(\dfrac{1}{|\nabla f|}\,\nabla f\right) = 0.$

Cf. Theorem 8.13 in [Ton], p. 109. The proof is straightforward.

Using the identity $\text{div}(u\,Y) = u\,\text{div}(Y) + Y(u)$ [for any $u \in C^\infty(M)$ and $Y \in \mathfrak{X}/(M)$],

$$\text{div}\left\{\frac{1}{|\nabla f|}\,\nabla f\right\} = \frac{1}{|\nabla u|}\,\text{div}\big(\nabla f\big) + (\nabla f)\Big(\frac{1}{|\nabla f|}\Big);$$

hence [by $\Delta f = -\text{div}(\nabla f)$],

$$\text{div}\left\{\frac{1}{|\nabla f|}\,\nabla f\right\} = -\frac{1}{|\nabla f|}\,\Delta f - \frac{1}{|\nabla f|^2}\,(\nabla f)\big(|\nabla f|\big). \tag{4.78}$$

We have previously shown that [by (4.78)]

$$\kappa(Z) = \frac{1}{|\nabla f|} \left\{ \Delta f + Z(Zf) \right\}$$

hence

$$\kappa(Z) = \frac{1}{|\nabla f|} Z(Zf) - \text{div}\left\{ \frac{1}{|\nabla f|} \nabla f \right\} - \frac{1}{|\nabla f|^2} (\nabla f)(|\nabla f|)$$

or

$$\kappa(Z) = -\text{div} \left\{ \frac{1}{|\nabla f|} \nabla f \right\}. \tag{4.79}$$

Exercise 4.60 Show that $\mathscr{F}$ is totally geodesic $\Longleftrightarrow$ $\text{Hess}(f)(X, X') = 0$ for any $X, X' \in \mathfrak{X}_{\mathscr{F}}$. □

The particular case when f is a harmonic function may be described as follows.

Corollary 4.61 *Let $\mathscr{F}$ be the foliation of the Riemannian manifold $M = V^{n+1} \setminus$ Crit(f) by the level hypersurfaces of a harmonic function $f : V^{n+1} \to \mathbb{R}$. Then*
i) *$\mathscr{F}$ is Riemannian* $\Longleftrightarrow$ $\text{Hess}(f)(X, \nabla f) = 0$ *for any $X \in \mathfrak{X}_{\mathscr{F}}$.*
ii) *All leaves of $\mathscr{F}$ are minimal* $\Longleftrightarrow$ $\text{Hess}(f)(\nabla f, \nabla f) = 0$.
iii) *$\mathscr{F}$ is Riemannian and harmonic* $\Longleftrightarrow$ *$|\nabla f|$ is constant on V^{n+1}.*

Example 4.62 Let $\Omega \subset \mathbb{C}$ be a domain, and $f : \Omega \to \mathbb{C}$ a holomorphic function. Let $f = u + iv$ be the real and imaginary parts of f. Let $M = \Omega \setminus \left[\text{Crit}(u) \cup \text{Crit}(v)\right]$, thought of as endowed with the Euclidean metric g of $\mathbb{R}^2$, and let $\mathscr{F}$ and $\mathscr{F}^\perp$ be the foliations on M by level curves of u and v, respectively. By the Cauchy-Riemann equations $u_x = v_y$ and $u_y = -v_x$

$$g\big(\nabla u, \nabla v\big) = u_x \, v_x + u_y \, v_y = v_y \, v_x - v_x \, v_y = 0$$

so that $\mathscr{F}$ and $\mathscr{F}^\perp$ are two orthogonal foliations of M. The mean curvature of $\mathscr{F}$ is

$$\kappa(Z) = \frac{1}{|\nabla u|^2} \, g\big(\nabla u, \nabla(|\nabla u|)\big).$$

By taking into account

$$\nabla u = u_x \frac{\partial}{\partial x} + u_y \frac{\partial}{\partial y}, \quad |\nabla u|^2 = u_x^2 + u_y^2, \quad \nabla\big(|\nabla u|^2\big) = 2\,|\nabla u|\,\nabla\big(|\nabla u|\big),$$

$$\nabla\big(|\nabla u|^2\big) = \nabla\big(u_x^2 + u_y^2\big) = 2\,u_x \left\{ u_{xx} \frac{\partial}{\partial x} + u_{xy} \frac{\partial}{\partial y} \right\} + 2\,u_y \left\{ u_{yx} \frac{\partial}{\partial x} + u_{yy} \frac{\partial}{\partial y} \right\},$$

it follows that

$$\nabla\big(|\nabla u|\big) = \frac{1}{2\,|\nabla u|} \, \nabla\big(|\nabla u|^2\big).$$

Therefore (by $u_{xx} + u_{yy} = 0$),

$$\kappa(Z) = \frac{u_x^2\, u_{xx} + 2u_x u_y u_{xy} + u_y^2\, u_{yy}}{\left(u_x^2 + u_y^2\right)^{3/2}}.$$

Cf. Jerrard, [Jer], and Jerrard and Rubel, [JeRu], for the study of the curvature of level curves of a harmonic function. Further properties of the (curvature of) level curves of harmonic functions were studied, as related to questions of harmonic analysis, by Flatto and Newman and Shapiro, [FNS], and Ortel and Schneider, [OrSc]. By a result in [OrSc], if $\Omega \subset \mathbb{C}$ is an open set and $u : \Omega \to \mathbb{R}$ is a harmonic function, then the curvature of the level curves of u can be strictly maximal or strictly minimal at a point of Ω. If however Ω is a doubly connected domain bounded by analytic convex Jordan curves, and if u is a harmonic measure of Ω with respect to the outer boundary of Ω, then the minimal curvature of the level curves of u is attained on the boundary of Ω. Cf. Theorems 2.1 and 3.1 in [OrSc], pp. 400–401, and p. 402. □

4.8 De Rham-Hodge Theory for Riemannian Foliations

4.8.1 Alaoui–Hector–Sergiescu Foliated De Rham Decomposition Theorem

Let M be a compact oriented C^∞ manifold and let $\mathscr{F}$ be a transversally oriented Riemannian foliation of M. Let $P = T(\mathscr{F})$ and $Q = \nu(\mathscr{F})$ be the tangent and transverse bundles, so that the sequence of vector bundles and vector bundle morphisms $0 \to P \hookrightarrow T(M) \xrightarrow{\pi} Q \to 0$ is exact. Let $\Omega_B^\bullet(\mathscr{F})$ denote the graded $\Omega_B^0(\mathscr{F})$-module of basic differential forms on $(M, \mathscr{F})$ and let $d_B \equiv d\big|_{\Omega_B^\bullet(\mathscr{F})}$. Through the present section we report briefly on the classical treatment (in the spirit of PDEs theory) by Kamber and Tondeur (cf. [KaTo16]) to the problem of finding a foliated analog to the de Rham decomposition theorem, holding for $\left(\Omega_B^\bullet(\mathscr{F}),\ d_B\right)$ and leading to the unique representation of basic cohomology classes by Δ_B-harmonic forms. The geometric context is that one endows M with a bundle-like Riemannian metric g such that the corresponding mean curvature form κ is basic, i.e., $\kappa \in \Omega_B^1(\mathscr{F})$. Let $\langle\,\cdot\,,\,\cdot\,\rangle$ be the ordinary L^2 inner product on $\Omega^\bullet(M)$, and let $\langle\,\cdot\,,\,\cdot\,\rangle_B$ be the restriction of $\langle\,\cdot\,,\,\cdot\,\rangle$ to $\Omega_B^\bullet(\mathscr{F})$. Let $\delta_B : \Omega_B^\bullet(\mathscr{F}) \to \Omega_B^{\bullet-1}(\mathscr{F})$ be the formal adjoint of d_B with respect to $\langle\,\cdot\,,\,\cdot\,\rangle_B$, and let $\Delta_B \equiv \delta_B \circ d_B + d_B \circ \delta_B$ be the basic Laplacian. Next, let us set

$$\mathscr{H}_B^r = \operatorname{Ker}\left\{\Delta_B\ :\ \Omega_B^r(\mathscr{F}) \to \Omega_B^r(\mathscr{F})\right\}$$

(the space of Δ_B-harmonic basic harmonic r-forms). The main result in [KaTo16] is

Theorem 4.63 *Let $\mathscr{F}$ be a transversally oriented Riemannian foliation on a compact orientable manifold $(M,\ g)$. Let us assume that g is bundle-like and $\kappa \in \Omega_B^1(\mathscr{F})$. Then*

$$\Omega^r_B(\mathscr{F}) \simeq \left\{d_B\, \Omega^{r-1}_B(\mathscr{F})\right\} \oplus \left\{\delta_B\, \Omega^{r+1}_B(\mathscr{F})\right\} \oplus \mathscr{H}^r_B\,. \tag{4.80}$$

Moreover, the terms in the right-hand side of (4.80) *are mutually orthogonal and* $\mathscr{H}^r_B$ *is finite dimensional.*

Here are a few historical remarks related to Theorem 4.63, a result that may be regarded as a corner stone to the theory of Riemannian foliations of compact manifolds. Theorem 4.63 was used through earlier works by F.W. Kamber and Ph. Tondeur (cf. [KaTo9, KaTo11], and [KaTo13]). Theorem 4.63 was first announced by A. Alaoui and G. Hector (cf. [AlHe1]) and a proof due to the same authors was published in [AlHe2], except for the finite dimensionality of the spaces $\mathscr{H}^r_B$. The statement $\dim_{\mathbb{R}} \mathscr{H}^r_B < \infty$ in Theorem 4.63 was actually proved by A. Alaoui and V. Sergiescu and G. Hector ([ASH]). The proofs in [AlHe2, ASH] rely on P. Molino's structure theorems for Riemannian foliations (cf. [Mol0]) and a rather involved spectral sequences argument relating the geometry of the foliation $\mathscr{F}$ to that of the foliation of $B^1_T(\mathscr{F})$ by closures of leaves of the lifted foliation $\mathscr{F}^1_T$. P. Molino's structure theorems (cf. *op. cit.*) aren't described in the present book mainly because they already appeared in book form (cf. P. Molino's book [Mol]), and also because of our manifest interest in a closer examination of certain PDEs aspects in the theory of Riemannian foliations, and Riemannian geometry in general. In turn, the proof to Theorem 4.63 given in [KaTo16] is a beautiful excursion into the theory of PDEs, and relies essentially on the construction of a strongly elliptic operator on $\Omega^\bullet(M)$ whose restriction to $\Omega^\bullet_B(\mathscr{F})$ is the basic Laplacian Δ_B. Said elliptic extension is $\Delta - \tilde{\eta}$, where Δ is the Laplace-Beltrami operator of the Riemannian manifold (M, g) and $\tilde{\eta}$ is a differential operator of order ≤ 1, that preserves the degrees of differential forms, whose construction we recall briefly in the sequel. It should be mentioned that $\Delta - \tilde{\eta}$ is not self-adjoint, in general.

4.8.2 Exterior Differential Calculus on $\Omega^\bullet(M)$ Vversus $\Omega^\bullet_B(\mathscr{F})$

Let $\nu \in \Omega^q_B(\mathscr{F})$ be the transverse volume form of the given (codimension q) Riemannian foliation $\mathscr{F}$. Let $*$ be the Hodge start with respect to the Riemannian metric g. Let $\chi_{\mathscr{F}} = * \nu \in \Omega^p(M)$ [with $p = n - q$ and $\dim(M) = n$] be the characteristic form of $\mathscr{F}$, so that (with the appropriate choice of transverse and tangential orientations of $\mathscr{F}$, compatible to an orientation of M) the Riemannian volume form is $d\,\mathrm{v}_g = \nu \wedge \chi_{\mathscr{F}}$. We need to recall the decreasing filtration

$$F^r\Omega^m = \Big\{\omega \in \Omega^m(M) \ : \ \mathrm{i}(X_1) \cdots \mathrm{i}(X_{m-r+1})\,\omega = 0, \tag{4.81}$$

$$X_j \in C^\infty(P), \quad 1 \leq j \leq m - r + 1\Big\}.$$

Definition 4.64 Let $r \in \mathbb{Z}_+$ with $0 \le r \le n$. An element of $F^{r+1}\Omega^{p+r}$ is called a *$\mathscr{F}$-trivial* $(p+r)$-form. □

Next, let us recall the Rummler formula

$$d\chi_{\mathscr{F}} + \kappa \wedge \chi_{\mathscr{F}} = \varphi_0 \tag{4.82}$$

where φ_0 is $\mathscr{F}$-trivial, i.e., $\varphi_0 \in F^2\Omega^{p+1}$. We emphasize that φ_0 is actually defined by (4.82) and the information furnished by (4.82) is only that the differential form in the left-hand side of (4.82) is congruent to zero mod $\mathscr{F}$-trivial forms. We also recall

Lemma 4.65 *If g is a bundle-like metric with $\kappa \in \Omega^1_B(\mathscr{F})$, then $d\kappa = 0$.*

Cf. e.g., Corollary A.6 in the Appendix to [KaTo16], p. 430. Next, let us consider the star operator

$$\hat{*} : \Omega^r_B(\mathscr{F}) \to \Omega^{q-r}_B(\mathscr{F})$$

associated to g_Q. Then

$$\alpha \wedge \hat{*}\alpha = \nu\,, \quad \alpha \in \Omega^r_B(\mathscr{F}).$$

The quantitative relationship between the star operators $\hat{*}$ and $*$ is

$$\hat{*}\alpha = (-1)^{p(q-r)} * \left(\alpha \wedge \chi_{\mathscr{F}}\right), \quad *\alpha = \left(\hat{*}\alpha\right) \wedge \chi_{\mathscr{F}}\,, \quad \alpha \in \Omega^r_B(\mathscr{F}). \tag{4.83}$$

The L^2 inner product $\langle\cdot\,,\,\cdot\rangle$ on $\Omega^r(M)$ restricts to $\Omega^r_B(\mathscr{F})$ as

$$\langle\alpha\,,\,\beta\rangle_B = \int_M \alpha \wedge \left(\hat{*}\beta\right) \wedge \chi_{\mathscr{F}}\,, \quad \alpha,\,\beta \in \Omega^r_B(\mathscr{F}). \tag{4.84}$$

Let $\tilde{\gamma}$ be the bounded linear operator

$$\tilde{\gamma} : \Omega^r(M) \to \Omega^{p+q-r+1}(M), \quad \tilde{\gamma}(\omega) = (-1)^{(q+1)(p+r)+1} * \left(\omega \wedge \chi_{\mathscr{F}}\right) \wedge \varphi_0\,,$$

where $\varphi_0 := d\chi_{\mathscr{F}} + \kappa \wedge \chi_{\mathscr{F}}$. Let us set

$$\gamma := \tilde{\gamma}\,\big|_{\Omega^r_B(\mathscr{F})}\,.$$

Then [by (4.83)]

$$\gamma(\alpha) = (-1)^{(p+1)(r+1)+qr} \left(\hat{*}\alpha\right) \wedge \varphi_0\,, \quad \alpha \in \Omega^r_B(\mathscr{F}). \tag{4.85}$$

Note that

$$\hat{*}\alpha \in F^{q-r}\,, \quad \varphi_0 \in F^2 \Longrightarrow \gamma(\alpha) \in F^{q-r+2}\Omega^{p+q-r+1}\,,$$

i.e., $\gamma(\alpha)$ is $\mathscr{F}$-trivial. Let $\delta : \Omega^r(M) \to \Omega^{r-1}(M)$ be the formal adjoint of d, i.e.,

$$\delta\,\alpha = (-1)^{n(r+1)+1} \;*\; d \;*\; \alpha\,, \quad \alpha \in \Omega^r(M).$$

The quantitative relationship between δ_B and δ is

$$\delta\alpha = \delta_B\alpha + *\,\gamma\,(\alpha), \quad \alpha \in \Omega^r_B(\mathscr{F}), \tag{4.86}$$

and $*\,\gamma(\alpha)$ is orthogonal to $\Omega^\bullet_B(\mathscr{F})$. As a corollary to (4.86) one may easily show that

$$\langle \delta\,\alpha\,,\,\beta\rangle = \langle \delta_B\,\alpha\,,\,\beta\rangle_B\,, \quad \alpha \in \Omega^r_B(\mathscr{F}), \quad \beta \in \Omega^{r-1}_B(\mathscr{F}),$$

i.e., δ_B is the formal adjoint of d on basic forms. By applying (4.86) together with $\delta^2 = 0$ one obtains

$$\gamma(\delta_B\,\alpha) = (-1)^{(n-r)r+1} \;*\; \delta\;*\,\gamma\,(\alpha) = (-1)^{n-r+1}\,d\,\gamma(\alpha). \tag{4.87}$$

Next, we may establish the quantitative relationship between the basic Laplacian $\Delta_B \equiv \delta_B\,d_B + d_B\,\delta_B$ and the Laplacian $\Delta \equiv \delta\,d + d\,\delta$ restricted to $\Omega^\bullet_B(\mathscr{F})$. Indeed (4.86) and (4.87) imply

$$\Delta\alpha = \Delta_B\alpha + \eta(\alpha), \quad \alpha \in \Omega^r_B(\mathscr{F}), \tag{4.88}$$

$$\eta(\alpha) \equiv *\,\gamma\,(d_B\,\alpha) + d\;*\;\gamma\,(\alpha).$$

It should be observed that

$$\eta = \tilde{\eta}\big|_{\Omega^r_B(\mathscr{F})}\,,$$

$$\tilde{\eta} : \Omega^r(M) \to \Omega^r(M), \quad \tilde{\eta}(\omega) \equiv *\,\tilde{\gamma}\,(d\,\omega) + d\;*\;\tilde{\gamma}\,(\omega).$$

Of course the meaning of (4.88) is that $\Delta - \tilde{\eta} : \Omega^\bullet(M) \to \Omega^\bullet(M)$ is an extension of $\Delta_B : \Omega^\bullet_B(\mathscr{F}) \to \Omega^\bullet_B(\mathscr{F})$. As Δ is strongly elliptic of order 2 and $\tilde{\eta}$ has lower order, the differential operator $\Delta - \tilde{\eta}$ is still strongly elliptic. Classical results in elliptic theory applied to $\Delta - \tilde{\eta}$ will lead to the seek for results about Δ_B.

Applying (4.86) twice, together with

$$\langle *\,\gamma\,(\alpha)\,,\,*\,\gamma\,(\beta)\rangle = \langle \gamma\,(\alpha),\,\gamma\,(\beta)\rangle = \langle (\gamma^*\,\gamma)\,(\alpha),\,\beta\rangle$$

leads to

$$\langle \Delta\alpha,\,\beta\rangle = \langle \Delta_B\,\alpha,\,\beta\rangle_B + \langle (\gamma^*\,\gamma)(\alpha),\,\beta\rangle. \tag{4.89}$$

As a corollary to (4.89) [by applying (4.88) and (4.89)],

$$\langle \eta\,(\alpha),\,\beta\rangle = \langle (\gamma^*\,\gamma)(\alpha),\,\beta\rangle, \quad \alpha,\,\beta \in \Omega_B(\mathscr{F}),$$

and $\eta(\alpha) - (\gamma^*\,\gamma)(\alpha)$ is orthogonal to $\Omega^r_B(\mathscr{F})$.

4.8.3 Sobolev Chain on $\Omega_B^\bullet(\mathscr{F})$

Let $C_B^\infty(\Lambda^r Q^*)$ consist of all $\beta \in C^\infty(\Lambda^r Q^*)$ such that $\mathscr{L}_X\beta = 0$ for any $X \in \mathfrak{X}_{\mathscr{F}}$. For every $\alpha \in \Omega_B^\bullet(\mathscr{F}) \simeq C_B^\infty(\Lambda^\bullet Q^*)$, let $\nabla^k\alpha$ be the k-th covariant derivative of α, defined in terms of the canonical connection

$$\nabla_Y s = \begin{cases} \mathring{\nabla}_X s, & X \in \mathfrak{X}_{\mathscr{F}}, \\ \pi \nabla^g_Y \sigma_g\, s, & Y \in C^\infty(P^\perp), \end{cases}$$

for any $Y \in \mathfrak{X}(M)$ and any $s \in C^\infty(Q)$. The Sobolev norms on $\Omega_B^\bullet(\mathscr{F})$ are defined by

$$\|\alpha\|_{B,s} = \Big(\sum_{k=0}^{s} \|\nabla^k\alpha\|_B^2\Big)^{1/2}, \quad s \geq 0,$$

where $\|\cdot\|_B$ is the norm associated to the L^2 inner product (4.84). Let $H_s(\Omega_B^\bullet)$ be the completion of $\Omega_B^\bullet(\mathscr{F})$ in the norm $\|\cdot\|_{B,s}$. The diagram

$$\begin{array}{ccc} H_s(\Omega_B^\bullet) & \hookrightarrow & H_s(\Lambda^\bullet Q^*) \\ \uparrow & & \uparrow \\ \Omega_B^\bullet(\mathscr{F}) & \hookrightarrow & C^\infty(\Lambda^\bullet Q^*) \end{array}$$

is commutative[17], hence $H_s(\Omega_B^\bullet)$ is isometric to the closure of $\Omega_B^\bullet(\mathscr{F})$ in $H_s(\Lambda^\bullet Q^*)$. The bundle $\Lambda^\bullet Q^*$ embeds in $\Lambda^\bullet T^*(M)$ as the subbundle of all differential forms ω on M such that $\mathfrak{X}_{\mathscr{F}} \rfloor \omega = 0$. The Sobolev norms on $H_s(\Omega^\bullet(M))$ descent to norms on $H_s(\Lambda^\bullet Q^*)$ [$\subset H_s(\Omega^\bullet(M))$], turning out to be equivalent to $\|\cdot\|_s$, and that are denoted by the same symbol $\|\cdot\|_s$. Therefore, $H_s(\Omega_B^\bullet)$ [$\subset H_s(\Omega^\bullet(M))$] is topologically isomorphic to the closure of $\Omega_B^\bullet(\mathscr{F})$ in $H_s(\Omega^\bullet(M))$, i.e., there exist positive constants $C > 0$ and C' such that

$$\|\alpha\|_{B,s} \leq C\,\|\alpha\|_s, \quad \|\alpha\|_s \leq C'\,\|\alpha\|_{B,s}, \quad \alpha \in \Omega_B^\bullet(\mathscr{F}), \quad s \geq 0. \tag{4.90}$$

In particular, the diagram

$$\begin{array}{ccccc} H_s(\Omega_B^\bullet) & \hookrightarrow & H_s(\Lambda^\bullet Q^*) & \hookrightarrow & H_s(\Omega^\bullet(M)) \\ \uparrow & & \uparrow & & \uparrow \\ H_{s+t}(\Omega_B^\bullet) & \hookrightarrow & H_{s+t}(\Lambda^\bullet Q^*) & \hookrightarrow & H_{s+t}(\Omega^\bullet(M)) \end{array}$$

---

[17] The vertical arrows are the natural injections.

is commutative. A little extra work [relying on (4.90) and the Rellich lemma for $\Lambda^\bullet Q^*$] shows that the "Relich lemma" holds for the Sobolev chain $H_a(\Omega_B^\bullet)$, $s \geq 0$, i.e.,

Lemma 4.66 *The inclusions*

$$H_{s+t}(\Omega_B^\bullet) \hookrightarrow H_s(\Omega_B), \quad s \geq 0, \ \ t > 0,$$

are compact and dense.

Cf. Proposition 4.5 in [KaTo16], p. 420. The Bott (partial) connection $\mathring{\nabla}$ may be thought of as the first-order differential operator $\mathring{\nabla} : C^\infty\big(\Lambda^\bullet Q^*\big) \to C^\infty\big(\mathrm{Hom}\,(P, \ \Lambda^\bullet Q^*)\big)$. As such, it extends to every $H_s\big(\Lambda^\bullet Q^*\big)$ and gives continuous mappings

$$\mathring{\nabla}_s : H_s\big(\Lambda^\bullet Q^*\big) \to H_{s-1}\big(\mathrm{Hom}(P, \ \Lambda^\bullet Q^*)\big).$$

The "Sobolev embedding theorem" holds for the Sobolev chain $H_a(\Omega_B^\bullet)$, $s \geq 0$, i.e.,

Lemma 4.67 *There exist continuous injections*

$$\alpha_s : \mathrm{Ker}\big(\mathring{\nabla}_{s+t}\big) \hookrightarrow \Omega_{B,\,s} := \mathrm{Ker}\Big\{\mathring{\nabla} \ : \ C^s\big(\Lambda^\bullet \ Q^*\big) \to C^{s-1}\big(\mathrm{Hom}(P, \ \Lambda^\bullet Q^*)\big)\Big\},$$

$$s > 0, \quad t \geq t_0 = \left[\frac{n}{2}\right] + 1,$$

such that the diagram

$$\begin{array}{ccccccc}
 & & & H_{s+t}\big(\Omega_B^\bullet\big) & & & \\
 & & & \downarrow & & & \\
\to & \Omega_{B,\,s+t} & \to & \mathrm{Ker}\big(\mathring{\nabla}_{s+t}\big) & \xrightarrow{\alpha_s} & \Omega_{B,\,s} & \to \\
 & \downarrow & & \downarrow & & \downarrow & \\
\to & C^{s+t}\big(\Lambda^\bullet Q^*\big) & \to & H_{s+t}\big(\Lambda^\bullet Q^*\big) & \xrightarrow{\alpha_s} & C^s\big(\Lambda^\bullet Q^*\big) & \to \\
 & \downarrow \mathring{\nabla} & & \downarrow \mathring{\nabla}_{s+t} & & \downarrow \mathring{\nabla} & \\
\to & C^{s+t-1}\big(\mathrm{Hom}(P, \ \Lambda \, Q^*)\big) & \to & H_{s+t-1}\big(\mathrm{Hom}(P, \ \Lambda \, Q^*)\big) & \xrightarrow{\alpha_s} & C^{s-1}\big(\mathrm{Hom}(P, \ \Lambda Q^*)\big) & \to
\end{array}$$

is commutative.

Cf. Proposition 4.12 in [KaTo16], p. 422.

4.8.4 Coercivity and Weak Solvability

Let Φ_B be the bilinear form on $\Omega_B^\bullet(\mathscr{F})$ given by

$$\Phi_B(\alpha, \beta) = \langle \Delta_B \alpha, \beta \rangle_B .$$

Φ_B is symmetric [as Δ_B is formally self-adjoint] and extends to a symmetric bilinear form on $H_1(\Omega_B^\bullet)$.

Lemma 4.68 Φ_B *is coercive, i.e., there exist constants* $C_B > 0$ *and* $\lambda_{B,0} > 0$ *such that*

$$C_B^2 \, \|v\|_{B,1}^2 \leq \Phi_B(v, v) + \lambda_{B,0} \, \|v\|_{B,0}^2 , \quad v \in H_1(\Omega_B^\bullet).$$

Cf. Proposition 5.2 in [KaTo16], p. 423. To sate the weak solvability lemma we need some notation. Let $K(\Phi_B)$ denote the null space of Φ_B, i.e.,

$$K(\Phi_B) = \left\{ v \in H_1(\Omega_B^\bullet) \ : \ \Phi_B(v, w) = 0, \quad w \in H_1(\Omega_B^\bullet) \right\}.$$

Also, for every $\lambda > 0$ we consider the symmetric bilinear form $\Phi_{B,\lambda}$ on $H_1(\Omega_B^\bullet)$ defined by

$$\Phi_{B,\lambda}(v, w) = \Phi_B(v, w) + \lambda \, \langle v, w \rangle_{B,0} , \quad v, w \in H_1(\Omega_B^\bullet).$$

Lemma 4.69 i) *For any* $v \in H_0(\Omega_B^\bullet)$ *and any* $\lambda \geq \lambda_{B,0}$ *there is a unique solution* $u \in H_1(\Omega_B^\bullet)$ *to*

$$\Phi_{B,\lambda}(u, y) = \langle v, y \rangle_{B,0} , \quad y \in H_1(\Omega_B^\bullet).$$

ii) *The equation*

$$\Phi_B(u, y) = \langle v, y \rangle_{B,0} , \quad y \in H_1(\Omega_B^\bullet),$$

has a solution $u \in H_1(\Omega_B^\bullet) \iff \langle v, y \rangle_{B,0} = 0$ *for every* $y \in K(\Phi_B)$.

iii) dim $K(\Phi_B) < \infty$.

Cf. Proposition 5.8, [KaTo16], p. 424. The unique solution $u \in H_1(\Omega_B^\bullet)$ for a given $v \in H_0(\Omega_B^\bullet)$ [as in statement (i) of Lemma 4.69 above] prompts the *Green operator*

$$G'_{B,\lambda} \ : \ H_0(\Omega_B^\bullet) \to H_1(\Omega_B^\bullet), \quad \lambda \geq \lambda_{B,0} ,$$

$$\Phi_{B,\lambda}\big(G'_{B,\lambda}(v), \, y\big) = \langle v, y \rangle_{B,0} , \quad y \in H_1(\Omega_B^\bullet).$$

Let $G_{B,\lambda} : H_0(\Omega_B^\bullet) \to H_0(\Omega_B^\bullet)$ be the composition

$$H_0(\Omega_B^\bullet) \xrightarrow{G'_{B,\lambda}} H_1(\Omega_B^\bullet) \hookrightarrow H_0(\Omega_B^\bullet).$$

Then $G_{B,\lambda}$ is a compact, self-adjoint, operator. Next, let us set

$$\Psi_{B,\lambda} := \mathrm{id} - \lambda\, G_{B,\lambda} \; : \; H_0(\Omega_B^\bullet) \to H_0(\Omega_B^\bullet).$$

Then $\Psi_{B,\lambda}$ is a self-adjoint, Fredholm, operator and $\mathrm{Ker}(\Psi_{B,\lambda}) = K(\Phi_B)$. The solution $u \in H_1(\Omega_B^\bullet)$ furnished by statement (ii) in Lemma 4.69 is characterized by

$$\Psi_{B,\lambda}(u) = G'_{B,\lambda}(v).$$

4.8.5 *Regularity*

Definition 4.70 Let $v \in H_0(\Omega_B^\bullet)$. A form $u \in H_1(\Omega_B^\bullet)$ is said to be a *weak solution* to $\Delta_B u = v$ if

$$\Phi_B(u, y) = \langle v, y\rangle_{B,0}, \quad y \in H_1(\Omega_B^\bullet). \tag{4.91}$$

□

As argued in the preceding section, a weak solution u to $\Delta_B u = v$ exists if and only if v is orthogonal to $K(\Phi_B)$ in $H_0(\Omega_B^\bullet)$. We may now state the relevant regularity result.

Lemma 4.71 *If $v = \beta \in \Omega_B^\bullet(\mathscr{F})$ is a C^∞ basic form such that $v \perp K(\Phi_B)$, then a weak solution $u \in H^1(\Omega_B^\bullet)$ of $\Delta_B u = \beta$ is a C^∞ basic form $u = \alpha \in \Omega_B^\bullet(\mathscr{F})$ and $\Delta_B\alpha = \beta$.*

Cf. Theorem 6.2 in [KaTo16], p. 425. The main ingredient in the proof of Lemma 4.71 is to show that a weak solution to $\Delta_B u = v$ prompts a weak solution to $(\Delta - \tilde{\eta})u = v$ and use the hypoellipticity of $\Delta - \tilde{\eta}$.

4.8.6 *Proof of Alaoui–Hector–Sergiescu Theorem*

The spaces of basic harmonic forms are finite dimensional as a consequence of the weak solvability Lemma 4.69 and the regularity Lemma 4.71. To decomposition in Theorem 4.63 follows from the orthogonal decomposition

$$\Omega_B^r(\mathscr{F}) = \{\Delta_B\, \Omega_B^r(\mathscr{F})\} \oplus \mathscr{H}_B^r\,. \tag{4.92}$$

The proof of (4.92) follows the arguments in the classical elliptic case (cf. e.g., [War]). As $\mathscr{H}_B^r$ is finite dimensional

$$\Omega_B^r(\mathscr{F}) = (\mathscr{H}_B^r)^\perp \oplus \mathscr{H}_B^r$$

and all we need to check is $\left(\mathscr{H}_B^r\right)^\perp = \Delta_B\, \Omega_B^r(\mathscr{F})$. Note that

$$\langle \Delta_B \alpha\,,\, \beta\rangle = \langle \alpha\,,\, \Delta_B \beta\rangle\,, \quad \alpha \in \Omega_B^r(\mathscr{F}), \quad \beta \in \mathscr{H}_B^r\,,$$

yields $\Delta_B\, \Omega_B^r(\mathscr{F}) \subset \left(\mathscr{H}_B^r\right)^\perp$. Let us prove the opposite inclusion $\left(\mathscr{H}_B^r\right)^\perp \subset \Delta_B$ $\Omega_B^r(\mathscr{F})$. By the regularity Lemma 4.71 the Green operator G_B

$$H_0\big(\Omega_B^\bullet\big) \to H_1\big(\Omega_B^\bullet\big) \hookrightarrow H_0\big(\Omega_B^\bullet\big)$$

induces a map $G_B : \Omega_B^\bullet(\mathscr{F}) \to \Omega_B^\bullet(\mathscr{F})$ hence $G_B\big[\Omega_B^\bullet(\mathscr{F})\big] \subset \mathscr{H}_B^\perp$ and the map $G_B : \Omega_B^\bullet(\mathscr{F}) \to \mathscr{H}_B^\perp$ has kernel $\mathscr{H}_B$. Indeed, let $\pi_B : H_0\big(\Omega_B^\bullet\big) \to K(\Phi_B)$ be the orthogonal projection with respect to $\langle\,\cdot\,,\,\cdot\,\rangle_{B,0}$. Then, for every $\alpha \in \Omega_B^\bullet(\mathscr{F})$

$$\Delta_B\big(G_B(\alpha)\big) = \delta_B\, d_B\, G_B\, \alpha + d_B\, \delta_B\, G_B\, \alpha = \alpha - \pi_B(\alpha).$$

Yet for $\alpha \in \big(\mathscr{H}_B\big)^\perp$ one has $\pi_B(\alpha) = 0$ hence $\alpha \in \Delta_B\, \Omega_B^\bullet(\mathscr{F})$. Q.e.d.

4.9 Riemannian Foliations and Curvature Groups

The purpose of the present section is to report on results by S. Dragomir and R. Petit (cf. [DrPe]) on curvature groups $H_B^\bullet(M,\, \mathscr{F},\, \Gamma)$ associated to a foliated Riemannian manifold $(M,\, g,\, \mathscr{F})$, that generalize I. Vaisman's curvature groups (cf. [Vai]) of a Riemannian manifold of constant sectional curvature. A purpose of [DrPe] is to study the relationship between the groups $H_B^\bullet(M,\, \mathscr{F},\, \Gamma)$ and the basic cohomology groups of $(M,\, \mathscr{F})$. A result in [DrPe] is that $H_B^\bullet(M,\, \mathscr{F},\, \Gamma) = 0$ for any Einstein foliation $\mathscr{F}$ of nonzero transverse scalar curvature.

Building on work by K. Kodaira and D.C. Spencer (cf. [KoSp]) I. Vaisman announced (cf. [Vai0]) in 1967 the remarkable finding that, to each symmetric linear connection Γ on a C^∞ manifold M, one may associate curvature groups $H^\bullet(M,\, \Gamma)$ and curvature numbers, similar to the de Rham cohomology and Betti numbers of a compact manifold. The most general setting consists in considering a connection Γ in the principal bundle $G \to P \to M$ over the n-dimensional manifold M and a linear representation ρ of the structure group G on a finite-dimensional vector space V. Next, let $\mathscr{S}^r$ be the $C^\infty(M)$-module of all tensor r-forms of type $(V,\, \rho)$ on M, and let $\mathscr{D}^r = \big\{DDT\, :\, T \in \mathscr{S}^{r-2}\big\}$ (a submodule of $\mathscr{S}^r$). Here D denotes the covariant derivative with respect to Γ. By following the ideas in [KoSp], one may then consider $\mathscr{J}^r = \mathscr{S}^r \times \mathscr{S}^{r+1}$ and $D^r : \mathscr{J}^r \to \mathscr{J}^{r+1}$ given by

$$D^r(T,\, S) = (DT - S,\; DDT - DS), \quad (S,\, T) \in \mathscr{J}^r\,, \tag{4.93}$$

and observe that

$$\mathscr{J}^\bullet = \Big(\bigoplus_{r=0}^{n} \mathscr{J}^r\,,\; D^\bullet\Big)$$

is an acyclic complex (cf. also [Vai0]). On the other hand, if $\mathscr{K}^r = \mathscr{J}^r \times \mathscr{D}^{r+1}$, then

$$\mathscr{K}^\bullet = \Big(\bigoplus_{r=0}^{n} \mathscr{K}^r\,,\; D^\bullet\Big)$$

is a subcompex of $\mathscr{J}^\bullet$ which is not acyclic in general.

Definition 4.72 The *curvature groups* of $(M,\ \Gamma)$ are the cohomology groups of $\mathscr{K}^\bullet$

$$H^r(M,\ \Gamma,\ \rho) = H^r\big(\mathscr{K}^\bullet\big), \quad 0 \le r \le n.$$

□

Cf. I. Vaisman, [Vai], p. 322. The case Γ is the Levi-Civita connection of a space form is studied in [Vai]. There $G = \mathrm{GL}(n, \mathbb{R})$, $P = F(M)$ (the principal bundle of linear frames tangent to M), and $V = \mathbb{R}^n$. If M is a space form of (constant) sectional curvature c, then $\mathscr{K}^r \simeq L^r$ (an isomorphism of modules) for any $1 \le r \le n-1$, where $L^r = \mathscr{J}^r \times \Omega^r(M)$. Under this isomorphism the cochain maps (4.93) go over to

$$\begin{aligned} D^r : L^t \to L^{r+1}\,, \quad & D^r(\lambda,\ \alpha) \\ = \big(D\lambda - \theta \wedge \alpha\,,\ d\alpha - c\,\epsilon_i \wedge \lambda^i\big), \quad & (\lambda,\ \alpha) \in L^r, \end{aligned} \tag{4.94}$$

where θ is the (tensor form associated to the) canonical 1-form on $F(M)$. The last term in (4.94) is "morally" $DD\lambda$. This is vector-valued, yet roughly speaking the second covariant derivative "equals" curvature, and the curvature 2-form of a space form is $\Omega_i^j = c\,\epsilon_i \wedge dx^j$, where $\epsilon_i = g_{ik}\,dx^k$. By a result of I. Vaisman (cf. [Vai], p. 337)

$$H^r(M,\ \Gamma,\ \rho) \simeq H^r(M,\ \mathscr{S})$$

(a group isomorphism) where $\mathscr{S}$ is the sheaf of all local solutions to a first-order partial differential system of the form

$$\nabla_j X^i = f\,\delta^i{}_j\,, \quad 1 \le i,\ j \le n, \tag{4.95}$$

for some C^∞ function f.

Definition 4.73 Tangent vector fields X^i satisfying (4.95) are called *s-fields*. □

Cf. [Vai], p. 336. S.I. Goldberg and Petridis studied (cf. [GoPe]) the general case of a semi-Riemannian manifold M equipped with the cochain complex

$$\tilde{L}^\bullet = \Big(\bigoplus_{r=0}^{n} \tilde{L}^r\,,\ D^\bullet\Big), \quad \tilde{L}^r = \big\{(\lambda, \alpha) \in L^r \ :\ DD\lambda = 0\big\},$$

and $D^r : \tilde{L}^r \to \tilde{L}^{r+1}$ is the restriction of the cochain map (4.94) to $\tilde{L}^r$. The *curvature groups* are taken to be the cohomology groups of $\tilde{L}^\bullet$, i.e.,

$$H^r(M,\ \Gamma) = H^r\big(\tilde{L}^\bullet\big), \quad 0 \le r \le n.$$

Similar to I. Vaisman's result (cf. [Vai]) the curvature groups are isomorphic to the cohomology groups of M with coefficients in the sheaf $\tilde{\mathscr{S}}$ of all homothetic s-fields ($\tilde{\mathscr{S}}$ is the subsheaf of $\mathscr{S}$ consisting of all local solutions to systems of the form (4.95) with $f = c \in \mathbb{R}$), provided that $\tilde{L}^\bullet$ satisfies Poincaré lemma. The cochain complex $\tilde{L}^\bullet$ satisfies Poincaré lemma (i.e., the associated sequence of sheaves is exact at the level of stalks) when M is flat, yet said property may fail to hold on an arbitrary semi-Riemannian manifold.

One of the purposes of the present section is to build (by following S. Dragomir and R. Petit, [DrPe]) basic analogs $L^\bullet_B(M,\ \mathscr{F},\ \Gamma)$ and $\mathscr{K}^\bullet_B(M,\ \mathscr{F},\ \Gamma)$ to the chain complexes $\tilde{L}^\bullet$ and $\mathscr{K}^\bullet$, respectively, in [GoPe, Vai]. The construction in [DrPe] is general enough to encompass the case of an arbitrary projectable connection on a foliated principal bundle over a foliated manifold. However, to stay within the present chapter's theme, the exposition is limited to transverse Levi-Civita connections Γ of foliated Riemannian manifolds $(M,\ \mathscr{F},\ g)$. The corresponding cohomology groups $H^r\big(\mathscr{K}^\bullet_B(M,\ \mathscr{F},\ \Gamma)\big)$ and $H^r_B(M,\ \mathscr{F},\ \Gamma)$ [coinciding with $H^r(M,\ \Gamma)$ in [GoPe], p. 55, and $H^r(M,\ \omega,\ \rho)$ in [Vai], p 332, when $\mathscr{F}$ is the trivial foliation by points] are related, and related to the cohomology of M with coefficients in the sheaf of local basic solutions to partial differential systems of the form

$$\nabla s = f\,\pi \tag{4.96}$$

where $\pi : T(M) \to Q = \nu(\mathscr{F}) = T(M)/T(\mathscr{F})$ is the projection. Precisely

Theorem 4.74 *Let $\mathscr{F}$ be a Riemannian foliation modeled on a space form $N^q(c)$ of sectional curvature $c \in \mathbb{R}$. For each $1 \le r \le n + q^2$ there is a module monomorphism*

$$\Phi^r : \mathscr{K}^r(M,\ \mathscr{F},\ \Gamma) \to L^r(M,\ \mathscr{F})$$

of the form

$$\Phi^r(\varphi\,,\ DD\lambda) = (\varphi\,,\ c\,\alpha(\lambda)), \quad \varphi \in \mathscr{I}^r(M,\ \mathscr{F}), \ \ \lambda \in \mathscr{I}^{r-1}(M,\ \mathscr{F}),$$

where the form $\alpha(\lambda)$ is prescribed by Lemma 4.94. *Moreover,*

$$\Phi^{r+1}\, D^r(\varphi,\ \psi) = D^r\, \phi^r(\varphi,\ \psi) + c\,(0,\ \alpha(\varphi)), \tag{4.97}$$
$$(\varphi,\ \psi) \in \mathscr{K}^r(M,\ \mathscr{F},\ \Gamma).$$

Let

$$\mathscr{K}_0^r(M, \mathscr{F}, \Gamma) = \{(\varphi, \psi) \in \mathscr{K}^r(M, \mathscr{F}, \Gamma) \ : \ DD\varphi = 0\}.$$

Then

$$\mathscr{K}_{B,0}^\bullet(M, \mathscr{F}, \Gamma) := \Big(\bigoplus_{r=0}^{q(q+1)} \mathscr{K}_{B,0}^r(M, \mathscr{F}, \Gamma), \ D_B^\bullet \Big)$$

is a cochain subcomplex of $\mathscr{K}_B^\bullet(M, \mathscr{F}, \Gamma)$. *If* $c \neq 0$ *the map* Φ_r *defines a module isomorphism*

$$\mathscr{K}_B^r(M, \mathscr{F}, \Gamma) \simeq L_B^r(M, \mathscr{F}). \tag{4.98}$$

In particular

$$\mathscr{K}_{B,0}^r(M, \mathscr{F}, \Gamma) \simeq L_B^r(M, \mathscr{F}, \Gamma)$$

and the cohomology groups of the complex $\mathscr{K}_{B,0}^\bullet(M, \mathscr{F}, \Gamma)$ *are isomorphic to the basic curvature groups* $H_B^r(M, \mathscr{F}, \Gamma)$.

Equation (4.96) is the foliated analog to (4.95) [and (4.96) has no local basic solutions unless f is a basic function]. By a result in [GoPe] the curvature groups $H^r(\tilde{L}^\bullet)$ are related to the de Rham cohomology groups of M [provided that (4.95) admits local solutions for each $f = c \in \mathbb{R} \setminus \{0\}$, and has no solutions for $f = 0$]. The proof relies on the fact that the de Rham complex satisfies the Poincaré lemma. The foliated analog to the de Rham cohomology is B. Reinhart's basic cohomology (intuitively, the "de Rham cohomology" of the leaf space $M/\mathscr{F}$). Yet, in general, the Poincaré lemma holds for the basic complex only in degree 1. The corresponding result for $H_B^\bullet(M, \mathscr{F}, \Gamma)$ is then more restrictive and requires assumptions both on the existence of basic solutions to (4.96) and on the structure of the basic complex (4.100). Leaving the explanation of notations momentarily aside, we may state

Theorem 4.75 *Let* $(M, \mathscr{F}, g)$ *be a foliated Riemannian manifold with* $\mathscr{F}$ *Riemannian and* g *bundle-like. Let* $\mathscr{S}$ *be the sheaf of all local basic sections* s *in* $Q = \nu(\mathscr{F})$ *such that* $\nabla s = c\,\pi$ *for some* $c \in \mathbb{R}$. *Let* $\mathscr{L}_B^r$ *be the sheaf on* M *associated to the module* $L_B^r(M, \mathscr{F}, \Gamma)$, *i.e.,*

$$\mathscr{L}_B^r(U) = \Big\{(\varphi, \beta) \in \mathscr{I}_B^r(U_T, \mathscr{F}_U) \times \Omega_B^r(\mathscr{F}_U) \ : \ DD\varphi = 0\Big\},$$

for any open subset $U \subset M$ *and any* $0 \le r \le q$. *If the basic Vaisman complex* $L_B^\bullet(M, \mathscr{F}, \Gamma)$ *satisfies the Poincaré lemma, then*

$$0 \to \mathscr{S} \xrightarrow{j} \mathscr{L}_B^0 \xrightarrow{D^0} \mathscr{L}_B^1 \xrightarrow{D^1} \cdots \xrightarrow{D^{q-1}} \mathscr{L}_B^0 \xrightarrow{D^0} \mathscr{L}_B^q \xrightarrow{D^q} 0 \tag{4.99}$$

is an exact sequence of sheaves. In particular $0 \to \mathscr{S} \to \mathscr{L}_B^\bullet$ *is a fine resolution of* $\mathscr{S}$, *so that the basic curvature groups of* $(M, \mathscr{F}, g)$ *are isomorphic to the cohomology groups of* M *with coefficients in* $\mathscr{S}$.

Cf. Theorem 2.2 in [DrPe], p. 5. The integrability conditions to (4.95) involve curvature (of the given Riemannian metric) and then a vanishing result holds (for the curvature groups $H^r(\tilde{L}^\bullet)$, cf. [GoPe], p. 552) under an assumption of nondegeneracy on the Ricci tensor. A foliated analog may then be established in terms of transverse Ricci curvature (and turns out to be related to the geometry of infinitesimal conformal transformations of a Riemannian foliation, as developed by F. Kamber and Ph. Tondeur, [KaTo7], and Ph. Tondeur and G. Toth, [ToTo]. Precisely, under the assumptions of Theorem 4.75 we may state

Theorem 4.76 *If S^∇ is everywhere non-degenerate, then the basic curvature groups of $(M, \mathscr{F}, g)$ vanish. In particular $H_B^\bullet(M, \mathscr{F}, \Gamma) = 0$ for any Einstein foliation $\mathscr{F}$ with $c^\nabla \neq 0$. Hence, an Einstein foliation with at least one nonzero basic curvature group is transversally Ricci flat.*

Similarly, the counterpart in [DrPe] to Vaisman's calculations involving exterior differential systems on a space form (cf. [Vai]) is to assume that the transverse geometry of the given Riemannian foliation is modeled on a space form.

4.9.1 The Basic Vaisman Complex

Let $(M, \mathscr{F}, g)$ be a n-dimensional foliated Riemannian manifold, where $\mathscr{F}$ is a transversally oriented Riemanian foliation of codimension $\operatorname{cod}(\mathscr{F}) = q$, and g is a bundle-like Riemnanian metric. Let $Q = \nu(\mathscr{F})$ be the transverse bundle, and let $\pi : T(M) \to Q$ be the projection. We recall the basic complex of $(M, \mathscr{F})$

$$\Omega_B^0(\mathscr{F}) \xrightarrow{d_B} \Omega_B^1(\mathscr{F}) \xrightarrow{d_B} \cdots \xrightarrow{d_B} \Omega_B^q(\mathscr{F}) \xrightarrow{d_B} 0 \tag{4.100}$$

and the basic cohomology of $(M, \mathscr{F})$, i.e., the cohomology of (4.100). By a classical result of A. El Kacimi-Alaoui and V. Sergiescu and V. Hector (cf. [ESH]) the basic cohomology groups are finite dimensional, provided M is compact. Infinite-dimensional examples may be built (cf. [Sch]) when the given foliation is not Riemannian (even if M is compact).

Let $\mathrm{GL}(q, \mathbb{R}) \to B_T^1(\mathscr{F}) \xrightarrow{p_T^1} M$ be the principal bundle of all transverse frames, and let

$$\theta_T^1 \in C^\infty\big(T^*(B_T^1(\mathscr{F})) \otimes \mathbb{R}^q\big)$$

be the canonical 1-form. As customary in the theory of Riemannian foliations, we endow $B_T^1 = B_T^1(\mathscr{F})$ with the lifted foliation $\mathscr{F}_T^1$, i.e., the foliation tangent to

$$P_T^1 = \left\{X \in T\big(B_T^1\big) \; : \; X \rfloor \theta_T^1 = 0, \quad X \rfloor d\,\theta_T^1 = 0\right\}.$$

The leaves of $\mathscr{F}_T^1$ are Galois coverings of the leaves of $\mathscr{F}$, i.e.,

i) P_T^1 is invariant by right translations,
ii) $P_T^1 \cap \mathrm{Ker}(dp_T^1) = (0)$,
iii) $(dp_T^1)P_T^1 = (p_T^1)^{-1}\, T(\mathscr{F})$.
Also $(B_T^1,\, p_T^1,\, M,\, \mathrm{GL}(q,\, \mathbb{R}),\, \mathscr{F}_T^1)$ is a foliated principal bundle in the sense of P. Molino, [Mol].
Let

$$\mathscr{I}^r(M,\, \mathscr{F}) \subset C^\infty\big(\Lambda^r T^*\big(B_T^1(\mathscr{F})\big) \otimes \mathbb{R}^q\big), \quad \mathscr{I}_B^r(M,\, \mathscr{F}) \subset \Omega_B^r(\mathscr{F}_T^1) \otimes \mathbb{R}^q,$$

be the module of all $\mathbb{R}^q$-valued tensorial r-forms, respectively, tensorial basic r-forms, of type $(\rho,\, \mathbb{R}^q)$ on $B_T^1(\mathscr{F})$. Here ρ is the natural representation of $\mathrm{GL}(q,\, \mathbb{R})$ on $\mathbb{R}^q$. Cf. also [Ton], p. 59. Then $\theta_T^1 \in \mathscr{I}_B^1(M, \mathscr{F})$. We shall make use of the $\Omega^0(M)$-modules

$$L^r(M,\, \mathscr{F}) = \mathscr{I}^r(M,\, \mathscr{F}) \times \Omega^r(M), \quad 0 \le r \le n.$$

Let $P^\perp$ be the g-orthogonal complement of $P = T(\mathscr{F})$. Let ∇ be the canonical adapted connection of $(M,\, \mathscr{F},\, g)$. Let

$$\omega \in C^\infty\Big(T^*\big(B_T^1(\mathscr{F})\big) \otimes \mathfrak{gl}(q,\, \mathbb{R})\Big)$$

be the connection 1-form associated to ∇. Then ω is adapted, i.e., $T(\mathscr{F}_T^1) \subset \Gamma = \mathrm{Ker}(\Omega)$. The canonical adapted connection is symmetric, i.e., the torsion $\Sigma_T^1 = d\theta_T^1 + \omega \wedge \theta_T^1$ vanishes. Next, let us consider the $\Omega^0(M)$-modules

$$L^r(M,\, \mathscr{F},\, \Gamma) = \big\{(\varphi,\, \alpha) \in L^r(M,\, \mathscr{F}) \;:\; DD\varphi = 0\big\}, \quad 0 \le r \le n.$$

Here $D\varphi$ is the covariant derivative of φ with respect to Γ. $L^r(M,\, \mathscr{F},\, \Gamma)$ is a submodule of $L^r(M,\, \mathscr{F})$.

To each local foliated chart $(U,\, x^\alpha,\, y^i) \in \mathscr{F}$ one may associate the section

$$s_U : U \to B_T^1(\mathscr{F}), \quad s_U(x) : \mathbb{R}^q \to Q_x, \quad x \in U,$$

$$s_U(x)\, e_i = \pi_x \left(\frac{\partial}{\partial y^i}\right)_x, \quad 1 \le i \le q.$$

Here $\{e_i \;:\; 1 \le i \le q\} \subset \mathbb{R}^q$ is the canonical basis. s_U is foliate, i.e.,

$$(d\, s_U)\, T(\mathscr{F}) = s_U^{-1}\, T(\mathscr{F}_T^1).$$

For every tensorial r-form φ on $B_T^1(\mathscr{F})$

$$(D\varphi)_U = d\,\varphi_U + \omega_U \wedge \varphi_U\,, \tag{4.101}$$

$$\omega_U := s_U^* \omega, \quad \varphi_U := s_U^* \varphi, \quad (D\varphi)_U = s_U^* D\varphi.$$

Hence, $\Sigma_T^1 = D\theta_T^1$. As another consequence of (4.101) covariant derivatives with respect to Γ preserve both tensorial type and basic forms. In particular $\mathscr{I}_B^r(M, \mathscr{F})$ is a $\Omega_B^0(\mathscr{F})$-module.

Theorem 4.77 i) *For any* $\psi \in \Omega^{r-1}(M)$ *and any* $\alpha \in \Omega^r(M)$

$$\left(\theta_T^1 \wedge (p_T^1)^* \psi, \alpha\right) \in L^r(M, \mathscr{F}, \Gamma), \quad 1 \le r \le n. \tag{4.102}$$

Let us set

$$D^r(\varphi, \alpha) = \left(D\varphi - \theta_T^1 \wedge (p_T^1)^* \alpha, d\alpha\right), \quad 0 \le r \le n, \tag{4.103}$$

for any $(\varphi, \alpha) \in L^r(M, \mathscr{F}, \Gamma)$. *Then* ii)

$$D^r L^r(M, \mathscr{F}, \Gamma) \subset L^{r+1}(M, \mathscr{F}, \Gamma), \quad D^{r+1} \circ D^r = 0,$$

i.e.,

$$L^\bullet(M, \mathscr{F}, \Gamma) = \left(\bigoplus_{r=0}^n L^r(M, \mathscr{F}, \Gamma), D^\bullet\right)$$

is a cochain complex. Moreover, let us set

$$L_B^r(M, \mathscr{F}) = \mathscr{I}_B^r(M, \mathscr{F}) \times \Omega_B^r(\mathscr{F}),$$

$$L_B^r(M, \mathscr{F}, \Gamma) = L_B^r(M, \mathscr{F}) \cap L^r(M, \mathscr{F}, \Gamma), \quad 0 \le r \le q.$$

Then iii) $L_B^r(M, \mathscr{F}, \Gamma) = L_B^r(M, \mathscr{F})$ *for* $r \in \{q-1, q\}$. *Finally,* iv)

$$D^r L_B^r\left(M, \mathscr{F}, \Gamma\right) \subset L_B^{r+1}\left(M, \mathscr{F}, \Gamma\right)$$

hence

$$L_B^\bullet(M, \mathscr{F}, \Gamma) = \left(\bigoplus_{r=0}^n L_B^r(M, \mathscr{F}, \Gamma), D_B^\bullet\right)$$

is a cochain subcomplex of $L^\bullet(M, \mathscr{F}, \Gamma)$.

Here D_B^r is the restriction of D^r to $L_B^r(M, \mathscr{F}, \Gamma)$.

Definition 4.78 $L^\bullet(M, \mathscr{F}, \Gamma)$ is the *Vaisman complex*, while $L_B^\bullet(M, \mathscr{F}, \Gamma)$ is the *basic Vaisman complex*. □

Definition 4.79 The cohomology groups

$$H^r(M, \mathscr{F}, \Gamma) = H^r\left(L^\bullet(M, \mathscr{F}, \Gamma)\right), \quad 0 \le r \le n,$$

$$H_B^r(M, \mathcal{F}, \Gamma) = H^r\big(L_B^\bullet(M, \mathcal{F}, \Gamma)\big), \quad 0 \le r \le q,$$

are, respectively, the *curvature groups* and the *basic curvature groups* of $(M, \mathcal{F}, g)$. □

The construction of the (basic) curvature groups generalizes easily to arbitrary symmetric adapted connections in $\mathrm{GL}(q, \mathbb{R}) \to B_T^1(\mathcal{F}) \to M$. When $\mathcal{F}$ is the trivial foliation by points ($q = n$) and Γ is the Levi-Civita connection of (M, g), the basic Vaisman complex is the complex $\tilde{L}^\bullet$ build by S.I. Goldberg and N.C. Petridis (cf. [GoPe], pp. 550–551) while $H^r(M, \mathcal{F}, \Gamma)$ is its cohomology $H^r\big(\tilde{L}^\bullet\big)$.

Only (4.102) in Theorem 4.77 needs to be justified. For any $\varphi \in \mathcal{I}^r(M, \mathcal{F})$ we set $\varphi = \varphi^i \otimes e_i$. Then

$$\big(D(\varphi \wedge (p_T^1)^* \psi)\big)^i = (D\varphi^i) \wedge (p_T^1)^* \psi + (-1)^r \varphi^i \wedge (p_T^1)^* \, d\psi \tag{4.104}$$

for any scalar r-form $\psi \in \Omega^r(M)$. Now (4.102) follows from (4.104) together with $\Sigma_T^1 = 0$.

To understand the statement of Theorem 4.75 one needs a bit of notation, as follows. Let $C_B^\infty(Q)$ denote the space of all $s \in C^\infty(Q)$ such that $\mathring{\nabla}_X s = 0$ for any $X \in \mathfrak{X}_{\mathcal{F}}$. Let $V(\mathcal{F})$ denote the module of all infinitesimal automorphisms of $(M, \mathcal{F})$, i.e., the set of all $Y \in \mathfrak{X}(M)$ such that $[X, Y] \in \mathfrak{X}(M)$ for any $X \in \mathfrak{X}_{\mathcal{F}}$. The map $Y \in V(\mathcal{F}) \mapsto \pi\, Y \in C_B^\infty(Q)$ is an isomorphism of $\Omega_B^0(\mathcal{F})$-modules. Moreover, if $U \subset M$ is an open subset, let $\mathcal{F}_U$ be the foliation induced on U by $\mathcal{F}$, and let $U_T := (p_T^1)^{-1}(U) \subset B_T^1(\mathcal{F})$. Also $j : \mathcal{S} \to \mathcal{L}_B^0$ is the sheaf morphism built as follows. Let $U \subset M$ be an open set, and let us set

$$j_U(s) = \Big(f_s\,,\ \frac{1}{q}\,\mathrm{div}_B(s)\Big), \quad s \in C_B^\infty(U, Q), \tag{4.105}$$

$$f_s : U_T \to \mathbb{R}, \quad f_s(z) = z^{-1}\big(s(x)\big), \quad z \in U_T\,, \quad x = p_T^1(z) \in U.$$

Cf. (9.24) in [Ton], p. 126, for the basic divergence operator $\mathbf{div}_B$ (the needed details are also recalled in the next section).

Lemma 4.80 *f_s is a $\mathbb{R}^q$-valued tensorial 0-form of type $(\rho, \mathbb{R}^q)$, and f_s is basic with respect to the lifted foliation $\mathcal{F}_{T, U_T}^1$ induced by $\mathcal{F}_T^1$ on U_T, i.e., $f_s \in \mathcal{I}_B^0(U_T\,, \mathcal{F}_U)$.*

Cf. Lemma 2.3 in [DrPe], p. 6.

Proof of Lemma 4.80. The fact that

$$f_s \circ R_a = \rho(a^{-1})\, f_s\,, \quad a \in \mathrm{GL}(q, \mathbb{R}), \tag{4.106}$$

is both elementary and follows within the general theory of associated bundles (cf. e.g., R. Crittenden, [Cri]). To check that f_s is basic, let $Y \in V(\mathcal{F})$ be the unique infinitesimal automorphism of $(M, \mathcal{F})$ such that $\pi\, Y = s$. Then $f_s = \theta_T^1\big(Y^\uparrow\big)$, where

$Y^\uparrow \in \mathfrak{X}\big(B^1_T(\mathscr{F})\big)$ is the horizontal lift of Y with respect to Γ. Let $\tilde{X} \in P^1_T \subset \Gamma$. There is a unique $X \in \mathfrak{X}_{\mathscr{F}}$ such that $X^\uparrow = \tilde{X}$. By a classical result [in the theory of connections on principal bundles, cf. e.g., Proposition 1.3 in [KoNo], Vol. I, p. 65] $\big[X^\uparrow, Y^\uparrow\big] - [X, Y]^\uparrow \in \operatorname{Ker}(dp^1_T)$ hence

$$\tilde{X}(f_s) = 2\,(d\theta^1_T)(\tilde{X}, Y^\uparrow) + \theta^1_T\big([\tilde{X}, Y^\uparrow]\big) = 0.$$

Q.e.d. Next, if $\nabla s = c\,\pi$, then $c = (1/q)\operatorname{div}_B(s)$ so that $j_U(s)$ is well defined. To prove Theorem 4.75 one only needs to check exactness at

$$\big(\mathscr{L}^0_B\big)_x = \lim_{\substack{\longrightarrow \\ U \ni x}} \mathscr{L}^0_B(U)$$

in the sequence of stalks

$$0 \to \big(\mathscr{L}^0_B\big)_x \to \big(\mathscr{L}^1_B\big)_x, \quad x \in M.$$

Let $\ell \in \operatorname{Ker}\big[\big(D^0_B\big)_x\big]$ and let $(f, \tilde{u}) \in \mathscr{L}^0_B(U)$ be a representative of ℓ defined on the simply connected domain of a local foliated chart $(U, x^\alpha, y^i) \in \mathscr{F}$. Here $\tilde{u} = u \circ p^1_T$. Then $\big(Df - \tilde{u}\,\theta^1_T, d_B u\big)$ is a representative of the null germ, so that $u = c$ for some $c \in \mathbb{R}$ and $Df = c\,\theta^1_T$ on U_T. Let $s \in C^\infty\big(U, Q\big)$ be the section

$$s(x) = z\big(f(z)\big), \quad x \in U,$$

where $z \in (p^1_T)^{-1}(x)$ is an arbitrary transverse frame at x. By (4.106), it follows that $s(x)$ is well defined, and $s : U \to Q$ is the unique section in the transverse bundle such that $f_s = f$. Let Γ^i_{jk} be the local coefficients of the canonical adapted connection, i.e.,

$$\nabla_{\partial/\partial y^j}\,\pi\left(\frac{\partial}{\partial y^k}\right) = \Gamma^i_{jk}\,\pi\left(\frac{\partial}{\partial y^i}\right).$$

Let $(U_T, x^\alpha, y^i, X^i_j)$ are the naturally induced local coordinates on $B^1_T(\mathscr{F})$, and let us set

$$Y_k = \frac{\partial}{\partial y^k} - \big(\Gamma^i_{k\ell} \circ p^1_T\big)\,X^\ell_j\,\frac{\partial}{\partial X^i_j}, \quad 1 \le k \le q,$$

so that

$$\left\{\frac{\partial}{\partial x^\alpha}, \; Y_i \,:\, 1 \le \alpha \le n-q, \; 1 \le i \le q\right\} \subset C^\infty\big(U_T, \Gamma\big)$$

is a local frame in Γ. Then $Df = c\,\theta^1_T$ evaluated on $\partial/\partial x^\alpha$ is actually equivalent to f being a basic function. We may locally write $s = s^i\,\pi\,(\partial/\partial y^i)$ on U. For arbitrary $x \in U$ we may pick up the transverse frame $z_0(e_i) = \pi_x\,\big(\partial/\partial y^i\big)_x$, so that

$$X^i_j(z_0) = \delta^i_j\,, \quad s^i(x) = f^i(z_0)\,\pi_x\,(\partial/\partial y^i)_x\,, \quad f = f^i e_i\,.$$

Evaluation of $(Df)_{z_0} = c\,\theta^1_{T,\,z_0}$ on $Y_{k,\,z_0}$ is then equivalent to $\nabla s = c\,\pi$.

4.9.2 Infinitesimal Conformal Transformations

Let $\mathscr{C}(Q,\,g_Q)$ be the affine space of all connections $D : \Omega^0(Q) \to \Omega^1(Q)$ parallelizing g_Q, i.e., $Dg_Q = 0$. If ∇ is the canonical adapted connection, then $\nabla \in \mathscr{C}(Q,\,g_Q)$. Let R^∇ be the curvature tensor field of ∇. Let $\{s_i : 1 \le i \le q\} \subset C^\infty(U,\,Q)$ be a (local) g_Q-orthonormal frame, and let us set

$$S^\nabla(r,\,s) = \sum_{i=1}^{q} g_Q\big(R^\nabla\big(\sigma(r),\,\sigma(s_i)\big)s,\,s\big), \quad r,\,s \in C^\infty(Q) \tag{4.107}$$

on U.

Definition 4.81 The vector bundle morphism

$$\rho^\nabla : Q \to Q, \quad g_Q\big(\rho^Q r,\,s\big) = S^\nabla(r,\,s),$$

is the *transverse Ricci operator* and $c^\nabla = \mathrm{Trace}(\rho^\nabla)$ is the *transverse scalar curvature*. The pair $(\mathscr{F},\,g_Q)$ is an *Einstein foliation* if $S^\nabla = (c^\nabla/q)\,g_Q$. □

Cf. [KaTo7], pp. 527–528. The transverse orientation of $\mathscr{F}$ together with g_Q give rise to a transversal volume form $\nu \in \Omega^q_B(\mathscr{F})$ (such that $d\nu = 0$). The volume form along the leaves is $\chi_{\mathscr{F}} = *\,\nu$ (the characteristic form) where $* : \Omega^q(M) \to \Omega^{n-q}(M)$ is the Hodge star operator associated to g. For any infinitesimal automorphism $Y \in V(\mathscr{F})$, its basic divergence $\mathrm{div}_B(Y) \in \Omega^0_B(\mathscr{F})$ is given by

$$\mathscr{L}_Y \nu = \mathrm{div}_B(Y)\,\nu\,,$$

where $\mathscr{L}_Y$ is the Lie derivative. For any transversal vector field $s = \pi(Y) \in C^\infty_B(Q)$, the basic divergence of Y depends only on s, and one often writes $\mathrm{div}_B(s)$ [instead of $\mathrm{div}_B(Y)$]. When $\mathscr{F}$ is harmonic, the basic divergence theorem is

$$\int_M \mathrm{div}_B\big(\pi\,Y\big)\,\nu \wedge \chi_{\mathscr{F}} = 0$$

for any $Y \in V(\mathscr{F})$ of compact support (cf. e.g., Theorem C in [KTT], p. 262). A version of the basis divergence theorem [containing explicitly the mean curvature form $\kappa \in \Omega^1(M)$] is also available for the case where $\mathscr{F}$ is not necessarily harmonic (cf. e.g., (5) in [BaDr], p. 124). Let us set

$$(\delta^*\eta)(r, s) = \frac{1}{2}\Big\{\big(\nabla_{\sigma(r)}\eta\big)\, s + \big(\nabla_{\sigma(s)}\eta\big)\, r\Big\}, \quad \eta \in C^\infty(Q^*), \quad r, s \in C^\infty(Q).$$

Let $C^\infty_B(Q^*)$ consist of all $\eta \in C^\infty(Q^*)$ such that $\mathscr{L}_X \eta = 0$ for any $X \in \mathfrak{X}_{\mathscr{F}}$ (the Lie derivative is once again defined in terms of the Bott connection $\overset{\circ}{\nabla}$).

Definition 4.82 An infinitesimal automorphism $Y \in V(\mathscr{F})$ is said to be *transversally conformal* if

$$\delta^* \eta = -\frac{1}{q}\,(\delta_B\, \alpha)\, g_Q\,. \tag{4.108}$$

□

Here $\eta \in C^\infty_B(Q^*)$ is the g_Q-dual of $\pi\, Y$, i.e.,

$$\eta(r) = g_Q\big(\pi\, Y,\, r\big), \quad r \in C^\infty(Q).$$

Also $\alpha := \eta \circ \pi$ and $\delta_B : \Omega^1_B(\mathscr{F}) \to \Omega^0_B(\mathscr{F})$ is the formal adjoint of $d_B : \Omega^0_B(\mathscr{F}) \to \Omega^1_B(\mathscr{F})$.

Lemma 4.83 *Let $f \in C^\infty(M)$ and let $s \in C^\infty_B(Q)$ be a basic solution to $\nabla s = f\,\pi$. Then*

i) $f \in \Omega^0_B(\mathscr{F})$,

ii) $Y := \sigma(s) \in V(\mathscr{F})$ *is a transversally conformal infinitesimal automorphism.*

Proof Let $(U, x^\alpha, y^i) \in \mathscr{F}$ be a foliate chart. Then

$$f = \frac{1}{q}\,(g_Q)^{ij}\, g_Q\Big(\nabla_{\partial/\partial y^i}\, s,\ \pi\,\Big(\frac{\partial}{\partial y^j}\Big)\Big) \in \Omega^0_B(\mathscr{F}_U).$$

On the other hand, for any $Z,\, W \in T(\mathscr{F})^\perp$

$$\mathscr{L}_Y \pi\, Z = \pi\, \nabla^g_Y Z - \nabla_Z s\,,$$

$$\big(\mathscr{L}_Y g_Q\big)(\pi\, Z,\, \pi\, W) = g_Q(\nabla_Z s,\, \pi\, W) + g_Q(\pi Z,\, \nabla_W s),$$

hence $\mathscr{L}_Y g_Q = 2\, f\, g_Q$, which is equivalent to (4.108). Q.e.d.

If $s = s^i\, \pi\, (\partial/\partial y^i)$ on U with $s^i \in \Omega^0_B(\mathscr{F}_U)$, then the integrability conditions for $\nabla s = f\,\pi$ are

$$s^k\,\big(R^\nabla\big)^\ell_{ijk} = \delta^\ell_j\, f_i - \delta^\ell_i\, f_j\,, \quad f_i := \frac{\partial f}{\partial y^i}\,. \tag{4.109}$$

Let us set

$$\big(S^\nabla\big)_{ij} = S^\nabla\big(\pi\, \partial_i\,,\, \pi\, \partial_j\big), \quad \partial_i \equiv \frac{\partial}{\partial y^i}\,.$$

Then [by (4.107)]

$$\big(S^\nabla\big)_{k\ell} = (g_Q)^{ij}\,\big(R^\nabla\big)^m_{kij}\,(g_Q)_{m\ell}\,. \tag{4.110}$$

Let us contract with $(g_Q)_{m\ell}$ in (4.109)

$$s^k \left(R^\nabla\right)^\ell_{ijk} (g_Q)_{m\ell} = (g_Q)_{mj}\, f_i - (g_Q)_{mi}\, f_j$$

and observe that the left-hand side is skew-symmetric in k and m (because of $\nabla g_Q = 0$). Contraction with $(g_Q)^{jm}$ then leads [by (4.110)] to

$$s^j \left(S^\nabla\right)_{ij} = -(q-1)\, f_i\,. \tag{4.111}$$

Next we may substitute from (4.111) into (4.109) so that to get

$$s^k \left(W^D\right)^\ell_{ijk} = 0. \tag{4.112}$$

Here $\left(W^D\right)^\ell_{ijk}$ are the components of the *transverse Weyl* tensor field

$$W^\nabla(Y,\, Z)r = R^\nabla(Y,\, Z)r - \frac{1}{q-1}\left\{S^\nabla(\pi Z,\, r)\,\pi Y - S^\nabla(\pi Y,\, r)\,\pi Z\right\}$$

with respect to $(U,\, x^\alpha,\, y^i)$. Therefore, (4.112) are necessary conditions for the existence of a solution to $\nabla s = f\,\pi$. The conditions are satisfied when $(\mathscr{F},\, g_Q)$ is transversally projectively flat, i.e., $W^\nabla = 0$. In particular, if $f = c$ with $c \in \mathbb{R}$, then (4.109) becomes

$$s^k \left(R^\nabla\right)^\ell_{ijk} = 0, \tag{4.113}$$

which is satisfied iff $(\mathscr{F},\, g_Q)$ is transversally Ricci flat [by (4.111)]. If the transverse Ricci tensor S^∇ is non-degenerate at every point of M, then the equation $\nabla s = c\,\pi$ has no solutions $s \in C^\infty_B(Q)$.

4.9.3 The Relationship to Basic Cohomology

Let $\Sigma^r(M,\, \mathscr{F})$ be the $\Omega^0(M)$-module of all differential r-forms $\theta^1_T \wedge (p^1_T)^*\psi$ with $\psi \in \Omega^{r-1}(M)$. Also, let us set

$$\Sigma^r_B(M,\, \mathscr{F}) = \left\{\theta^1_T \wedge (p^1_T)^*\psi \;:\; \psi \in \Omega^{r-1}_B(\mathscr{F})\right\} \subset \Omega^r_B(\mathscr{F}^1_T).$$

Lemma 4.84 i) *One has*

$$D\,\Sigma^r(M,\, \mathscr{F}) \subset \Sigma^{r+1}(M,\, \mathscr{F}), \quad D\,\Sigma^r_B(M,\, \mathscr{F}) \subset \Sigma^{r+1}_B(M,\, \mathscr{F}). \tag{4.114}$$

Let us set $\nabla^r \equiv D\big|_{\Sigma^r(M,\, \mathscr{F})}$ *and* $\nabla^r_B \equiv \nabla^r\big|_{\Sigma^r_B(M,\, \mathscr{F})}$. *Then* ii)

$$\Sigma^\bullet(M,\mathscr{F}) = \Big(\bigoplus_{r=0}^{n+1} \Sigma^r(M,\mathscr{F}),\ \nabla^\bullet\Big), \tag{4.115}$$

$$\Sigma_B^\bullet(M,\mathscr{F}) = \Big(\bigoplus_{r=0}^{n+1} \Sigma_B^r(M,\mathscr{F}),\ \nabla_B^\bullet\Big), \tag{4.116}$$

are cochain complexes. Let

$$H^1(\Sigma) = \mathrm{Ker}\big(\nabla^1\big), \quad H^1(\Sigma_B) = \mathrm{Ker}(\nabla_B^1),$$

$$H^r(\Sigma) = \frac{\mathrm{Ker}(\nabla^r)}{\nabla^{r-1}\,\Sigma^{r-1}(M,\mathscr{F})}, \quad H^r(\Sigma_B) = \frac{\mathrm{Ker}(\nabla_B^r)}{\nabla_B^{r-1}\,\Sigma_B^{r-1}(M,\mathscr{F})}, \quad r \geq 2,$$

be the cohomology groups of the complexes (4.115)–(4.116). *Then* iii) *the natural map $H^2(\Sigma_B) \to H^2(\Sigma)$ is a group monomorphism.*

Proof As Γ is symmetric

$$D\big(\theta_T^1 \wedge (p_T^1)^*\,\psi\big) = -\theta_T^1 \wedge (p_T^1)^*\,(d\psi)$$

[thus yielding (4.114)] and $D^{r+1} \circ D^r = 0$. To check the last statement in Lemma 4.84, let $[\omega]_B$ be the class of $\omega \in \mathrm{Ker}\big(\nabla_B^2\big)$ in $H^2(\Sigma_B)$, and let $[\omega]$ be the corresponding class in $H^2(\Sigma)$. If $[\omega] = 0$, then $\omega = -\theta_T^1 \wedge d(f \circ p_T^1)$ for some $f \in \Omega^0(M)$. On the other hand, $\omega \in \Omega_B^2(\mathscr{F}_T^1)$. Therefore, for any $X \in \mathfrak{X}_{\mathscr{F}}$

$$0 = X^\uparrow \rfloor\, \omega = -\frac{1}{2}\Big\{\theta_T^1(X^\uparrow)\,(p_T^1)^*\,df - \big[X(f)\circ p_T^1\big]\,\theta_T^1\Big\}$$

hence $f \in \Omega_B^0(\mathscr{F})$ (yielding $[\omega]_B = 0$). Q.e.d.

Lemma 4.85 *The basic cohomology groups $H_B^r(\mathscr{F})$ of a Riemannian foliation $\mathscr{F}$ with* $\mathrm{cod}(\mathscr{F}) = q \geq 2$ *are isomorphic to the groups $H^{r+1}(\Sigma_B)$ with $r \in \{0, 1, \cdots, q\}$.*

Proof Let $\mathscr{M}^r : \Omega^r(M) \to \Sigma^{r+1}(M,\mathscr{F})$ be exterior multiplication with θ_T^1, i.e.,

$$\mathscr{M}^r\psi = \theta_T^1 \wedge \big(p_T^1\big)^*\,\psi, \quad \psi \in \Omega^r(M).$$

As $q \geq 2$

$$\mathrm{Ker}\big(\mathscr{M}^r\big) = \big\{\psi \in \Omega^r(M)\ :\ \sigma(s)\rfloor\,\psi = 0,\ \ s \in C^\infty(Q)\big\}. \tag{4.117}$$

Let $\mathscr{M}_B^r := \mathscr{M}^r\big|_{\Omega_B^r(\mathscr{F})}$. Then [by (4.117)] $\mathrm{Ker}\big(\mathscr{M}_B^r\big) = (0)$ so that $\mathscr{M}_B^r$ is an isomorphism of modules $\Omega_B^r(\mathscr{F}) \simeq \Sigma_B^{r+1}(M,\mathscr{F})$. Also $\mathscr{M}_B^r$ maps d_B-cycles into ∇_B^{r+1}-cycles, and d_B-boundaries into ∇_B^r-boundaries. Q.e.d.

Let $\mathfrak{S}^r_B$ be the sheaf associated to the module $\Sigma^r_B(M, \mathscr{F})$, i.e.,

$$\mathfrak{S}^r_B(U) = \left\{\theta^1_T \wedge (p^1_T)^*\psi \; : \; \psi \in \Omega^{r-1}_B(\mathscr{F}_U)\right\}$$

for any open set $U \subset M$.

Theorem 4.86 *Let us assume that the equation $\nabla s = c\,\pi$ admits local basic solutions for every $c \in \mathbb{R} \setminus \{0\}$, and admits no nontrivial basic solution for $c = 0$. Let $\nabla : \mathscr{S} \to \mathfrak{S}^1_B$ be the sheaf morphism*

$$\nabla^U : \mathscr{S}(U) \to \mathfrak{S}^1_B(U), \quad \nabla^U s = \frac{1}{q}\operatorname{div}_B(s)\,\theta^1_T, \quad s \in \mathscr{S}(U), \;\; U \subset M.$$

Then the sheaf sequence

$$0 \to \mathscr{S} \xrightarrow{\nabla} \mathfrak{S}^1_B \xrightarrow{\nabla^1_B} \mathfrak{S}^2_B \xrightarrow{\nabla^2_B} \mathfrak{S}^3_B \tag{4.118}$$

is exact at $\mathscr{S}$ and $\mathscr{S}^r_B$, $r \in \{1, 2\}$. If additionally the basic complex (4.100) of $(M, \mathscr{F})$ satisfies the Poicaré lemma, then

$$0 \to \mathscr{S} \xrightarrow{\nabla} \mathfrak{S}^1_B \xrightarrow{\nabla^1_B} \cdots \xrightarrow{\nabla^q_B} \mathfrak{S}^{q+1}_B \xrightarrow{\nabla^{q+1}_B} 0 \tag{4.119}$$

is exact. In particular $0 \to \mathscr{S} \to \mathfrak{S}^\bullet_B$ is a fine resolution of $\mathscr{S}$ so that $H^{r+1}(\Sigma_B) \simeq H^r(M, \mathscr{S})$ (a group isomorphism) for any $0 \le r \le q$. Moreover, if $q \ge 2$ and the basic Vaisman complex of $(M, \mathscr{F}, g)$ satisfies the Poincaré lemma, then the basic curvature groups are isomorphic to the basic cohomology groups of $(M, \mathscr{F})$.

Proof Let $x \in M$ and $\ell \in \operatorname{Ker}(\nabla_x)$. If $s \in \mathscr{S}(U)$ is a representative of ℓ, then $\operatorname{div}_B(s) = 0$ on U. Yet $\nabla s = c\,\pi$ for some $c \in \mathbb{R}$ yields

$$c = \frac{1}{q}\operatorname{div}_B(s)$$

hence $\nabla s = 0$ on U and then $s = 0$, proving exactness at $\mathscr{S}_x$. Next, let us consider the sequence of stalks

$$\mathscr{S}_x \to \left(\mathfrak{S}^1_B\right)_x \to \left(\mathfrak{S}^2_B\right)_x$$

and let us check exactness at the term $\left(\mathfrak{S}^1_B\right)_x$. Let $\ell \in \operatorname{Ker}\left[\left(\nabla^1_B\right)_x\right]$. Then there is a connected open neighborhood $U \subset M$ of x and a basic function $f \in \Omega^0_B(\mathscr{F}_U)$ such that $(f \circ p^1_T)\,\theta^1_T$ is a representative of ℓ and

$$0 = \nabla^1_B\left((f \circ p^1_T)\,\theta^1_T\right) = -\mathscr{M}^2_B(d_B f)$$

on U. Therefore, $d_B f = 0$, i.e., $f = c$ for some $c \in \mathbb{R}$. Again by the assumption in Theorem 4.86 there is a neighborhood $V \subset M$ of x and a basic section

$s \in C^\infty(V, Q)$ such that $\nabla s = c\pi$. Then $c = (1/q)\,\mathrm{div}_B(s)$ so that $\nabla^V s$ and $c\,\theta_T^1$ agree on $(p_T^1)^{-1}(W)$ for some open subset $W \subset M$ such that $x \in W \subset U \cap V$. We may conclude that $\nabla^W s$ is a representative of ℓ, i.e., $\ell \in \nabla_x\big(\mathcal{S}_x\big)$. Conversely, if $\ell \in \nabla_x\big(\mathcal{S}_x\big)$, then ℓ admits a representative of the form $\nabla^U s$ for some $s \in \mathcal{S}(U)$. Finally, $\nabla_x^1 \ell = 0$ because $\mathrm{div}_B(s) \in \mathbb{R}$ and then

$$\nabla^1\big(\nabla^U s\big) = \frac{1}{q}\,\mathrm{div}_B(s)\,\Sigma_T^1 = 0.$$

Moreover, exactness at $\big(\mathfrak{S}_B^2\big)_x$ follows from the existence of a canonical injection $H_B^1(\mathcal{F}) \to H^1(M, \mathbb{R})$. To end the proof of Theorem 4.86 one needs to check the exactness at $\big(\mathfrak{S}_B^r\big)_x$ in the sequence

$$\big(\mathfrak{S}_B^{r-1}\big)_x \to \big(\mathfrak{S}_B^r\big)_x \to \big(\mathfrak{S}_B^{r+1}\big)_x, \quad r \geq 3, \quad x \in M.$$

Let $\ell \in \mathrm{Ker}\big(\nabla_x^r\big)$. There is an open set $U \subset M$ such that $x \in U$ and there is a basic form $\psi \in \Omega_B^{r-1}(\mathcal{F}_U)$ such that $\theta_T^1 \wedge (p_T^1)^*\psi$ is a representative of ℓ and

$$\nabla^r\big(\theta_T^1 \wedge (p_T^1)^*\psi\big) = 0$$

on U. Hence, $d_B\psi = 0$ so that [by the Poincaré Lemma for the complex (4.100)] there is an open set $V \subset U$ with $x \in V$, and there is $\alpha \in \Omega_B^{r-2}\big(\mathcal{F}_V\big)$ such that $\psi = d_B\alpha$. Thus, $-\nabla^{r-1}\big(\theta_T^1 \wedge (p_T^1)^*\alpha\big)$ is a representative of ℓ, so that $\ell \in \nabla_x^{r-1}\big[\big(\mathfrak{S}_B^{r-1}\big)_x\big]$. Finally, we may apply Lemma 4.85 and Theorem 4.75 to conclude that $H_B^r(\mathcal{F}) \simeq H_B^r(M, \mathcal{F}, \Gamma)$. Q.e.d.

Example 4.87 (*Curvature groups of Riemannian foliations by level hypersurfaces*) Let $r : V^{n+1} \to \mathbb{R}$ be a C^∞ function on a Riemannian manifold (V^{n+1}, g). Removing the set of critical points of r one obtains a foliation $\mathcal{F}$ of $M = V^{n+1} \setminus \mathrm{Crit}(r)$ by level hypersurfaces of r. Let us assume that $\mathcal{F}$ is Riemannian and g is bundle-like. In particular, if

$$Z := \frac{1}{|\nabla r|}\,\nabla r,$$

then $[X, Z] \in \mathfrak{X}_{\mathcal{F}}$ for every $X \in \mathfrak{X}_{\mathcal{F}}$. Also $\nabla_Y^g Z \in \mathfrak{X}_{\mathcal{F}}$ for any $Y \in \mathfrak{X}(M)$. Thus,

$$\nabla_Y s = Y(f)\,\pi\,Z$$

for any C^∞ section $s = f\,\pi\,Z$ in Q. If s is a basic solution to $\nabla s = c\pi$ for some $c \in \mathbb{R}$, then $f \in \Omega_B^0(\mathcal{F})$ and $Z(f) = c$. Let $(U, x^1, \cdots, x^{n-1}, t)$ be a local coordinate system on M such that $r(x, t) = t - \phi(x)$ on U. Then $\mathfrak{X}_{\mathcal{F}_U}$ is the span of

$$X_i \equiv \frac{\partial}{\partial x^i} + \phi_{x^i}\,\frac{\partial}{\partial t}, \quad 1 \leq i \leq n-1,$$

hence

$$d\,f = \frac{c}{|\nabla r|}\left\{d\,t - \phi_{x^i}\,d\,x^i\right\} \tag{4.120}$$

or $\nabla f = c\,Z$. Consequently, the equation $\nabla s = c\,\pi$ admits local basic solutions for each $c \neq 0$, while $f = 0$ when $c = 0$. We may state

Corollary 4.88 *Let $\mathscr{F}$ be the foliation of $M = V^{n+1} \setminus \mathrm{Crit}(r)$ by level hypersurfaces of $r \in C^\infty(V^{n+1})$. If $\mathscr{F}$ is Riemannian and g is bundle-like, then the basic Vaisman complex of $(M, \mathscr{F}, g)$ satisfies the Poincaré lemma. Consequently,*

$$H^1_B(M, \mathscr{F}, \Gamma) \simeq H^1(M, \mathscr{S}) \simeq H^2(\Sigma_B). \tag{4.121}$$

Proof As $q = 1$ the basic Vaisman complex is

$$L^0_B(M, \mathscr{F}, \Gamma) \xrightarrow{D^0_B} L^1_B(M, \mathscr{F}, \Gamma) \xrightarrow{D^1_B} 0$$

and

$$L^s_B(M, \mathscr{F}, \Gamma) = L^s_B(M, \mathscr{F}), \quad s \in \{0, 1\}.$$

Let $\varphi \in \mathscr{I}^1_B(M, \mathscr{F})$ and $\alpha \in \Omega^1_B(\mathscr{F})$ such that

$$D\varphi - \theta^1_T \wedge (p^1_T)^*\alpha = 0, \quad d\alpha = 0.$$

There is a natural injection $H^1_B(\mathscr{F}) \to H^1(M, \mathbb{R})$ hence the basic complex

$$\Omega^0_B(\mathscr{F}) \xrightarrow{d_B} \Omega^1_B(\mathscr{F}) \xrightarrow{d_B} 0$$

satisfies the Poincaré lemma. Thus, for each $x_0 \in M$, there is an open neighborhood $U \subset M$ of x_0 such that $H^1_B(\mathscr{F}_U) = 0$. Let $f \in \Omega^0_B(\mathscr{F}_U)$ such that $\alpha = d_B f$. We wish to solve (locally) for $\psi \in \mathscr{I}^0_B(M, \mathscr{F})$ in the equation

$$\varphi = D\psi - \left((p^1_T)^* f\right)\theta^1_T\,. \tag{4.122}$$

Let $(U, x^\alpha, y) \in \mathscr{F}$ be a foliated chart and let $\left((p^1_T)^{-1}(U), \tilde{x}^\alpha, \tilde{y}, t\right)$ be the induced local coordinates on $B^1_T(\mathscr{F})$, i.e., $\tilde{x}^\alpha = x^\alpha \circ p^1_T$ and $\tilde{y} = y \circ p^1_T$ while $t : (p^1_T)^{-1}(U) \to \mathbb{R}$ (the fiber coordinate) is given by

$$z(1) = t(z)\left(\frac{\partial}{\partial y}\right)_x, \quad z \in (p^1_T)^{-1}(U), \quad x = p^1_T(z).$$

Let $\gamma \in \Omega^0_B(\mathscr{F}_U)$ be determined by

$$\nabla_{\partial/\partial y}\,\pi\left(\frac{\partial}{\partial y}\right) = \gamma\,\pi\left(\frac{\partial}{\partial y}\right)$$

(the Christoffel symbol of the adapted connection ∇). The horizontal distribution Γ in $\mathbb{R}^* \to B^1_T(\mathscr{F}) \to M$ [with $\mathbb{R}^* = \mathbb{R} \setminus \{0\}$] is locally the span of

$$\left\{\frac{\partial}{\partial \tilde{x}^\alpha}, \ \frac{\partial}{\partial \tilde{y}} - \left((p^1_T)^* \gamma\right) t \frac{\partial}{\partial t} \ : \ 1 \le \alpha \le n-1\right\}.$$

Hence, the canonical and connection 1-forms θ^1_T, $\omega \in C^\infty\big(T^*(B^1_T(\mathscr{F}))\big)$ (both scalar valued, as $q = 1$) admit the local expressions

$$\theta^1_T = \frac{1}{t}\, d\,\tilde{y}, \quad \omega = \frac{1}{t}\left\{d\,t + \left((p^1_T)^* \gamma\right) t\, d\,\tilde{y}\right\}.$$

The last ingredient one needs for solving (4.122) is the section $s_U \in C^\infty\big(U, \ B^1_T(\mathscr{F})\big)$ given by

$$s_U(x)\, 1 = \pi_x \left(\frac{\partial}{\partial y}\right)_x, \quad x \in U.$$

Indeed, as

$$s^*_U\, d\,\tilde{y} = d\,y, \quad s^*_U\, d\,t = 0,$$

equation (4.122) is equivalent to

$$\rho_U\, d\,y = (D\psi)_U - f\, d\,y \tag{4.123}$$

where $\psi = \rho\, d\,\tilde{y}$ for some $\rho \in \Omega^0_B(\mathscr{F}^1_T)$ and $\rho_U = \rho \circ s_U$. As s_U is a foliate section, the function ρ_U is basic for $\mathscr{F}_U$. Moreover [by (4.101) together with $t \circ s_U = 1$], Equation (4.123) becomes

$$\rho_U\, d\,y = d\,\psi_U + \big(\psi_U\, \gamma - f\big)\, d\,y. \tag{4.124}$$

Let us observe that

$$\gamma = \frac{1}{u}\, u_y, \quad u := t \circ \pi(Z) \in \Omega^0_B\big(\mathscr{F}_U\big).$$

Hence, (4.124) is equivalent to the ODE

$$\psi'_U + \gamma\, \psi_U = f + \rho_U$$

or

$$\left(\frac{\psi_U}{u}\right)' = \frac{1}{u}\,(f + \rho_U),$$

which may be solved for ψ_U. Then (4.121) follows from Theorem 4.75 and the first part of Theorem 4.86. Q.e.d. □

Example 4.89 (*Curvature groups of Killing flows*) Let X be a nonsingular Killing vector field on a closed Riemannian manifold (M, g), i.e., M is compact and $\partial M = \emptyset$. Let us assume that $|X| = 1$ and consider the foliation $\mathscr{F}$ of M by the orbits of X. If G is the closure in M of $\{\exp(tX) \; : \; t \in \mathbb{R}\}$ (the 1-parameter group got by integrating X), then G is a compact abelian group and $\Omega^\bullet_B(\mathscr{F}) \subset \Omega^\bullet(M)^G$ (the G-invariant differential forms on M). Also

$$0 \to \Omega^r_B(\mathscr{F}) \overset{i}{\hookrightarrow} \Omega^r(M)^G \xrightarrow{i_X} \Omega^{r-1}_B(\mathscr{F}) \to 0 \tag{4.125}$$

is a short exact sequence in all degrees r. We may state

Corollary 4.90 *Let $\chi_{\mathscr{F}} \in \Omega^1(M)$ be the characteristic form of $(M, \mathscr{F})$, i.e., $\chi_{\mathscr{F}}(Y) = g(X, Y)$ for any $Y \in \mathfrak{X}(M)$. Let*

$$\delta : H^{r-1}_B(\mathscr{F}) \to H^{r+1}_B(\mathscr{F}), \quad \delta\,[\omega] = \left[d\,\chi_{\mathscr{F}}\right] \cdot [\omega], \quad [\omega \in H^{r-1}_B(\mathscr{F}).$$

If $\delta = 0$, then the basic complex of $(M, \mathscr{F})$ satisfies the Poincaré lemma. Let $x_0 \in M$ and let us consider a system of vector fields $\{Z_i \; : \; 1 \leq i \leq n-1\} \subset \mathfrak{X}(U)$ defined on a neighborhood $U \subset M$ of x_0 such that $\{Z_{i,x} \; : \; 1 \leq i \leq n-1\}$ are linearly independent and $g(X, Z_i)_x = 0$ at each $x \in U$. Let us assume that
 i) *$T(\mathscr{F})^\perp$ is involutive,*
 ii) *for every $c \in \mathbb{R}^*$ the system of equations*

$$Z_i(f^j) + \gamma^j_{ik}\, f^k = c\,\delta^j_i\,, \tag{4.126}$$

admits a solution $(f^1, \cdots, f^{n-1}) \in \Omega^0_B(\mathscr{F})^{n-1}$ defined in some neigborhood $V \subset U$ of x_0 while the system (4.126) *with $c = 0$ admits only the trivial solution. Here $\gamma^i_{jk} \in \Omega^0_B(\mathscr{F}_U)$ are given by*

$$\nabla_{Z_j}\, \pi\, Z_k = \gamma^i_{jk}\, \pi\, Z_k\,, \quad 1 \leq i,\, j,\, k \leq n-1.$$

Then

$$H^{r+1}(\Sigma_B) \simeq H^r(M, S), \quad 0 \leq r \leq n-1.$$

Proof By (4.125) together with

$$H^r\big(\Omega^\bullet(M)^G\big) \simeq H^r(M, \mathbb{R})$$

there is a long exact sequence of cohomology groups

$$\cdots \to H^{r-2}_B(\mathscr{F}) \xrightarrow{\delta} H^r_B(\mathscr{F}) \xrightarrow{i_*} H^r(M, \mathbb{R}) \xrightarrow{(i_X)_*} H^{r-1}_B(\mathscr{F}) \xrightarrow{\delta} \cdots$$

with δ as a connecting homomorphism. Therefore, when $\delta = 0$, the homomorphism $i_* : H^r_B(\mathscr{F}) \to H^r(M, \mathbb{R})$ is injective and the Poincaré lemma for the de Rham complex (i.e., $H^r(U, \mathbb{R}) = 0$ for some neighborhood $U \subset M$ of x_0) yields the first statement in Corollary 4.90. For instance, if $T(\mathscr{F})^{\perp}$ is involutive, then (by Theorem 10.6 in [Ton], pp. 133–134) $d\,\chi_{\mathscr{F}} = 0$ is equivalent to the requirement that the orbits of X be geodesics of $(M,\ g)$ (which is of course true because X is a unit Killing vector field) and then $\delta = 0$. A system of vector fields $\{Z_i\ :\ 1 \leq i \leq n-1\}$ as above is a local frame of $T(\mathscr{F})^{\perp}$, and each basic section $s \in C^{\infty}\big(Q\big)$ is locally represented as $s = d^i\,\pi\,Z_i$ for some $f^i \in \Omega^0_B(\mathscr{F}_U)$. Then (4.126) is equivalent to $\nabla s = c\,\pi$, hence we may apply Theorem 4.86 (yielding the second statement in Corollary 4.90).

Corollary 4.91 *Let* $(\mathscr{F},\ \mathscr{F}^{\perp})$ *be a pair of orthogonal foliations of codimensions* $\operatorname{cod}(\mathscr{F}) = n-1$ *and* $\operatorname{cod}(\mathscr{F}^{\perp}) = 1$ *on a closed n-dimensional* $(n \geq 3)$ *Riemannian manifold* M*. If* $\mathscr{F}$ *is a nonsingular Killing flow whose basic Vaisman complex satisfies the Poincaré lemma, then*

$$H^r(M,\ \mathscr{F},\ \Gamma) \simeq H^r_B(\mathscr{F}),\quad 0 \leq r \leq n-1.$$

If in addition $\mathscr{F}$ *is Einstein, then* $\mathscr{F}$ *is transversally Ricci flat.*

Proof The first statement follows from Theorem 4.86 (as $q = n-1 \geq 2$). In particular

$$H^{n-1}(M,\ \mathscr{F},\ \Gamma) \simeq H^{n-1}_B(\mathscr{F}),\quad H^{n-1}_B(\mathscr{F}) \neq 0,$$

by Theorem 10.17 in [Ton], p. 141. Finally (by Theorem 4.76),

$$S^{\nabla} = \frac{c^{\nabla}}{n-1}\, g_Q \Longrightarrow c^{\nabla} = 0.$$

Q.e.d. □

4.9.4 Basic Curvature Groups of Riemannian Foliations Modeled on Space Forms

Let $(M,\ \mathscr{F},\ g)$ be a foliated Riemannian manifold and let Γ be the canonical adapted connection in $\mathrm{GL}(q,\ \mathbb{R}) \to B^1_T(\mathscr{F}) \to M$. By following [Vai], we set

$$\mathscr{J}^r(M,\ \mathscr{F}) = \mathscr{I}^r(M,\ \mathscr{F}) \times \mathscr{I}^{r+1}(M, \mathscr{F}),$$

$$\mathscr{D}^r(M,\ \mathscr{F},\ \Gamma) = \big\{DD\lambda\ :\ \lambda \in \mathscr{J}^{r-2}(M,\ \mathscr{F})\big\},$$

$$\mathscr{K}^r(M,\ \mathscr{F},\ \Gamma) = \mathscr{I}^r(M,\ \mathscr{F}) \times \mathscr{D}^{r+1}(M,\ \mathscr{F}\,, \Gamma),$$

and consider

$$D^r\ :\ \mathscr{J}^r(M,\ \mathscr{F}) \to \mathscr{J}^{r+1}(M,\ \mathscr{F}),$$

$$D^r(\varphi, \psi) = (D\varphi - \psi, \ DD\varphi - D\psi), \quad (\varphi, \psi) \in \mathscr{J}^r(M, \mathscr{F}).$$

Proposition 4.92 *$\mathscr{J}^\bullet(M, \mathscr{F}, \Gamma) := \left(\bigoplus_{r=0}^{n+q^2} \mathscr{J}^r(M, \mathscr{F}), \ D^\bullet\right)$ is an acyclic cochain complex. Also $\mathscr{K}^\bullet(M, \mathscr{F}, \Gamma) := \left(\bigoplus_{r=0}^{n+q^2} \mathscr{K}^r(M, \mathscr{F}), \ D^\bullet\right)$ is a cochain subcomplex of $\mathscr{J}^\bullet(M, \mathscr{F}, \Gamma)$, in general not acyclic.*

The proof is straightforward, and left as an exercise to the reader. Let then $H^r\big(\mathscr{K}^\bullet(M, \mathscr{F}, \Gamma)\big)$ be the cohomology groups of $\mathscr{K}^\bullet(M, \mathscr{F}, \Gamma)$. When $\mathscr{F}$ is the trivial foliation by points, the complex $\mathscr{J}^\bullet(M, \mathscr{F})$ is $\mathscr{J}^\bullet$, and $H^r\big(\mathscr{K}^\bullet(M, \mathscr{F}, \Gamma)\big)$ are the curvature groups of type ρ of Γ, as in [Vai], p. 332 (also recalled earlier in this section). Next, let us set

$$\mathscr{J}^r B(M, \mathscr{F}) = \mathscr{I}_B^r(M, \mathscr{F}) \times \mathscr{I}_B^{r+1}(M, \mathscr{F}),$$

$$\mathscr{D}_B^r(M, \mathscr{F}, \Gamma) = \left\{DD\lambda \ : \ \lambda \in \mathscr{I}_B^{r-2}(M, \mathscr{F})\right\},$$

$$\mathscr{K}_B^r(M, \mathscr{F}, \Gamma) = \mathscr{I}_B^r(M, \mathscr{F}) \times \mathscr{D}_B^{r+1}(M, \mathscr{F}, \Gamma),$$

and consider

$$D^r : \mathscr{J}^r(M, \mathscr{F}) \to \mathscr{J}^{r+1}(M, \mathscr{F}), \quad D_B^r := D^r\big|_{\mathscr{J}_B^r(M, \mathscr{F})}.$$

Proposition 4.93 *Let $(M, \mathscr{F}, g)$ be a foliated Riemannian manifold and let Γ be the canonical adapted connection. Then $\mathscr{J}_B^\bullet(M, \mathscr{F}, \Gamma) := \left(\bigoplus_{r=0}^{q(q+1)} \mathscr{J}_B^r(M, \mathscr{F}), \ D_B^\bullet\right)$ is an acyclic cochain complex. Also $\mathscr{K}_B^\bullet(M, \mathscr{F}, \Gamma) := \left(\bigoplus_{r=0}^{q(q+1)} \mathscr{K}_B^r(M, \mathscr{F}), \ D_B^\bullet\right)$ is a cochain subcomplex of $\mathscr{J}_B^\bullet(M, \mathscr{F}, \Gamma)$, in general not acyclic.*

Proof One has $D_B^{r+1} \circ D_B^r = 0$ by Proposition 4.92. Also, if $D_B^r(\varphi, \psi) = 0$, i.e., $D\varphi = \psi$, then let $\lambda \in \mathscr{J}_B^{r-1}(\mathscr{F})$ be an arbitrary basic tensor $(r-1)$-form and let us set $\mu = D\lambda - \varphi \in \mathscr{J}_B^r(\mathscr{F})$, so that $D_B^{r-1}(\lambda, \mu) = (\varphi, \psi)$ and one may conclude that $H^r\big(\mathscr{J}_B^\bullet(M, \mathscr{F}, \Gamma)\big) = 0$.

By a result of Vaisman (cf. [Vai], p. 333) there are natural homomorphisms $H^{r-1}(M, \mathbb{R}) \to H^r(M, \Gamma, \rho)$ of the de Rham cohomology groups of a Riemannian manifold M into the curvature groups of its Levi-Civita connection Γ. The cohomology groups $H^r\big(\mathscr{K}_B^\bullet(M, \mathscr{F}, \Gamma)\big)$ may be thought of as the curvature groups of the leaf space $M/\mathscr{F}$. As well known, in general $M/\mathscr{F}$ is not a manifold but only a topological space, and its topology (the quotient topology) may even fail to be Hausdorff. In the spirit of foliation theory, one is then led to work with geometric objets on M (thought of in this context as a desingularization of $M/\mathscr{F}$) though possibly invariant under sliding along the leaves. As an analog to the quoted result in [Vai], let $h^r : \Omega^{r-1}(M) \to \mathscr{I}^r(M, \mathscr{F})$ be the monomorphism

$$h^r(\alpha) = \left(\left((p_T^1)^* \alpha\right) \wedge \theta_T^1 ,\ 0\right), \quad \alpha \in \Omega^{r-1}(M).$$

Then $D^r \circ h^r = h^{r+1} \circ d$ and $h^r\left[\Omega^{r-1}(M)\right] \subset \mathscr{K}^r(M, \mathscr{F}, \Gamma)$, hence there is a natural homomorphism

$$\left(h^r\right)^* : H^{r-1}(M, \mathbb{R}) \to H^r\left(\mathscr{K}^\bullet(M, \mathscr{F}, \Gamma)\right).$$

Moreover, $h_B^r := h^r\big|_{\Omega_B^{r-1}(\mathscr{F})}$ is $\mathscr{K}_N^r(M, \mathscr{F}, \Gamma))$-valued [as $\theta_T^1 \in \mathscr{I}_B^1(M, \mathscr{F})$]. Also h_B^r is an injection of $\Omega_B^{r-1}(\mathscr{F})$ into $\mathscr{K}_B^r(M, \mathscr{F}, \Gamma)$ and there is a natural homomorphism

$$\left(h_B^r\right)^* : H_B^{r-1}(\mathscr{F}) \to H^r\left(\mathscr{K}_B^\bullet(M, \mathscr{F}, \Gamma)\right).$$

Let N be a q-dimensional Riemannian manifold and let $\Gamma(N)$ be the pseudogroup of all local isometries of N. Following Haefliger's (by now classical) viewpoint (cf. [Hae]), one may think of $\mathscr{F}$ as a foliation whose transverse geometry is modeled on N, i.e., there is an open covering $\{U_i\}_{i\in I}$ of M and a family of C^∞ submersions $\{f_i : U_i \to N\}_{i\in I}$ such that 1) $f_j = \gamma_{ji} \circ f_i$ for some $\gamma_{ji} \in \Gamma(N)$ and any $i,\ j \in I$ with $U_i \cap U_j \neq \emptyset$, and 2) $\mathscr{F}_{U_i}$ is the simple foliation tangent to the fibers of $f_i : U_i \to N$.

Proof of Theorem 4.74. When the transverse geometry of $\mathscr{F}$ is modeled on $N = N^q(c)$, a precise description of the modules $\mathscr{D}^r(M, \mathscr{F}, \Gamma)$ is already available. Indeed, for each $\varphi = \varphi^i \otimes e_i \in \mathscr{I}^r(M, \mathscr{F})$

$$DD\varphi^i = \Omega^i{}_j \otimes \varphi^j , \quad 1 \le i \le q, \tag{4.127}$$

where $\Omega = D\omega \in \mathscr{I}^2(M, \mathscr{F})$ is the curvature 2-form of Γ. Yet ω is the transverse Levi-Civita connection, hence it is projectable, i.e., $\omega \in \Omega_B^1\left(\mathscr{F}_T^1\right)$ (cf. e.g., [Mol], p. 81). In particular $T(\mathscr{F}_T^1) \rfloor \Omega = 0$. By (4.127)

$$\mathscr{D}^{r+1}(M, \mathscr{F}, \Gamma) = \left\{\Omega \wedge \varphi \ : \ \varphi \in \mathscr{I}^{r-1}(M, \mathscr{F})\right\}.$$

On the other hand,

$$2\,\Omega\left(X^\uparrow ,\ Y^\uparrow\right)_z z^{-1}(s_x) = z^{-1}\left(R^\nabla(X, Y)s\right)_x ,$$

$$X,\ Y \in \mathfrak{X}(M), \quad s \in C^\infty(Q), \quad z \in (p_T^1)^{-1}(x), \quad x \in M,$$

so that (locally)

$$\Omega^i{}_j = \frac{1}{2}\,(R^\nabla)^s_{k\ell m}\ X^m_j\ Y^i_s\ d\,y^k \wedge d\,y^\ell , \tag{4.128}$$

where $\left[Y^i_j(z)\right] = \left[X)_j(z)\right]^{-1}$ for any $z \in U_T$. By the assumption in Theorem 4.74

$$(R^\nabla)^\ell_{ijk} = c\{(g_Q)_{kj}\,\delta_i^\ell - (g_Q)_{ki}\,\delta_j^\ell\}$$

hence (4.127) becomes

$$\Omega^i{}_j = c\, Y^i_k\, X^m_j\, \left[(g_Q)_{m\ell} \circ p^1_T\right] d\,y^k \wedge d\,y^\ell\,. \tag{4.129}$$

We shall need the following.

Lemma 4.94 *Let $\mathscr{F}$ be a Riemannian foliation of M modeled on the space form $N^q(c)$. For each $\varphi \in \mathscr{I}^{r-1}(M,\,\mathscr{F})$ there is a scalar r-form $\alpha(\varphi) \in \Omega^r(M)$ such that*

$$\mathscr{D}^{r+1}(M,\,\mathscr{F},\,\Gamma) = \left\{c\,\theta^1_T \wedge (p^1_T)^*\alpha(\varphi)\ :\ \varphi \in \mathscr{I}^{r-1}(M,\,\mathscr{F})\right\}.$$

The form $\alpha(\varphi)$ is given by (4.131).

Proof For each $\varphi \in \mathscr{I}^{r-1}(M,\,\mathscr{F})$ let $\alpha^i(\varphi) \in \Omega^{r-1}(M)$ be defined by

$$\alpha^i(\varphi)_x\big(v_1\,,\ \cdots\,,\ v_{r-1}\big) = X^i_j(z)\,\varphi^j_z\big(w_1\,,\ \cdots\,,\ w_{r-1}\big),\quad v_i \in T_x(M),\quad x \in M,$$

where $z \in (p^1_T)^{-1}(x)$ and $w_i \in \Gamma_z$ satisfy $(d_z p^1_T) w_i = v_i$, i.e., w_i is the horizontal lift of v_i. If $z' = za$ with $a \in \mathrm{GL}(q,\,\mathbb{R})$, then

$$X^k_j(z') = a^i_j\, X^k_i(z).$$

Also, if $w'_i \in \Gamma_{z'}$ satisfies $(d_{z'} p^1_T) w'_i = v_i$, then

$$\varphi_{z'}\big(w'_1\,,\ \cdots\,,\ w'_{r-1}\big) = \rho(a^{-1})\,\varphi_z\big(w_1\,,\ \cdots\,,\ w_{r-1}\big)$$

merely because φ is a tensor r-form [of type $\left(\rho\,,\ \mathbb{R}^q\right)$]. It follows that the map

$$z \longmapsto X^i_j(z)\ \varphi^j_z\big(w_1\,,\ \cdots\,,\ w_{r-1}\big)$$

is constant on the fibers of p^1_T, i.e., $\alpha^i(\varphi)_x(v_1\,,\ \cdots\,,\ v_{r-1})$ is well defined. Note that

$$\theta^1_T = \big(Y^j_i\ d\,y^i\big) \otimes e_j\,.$$

Then [by (4.129)]

$$\mathscr{D}^{r+1}(M,\,\mathscr{F},\,\Gamma) = \left\{\big(\Omega^i{}_j \wedge \varphi^j\big) \otimes e_i\ :\ \varphi \in \mathscr{I}^{r-1}(M,\,\mathscr{F})\right\} =$$

$$= \left\{c\ Y^i_k\ X^m_j\ \left[(g_Q)_{m\ell} \circ p^1_T\right]\,\big(d\,y^k \wedge d\,y^\ell \wedge \varphi^j\big) \otimes e_i\ :\ \varphi \in \mathscr{I}^{r-1}(M,\,\mathscr{F})\right\} =$$

$$= \left\{\theta^1_T \wedge \left[c\,X^k_j\,[(g_Q)_{k\ell} \circ p^1_T]\,d\,y^\ell \wedge \varphi^j\right]\ :\ \varphi \in \mathscr{I}^{r-1}(M,\,\mathscr{F})\right\}$$

that is

$$\mathscr{D}^{r+1}(M, \mathscr{F}, \Gamma) = \left\{ c\, \theta_T^1 \wedge (p_T^1)^*(\epsilon_i \wedge \alpha^i(\varphi)) \ : \ \varphi \in \mathscr{I}^{r-1}(M, \mathscr{F}) \right\} \quad (4.130)$$

where $\epsilon_i = g_{ij}\, d\, y^j$. Lemma 4.94 is the foliated counterpart of the main observation in [Vai], p. 334.

Let us go back to the proof of Theorem 4.74. Let us consider the map

$$\Phi^r : \mathscr{I}^r(M, \mathscr{F}) \times \left[\theta_T^1 \wedge (p_T^1)^*\, \Omega^{r-1}(M) \right] \to L^r(M, \mathscr{F}),$$

$$\Phi^r\left(\varphi,\ \theta_T^1 \wedge (p_T^1)^*\varphi\right) = (\varphi, \alpha), \quad \varphi \in \mathscr{I}^r(M, \mathscr{F}), \quad \alpha \in \Omega^{r-1}(M).$$

Note that

$$\alpha(\varphi) = \epsilon_i \wedge \alpha^i(\varphi) \quad (4.131)$$

is a globally defined scalar r-form on M, so that the restriction of Φ^r to $\mathscr{K}^r(M, \mathscr{F}, \Gamma)$ is the module monomorphism $\mathscr{K}^r(M, \mathscr{F}, \Gamma) \hookrightarrow L^r(M, \mathscr{F})$ seek for.

For any $\varphi \in \mathscr{I}^r(M, \mathscr{F})$ and any $\lambda \in \mathscr{I}^{r-1}(M, \mathscr{F})$

$$\begin{aligned} D^r(\varphi,\ DD\lambda) =& D^r(\varphi,\ c\,\theta_T^1 \wedge (p_T^1)^*\, \alpha(\lambda)) \\ &= (D\varphi - c\,\theta_T^1 \wedge (p_T^1)^*\, \alpha(\lambda),\ DD\varphi + c\,\theta_T^1 \wedge (p_T^1)^*\, d\,\alpha(\lambda)) \end{aligned}$$

because of $D\theta_T^1 = 0$. Finally, we may apply Φ^{r+1} to both members, so that to derive (4.97).

To show that $\mathscr{K}_0^\bullet(M, \mathscr{F}, \Gamma)$ is a subcomplex of $\mathscr{K}^\bullet(M, \mathscr{F}, \Gamma)$ it suffices to check that

$$D^r\, \mathscr{K}_0^r(M, \mathscr{F}, \Gamma) \subset \mathscr{K}_0^{r+1}(M, \mathscr{F}, \Gamma).$$

Indeed, if $(\varphi,\ DD\lambda) \in \mathscr{K}_0^r(M, \mathscr{F}, \Gamma)$, then

$$D^r(\varphi,\ DD\lambda) = \left(\tilde{\varphi},\ DD\tilde{\lambda}\right),$$

$$\tilde{\varphi} := D\varphi - DD\lambda, \quad \tilde{\lambda} := \varphi - D\lambda,$$

and

$$DD\tilde{\varphi} = D\left(D^2\varphi\right) - D^2\left(c\,\theta_T^1 \wedge (p_T^1)^*\, \alpha(\lambda)\right) = -c\,\theta_T^1 \wedge (p_T^1)^*\, d^2\, \alpha(\lambda) = 0.$$

To prove (4.98) let $\alpha \in \Omega_B^r(\mathscr{F})$. Locally [with respect to a foliated chart $(U, x^\alpha, y^i) \in \mathscr{F}$]

$$\alpha = \alpha_{i_1 \cdots i_r}\, d\, y^{i_1} \wedge \cdots \wedge d\, y^{i_r}, \quad \alpha_{i_1 \cdots i_r} \in \Omega_B^0(\mathscr{F}_U).$$

Let

$$\lambda = \lambda^i \otimes e_i \in C^\infty\big(\Lambda^{r-1} T^*(B_T^1(\mathscr{F})) \otimes \mathbb{R}^q\big),$$

$$\lambda^i = \frac{1}{c}\, Y_j^i\big[(g_Q)^{jk} \circ p_T^1\big]\, \big(\alpha_{k\ell_1\cdots\ell_{r-1}} \circ p_T^1\big)\, d\, y^{\ell_1} \wedge \cdots \wedge d\, y^{\ell_{r-1}}\,. \tag{4.132}$$

Then λ is globally defined, basic with respect to the lifted foliation $\mathscr{F}_T^1$, and horizontal, i.e.,

$$\operatorname{Ker}\big(d\, p_T^1\big) \rfloor \lambda = 0.$$

Also

$$Y_j^i(z\, a) = \big(a^{-1}\big)_k^i\, Y_j^k(z) \Longrightarrow R_a^* \lambda = \rho(a^{-1})\, \lambda$$

for any $a \in \mathrm{GL}(q, \mathbb{R})$, so that λ is a tensorial $(r-1)$-form of type $(\rho\, , \mathbb{R}^q)$ on $B_T^1(\mathscr{F})$, i.e., $\lambda \in \mathscr{I}^{r-1}(M, \mathscr{F})$. Next, a calculation relying on (4.132) shows that $c\,\alpha(\lambda) = \alpha$ so that $\Phi^r : \mathscr{K}_B^r(M, \mathscr{F}, \Gamma) \to L_B^r(M, \mathscr{F})$ is surjective. Let

$$\big(\Phi^r\big)^* : H^r\big(\mathscr{K}_{B,0}^\bullet(M, \mathscr{F}, \Gamma)\big) \to H_B^r(M, \mathscr{F}, \Gamma)$$

be the induced map on cohomology, i.e., $\Phi^r[\varphi, \psi] = [\Phi^r(\varphi, \psi)]$ for any $(\varphi, \psi) \in \operatorname{Ker}(D^r) \subset \mathscr{K}_{B,0}^r(M, \mathscr{F}, \Gamma)$ [square brackets denote cohomology classes]. By (4.97) (together with $DD\varphi = 0$) the definition of $\Phi^r[\varphi, \psi]$ doesn't depend on the choice of representative (φ, ψ). It remains to be shown that $(\Phi^r)^*$ is injective. If $\psi = DD\lambda \in \mathscr{D}_B^{r+1}(M, \mathscr{F}, \Gamma)$, then

$$(\Phi^r)^* [\varphi, DD\lambda] = 0 \Longrightarrow \varphi = D\chi - \theta_T^1 \wedge (p_T^1)^* \beta\,, \quad c\,\alpha(\lambda) = d\,\beta,$$

for some $\chi \in \mathscr{I}^{r-1}(M, \mathscr{F})$ and some $\beta \in \Omega_B^{r-1}(\mathscr{F})$. Then

$$[\varphi\, , DD\lambda] = \big[D_B^{r-1}(\chi\, , \gamma)\big], \quad \gamma := \theta_T^1 \wedge (p_T^1)^* \beta\,.$$

Example 4.95 (*Curvature groups of foliations defined by suspension*) Let (T, g_T) be a q-dimensional connected Riemannian manifold, and let $H = \operatorname{Isom}(T, g_T)$ be the group of all global isometries of (T, g_T). Let $(M, \mathscr{F})$ be a n-dimensional foliated manifold defined by suspension of a group homomorphism $h : \pi_1(B, b_0) \to H$, where $\pi_1(B, b_0)$ is the first homotopy group of an $(n-q)$-dimensional connected Riemannian manifold (B, g_B), i.e., $M = \tilde{M}\big/\pi_1(B, b_0)$ with $\tilde{M} = T \times \hat{B}$ and $\hat{B}$ is the universal covering manifold of B. Here one uses the natural action of $\pi_1(B, b_0)$ on $\tilde{M}$, as induced by h, i.e.,

$$R_{[\gamma]}\big(t, \hat{b}\big) = \big(h([\gamma]^{-1})(t), \hat{b} \cdot [\gamma]\big),$$

$$[\gamma] \in \pi_1(B, b_0), \quad (t, \hat{b}) \in \tilde{M}.$$

Let $\tilde{\mathcal{F}}$ be the simple foliation of $\tilde{M}$ determined by the submersion $\text{pr}_1 : T \times \hat{B} \to T$ (the first projection). Each $R_{[\gamma]}$ is a foliated map of $(\tilde{M}, \tilde{\mathcal{F}})$ in itself, hence $\tilde{\mathcal{F}}$ descends to a foliation $\mathcal{F}$ of M such that the projection $\pi : \tilde{M} \to M$ is a foliated étale map. The induced map on basic cohomology

$$\pi^* : H^r_B(\mathcal{F}) \to H^r_B(\tilde{\mathcal{F}}), \quad 0 \le r \le q,$$

is injective, hence the Poincaré lemma for the basic complex of $(M, \mathcal{F})$ follows from the Poincaré lemma for the de Rham complex of T. Let $g_{\hat{B}} = \hat{p}^* g_B$, where $\hat{p} : \hat{B} \to B$ is the projection, and let

$$\tilde{g} = \text{pr}_1^* g_T + \text{pr}_2^* g_{\hat{B}}$$

be the product metric on $\tilde{M}$. Right translations $R_{[\gamma]}$ are isometries of $(\tilde{M}, \tilde{g})$, hence $\tilde{g}$ descends to a bundle-like metric g on $(M, \mathcal{F})$ such that $\pi^* g = \tilde{g}$. Let Γ_g and $\Gamma_{\tilde{g}}$ be the transverse Levi-Civita connections in $B^1_T(\mathcal{F}) \to M$ and $B^1_T(\tilde{\mathcal{F}}) \to \tilde{M}$, respectively. Then

$$H^r_B\big(M, \mathcal{F}, \Gamma_g\big) \simeq H^r_B\big(\tilde{M}, \tilde{\mathcal{F}}, \Gamma_{\tilde{g}}\big), \quad 0 \le r \le q. \tag{4.133}$$

Let $p : M \to B$ be given by $p\big(\pi(t, \hat{b})\big) = \hat{p}\big(\hat{b}\big)$, so that (M, p, B, T, G) is a locally trivial smooth bundle with standard fiber T and structure group $G = h\big(\pi_1(B, b_0)\big)$. Each fiber of p is a connected total transversal in $(M, \mathcal{F})$. Let $t_0 \in T$ be a fixed point of G and let us consider the leaf

$$L_0 = \big\{\pi(t_0, \hat{b}) \; : \; \hat{b} \in \hat{B}\big\} \in M/\mathcal{F}.$$

Then L_0 meets each fiber of p in exactly one point, and we may set

$$\{x_0\} = L_0 \cap T_0, \quad T_0 := p^{-1}(b_0).$$

Let $\hat{b}_0 \in \hat{B}$ be the homotopy class of the constant map $[0, 1] \to \{b_0\}$. If γ is a loop at x_0 in L_0, then $p \circ \gamma$ is a loop at b_0 in B, which lifts to a path $\hat{\gamma} : [0, 1] \to \hat{B}$ such that $\hat{\gamma}(0) = \hat{b}_0$. Let $h_{\hat{\gamma}}$ denote the sliding along the leaves of $\tilde{\mathcal{F}}$ following $\hat{\gamma}$ [the germ at $(t_0, \hat{b}_0)$ of a diffeomorphism of a neighborhood of $(t_0, \hat{\gamma}(0))$ in $T \times \big\{\hat{\gamma}(0)\big\}$ onto a neighborhood of $\big(t_0, \hat{\gamma}(1)\big)$ in $T \times \big\{\hat{\gamma}(1)\big\}$]. Then $\text{pr} \circ h_{\hat{\gamma}} \circ i = h([\gamma])$, which is an isometry of g_T and then a Γ_{g_T}-preserving map [Γ_{g_T} is the Levi-Civita connection of g_T regarded as a distribution on $B^1(T)$]. In particular $h([\gamma])$ preserves the ordinary curvature groups $H^\bullet(T, \Gamma_{g_T})$. On the other hand,

$$i : T \to \tilde{M}, \quad i(t) = (t, \hat{b}_0), \quad t \in T,$$

is a totally geodesic isometric immersion of (T, g_T) into $(\tilde{M}, \tilde{g})$, hence

$$H^r(T,\ \Gamma_{g_T}) \simeq H^r_B(\tilde{M},\ \tilde{\mathscr{F}},\ \Gamma_{\tilde{g}}),\quad 0 \le r \le q. \tag{4.134}$$

Then [by (4.133)–(4.134)] the basic curvature groups of $(M,\ \mathscr{F},\ g)$ are G-invariant and the basic Vaisman complex satisfies the Poincaré lemma whenever the Goldberg-Petridis complex of $(T,\ g_T)$ does [e.g., when $(T,\ g_T)$ is locally Euclidean]. □

Izu Vaisman (born on June 22, 1938 in Iaşi, Romania) is a Romanian mathematician.

Izu Vaisman (born 1938)

V. earned a Master of Science in Mathematics in 1959 and a Ph.D. in Mathematics in 1965, and was awarded a Doctorate *Honoris Causa* in 1999, at the University *Al.I. Cuza* of Iaşi, Romania. **V.** was an assistant professor (1964–1969) and an associate professor (1969–1976) at the University *Al.I. Cuza* of Iaşi, Romania. **V.** was a professor (1976–2006) at the University of Haifa, and is a professor emeritus since 2006. **V.** served as Chairman of the Department of Mathematics of University of Haifa (1982–1984 and 1999–2001). **V.** had crucial contributions to the field of Differential Geometry, and in particular Foliation Theory. **V.** authored many wonderful books, that both capture the state of the art of the topic mathematics research at the time of their publication, and are indeed suitable for learning. The proportions of the influence of **V.** on his contemporaries and on his students are hard to overestimate, and **V.**'s scientific creation and life, both as a member of the mathematical community and as a human being, are bound to attract the attention of the historians of mathematics in the near future. **V.**'s book *Symplectic Geometry and Secondary Characteristic Classes* [Progress in Mathematics, Birkhäuser, 1987, 232 pp.] grew out of his study of the Maslov class, which is a fundamental invariant in asymptotic analysis of partial differential equations of quantum physics. One of the many interpretations of this class had been given earlier by F. Kamber and Ph. Tondeur, exhibiting the Maslov class as a secondary characteristic class of a complex trivial vector bundle endowed with a real reduction of its structure group. Accordingly, **V.** attempted the exposition of the whole range of secondary characteristic classes involved in this interpretation.

4.10 Foliated Harmonic Maps

4.10.1 Lichnerowicz Type Theorem on Transversally Holomorphic Maps

The purpose of the present section is to report on the result by E. Barletta and S. Dragomir (cf. [BaDr]) that any transversally holomorphic foliated map φ :

$(M, \mathscr{F}) \to (M', \mathscr{F}')$ of Kählerian foliations with $\mathscr{F}$ harmonic, is a transversally harmonic map and an absolute minimum of the energy functional $E_T(\varphi) = \frac{1}{2}\int_M \|d_T\varphi\|^2\ \mu$ in its foliated homotopy class.

Let $(M, \mathscr{F})$ and $(M', \mathscr{F}')$ be two foliated manifolds and $\varphi : M \to M'$ a foliated map. When $\mathscr{F}$ and $\mathscr{F}'$ are Riemannian foliations one associates to φ a concept of tension field $\tau(\varphi) \in C^\infty\big(\varphi^{-1}\nu(\mathscr{F}')\big)$, coinciding with the ordinary tension field[18] of φ when M, M' are Riemannian manifolds and $\mathscr{F}$, $\mathscr{F}'$ are the trivial foliations by points. A calculation based on techniques introduced by F. Kamber and Ph. Tondeur (cf. [KaTo16]) will lead to the expression (4.141) of $\tau(\varphi)$. It is noteworthy that (4.141) contains explicitly the mean curvature $\kappa \in \Omega^1_B(\mathscr{F})$ of the Riemannian foliation $\mathscr{F}$. A result in [BaDr] is that φ is transversally harmonic if and only if φ satisfies the PDEs system (4.142), in which the principal part is the basic Laplacian Δ_B of $\mathscr{F}$, a strongly transversally elliptic operator (cf. A. El Kacimi-Alaoui, [KaAl], p. 91). The main result in [BaDr] is that a transversally holomorphic foliated map φ of $(M, \mathscr{F})$ into $(M', \mathscr{F}')$, with $\mathscr{F}$, $\mathscr{F}'$ Kählerian and $\mathscr{F}$ harmonic, is transversally harmonic map (cf. Theorem 4.98 below). An application of the result regards the holomorphic maps of Vaisman (or generalized Hopf) manifolds.[19]

4.10.1.1 Transversally Harmonic Maps

Let M be a real m-dimensional C^∞ manifold, and let $\mathscr{F}$ be a codimension q foliation of M. Let $T(\mathscr{F})$ be the tangent bundle of $\mathscr{F}$ and let $\nu(\mathscr{F}) = T(M)/T(\mathscr{F})$ be the transverse bundle. Let $\pi : T(M) \to \nu(\mathscr{F})$ be the projection. Let us assume that $\mathscr{F}$ is a Riemannian foliation, i.e., $\mathscr{F}$ is a Γ-foliation (in the sense of [Hae]), where Γ is the pseudogroup of all local isometries of some model Riemannian manifold. Equivalently (cf. e.g., [Mol], p. 78) $\nu(\mathscr{F})$ is equipped with a holonomy invariant metric $g_{\nu(\mathscr{F})}$. Let g be a bundle-like Riemannian metric on M inducing $g_{\nu(\mathscr{F})}$. Assume $\mathscr{F}$ to be transversally oriented and let $\nu \in \Omega^q_B(\mathscr{F})$ be the transversal volume element ($d_B\nu = 0$). Let $\chi_{\mathscr{F}}$ be the characteristic p-form, i.e., $\chi_{\mathscr{F}} = *\,\nu$ where $* : \Omega^q(M) \to \Omega^{m-q}(M)$ is the Hodge star operator of (M, g) (here $p = m - q$). The Riemannian volume form $\mu \in \Omega^m(M)$ is then given by $\mu = \nu \wedge \chi_{\mathscr{F}}$. Let $(M', \mathscr{F}')$ be another foliated manifold, and let $\varphi : M \to M'$ be a foliated map of $(M, \mathscr{F})$ into $(M', \mathscr{F}')$, i.e.,

$$(d_x\varphi)T(\mathscr{F})_x \subset T(\mathscr{F}')_{\varphi(x)}, \quad x \in M.$$

The differential $d\varphi$ gives rise naturally to a section

$$d_T\varphi \in C^\infty\big(\nu(\mathscr{F})^* \otimes \varphi^{-1}\,\nu(\mathscr{F}')\big).$$

---

[18] Cf. e.g., S. Dragomir and J.C. Wood, [DrWo], p. 107.

[19] In the sense of S. Dragomir and L. Ornea, [DrOr], p. 33.

If both $\mathscr{F}$ and $\mathscr{F}'$ are Riemannian foliations, then their metrics g_Q and $g_{Q'}$ [where $Q = \nu(\mathscr{F})$ and $Q' = \nu(\mathscr{F}')$] induce a metric on $\nu(\mathscr{F})^* \otimes \varphi^{-1}\,\nu(\mathscr{F}')$ and one may consider the basic function

$$e_T(\varphi) \in \Omega^0_B(\mathscr{F}), \quad e_T(\varphi) = \frac{1}{2}\,\|d_T\,\varphi\|^2,$$

cf. [KAGo], p. 18. Let us assume M to be orientable and closed (i.e., M is compact and $\partial M = \emptyset$) and let us set

$$E_T(\varphi) = \int_M e_T(\varphi)\,\mu\,.$$

Definition 4.96 We say φ is *transversally harmonic* if it is a critical point of E_T for any variation of φ by foliated maps. □

Cf. also J.Y. Chiang, [Chi].

Definition 4.97 Let $\varphi : M \to M'$ be a foliated map. The *foliated homotopy class* $[\varphi]$ of φ consists of all C^∞ maps that are smoothly homotopic to φ through foliated maps. □

We shall establish

Theorem 4.98 *Let $(M, \mathscr{F})$ and $(M', \mathscr{F}')$ be two foliated manifolds, where M is closed and $\mathscr{F}$, $\mathscr{F}'$ are Kählerian foliations. Let $\varphi : M \to M'$ be a transversally holomorphic (anti-holomorphic) foliated map. The following statements are equivalent:*

(i) *φ is transversally harmonic.*

(ii) $\text{grad}_B\,\varphi^\alpha \in \text{Ker}(\kappa)$, $\varphi^\alpha := y'^\alpha \circ \varphi$, *for any foliated chart (U', x'^s, y'^α) on $(M', \mathscr{F}')$. If additionally $\mathscr{F}$ is harmonic (i.e., $\kappa = 0$), then φ is (a transversally harmonic map and) an absolute minimum for the energy E_T in its foliated homotopy class.*

Definition 4.99 A foliation $\mathscr{F}$ of M is *transversally holomorphic* if it is equipped with a holonomy invariant complex structure $J : \nu(\mathscr{F}) \to \nu(\mathscr{F})$. □

Definition 4.100 A Riemannian foliation $(\mathscr{F}, g_{\nu(\mathscr{F})})$ of M is (*transversally*) *Kählerian* if it is transversally holomorphic and the complex structure J in the normal bundle satisfies

(1) $g_{\nu(\mathscr{F})}(Jr,\ Js) = g_{\nu(\mathscr{F})}(r,\ s)$ for any $r,\ s \in C^\infty(\nu(\mathscr{F}))$,

(2) $\nabla J = 0$. □

Here ∇ is the unique Riemannian ($\nabla g_{\nu(\mathscr{F})} = 0$) torsion-free adapted connection in $\nu(\mathscr{F})$. As presented in Definition 4.100, the notion is due to S. Nishikawa and Ph. Tondeur (cf. [NiTo]) who studied the Lie algebra of transversally holomorphic fields for harmonic Kähler foliations of closed orientable manifolds of constant transversal scalar curvature. Cf. also [KaAl], p. 95. Note that

$$\Omega(r, s) = g_{\nu(\mathscr{F})}(r, Js), \quad r, s \in C^\infty(\nu(\mathscr{F})),$$

determines a closed basic 2-forms $\Omega_T \in \Omega^2_B(\mathscr{F})$.

Definition 4.101 A foliated map φ of $(M, \mathscr{F})$ into $(M', \mathscr{F}')$, with $\mathscr{F}$ and $\mathscr{F}'$ transversally holomorphic, is said to be *transversally holomorphic* (respectively, *anti-holomorphic*) if $(d_T\varphi) \circ J = J' \circ (d_T\varphi)$ [respectively, $(d_T\varphi) \circ J = -J' \circ (d_T\varphi)$].

4.10.1.2 Euler-Lagrange Equations

Let $V(\mathscr{F})$ be the Lie algebra of all foliated vector fields

$$V(\mathscr{F}) = \left\{ Y \in \mathfrak{X}(M) \ : \ [X, Y] \in \mathfrak{X}_{\mathscr{F}} , \quad \forall\, X \in \mathfrak{X}_{\mathscr{F}} \right\}$$

and let us set $\ell(\mathscr{F}) := V(\mathscr{F})/\mathfrak{X}_{\mathscr{F}}$.

Definition 4.102 An element $s \in \ell(\mathscr{F})$ is a *transverse vector field.* □

Definition 4.103 The *basic divergence* of a foliated vector field $Y \in V(\mathscr{F})$ is the basic function $\operatorname{div}_B(Y) \in \Omega^0_B(\mathscr{F})$ determined by

$$\mathscr{L}_Y \nu = \left(\operatorname{div}_B Y\right) \nu. \tag{4.135}$$

□

Exercise 4.104 Show that

$$\pi\, Y = \pi\, Y' \Longrightarrow \operatorname{div}_B Y = \operatorname{div}_B Y',$$

i.e., the basic divergence $\operatorname{div}_B Y$ depends only on the transverse vector field $\pi\, Y$ associated to Y. □

Let $M^\sharp \to M$ be the principal SO(q)-bundle of all oriented orthonormal transverse frames. If $z : \mathbb{R}^q \to \nu(\mathscr{F})_x$ is a transverse frame adapted to $M^\sharp$, the transverse volume element $\nu \in \Omega^q_B(\mathscr{F})$ is given by

$$\nu_x(v_1 , \cdots , v_q) = \det\left[(g_Q)_x\left(\pi_x\, v_i , z(e_j)\right)\right]_{1 \le i,\, j \le q} ,$$

$$v_i \in T_x(M), \quad 1 \le i \le q,$$

where $\{e_i \ : \ 1 \le i \le q\} \subset \mathbb{R}^q$ is the canonical basis. Let $(U, x^s , y^i) \in \mathscr{F}$ be a foliated chart, and let us set

$$z(e_i) = \lambda^j_i\, \pi_x\left(\frac{\partial}{\partial y^j}\right)_x , \quad \lambda = \left[\lambda^j_i\right] \in \mathrm{GL}(q, \mathbb{R}).$$

The function

$$c : M \to \{\pm 1\}, \quad c(x) = \text{sign} \det(\lambda), \quad x \in M,$$

is well defined [i.e., $c(x)$ doesn't depend on the choice of adapted transverse frame at x]. Then

$$\nu = c\, q!\, \sqrt{\mathfrak{g}}\, d\, y^1 \wedge \cdots \wedge d\, y^q \tag{4.136}$$

where

$$\mathfrak{g} = \det\left[g_{ij}\right], \quad g_{ij} = g_Q\Big(\frac{\partial}{\partial y^i}, \frac{\partial}{\partial y^j}\Big), \quad 1 \le i, j \le q.$$

Then [by (4.135)–(4.136)]

$$\text{div}_B\Big(Y^i(y)\, \frac{\partial}{\partial y^i}\Big) = \frac{1}{\sqrt{\mathfrak{g}}}\, \frac{\partial}{\partial y^i}\Big(\sqrt{\mathfrak{g}}\, Y^i(y)\Big). \tag{4.137}$$

We shall need the star operator

$$\hat{*} : \Omega^k_B(\mathscr{F}) \to \Omega^{q-k}_B(\mathscr{F})$$

associated to g_Q (cf. e.g., Section 4 in [KaTo11]), characterized by the property

$$\alpha \wedge \hat{*}\, \alpha = \nu\,, \quad \alpha \in \Omega^k_B(\mathscr{F}).$$

The L^2 inner product on $\Omega^k(M)$ restricts on basic forms to the expression

$$\langle \alpha\,,\, \beta \rangle_B = \int_M \alpha \wedge (\hat{*}\, \beta) \wedge \chi_{\mathscr{F}}\,.$$

With respect to this inner product one may define the formal adjoint

$$\delta_B : \Omega^k_B(\mathscr{F}) \to \Omega^{k-1}_B(\mathscr{F})$$

of d_B and the basic Laplacian

$$\Delta_B = \delta_B\, d_B + d_B\, \delta_B\,.$$

Let $P^\perp$ be the orthogonal complement of $P = T(\mathscr{F})$ with respect to g, and let $\pi_P : T(M) \to P$ be the projection with respect to the decomposition $T(M) = P \oplus P^\perp$. Let

$$W_Z : P \to P, \quad W_Z X = -\pi_P\, \nabla^g_X Z\,, \quad X \in C^\infty(P), \quad Z \in C^\infty(P^\perp),$$

be the Weingarten operator, and $\kappa \in \Omega^1_B(\mathscr{F})$ the mean curvature

$$\kappa(Z) = \text{Trace}\, W_Z\,, \quad \kappa(X) = 0.$$

By the identity (2.6) in [KaTo16], p. 418

$$\delta_B\, \alpha = (-1)^{q(k+1)+1}\, \hat{*}\, \big(d_B - \kappa\, \wedge\, \big)\, \hat{*}\alpha\,, \quad \alpha \in \Omega_B^k(\mathscr{F}). \tag{4.138}$$

Let $\tau \in C^\infty(P)$ be determined by

$$\kappa(Z) = g_P\big(\tau\,,\, \pi\, Z\big), \quad Z \in C^\infty\big(P^\perp\big).$$

We shall make use of the following result[20]

$$\int_M \big(\mathrm{div}_B\, Y\big)\, \mu = \int_M g_Q\big(\tau\,,\, \pi\, Y\big)\, \mu. \tag{4.139}$$

For harmonic foliations this is the *transversal divergence theorem*, cf. Theorem C in [KTT], p. 262. Let $\sigma\, :\, Q \to P^\perp$ be the natural bundle isomorphism. If $\eta \in C^\infty(Q^*)$ let $\eta_T \in \Omega^1(M)$ be given by $\eta_T(Y) = \eta\big(\pi\, Y\big)$. Moreover, let $\eta \in C_B^\infty(Q^*)$, i.e., $\mathscr{L}_X \eta = 0$ for any $X \in \mathfrak{X}_{\mathscr{F}}$. The mapping $\eta \longmapsto \eta_T$ is an isomorphism $C_B^\infty(Q^*) \simeq \Omega_B^1(\mathscr{F})$. For any basic function $f \in \Omega_B^1(\mathscr{F})$ let $\eta_f \in C_B^\infty(Q^*)$ be determined by $(\eta_f)_T = d_B f$.

Definition 4.105 The *basic gradient* of f is

$$\mathrm{grad}_B\, f := \sigma\big(s_f\big) \in C^\infty\big(T(\mathscr{F})^\perp\big),$$

$$s_f \in C^\infty\big(\nu(\mathscr{F})\big), \quad g_Q\big(s_f\,,\, r\big) = \eta_f(r), \quad r \in C^\infty(Q).$$

□

A calculation relying on (4.138) leads to

$$\Delta_B f = \frac{1}{\sqrt{\mathfrak{g}}}\, \frac{\partial}{\partial y^i}\Big(\sqrt{\mathfrak{g}}\, g^{ij}\, \frac{\partial f}{\partial y^j}\Big) - \kappa\big(\mathrm{grad}_B\, f\big). \tag{4.140}$$

Let ∇ and ∇' be the transverse Levi-Civita connections of $(M,\, \mathscr{F},\, g)$ and $(M',\, \mathscr{F}',\, g')$. Let $\varphi^{-1}\nabla'$ be the pullback of ∇' by φ [a connection in the pullback bundle $\varphi^{-1}\nu(\mathscr{F}') \to M$]. Let ∇^φ be the connection in $\nu(\mathscr{F})^* \otimes \varphi^{-1}\nu(\mathscr{F}')$ determined by

$$\nabla^\varphi(\omega \otimes s) = (\nabla\omega) \otimes s + \omega \otimes \varphi^{-1}(\nabla')s,$$

$$\omega \in C^\infty\big(\nu(\mathscr{F})^*\big), \quad s \in C^\infty\big(\varphi^{-1}\nu(\mathscr{F}')\big).$$

[20] Holding for any transversally oriented Riemannian foliation $\mathscr{F}$ on a closed oriented Riemannian manifold $(M,\, g)$.

Let B_φ be the $\varphi^{-1}\nu(\mathcal{F}')$-valued bilinear form on $\nu(\mathcal{F})$ determined by $\nabla^\varphi d_T\varphi$ [*via* the isomorphism $C_B^\infty(\nu(\mathcal{F})^*) \simeq \Omega_B^1(\mathcal{F})$].

Definition 4.106 The *tension field* of φ is $\tau(\varphi) := \mathrm{Trace}(B_\varphi)$. □

Let $\Gamma'^\alpha_{\beta\gamma}$ be the coefficients of ∇' with respect to a foliated chart $(U', x'^r, y'^\alpha) \in \mathcal{F}'$ [with $1 \le r \le p'$ and $1 \le \alpha \le q'$, $p' + q' = n' = \dim(M')$]. Then $\tau(\varphi)$ may be locally represented as

$$\tau(\varphi) = \tau^\alpha(\varphi)\, Y_\alpha\,,$$

$$\tau^\alpha(\varphi) = \Delta_B\varphi^\alpha + \kappa\big(\mathrm{grad}_B\,\varphi^\alpha\big) + g^{ij}\,\Gamma'^\alpha_{\beta\gamma}\,\frac{\partial\varphi^\beta}{\partial y^i}\,\frac{\partial\varphi^\gamma}{\partial y^j}\,. \tag{4.141}$$

Here

$$Y_\alpha(x) = \pi'_{\varphi(x)}\left(\frac{\partial}{\partial y'^\alpha}\right)_{\varphi(x)}, \quad x \in U,$$

are the natural lifts of $\pi'(\partial/\partial y'^\alpha)$ to $\varphi^{-1}\nu(\mathcal{F}')$. We may now derive the Euler-Lagrange equations of the variational principle $\delta\int_M e_T(\varphi)\,\mu = 0$. To this end, let $\{\varphi_t\}_{|t|<\epsilon}$ be a smooth 1-parameter variation of $\varphi_0 = \varphi$ through foliated maps $\varphi_t : M \to M'$. Then

$$\frac{d}{dt}\,e_T(\varphi_t) = g^{ij}\left\{\frac{\partial^2\varphi_t^\alpha}{\partial y^i\,\partial t}\,\frac{\partial\varphi_t^\beta}{\partial y^j}\,\big(g'_{\alpha\beta}\circ\varphi_t\big) + \frac{1}{2}\,\frac{\partial\varphi_t^\alpha}{\partial y^i}\,\frac{\partial\varphi_t^\beta}{\partial y^j}\,\frac{\partial g'_{\alpha\beta}}{\partial y'^\gamma}\,\frac{\partial\varphi_t^\gamma}{\partial t}\right\}.$$

Let us set

$$Y_t = Y_t^j\,\frac{\partial}{\partial y^j}\,, \quad Y_t^j := g^{ij}\,\frac{\partial\varphi_t^\alpha}{\partial y^i}\,\frac{\partial\varphi_t^\beta}{\partial t}\,\big(g'_{\alpha\beta}\circ\varphi_t\big).$$

Then [by (4.137) and (4.140)]

$$\mathrm{div}_B\,Y_t = \frac{d}{dt}\,e_T(\varphi_t) + \left\{\Delta_B\varphi_t^\alpha + \kappa(\mathrm{grad}_B\,\varphi_t^\alpha)\right\}\frac{\partial\varphi_t^\beta}{\partial t}\,\big(g'_{\alpha\beta}\circ\varphi_t\big) +$$

$$+\, g^{ij}\left[\frac{\partial g'_{\alpha\lambda}}{\partial y'^\beta} - \frac{1}{2}\,\frac{\partial g'_{\alpha\beta}}{\partial y'^\lambda}\right]\frac{\partial\varphi_t^\alpha}{\partial y^i}\,\frac{\partial\varphi_t^\beta}{\partial y^j}\,\frac{\partial\varphi_t^\lambda}{\partial t}\,.$$

Let us integrate over M and use (4.139) together with the identity

$$g^{ij}\left[\frac{\partial g'_{\alpha\lambda}}{\partial y'^\beta} - \frac{1}{2}\,\frac{\partial g'_{\alpha\beta}}{\partial y'^\lambda}\right]\frac{\partial\varphi_t^\alpha}{\partial y^i}\,\frac{\partial\varphi_t^\beta}{\partial y^j} = g^{ij}\,g'_{\alpha\lambda}\,\Gamma'^\alpha_{\beta\gamma}\,\frac{\partial\varphi_t^\beta}{\partial y^i}\,\frac{\partial\varphi_t^\gamma}{\partial y^j}\,.$$

We obtain [by also using (4.141)]

$$\int_M g_Q(\tau,\,\pi Y_t)\,\mu = \frac{d}{dt}\left\{E_T(\varphi_t)\right\} + \int_M \big(g'_{\alpha\beta}\circ\varphi_t\big)\,\frac{\partial\varphi_t^\alpha}{\partial t}\,\tau^\beta(\varphi_t)\,\mu\,.$$

Finally [by $\{d\,E_T(\varphi_t)/d\,t\}_{t=0} = 0$],

$$\int_M (g'_{\alpha\beta} \circ \varphi_t) \left(\frac{\partial \varphi_t^\alpha}{\partial t}\right)_{t=0} \left[\tau^\beta(\varphi) - \frac{\partial \varphi^\beta}{\partial y^i}\, \tau^i\right] \mu = 0$$

where τ^i are determined by $\tau = \tau^i\, \pi\,(\partial/\partial y^i)$. Locally, one may always produce a variation $\varphi_t^\alpha(y)$ such that

$$\varphi_0^\alpha(y) = \varphi^\alpha(y), \qquad \left\{\frac{\partial \varphi_t^\alpha}{\partial t}\right\}_{t=0} = \tau^\alpha(\varphi) - \tau^i\, \frac{\partial \varphi^\alpha}{\partial y^i}\,.$$

Hence, φ is transversally harmonic if and only if

$$\tau^\alpha(\varphi) = \tau^i\, \frac{\partial \varphi^\alpha}{\partial y^i},$$

i.e.,

$$\Delta_B \varphi^\alpha + g^{ij}\left(\Gamma'^\alpha_{\beta\gamma} \circ \varphi\right) \frac{\partial \varphi^\beta}{\partial y^i}\, \frac{\partial \varphi^\gamma}{\partial y^j} = 0 \tag{4.142}$$

The reader should observe the cancellation of the mean curvature terms [because of $\kappa(\text{grad}_B\, f) = \tau^i\,(\partial f/\partial y^i)$]. It is noteworthy that the derivation of (4.142) required no additional assumptions on $\mathscr{F}$ [such as the customary assumption that $\mathscr{F}$ be harmonic].

4.10.1.3 Transversally Holomorphic Maps

Let $(\mathscr{F},\, J)$ and $(\mathscr{F}',\, J')$ be transversally holomorphic foliations of M and M', respectively. Let $J^{\mathbb{C}}$ be the extension of J to $\nu(\mathscr{F}) \otimes \mathbb{C}$ [by $\mathbb{C}$-linearity], and let

$$\nu^+(\mathscr{F}) = \text{Eigen}\big(J^{\mathbb{C}};\, +i\big), \quad \nu^-(\mathscr{F}) = \overline{\nu^+(\mathscr{F})}\,,$$

be, respectively, the eigenbundles corresponding to the eigenvalues $\text{Spec}(J^{\mathbb{C}}) = \{\pm i\}$. Let

$$\pi'_\pm : \nu(\mathscr{F}') \otimes \mathbb{C} \to \nu^\pm(\mathscr{F}')$$

be the projections associated to the direct sum decomposition

$$\nu(\mathscr{F}') \otimes \mathbb{C} = \nu^+(\mathscr{F}') \oplus \nu^-(\mathscr{F}').$$

Let $\varphi : M \to M'$ be a foliated map. Let

$$\partial_T \varphi := \pi'_+ \circ (d_T\, \varphi)^{\mathbb{C}}\Big|_{\nu^+(\mathscr{F})}, \qquad \overline{\partial}_T \varphi := \pi'_- \circ (d_T\, \varphi)^{\mathbb{C}}\Big|_{\nu^+(\mathscr{F})}\,.$$

Here $(d_T\varphi)^{\mathbb{C}} : \nu(\mathscr{F}) \otimes \mathbb{C} \to \nu(\mathscr{F}') \otimes \mathbb{C}$ denotes the $\mathbb{C}$-linear extension of $d_T\varphi : \nu(\mathscr{F}) \to \nu(\mathscr{F}')$. The following commutative diagram sums up the mappings introduced so far.

$$
\begin{array}{ccc}
\nu^+(\mathscr{F}) & \stackrel{\overline{\partial}_T\varphi}{\longrightarrow} & \nu^-(\mathscr{F}') \\
\cap & & \uparrow \pi'_- \\
\nu(\mathscr{F}) \otimes \mathbb{C} & \stackrel{(d_T\varphi)^{\mathbb{C}}}{\longrightarrow} & \nu(\mathscr{F}') \otimes \mathbb{C} \\
\cup & & \downarrow \pi'_+ \\
\nu^+(\mathscr{F}) & \stackrel{\partial_T\varphi}{\longrightarrow} & \nu^+(\mathscr{F}')
\end{array}
$$

From now on we shall not distinguish notationally between J and $d_T\varphi$, and their $\mathbb{C}$-linear extensions $J^{\mathbb{C}}$ and $(d_T\varphi)^{\mathbb{C}}$.

Definition 4.107 The foliated map φ is *transversally holomorphic* (respectively, *transversally anti-holomorphic*) if

$$(d_T\,\varphi)\,\nu^+(\mathscr{F}) \subset \varphi^{-1}\,\nu^+(\mathscr{F}').$$

[respectively, if

$$(d_T\,\varphi)\,\nu^+(\mathscr{F}) \subset \varphi^{-1}\,\nu^+(\mathscr{F}')].$$

□

By the very definitions

Lemma 4.108 *Let $\varphi : M \to M'$ be a foliated map of $(M, \mathscr{F}, J)$ into $(M', \mathscr{F}', J')$. The following statements are equivalent:*

(i) *φ is transversally holomorphic.*

(ii) $\overline{\partial}_T\varphi = 0$.

Lemma 4.109 *Let $\varphi : M \to M'$ be a foliated map of $(M, \mathscr{F}, J)$ into $(M', \mathscr{F}', J')$. The following statements are equivalent:*

(i) *φ is transversally anti-holomorphic.*

(ii) $\partial_T\varphi = 0$.

Let us assume from now on that $\mathscr{F}$ and $\mathscr{F}'$ admit holonomy invariant bundle metrics g_Q and $g_{Q'}$ that are compatible with the transverse complex structures J and J'. Next, let us set

$$k_+(\varphi) := \|\partial_T\,\varphi\|^2, \quad k_-(\varphi) := \|\overline{\partial}_T\,\varphi\|^2.$$

Then $e_T(\varphi) = k_+(\varphi) + k_-(\varphi)$. We may state

Theorem 4.110 *Let $(M, \mathscr{F})$ and $(M', \mathscr{F}')$ be two foliated manifolds, with M compact, and let $\varphi : M \to M'$ be a foliated map. Let us assume that $\mathscr{F}$ and $\mathscr{F}'$ are Kählerian foliations and that $\mathscr{F}$ is harmonic ($\kappa = 0$). Let us set*

$$k(\varphi) := k_+(\varphi) - k_-(\varphi), \qquad K(\varphi) := \int_M k(\varphi)\,\mu.$$

Then $t \longmapsto K(\varphi_t)$ is a constant map for any smooth 1-parameter variation $\{\varphi_t\}_{|t|<\epsilon}$ of $\varphi_0 = \varphi$ through foliated maps.

Let $\varphi^{-1}\Omega' \in C^\infty_B\big(\Lambda^2\,\nu(\mathscr{F})^*\big)$ be determined by $\big(\varphi^{-1}\Omega'\big)_T = \varphi^*\,\Omega'_T$ [*via* the isomorphism $C^\infty_B\big(\Lambda^2\,\nu(\mathscr{F})^*\big) \simeq \Omega^2_B(\mathscr{F})$]. A calculation shows that

$$2\,k(\varphi) = \big\langle \Omega\,,\ \varphi^{-1}\,\Omega'\big\rangle.$$

The map $\alpha \longmapsto \alpha_T$ is an isometry (cf. Appendix A to this section). Hence,

$$K(\varphi) = \frac{1}{2}\int_M \big\langle \Omega_T\,,\ \varphi^*\,\Omega'_T\big\rangle\,\mu = \frac{1}{2}\int_M \big(\varphi^*\Omega'_T\big)\wedge *\,\Omega_T\,.$$

Let $\alpha \in \Omega^1_B(\mathscr{F})$ be defined by

$$\alpha = \int_0^1 \varphi_t^*\Big(\frac{\partial \varphi_t}{\partial t}\,\rfloor\,\Omega'_T\Big)\,d\,t.$$

Then (by (16.11) in [DrWo], as $d\Omega'_T = 0$)

$$d\alpha = \varphi_1^*\,\Omega'_T - \varphi_0^*\,\Omega'_T$$

hence (by the theorem of Stokes)

$$K(\varphi_1) - K(\varphi_0) = \frac{1}{2}\int_M \alpha \wedge *\,\Omega_T\,.$$

Next (by (2.4) in [KaTo16])

$$*\,\Omega_T = \big(\hat{*}\,\Omega_T\big)\wedge \chi_{\mathscr{F}} = \frac{1}{(n-1)!}\,\Omega_T^{n-1}\wedge \chi_{\mathscr{F}}$$

with $q = 2n$. By Rummler's formula (cf. also (2.2) in [KaTo16])

$$d\,\chi_{\mathscr{F}} + \kappa \wedge \chi_{\mathscr{F}} = \beta$$

for some $\beta \in F^2\Omega^3$. Finally (as $\mathscr{F}$ was assumed to be harmonic),

$$K(\varphi_1) - K(\varphi_0) = \frac{1}{2(n-1)!} \int_M \alpha \wedge \Omega_T^{n-1} \wedge \beta .$$

The meaning of $\beta \in F^2\Omega^3$ (cf. also (2.1) in [KaTo16]) is that, for any $(U, x^s, y^i) \in \mathscr{F}$, β contains only monomials of the form $dx^s \wedge dy^j \wedge dy^k$ and $dy^i \wedge dy^j \wedge dy^k$. Then each monomial in $\alpha \wedge \Omega_T^{n-1} \wedge \beta$ has degree at least $q+1$ in the dy^i's, so $\alpha \wedge \Omega_T^{n-1} \wedge \beta = 0$. Q.e.d.

Let (M, J, g) be a Hermitian manifold, where J is its complex structure and g the Hermitian metric. Let $\Phi \in \Omega^2(M)$ be given by $\Phi(X, Y) = g(X, JY)$ for any $X, Y \in \mathfrak{X}(M)$.

Definition 4.111 (M, J, g) is said to be a *locally conformal Kähler* (l.c.K.) manifold if there is a closed 1-form $\omega \in \Omega^1(M)$ (the *Lee form*) such that $d\Phi = \omega \wedge \Phi$. □

Cf. P. Libermann, [Lib].

Definition 4.112 A *Vaisman manifold* is a l.c.K. manifold whose Lee form ω satisfies $\omega_x \neq 0$ for every $x \in M$, and $\nabla^g \omega = 0$. □

Cf. [DrOr]. For instance let $\mathbb{C}H^n(\lambda) = \big(\mathbb{C}^n \setminus \{0\}\big)/G_\lambda$ be the complex Hopf manifold (cf. e.g., [KoNo], Vol. II, p. 137) where G_λ is the discrete group generated by the transformation

$$\mathbb{C}^n \setminus \{0\} \to \mathbb{C}^n \setminus \{0\}, \quad z \longmapsto \lambda z, \quad \lambda \in \mathbb{C} \setminus S^1 ,$$

(G_λ acts freely on $\mathbb{C}^b \setminus \{0\}$ as a properly discontinuous group of holomorphic transformations). Let g_0 be the Boothby metric on $\mathbb{C}^n H(\lambda)$, i.e.,

$$g_0 = \big|z\big|^{-2}\, \delta_{jk}\, dz^j \otimes d\overline{z}^k$$

with respect to a local complex coordinate system (z^j) on $\mathbb{C}^n H(\lambda)$. Then $(\mathbb{C}H^n(\lambda), g_0)$ is a Vaisman manifold with the Lee form $\omega_0 = d\,\log |z|^2$ (cf. e.g., [DrOr], p. 21). Vaisman manifolds were introduced in [Vai2] and called *generalized Hopf manifolds* there. However (as pointed out in [DrOr], p. 33) the term generalized Hopf manifold was already in use for other generalizations of complex Hopf manifolds, cf. e.g., [BrVe]. When $\omega = 0$, g is a Kählerian metric, so the class of l.c.K. manifolds contains that of Kählerian manifolds. Nevertheless, when $\omega \neq 0$ it is questionable whether M admits any globally defined Kählerian metric (e.g., every compact Vaisman manifold has an odd first Betti number[21], cf. [DrOr], p. 8).

Let M be a Vaisman manifold, and let $B := \omega^\sharp \in \mathfrak{X}(M)$, where $\sharp$ denotes raising of indices with the metric g.

Definition 4.113 B is called the *Lee field* of M, and $A := JB$ is the *anti-Lee field.* □

[21] While the odd order Betti numbers of a compact Kählerian manifold are known to be even, cf. e.g., [Gol].

As $[A, B] = 0$ the distribution $\mathbb{R}B \oplus \mathbb{R}A$ is completely integrable and gives rise to a foliation $\mathscr{F}$ of M by Riemannian surfaces (the *vertical foliation* of M). By a result in [Vai2], each leaf of $\mathscr{F}$ is complex parallelizable, and a totally geodesic and locally flat submanifold of M. In particular $\mathscr{F}$ is harmonic. Also, by a result of K. Tsukada (cf. Theorem 2.2 in [Tsu1], p. 4) $\mathscr{F}$ is transversally Kählerian, and every holomorphic map $\varphi : M \to M'$ of compact Vaisman manifolds is foliated (i.e., maps the leaves of $\mathscr{F}$ to leaves of the vertical foliation of the target Vaisman manifold M'). A slight modification of the proof of Theorem 2.5 in [Tsu2], p. 4, shows that the statement remains true when φ is anti-holomorphic. As an application of Theorem 4.98 above one obtains

Corollary 4.114 *Let $\varphi : M \to M'$ be a holomorphic (anti-holomorphic) map of compact Vaisman manifolds. Then φ is transversally harmonic (with respect to the vertical foliations of $\mathscr{F}$ and $\mathscr{F}'$) and an absolute minimum of E_T in $[\varphi]$.*

By a result of A. Lichnerowicz (cf. [Lic]) any holomorphic map $\varphi : M \to M'$ of compact Kähler manifolds is harmonic and an absolute minimum in its homotopy class. When M is a l.c.K. manifold A. Lichnerowicz's theorem doesn't hold, in general. Indeed, by a result of I. Vaisman (cf. [Vai3]) φ is harmonic if and only if the Lee field of M lies in $\mathrm{Ker}(d\varphi)$. Corollary 4.114 is therefore a l.c.K. analog to Lichnerowicz's theorem, for $\pm$ holomorphic maps between compact Vaisman manifolds.

Let $(M, \mathscr{F})$ be a foliated manifold, and let $(g_{\nu(\mathscr{F})}, J)$ be a transverse Hermitian structure.

Definition 4.115 $\mathscr{F}$ is a (*transversally*) *locally conformal Kähler* (*l.c.K.*) foliation if there is a globally defined closed basic 1-form $\omega \in \Omega^1_B(\mathscr{F})$ (the *basic Lee form*) such that $d_B \Omega_T = \omega \wedge \Omega_T$. □

For instance, the simple foliation defined by the C^∞ submersion $f : M \to N$ of M onto a l.c.K. manifold N is transversally l.c.K., in the sense of Definition 4.115.

Let $(\mathscr{F}, g_Q, J)$ be a transversally l.c.K. manifold, and let ω be its basic Lee form. As $d_B\omega = 0$ (by Poincaré's lemma) there is an open cover $\{U_i\}_{i\in I}$ of M and a family $\{f_i\}_{i\in I}$ of C^∞ functions $f_i : U_i \to \mathbb{R}$ such that $\omega = df_i$ on U_i. In particular, as ω is basic, the functions f_i must be basic, as well, i.e., $f_i \in \Omega^0_B(\mathscr{F}_{U_i})$. Let $j_i : U_i \to M$ be the inclusion. Then $g_i := \exp(-f_i)\, j_i^*\, g_Q$ is holonomy invariant anad compatible with J. Moreover, if $\Omega_i(r, s) := g_i(r, Js)$, then

$$\left(\Omega_i\right)_T \in \Omega^2_B(\mathscr{F}), \quad d_B(\Omega_i)_T = 0,$$

hence $\mathscr{F}_{U_i}$ is Kählerian.

Definition 4.116 The transverse vector field $B \in \ell(\mathscr{F})$ determined by

$$g_Q(B, s) = \omega(Y_s), \quad s \in C^\infty(Q), \quad Y_s \in \mathfrak{X}(M), \quad \pi Y_s = s,$$

is the *transverse Lee field* of $\mathscr{F}$. □

Let $(M, \mathscr{F})$, $(M', \mathscr{F}')$ be two foliated manifolds and let $\varphi : M \to M'$ be a foliated map. Let g_Q be a holonomy invariant bundle metric in $Q = \nu(\mathscr{F})$, and let ∇, ∇', by two adapted connections in Q and $Q' = \nu(\mathscr{F}')$. The pair (∇, ∇') induces a connection ∇^{φ} in the vector bundle $\nu(\mathscr{F})^* \otimes \varphi^{-1}\nu(\mathscr{F}') \to M$. Let $(U, x^s, y^i) \in \mathscr{F}$ and $(U', x'^r, y'^{\alpha}) \in \mathscr{F}'$ be two foliated charts, chosen such that $\varphi(U) \subset U'$. Moreover, let $\nabla^{\varphi}_j \varphi^{\alpha}_i$ be the C^{∞} functions defined by

$$\nabla^{\varphi}_{\partial/\partial y^j} d_T \varphi = \left(\nabla^{\varphi}_j \varphi^{\alpha}_i\right) \lambda^i \otimes Y_{\alpha}$$

where $\lambda^i \in C^{\infty}_B(Q^*)$ and $(\lambda^i)_T = d\, y^i$. Finally, let

$$\tau(\varphi;\, g_Q,\, \nabla,\, \nabla') = \tau^{\alpha}(\varphi;\, g_Q,\, \nabla,\, \nabla')\, Y_{\alpha} \in C^{\infty}\big(U,\, \varphi^{-1}\,\nu(\mathscr{F})\big),$$

$$\tau^{\alpha}(\varphi;\, g_Q,\, \nabla,\, \nabla') = g^{ij}\left(\nabla^{\varphi}_j \varphi^{\alpha}_i\right).$$

Note that, give a holonomy invariant bundle metric $g_{Q'}$ on Q', if ∇ and ∇' are the transverse Levi-Civita connections of $(\mathscr{F}, g_Q)$ and $(\mathscr{F}', g_{Q'})$, then $\tau(\varphi;\, g_Q,\, \nabla,\, \nabla') = \tau(\varphi)$. We may state

Proposition 4.117 *Let $(\mathscr{F}, J)$ and $(\mathscr{F}', J')$ be two transversally holomorphic foliations, respectively, on M and M'. Let ∇ and ∇' be almost complex (i.e., $\nabla J = 0$ and $\nabla' J' = 0$) adapted torsion-free connections, respectively, in Q and Q'. Let g_Q be a holonomy invariant bundle metric in Q, compatible with J. If $\varphi : M \to M'$ is a transversally holomorphic foliated map, then*

$$\tau(\varphi;\, g_Q,\, \nabla,\, \nabla') = 0.$$

The proof of Proposition 4.117 is straightforward (a few details are provided in Appendix B to this section). Proposition 4.117 together with (4.141)–(4.142) implies the statement (i) $\Longleftrightarrow$ (ii) in Theorem 4.98. To prove the last statement in Theorem 4.98 let us set

$$E_{\pm}(\varphi) = \int_M k_{\pm}(\varphi)\, \mu.$$

Then a foliated map φ is transversally holomorphic (respectively, anti-holomorphic) if and only if $E_-(\varphi) = 0$ [respectively, $E_+(\varphi) = 0$]. Let $\psi : M \to M'$ be a foliated map smoothly homotopic to φ. Then

$$E_T(\psi) = E_+(\psi) + E_-(\psi) \geq \left|E_+(\psi) - E_-(\psi)\right| = |K(\psi)| = |K(\varphi)|$$

where the last equality is got by picking up a smooth homotopy φ_t, by foliated maps, with $\varphi_1 = \psi$ and $\varphi_0 = \varphi$, followed by applying Theorem 4.110. Finally,

$$E_T(\psi) \geq \left|E_+(\varphi) - E_-(\varphi)\right| \geq \begin{cases} E_+(\varphi) & \text{if } \varphi \text{ is transversally holomorphic,} \\ \\ E_-(\varphi) & \text{if } \varphi \text{ is transversally anti-holomorphic.} \end{cases}$$

Q.e.d.

Let $\mathscr{F}$ be a l.c.K. foliation of M. The transverse Levi-Civita connection of the (local) bundle metrics g_i glue up to a (globally defined) torsion-free adapted connection D in Q expressed by

$$D_X s = \begin{cases} \pi\,[X,\,\sigma(s)] & \text{if } X \in \mathfrak{X}_{\mathscr{F}}\,, \\ \\ \nabla_X s - \frac{1}{2}\left\{\omega(X)\,s + \omega\big(\sigma(s)\big)\,X - g_Q\big(\pi\,X,\ s\big)\ B\right\} & \text{if } X \in C^\infty(P^\perp), \end{cases}$$

(D is the *transverse Weyl connection*). Clearly $DJ = 0$. The statement (i) $\Longleftrightarrow$ (ii) in Theorem 4.98 admits the following generalization.

Proposition 4.118 *Let $\varphi : M \to M'$ be a transversally holomorphic foliated map of $(M,\ \mathscr{F})$ into $(M',\ \mathscr{F}')$, where $\mathscr{F}$ is a l.c.K. foliation of basic Lee form ω, while $\mathscr{F}'$ is a Kählerian foliation. The following statements are equivalent:*

(i) *φ is transversally harmonic.*

(ii) $\text{grad}_B\, \varphi^\alpha \in \text{Ker}\big(\kappa - \omega\big)$ *for any $1 \leq \alpha \leq q'$, and any foliated chart $(U',\ x'^r,\ y'^\alpha) \in \mathscr{F}'$.*

The proof of Proposition 4.118 follows from

$$\tau^\alpha\big(\varphi\,;\ g_Q\,,\ D,\ \nabla'\big) = \tau^\alpha(\varphi) - \omega\big(\text{grad}_B\, \varphi^\alpha\big)$$

together with (4.141)–(4.142).

4.10.1.4 Appendix A

Let $(\mathscr{F},\ g_Q)$ be a Riemannian foliation of M and let g be a bundle-like metric inducing g_Q. Let $C^\infty_B\big(\Lambda^k\, Q^*\big)$ consist of all sections $\alpha \in C^\infty\big(\Lambda^k\, Q^*\big)$ such that $\mathscr{L}_X \alpha = 0$ for any $X \in \mathfrak{X}_{\mathscr{F}}$. For any $\alpha \in C^\infty\big(\Lambda^k\, Q^*\big)$ let $\alpha_T \in \Omega^k(M)$ be the k-form given by

$$\alpha_T(Y_1\,,\ \cdots\,,\ Y_k) := \alpha\big(\pi\, Y_1\,,\ \cdots\,,\ \pi\, Y_k\big), \quad Y_i \in \mathfrak{X}(M), \quad 1 \leq i \leq k.$$

Then $\alpha \longmapsto \alpha_T$ is an isomorphism $C^\infty_B(\Lambda^k\, Q^*) \simeq \Omega^k_B(\mathscr{F})$. Let g^* be the bundle metric induced by g in $\Lambda^k\, T^*(M) \to M$. Also, let g^*_Q be the metric induced by g_Q in $\Lambda^k\, Q^* \to M$.

Lemma 4.119 *$\alpha \to \alpha_T$ is an isometry, i.e., $g^*(\alpha_T\,,\ \beta_T) = g^*_Q(\alpha\,,\ \beta)$ for any $\alpha,\ \beta \in C^\infty_B(\Lambda^k Q^*)$.*

Proof Let $(U, x^\alpha, y^j) \in \mathscr{F}$ be a foliated chart and let

$$\begin{pmatrix} g_{\alpha\beta} & g_{\alpha, j+p} \\ g_{i+p, \beta} & g_{i+p, j+p} \end{pmatrix}$$

be the local components of g. Then $g_{\alpha\beta}$ are the components of the leafwise metric induced by g, hence $\left[g_{\alpha\beta}\right]$ is invertible. Let us set

$$\left[g^{\alpha\beta}\right] = \left[g_{\alpha\beta}\right]^{-1}.$$

A calculation shows that the $P^\perp$ component of $\partial/\partial y^i$ is given by

$$\left(\frac{\partial}{\partial y^i}\right)^\perp = g^{\alpha\beta}\, g_{i+p, \beta}\, \frac{\partial}{\partial x^\alpha} - \frac{\partial}{\partial y^i}.$$

Then

$$(g_Q)_{ij} = g_Q\Big(\pi\, \frac{\partial}{\partial y^i},\ \pi\, \frac{\partial}{\partial y^j}\Big) = g\Big(\Big(\frac{\partial}{\partial y^i}\Big)^\perp,\ \Big(\frac{\partial}{\partial y^j}\Big)^\perp\Big)$$

yields

$$(g_Q)_{ij} = g_{i+p, j+p} - g^{\lambda\mu}\, g_{i+p, \lambda}\, g_{j+p, \mu}. \tag{4.143}$$

For simplicity, we look at the case $k = 2$. Let $\lambda^i \in C^\infty_B(Q^*)$ determined by $(\lambda_i)_T = dy^i$. Every $\alpha \in C^\infty_B\left(\Lambda^2 Q^*\right)$ may be represented as

$$\alpha = \alpha_{ij}(y)\, \lambda_i \wedge \lambda^j$$

so that

$$g_Q^*(\alpha, \beta) = \frac{1}{2}\, \alpha_{ij}\, \beta_{k\ell} \left\{(g_Q)^{ik}\, (g_Q)^{j\ell} - (g_Q)^{i\ell}\, (g_Q)^{jk}\right\}, \quad \left[(g_Q)^{ij}\right] = \left[(g_Q)_{ij}\right]^{-1}.$$

Moreover, let us set

$$\begin{pmatrix} \tilde{g}^{\alpha\beta} & g^{\alpha, j+p} \\ g^{i+p, \beta} & g^{i+p, j+p} \end{pmatrix} = \begin{pmatrix} g_{\alpha\beta} & g_{\alpha, j+p} \\ g_{i+p, \beta} & g_{i+p, j+p} \end{pmatrix}^{-1}.$$

Then

$$g^*(\alpha_T, \beta_T) = \frac{1}{2}\, \alpha_{ij}\, \beta_{k\ell} \left\{g^{i+p, k+p}\, g^{j+p, \ell+p} - g^{i+p, \ell+p}\, g^{j+p, k+p}\right\}.$$

To end the proof it suffices to show that

$$(g_Q)^{ij} = g^{i+p,\, j+p} \, .$$

Indeed [by (4.143)],

$$g^{i+p,\, j+p}\,(g_Q)_{jk} = g^{i+p,\, j+p}\left\{g_{j+p,\, k+p} - g^{\lambda\mu}\, g_{j+p,\, \lambda}\, g_{k+p,\, \mu}\right\} =$$

$$= \delta^i_k - g^{i+p,\, \alpha}\, g_{\alpha,\, k+p} - g^{\lambda\mu}\, g^{i+p,\, j+p}\, g_{j+p,\, \lambda}\, g_{k+p,\, \mu} =$$

$$= \delta^i_k - g^{i+p,\, \alpha}\, g_{\alpha,\, k+p} - g^{\lambda\mu}\left[- g^{i+p,\, \beta}\, g_{\beta\lambda}\right] g_{k+p,\, \mu} = \delta^i_k \, .$$

Q.e.d.

4.10.1.5 Appendix B

A transversally holomorphic foliation $\mathcal{F}$ may be thought of (cf. A. Haefliger, [Hae]) as the synthetic object $\left\{U_i\,,\quad f_i\,:\,U_i \to N,\quad \gamma_{ij} \in \Gamma\right\}$ where
(i) $\{U_i\}$ is an open cover of M,
(ii) N is a complex manifold,
(iii) Γ is the pseudogroup of all local complex analytic automorphisms of N,
(iv) $f_i : U_i \to N$ is a C^∞ submersion,
(v) $\gamma_{ij} \in \Gamma$, such that $\gamma_{ij} \circ f_j = f_i$ on $U_i \cap U_j$. Each f_i is referred to as a *local defining submersion* of $\mathcal{F}$.

Let $\mathcal{F}$ be a transversally holomorphic foliation, and let $f : U \to N$ be a local defining submersion of $\mathcal{F}$, and let $\{w^j\}$ be a local complex coordinate system on N. Let us set $z^j = w^j \circ f$, $1 \le j \le n$, and let $\theta^j \in C^\infty_B(Q^* \otimes \mathbb{C})$ such that $(\theta^j)_T = dz^j$. Also let $\alpha_j \in C^\infty_B(Q \otimes \mathbb{C})$ be given by

$$\alpha_{j,x} = (d_T f)^{-1}_x \left(\frac{\partial}{\partial w^j}\right)_{f(x)}, \qquad x \in U.$$

Thinking of $\mathcal{F}$ as determined by the data $\{U_i\,,\ f_i\,,\ \gamma_{ij}\}$, the (holonomy invariant) complex structure J on Q is related to the complex structure J_N on N by

$$(d_T f)_x \circ J_x = (J_N)_{f(x)} \circ (d_T f)_x\,, \quad x \in U,$$

hence α_j is a local frame of $\nu^+(\mathcal{F})$, defined on the open set U. Also

$$\theta^j(\alpha_k) = \delta^j_k\,, \quad \theta^j(\alpha_{\overline{k}}) = 0, \quad \alpha_{\overline{k}} := \overline{\alpha_k}\,.$$

Let φ be a foliated map of $(M,\ \mathcal{F})$ into $(M'\,,\ \mathcal{F}')$, with $\mathcal{F}$ and $\mathcal{F}'$ transversally holomorphic. Let

$$Y_\beta = \varphi^{-1}\,\alpha'_\beta\,, \quad 1 \le \beta \le n'\,,$$

be the natural lifts to $\varphi^{-1} Q' \otimes \mathbb{C}$ of α'_β, determined by a choice of local defining submersion $f' : U' \to N'$ for $(M', \mathscr{F}')$. Then

$$\partial_T \varphi = \varphi^\beta_j \, \theta^j \times Y_\beta \,, \quad \overline{\partial}_T \varphi = \varphi^\beta_{\overline{j}} \, \theta^{\overline{j}} \otimes Y_\beta \,,$$

for some C^∞ functions φ^β_j, $\varphi^\beta_{\overline{j}} : U \to \mathbb{C}$. If $\varphi^\beta = z'^\beta \circ \varphi$ and $Z_j \in C^\infty\big(T(M) \otimes \mathbb{C}\big)$ is a complex tangent vector field such that $\pi \, Z_j = \alpha_j$, then

$$\varphi^\beta_j = Z_j\big(\varphi^\beta\big), \quad \varphi^\beta_{\overline{j}} = Z_{\overline{j}}(\varphi^\beta), \quad Z_{\overline{j}} := \overline{Z_j} \,.$$

Let ∇ be an adapted connection in Q and let Γ^A_{BC} be determined by

$$\nabla_{Z_B} \alpha_C = \Gamma^A_{BC} \, \alpha_A$$

where the indices $A, \ B, \ C$ range in $\{1, \ \cdots, \ n, \ \overline{1}, \ \cdots, \ \overline{n}\}$. Then $\nabla J = 0$ implies $\Gamma^{\overline{k}}_{ij} = 0$ and $\Gamma^{\overline{k}}_{\overline{i}j} = 0$. Moreover,

$$\Gamma^{\overline{i}}_{j\overline{k}} = 0 \tag{4.144}$$

as a mere consequence of the torsion-free assumption. Indeed $T_\nabla(Z_j \, , \, Z_{\overline{k}}) = 0$ may be written as

$$\Gamma^{\overline{i}}_{j\overline{k}} \alpha_{\overline{i}} - \Gamma^i_{\overline{k}j} \alpha_i = \pi \left[Z_j \, , \, Z_{\overline{k}}\right]$$

or (by applying $\theta^{\overline{i}}$)

$$\Gamma^{\overline{i}}_{j\overline{k}} = \big(d \, \overline{z}^i\big) \left[Z_j \, , \, Z_{\overline{k}}\right]. \tag{4.145}$$

Next, let us apply $(d_T f)_x$ to $\pi_x \, Z_{j,x} = \alpha_{j,x}$, $x \in U$, so that to obtain $(d_x f) Z_{j,x} = \big(\partial/\partial w^j\big)_{f(x)}$ and then

$$Z_{\overline{j}}\big(\overline{z}^i\big) = Z_{\overline{j}}\big(\overline{w}^i \circ f\big) = \big[(d \, f) Z_{\overline{j}}\big]\big(\overline{w}^i\big) = \delta^i_j \,.$$

Finally,

$$0 = 2 \, \big(d^2 \, \overline{z}^j\big)\big(Z_j \, , \, Z_{\overline{k}}\big) = -(d\overline{z}^j)\big[Z_j \, , \, Z_{\overline{k}}\big]$$

and (4.145) leads to (4.144). Similar to ∇, the only "surviving" coefficients of ∇' are $\Gamma'^\alpha_{\beta\gamma}$. Let φ be transversally holomorphic (i.e., $\varphi^{\overline{\beta}}_{\overline{j}} = 0$). Then

$$d_T \varphi = \varphi^\beta_j \, \theta^j \otimes Y_\beta + \varphi^{\overline{\beta}}_{\overline{j}} \, \theta^{\overline{j}} \otimes Y_{\overline{\beta}} \,.$$

Moreover,

$$\nabla_{Z_j} \theta^i = -\Gamma^i_{jk} \, \theta^k \,, \quad \nabla_{Z_j} \theta^{\overline{i}} = 0,$$

$$(\varphi^{-1}\nabla')_{Z_j} Y_\gamma = \varphi_j^\beta \, \Gamma'^\alpha_{\beta\gamma} \, Y_\alpha \, , \quad (\varphi^{-1}\nabla')_{Z_j} Y_{\overline{\gamma}} = 0.$$

The proof of the first two identities is straightforward. As to the last two, it suffices to choose[22] $Z_j \in V(\mathscr{F}) \otimes \mathbb{C}$. These identities lead to

$$\nabla^\varphi_{Z_j} d_T\varphi = \left\{ Z_j\left(\varphi_k^\alpha\right) - \varphi_i^\alpha \, \Gamma^i_{jk} + \varphi_j^\beta \varphi_k^\gamma \Gamma'^\alpha_{\beta\gamma} \, \theta^k \otimes Y_\alpha + Z_j\left(\varphi_{\overline{k}}^{\overline{\alpha}}\right) \theta^{\overline{k}} \otimes Y_{\overline{\alpha}} \, . \right.$$

Let us set $X_{j\overline{k}} := \left[Z_j \, , \, Z_{\overline{k}}\right]$ for the sake of brevity. As shown above $X_{j\overline{k}} \in C^\infty(P \otimes \mathbb{C})$. Also, the calculation

$$X(\varphi^\alpha) = X\left(z'^\alpha \circ \varphi\right) = \left[\left(d_{\varphi(x)} f'\right) (d_x\varphi) X\right](w'^\alpha) = 0$$

for any[23] $X \in P_x$, shows that φ^α are (local) basic functions on $(U, \, \mathscr{F}_U)$. Finally,

$$\left(\nabla^\varphi_{Z_j} d_T\varphi\right) Z_{\overline{k}} = Z_j\left(\varphi_{\overline{k}}^{\overline{\alpha}}\right) Y_{\overline{\alpha}} = \left\{ X_{j\overline{k}}(\varphi^{\overline{\alpha}}) + Z_{\overline{k}}\left(\varphi_j^{\overline{\alpha}}\right) \right\} Y_{\overline{\alpha}} = 0$$

[as $\varphi^{\overline{\alpha}} \in \Omega^0_B(\mathscr{F}_U)$ and $\varphi_j^{\overline{\alpha}} = 0$], i.e.,

$$\tau^\alpha\left(\varphi \, ; \, g_Q \, , \, \nabla \, , \, \nabla'\right) = 0.$$

Q.e.d.

4.10.2 $(\mathscr{F}, \mathscr{G})$-Harmonic Maps of Foliated Riemannian Manifolds

The purpose of the present section is to report on the results by S. Dragomir and A. Tommasoli, [DrTom]. We study $(\mathscr{F}, \mathscr{G})$-harmonic maps, i.e., critical points $\phi \in C^\infty(M, N)$ of the functional

$$E_T(\phi) = \frac{1}{2} \int_M \left\| d_T\varphi \right\|^2 d \, \mathrm{v}_g$$

with respect to variations through foliated maps. In particular we study $(\mathscr{F}, \mathscr{G})$-harmonic morphisms, i.e., smooth foliated maps preserving the basic Laplace equation $\Delta_B u = 0$. We shall show that CR maps of compact Sasakian manifolds preserving the Reeb flows are weakly stable $(\mathscr{F}, \mathscr{G})$-harmonic maps. We study $(\mathscr{F}, \mathscr{G}_0)$-harmonic maps into spheres and give foliated analogs to B. Solomon's (cf. [Sol]) results.

---

[22] This is possible because f induces an isomorphism $C^\infty_B(U, \, Q) \simeq \mathfrak{X}(N)$, cf. [Mol], p. 37.

[23] Because φ is foliate and $\mathrm{Ker}(d \, f') = P'$.

J.J. Konderak and R.A. Wolak (cf. [KoWo1]-[KoWo2]) defined transversally harmonic maps as smooth maps $\phi : M \to N$ of foliated Riemannian manifolds satisfying a condition similar to the vanishing of the tension tensor in Riemannian geometry. They used the canonical adapted connections to build a foliated analog $\beta_T(\phi)$ to the second fundamental form of ϕ and computed the trace

$$\tau_T(\phi) \equiv \text{Trace}_{g_Q} \beta_T(\phi)$$

with respect to the given holonomy invariant Riemannian bundle metric g_Q in the transverse bundle. Then ϕ is transversally harmonic if $\tau_T(\phi) = 0$. However such maps do not, in general, extremize the energy functional (4.146), but rather the integral of the energy density divided by the volume of each fiber (cf. Theorem 2.3 in [KoWo2]). In the present section we deal with a different, perhaps more natural, notion of a transversally harmonic map between foliated Riemannian manifolds, referred to as a $(\mathscr{F}, \mathscr{G})$-*harmonic map* so that to avoid the confusion with the work by J.J. Konderak and R.A. Wolak (cf. *op. cit.*) and also with the works by Y-J. Chiang and R.A. Wolak, [ChWo], H.K. Pak and J.H. Park, [PaPa], M.J. Jung and S.D. Jung, [JuJu].

In Sect. 4.10.2.1 we start from the natural generalization of the energy density of a smooth foliated, i.e., the *transverse energy density*

$$e_T(\phi) = \frac{1}{2} \left\| d_T \phi \right\|^2 \in \Omega^0_B(\mathscr{F}), \quad \phi \in C^\infty(\mathscr{F}, \mathscr{G}),$$

cf. A. El Kacimi Alaoui and E.G. Gomez, [KAGo]. This coincides with the ordinary energy density of ϕ when $\mathscr{F}$ and $\mathscr{G}$ are the trivial foliations by points. A smooth foliated map $\phi : M \to N$ is said to be a $(\mathscr{F}, \mathscr{G})$-*harmonic map* if ϕ is a critical point of the energy functional

$$E_T\big(\mathscr{F}, \mathscr{G}\big) \to \mathbb{R}, \quad E_T(\phi) = \int_M e_T(\phi) \, d\, \text{v}_g \,. \tag{4.146}$$

The Euler-Lagrange equations of the variational principle $\delta\, E_T(\phi) = 0$ are

$$\text{Trace}_{g_Q} \beta_T(\phi) - \phi_* \, \tau_{\mathscr{F}} = 0 \tag{4.147}$$

where $\tau_{\mathscr{F}}$ is the mean curvature of the foliation $\mathscr{F}$. Therefore, transverse harmonicity as proposed in [KoWo1] coincides with $(\mathscr{F}, \mathscr{G})$-harmonicity as based on the action (4.146) provided $\mathscr{F}$ is harmonic (i.e., all leaves of $\mathscr{F}$ are minimal submanifolds of M). Equations (4.147) turn out to be a second-order nonlinear system of PDEs whose principal part is the basic Laplacian Δ_B of $(M, \mathscr{F}, g)$.

In Sect. 4.10.2.2 we start a study of $(\mathscr{F}, \mathscr{G})$-harmonic morphisms [that are to $(\mathscr{F}, \mathscr{G})$-harmonic maps what harmonic morphisms are to ordinary harmonic maps]. The basic Laplacian Δ_B is not elliptic (in fact Δ_B is only transversally elliptic, in the sense of [KaAl]) yet the Laplace-Beltrami operator Δ of (M, g) is an elliptic extension [from basic functions $\Omega_B^0(\mathscr{F})$ to arbitrary smooth functions $C^\infty(M)$] of Δ_B. Therefore, positive solutions to $\Delta_B u = 0$ satisfy the Harnack inequality and consequently a class of $(\mathscr{F}, \mathscr{G})$-harmonic morphisms (with values in a Riemannian manifold N thought of as endowed with the trivial foliation $\mathscr{G} = \mathscr{G}_0$) may be shown to consist of open maps. This is a foliated analog to a result by B. Fuglede, [Fug]. Maps $\phi^* : H_B^1(\mathscr{G}) \to H_B^1(\mathscr{F})$ among basic cohomology groups, induced by $(\mathscr{F}, \mathscr{G})$-harmonic morphisms $\phi : M \to N$, are investigated as an application of Park and Richardson's Hodge-de Rham decomposition theorem (cf. [PaRi]) for the basic complex $\big(\Omega_B^\bullet(\mathscr{F}),\, d_B\big)$.

In Sect. 4.10.2.3 we derive the second variation formula for the transverse energy and start a study of stability of $(\mathscr{F}, \mathscr{G})$-harmonic maps.

As an application of the techniques developed in the present section we continue the study of foliated maps of trasnversally Kählerian foliations, as motivated by the lack of a l.c.K. analog to Lichnerowicz's theorem, cf. our earlier discussion in this chapter. Very much the same pathology is known to occur in CR geometry. Indeed CR maps of strictly pseudoconvex CR manifolds fail (by a result of H. Urakawa, [Ura]) to be harmonic (even if the chosen metrics are Sasakian, which is perhaps the closest odd dimensional analog to Kählerian metrics). Both problems find a solution within foliation theory (cf. Theorem 4.167 in Sect. 4.7.2.4) and the transversally elliptic nonlinear system (4.183) appears as a natural replacement of the ordinary harmonic maps equations.

In Sect. 4.10.2.5 we look at the topology of $(\mathscr{F}, \mathscr{G})$-harmonic maps into spheres and establish a foliated analog to a result by B. Solomon, [Sol].

4.10.2.1 $(\mathscr{F}, \mathscr{G})$-Harmonic maps

Let M be a real n-dimensional, connected, C^∞ manifold. Let

$$\mathscr{F} = \big(\{U_i\}_{i\in I}\,,\; M_0\,,\; \Gamma,\; \{f_i\}_{i\in I}\big)$$

be a codimension q foliation of M, where (i) $\{U_i\}_{i\in I}$ is an open cover of M, (ii) M_0 is a real q-dimensional C^∞ manifold, (iii) Γ is a pseudogroup of local transformations of M_0, and (iv) each $f_i : U_i \to M_0$ is a C^∞ submersion such that for every pair of indices $i, j \in I$ with $U_{ji} = U_i \cap U_j \neq \emptyset$ there is $\gamma_{ji} \in \Gamma$ such that $f_j = \gamma_{ji} \circ f_i$ on U_{ji}. By following A. Haefliger (cf. [Hae]), a foliation $\mathscr{F}$ within this approach is also referred to as a Γ-foliation. Let $x \in M$ and let $i \in I$ such that $x \in U_i$. We set as customary $P_x = \mathrm{Ker}(d_x f_i)$ and denote the resulting completely integrable distribution by $P = T(\mathscr{F})$. Our main object of study are foliated maps from $(M, \mathscr{F})$ into another foliated manifold $(N, \mathscr{G})$, i.e., C^∞ maps $\phi : M \to N$ such that $\phi_* T(\mathscr{F}) \subset \phi^{-1} T(\mathscr{G})$. Throughout $\phi^{-1}E \to M$ denotes the pullback of the bundle $E \to N$ by the map $\phi : M \to N$. A foliated map $\phi : M \to N$ sends leaves of $\mathscr{F}$ onto leaves of $\mathscr{G}$. As such $\phi : M \to N$ induces a map of leaf spaces $\overline{\phi} : M/\mathscr{F} \to N/\mathscr{G}$. In general $M/\mathscr{F}$ has a rather poor structure (a topological space, often non Hausdorff) and $\overline{\phi}$ is only continuous. As customary in foliation theory, calculations of sorts will be conducted on M rather than on $M/\mathcal{F}$, in terms of geometric objects which are invariant under sliding along the leaves.

Let $Q = \nu(\mathcal{F}) = T(M)/T(\mathcal{F})$ be the transverse bundle and let $\pi : Q \to M$ be the projection. We recall that a foliation $\mathcal{F}$ of M is transversally Riemannian if it is endowed with a transverse O(q)-structure. This is equivalent to $\mathcal{F}$ being a $\Gamma(M_0, g_0)$-foliation, where $\Gamma(M_0, g_0)$ is the pseudogroup of all local isometries of the (model) Riemannian manifold (M_0, g_0). In terms of first-order geometric structures this is also equivalent to the existence of a Riemannian bundle metric g_Q on Q that is invariant under sliding along the leaves

$$\stackrel{\circ}{\nabla}_X g_Q = 0, \quad X \in \mathfrak{X}_{\mathcal{F}}, \tag{4.148}$$

where $\stackrel{\circ}{\nabla}$ is the Bott (partial) connection of $(M, \mathcal{F})$. Indeed, if $\mathcal{F}$ is a $\Gamma(M_0, g_0)$ foliation one may set

$$(g_T)_i = f_i^* g_0, \quad i \in I,$$

and observe that $\gamma_{ji} \in \Gamma(M_0, g_0)$ yields $(g_T)_i = (g_T)_j$ on U_{ji}, so that g_T is a globally defined transverse metric on $(M, \mathcal{F})$. Then one may consider $g_Q(r, s) := g_T(Y, Z)$ for any $r, s \in C^\infty(Q)$, where $Y, Z \in \mathfrak{X}(M)$ are such that $\pi Y = r$ and $\pi Z = s$.

Let g be a Riemannian metric on M and let $\sigma_g : Q \to T(\mathcal{F})^\perp$ be the natural vector bundle isomorphism. The metric g induces a bundle metric on Q

$$g_Q(r, s) := g\big(\sigma_g(r), \sigma_g(s)\big), \quad r, s \in Q, \tag{4.149}$$

yet g_Q may fail to satisfy (4.148), and (4.149) is indeed a transverse Riemannian structure provided that g is bundle-like. Let $(M, \mathcal{F}, g)$ and $(N, \mathcal{G}, h)$ be two foliated Riemannian manifolds, where $\mathcal{F}$ and $\mathcal{G}$ are Riemannian foliations, and the Riemannian metrics g and h are bundle-like. Let $C^\infty(\mathcal{F}, \mathcal{G})$ denote the space of all C^∞ foliated maps of $(M, \mathcal{F})$ into $(N, \mathcal{G})$. For every $\phi \in C^\infty(\mathcal{F}, \mathcal{G})$ the *transverse differential* of ϕ is the vector bundle morphism

$$d_T\phi : \nu(\mathcal{F}) \to \phi^{-1}\, \nu(\mathcal{G}),$$

$$(d_T\phi)_x s = \pi'_{\phi(x)}\, (d_x\phi) Y, \quad s \in Q_x, \quad Y \in T_x(M), \quad \pi_x(Y) = s,$$

where $\pi' : T(N) \to \nu(\mathcal{G})$ is the projection. Let $\{s_i : 1 \le i \le q\}$ be a local g_Q-orthonormal frame in Q, defined on the open set $U \subset M$. The norm $\|d_T\phi\| : M \to [0, +\infty)$ is given by

$$\|d_T\phi\|_x^2 = \sum_{i=1}^{q} h_{\nu(\mathcal{G}), \phi(x)}\Big(\big(d_T\phi\big)_x s_{i,x}, \big(d_T\phi\big)_s s_{i,x}\Big), \quad x \in U. \tag{4.150}$$

Here $h_{\nu(\mathscr{G})}$ is the bundle metric induced by h on $\nu(\mathscr{G})$. Note that (4.150) is an analog to the Hilbert-Schmidt norm of $d\phi$ (that we shall recall shortly). Let M be oriented, and let

$$d\,\mathrm{v}_g = \sqrt{\mathfrak{g}}\,dx^1 \wedge \cdots \wedge x^n$$

be the canonical volume form on (M, g), locally expressed with respect to a local coordinate system (U, x^i).

Definition 4.120 Let $\Omega \subset\subset M$ be a relatively compact domain. The functional

$$E_{T,\Omega} : C^\infty(\mathscr{F}, \mathscr{G}) \to [0, +\infty), \quad E_{T,\Omega}(\phi) = \frac{1}{2}\int_M \|d_T\phi\|^2 \, d\,\mathrm{v}_g \, ,$$

is the *transverse energy* integral. □

When M is compact we write simply $E_{T,M} = E_T$. The following notion is central to the present section.

Definition 4.121 A foliated map $\phi \in C^\infty(\mathscr{F}, \mathscr{G})$ is said to be $(\mathscr{F}, \mathscr{G})$*-harmonic* if

$$\frac{d}{dt}\Big\{E_{T,\Omega}(\phi_t)\Big\}_{t=0} = 0$$

for any relatively compact domain $\Omega \subset\subset M$ and any smooth 1-parameter variation $\{\phi_t\}_{|t|<\epsilon} \subset C^\infty(\mathscr{F}, \mathscr{G})$ of $\phi = \phi_0$, supported in Ω, i.e., $\mathrm{Supp}(V) \subset \Omega$, where

$$V \in C^\infty\big(\phi^{-1}T(N)\big), \quad V_x = \frac{\partial \phi_t}{\partial t}(x), \quad x \in M.$$

□

Before proceeding with the study of $(\mathscr{F}, \mathscr{G})$-harmonic maps, we ought to recall a few elements of (ordinary) harmonic maps theory. Let (M, g) and (N, h) be Riemannian manifolds, with M compact and orientable. Let $\{E_j : 1 \le j \le n\}$ be a local g-orthonormal frame of $T(M)$, defined on the open subset $U \subset M$. For every C^∞ map $\phi : M \to N$, the Hilbert-Schmidt norm $\|d\phi\| : M \to [0, +\infty)$ of $d\phi$ is given by

$$\|dphi\|_x = \sum_{j=1}^n h_{\phi(x)}\big((d_x\phi)E_{j,x}\, , \, (d_x\phi)E_{j,x}\big), \quad x \in U.$$

The energy integral is

$$E : C^\infty(M, N) \to \mathbb{R}, \quad E(\phi) = \frac{1}{2}\int_M \|d\phi\|^2 \, d\,\mathrm{v}_g \, .$$

A C^∞ map $\phi : M \to N$ is *harmonic* if $\big\{E(\phi_t)/dt\big\}_{t=0} = 0$ for any smooth 1-parameter variation $\{\phi_t\}_{|t|<\epsilon} \subset C^\infty(M, N)$ of $\phi_0 = \phi$. A C^∞ map $\phi : M \to N$ is a *harmonic morphism* if for every harmonic function $v : V \to \mathbb{R}$ (i.e., $\Delta_h v = 0$ in V) defined on the open set $V \subset N$, one has $\Delta_g(v \circ \phi) = 0$ in $\phi^{-1}(V)$. Cf. (4.12) in [EeLe1], p. 15. Throughout Δ_g is the Laplace-Beltrami operator of the Riemannian manifold (M, g). Let $\dim(M) = n \geq \dim(N)$. The Fuglede-Ishihara theorem (cf. e.g., Theorem 4.2.2 in [BaWo], p. 108) asserts that a smooth map is a harmonic morphism if and only if ϕ is both a harmonic map and a horizontally weakly conformal map. For a discussion of horizontal weak conformality the reader may see the wonderful monograph by P. Baird and J.C. Wood (cf. [BaWo], pp. 45–48). Nonconstant harmonic morphisms are open mappings (cf. [Fug] or Theorem 4.3.7 in [BaWo]). A useful characterization of a harmonic morphism $\phi : M \to N$ is that ϕ pulls back harmonic 1-forms into harmonic 1-forms.

Let $(M, \mathscr{F}, g)$ and $(N, \mathscr{G}, h)$ be foliated Riemannian manifolds.

Definition 4.122 A foliated map $\phi \in C^\infty(\mathscr{F}, \mathscr{G})$ is said to be a $(\mathscr{F}, \mathscr{G})$-*harmonic morphism* if for any basic function $f \in \Omega^0_B(\mathscr{G}_V)$ such that $\Delta'_B f = 0$ in $V \subset N$, it follows that $\Delta_B(f \circ \phi) = 0$ in $U = \phi^{-1}(V)$. □

Here Δ_B and Δ'_B are, respectively, the basic Laplacians of $(M, \mathscr{F}, g)$ and $(N, \mathscr{G}, h)$. For every foliated map $\phi : M \to N$ one has

$$\phi^* \, \Omega^0_B(\mathscr{G}_V) \subset \Omega^0_B(\mathscr{F}_U)$$

so that the notion of a $(\mathscr{F}, \mathscr{G})$-harmonic morphism is well defined. Here $\phi^*(f) := \phi \circ f$. Also $\mathscr{F}_U$ and $\mathscr{G}_V$ are the foliations induced on U and V by $\mathscr{F}$ and $\mathscr{G}$. It was a natural open question at the time [DrTom] was written whether a $(\mathscr{F}, \mathscr{G})$-harmonic morphism is a $(\mathscr{F}, \mathscr{G})$-harmonic map. The partial answer provided in [DrTom] is presented in Sect. 4.10.2.2.

Let $(M, \mathscr{F})$ be an orientable foliated manifold, where $\mathscr{F}$ is a transversally oriented foliation of codimension q. Let $\nu \in \Omega^q_B(\mathscr{F})$ be the transversal volume form ($d_B \nu = 0$). Let us assume that $\mathscr{F}$ is Riemannian, and let g be a bundle-like Riemannian metric on M. The characteristic form is the $(n - q)$-form $\chi_{\mathscr{F}} = * \nu$ where $* : \Omega^q(M) \to \Omega^{n-q}(M)$ is the Hodge star operator of (M, g). The Riemannian volume form $d\,\mathrm{v}_g \in \Omega^n(M)$ is then given[24] by $d\,\mathrm{v}_g = \nu \wedge \chi_{\mathscr{F}}$. Let $(N, \mathscr{G}, h)$ be another foliated Riemannian manifold, with $\mathscr{G}$ Riemannian and h bundle-like. Let $\Omega \subset\subset M$ be a relatively compact domain. Our task in this section is to derive the Euler-Lagrange equations of the variational principle $\delta\, E_{T,\Omega}(\phi) = 0$. To this end et $\phi \in C^\infty(\mathscr{F}, \mathscr{G})$ be a foliated map and let $\{\phi_t\}_{|t|<\epsilon} \subset C^\infty(\mathscr{F}, \mathscr{G})$ be a smooth 1-parameter variation of $\phi = \phi_0$ through foliated maps. Next, let

$$\Phi : \tilde{M} := M \times (-\epsilon, \epsilon) \to N, \quad \Phi(x, t) = \phi_t(x), \quad (x, t) \in \tilde{M}.$$

For each $X \in \mathfrak{X}(M)$ we denote by $\tilde{X} \in \mathfrak{X}(\tilde{M})$ the tangent vector field

$$\tilde{X}_{(x,t)} = (d_x \alpha_t) X_x\,, \quad (x, t) \in \tilde{M},$$

[24] With an appropriate choice of the orientation of M and the tangential and transverse orientations of $\mathscr{F}$.

$$\alpha_t : M \to \tilde{M}, \quad \alpha_t(x) = (x, t).$$

Also we set

$$\tilde{P}_{(x,t)} = (d\,\alpha_t)\,T(\mathscr{F})_x\,, \quad (x, t) \in \tilde{M}. \tag{4.151}$$

If $\{X_\alpha : 1 \le \alpha \le p\}$ is a local frame of $T(\mathscr{F})$ (here $p := n - q$) defined on the open set $U \subset M$, then $\{\tilde{X}_\alpha : 1 \le \alpha \le p\}$ is a local frame of $\tilde{P}$ defined on $\tilde{U} = U \times (-\epsilon, \epsilon)$, hence $\tilde{P}$ is a smooth involutive distribution on $\tilde{M}$. By the classical Frobenius theorem $\tilde{P}$ is completely integrable. Let $\tilde{F}$ be the corresponding foliation of $\tilde{M}$, such that $T(\tilde{\mathscr{F}}) = \tilde{P}$. In particular $\alpha_t \in C^\infty(\mathscr{F}, \mathscr{G})$, i.e., $\alpha_t : M \to \tilde{M}$ is a foliated map for each $|t| < \epsilon$. Similarly, if $s \in C^\infty(\nu(\mathscr{F})$ is a section in the transverse bundle, we denote by $\tilde{s} \in C^\infty\big(\nu(\tilde{\mathscr{F}})\big)$ the section

$$\tilde{s}(x, t) = \big(d_T\,\alpha_t\big)_x\,s(x), \quad (x, t) \in \tilde{M}.$$

Also $\phi_*\,s \in C^\infty\big(\phi^{-1}\,\nu(\mathscr{G})\big)$ denotes the section in the pullback bundle $\phi^{-1}\nu(\mathscr{G}) \to M$ given by

$$(\phi_*\,s)(x) = (d_T\phi)_x\,s(x), \quad x \in M\,.$$

Let ∇ be the canonical adapted connection of $(M, \mathscr{F}, g)$

$$\nabla_X s = \begin{cases} \mathring{\nabla}_X s & \text{if } X \in T(\mathscr{F}), \\ \pi\nabla^g_X\,\sigma_g(s) & \text{if } X \in T(\mathscr{F})^\perp, \end{cases} \qquad s \in C^\infty(Q), \tag{4.152}$$

[∇^g is the Levi-Civita connection of (M, g)]. Then $\nabla \in \mathscr{C}\big(\nu(\mathscr{F}),\ g_{\nu(\mathscr{F})}\big)$. Throughout, if (E, g) is a Riemannian vector bundle, $\mathscr{C}(E)$ denotes the affine space of all connections in E, while $\mathscr{C}(E, g)$ is the affine subspace of all $D \in \mathscr{C}(E)$ such that $Dg = 0$. Let $\nabla' \in \mathscr{C}\big(\nu(\mathscr{G}),\ h_{\nu(\mathscr{G})}\big)$ be the canonical adapted connection of $(N, \mathscr{G}, h)$ and let

$$D^\phi = \phi^{-1}\nabla' \in \mathscr{C}\big(\phi^{-1}\,\nu(\mathscr{G})\big)$$

be the pullback of ∇', i.e., the connection induced by ∇' pn $\phi^{-1}\nu(\mathscr{G}) \to M$. Let

$$h^\phi_{\nu(\mathscr{G})} = \phi^{-1}\,h_{\nu(\mathscr{G})}$$

be the pullback of $h_{\nu(\mathscr{G})}$ by ϕ [so that $\big(\phi^{-1}\,\nu(\mathscr{G}),\ h^\phi_{\nu(\mathscr{G})}\big)$ is a Riemannian vector bundle]. It is immediate that

(i) the bundle metric $h^\phi_{\nu(\mathscr{G})}$ is parallel with respect to D^ϕ, i.e.,

$$D^\phi \in \mathscr{C}\big(\phi^{-1}\,\nu(\mathscr{G}),\ h^\phi_{\nu(\mathscr{G})}\big),$$

and

(ii) Φ is a foliated map of $(\tilde{M}, \tilde{\mathscr{F}})$ into $(N, \mathscr{G})$, i.e., $\Phi \in C^\infty(\tilde{F}, \mathscr{G})$. Moreover [by (4.150)],

$$\|d_T \phi_t\|^2 = \sum_{j=1}^{q} h^{\Phi}_{\nu(\mathscr{G})}\big(\Phi_* \tilde{s}_j , \Phi_* \tilde{s}_j\big) \circ \alpha_t , \quad |t| < \epsilon , \tag{4.153}$$

on U. Let us set

$$f(x, t) = \frac{1}{2} \, \|d_T \phi_t\|_x^2, \quad (x, t) \in \tilde{M}.$$

Then [by (4.153)]

$$\frac{\partial f}{\partial t} = \frac{1}{2} \sum_j \frac{\partial}{\partial} \Big[h^{\Phi}_{\nu(\mathscr{G})}\big(\Phi_* \tilde{s}_j , \Phi_* \tilde{s}_j\big)\Big] = \sum_j h^{\Phi}_{\nu(\mathscr{G})}\Big(D^{\Phi}_{\partial/\partial t} \Phi_* \tilde{s}_j , \Phi_* \tilde{s}_j\Big)$$

on $\tilde{U}$.

Lemma 4.123 *Let $\mathbb{V} \in C^\infty\big(\Phi^{-1} T(N)\big)$ and $\mathbf{v} \in C^\infty\big(\Phi^1, \nu(\mathscr{G})\big)$ be given by*

$$\mathbb{V}_{(x,t)} = \big(d_{(x,t)}\Phi\big) \left(\frac{\partial}{\partial t}\right)_{(x,t)}, \quad \mathbf{v}(x, t) = \pi'_{\Phi(x,t)} \mathbb{V}_{(x,t)}, \quad (x, t) \in \tilde{M},$$

where $\pi' : T(N) \to \nu(\mathscr{G})$ is the projection. Then

$$D^{\Phi}_{\partial/\partial t} \, \Phi^* \tilde{s} = D^{\Phi}_{\widetilde{\sigma_g(s)}} \mathbf{v} \tag{4.154}$$

for any $s \in C^\infty\big(\nu(\mathscr{F})\big)$.

The proof of (4.154) is a calculation in local foliated coordinates. For this and further calculations, we need to adopt the following conventions, as to the ranges of the various indices. Let $\big(U, x^\alpha , y^j\big)$ and $\big(\mathscr{U}, X^a , Y^u\big)$ be local coordinate systems adapted to $(M, \mathscr{F})$ and $(N, \mathscr{G})$, respectively, such that $\phi(U) \subset \mathscr{U}$. The range of indices is

$$\alpha , \beta , \cdots \in \{1, \cdots , p\}, \quad i, j, \cdots \in \{1, \cdots , q\},$$

$$a, b, \cdots \in \{1, \cdots , p'\}, \quad u, v, \cdots \in \{1, \cdots , q'\}.$$

Then $\big\{s_j = \pi \, \big(\partial/\partial y^j\big) \, : 1 \le j \le q\big\}$ is a local frame in Q. Given $s \in C^\infty(Q)$ locally represented as $s = f^j s_j$ for some $f^i \in C^\infty(U)$, one has

$$\tilde{s} = \big(f^j \circ p\big) \tilde{s}_j \quad \text{on} \quad \tilde{U} = U \times (-\epsilon , \epsilon),$$

where $p : \tilde{M} \to M$ is the projection. Also

$$\left(\Phi_*\tilde{s}\right)(x,\, t) = f^j(x)\left(d_T\Phi\right)_{(x,t)}\tilde{\pi}_{(x,t)}(d_x\alpha_t)\left(\frac{\partial}{\partial y^j}\right)_x$$

where $\tilde{\pi} : T(\tilde{M}) \to \nu(\tilde{\mathscr{F}})$ is the projection. Thus,

$$\left(\Phi_*\,\tilde{s}\right)(x,\, t) = \pi'_{\Phi(x,t)}\left(d_{(x,t)}\Phi\right)\tilde{Y}_{(x,t)}$$

where $Y := f^j\,\partial/\partial y^j \in \mathfrak{X}(U)$. Finally,

$$\Phi_*\,\tilde{s} = \tilde{Y}\left(\Phi^{p'+u}\right)\pi'^{\Phi}\left(\frac{\partial}{\partial Y^u}\right)^{\Phi}. \tag{4.155}$$

As to the notations in (4.155), for every tangent vector field W on N its *natural lift* is

$$W^{\Phi} \in C^{\infty}\left(\Phi^{-1}T(N)\right), \quad W^{\Phi}(x,\, t) = W_{\Phi(x,t)}\,, \quad (x,\, t) \in \tilde{M}.$$

Also $\Phi^{p'+u} = Y^u \circ \Phi$ and $\pi'^{\Phi} : \Phi^{-1}T(N) \to \Phi^{-1}\nu(\mathscr{G})$ is the pullback by Φ of the projection$\pi' : T(N) \to \nu(\mathscr{G})$. The canonical adapted connection $\nabla' \in \mathscr{C}\big(\nu(\mathscr{G}),\, h_{\nu(\mathscr{G})}\big)$ is locally expressed by

$$\nabla'_{\partial/\partial X^a}\pi'\left(\frac{\partial}{\partial Y^u}\right) = 0, \quad \nabla'_{\partial/\partial Y^u}\pi'\left(\frac{\partial}{\partial Y^v}\right) = \Gamma'^{w}_{uv}\,\pi'\left(\frac{\partial}{\partial Y^w}\right).$$

Consequently, its pullback $D^{\Phi} = \Phi^{-1}\nabla'$ reads

$$D^{\Phi}_Z\pi'^{\phi}\left(\frac{\partial}{\partial Y^v}\right)^{\Phi} = Z\left(\Phi^{p'+u}\right)\left(\Gamma'^{w}_{uv}\circ\Phi\right)\pi'^{\Phi}\left(\frac{\partial}{\partial Y^w}\right)^{\Phi} \tag{4.156}$$

for any $Z \in \mathfrak{X}(\tilde{M})$. Next, we need the local expression of the isomorphism

$$\sigma_g : Q \simeq T(\mathscr{F})^{\perp}, \quad \sigma_g(s) = f^j\left(\frac{\partial}{\partial y^j}\right)^{\perp}, \quad s = f^j\,s_j\,.$$

To this end let

$$g : \begin{pmatrix} g_{\alpha\beta} & g_{\alpha,\,p+j} \\ g_{p+i,\,\beta} & g_{p+i,\,p+j} \end{pmatrix}$$

be the local components of g with respect to the local coordinate system $(U,\, x^{\alpha}\,,\, y^i)$. We set

$$(g_Q)_{ij} = g_Q\left(\pi\left(\frac{\partial}{\partial y^i}\right),\, \pi\left(\frac{\partial}{\partial y^j}\right)\right) = g\left(\left(\frac{\partial}{\partial y^i}\right)^{\perp},\, \left(\frac{\partial}{\partial y^i}\right)^{\perp}\right)$$

for any $1 \leq i,\, j \leq q$. If $P = T(\mathscr{F})$ and g_P is the leafwise metric induced by g on P, then $g_{\alpha\beta} = g_P\big(\partial/\partial x^{\alpha}\,,\, \partial/\partial x^{\beta}\big)$. In particular $\left[g_{\alpha\beta}\right]$ is positive definite. Let us set

$[g^{\alpha\beta}] = [g_{\alpha\beta}]^{-1}$. A calculation then shows

$$\left(\frac{\partial}{\partial y^i}\right)^{\perp} = \frac{\partial}{\partial y^i} - g^{\alpha\beta}\, g_{p+i,\,\beta}\, \frac{\partial}{\partial x^\alpha}\,. \tag{4.157}$$

At this point we may prove (4.154). Note that $[\partial/\partial t,\ \tilde{Y}] = 0$. On one hand [by (4.155)],

$$D^{\Phi}_{\partial/\partial t}\, \Phi_* \tilde{s} = \left\{\tilde{Y}\left(\frac{\partial \Phi^{p'+w}}{\partial t}\right) + \tilde{Y}(\Phi^{p'+u})\, \frac{\partial \Phi^{p'+v}}{\partial t}\, \left(\Gamma'^{w}_{vu} \circ \Phi\right)\right\} \pi'^{\Phi} \left(\frac{\partial}{\partial Y^w}\right)^{\Phi}. \tag{4.158}$$

On the other hand, let $\mathbf{v} = \pi'^{\Phi}\, \mathbb{V}$ so that [by $\pi'\, \partial/\partial X^a = 0$]

$$\mathbf{v} = \frac{\partial \Phi^{p'+u}}{\partial t}\, \pi'^{\Phi} \left(\frac{\partial}{\partial Y^u}\right)^{\Phi}. \tag{4.159}$$

Consequently [by (4.157)],

$$D^{\Phi}_{\widetilde{\sigma_g(s)}}\, \mathbf{v} = f^i \left\{\tilde{\partial}_{p+i} \left(\frac{\partial \Phi^{p'+u}}{\partial t}\right) \pi'^{\Phi} \left(\frac{\partial}{\partial Y^u}\right)^{\Phi} + \frac{\partial \Phi^{p'+u}}{\partial t}\, D^{\Phi}_{\tilde{\partial}_{p+i}}\, \pi'^{\Phi} \left(\frac{\partial}{\partial Y^u}\right)^{\Phi} + \right.$$

$$\left. - g^{\alpha\beta}\, g_{p+i,\,\beta}\, \tilde{\partial}_\alpha \left(\frac{\partial \Phi^{p'+u}}{\partial t}\right) \pi'^{\Phi} \left(\frac{\partial}{\partial Y^u}\right)^{\Phi} - g^{\alpha\beta}\, g_{p+i,\,\beta}\, \frac{\partial \Phi^{p'+u}}{\partial t}\, D^{\Phi}_{\tilde{\partial}_\alpha}\, \pi'^{\Phi} \left(\frac{\partial}{\partial Y^u}\right)^{\Phi} \right\}$$

where we have set $\partial_\alpha = \partial/\partial x^\alpha$ and $\partial_{p+i} = \partial/\partial y^i$. The last two terms vanish due to the fact that Φ is a foliated map [cf. also (4.156) with $Z = \tilde{\partial}_\alpha$]. Indeed, as $\Phi \in C^\infty(\tilde{\mathscr{F}},\, \mathscr{G})$

$$T(\mathscr{G})_{(x,t)} \ni \left(d_{(x,t)}\Phi\right) \left(d_x\, \alpha_t\right) \left(\partial_\alpha\right)_x = \tilde{\partial}_\alpha \left(\Phi^a\right)_{(x,t)} \left(\frac{\partial}{\partial X^a}\right)_{\Phi(x,t)} + \tilde{\partial}_\alpha \left(\Phi^{p'+u}\right)_{(x,t)} \left(\frac{\partial}{\partial Y^u}\right)_{\Phi(x,t)}$$

hence

$$\tilde{\partial}_\alpha \left(\Phi^{p'+u}\right) = 0. \tag{4.160}$$

Moreover [by (4.156) and (4.160)],

$$D^{\Phi}_{\widetilde{\sigma_g(s)}}\, \mathbf{v} = \left\{\tilde{Y}\left(\frac{\partial \Phi^{p'+w}}{\partial t}\right) + \frac{\partial \Phi^{p'+u}}{\partial t}\, \tilde{Y}(\Phi^{p'+v})\left(\Gamma'^{w}_{uv} \circ \Phi\right)\right\} \left(\frac{\partial}{\partial Y^w}\right)^{\Phi}. \tag{4.161}$$

Finally (by Proposition 5.5 in [Ton], p. 49), the torsion

$$T_{\nabla'}(Z,\ W) = \nabla'_Z \pi' W - \nabla'_W \pi' Z - \pi'[Z,\ W], \quad Z,\ W \in \mathfrak{X}(N),$$

vanishes (i.e., $T_{\nabla'} = 0$) hence $\Gamma'^{u}_{vw} = \Gamma'^{u}_{wv}$ and then (4.158) and (4.161) yield (4.154). Lemma 4.123 is proved.

Let us set $\pi'^{\Phi} = \pi' \circ \Phi$. By (4.154) and $D^{\Phi} h^{\Phi}_{\nu(\mathscr{G})} = 0$

$$\frac{\partial f}{\partial t} = \sum_j \left\{ \widetilde{\sigma_g(s_j)} \left[h^{\Phi}_{\nu(\mathscr{G})}\left(\pi'^{\Phi}\mathbb{V},\ \Phi_* \tilde{s}_j\right)\right] - h^{\Phi}_{\nu(\mathscr{G})}\left(\pi'^{\Phi}\mathbb{V},\ D^{\Phi}_{\widetilde{\sigma_g(s_j)}} \Phi_* \tilde{s}_j\right)\right\}. \tag{4.162}$$

Lemma 4.124 *For every $|t| < \epsilon$ let $r_t \in C^{\infty}\big(\nu(\mathscr{F})\big)$ be determined by*

$$h^{\Phi}_{\nu(\mathscr{G})}\left(\pi'^{\Phi}\mathbb{V},\ \Phi_* \tilde{s}\right) \circ \alpha_t = g_{\nu(\mathscr{F})}\left(r_t\,,\ s\right), \quad s \in C^{\infty}\big(\nu(\mathscr{F})\big). \tag{4.163}$$

Then $r_t \in \ell(M,\ \mathscr{F})$, i.e., r_t is a transverse vector field for any $|t| < \epsilon$.

Here $\ell(\mathscr{F}) = \ell(M,\ \mathscr{F}) = \pi\ V(M,\ \mathscr{F})$ and $V(\mathscr{F}) \equiv V(M,\ \mathscr{F})$ is the set of all infinitesimal automorphisms of $\mathscr{F}$, i.e., all tangent vector fields $Y \in \mathfrak{X}(M)$ such that $[X,\ Y] \in \mathfrak{X}_{\mathscr{F}}$ for any $X \in \mathfrak{X}_{\mathscr{F}}$.

Proof of Lemma 4.124. Let $s \in C^{\infty}\big(\nu(\mathscr{F})\big)$ be locally represented as $s = f^i\, \pi\, \partial/\partial^i$. Then $s \in \ell(\mathscr{F})$ if and only if f^i are basic functions. By (4.155) and (4.159)

$$r_t^i = g_Q^{ij} \left\{ \frac{\partial \Phi^{p'+u}}{\partial t}\ \tilde{\partial}_{p+j}\left(\Phi^{p'+v}\right) \left[\left(h_{\nu(\mathscr{G})}\right)_{uv} \circ \Phi \right] \right\} \circ \alpha_t \tag{4.164}$$

hence $r_t^i \in \Omega^0_B(\mathscr{F})$ for any $|t| < \epsilon$. Q.e.d.

Next [by (4.163) and $\nabla g_{\nu(\mathscr{F})} = 0$],

$$\widetilde{\sigma_g(s_j)} \left[h^{\Phi}_{\nu(\mathscr{G})}\left(\pi'^{\Phi}\mathbb{V},\ \Phi_* \tilde{s}_j\right)\right] \circ \alpha_t = \sigma_g(s_j)\left[g_{\nu(\mathscr{F})}\left(r_t\,,\ s_j\right)\right] =$$

$$= g_{\nu(\mathscr{F})}\left(\nabla_{\sigma_g(s_j)} r_t\,,\ s_j\right) + g_{\nu(\mathscr{F})}\left(r_t\,,\ \nabla_{\sigma_g(s_j)} s_j\right).$$

The term $\sum_j g_{\nu(\mathscr{F})}\left(\nabla_{\sigma_g(s_j)} r_t\,,\ s_j\right)$ may be recognized as a basic divergence. Indeed, let $SO(M,\ \mathscr{F}) \to M$ be the SO(q)-principal bundle of al oriented orthonormal transverse frames. If $z : \mathbb{R}^q \to \nu(\mathscr{F})_x$ is a transverse frame adapted to $SO(M,\ \mathscr{F})$ the transverse volume form $\nu \in \Omega^q_B(\mathscr{F})$ is given by

$$\nu_x\left(v_1\,,\ \cdots\,,\ v_q\right) = \det\left[\left(g_{\nu(\mathscr{F})}\right)_x \left(\pi_x\, v_i\,,\ z(e_j)\right)\right]$$

for any $v_i \in T_x(M)$, where $\{e_1\,,\ \cdots\,,\ e_q\} \subset \mathbb{R}^q$ is the canonical linear basis. Let $\left(U,\ x^{\alpha}\,,\ y^i\right)$ be a foliated local chart of $(M,\ \mathscr{F})$ such that $x \in U$. Then

$$z(e_j) = a^i_j\, \pi_x \left(\frac{\partial}{\partial y^i}\right)_x, \quad 1 \le j \le q,$$

for some $a = \left[a^i_j\right] \in \mathrm{GL}(q,\ \mathbb{R})$. Next, we may consider the function $c : M \to \{\pm 1\}$ defined by $c(x) = \operatorname{sign}\left(\det(a)\right)$ for any $x \in M$. The definition of $c(x)$ doesn't depend on the choice of the transverse frame $z \in SO(M, \mathcal{F})_x$. A calculation shows that

$$\nu = c\, q! \sqrt{G}\, d y^1 \wedge \cdots \wedge d y^q . \tag{4.165}$$

Here

$$G = \det \left[(g_Q)_{ij} \right], \quad (g_Q)_{ij} = g_Q\Big(\pi \frac{\partial}{\partial y^i}, \pi \frac{\partial}{\partial y^j}\Big).$$

Definition 4.125 The *basic divergence* of an infinitesimal automorphism $Y \in V(\mathcal{F})$ is the basic function

$$\operatorname{div}_B(Y) \in \Omega^0_B(\mathcal{F}), \quad \mathcal{L}_Y \nu = \operatorname{div}_B(Y)\, \nu. \tag{4.166}$$

□

Here $\mathcal{L}_Y$ denotes the Lie derivative. It should be observed that $\operatorname{div}_B(Y)$ depends only on the transverse vector field $s = \pi(Y) \in \ell(M, \mathcal{F})$ [and one often writes $\operatorname{div}_B(s)$ rather than $\operatorname{div}_B(Y)$]. A calculation based on [based on (4.165)-(4.166)] shows that

$$\operatorname{div}_B(Y) = \frac{1}{\sqrt{G}} \frac{\partial}{\partial y^i}\Big(\sqrt{G}\, Y^i\Big) \tag{4.167}$$

for any $Y \in V(\mathcal{F})$ such that locally $\pi(Y) = Y^i\, \pi\, \partial/\partial y^i$ [note that $Y^i \in \Omega^0_B(\mathcal{F}_U)$]. As previously announced

Lemma 4.126 *Let $\{s_j : 1 \le j \le q\}$ be a local g_Q-orthonormal frame in $Q = \nu(\mathcal{F}) \to M$, defined on the open set $U \subset M$. Then*

$$\sum_{j=1}^{q} g_Q\big(\nabla_{\sigma_g(s_j)} r_t\, , s_j\big) = \operatorname{div}_B(r_t) \tag{4.168}$$

on U for any $|t| < \epsilon$.

Proof One has [by (4.179)]

$$\Gamma^i_{ij} = \frac{1}{2G} \frac{\partial G}{\partial y^j}$$

so that

$$(g_Q)^{ij}\, g_Q\big(\nabla_{(\partial_{p+i})^\perp} r_t\, , \pi\, \partial_{p+j}\big) = \frac{\partial r^i_t}{\partial y^i} + \Gamma^i_{ij}\, r^j_t = \operatorname{div}_B\big(r^i_t\, \partial_{p+i}\big).$$

Q.e.d.

By (4.163) and (4.168)

$$\sum_j \widetilde{\sigma_g(s_j)}\Big[h^{\Phi}_{\nu(\mathscr{G})}\Big(\pi'^{\Phi}\mathbb{V},\ \Phi_*\tilde{s}_j\Big)\Big]\circ\alpha_t = \mathrm{div}_B(r_t) + \sum_j h^{\Phi}_{\nu(\mathscr{G})}\Big(\pi'^{\Phi}\mathbb{V},\ \Phi_*\widetilde{\nabla_{\sigma_g(s_j)} s_j}\Big)$$

hence (4.162) becomes

$$\frac{\partial f}{\partial t}\circ\alpha_t = \mathrm{div}_B(r_t) - h^{\Phi}_{\nu(\mathscr{G})}\Big(\pi'^{\Phi}\mathbb{V},\ \sum_j\Big[D^{\Phi}_{\widetilde{\sigma_g(s_j)}}\Phi_*\tilde{s}_j - \Phi_*\widetilde{\nabla_{\sigma_g(s_j)} s_j}\Big]\Big)\circ\alpha_t\,.$$

Let us integrate over Ω and use

$$\int_M \mathrm{div}_B(s)\,d\,\mathrm{v}_g = \int_M g_Q\big(\tau_{\mathscr{F}}\,,\ s\big)\,d\,\mathrm{v}_g \tag{4.169}$$

for any compactly supported $s \in \ell(\mathscr{F})$.

Theorem 4.127 *Let $\phi \in C^\infty(\mathscr{F},\mathscr{G})$ be a foliated map. For every relatively compact domain $\Omega \subset\subset M$, and any smooth* 1-.*parameter variation $\Phi \in C^\infty\big(\tilde{\mathscr{F}},\mathscr{G}\big)$ of $\phi = \phi_0$ supported in Ω*

$$\frac{d}{dt}\Big\{E_{T,\Omega}(\phi_t)\Big\}_{t=0} = -\int_\Omega h^{\phi}_{\nu(\mathscr{G})}\big(\pi'^{\phi}V,\ \tau_T(\phi) - \phi_*\,\tau_{\mathscr{F}}\big)\,d\,\mathrm{v}_g \tag{4.170}$$

where

$$\tau_T(\phi) \equiv \mathrm{Trace}_{g_{\nu(\mathscr{F})}}\,\beta_T(\phi) \in C^\infty\big(\phi^{-1}\nu(\mathscr{F})\big),$$

$$\beta_T(\phi)(r,\ s) = D^{\phi}_{\sigma_g(r)}\,\phi_*\,s - \phi_*\nabla_{\sigma_g(r)}\,s\,,\quad r,\ s \in C^\infty\big(\nu(\mathscr{F})\big).$$

Consequently, $\phi : M \to N$ is $(\mathscr{F},\mathscr{G})$-harmonic if and only if

$$\tau_T(\phi) - \phi_*\,\tau_{\mathscr{F}} = 0. \tag{4.171}$$

The section $\tau_T(\phi)$ is locally given by

$$\tau_T(\phi) = \Big\{\frac{1}{\sqrt{G}}\frac{\partial}{\partial y^i}\Big(\sqrt{G}\,(g_Q)^{ij}\,\frac{\partial\phi^{p'+w}}{\partial y^j}\Big) + (g_Q)^{ij}\,\frac{\partial\phi^{p'+u}}{\partial y^i}\,\frac{\partial\phi^{p'+v}}{\partial y^j}\,\big(\Gamma'^{w}_{uv}\circ\phi\big)\Big\}\,\pi'^{\phi}\Big(\frac{\partial}{\partial Y^w}\Big)^{\phi} \tag{4.172}$$

for any local foliated charts $(U,\ x^\alpha,\ y^i) \in \mathscr{F}$ and $(\mathscr{U},\ X^a,\ Y^u) \in \mathscr{G}$ such that $\phi(U) \subset \mathscr{U}$.

Definition 4.128 The identity (4.170) is the *first variation* formula for the transverse energy $E_{T,\Omega} : C^\infty(\mathscr{F},\mathscr{G}) \to \mathbb{R}$. Equation (4.171) is the $(\mathscr{F},\mathscr{G})$-*harmonic map system.* □

As well as the transverse formula (4.169), the $(\mathscr{F},\mathscr{G})$-harmonic map system contains explicitly the mean curvature vector

$$\tau_{\mathscr{F}} \in C^\infty\big(\nu(\mathscr{F})\big),\quad g_Q\big(\tau_{\mathscr{F}}\,,\ \pi\,Z\big) = \kappa(Z),\quad Z \in T(\mathscr{F})^\perp,$$

of the Riemannian foliation $\mathscr{F}$. Here $\kappa \in \Omega^1(M)$ is the mean curvature of $\mathscr{F}$. We recall that one sets

$$\alpha(X,\, X') = \pi\, \nabla^g_X X'\,, \quad X,\, X' \in \mathfrak{X}_{\mathscr{F}}\,,$$

$$W(Z): T(\mathscr{F}) \to T(\mathscr{F}), \quad g_Q\big(\alpha(X,\, X'),\, \pi\, Z\big) = g_{T(\mathscr{F})}\big(W(Z)X,\, X'\big), \quad Z \in C^\infty\big(T(\mathscr{F})^\perp\big).$$

Finally, $\kappa \in \Omega^1(M)$ is given by

$$\kappa(X) = 0, \quad \kappa(Z) = \text{Trace}\, W(Z), \quad X \in \mathfrak{X}_{\mathscr{F}}\,, \quad Z \in C^\infty\big(T(\mathscr{F})^\perp\big).$$

Definition 4.129 The bilinear form $\beta_T(\phi) \in C^\infty\big(Q^* \otimes Q^*\big)$ is the *transverse second fundamental form* of ϕ and its trace $\tau_T(\phi)$ with respect to g_Q is the *transverse tension* field of ϕ. □

The proof of Theorem 4.127 may be completed applying the following identities

$$\tilde{Y}\big(\Phi^{p'+u}\big) \circ \alpha_0 = Y\big(\phi^{p'+u}\big), \tag{4.173}$$

$$\tilde{Z}\big(\tilde{Y}\, \Phi^{p'+u}\big) \circ \alpha_0 = Z\big(Y\, \phi^{'+u}\big), \tag{4.174}$$

$$\big(\Phi_*\, \tilde{s}\big) \circ \alpha_0 = \phi_*\, s, \tag{4.175}$$

$$\big(D^\Phi_{\tilde{Y}}\, \Phi_*\, \tilde{s}\big) \circ \alpha_0 = D^\phi_Y\, \phi_*\, s, \tag{4.176}$$

for any $Y,\, Z \in \mathfrak{X}(M)$ and any $s \in C^\infty(Q)$. At its turn, the proof of (4.173)-(4.175) is straightforward. The identities (4.173)-(4.175) and (4.156) then yield (4.176). To prove (4.172) we compute

$$\tau_T(\phi) = \big(g_Q\big)^{ij}\, \beta_T(\phi)\big(\pi\, \partial_{p+i}\,,\, \pi\, \partial_{p+j}\big). \tag{4.177}$$

One has [by (4.157)]

$$\beta_T(\phi)\big(\pi\, \partial_{p+i}\,,\, \pi\, \partial_{p+j}\big) = \left\{\frac{\partial^2 \phi^{p'+w}}{\partial y^i\, \partial y^j} + \frac{\partial \phi^{p'+u}}{\partial y^i}\, \frac{\partial \phi^{p'+v}}{\partial y^j}\, \big(\Gamma'^w_{uv} \circ \phi\big) - \Gamma^k_{ij}\, \frac{\partial \phi^{p'+w}}{\partial y^k}\right\}\Big(\pi'\, \frac{\partial}{\partial Y^w}\Big)^\phi.$$

Here Γ^i_{jk} are local coefficients of the canonical adapted connection $\nabla \in \mathscr{C}(Q,\, g_Q)$, i.e.,

$$\nabla_{(\partial/\partial y^j)^\perp}\, \pi\, \frac{\partial}{\partial y^k} = \Gamma^i_{jk}\, \pi\, \frac{\partial}{\partial y^j}\,.$$

As $\mathscr{L}_X\, g_Q = 0$ for any $X \in \mathfrak{X}_{\mathscr{F}}$, it follows that $\big(g_Q\big)_{ij} \in \Omega^0_B\big(\mathscr{F}_U\big)$. Also $\nabla g_Q = 0$ implies

$$\Gamma^i_{jk} = \left(g_Q\right)^{i\ell} \Gamma_{jk\ell}\,, \quad \Gamma_{ijk} = \frac{1}{2}\left\{\frac{\partial \left(g_Q\right)_{ik}}{\partial y^j} + \frac{\partial \left(g_Q\right)_{jk}}{\partial y^i} - \frac{\partial \left(g_Q\right)_{ij}}{\partial y^k}\right\},$$

hence

$$-\left(g_Q\right)^{jk}\Gamma^i_{jk} = \frac{\partial \left(g_Q\right)^{ij}}{\partial y^j} + \frac{1}{2}\left(g_Q\right)^{ij}\left(g_Q\right)^{k\ell}\frac{\partial \left(g_Q\right)_{k\ell}}{\partial y^j}\,. \tag{4.178}$$

Finally, (4.172) follows from (4.177) to (4.178) and the identity

$$\frac{1}{G}\frac{\partial G}{\partial y^i} = \left(g_Q\right)^{jk}\frac{\partial \left(g_Q\right)_{jk}}{\partial y^i}\,. \tag{4.179}$$

Let us briefly examine the structure of the $(\mathscr{F},\,\mathscr{G})$-harmonic map system (4.171). Let $(M,\,\mathscr{F},\,g)$ be a compact orientable foliated Riemannian manifold where $\mathscr{F}$ is transversally oriented Riemannian foliation and g is bundle-like. Let $\hat{ast}$: $\Omega^j_B(\mathscr{F}) \to \Omega^{q-j}_B(\mathscr{F})$ be the star operator associeted to g_Q, i.e.,

$$\omega \wedge \hat{*}\,\omega = \nu, \quad \omega \in \Omega^j_B(\mathscr{F}).$$

The ordinary L^2 inner product on $\Omega^j(M)$ restricts on basic forms to

$$\langle \alpha\,,\,\beta\rangle = \int_M \alpha \wedge \left(\hat{*}\,\beta\right) \wedge \chi_{\mathscr{F}}\,.$$

Let

$$\delta_B : \Omega^j_B(\mathscr{F}) \to \Omega^{q-j}_B(\mathscr{F})$$

be the formal adjoint of d_B with respect to $\langle\ ,\ \rangle$.

Definition 4.130 The *basic Laplacian* is

$$\Delta_B = \delta_B\, d_B + d_B\, \delta_b : \Omega^j_B(\mathscr{F}) \to \Omega^j_B(\mathscr{F})\,, \quad 0 \le j \le q. \tag{4.180}$$

□

The adjoint δ_B may be expressed as

$$\delta_B\,\omega = (-1)^{q(j+1)+1}\,\hat{*}\left(d_B\,\hat{*}\,\omega - \kappa \wedge \hat{*}\,\omega\right), \quad \omega \in \Omega^j_B(\mathscr{F}). \tag{4.181}$$

It should be emphasized that (4.181) requires the additional assumption $\kappa \in \Omega^1_B(\mathscr{F})$. As a consequence of (4.181) the basic Laplacian (on functions) is locally given by

$$\Delta_B u = \frac{1}{\sqrt{G}}\frac{\partial}{\partial y^i}\left\{\sqrt{G}\left(g_Q\right)^{ij}\frac{\partial u}{\partial y^j}\right\} - \left(g_Q\right)^{ij}\frac{\partial u}{\partial y^i}\kappa_j\,, \quad u \in \Omega^0_B(\mathscr{F}). \tag{4.182}$$

Here $\kappa_i = \kappa\big(\partial/\partial y^i\big)$. Locally $\tau_{\mathscr{F}} = \kappa^i \, \pi\big(\partial/\partial y^i\big)$ and $\kappa^i = \big(g_Q\big)^{ij} \, \kappa_j$. The restriction of Δ_B to a transversal manifold is an elliptic operator so that Δ_B is transversally elliptic (in the sense of [KaAl], p. 58). Next [by (4.172) and (4.182)], the $(\mathscr{F}, \mathscr{G})$-harmonic map system may be written

$$\Delta_B \phi^{p'+w} + \big(g_Q\big)^{ij} \frac{\partial \phi^{p'+u}}{\partial y^i} \frac{\partial \phi^{p'+v}}{\partial y^j} \big(\Gamma'^{w}_{uv} \circ \phi\big) = 0, \tag{4.183}$$

a nonlinear transversally elliptic system of variational origin.

Example 4.131 (*$(\mathscr{F}, \mathscr{G})$-harmonic maps of linear foliations*) Let $T^2 = S^1 \times S^1$ be the 2-torus carrying the linear foliation $\mathscr{F}_\lambda$ of gradient λ_2/λ_1 [with $\lambda = (\lambda_1 , \lambda_2) \in \mathbb{R}^2$, $\lambda \neq 0$], i.e., $\mathscr{F}_\lambda$ is the flow of T^2 spanned by $X_\lambda = \lambda_1 S_1 + \lambda_2 S_2$ where $S_A \in \mathfrak{X}(T^2)$ are the (globally defined) vector fields locally given by $S_A = \partial/\partial\theta^A$ on $U \subset T^2$ for each coordinate system

$$\chi = (\theta^1 , \theta^2) : U \to \big(\theta_0^1 - \pi , \theta_0^1 + \pi\big) \times \big(\theta_0^2 - \pi , \theta_0^2 + \pi\big) \subset \mathbb{R}^2, \quad \chi^{-1}(\alpha) = \big(e^{i\alpha_1} , e^{i\alpha_2}\big),$$

for some $\|\theta_0\| < \pi$ and any $\|\alpha - \theta_0\| < \pi$. Here $\|\alpha\| = \max\big\{|\alpha_1\| , |\alpha_2|\big\}$. If g is the natural flat metric on T^2, i.e., $g\big(S_A , S_B\big) = \delta_{AB}$, then the induced bundle metric $g_\nu\big(\mathscr{F}_\lambda\big)$ is invariant under sliding along the curves

$$\gamma_\theta : \mathbb{R} \to T^2, \quad \gamma_\theta(t) = \chi^{-1}(\theta + t\lambda), \quad t \in \mathbb{R}, \quad \|\theta\| < \pi .$$

Indeed, as $Z_\lambda = \|\lambda\|^{-1}\big(\lambda_2 S_1 - \lambda_1 S_2\big)$ spans $T\big(\mathscr{F}_\lambda\big)^\perp$, the bundle isomorphism $\sigma_g : \nu\big(\mathscr{F}_\lambda\big) \to T(\mathscr{F})^\perp$ is given by $\sigma_g\big(s_\lambda\big) = Z_\lambda$ where $s_\lambda = \pi\big(Z_\lambda\big)$. Here $\|\lambda\|^2 = \lambda_1^2 + \lambda_2^2$. Therefore, $g_{\nu(\mathscr{F}_\lambda)}\big(s_\lambda , s_\lambda\big) = 1$ and $\mathring{\nabla}_{X_\lambda} s_\lambda = 0$ so that $\mathring{\nabla}_{X_\lambda} g_{\nu(\mathscr{F}_\lambda)} = 0$, i.e., $\big(\mathscr{F}_\lambda , g_{\nu(\mathscr{F}_\lambda)}\big)$ is a Riemannian flow and g is bundle-like. Since $\nabla^g_{X_\lambda} X_\lambda = 0$ the leaves of $\mathscr{F}_\lambda$ are geodesics of $\big(T^2, g\big)$. In particular $\tau_{\mathscr{F}} = 0$. If $\phi \in C^\infty\big(\mathscr{F}_\lambda, \mathscr{G}\big)$ is a foliated map $\pi : T^2 \to N$ into a foliated Riemannian manifold $(N, \mathscr{G}, h)$, then

$$\tau_T(\phi) = D^\phi_{Z_\lambda} \phi_* s_\lambda - \phi_* \nabla_{Z_\lambda} s_\lambda = \Big\{Z_\lambda^2(\phi^{p'+w}) + (\gamma'^{w}_{uv} \circ \phi) Z_\lambda(\phi^{p'+u}) Z_\lambda(\phi^{p'+v})\Big\}\Big(\frac{\partial}{\partial Y^w}\Big)^\phi$$

on $\phi^{-1}(\mathscr{U}($. Let $U, \chi)$ be a local coordinate system on T^2 as above with $U \cap \phi^{-1}(\mathscr{U}) \neq \emptyset$. If $\mathscr{F}_\lambda$ is a linear foliation of irrational gradient $\lambda_2/\lambda_1 \in \mathbb{R} \setminus \mathbb{Q}$, then the leaf $\gamma_\theta(\mathbb{R})$ is dense in T^2 (and diffeomorphic to $\mathbb{R}$) hence $\Omega^0_B\big(\mathscr{F}_\lambda\big) = \mathbb{R}$. Consequently [by (4.183)], each $(\mathscr{F}, \mathscr{G})$-harmonic map $\phi : T^2 \to N$ is locally given by $\phi(\theta) = \big(\psi(\theta), c\big)$ where $\psi : U \to \mathbb{R}^{p'}$ and $c \in \mathbb{R}^{q'}$ are, respectively, an arbitrary C^∞ function and an arbitrary constant. Let us look now at the opposite case where $\mathscr{F}_\lambda$ has rational gradient $\lambda_2/\lambda_1 \in \mathbb{Q}$. Then $\mathscr{F}_\lambda$ is simple and has compact leaves (diffeomorphic to S^1). The change of local coordinate $x = \lambda_1\theta^1 + \lambda_2\theta^2$, $y = \lambda_2\theta^1 - \lambda_1\theta^2$ furnishes a local foliated chart $(U, x, y) \in \mathscr{F}_\lambda$. Indeed $\pi(\partial/\partial x) = 0$ and $\pi(\partial/\partial y = \|\lambda\|^{-1} s_\lambda$ so that y is a transverse local coordinate. In the new coordinates $Z_\lambda = \|\lambda\| \partial/\partial y$ and $\phi^{p'+u}$ are functions of y alone so that [by (4.171)] $\phi : T^2 \to N$

is $(\mathscr{F}, \mathscr{G})$-harmonic if and only if

$$\frac{\partial^2 \phi^{p'+w}}{\partial y^2} + \left(\Gamma'^{w}_{uv} \circ \phi\right) \frac{\partial \phi^{p'+u}}{\partial y} \frac{\partial \phi^{p'+v}}{\partial y} = 0.$$

If for instance $N = T^2$ and $\mathscr{G} = \mathscr{F}_\mu$ (with $\mu \in \mathbb{R}^2 \setminus \{0\}$), then $\phi^2(y) = ay + b$ for some $a, b \in \mathbb{R}$. □

4.10.2.2 $(\mathscr{F}, \mathscr{G})$-Harmonic morphisms

If $\nabla \in \mathscr{C}(Q, g_Q)$ is the canonical adapted connection we set

$$(\nabla_X \omega)s = X(\omega(s)) - \omega(\nabla_X s), \quad \omega \in C^\infty(Q^*), \quad s \in C^\infty(Q).$$

In particular for each $u \in \Omega^0_B(\mathscr{F})$

$$\nabla_X \left((d_B u) \circ \sigma_g \right) = 0, \quad X \in \mathfrak{X}_{\mathscr{F}},$$

$$\nabla_Y \left((d_B u) \circ \sigma_g \right) = (\nabla^g_Y d_B u) \circ \sigma_g, \quad Y \in C^\infty(T(\mathscr{F})^\perp).$$

We adopt the notation

$$u_{i,j} = \left(\nabla_{\partial/\partial y^i} \left((d_B u) \sigma_g \right) \pi \left(\frac{\partial}{\partial y^i} \right) = \frac{\partial^2 u}{\partial y^i \partial y^j} - \Gamma^k_{ij} \frac{\partial u}{\partial y^k}, \quad u \in \Omega^0_B(\mathscr{F}).$$

For any $\phi \in C^\infty(\mathscr{F}, \mathscr{G})$ and $f \in \Omega^0_B(\mathscr{G})$

$$\Delta_B(f \circ \phi) = \left(\frac{\partial f}{\partial Y^u} \circ \phi \right) \Delta_B \phi^{p'+u} + (g_Q)^{ij} \frac{\partial \phi^{p'+u}}{\partial y^i} \frac{\partial \phi^{p'+v}}{\partial y^j} \left(f_{u,v} + \Gamma'^{w}_{uv} \frac{\partial f}{\partial Y^w} \right) \circ \phi. \tag{4.184}$$

Theorem 4.132 *Let $\mathscr{G}_0$ be the trivial foliation by points on N. Then any $(\mathscr{F}, \mathscr{G}_0)$-harmonic morphism $\phi : M \to N$ of $(M, \mathscr{F}, g)$ into $(N, \mathscr{G}_0, h)$ is a $(\mathscr{F}, \mathscr{G}_0)$-harmonic map.*

(i) *If* $\dim N) > \operatorname{cod}(\mathscr{F})$, *then there are no $(\mathscr{F}, \mathscr{G}_0)$-harmonic morphisms of M into N, except for the constant maps.*

(ii) *If* $\dim(N) \leq \operatorname{cod}(\mathscr{F})$, *then, for each $(\mathscr{F}, \mathscr{G}_0)$-harmonic morphism, $\phi : M \to N$*

$$\phi^* h = \Lambda_\phi \, g \quad on \quad \mathscr{H} = \operatorname{Ker}(d\phi)^\perp \tag{4.185}$$

where $\Lambda_\phi = (1/\dim(N)) \, \|d_T \phi\|^2$. Also $\phi : M \setminus C(\phi) \to N$ is a C^∞ submersion where $C(\phi) = \{x \in M : d_x \phi = 0\}$. Moreover, each leaf of the foliation induced *by $\mathcal{F}$ on $M \setminus C(\phi)$ is a submanifold of some connected component of a fiber of $\phi : M \setminus C(\phi) \to N$. In particular, if* $\dim(N) = \operatorname{cod}(\mathcal{F})$ *and* $C(\phi) =$, *then $\mathcal{F}$ is the simple foliation determined by the submersion $\phi : M \to N$.*

Proof By a result of Ishihara, [Ish], for every point $y_0 \in N$ and every system of constants C_u, $C_{u,v} \in \mathbb{R}$ such that $C_{uv} = C_{vu}$, there is a normal coordinate system $(\mathcal{U}, Y^u)$ centered at $y_0 \in \mathcal{U} \subset N$ and a harmonic function $f : \mathcal{U} \to \mathbb{R}$ such that $f_u(y_0) = C_u$ and $f_{u,v}(y_0) = C_{uv}$ for any $1 \leq u, v \leq n'$. This is commonly referred to as *Ishihara's lemma* (cf. again [Ish]). Here $f_u = \partial f / \partial Y^u$ and $f_{u,v}$ are the local components of the covariant derivative of df. Of course the canonical adapted connection of $\mathcal{G}_0$ is the Levi-Civita connection of the background Riemannian metric. Let $x_0 \in M$ and $y_0 := \phi(x_0)$. Let us apply Ishihara's lemma with

$$C_u = \delta_{u,u_0}, \quad X_{uv} = 0, \quad u_0 \in \{1, \cdots, n'\},$$

so that to produce a normal coordinate system $(\mathcal{U}, Y^u)$ at y_0 such that $\phi(U) \subset \mathcal{U}$ and a harmonic function $f : \mathcal{U} \to \mathbb{R}$ with $f_u(y_0) = \delta_{u u_0}$ and $f_{u,v}(y_0) = 0$. As $\phi : M \to N$ is a $(\mathcal{F}, \mathcal{G}_0)$-harmonic morphism

$$\Delta_B(f \circ \phi) = 0 \quad \text{on} \quad U;$$

hence [by (4.184) at the point x_0 with $p' = 0$],

$$\left(\Delta_B \phi^{u_0}\right)(x_0) + \left(g_Q\right)^{ij}(x_0) \frac{\partial \phi^v}{\partial y^i}(x_0) \frac{\partial \phi^w}{\partial y^j}(x_0) \, \Gamma'^{u_0}_{vw}(y_0) = 0,$$

i.e., ϕ is a $(\mathcal{F}, \mathcal{G}_0)$-harmonic map. Next, let $C_{uv} \in \mathbb{R}$ be a system of constants such that

$$C_{uv} = C_{vu}, \quad \sum_{u=1}^{n'} C_{uu} = 0.$$

By Ishihara's lemma there is a harmonic function $f : \mathcal{U} \to \mathbb{R}$ defined on a normal coordinate neighborhood $\mathcal{U} \subset N$ centered at $y_0 = \phi(x_0)$ such that

$$f_u(y_0) = 0, \quad f_{u,v}(y_0) = C_{uv}.$$

Again by (4.184)

$$0 = \Delta_B(f \circ \phi)(x_0) = \left(g_Q\right)^{ij}(x_0) \frac{\partial \phi^u}{\partial y^i}(x_0) \frac{\partial \phi^v}{\partial y^j}(x_0) \, C_{uv}.$$

Let us set

$$X^{uv} = \left(g_Q\right)^{ij} \frac{\partial \phi^u}{\partial y^i} \frac{\partial \phi^v}{\partial y^j}.$$

Then $C_{uv} X^{uv}(x_0) = 0$ or

$$\sum_{u \neq v} C_{uv} X^{uv}(x_0) + \sum_u \left\{X^{uu}(x_0) - X^{11}(x_0)\right\} C_{uu} = 0. \tag{4.186}$$

Now let us choose $C_{uv} = 0$ for any $u \neq v$ and

$$C_{uv} = \begin{cases} 1 & \text{if } u = u_0 \,, \\ -1 & \text{if } u = 1, \quad u_0 \in \{2, \cdots, n'\}, \\ 0 & \text{otherwise.} \end{cases}$$

Then [by (4.186)] $X^{u_0 u u_0}(x_0) - X^{11}(x_0) = 0$, that is

$$X^{11}(x_0) = \cdots = X^{n' n'}(x_0)$$

and (4.186) becomes

$$\sum_{u \neq v} C_{uv} X^{uv}(x_0) = 0.$$

At this point we fix $u_0 \,, v_0 \in \{1, \cdots, n'\}$ with $u_0 \neq v_0$ and choose

$$C_{uv} = \begin{cases} 1 & \text{if } u = u_0 \text{ and } v = v_0 \,, \\ 0 & \text{otherwise,} \end{cases}$$

so that to yield $X^{u_0 v_0}(x_0) = 0$. We obtained $X^{uv}(x_0) = 0$ for any $u \neq v$. Next, let us set

$$\Lambda_\phi = X^{11} = \left(g_Q\right)^{ij} \frac{\partial \phi^1}{\partial y^i} \frac{\partial \phi^1}{\partial y^j} \in C^\infty(U).$$

Then

$$\left(g_Q\right)^{ij}(x_0) \frac{\partial \phi^u}{\partial y^i}(x_0) \frac{\partial \phi^v}{\partial y^j}(x_0) = \Lambda_\phi(x_0)\, \delta^{uv} \,. \tag{4.187}$$

Let us contract u and v in (4.187) and use $h_{v(\mathscr{G}_0)} = h$ and $h_{uv}(y_0) = \delta_{uv}$. We get

$$\Lambda_\phi = \frac{1}{n'} \left\| d_T \, \phi \right\|^2_{x_0}$$

so that Λ_ϕ is a globally defined C^∞ function on M and $\Lambda_\phi(x) \geq 0$ for any $x \in M$. As $\phi \in C^\infty(\mathscr{F}, \mathscr{G}_0)$, the functions $\phi^u = Y^u \circ \phi$ are basic (with respect to $\mathscr{F}_U$). Consequently, if $\Lambda_\phi = 0$, then ϕ is a constant map. Clearly $C(\phi)$ is the zero set of Λ_ϕ. Let us assume that there is $x_0 \in M$ such that $\Lambda_\phi(x_0) \neq 0$. As $\left[\left(g_Q\right)^{ij}(x_0)\right]$ is a positive definite matrix, there is $a = \left[a^i_j\right] \in \mathrm{GL}(q, \mathbb{R})$ such that

$$\left(g_Q\right)^{ij}(x_0) = \sum_{k=1}^{q} a_k^i\, a_k^j\,,$$

Let $\{W^1, \cdots, W'\} \subset \mathbb{R}^q$ be given by

$$W^u = \sum_{k=1}^{q} a_k^i\, \frac{\partial \phi^u}{\partial y^i}(x_0)\, e_k\,, \quad 1 \le u \le n'\,.$$

Then we may write (4.187) as

$$\left(W^u\,,\, W^v\right) = \Lambda_\phi(x_0)\, \delta^{uv}$$

where $(\cdot\,,\,\cdot)$ is the Euclidean inner product in $\mathbb{R}^q$. Hence, the vectors $\{W^u \,:\, 1 \le u \le n'\}$ are nonzero and mutually orthogonal. Therefore, $n' \le q$ and the $n' \times q$ matrix

$$\left(W^1, \cdots, W^{n'}\right)^t = \left[\frac{\partial \phi^u}{\partial y^i}(x_0)\right] \cdot a$$

has rank n' (a dot denotes ordinary matrix product). In particular $d_{x_0}\phi$ has rank n' and we may conclude that $\phi : M \setminus C(\phi) \to N$ is a C^∞ submersion on some open neighborhood of x_0. Next, the Riemannian gradient of ϕ^u is (in the chosen local coordinates)

$$\nabla \phi^u = \left(g_Q\right)^{ij} \frac{\partial \phi^u}{\partial y^j} \frac{\partial}{\partial y^i}$$

so that

Claim. *The vectors* $\left\{\left(\nabla \phi^u\right)_{x_0} \,:\, 1 \le u \le n'\right\}$ *are linearly independent.*

Proof Indeed, if $\alpha_u \left(\nabla \phi^u\right)_{x_0} = 0$ with $\alpha_u \in \mathbb{R}$, then

$$\alpha_u \left(g_Q\right)^{ij}(x_0)\, \frac{\partial \phi^u}{\partial y^j}(x_0) = 0 \Longrightarrow \alpha_u = 0$$

as $\operatorname{rank}\left[\partial \phi^u / \partial y^j\right]_{x_0} = n'$. Q.e.d.

Let $\mathscr{H}_{x_0}$ be the linear span of $\left\{\left(\nabla \phi^u\right)_{x_0} \,:\, 1 \le u \le n'\right\}$ over $\mathbb{R}$.

Claim. *The definition of* $\mathscr{H}_{x_0}$ *doesn't depend on the choice of normal local coordinates centered at* $\phi(x_0)$.

Proof Indeed, let $z,\, \overline{z} : \mathbb{R}^{n'} \to T_{\phi(x_0)}(N)$ be two linear frames tangent to N at $\phi(x_0)$, so that $\overline{z}(e_v) = a_v^u\, z(e_u)$ for some $\left[a_v^u\right] \in \mathrm{GL}(n', \mathbb{R})$. Let $N_r = \left\{w \in T_{\phi(x_0)}(N) \,:\, \|w\| < r\right\}$, $r > 0$, such that the exponential map $\exp_{\phi(x_0)} : N_r \to N$ is a C^∞ diffeomorphism on the image $\mathscr{U} = \exp_{\phi(x_0)}\left(N_r\right)$. If $\chi = \left(Y^1, \cdots, Y^{n'}\right)$ and $\overline{\chi} = \left(\overline{Y}^1, \cdots, \overline{Y}^{n'}\right)$ are the corresponding local charts, i.e.,

$$\chi = \left(z \circ \exp_{\phi(x_0)}\right)^{-1}, \quad \overline{\chi} = \left(\overline{z} \circ \exp_{\phi(x_0)}\right)^{-1},$$

then $\nabla \overline{\phi}^u = a^u_v \nabla \phi^v$ where $\overline{\phi}^u = \overline{Y}^u \circ \phi$. Moreover,

$$\begin{aligned}(d_{x_0}\phi)(\nabla\phi^u)_{x_0} =& (\nabla\phi^u)_{x_0}(\phi^v)\left(\frac{\partial}{\partial Y^v}\right)_{\phi(x_0)} = g\left(\nabla\phi^u, \nabla\phi^v\right)_{x_0}\left(\frac{\partial}{\partial Y^v}\right)_{\phi(x_0)} \\ =& \Lambda_\phi(x_0)\left(\frac{\partial}{\partial Y^u}\right)_{\phi(x_0)}\end{aligned}$$

hence

$$\left(\phi^* h\right)\left(\nabla\phi^u, \nabla\phi^v\right)_{x_0} = \Lambda(x_0)^2 \delta^{uv} = \Lambda_\phi(x_0)\, g\left(\nabla\phi^u, \nabla\phi^v\right)_{x_0},$$

i.e., $(\phi^* h)_{x_0} = \Lambda_\phi(x_0)\, g_{x_0}$ holds on $\mathscr{H}_{x_0}$ as announced. By $(d\phi)T(\mathscr{F}) = (0)$ and a dimension argument, it follows that $\mathscr{H}^\perp = \operatorname{Ker}(d\phi)$ and $T(\mathscr{F}) \subset \operatorname{Ker}(d\phi)$. Theorem 4.132 is proved.

A version of Theorem 4.132 where the target manifold N carries an arbitrary Riemannian foliation $\mathscr{G} \neq \mathscr{G}_0$ would require a foliated version of Ishihara's lemma. The problem is open.

Let us discuss the realtionship of $(\mathscr{F}, \mathscr{G})$-harmonic morphisms to basic cohomology. Every foliated map $\phi \in C^\infty(\mathscr{F}, \mathscr{G})$ induces a map $\phi^* : \Omega^1_B(\mathscr{G}) \to \Omega^1_B(\mathscr{F})$ such that $\phi^* \circ d'_B = d_B \circ \phi^*$. Hence, ϕ^* descends to a map $\phi^* : H^1_B(\mathscr{G}) \to H^1_B(\mathscr{F})$ between the first basic cohomology groups. Here $d'_B : \Omega^\bullet_B(\mathscr{G}) \to \Omega^\bullet_B(\mathscr{G})$ is the basic complex of the foliated manifold $(N, \mathscr{G})$.

Definition 4.133 A $(\mathscr{F}, \mathscr{G})$-harmonic morphism is said to be *submersive* if $\phi : M \setminus C(\phi) \to N$ is a submersion. □

By Theorem 4.132 every $(\mathscr{F}, \mathscr{G})$-harmonic morphism is submersive provided that $\mathscr{G} = \mathscr{G}_0$. We shall show that

Theorem 4.134 *Let $(M, \mathscr{F}, g)$ be a compact, orientable, foliated Riemannian manifold, where $\mathscr{F}$ is transversally oriented and Riemannian, and g is bundle-like. Then, for any surjective submersive $(\mathscr{F}, \mathscr{G})$-harmonic morphism $\phi : M \to N$ of $(M, \mathscr{F}, g)$ into another compact, orientable, foliated Riemannian manifold $(N, \mathscr{G}, h)$(with $\mathscr{G}$ transversally oriented and Riemannian, and h bundle-like), the induced map on basic cohomology $\phi^* : H^1_B(\mathscr{G}) \to H^1_B(\mathscr{F})$ is injective. In particular* $\dim_{\mathbb{R}} \mathscr{H}^1_B(\mathscr{F}) \geq \dim_{\mathbb{R}} \mathscr{H}^1_B(\mathscr{G})$.

Here

$$\mathscr{H}^\bullet_B(\mathscr{F}) = \left\{\omega \in \Omega^\bullet_B(\mathscr{F}) \,:\, \Delta_B \omega = 0\right\}.$$

Lemma 4.135 *A foliated map $\phi \in C^\infty(\mathscr{F}, \mathscr{G})$ is a $(\mathscr{F}, \mathscr{G})$-harmonic morphism if and only if $\phi^*\omega \in \Omega^1_B(\mathscr{F})$ is Δ_B-harmonic for each Δ'_B-harmonic form $\omega \in \Omega^1_B(\mathscr{G})$.*

Proof Let us assume that ϕ is a $(\mathscr{F}, \mathscr{G})$-harmonic morphism and let us show $\phi^* \mathscr{H}^1_B(\mathscr{G}) \subset \mathscr{H}^1_B(\mathscr{F})$. Let $\omega \in \mathscr{H}^1_B(\mathscr{G})$. The identity

$$\left(\Delta'_B\omega\,,\,\omega\right)_{L^2} = \left\|d'_B\,\omega\right\|^2 + \left\|\delta'_B\,\omega\right\|^2$$

implies that ω is d'_B-closed. In particular $d\omega = 0$ hence (by Poincaré lemma) for each $y_0 \in N$ there is aneighborhood $\mathscr{U} \subset N$ of y_0 and a C^∞ function $f : \mathscr{U} \to \mathbb{R}$ such that $df = \omega$. As $\omega \in \Omega^1_B(\mathscr{G})$, it follows that $f \in \Omega^0_B(\mathscr{G}_{\mathscr{U}})$. Yet ω is also δ'_B-closed, hence f is Δ'_B-harmonic. Then

$$d_B\big(\phi^*\omega\big) = d^2_B(f\circ\phi) = 0, \quad \delta_B(\phi^*\omega) = \delta_B\,d_B(f\circ\phi) = \Delta_B(f\circ\phi) = 0,$$

because ϕ is a $(\mathscr{F}, \mathscr{G})$-harmonic morphism. Viceversa, let $f : \mathscr{U} \to \mathbb{R}$ be a local Δ'_B-harmonic function and let us set $\omega = d'_B f \in \Omega^1_B(\mathscr{G})$. Then

$$\Delta'_B\omega = d'_B\,\delta'_B\,d'_B\,f = d'_B\,\Delta'_B f = 0,$$

i.e., $\omega \in \mathscr{H}^1_N(\mathscr{G})$. Our assumption is that $\phi^* \mathscr{H}^1_B(\mathscr{G}) \subset \mathscr{H}^1_B(\mathscr{F})$ hence $\phi^*\omega$ must be δ_B-closed. Finally,

$$0 = \delta_B\big(\phi^*\omega\big) = \delta_B\,d_B\big(f\circ\phi\big)$$

so that ϕ is a $(\mathscr{F}, \mathscr{G})$-harmonic morphism. Q.e.d.

To prove Theorem 4.134 let $\gamma \in H^1_B(\mathscr{G})$ such that $\phi^*\gamma = 0$. As N is a compact orientable manifold, and $\mathscr{G}$ is transversally oriented, we may use de Rham-Hodge theory for Riemannian foliations (cf. e.g., [PaRi]) to conclude that γ has a Δ'_B-harmonic representative $\omega \in \mathscr{H}^1_B(\mathscr{G})$. Precisely, the existence of a Δ'_B-harmonic representative for γ follows from Theorem 3.2. in [PaRi], p. 1268 (implying Theorem 12.30 in [Ton]). As $\phi : M \to N$ is a $(\mathscr{F}, \mathscr{G})$-harmonic morphism, we may apply Lemma 4.135 to conclude that $\phi^*\omega$ is a Δ_B-harmonic representative of $\phi^*\gamma \in H^1_B(\mathscr{F})$. By Theorem 3.2 in [PaRi] the assumption $\phi^*\gamma = 0$ yields $\phi^*\omega = 0$. As $\phi : M \setminus C(\phi) \to N$ is a submersion, it follows that $\omega_{\phi(x)} = 0$ for any $x \in M \setminus C(\phi)$. By Sard's theorem the set of all regular values of ϕ is dense in N hence (by continuity) $\omega_y = 0$ for any $y \in N$. Q.e.d.

The spaces $\mathscr{H}^\bullet_B(\mathscr{F})$ and $\mathscr{H}^\bullet_B(\mathscr{G})$ are finite dimensional (again by Theorem 3.2 in [PaRi]). The last statement in Theorem 4.134 then follows from Theorem 12.30 in [Ton].

Example 4.136 (*Morphisms of Killing flows*) Let (M, g) be a compact Riemannian manifold, and let T be a unit Killing vector field on M. Let $\mathscr{F}$ be the Riemannian foliation of M by orbits of T. We denote by G the closure of $\{\exp(t\,T) : t \in \mathbb{R}\}$ in $\mathrm{Isom}(M, g)$. G is a compact Abelian group, hence a torus. Every basic form on $(M, \mathscr{F})$ is then G-invariant, i.e.,

$$\iota : \Omega^r_B(\mathscr{F}) \subset \Omega^r(M)^G, \quad 0 \le r \le n-1.$$

Let $\phi \in C^\infty(\mathscr{F}, \mathscr{G})$ be a foliated map of $(M, \mathscr{F})$ into $(N, \mathscr{G})$, where $\mathscr{G}$ is the foliation of N by orbits of a unit Killing vector field $S \in \mathfrak{X}(N)$. Then

$$(d_x\phi)\, T_x = \lambda(x)\, S_{\phi(x)}$$

for some $\lambda \in C^\infty(M)$ and any $x \in M$.

Definition 4.137 The function $\lambda = (\phi^* h)(T, T)$ is the *square dilation* of ϕ. □

Claim. *The diagram*

$$\begin{array}{ccccccc} H^r_B(\mathscr{F}) & \xrightarrow{\iota_*} & H^r(M, \mathbb{R}) & \xrightarrow{i(T)_*} & H^{r-1}_B(\mathscr{F}) & \xrightarrow{\delta_{\mathscr{F}}} & H^{r+1}_B(\mathscr{F}) \\ \uparrow & & \uparrow & & \uparrow & & \uparrow \\ H^r_B(\mathscr{G}) & \xrightarrow{\iota_*} & H^r(N, \mathbb{R}) & \xrightarrow{i(S)_*} & H^{r-1}_B(\mathscr{G}) & \xrightarrow{\delta_{\mathscr{G}}} & H^{r+1}_B(\mathscr{G}) \end{array} \tag{4.188}$$

has exact lines and the vertical arrows are induced by ϕ.

Proof The first claim follows from Theorem 10.13 in [Ton], p. 139, as the short exact sequence of complexes

$$0 \to \Omega^\bullet_B(\mathscr{F}) \xhookrightarrow{\iota} \Omega^\bullet(M)^G \xrightarrow{i(T)} \Omega^{\bullet -1}_B(\mathscr{F}) \to 0$$

is exact and

$$H^r\big(\Omega^\bullet(M)^G\big) \simeq H^r(M, \mathbb{R})$$

(a group isomorphism). Moreover, the pullback of basic forms by the foliated map ϕ induces maps on cohomology $\phi^* : H^r_B(\mathscr{G}) \to H^r_B(\mathscr{F})$ for every $r \geq 0$. Next, for any $\omega \in \Omega^r(N)$

$$i(T)\, \phi^*\, \omega = \lambda\, \phi^*\, i(S)\, \omega. \tag{4.189}$$

We recall (cf. e.g., formula (10.15) in [Ton], p. 140) that the connecting homomorphism

$$\delta_{\mathscr{F}} : H^{r-1}_B(\mathscr{F}) \to H^{r+1}_B(\mathscr{F})$$

is given by

$$\delta_{\mathscr{F}}\, [\alpha]_B = \big[(d\, \chi_{\mathscr{F}}) \wedge \alpha\big]_B$$

for any $\alpha \in \Omega^{r-1}_B(\mathscr{F})$ such that $d_B\alpha = 0$. Here $[\alpha]_B$ is the basic cohomology class of α. Then

$$\big(\delta_{\mathscr{F}}\, \phi^* - \phi^*\, \delta_{\mathscr{G}}\big)\, [\beta]_B = \big[d\big(\chi_{\mathscr{F}} - \phi^*\, \chi_{\mathscr{G}}\big) \wedge \phi^*\, \beta\big]_B$$

for any $\beta \in \Omega^{r-1}_B(\mathscr{G})$ with $d'_B\beta = 0$. Let us set

$$\alpha = \left(\chi_{\mathscr{F}} - \phi^* \chi_{\mathscr{G}}\right) \wedge \phi^* \beta \in \Omega^{r+1}(M).$$

As $\chi_{\mathscr{F}}(T) = 1$

$$i(T)\,\alpha = (1 - \lambda)\,\phi^*\,\beta. \tag{4.190}$$

As T is Killing one has (by Theorem 10.6 in [Ton], pp. 133–134) $\mathscr{L}_T \chi_{\mathscr{F}} = 0$. Thus (by Cartan's formula), $i(T) d\,\chi_{\mathscr{F}} = 0$. Similarly [by (4.189)],

$$i(T)\,\phi^*\,d\,\chi_{\mathscr{G}} = 0.$$

As $i(T)_*\,\phi^*\,\beta = 0$, it follows that $i(T)\,d\,\alpha = 0$, thus yielding

Proposition 4.138 *Let $\phi : M \to N$ be a surjective, submersive $(\mathscr{F}, \mathscr{G})$-harmonic morphism of Killing flows $(M, \mathscr{F}, g)$ and $(N, \mathscr{G}, h)$, of square dilation λ. If $\lambda = 1$, then the squares in* (4.188) *are commutative and*

(i) Ker$\left\{\phi^* : H^2(N, \mathbb{R}) \to H^2(M, \mathbb{R})\right\} \subset \iota_*\, H^2_B(\mathscr{G})$. *In particular, if N is a real surface, then $\phi^* : H^2(N, \mathbb{R}) \to H^2(M, \mathbb{R})$ is injective. Let us assume that the Betti numbers $b_2(M)$ and $b_3(N)$ vanish. Then*

(ii) $\phi^* : H^3_B(N, \mathscr{G}) \to H^3_B(M, \mathscr{F})$ *is injective.*

Proof The commutativity of the middle square in (4.188) follows from (4.189). Next [by (4.190)], the $(r+1)$-form α is basic, so that

$$d\left(\chi_{\mathscr{F}} - \phi^*\,\chi_{\mathscr{G}}\right) \wedge \phi^*\,\beta = d_B\,\alpha$$

hence the right-hand square in (4.188) is commutative. To prove statement (i) let $[\omega]$ be a de Rham cohomology class in Ker $\left[i(S)_*\right] = \text{Im}\left(\iota_*\right)$. To prove statement (ii) let $[\gamma]_B \in \text{Ker}\left(\phi^*\right) \cap H^3_B(\mathscr{G})$. As $b_3(N) = 0$ the sequence

$$H^1_B(\mathscr{G}) \xrightarrow{\delta_{\mathscr{G}}} H^3_B(\mathscr{G}) \to 0$$

is exact. Hence, there is $[\beta]_B \in H^1_B(\mathscr{G})$ such that $\delta_{\mathscr{G}}\,[\beta]_B = [\gamma]_B$. Thus,

$$0 = \phi^*\,[\gamma]_B = \phi^*\,\delta_{\mathscr{G}}\,[\beta]_B = \delta_{\mathscr{F}}\,\phi^*\,[\beta]_B,$$

i.e., [by $b_2(M) = 0$],

$$\phi^*\,[\beta]_B \in \text{Ker}\,(\delta_{\mathscr{F}}) = \text{Im}\,[i(T)_*] = 0.$$

Q.e.d. □

Our next task is to prove the following.

Theorem 4.139 *Let $\phi \in C^\infty\left(\mathscr{F}, \mathscr{G}_0\right)$ be a nonconstant $\left(\mathscr{F}, \mathscr{G}_0\right)$-harmonic morphism. Then ϕ is an open map, i.e., ϕ maps open subsets of M into open subsets of N. Consequently, if M is compact and N is connected, then N is compact and ϕ is surjective.*

This is a foliated analog to B. Fuglede's result in [Fug]. The proof is by contradiction. Let $x_0 \in M$ and let $U \subset M$ be an open connected neighborhood of x_0 such that $\phi(U)$ is not an open neighborhood of $y_o = \phi(x_0)$. We also assume that U is a relatively compact subset of a local neighborhood $\varphi : D \to \mathbb{R}^n$, i.e., $\overline{U}$ is relatively compact in D. Here $n = \dim_{\mathbb{R}}(M)$. The there is a sequence $y_\nu \in N \setminus \phi(U)$ such that $y_\nu \to y_0$ as $\nu \to \infty$. Moreover, let $\Gamma(x, y)$ be a positive Green function on a neighborhood $\mathscr{U} \subset N$ of y_0, i.e., for any $y \in \mathscr{U}$ one has $\Gamma(\cdot, y) > 0$ and

$$\Delta' \Gamma(\cdot, y) = 0 \quad \text{in} \quad \mathscr{U} \setminus \{y\}.$$

Here Δ' is the Laplace-Beltrami operator of (N, h). As well known $\Gamma(x, y) \to \infty$ when (x, y) tends to the diagonal $\mathscr{U} \times \mathscr{U}$. We may assume without loss of generality that $U \subseteq \phi^{-1}(\mathscr{U})$. Let us set

$$v_\nu(x) = \Gamma\big(x, y_\nu\big), \quad x \in \mathscr{U} \setminus \{y_\nu\},$$

so that v_ν is positive and

$$\Delta' v_\nu = 0 \quad \text{in} \quad \mathscr{U} \setminus \{y_\nu\} \supset \phi(U).$$

Of course $\mathscr{G} = \mathscr{G}_0$ yields $\Delta'_B = \Delta'$ hence the function $u_\nu = v_\nu \circ \phi$ is positive and

$$\Delta_B u_\nu = 0 \quad \text{in} \quad \phi^{-1}\big(\mathscr{U} \setminus \{y_\nu\}\big)$$

and in particular in U. Yet

$$u_\nu(x_0) = v_\nu(y_0) = \Gamma\big(y_0, y_\nu\big) \to \infty, \quad \nu \to \infty,$$

hence there is a compact subset $K \subset U$ such that $x_0 \in K$ and the sequence

$$\Big\{ \sup \big\{ u_\nu(x) : x \in K \big\} \Big\}_{\nu \geq 1}$$

is bounded. Let Δ be the Laplace-Beltrami operator of (M, g). Δ is an extension of the basic Laplacian Δ_B [from $\Omega^0_B(\mathscr{F})$ to $C^\infty(M)$] hence $\Delta u_\nu = 0$ in U. The reminder of the proof of Theorem 4.139 requires the Harnack inequality. Let us set

$$\varphi = \big(x^1, \cdots, x^n\big), \quad \big[g^{AB}\big] = \big[g_{AB}\big]^{-1}, \quad g_{AB} = g\Big(\frac{\partial}{\partial x^A}, \frac{\partial}{\partial x^B}\Big) \circ \varphi^{-1}.$$

Lemma 4.140 *Let $\Omega = \varphi(U) \subset \mathbb{R}^n$ and let $f \in C^\infty(\Omega)$ satisfy $f \geq 0$ in Ω and $\Delta(f \circ \varphi) = 0$ in U. Then, for any Euclidean ball $B_{4R}(y) \subset \Omega$,*

$$\sup_{x \in B_R(y)} f(x) \leq C \inf_{x \in B_R(y)} f(x) \tag{4.191}$$

where $C = C(n, \Lambda/\lambda, \nu R) > 0$ *is a constant depending only on* n, Λ/λ *and* νR *with*

$$\Lambda = m \max_{A,\,B} \sup_{x\in\overline{\Omega}} \left|g^{AB}(x)\right|, \quad \nu = \lambda^{-1}\sqrt{n} \sup_{x\in\overline{\Omega}} \left|b^A(x)\right|. \tag{4.192}$$

Here

$$b^A = g^{AB}\frac{\partial \log \sqrt{G}}{\partial x^B}, \quad G = \det\left[g_{AB}\right], \quad \lambda = \inf_{x\in\overline{\Omega}} \lambda_1(x) > 0,$$

where $\lambda_1(x)$ is the smallest eigenvalue of the matrix $\left[g^{AB}(x)\right]$.

Proof of Lemma 4.140. Under the assumptions of the lemma

$$\sum_{A,\,B=1}^{n} \frac{\partial}{\partial x^A}\left(g^{AB}(x)\frac{\partial f}{\partial x^B}\right) + b^A(x)\frac{\partial f}{\partial x^A} = 0 \quad \text{in} \quad \Omega.$$

Next [with the definitions in (4.192)]

$$g^{AB}(x)\,\xi_A\,\xi_B \geq \lambda\,|\xi|^2, \quad \xi \in \mathbb{R}^n,$$

$$\sum_{A,\,B} \left|g^{AB}(x)\right|^2 \leq \Lambda^2, \quad \lambda^{-2}\sum_{A} \left|b^A(x)\right|^2 \leq \nu^2, \quad x \in \Omega.$$

Also Ω is bounded hence $f \in W^{1,2}(\Omega)$. Therefore, one may apply Theorem 8.20 in [GiTr], p. 199, to conclude that (4.191) holds.

Let us go back to the proof of Theorem 4.139. As $\varphi(K)$ is a compact set there exist $s \geq 1$ and $x_i \in \varphi(K)$ and $R_i > 0$ such that

$$\varphi(K) \subset \bigcup_{i=1}^{s} B_{4R_i}(x_i) \subset \Omega.$$

Let $f_\nu = u_\nu \circ \varphi^{-1}$. Then [by (4.191)]

$$\sup\left\{u_\nu(x)\,:\,x\in K\right\} \leq \sup\left\{f_\nu(x)\,:\,x\in\bigcup_{i=1}^{s} B_{R_i}(x_i)\right\} \leq C\max\left\{\inf_{x\in B_{R_i}(x_i)} f_\nu(x)\,:\,1\leq i\leq s\right\}.$$

As $\phi : M \to N$ is nonconstant $K \setminus \phi^{-1}(y_0) \neq \emptyset$. Also as M is connected ϕ cannot be locally constant. Therefore, (relabeling x_i and R_i if necessary) one may assume

$$\sup\left\{u_\nu(x)\,:\,x\in K\right\} \leq C \inf_{x\in B_{R_1}(x_1)} f_\nu(x),$$

$$B_{R_1}(x_1) \cap \varphi\left[K \setminus \phi^{-1}(y_0)\right] \neq \emptyset.$$

Indeed if $\varphi^{-1}\big[B_{R_1}(x_1)\big] \subseteq \phi^{-1}(y_0)$, then, for each $x \in \varphi^{-1}\big[B_{R_1}(x_1)\big]$, one has $u_\nu(x) = v_\nu(y_0) = u_\nu(x_0)$, a contradiction. Thus,

$$\sup_{x \in K} u_\nu(x) \leq C \inf \Big\{ f_\nu(x) \,:\, x \in B_{R_1}(x_1) \cap \varphi\big[K \setminus \phi^{-1}(y_0)\big] \Big\}. \tag{4.193}$$

Finally,

$$x \in K \setminus \phi^{-1}(y_0) \Longrightarrow u_\nu(x) = v_\nu\big(\phi(x)\big) = \Gamma\big(\phi(x),\, y_\nu\big) \to \Gamma\big(\phi(x),\, y_0\big) < \infty, \quad \nu \to \infty,$$

and the fact that $\big\{ \sup_{x \in K} u_\nu(x) \big\}_{\nu \geq 1}$ is unbounded contradicts (4.193). To complete the proof of Theorem 4.139 note that $\phi(M)$ is compact (as the image of a compact set) and in particular closed. Yet (by the first statement in Theorem 4.139) $\phi(M)$ is also open so that (as N is connected) $\phi(M) = N$. As a consequence of Theorems 4.134 and 4.139 one obtains

Corollary 4.141 *Let M, $\mathscr{F}$, g) be a compact orientable foliated Riemannian manifold where $\mathscr{F}$ is a transversally oriented Riemannian foliation and g is bundle-like. Let $\phi \in C^\infty\big(\mathscr{F},\, \mathscr{G}_0\big)$ be a foliated map into a Riemannian manifold N thought of as equipped with the trivial foliation by points $\mathscr{G}_0$. If $\phi : M \to N$ is a nonconstant $\big(\mathscr{F},\, \mathscr{G}_0\big)$-harmonic morphism, then*

(i) *N is compact,*

(ii) *$\phi^* : H^1(N,\, \mathbb{R}) \to H^1_B(\mathscr{F})$ is injective,*

iii) *$\dim_{\mathbb{R}} \mathscr{H}^1_B(\mathscr{F}) \geq b_1(N)$.*

Here $b_1(N)$ is the first Betti number of N. To replace $\mathscr{G}_0$ in Theorem 4.139 by an arbitrary Riemannian foliation $\mathscr{G}$ one should produce estimates from below on a fundamental solution $\Gamma_B(x,\, y)$ to the basic Laplacian [describing the behavior of $\Gamma_B(x,\, y)$ as $y \to x$]. This is an open problem.

Let $(M,\, g)$ and $(N,\, h)$ be two Riemannian manifolds and $x \in M$.

Definition 4.142 A C^∞ map $\phi : M \to N$ is *horizontally weakly conformal* at x if either $d_x\phi = 0$ or

$$h_{\phi(x)}\big((d_x\phi)v\,,\, (d_x\phi)w\big) = \Lambda(x)\, g_x(v\,,\, w)$$

for some $\Lambda(x) \in \mathbb{R} \setminus \{0\}$ and any $v,\, w \in \mathscr{H}_x = \operatorname{Ker}\big(d_x\phi\big)^\perp \subset T_x(M)$. □

Let us consider $(M,\, \mathscr{F},\, g)$ and $(N,\, \mathscr{G}_0\,,\, h)$ where $\mathscr{F}$ is transversally oriented and Riemannian, and g is bundle-like. If $\phi \in C^\infty\big(\mathscr{F},\, \mathscr{G}_0\big)$ is a nonconstant $\big(\mathscr{F},\, \mathscr{G}_0\big)$-harmonic morphism, then (by Theorem 4.134) ϕ is horizontally weakly conformal and a $\big(\mathscr{F},\, \mathscr{G}_0\big)$-harmonic map. The following result deals with the converse.

Theorem 4.143 *Let $\phi \in C^\infty(\mathscr{F},\, \mathscr{G})$ be a foliated map of foliated Riemannian manifolds where both $\mathscr{F}$ and $\mathscr{G}$ are transversally oriented and Riemannian while g and h are bundle-like. If $\phi : M \to N$ is horizontally weakly conformal of square dilation $\Lambda \in C^\infty(M)$ and a $(\mathscr{F},\, \mathscr{G})$-harmonic map, then*

$$\Delta_B(f \circ \phi) = \Lambda\big[\Delta'_B f + \sigma_h\big(\tau_{\mathcal{G}}\big) f\big] \circ \phi \tag{4.194}$$

for any C^∞ basic function $f : \mathcal{U} \subseteq N \to \mathbb{R}$. In particular if the foliation $\mathcal{G}$ is harmonic, then ϕ is a $(\mathcal{F}, \mathcal{G})$-harmonic morphism.

Here $\sigma_h : \nu(\mathcal{G}) \to T(\mathcal{G})^\perp$ is the natural vector bundle isomorphism and $\tau_{\mathcal{G}} \in C^\infty\big(\nu(\mathcal{G})\big)$ is the mean curvature vector of $\mathcal{G}$ in (N, h). A crucial ingredient in the proof of Theorem 4.143 is the identity

$$\Delta_B f = \sum_{i=1}^{q} \Big\{\sigma_g\big(r_i\big)\,\sigma_g\big(r_i\big) - \sigma_g \nabla_{\sigma_g(r_i)} r_i\Big\}\, f - \sigma_g\big(\tau_{\mathcal{F}}\big)\, f. \tag{4.195}$$

A proof of (4.195) is given at the end of Sect. 4.7.2.2. Let $\big(U,\, x^\alpha,\, y^i\big) \in \mathcal{F}$ and $\big(\mathcal{U},\, X^a,\, Y^u\big) \in \mathcal{G}$ be local foliated charts such that $\phi(U) \subseteq \mathcal{U}$. Let us rewrite (4.184) as

$$\Delta_B(f \circ \phi) = \Big[\tau_T(\phi)^w - \big(g_Q\big)^{ij}\, \frac{\partial\, \phi^{p'+w}}{\partial\, y^i}\, \kappa_j\Big] \Big(\frac{\partial f}{\partial\, Y^w}\Big)^{\phi} + \big(g_Q\big)^{ij}\, \frac{\partial\, \phi^{p'+u}}{\partial\, y^i}\, \frac{\partial\, \phi^{p'+v}}{\partial\, y^i}\, f_{u,v} \tag{4.196}$$

where

$$f_{u,v} = \frac{\partial^2 f}{\partial Y^u\, \partial Y^v} - \Gamma'^{w}_{uv}\, \frac{\partial\, f}{\partial\, Y^w}, \quad f \in \Omega^0_B(\mathcal{G}).$$

If ϕ is $(\mathcal{F}, \mathcal{G})$-harmonic $\tau_T(\phi) = \phi_*\, \tau_{\mathcal{F}}$. Also if ϕ is horizontally weakly conformal with square dilation Λ, then (by Lemma 2.4.4 in [BaWo], p. 47)

$$g^{AB}\Big(\frac{\partial\, \phi^M}{\partial\, x^A}\, \frac{\partial\, \phi^N}{\partial\, x^B}\Big) = \Lambda\, h^{MN}, \tag{4.197}$$

$$\big[g^{AB}\big] = \big[g_{AB}\big]^{-1}, \quad g_{AB} = g\Big(\frac{\partial}{\partial\, x^A},\, \frac{\partial}{\partial\, x^B}\Big), \quad 1 \le A,\, B \le n, \quad 1 \le M,\, N \le n'.$$

We also set $\big[g^{\alpha\beta}\big] = \big[g_{\alpha\beta}\big]^{-1}$ (the leafwise cometric relative to $\mathcal{F}$). Then (4.157) yields

$$g^{p+i,\, p+j} = \big(g_Q\big)^{ij}. \tag{4.198}$$

Next [by (4.197) for $M = p' + u$ and $N = p' + v$],

$$\big(g_Q\big)^{ij}\, \frac{\partial\, \phi^{p'+u}}{\partial\, y^i}\, \frac{\partial\, \phi^{p'+v}}{\partial\, y^j} = \Lambda\, h^{p'+u,\, p'+v}$$

[as $\phi^{p'+u} \in \Omega^0_B\big(\mathcal{F}_U\big)$ for any $1 \le u \le q'$] and (4.196) becomes

$$\Delta_B(f \circ \phi) = \Lambda\, \big(h_{\nu(\mathcal{G})}\big)^{uv}\, f_{u,v}. \tag{4.199}$$

Yet [by (4.195)],

$$\Delta'_B f = \sum_{u=1}^{q'} \left\{ \sigma_h(\gamma_u)^2 - \sigma_h \nabla'_{\sigma_h(\gamma_u)} \gamma_u \right\} f - \sigma_h(\tau_{\mathscr{G}}) f$$

on $\mathscr{U}$ for any local $h_{\nu(\mathscr{G})}$-orthonormal frame $\{\gamma_u : 1 \le u \le q'\}$ (defined on the open set $\mathscr{U} \subset N$) in the transverse bundle $\nu(\mathscr{G}) \to N$. Thus,

$$(h_{\nu(\mathscr{G})})^{uv} f_{u,v} = \Delta'_B f + \sigma_h(\tau_{\mathscr{G}}) f$$

so that (4.199) yields (4.194).

We anticipated the useful local expression (4.195) of Δ_B [alternative to (4.182)]. To prove (4.195) let $\{r_i : 1 \le i \le q\}$ be a local g_Q-orthonormal frame of Q defined on $U \subset M$. As

$$\sigma_g(r) f = g_Q(\pi \nabla f, r), \quad r \in C^\infty(\nu(\mathscr{F})), \quad f \in C^\infty(M),$$

we may conduct the calculation (as $\Delta g_Q = 0$)

$$\sum_i \left\{ \sigma_g(r_i) \sigma_g(r_i) - \sigma_g \nabla_{\sigma_g(r_i)} r_i \right\} f = \sum_i \left\{ \sigma_g(r_i)[g_Q(\pi \nabla f, r_i)] - g_Q(\pi \nabla f, \nabla_{\sigma_g(r_i)} r_i) \right\} =$$

$$= \sum_i g_Q(\nabla_{\sigma_g(r_i)} \pi \nabla f, r_i) = (g_Q)^{ij} g_Q(\nabla_{(\partial_{p+i})^\perp} \pi \nabla f, \pi \partial_{p+j})$$

and

$$\pi \nabla f = (g_Q)^{ij} \frac{\partial f}{\partial y^i} \pi \partial_{p+i}, \quad f \in \Omega^0_B(\mathscr{F});$$

hence [by (4.167) and (4.182)],

$$\sum_i \left\{ \sigma_g(r_i) \sigma_g(r_i) - \sigma_g \nabla_{\sigma_g(r_i)} r_i \right\} f = \mathrm{div}_B[(\pi \nabla f)^i \partial_{p+i}] =$$

$$= \mathrm{div}_B\left[(g_Q)^{ij} \frac{\partial f}{\partial y^j} \partial_{p+i} \right] = \Delta_B f + (g_Q)^{ij} \frac{\partial f}{\partial y^i} \kappa_j$$

implying that (4.195) holds on U for any $f \in \Omega^0_B(\mathscr{F})$. Let $\mathscr{D}(\Delta) = C^\infty(M)$ and $\mathscr{D}(\Delta_B) = \Omega^0_B(\mathscr{F})$ be the domains of Δ and Δ_B, respectively. As a byproduct of (4.195) $\Delta_B \subset \Delta$, i.e., the Laplace-Beltrami operator of (M, g) is an elliptic extension of the basic Laplacian (on functions).

4.10.2.3 Stability of $(\mathcal{F}, \mathcal{G})$-harmonic maps

Through this section $\mathcal{F}$ is a transversally oriented Riemannian foliation and g is a bundle-like Riemannian metric on $(M, \mathcal{F})$. We shall establish

Theorem 4.144 *Let $(M, \mathcal{F}, g)$ be a compact orientable foliated Riemannian manifold. Let $\phi \in C^\infty(\mathcal{F}, \mathcal{G})$ be a $(\mathcal{F}, \mathcal{G})$-harmonic map into another foliated Riemannian manifold $(N, \mathcal{G}, h)$. Let $\{\phi_{s,t} \in C^\infty(\mathcal{F}, \mathcal{G}) : -\epsilon < s, t < \epsilon\}$ be a smooth 2-parameter variation of $\phi_{0,0} = \phi$ through foliated maps. We set $\tilde{M} \equiv M \times (-\epsilon, \epsilon) \times (-\epsilon, \epsilon)$ and*

$$\Phi : \tilde{M} \to N, \quad \Phi(x, s, t) = \phi_{s,t}(x), \quad |s| < \epsilon, \quad |t| < \epsilon,$$

$$V(x) = \big(d_{(x,0,0)}\Phi\big)\Big(\frac{\partial}{\partial s}\Big)_{x,0,0}, \quad W(x) = \big(d_{(x,0,0)}\Phi\big)\Big(\frac{\partial}{\partial t}\Big)_{x,0,0},$$

for any $x \in M$ so that $V, W \in C^\infty\big(\phi^{-1}T(N)\big)$. Then

$$\frac{\partial^2 E_T(\phi_{s,t})}{\partial s\, \partial t}(0, 0) = \int_M h^\phi_{\nu(\mathcal{G})}\Big(\Big[J_\phi + D_{\sigma_g(\tau_{\mathcal{F}})}\Big]\pi'^\phi V, \pi'^\phi W\Big)\, d\,v_g . \tag{4.200}$$

Here $J_\phi : C^\infty_B\big(\phi^{-1}\nu(\mathcal{G})\big) \to C^\infty_B\big(\phi^{-1}\nu(\mathcal{G})\big)$ is locally given by

$$J_\phi v = -\sum_{i=1}^{q}\Big\{D^\phi_{\sigma_g(r_i)}\, D^\phi_{\sigma_g(r_i)} v - D^\phi_{\sigma_g[\nabla_{\sigma_g(r_i)} r_i]} v + R'^\phi\big(v, \phi_* r_i\big)\phi_* r_i\Big\}$$

for any $v \in C^\infty_B\big(\phi^{-1}\nu(\mathcal{G})\big)$ and any local g_Q-orthonormal frame $\{r_i : 1 \le i \le q\}$ in Q. Also $R'^\phi \in C^\infty\big(\bigotimes^3 \phi^{-1}\nu(\mathcal{G})^*\big)$ is naturally induced by the curvature tensor field R' of the canonical adapted connection $\nabla' \in \mathcal{C}\big(\nu(\mathcal{G}), h_{\nu(\mathcal{G})}\big)$.

Definition 4.145 We refer to

$$J_T(\phi) \equiv J_\phi + D^\phi_{\sigma_g(\tau_{\mathcal{F}})}$$

as the *transverse Jacobi operator* while (4.200) is the *second variation formula* for the transverse energy integral. □

Also $C^\infty_B\big(\phi^{-1}\nu(\mathcal{G})\big)$ denotes the space of all $v \in C^\infty\big(\phi^{-1}\nu(\mathcal{G})\big)$ such that $D^\phi_X v = 0$ for any $X \in \mathfrak{X}_{\mathcal{F}}$. For each $(s, t) \in (-\epsilon, \epsilon)^2$ we consider the injection $\alpha_{s,t} : M \to \tilde{M}$ given by $\alpha_{s,t}(x) = (x, s, t)$ and set $\tilde{P}_{(x,s,t)} = \big(d_x\alpha_{s,t}\big)T(\mathcal{F})_x$, $x \in M$. Then $\tilde{P}$ in an involutive smooth distribution of rank p. By the Frobenius theorem $\tilde{P}$ is completely integrable. Let $\tilde{\mathcal{F}}$ be the corresponding foliation of $\tilde{M}$ so that $T(\tilde{\mathcal{F}}) = \tilde{P}$ (in particular $\tilde{\mathcal{F}}$ is the pullback of $\mathcal{F}$, i.e., $\tilde{\mathcal{F}} = \pi_1^*\mathcal{F}$ where $\pi_1 : \tilde{M} \to M$ is the natural projection). Note that Φ is a foliated map, i.e., $\Phi \in C^\infty(\tilde{\mathcal{F}}, \mathcal{G})$, hence it induces the bundle morphism $d_T\Phi : \nu(\tilde{\mathscr{F}}) \to \Phi^{-1}\nu(\mathscr{G})$. If $\mathbf{r} \in C^\infty\big(\nu(\tilde{\mathscr{F}}\big)$ we define a section $\Phi_*\mathbf{r}$ in the pullback bundle $\Phi^{-1}\nu(\mathscr{G}) \to \tilde{M}$ by setting

$$\big(\Phi_*\mathbf{r}\big)(x,\, s,\, t) = \big(d_T\,\Phi\big)_{(x,s,t)}\,\mathbf{r}(x,\, s,\, t), \quad (x,\, s,\, t) \in \tilde{M}.$$

We wish to compute $\left\{\dfrac{\partial^2 E_T\big(\phi_{s,t}\big)}{\partial s\,\partial t}\right\}_{s=t=0}$ where

$$E_T\big(\phi_{s,t}\big) = \frac{1}{2}\int_M \big\|d_T\,\phi_{s,t}\big\|^2\, d\,v_g\,.$$

Note that $\alpha_{s,t} : M \to \tilde{M}$ is a foliated map, i.e.,

$$\alpha_{s,t} \in C^\infty\big(\mathscr{F},\, \tilde{\mathscr{F}}\big), \quad |s| < \epsilon, \quad |t| < \epsilon.$$

For every section $r \in C^\infty\big(\nu(\mathscr{F})\big)$ we set

$$\tilde{r} \in C^\infty\big(\nu(\tilde{\mathscr{F}})\big), \quad \tilde{r}(x,\, s,\, t) = \big(d_T\,\alpha_{s,t}\big)_x\, r(x).$$

Also let $w^\phi = w \circ \phi \in C^\infty\big(\phi^{-1}\,\nu(\mathscr{G})\big)$ be the natural lift of $w \in C^\infty\big(\nu(\mathscr{G})\big)$. Let $\{r_i \,:\, 1 \le i \le q\}$ be a local g_Q-orthonormal frame of $Q \to M$, defined on the open set $U \subset M$. If

$$f \in C^\infty\big(\tilde{M}\big), \quad f(x,\, s,\, t) = \big\|d_T\,\phi_{s,t}\big\|^2(x),$$

then

$$f(x,\, s,\, t) = \sum_i h^\Phi_{\nu(\mathscr{G})}\big(\Phi_*\,\tilde{r}_i\,,\, \Phi_*\,\tilde{r}_i\big)_{(x,\,s,\,t)}\,, \quad x \in U,$$

where $h^\Phi_{\nu(\mathscr{G})} = \Phi^{-1}\,h_{\nu(\mathscr{G})}$ is the pullback of $h_{\nu(\mathscr{G})}$ by Φ [a Riemannian bundle metric on $\Phi^{-1}\,\nu(\mathscr{G}) \to \tilde{M}$]. Then

$$\frac{\partial f}{\partial t} = 2\sum_i h^\Phi_{\nu(\mathscr{G})}\big(D^\Phi_{\partial/\partial t}\,\Phi_*\tilde{r}_i\,,\, \Phi_*\tilde{r}_i\big) \quad \text{on} \quad \tilde{U} \equiv U \times (-\epsilon,\, \epsilon)^2\,,$$

where $D^\Phi = \Phi^{-1}\nabla' \in \mathscr{C}\big(\Phi^{-1}\nu(\mathscr{G}),\, h^\Phi_{\nu(\mathscr{G})}\big)$ is the pullback of the canonical adapted connection ∇'. Let us consider the sections

$$\mathbb{V},\, \mathbb{W} \in C^\infty\big(\Phi^{-1}T(N)\big)\,,$$

$$\mathbb{V}(x,\, s,\, t)\big(d_{(x,s,t)}\,\Phi\big)\left(\frac{\partial}{\partial s}\right)_{(x,\,s,\,t)}, \quad \mathbb{W}(x,\, s,\, t)\big(d_{(x,s,t)}\,\Phi\big)\left(\frac{\partial}{\partial t}\right)_{(x,\,s,\,t)}, \quad (x,\, s,\, t) \in \tilde{M}.$$

We observe that

$$D^{\Phi}_{\partial/\partial t}\,\Phi_* \tilde{r} = D_{\widetilde{\sigma_g(r)}}\,\pi'\mathbb{W}, \quad r \in C^{\infty}\big(\nu(\mathscr{F})\big). \tag{4.201}$$

Given $Y \in \mathfrak{X})M)$ we define

$$\tilde{Y} \in \mathfrak{X}(\tilde{M}), \quad \tilde{Y}_{(x,s,t)} = \big(d_x\,\alpha_{s,t}\big)Y_x.$$

Then [by (4.201)]

$$\frac{\partial f}{\partial t} = 2\sum_i \Big\{\widetilde{\sigma_g(r_i)}\Big[h^{\Phi}_{\nu(\mathscr{G})}\big(\pi'^{\Phi}\,\mathbb{W},\ \Phi_*\tilde{r}_i\big)\Big] - h^{\Phi}_{\nu(\mathscr{G})}\big(\pi'^{\Phi}\,\mathbb{W},\ D^{\Phi}_{\widetilde{\sigma_g(r_i)}}\,\Phi_*\tilde{r}_i\big)\Big\}. \tag{4.202}$$

We need the following:

Lemma 4.146 *For each $(s, t) \in (-\epsilon, \epsilon)^2$ let $r_{s,t}1C^{\infty}\big(\nu(\mathscr{F})\big)$ be the section determined by*

$$h^{\Phi}_{\nu(\mathscr{G})}\big(\pi'^{\Phi}\,\mathbb{W},\ \Phi_*\tilde{u}\big)\circ\alpha_{s,t} = g_Q\big(r_{s,t},\ u\big), \quad u \in C^{\infty}(Q). \tag{4.203}$$

Then $r_{s,t}$ is a transverse vector field on $(M, \mathscr{F})$, i.e., $r_{s,t} \in \ell(\mathscr{F})$.

By (4.203) and $\nabla g_{\nu(\mathscr{F})} = 0$

$$\widetilde{\sigma_g(r_i)}\Big[h^{\Phi}_{\nu(\mathscr{G})}\Big(\pi'^{\Phi}\,\mathbb{W},\ \Phi_*\tilde{r}_i\Big)\Big]\circ\alpha_{s,t} = \sigma_g(r_i)\big[g_{\nu(\mathscr{F})}\big(r_{s,t},\ r_i\big)\big] =$$

$$= g_{\nu(\mathscr{F})}\big(\nabla_{\sigma_g(r_i)}r_{s,t},\ r_i\big) + g_{\nu(\mathscr{F})}\big(r_{s,t},\ \nabla_{\sigma_g(r_i)}r_i\big).$$

Lemma 4.147 *Let $\{r_i\ :\ 1 \le i \le q\}$ be a local g_Q-orthonormal frame in the normal bundle $\nu(\mathscr{F}) \to M$ defined on the open set $U \subseteq M$. Then*

$$\sum_{j=1}^{q} g_Q\big(\nabla_{\sigma_g(r_j)}\,r_{s,t},\ r_j\big) = \operatorname{div}_B\big(r_{s,t}\big) \tag{4.204}$$

on U for any $(s, t) \in (-\epsilon, \epsilon)^2$.

Thus [by (4.203)–(4.204)],

$$\sum_i \widetilde{\sigma_g(r_i)}\Big[h^{\phi}_{\nu(\mathscr{G})}\Big(\pi'^{\Phi}\,\mathbb{W},\ \Phi_*\tilde{r}_i\Big)\Big]\circ\alpha_{s,t} = \operatorname{div}_B\big(r_{s,t}\big) + \sum_i h^{\Phi}_{\nu(\mathscr{G})}\Big(\pi'^{\Phi}\,\mathbb{W},\ \Phi_*\,\widetilde{\nabla_{\sigma_g(r_i)}\,r_i}\Big)$$

so that (4.202) may be written

$$\frac{1}{2}\,\frac{\partial f}{\partial t}\circ\alpha_{s,t} = \operatorname{div}_B\big(r_{s,t}\big) - h^{\Phi}_{\nu(\mathscr{G})}\Big(\pi'^{\Phi}\,\mathbb{W},\ \sum_i\Big[D^{\Phi}_{\widetilde{\sigma_g(r_i)}}\,\Phi_*\tilde{r}_i - \Phi_*\,\widetilde{\nabla_{\sigma_g(r_i)}\,r_i}\Big]\Big)\circ\alpha_{s,t}\,.$$

Let us integrate over M and use the foliated analog (4.169) to Green's lemma. We obtain [by (4.203)]

$$\frac{1}{2}\int_M \left(\frac{\partial f}{\partial t}\circ \alpha_{s,t}\right) d\, v_g = -\int_M \left[h^{\Phi}_{\nu(\mathscr{G})}\left(\pi'^{\Phi}\,\mathbb{W},\ \mathrm{trace}_{g_{\nu(\mathscr{F})}} B(\Phi) - \Phi_*\,\tilde{\tau}_{\mathscr{F}}\right)\circ \alpha_{s,t}\right] d\, v_g \tag{4.205}$$

where

$$B(\Phi)(u,\, v) = D^{\Phi}_{\widetilde{\sigma_g(u)}}\,\Phi_*\,\tilde{v} - \Phi_*\widetilde{\nabla_{\sigma_g(u)} v}, \quad u,\ v \in C^\infty(Q).$$

Moreover,

$$\frac{1}{2}\int_M \frac{\partial^2 f}{\partial s\,\partial t}\, d\, v_g = -\int_M \left\{h^{\Phi}_{\nu(\mathscr{G})}\left(D^{\Phi}_{\partial/\partial s}\,\pi'^{\Phi}\,\mathbb{W},\ \mathrm{trace}_{g_{\nu(\mathscr{F})}} B(\Phi) - \Phi_*\,\tilde{\tau}_{\mathscr{F}}\right) + \right.$$

$$\left. + \, h^{\Phi}_{\nu(\mathscr{G})}\left(\pi'^{\Phi}\,\mathbb{W},\ D^{\Phi}_{\partial/\partial s}\left[\mathrm{trace}_{g_{\nu(\mathscr{F})}} B(\Phi) - \Phi_*\,\tilde{\tau}_{\mathscr{F}}\right]\right)\right\} d\, v_g .$$

The first term evaluated at $(s,\ t) = (0,\ 0)$ is

$$h^{\Phi}_{\nu(\mathscr{G})}\left(D^{\Phi}_{\partial/\partial s}\,\pi'^{\Phi}\,\mathbb{W},\ \mathrm{trace}_{g_{\nu(\mathscr{F})}} B(\Phi) - \Phi_*\,\tilde{\tau}_{\mathscr{F}}\right)\circ \alpha_{0,0} =$$

$$= h^{\phi}\left(\left[D^{\Phi}_{\partial/\partial s}\,\pi'^{\Phi}\,\mathbb{W}\right]\circ \alpha_{0,0},\ \tau_T(\phi) - \phi_*\,\tau_{\mathscr{F}}\right) = 0$$

as ϕ is $/\mathscr{F},\ \mathscr{G})$-harmonic. On the other hand [by (4.201)],

$$D^{\Phi}\left\{\mathrm{trace}_{g_{\nu(\mathscr{F})}} B(\Phi) - \Phi_*\,\tilde{\tau}_{\mathscr{F}}\right\} = \tag{4.206}$$

$$- D^{\Phi}_{\widetilde{\sigma_g(\tau_{\mathscr{F}})}}\,\pi'^{\Phi}\,\mathbb{V} + \sum_i \left\{D^{\Phi}_{\partial/\partial s}\, D^{\Phi}_{\widetilde{\sigma_g(r_i)}}\,\Phi_*\,\tilde{r}_i - D^{\Phi}_{\widetilde{\sigma_g[\nabla_{\sigma_g(r_i)} r_i]}}\,\pi'^{\Phi}\,\mathbb{V}\right\}.$$

Let R^{D^Φ} be the curvature tensor filed of D^Φ. Then (as $[\partial/\partial s,\ \tilde{Y}] = 0$ for any $Y \in \mathfrak{X}(M)$)

$$D^{\Phi}_{\partial/\partial s}\, D^{\Phi}_{\widetilde{\sigma_g(r_i)}}\,\Phi_*\,\tilde{r}_i = R^{D^\Phi}\left(\frac{\partial}{\partial s},\ \widetilde{\sigma_g(r_i)}\right)\Phi_*\,\tilde{r}_i + D^{\Phi}_{\widetilde{\sigma_g(r_i)}}\, D^{\Phi}_{\widetilde{\sigma_g(r_i)}}\,\pi'^{\Phi}\,\mathbb{V}$$

and Theorem 4.144 follows from (4.206) by evaluating at $s = t = 0$. Q.e.d.

Definition 4.148 The second-order differential operator

$$\Delta_\phi : C^\infty_B\left(\phi^{-1}\nu(\mathscr{G})\right) \to C^\infty_B\left(\phi^{-1}\nu(\mathscr{G})\right)$$

locally given by

$$\Delta_\phi = -\sum_{i=1}^{q} \left\{ D^\phi_{\sigma_g(r_i)} D^\phi_{\sigma_g(r_i)} - D^\phi_{\sigma_g[\nabla_{\sigma_g(r_i)}(r_i)]} \right\} + D^\phi_{\sigma_g(\tau_{\mathscr{F}})}$$

is called the *rough basic Laplacian.* □

Also we set

$$\mathscr{R}_\phi\, w = \sum_{i=1}^{q} R'^\phi\big(w,\ \phi_* r_i\big)\, \phi_* r_i\,, \quad w \in C^\infty\big(\phi^{-1}\nu(\mathscr{G})\big)$$

so that

$$J_T(\phi) = \Delta_\phi - \mathscr{R}_\phi. \tag{4.207}$$

This gives a bundle morphism $\mathscr{R}_\phi : \phi^{-1}(\mathscr{G}) \to \phi^{-1}(\mathscr{G})$ such that

$$h^\phi_{\nu(\mathscr{G})}\big(\mathscr{R}_\phi\, v,\ w\big) = h^\phi_{\nu(\mathscr{G})}\big(v,\ \mathscr{R}_\phi\, w\big), \quad v,\ w \in \phi^{-1}(\mathscr{G}).$$

Definition 4.149 The *basic Hessian* $H\big(E_T\big)$ of the transverse energy integral E_T at a critical point ϕ is

$$H\big(E_T\big)_\phi = \int_M h^\phi_{\nu(\mathscr{G})}\big(J_T(\phi)\, v,\ w\big)\, d\, v_g$$

for any $v,\ w \in C^\infty_B\big(\phi^{-1}\nu(\mathscr{G})\big)$. □

Definition 4.150 The *basic index* $\mathrm{ind}_B(\phi)$ of a $(\mathscr{F},\ \mathscr{G})$-harmonic map $\phi \in C^\infty(\mathscr{F},\ \mathscr{G})$ is the least upper bound of the dimensions $\dim_{\mathbb{R}}(F)$ as F runs over the subspaces of $C^\infty_B\big(\phi^{-1}\nu(\mathscr{G})\big)$ on which $H\big(E_T\big)_\phi$ is negative definite. □

Definition 4.151 The *basic nullity* $\mathrm{null}_B(\phi)$ of ϕ is the dimension of the space

$$\Big\{ v \in C^\infty_B\big(\phi^{-1}\nu(\mathscr{G})\big)\ :\ H\big(E_T\big)_\phi(v,\ w) = 0, \quad w \in C^\infty_B\big(\phi^{-1}\nu(\mathscr{G})\big) \Big\}.$$

□

Definition 4.152 A $(\mathscr{F}, \mathscr{G})$-harmonic map ϕ is *weakly stable* if $\mathrm{ind}_B(\phi) = 0$, i.e.,

$$H\big(E_T\big)_\phi(v,\ v) \geq 0, \quad v \in C^\infty_B\big(\phi^{-1}\nu(\mathscr{G})\big).$$

Otherwise ϕ is *unstable.* □

Theorem 4.153 *If the foliated Riemannian manifold $(N,\ \mathscr{G},\ h)$ has nonnegative transverse sectional curvature, then any $(\mathscr{F},\ \mathscr{G})$-harmonic map $\phi \in C^\infty(\mathscr{F},\ \mathscr{G})$ is weakly stable.*

Lemma 4.154 *For each $(\mathcal{F}, \mathcal{G})$-harmonic map $\phi \in C^\infty(\mathcal{F}, \mathcal{G})$ the rough basic Laplacian Δ_ϕ satisfies*

$$\int_M h^\phi_{\nu(\mathcal{G})}\big(\Delta_\phi\, v,\, w\big)\, d\, v_g = \int_M \Big(h^\phi_{\nu(\mathcal{G})}\Big)^*\big(D^\phi\, v,\, D^\phi\, w\big)\, d\, v_g \tag{4.208}$$

for any $v, w \in C^\infty_B\big(\phi^{-1}\nu(\mathcal{G})\big)$. In particular the transverse Jacobi operator $J_T(\phi)$: $C^\infty_B\big(\phi^{-1}\nu(\mathcal{G})\big) \to C^\infty_B\big(\phi^{-1}\nu(\mathcal{G})\big)$ is formally self-adjoint.

As to the notation in (4.208) one uses the map

$$D^\phi : C^\infty_B\big(\phi^{-1}\nu(\mathcal{G})\big) \to C^\infty_B\big(T^*(M) \otimes \phi^{-1}\nu(\mathcal{G})\big),$$

$$\big(D^\phi\, v\big)Y = D^\phi_Y\, v, \quad v \in C^\infty_B\big(\phi^{-1}\nu(\mathcal{G})\big), \quad Y \in \mathfrak{X}(M).$$

The space $C^\infty_B\big(T^*(M) \otimes \phi^{-1}\nu(\mathcal{G})\big)$ consists of all C^∞ sections φ in $T^*(M) \otimes \phi^{-1}\nu(\mathcal{G}) \to M$ such that $T(\mathcal{F}) \rfloor \varphi = 0$. Also $\big(h^\phi_{\nu(\mathcal{G})}\big)^*$ is the pointwise inner product induced by $h^\phi_{\nu(\mathcal{G})}$ on $C^\infty_B\big(T^*(M) \otimes \phi^{-1}\nu(\mathcal{G})\big)$, i.e., locally

$$\big(h^\phi_{\nu(\mathcal{G})}\big)^*\,(\varphi,\, \psi) = \sum_{i=1}^q h_{\nu(\mathcal{F})}\Big(\varphi\big(\sigma_g(r_i)\big),\, \psi\big(\sigma_g(r_i)\big)\Big).$$

Let

$$\big(D^\phi\big)^* : C^\infty_B\big(T^*(M) \otimes \phi^{-1}\nu(\mathcal{G})\big) \to C^\infty_B\big(\phi^{-1}\nu(\mathcal{G})\big),$$

$$\big\langle D^\phi\, v,\, \varphi\big\rangle_B = \big\langle v,\, \big(D^\phi\big)^*\varphi\big\rangle_B,$$

be the adjoint of D^ϕ. The next step is to establish a useful local expression of $\big(D^\phi\big)^*$. One has (by $D^\phi\, h^\phi_{\nu(\mathcal{G})} = 0$)

$$\big\langle v,\, \big(D^\phi\big)^*\varphi\big\rangle_B = \int_M \big(h^\phi_{\nu(\mathcal{G})}\big)^*\,\big(D^\phi\, v,\, \varphi\big)\, d\, v_g\,,$$

$$\big(h^\phi_{\nu(\mathcal{G})}\big)^*\,\big(D^\phi\, v,\, \varphi\big) = \sum_i h^\phi_{\nu(\mathcal{G})}\Big(D^\phi_{\sigma_g(r_i)}\, v,\, \varphi\, \sigma_g\big(r_i\big)\Big) =$$

$$= \sum_i \Big\{\sigma_g\big(r_i\big)\Big[h^\phi_{\nu(\mathcal{G})}\,\big(v,\, \varphi\, \sigma_g\big(r_i\big)\big)\Big] - h^\phi_{\nu(\mathcal{G})}\big(v,\, D^\phi_{\sigma_g(r_i)}\, \varphi\, \sigma_g\big(r_i\big)\big)\Big\}.$$

Next we consider the transverse vector field $r_v \in \ell(\mathcal{F})$ determined by

$$h^\phi_{\nu(\mathcal{G})}\big(v,\, \varphi\, \sigma_g(r)\big) = g_{\nu(\mathcal{F})}\big(r_v,\, r\big), \quad r \in C^\infty(\nu(\mathcal{F})).$$

Its basic divergence may be computed locally as

$$\operatorname{div}_B(r_v) = \sum_{i=1}^{q} g_{\nu(\mathscr{F})}(\nabla_{\sigma_g(r_i)} r_v,\, r_i) = \sum_i \left\{\sigma_g(r_i)\left[g_{\nu(\mathscr{F})}(r_v,\, r_i)\right] - g_{\nu(\mathscr{F})}(r_v,\, \nabla_{\sigma_g(r_i)} r_i)\right\}$$

hence

$$\sum_i \sigma_g(r_i)\left[h^{\phi}_{\nu(\mathscr{G})}(v,\, \varphi\,\sigma_g(r_i))\right] = \operatorname{div}_B(r_v) + h^{\phi}_{\nu(\mathscr{G})}\Big(v,\, \sum_i \varphi\,\sigma_g\,\nabla_{\sigma_g(r_i)} r_i\Big).$$

Consequently [by (4.169)],

$$\langle v,\, (D^{\phi})^{*}\varphi\rangle_B = \int_M \left\{h^{\phi}_{\nu(\mathscr{G})}\Big(v,\, \varphi\,\sigma_g(\tau_{\mathscr{F}}) - \operatorname{Trace}_{g_Q}(\mathscr{D}^{\phi}\varphi)\Big)\right\} d\,v_g$$

where the bilinear form $\mathscr{D}^{\phi}\varphi$ is given by

$$(\mathscr{D}^{\phi}\varphi)(r,\, s) = (D^{\phi}_{\sigma_g(r)}\varphi)s, \quad r,\, s \in C^{\infty}_B(Q),$$

$$(D^{\phi}_Y\varphi)r = D^{\phi}_Y\varphi\,\sigma_g(r) - \varphi\,\sigma_g\nabla_Y r, \quad Y \in \mathfrak{M}, \quad r \in C^{\infty}(Q).$$

We may conclude that

$$(D^{\phi})^{*}\varphi = -\operatorname{Trace}_{g_Q}(\mathscr{D}^{\phi}\varphi) + \varphi\,\sigma_g(\tau_{\mathscr{F}}), \tag{4.209}$$

where (locally)

$$\operatorname{Trac}_{g_Q}(\mathscr{D}^{\phi}\varphi) = \sum_{i=1}^{q}\left[D^{\phi}_{\sigma_g(r_i)}\varphi\,\sigma_g(r_i) - \varphi\,\sigma_g\nabla_{\sigma_g(r_i)} r_i\right].$$

In particular for $\varphi = D^{\phi}\,v$

$$\Delta_{\phi}\,v = (D^{\phi})^{*}\,D^{\phi}\,v, \quad v \in C^{\infty}_B(\phi^{-1}\nu(\mathscr{G})). \tag{4.210}$$

The identity (4.208) follows from (4.210). For any foliated Riemannian manifold $(M,\, \mathscr{F},\, g)$ let $x \in M$ and $S \subset Q_x$ a 2-dimensional subspace. Given a $g_{Q,x}$-orthonormal basis $\{r,\, s\} \subset S$ we set

$$k_T(S) = g_{Q,x}\Big(R_x(\sigma_{g,x}(r),\, \sigma_{g,x}(s))\,s\,,\, r\Big).$$

The definition of $k_T(S)$ (the *transverse sectional curvature* of S) doesn't depend on the choice of orthonormal basin in S. By (2.6) in [ToVa], p. 309, $k_T(S) \le 0$ if and only if the model Riemannian manifold M_0 has nonpositive sectional curvature.

At this point one may prove Theorem 4.153. Indeed if the model Riemannian manifold N_0 of $(N, \mathcal{G}, h)$ has nonpositive sectional curvature, then $\langle \mathcal{R}_\phi v, v\rangle_B \leq 0$ so that [by (4.208)]

$$H\big(E_T\big)_\phi(v, v) = \big\|D^\phi v\big\|^2 - \langle \mathcal{R}_\phi v, v\rangle_B \geq 0$$

for any $v \in C^\infty_B\big(\phi^{-1}\nu(\mathcal{G})\big)$. Here $\|\varphi\|^2 = \langle \varphi, \varphi\rangle_B$.

Let $\mathfrak{Har}(\mathcal{F}, \mathcal{G})$ be the set of all $(\mathcal{F}, \mathcal{G})$-harmonic maps $\phi \in C^\infty(\mathcal{F}, \mathcal{G})$. Given $\phi \in \mathfrak{Har}(\mathcal{F}, \mathcal{G})$ let $T_\phi\big(\mathfrak{Har}(\mathcal{F}, \mathcal{G})\big)$ consist of all infinitesimal variations $V \in T_\phi\big(C^\infty(M, N)\big)$ induced by some smooth 1-parameter variation $\{\phi_s\}_{|s|<\epsilon} \subset \mathfrak{Har}(\mathcal{F}, \mathcal{G})$ with $\phi_0 = \phi$.

Proposition 4.155 *For each* $\phi \in \mathfrak{Har}(\mathcal{F}, \mathcal{G})$

$$\pi'^\phi\, T_\phi\big(\mathfrak{Har}(\mathcal{F}, \mathcal{G})\big) \subset \operatorname{Ker}\big[J_T(\phi)\big]. \tag{4.211}$$

In particular $\dim_{\mathbb{R}}\big[\pi'^\phi\, T_\phi\big(\mathfrak{Har}(\mathcal{F}, \mathcal{G})\big)\big] \leq \operatorname{null}_B(\phi)$.

Proof Let $V \in T_\phi\big(\mathfrak{Har}(\mathcal{F}, \mathcal{G})\big)$ and $\{\phi_s\}_{|s|<\epsilon} \subset \mathfrak{Har}(\mathcal{F}, \mathcal{G})$ a smooth 1-parameter variation of ϕ inducing V. Let $\{\phi_{s,t}\}_{|t|<\epsilon} \subset C^\infty(\mathcal{F}, \mathcal{G})$ be an arbitrary 1-parameter variation of ϕ_s, i.e., $\phi_{s,0} = \phi_s$ for each $|s| < \epsilon$. As ϕ_s is $(\mathcal{F}, \mathcal{G})$-harmonic

$$\left\{\frac{\partial\, E_T\big(\phi_{s,t}\big)}{\partial\, t}\right\}_{t=0} = 0$$

so that

$$\int_M h^\phi_{\nu(\mathcal{G})}\Big(J_T(\phi)\, \pi'^\phi\, V,\, \pi'^\phi\, W\Big)\, d\, v_g = \frac{\partial^2}{\partial s\, \partial t}\Big\{E_T\big(\phi_{s,t}\big)\Big\}_{s=t=0} = 0$$

where W is the infinitesimal variation induced by $\{\phi_{0,t}\}_{|t|<\epsilon}$. Consequently, $J_T(\phi)\, \pi'^\phi\, V = 0$.

Example 4.156 (*Basic index and nullity of constant maps*) Let $y_0 \in N$ and $\phi : M \to N$ the constant map $\phi(x) = y_0$ for any $x \in M$. If $v \in C^\infty\big(\phi^{-1}\nu(\mathcal{G})\big)$, then $v(x) \in \nu(\mathcal{G})_{y_0}$ for any $x \in M$. Let $\{w_u : 1 \leq u \leq q'\}$ be a linear basis in $\nu(\mathcal{G})_{y_0}$ and let us set $v_u(x) = w_u$ for each $x \in M$ (so that $\{v_u : 1 \leq u \leq q'\}$ is a global frame in $\phi^{-1}\nu(\mathcal{G}) \to M$. Let $v \in C^\infty\big(\phi^{-1}\nu(\mathcal{G})\big)$ be represented as $v = f^u v_u$ for some $f^u \in C^\infty(M)$. next $\phi_* r_i = 0$ implies $\mathcal{R}_\phi\, v = 0$ hence $J_T(\phi)\, v = \Delta_\phi\, v$. Let $v \in C^\infty_B\big(\phi^{-1}\nu(\mathcal{G})\big)$ so that $f^u \in \Omega^0_B(\mathcal{F})$. Then $D^\phi\, v = \big(d\, f^u\big) \otimes v_u$ and we may conclude [by (4.195)] that $J_T(\phi)\, v = -\big(\Delta_B\, f^u\big) v_u$. It follows that

Proposition 4.157 *For any constant map* $\phi : M \to N$ *the spectrum* $\operatorname{Spec}\big[J_T(\phi)\big]$ *of the transverse Jacobi operator is the set of eigenvalues of* $-\Delta_B : \Omega^0_B(\mathcal{F}) \to \Omega^0_B(\mathcal{F})$ *counted* $\operatorname{cod}(\mathcal{G})$ *times. In particular* $\operatorname{Spec}\big[J_T(\phi)\big]$ *doesn't depend upon* $\{y_0\} = \phi(M) \subset N$. *Also* $\operatorname{ind}_B(\phi) = 0$ *and* $\operatorname{null}_B(\phi) = \operatorname{cod}(\mathcal{G})$.

□

Example 4.158 (*The identity mapping*) Let $(M, \mathscr{F}, g)$ be a compact foliated Riemannian manifold and $\phi = \text{id}$ the identity, i.e., $\phi(x) = x$. Clearly $\text{id} \in \mathfrak{Har}(\mathscr{F}, \mathscr{F})$. We say $(M, \mathscr{F}, g)$ is *stable* if $\phi = \text{id}$ in weakly stable (i.e., $\text{ind}_B(\text{id}) = 0$). Otherwise $(M, \mathscr{F}, g)$ is *unstable*. It is immediate that $\text{id}^{-1} Q = Q$ and $\text{id}_* = 1_{C^\infty(Q)}$ [the identity mapping of $C^\infty(Q)$]. Also $D^{\text{id}} = \nabla \in \mathscr{C}(Q, g_Q)$ hence $\mathscr{R}_{\text{id}} = S^\nabla$ where

Definition 4.159 The *transverse Ricci operator* S^∇ is locally given by

$$S^\nabla(v) = \sum_{i=1}^{q} R^\nabla(v, r_i) r_i, \quad r_i \in C^\infty(Q).$$

□

Here $R^\nabla \in C^\infty\left(\otimes^3 Q^*\right)$ is induced by the curvature tensor field of ∇. Next

$$J_T(\text{id}) = \overline{\Delta}_B - S^\nabla$$

where

Definition 4.160 The *rough basic Laplacian* $\overline{\Delta}_B = \delta_{\text{id}}$ is (locally) given by

$$\overline{\Delta}_B = -\sum_{i=1}^{q} \left\{ \nabla_{\sigma(r_i)} \nabla_{\sigma(r_i)} - \nabla_{\nabla_{\sigma(r_i)} r_i} \right\} + \nabla_{\sigma(\tau_{\mathscr{F}})}. \tag{4.212}$$

□

Cf. also e.g., [NiTo], p. 603. Then

$$H(E_T)_{\text{id}}(v, w) = \int_M \left\{ g_Q\left(\overline{\Delta}_B v, w\right) - s^\nabla(v, w) \right\} d\, v_g, \quad v, w \in C_B^\infty(Q).$$

Here

Definition 4.161 The *transverse Ricci curvature* is

$$s^\nabla(v, w) = g_Q\left(S^\nabla v, w\right), \quad v, w \in C_B^\infty(Q).$$

□

Definition 4.162 A Riemannian foliation $\mathscr{F}$ is *transversally Einstein* if $S^\nabla = (\lambda/2)\text{id}_Q$ for some $\lambda \in \mathbb{R}$. □

Cf. also [ToTo], p.232. We conjecture that a transversally Einstein foliation is stable if and only if the first eigenvalue $\lambda_1^{B,0}$ of Δ_B [acting on $\Omega_B^0(\mathscr{F})$] satisfies $\lambda_1^{B,0} \geq \lambda$, which is the foliated analog to Proposition 2.11 in [Smit], p. 234. □

A few comments are in order, as a follow up to Example 4.158. Another way to generalize harmonic map theory to the foliation context is due to F. Kamber and Ph. Tondeur, [KaTo6]. Their approach to the energy of a Riemannian foliation $(\mathscr{F}, g_Q)$ of a compact oriented Riemannian manifold (M, g) is

$$E(\mathscr{F}) = \frac{1}{2}\int_M g_Q(\pi \wedge *\pi) \tag{4.213}$$

where the projection $\pi T(M) \to Q$ is thought of as a Q-valued differential 1-form $\pi \in \Omega^1(M, Q)$. Of course the bundle metric g_Q induces a bilinear map of $Q \otimes Q$ onto the trivial line bundle over M [explaining the notation for the integrand in (4.213)] and the Riemannian metric g (so far unrelated to g_Q) is used to build the Hodge star operator, successively extended to Q-valued differential forms

$$* : \Omega^1(M, Q) \to \Omega^{n-1}(M, Q).$$

The energy (4.213) turns out to be trivial, i.e.,

$$E(\mathscr{F}) = \frac{q}{2}\,\text{Vol}(M)$$

when g_Q is induced by g. Yet given a family $\{\mathscr{F}_t\}_{|t|<\epsilon}$ of Riemannian foliations such that $\mathscr{F}_0 = \mathscr{F}$ in general the metric g fails to be bundle-like for $\mathscr{F}_t$ with $t \neq 0$. Every section $s \in C^\infty(Q)$ induces a *special variation* of $\mathscr{F}$. Indeed if

$$\mathscr{F} = \left(\{U_i\}_{i\in I}, \ \{f_i\}_{i\in I}, \ \{\gamma_{ji}\}_{i,j\in I}\right)$$

is a Riemannian foliation modeled on the Riemannian manifold M_0, then let us set

$$s_i \in C^\infty\left(U_i, \ f_i^{-1}T(M_0)\right), \quad s_i(x) = (d_x f_i)\,\sigma_g(s)_x, \quad x \in U_i,$$

$$f_i^t(x) = \exp_{f_i(t)}\left(t\, s_i(x)\right), \quad x \in U_i, \quad |t| < \epsilon,$$

with $\epsilon > 0$ sufficiently small (such that the exponential mapping for M_0 is well defined). Next we set

$$\mathscr{F}_t = \left(\{U_i\}_{i\in I}, \ \{f_i^t\}_{i\in I}, \ \{\gamma_{ji}\}_{i,j\in I}\right), \quad |t| < \epsilon,$$

and a calculation shows that

$$\frac{d}{dt}\left\{E(\mathscr{F}_t)\right\}_{t=0} = \int_M g_Q\left(s, \ \tau_{\mathscr{F}}\right) d\,v_g$$

where $\tau_{\mathscr{F}}$ is the mean curvature vector of $\mathscr{F}$ (cf. [KaTo7], p.529) so that $\mathscr{F}$ is harmonic if and only if $\mathscr{F}$ is a critical point of (4.213) with respect to special variations. The variational approach to harmonic foliations is related to our Example 4.158 as follows. Let $v,\ w \in C^\infty(Q)$ and consider the 2-parameter variation

$$\{\mathscr{F}_{s,t}\}_{-\epsilon s,\, t<\epsilon}\,, \quad \mathscr{F}_{s,t} = \left(\{U_i\}_{i\in I}\,,\ \{f_i^{s,t}\}_{i\in I}\,,\ \{\gamma_{ji}\}_{i,j\in I}\right),$$

$$f_i^{s,t}(x) = \exp_{f_i(x)}\left(s\, v_i(x) + t\, w_i(x)\right), \quad x \in U_i\,, \quad |s| < \epsilon, \quad |t| < \epsilon.$$

Then (cf. [KaTo7], pp. 536–538)

$$\frac{\partial^2}{\partial s\, \partial t}\left\{E\left(\mathscr{F}_{s,t}\right)\right\}_{s=t=0} = \int_M g_Q\left(J_\nabla\, v,\ w\right) d\, v_g$$

where

Definition 4.163 The *Jacobi operator* is

$$J_\nabla : C^\infty(Q) \to C^\infty(Q), \quad J_\nabla \equiv \overline{\Delta} - S^\nabla\,.$$

□

Here $\overline{\Delta}$ is the rough Laplacian on $C^\infty(Q)$, i.e., $\overline{\Delta} = d_\nabla^*\, d_\nabla$. Also $d_\nabla\, v = \nabla\, v \in \Omega^1(M,\ Q)$ and $d_\nabla^* : \Omega^1(M,\ Q) \to \Omega^0(M,\ Q)$ is the formal adjoint of d_∇ with respect to the L^2 inner product

$$\langle \alpha,\ \beta\rangle = \int_M g_Q^*(\alpha,\ \beta)\, d\, v_g\,, \quad \alpha,\ \beta \in \Omega^k(M,\ Q), \quad k \in \{0,\ 1\}.$$

We may show

Proposition 4.164 *Let* $\mathscr{D}\left(J_\nabla\right) = \Omega^0(M,\ Q)$ *and* $\mathscr{D}\left(J_T(\mathrm{id})\right) = C^\infty(Q)$ *be the domains of* J_∇ *and* $J_T(\mathrm{id})$*, respectively. Then* $J_T(\mathrm{id}) \subset J_\nabla$*, i.e., the Jacobi operator is an elliptic extension of the transverse Jacobi operator of the identity mapping.*

It should be observed that although $\overline{\Delta}_B$ and $J_T(\mathrm{id})$ are defined on the whole of $\Omega^0(M,\ Q)$ [unlike Δ_B which is only defined on basic functions (forms)] these operators are not elliptic: their ellipticity degenerates precisely along the leaves of $\mathscr{F}$. Indeed let $\mathrm{Diff}_k(E,\ F)$ denote the space of all linear operators $L : C^\infty(E) \to C^\infty(F)$, of order k, where E and F are vector bundles over M. Also we set $T'(M) = T^*(M) \setminus \{0\}$ and

$$\begin{aligned}\mathrm{Smbl}_k(E,\ F) =& \left\{\Sigma \in \mathrm{Hom}\left(\Pi^{-1}E,\ \Pi^{-1}F\right) : \Sigma(\rho\, \omega)\right.\\ &\left.= \rho^k\, \Sigma(\omega), \quad \omega \in T'(M), \quad \rho > 0\right\}\end{aligned}$$

where $\Pi : T'(M) \to M$ is the projection. Next let $\sigma_k : \mathrm{Diff}_k(E,\ F) \to \mathrm{Smbl}_k(E,\ F)$ be the linear map (2.1) in [Wel5], p. 115 (so that $\sigma_k(L)$ is the k-symbol of the differential operator L). Let $\omega \in T'(M)$ with $x = \Pi(\omega) \in M$ and $v \in Q_x$. We shall show that

$$\sigma_2\big[J_T(\mathrm{id})\big]_\omega = g_Q(\pi\,\nabla f,\, \pi\,\nabla f)_x\, v \tag{4.214}$$

where $f \in C^\infty(M)$ is chosen such that $(d\,f)_x = \omega$. To this end let $s \in C^\infty(Q)$ such that $s(x) = v$ and set $u = f - f(x)$. As $\mathscr{R}_{\mathrm{id}}$ has tensorial character $\mathscr{R}_{\mathrm{id}}\big(u^2\, s\big)_x = 0$ so that [by (4.207)]

$$\sigma_2\big[J_T(\mathrm{id})\big]_\omega (v) = -\frac{1}{2}\,\overline{\Delta}_B\big(u^2\, s\big)(x).$$

Then (4.214) follows from the identities

$$\Delta_B\big(h^2\big) = 2\,h\,\Delta_B\,h + 2\,\big\|(\nabla\,h)^\perp\big\|^2,$$

$$\overline{\Delta}_B(h\,s) = h\,\overline{\Delta}_B\,s - \big(\Delta_B\,h\big)\,s - 2\,\nabla_{(\nabla h)^\perp}\,s,$$

for any $h \in C^\infty(M)$. By (4.214) the linear map $\sigma_2\big[J_T(\mathrm{id})\big]_\omega : Q_x \to Q_x$ fails to be an isomorphism precisely when $(\nabla\,f)_X \in P_x$. This may be written locally (with respect to a foliated chart $\big(U,\, x^\alpha,\, y^i\big) \in \mathscr{F}$ at x)

$$g^{\alpha,\,p+j}(x)\,\omega_\alpha + g^{p+i,\,p+j}\,\omega_{p+i} = 0, \quad 1 \le j \le q, \tag{4.215}$$

where $\omega = \omega_\alpha\,\big(d\,x^\alpha\big)_x + \omega_{p+i}\,\big(d\,y^i\big)_x$. On the other hand, if

$$\delta\,x^\alpha = d\,x^\alpha + g^{\alpha\beta}\,g_{p+1,\,\beta}\,d\,y^i\,, \quad 1 \le \alpha \le p,$$

then $\big\{\delta\,x^\alpha\,,\, d\,y^i \;:\; 1 \le \alpha \le p,\;\; 1 \le i \le q\big\}$ is a local coframe adapted to the decomposition

$$T^*(M) = P^* \oplus \big(P^\perp\big)^*$$

and ω decomposes accordingly

$$\omega = \omega_\alpha\,\big(\delta\,x^\alpha\big)_x + \Big(\omega_{p+i} - g^{\alpha\beta}(x)\,g_{p+1,\,\beta}(x)\,\omega_\alpha\Big)\,\big(d\,y^i\big)_x\,.$$

Finally [by (4.198)], the condition (4.215) is equivalent to

$$\omega_{p+1} = g^{\alpha\beta}(x)\,g_{p+i,\,\beta}(x)\,\omega_\alpha,$$

i.e., $\omega \in P^*_x$.

To prove Proposition 4.164 let $s \in \Omega^0(M,\, Q)$ and $\omega = d_\nabla\,s$. Then [by (3.1) in [KaTo6], p.102]

$$d^*_\nabla\,\omega = -\sum_{A=1}^{n}\Big(\nabla_{E_A}\,\omega\Big)E_A$$

on U for any (local) g-orthonormal frame $\big\{E_A \;:\; 1 \le A \le n\big\}$ on $U \subset M$. Then

$$\overline{\Delta}\, s = -\sum_{A=1}^{n} \left\{ \nabla_{E_A} \nabla_{E_A} s - \nabla_{\nabla^g_{E_A} E_A} s \right\}$$

on U. Let $\{X_\alpha : 1 \leq \alpha \leq p\}$ (respectively, $\{r_i : 1 \leq i \leq q\}$) be a local g-orthonormal (respectively, g_Q-orthonormal) frame of P (respectively, of Q) on U. We may work with the frame $\{E_A\} = \{X_\alpha\, ,\, \sigma(r_i)\}$. Thus [by taking into account (4.152) and (4.212)],

$$\overline{\Delta}\, s = \overline{\Delta}_B\, s - \sum_{\alpha=1}^{p} \left\{ \mathring{\nabla}_{X_\alpha} \mathring{\nabla}_{X_\alpha} s - \mathring{\nabla}_{\pi^\perp \nabla^g_{X_\alpha} X_\alpha} s \right\} + \mathring{\nabla}_{\tau^\perp_{\mathscr{F}}} s \tag{4.216}$$

on U so that $\overline{\Delta}\, s = \overline{\Delta}_B\, s$ for any $s \in C^\infty_B(Q)$. In spite of the relationship to the stability of Riemannian foliations (as emphasized by Proposition 4.164) the approach in [KaTo6] appears as a chapter of the theory of harmonic maps in bundles whose total spaces carry Sasaki type metrics (cf. Konderak, [Kon]) staying to harmonic maps theory as the newer branch of harmonic vector fields (cf. e.g., G. Wiegmink, [Wie]). Establishing an explicit quantitative relationship between the harmonicity of $\mathscr{F}$ and that of the section determined by P (in the Grassmann bundle of p-planes tangent to M) is an open problem.

4.10.2.4 Transversally Holomorphic Maps of Kählerian Foliations

Let $\mathscr{F}$ be a trasversally oriented foliation of a Riemannian manifold $(M,\, g)$.

Definition 4.165 $\mathscr{F}$ is *transversally Kählerian* if

i) $\mathscr{F}$ is Riemannian and g is bundle-like,

ii) The transverse bundle $Q \to M$ carries a holonomy invariant complex structure $J_Q : Q \to Q$ [i.e., $\mathring{\nabla}_X J_Q = 0$ for any $X \in \mathfrak{X}_{\mathscr{F}}$] such that $g_Q(J_Q\, r,\, J_Q\, s) = g_Q(r,\, s)$ for any $r,\, s \in C^\infty(Q)$,

iii) $\nabla J_Q = 0$. □

Cf. S. Nishikawa and Ph. Tondeur, [NiTo], p. 599. If $\mathscr{F}$ is transversally Kählerian, then $\Omega_Q(r,\, s) = g_Q(r,\, J_Q\, s)$ determines a closed basic form $(\Omega_Q)_T \in \Omega^2_B(\mathscr{F})$ given by

$$(\Omega_Q)_T(Y,\, Z) = \Omega_Q(\pi\, Y,\, \pi\, Z), \quad Y,\, Z \in \mathfrak{X}(M).$$

Let $(N,\, \mathscr{G},\, h)$ be another foliated Riemannian manifold whose transverse bundle is equipped with a complex structure $J_{\nu(\mathscr{G})}$ such that $(\mathscr{G},\, J_{\nu(\mathscr{G})})$ is a transversally Kählerian foliation.

Definition 4.166 A foliated map $\phi \in C^\infty(\mathscr{F},\, \mathscr{G})$ is *transversally holomorphic* if

$$(d_T\, \phi) \circ J_Q = J^\phi_{\nu(\mathscr{G})} \circ (d_T\, \phi) \tag{4.217}$$

where $J^{\phi}_{\nu(\mathscr{G})} = J_{\nu(\mathscr{G})} \circ \phi$. □

By a result of E. Barletta et al. (cf. Theorem 1 in [BaDr], p. 122) a transversally holomorphic map is transversally harmonic if and only if $\operatorname{grad}_B(\phi^{p'+u}) \in \operatorname{Ker}(\kappa)$ for any $1 \le u \le q'$ and any foliated chart $(\mathscr{U}, X^a, Y^u) \in \mathscr{G}$. Here $\operatorname{grad}_B(f) \in C^\infty(P^\perp)$ is the *basic gradient* of the basic function $f \in \Omega^0_B(\mathscr{F})$. Precisely if $\eta \in C^\infty(Q^*)$ we set

$$\eta_T(Y, Z) = \eta(\pi\, Y, \pi\, Z), \quad Y, Z \in \mathfrak{X}(M)$$

so that $\eta \mapsto \eta_T$ is a linear isomorphism of $C^\infty_B(Q^*) = \{\eta \in C^\infty(Q^*) : \mathring{\nabla}\eta = 0\}$ onto $\Omega^1_B(\mathscr{F})$. Given $f \in \Omega^0_B(\mathscr{F})$ let $\eta_f \in C^\infty_B(Q^*)$ and $s_f \in C^\infty(Q)$ be determined by

$$(\eta_f)_T = d_B\, f, \quad g_Q(s_f\, , r) = \eta_f(r), \quad r \in C^\infty(Q).$$

Finally, $\operatorname{grad}_B(f) = \sigma_g(s_f)$.

Transversally Kählerian foliations abound. For instance the canonical S^1-action on the anti de Sitter space

$$H^{2n+1}_1 = \{z = (z_0\, , z_1\, , \dots, z_n) \in \mathbb{C}^{n+1} : -|z_0|^2 + \sum_{k=1}^{n} |z_k|^2 = -1\}$$

determines a fibration $H^{2n+1}_1 \to \mathbb{C}H^n$ (over the complex hyperbolic space $\mathbb{C}H^n$, [KoNo], Vol II, p. 282) which give rise to a (totally geodesic) transversally Kählerian foliation $\mathscr{F}$ of H^{2n+1}_1. Also the fibers of any principal S^1-bundle $P^{2n+1} \xrightarrow{p} \mathbb{C}P^n$ endowed with a connection 1-form $\eta \in \Omega^1(P^{2n+1})$ determine a transversally Kählerian foliation of P^{2n+1}. Here one endows P^{2n+1} with the Riemannian metric $g = p^*\, g_{FS} + \eta \otimes \eta$ on P^{2n+1}, where g_{FS} is the Fubini-Study metric on $\mathbb{C}P^n$.

As customary one extends J_Q (by $\mathbb{C}$-linearity) to a complex structure $J_Q^{\mathbb{C}}$ on $Q \otimes \mathbb{C}$ and sets $Q^{1,0} = \operatorname{Eigen}(J_Q^{\mathbb{C}}\,;\, i)$ (the eigenbundle of $J_Q^{\mathbb{C}}$ corresponding to the eigenvalue $i = \sqrt{-1}$) and $Q^{0,1} = \overline{Q^{1,0}}$. An overbar denotes complex conjugation. Let $\pi_{1,0} : Q \otimes \mathbb{C} \to Q^{1,0}$ be the projection associated to the decomposition $Q \otimes \mathbb{C} = Q^{1,0} \oplus Q^{0,1}$. For each foliated map $\phi \in C^\infty(\mathscr{F}, \mathscr{G})$ of transversally Kählerian foliations one defines the section $\partial_T\, \phi$ in the complex vector bundle $(Q^{1,0})^* \otimes \phi^{-1}\nu(\mathscr{G})^{1,0} \to M$ [respectively, $\overline{\partial}_T\, \phi$ in $(Q^{0,1})^* \otimes \phi^{-1}\nu(\mathscr{G})^{1,0} \to M$] as the restriction to $Q^{1,0}$ (respectively, to $Q^{0,1}$) of $\pi'_{1,0} \circ (d_T\, \phi)$. Then (4.217) is equivalent to $\overline{\partial}_T\, \phi = 0$. The Hermitian bundle metrics g_Q and $h_{\nu(\mathscr{G})}$ induce inner products on $C^\infty\left((Q^{1,0})^* \otimes \phi^{-1}\nu(\mathscr{G})^{1,0}\right)$ and $C^\infty\left((Q^{0,1})^* \otimes \phi^{-1}\nu(\mathscr{G})^{1,0}\right)$ so that we may set

$$k_{1,0}(\phi) = \left\|\partial_T\, \phi\right\|^2, \quad k_{0,1}(\phi) = \left\|\overline{\partial}_T\, \phi\right\|^2, \quad K(\phi) = \int_M \{k_{1,0}(\phi) - k_{0,1}(\phi)\}\, d\, v_g\, ,$$

for each $\phi \in C^\infty(\mathcal{F}, \mathcal{G})$. When $\mathcal{F} = \mathcal{F}_0$ and $\mathcal{G} = \mathcal{G}_0$ (the trivial foliations by points) the map $t \mapsto K(\phi_t)$ is constant for any smooth 1-parameter family $\{\phi_t\}_{|t|<\epsilon} \subset C^\infty(M, N)$ (cf. A. Lichnerowicz, [Lic]). A foliated analog of the result and its application to holomorphic maps of Vaisman manifolds and CR maps of Sasakian manifolds is demonstrated below. We obtain

Theorem 4.167 *Let $\{\phi_t\}_{|t|<\epsilon} \subset C^\infty(\mathcal{F}, \mathcal{G})$ be a smooth 1-parameter family of foliated maps of the foliated Riemannian manifolds $(M, \mathcal{F}, g)$ and $(N, \mathcal{G}, h)$, where M is compact and both $\mathcal{F}$ and $\mathcal{G}$ are transversally Kählerian foliations. Let $\{\alpha_t\}_{|t|<\epsilon} \subset \Omega^1(M)$ be given by*

$$\alpha_t(X)_x = \int_0^t \left(\Omega_{\nu(\mathcal{G})}\right)_{T,\, \phi_\tau(x)} \left(\mathbb{V}_{(x,\tau)},\, (d_x \phi_\tau) X_x\right) d\tau\,, \tag{4.218}$$

$$|t| < \epsilon\,, \quad X \in \mathfrak{X}(M), \quad x \in M.$$

Then

$$K(\phi_t) - K(\phi_0) = \frac{(-1)^p}{2(m-1)!} \int_M (\Omega_Q)_T^{m-1} \wedge \kappa \wedge \chi_{\mathcal{F}} \wedge \alpha_t \tag{4.219}$$

where $2m = \text{cod}(\mathcal{F})$ and $p = \dim(M) - 2m$. In particular any holomorphic map $\phi : M \to N$ of Vaisman manifolds, with M compact, is a weakly stable $(\mathcal{F}, \mathcal{G})$-harmonic map. Also any CR map $\phi : M \to N$ of Sasakian manifolds, with M compact and preserving the Reeb flows, is a weakly stable $(\mathcal{F}, \mathcal{G})$-harmonic map. In particular any foliated CR map $\phi : P^{2n+1} \to N$ into a Sasakian manifold N is $(\mathcal{F}, \mathcal{G})$-harmonic and $\text{ind}_B(\phi) = 0$.

Let (M, J, g) be a complex n-dimensional Hermitian manifold where J denotes the complex structure and g is the hermitian metric. We recall (cf. e.g., Theorem 1.1 in [DrOr], p. 1) that g is a locally conformal Kähler (l.c.K.) metric if there is a closed 1-form $\omega \in \Omega^1(M)$ (the Lee form) such that $d\,\Omega = \omega \wedge \Omega$ where $\Omega(X, Y) = g(X, JY)$ for any $X, Y \in \mathfrak{X}(M)$. A l.c.K. manifold (M, J, g) is a generalized Hopf, or Vaisman, manifold if $\nabla^g \omega = 0$ (cf. [DrOr], p. 33). The Lee field is $B = \omega^\sharp \in \mathfrak{X}(M)$ ($\sharp$ denotes raising of indices with respect to g). The anti-Lee field is then $A = -JB$. As $[A, B] = 0$ the distribution $\mathbb{R}B \oplus \mathbb{R}A$ is completely integrable and gives rise to a foliation $\mathcal{F}$ on M by Riemann surfaces (the vertical foliation). Let M be a compact Vaisman manifold. By a result of I. Vaisman [Vai2], each leaf of $\mathcal{F}$ is complex parallelizable and a totally geodesic and locally flat submanifold of M. In particular $\mathcal{F}$ is harmonic. Moreover, by a result of K. Tsukada (cf. Theorem 2.2 in [Tsu1], p. 4) $\mathcal{F}$ is transversally Kählerian and any holomorphic map $\phi : M \to N$ into another Vaisman manifold N, with M compact, is a foliated map. Also, by a result of E. Barletta et al. (cf. Corollary 1 in [BaDr], p. 128) any holomorphic map $\phi : M \to N$ of Vaisman manifolds is transversally harmonic with respect to the vertical foliations. It remains to be shown that $\text{ind}_B(\phi) = 0$. To this end we ought to prove (4.219). Let $Q' = \nu(\mathcal{G})$ and let $\phi^{-1}\,\Omega_{Q'} \in C_B^\infty(\Lambda^2 Q^*)$ be determined by

$\left(\phi^{-1}\,\Omega_{Q'}\right)_T = \phi^*\left(\Omega_{Q'}\right)_T$ [by exploiting the isomorphism $C_B^\infty\left(\Lambda^2\,Q^*\right) \simeq \Omega_B^2(\mathscr{F})$]. A calculation shows that

$$2\left\{k_{1,0}(\phi_t) - k_{0,1}(\phi_t)\right\} = g_Q^*\left(\Omega_Q\,,\ \phi_t^{-1}\,\Omega_{Q'}\right).$$

Hence, (as the map $\alpha \mapsto \alpha_T$ is an isometry)

$$K(\phi_\tau) = \frac{1}{2}\int_M g^*\left(\left(\Omega_Q\right)_T\,,\ \phi_\tau^*\left(\Omega_{Q'}\right)_T\right)d\,v_g = \frac{1}{2}\int_M \left[\phi^*\left(\Omega_{Q'}\right)_T\right] \wedge *\left(\Omega_Q\right)_T.$$

At this point we exploit a general result in [Lic], i.e., for any closed 2-form $\Omega \in \Omega^2(N)$ and any smooth 1-parameter family $\left\{\phi_t\right\}_{|t|<\epsilon} \subset C^\infty(M,\,N)$

$$d\left[\phi_t^*\left(\left(\mathbb{V}\circ\alpha_t\right)\lrcorner\,\Omega\right)\right] = \frac{d}{d\,t}\{\phi_t^*\Omega\},\quad |t|<\epsilon. \tag{4.220}$$

Cf. also (16.11) in [DrWo]. Indeed (4.220) for $\Omega = \left(\Omega_{Q'}\right)_T$ and integration on $[0,\ t]$ give $d\,\alpha_t = \phi_t^*\left(\Omega_{Q'}\right)_T - \phi_0^*\left(\Omega_{Q'}\right)_T$ so that (by the Stokes theorem)

$$K(\phi_t) - K(\phi_0) = \frac{1}{2}\int_M (d\,\alpha_t) \wedge *\left(\Omega_Q\right)_T = \frac{1}{2}\int_M \alpha_t \wedge d\left[\,*\left(\Omega_Q\right)_T\right].$$

At this point we need to recall Rummler's formula

$$d\,\chi_{\mathscr{F}} + \kappa \wedge \chi_{\mathscr{F}} = \varphi_0 \in F^2\Omega^{p+1}(M).$$

Here $F^r\Omega^s(M)$ consists of all differential s-forms $\eta \in \Omega^s(M)$ such that

$$i(X_1)\cdots i(X_{s-r+1})\eta = 0,\quad X_j \in \mathfrak{X}_{\mathscr{F}}\,,\quad 1 \le j \le s-r+1.$$

Then $F^r\Omega^s(M)$ is a (decreasing) filtration of the de Rham complex $\left\{\Omega^\bullet(M),\, d\right\}$. Its use in the present context may be stated loosely as follows. If $\eta \in F^r\Omega^s(M)$, then [locally with respect to a foliated chart $\left(U,\ x^\alpha,\ y^i\right) \in \mathscr{F}$] the s-form η contains exterior monomials of the form $d\,x^{\alpha_1} \wedge\cdots\wedge d\,x^{\alpha_k}$ of degree at most $k \le s-r$. Therefore, η contains exterior monomials of the form $d\,y^{i_1} \wedge\cdots\wedge d\,y^{i_\ell}$ of degree $\ell \ge r$. Next, by (2.4) in [KaTo16], p. 417, one has

$$*\left(\Omega_Q\right)_T = \left[\,\hat{*}\left(\Omega_Q\right)_T\,\right] \wedge \chi_{\mathscr{F}} = \frac{1}{(m-1)!}\left(\Omega_Q\right)_T^{m-1} \wedge \chi_{\mathscr{F}}$$

hence [as $d_B\left(\Omega_Q\right)_T = 0$]

$$\alpha_t \wedge d\left[\,*\left(\Omega_Q\right)_T\,\right] = \frac{1}{(m-1)!}\,\alpha_t \wedge \left(\Omega_Q\right)_T^{m-1} \wedge d\,\chi_{\mathscr{F}} = -\frac{1}{(m-1)!}\,\alpha_t \wedge \left(\Omega_Q\right)_T^{m-1} \wedge \kappa \wedge \chi_{\mathscr{F}}$$

by Rummler's formula [thus leading to (4.219)]. Indeed [by (4.218)], $\alpha_t(X) = 0$ for any $X \in \mathfrak{X}_{\mathcal{F}}$ hence locally $\alpha_t = \alpha_i(t)\, d\, y^i$ for some $\{\alpha_i(t)\}_{|t|<\epsilon} \subset \Omega^0_B(\mathcal{F}_U)$. Also $(\Omega_Q)^{m-1}_T \in \Omega^{2m-2}_B(\mathcal{F})$ and φ_0 contributes (again by Rummler's formula) exterior monomials of the form $d\, y^i \wedge d\, y^j$. Thus, $\alpha_t \wedge (\Omega_Q)^{m-1}_T \wedge \varphi_0$ has degree at least $2m + 1$ in the $d\, y^i$'s and therefore it must vanish.

The complex structure $J_Q : \nu(\mathcal{F}) \to \nu(\mathcal{F})$ (where $\mathcal{F}$ is the vertical foliation of the Vaisman manifold M) is given by $J_Q(s) = \pi(J\, Y)$ where $Y \in \mathfrak{X}(M)$ is a vector field such that $\pi(Y) = s$. As $\mathbb{R}B \oplus \mathbb{R}A$ is J-invariant, this is well defined. Consequently, the transverse differential of any holomorphic map $\phi : M \to N$ of Vaisman manifolds satisfies (4.217) so that $k_{0,1}(\phi) = 0$. Let $\{\phi_t\}_{|t|<\epsilon} \subset C^\infty(\mathcal{F}, \mathcal{G})$ be a smooth 1-parameter variation of ϕ (with $\phi_0 = \phi$). Then

$$E_T(\phi_t) = \int_M e_T(\phi)\, d\, v_g = \int_M \{k_{1,0}(\phi_t) + k_{0,1}(\phi_t)\}\, d\, v_g \le$$

$$\le \left| \int_M \{k_{1,0}(\phi_t) + k_{0,1}(\phi_t)\}\, d\, v_g \right| = |K(\phi_t)| = |K(\phi)| = E(\phi)$$

by (4.219) (as $\mathcal{F}$ is harmonic). Therefore, $\mathcal{E}''(0) \ge 0$ where we have set $\mathcal{E}(t) = E_T(\phi_t)$. On the other hand, let

$$\{\phi_{s,t}\}_{|s|<\epsilon,\, |t|<\epsilon} \subset C^\infty(\mathcal{F}, \mathcal{G}), \quad \phi_{s,t} = \phi_{s+t}\,, \quad -\epsilon < s,\, t < \epsilon.$$

Let $v \in C^\infty_B(\phi^{-1} Q')$ be induced by $\{\phi_t\}_{|t|<\epsilon}$ [as in (4.154), i.e., $v(X) = \mathbf{v}(x,\, 0)$ for any $x \in M$]. Finally,

$$0 \le \mathcal{E}''(0) = \left\{ \frac{\partial^2 E_T(\phi_{s,t})}{\partial s\, \partial t} \right\}_{s=t=0} = H(E_T)_\phi(v,\, v)$$

so that $\mathrm{ind}_B(\phi) = 0$.

To prove the CR statement in Theorem 4.167 we need to recall a few notations and conventions from [DrTo]. Let M be a $(2n + 1)$-dimensional manifold.

Definition 4.168 A complex subbundle $T_{1,0}(M) \subset T(M) \otimes \mathbb{C}$, of complex rank n, is a *CR structure* on M if

(i) $T_{1,0}(M)_x \cap T_{0,1}(M)_x = (0)$ for any $x \in M$,

(ii) $[Z,\, W] \in C^\infty(U,\, T_{1,0}(M))$ for any $Z,\, W \in C^\infty(U,\, T_{1,0}(M))$ and any open set $U \subset M$. □

Here $T_{0,1}(M) = \overline{T_{1,0}(M)}$. let $\{P_\alpha : 1 \le \alpha \le n\}$ be a local frame of $T_{1,0}(M)$ on $U \subset M$. CR structures are bundle-theoretic description of the *tangential Cauchy-Riemann* equations $\overline{P}_\alpha(u) = 0$, $1 \le \alpha \le n$. A solution $u \in C^1(U)$ to the tangential C-R equations is a *CR function*. Let $\mathrm{CR}^\infty(U)$ be the space of all CR functions of class C^∞ on U. Let $H(M) \subset T(M)$ be the real rank $2n$ subbundle defined by

$$H(M) = \text{Re}\left\{T_{1,0}(M) \oplus T_{0,1}(M)\right\}, \quad T_{0,1}(M) \equiv \overline{T_{1,0}(M)}\,.$$

The distribution $H(M)$ carries the complex structure J given by

$$J(Z + \overline{Z}) = i(Z - \overline{Z}),\ Z \in T_{1,0}(M), \quad i = \sqrt{-1}\,.$$

Let $H(M)^{\perp} \subset T^*(M)$ be the real line bundle defined by

$$H(M)^{\perp}_x = \left\{\omega \in T^*_x(M)\ :\ \text{Ker}(\omega) \supset H(M)_x\right\}, \quad x \in M.$$

Definition 4.169 A *pseudohermitian structure* on the CR manifold $(M,\ T_{1,0}(M))$ is a globally defined nowhere zero section $\theta \in C^{\infty}\big(H(M)^{\perp}\big)$. □

Definition 4.170 The *Levi form* is

$$G_\theta(X,\ Y) = (d\,\theta)(X,\ J\,Y), \quad X,\ Y \in H(M).$$

□

Definition 4.171 A CR manifold M is *strictly pseudoconvex* if G_θ is positive definite for some θ. □

If this is the case θ is a contact form, i.e., $\theta \wedge (d\,\theta)^n$ is a volume form on M. Also G_θ may be extended to a Riemannian metric g_θ (the *Webster metric*) on M by setting

$$g_\theta(X,\ Y) = G_\theta(X,\ Y), \quad g_\theta(X,\ T) = 1, \quad X,\ Y \in H(M),$$

and using the direct sum decomposition $T(M) = H(M) \oplus \mathbb{R}T$. Here

Definition 4.172 $T \in \mathfrak{X}(M)$ is the *Reeb vector field*, i.e., the unique tangent vector field on M determined by $\theta(T) = 1$ and $T \rfloor d\,\theta = 0$. □

Definition 4.173 A *CR map* is a smooth map $\phi : M \to N$ of CR manifolds such that

$$\phi_*\, T_{1,0}(M) \subset \phi^{-1} T_{1,0}(N).$$

□

It is a natural question whether a CR analog to A. Lichnerowicz's theorem holds (at least in the strictly pseudoconvex case). Precisely, given a CR map $\phi : M \to N$ of compact strictly pseudoconvex CR manifolds, is $\phi : M \to N$ harmonic with respect to the Webster metrics g_θ and g_Θ (for some choice of contact forms θ and Θ on M and N, respectively)? Such a result is ruled out in general by H. Urakawa's finding (cf. [Ura]) that a CR map $\phi : M \to N$ is a harmonic map of $(M,\ g_\theta)$ into $(N,\ g_\Theta)$ if and only if $T(\lambda) = 0$ where $\lambda \in C^{\infty}(M)$ is determined by $\phi^*\Theta = \lambda\,\theta$.

As subelliptic theory applies to CR geometry the way elliptic theory applies to Riemannian geometry (cf. S. Dragomir, [Dra5]) the concept of subelliptic harmonic map (due to J. Jost and C.-J. Xu, [JoXu]) appears to be the closest pseudohermitian analog to that of a harmonic map. Yet a CR map $\phi : M \to N$ above isn't subelliptic harmonic either, unless $\phi : M \to N$ is a *pseudohermitian map*, i.e., $\phi^*\Theta = c\,\theta$ for some $c \in \mathbb{R}$ (cf. Theorem 1.1 in [BaDrUr], p. 724). As in the case of holomorphic maps of l.c.K. manifolds the problem admits a surprising solution within foliation theory.

Definition 4.174 The *Reeb flow* is the flow $\mathscr{F}_\theta$ tangent to T, i.e., $T(\mathscr{F}_\theta) = \mathbb{R}T$. □

Then H. Urakawa's condition to the harmonicity of a CR map $\phi : M \to N$ reads $\lambda \in \Omega^0_B(\mathscr{F}_\theta)$. The Webster metric g_θ induces a Riemannian bundle metric $(g_\theta)_Q$ in the transverse bundle to the flow. At this point a number of natural questions may be asked. Which of the Reeb flows is Riemannian (transversally Kählerian)? Are CR maps of strictly pseudoconvex CR manifolds transversally holomorphic? To answer these questions is of course to prove Theorem 4.167. We need the following.

Lemma 4.175 *Let M be a strictly pseudoconvex CR manifold and θ a contact form on M such that G_θ is positive definite. Then*

$$\left(\mathring{\nabla}_T (g_\theta)_Q\right)(r,\, s) = 2\,A\big(\sigma(r),\, \sigma(s)\big) \tag{4.221}$$

for any $r, s \in C^\infty\big(\nu(\mathscr{F}_\theta)\big)$. Here $(A\,X,\, Y) = g_\theta(X,\, \tau\,Y)$ and τ is the pseudohermitian torsion of $(M,\, \theta)$. Consequently, $\big(\mathscr{F}_\theta,\, (g_\theta)_Q\big)$ is a Riemannian foliation and g_θ is a bundle-like Riemannian metric if and only if $\tau = 0$. Moreover,

$$\mathring{\nabla}_T\, J_Q = 2\pi \circ J \circ \tau \circ \sigma \tag{4.222}$$

where $J_Q : \nu(\mathscr{F}_\theta) \to \nu(\mathscr{F}_\theta)$ is the complex structure given by $J_Q\, s = \pi\; J\, Y$ for any $s \in C^\infty\big(\nu(\mathscr{F}_\theta)\big)$ and $Y \in \mathfrak{X}(M)$ such that $\pi\; Y = s$. Consequently, J_Q is holonomy invariant if and only if $\tau = 0$.

Here $\sigma : \nu(\mathscr{F}_\theta) \to H(M)$ is the natural isomorphism (induced by g_θ). Also τ (the *pseudohermitian torsion*) is the vector-valued 1-form on M given by $\tau(X) = \mathrm{Tor}(T,\, X)$ for any $X \in \mathfrak{M}$ where Tor is the torsion tensor field of the Tanaka-Webster connection of $(M,\, \theta)$, i.e., the unique connection ∇^θ on M such that

(i) $H(M)$ is parallel with respect to ∇^θ,

(ii) $\nabla^\theta\, g_\theta = 0$ and $\nabla^\theta\, J = 0$,

(iii) the torsion Tor of ∇^θ is *pure*, i.e., $\tau \circ J + J \circ \tau = 0$ and $\mathrm{Tor}(Z,\, W) = 0$ and $\mathrm{Tor}(Z,\, \overline{W}) = 2i\, G_\theta(Z,\, \overline{W})T$ for any $Z,\, W \in T_{1,0}(M)$.

It is customary to extend $J : H(M) \to H(M)$ to a field φ of $(1,\, 1)$-tensors on M by setting $\varphi = J$ on $H(M)$ and $\varphi(T) = 0$. Then $(\varphi,\, \xi,\, \eta,\, g)$ is a contact metric structure on M (in the sense of [Bla]) where $\xi = -T$, $\eta = -\theta$ and $g = g_\theta$. Moreover, $(\varphi,\, \xi,\, \eta)$ is normal (i.e., $[\varphi,\, \varphi] + 2\,(d\,\eta) \otimes \xi = 0$, cf. [Bla], p. 49) if and only if $\tau = 0$. If this is the case g is a Sasakian metric (in the sense of [Bla], p. 73). Summing up, a strictly pseudoconvex CR manifold endowed with a contact form θ such that the Levi form G_θ is positive definite is a Sasakian manifold (i.e., the Webster metric is Sasakian) if and only if (M, θ) has vanishing pseudohermitian torsion.

Let us prove Lemma 4.175. For any $r, s \in C^\infty(\nu(\mathscr{F}_\theta))$

$$\big(\mathring{\nabla}_T (g_\theta)_Q\big)(r, s) = T\big((g_\theta)_Q(r, s)\big) - (g_\theta)_Q\big(\mathring{\nabla}_T r, s\big) - (g_\theta)_Q\big(r, \mathring{\nabla}_T s\big) =$$

$$= T\big(g_\theta(\sigma(r), \sigma(s))\big) - g_\theta\big([T, \sigma(r)]^\perp, \sigma(s)\big) - g_\theta\big(\sigma(r), [T, \sigma(s)]^\perp\big)$$

where $Y^\perp = \Pi_H(Y)$ and $\Pi_H : T(M) \to H(M)$ is the natural projection (with respect to the decomposition $T(M) = H(M) \oplus \mathbb{R}T$). It should be observed that

$$\theta([T, X]) = -2(d\theta)(T, X) = 0, \quad X \in H(M),$$

[due to $\theta(X) = 0$ and $\theta(T) = 1$]. An equivalent formulation is that

$$\theta \circ \mathscr{L}_T = -2(T \rfloor d\theta) = 0.$$

Consequently, $[T, X] \in H(M)$ for any $X \in H(M)$. Therefore [as σ is $H(M)$-valued and $\nabla^\theta T = 0$],

$$\big(\mathring{\nabla}_T (g_\theta)_Q\big)(r, s) = \big(\mathscr{L}_T\, g_\theta\big)\big(\sigma(r), \sigma(s)\big) = T\big(g_\theta(\sigma(r), \sigma(s))\big) =$$

$$= -g_\theta\big(\nabla^\theta_T \sigma(r) - \mathrm{Tor}(T, \sigma(r)), \sigma(s)\big) - g_\theta\big(\sigma(t), \nabla^\theta_T \sigma(s) - \mathrm{Tor}(T, \sigma(s))\big) =$$

$$= \big(\nabla^\theta g_\theta\big)\big(\sigma(r), \sigma(s)\big) + 2\,A\big(\sigma(r), \sigma(s)\big)$$

(as A is symmetric, cf. Lemma 1.4 in [DrTo], p. 38. This proves (4.221). Moreover (by the very definition of $\mathring{\nabla}$),

$$\big(\mathring{\nabla}_T\, J_Q\big)s = \pi\big\{[T, J\sigma(s)] - J\,[T, \sigma(s)]\big\} =$$

$$= \pi\big\{\nabla^\theta_T\, J\sigma(s) - \mathrm{Tor}(T, J\sigma(s)) - J\,\nabla^\theta_T \sigma(s) + J\,\mathrm{Tor}(T, \sigma(s))\big\} =$$

$$= \pi\big(\nabla^\theta_T\, J\big)\sigma(s) + \pi(J \circ \tau - \tau \circ J)\sigma(s),$$

i.e., $\mathring{\nabla}_T\, J_Q = 2\pi \circ J \circ \tau \circ \sigma$ and (4.222) is proved.

Theorem 4.176 *Let M be a strictly pseudoconvex CR manifold and θ a contact form on M with G_θ positive definite. Then*

$$(g_\theta)_Q\big(J_Q r, J_Q s\big) = (g_\theta)_Q(r, s), \quad r, s \in \nu(\mathscr{F}_\theta), \quad \nabla_Y\, J_Q = 0, \quad Y \in H(M).$$

Consequently, if g_θ is a Sasakian metric, then $\big((g_\theta)_Q,\ J_Q\big)$ is a transverse Kählerian structure on $\big(M,\ \mathscr{F}_\theta\big)$. Moreover, for any CR map $\phi : M \to N$ of strictly pseudoconvex CR manifolds the following statements are equivalent:

(i) $\phi : M \to N$ preserves the Reeb flows, i.e., $\phi \in C^\infty\big(\mathscr{F}_\theta,\ \mathscr{F}_\Theta\big)$.

(ii) Either $(d\,\phi)T = 0$ [i.e., $\phi \in C^\infty\big(\mathscr{F}_\theta,\ \mathscr{G}_0\big)$] or there is $\lambda \in C^\infty(M)$ such that $\lambda \neq 0$ and $\phi : U \to N$ is an isometry of $\big(U,\ g_{\hat\theta}\big)$ into $\big(N,\ g_\Theta\big)$, where $U \subset M$ is the open set $U = M \setminus \{x \in M\ :\ \lambda(x) = 0\}$ and $\hat\theta = \lambda\,\theta$. Finally, if $\phi : M \to N$ is a CR map and $\phi \in C^\infty\big(\mathscr{F}_\theta,\ \mathscr{F}_\Theta\big)$, then ϕ is transversally holomorphic.

Proof For any $Y \in H(M)$ and $s \in C^\infty\big(\nu(\mathscr{F}_\theta)\big)$ [by (4.152)]

$$\big(\nabla_Y J_Q\big)s = \pi\big\{\nabla^{g_\theta}_Y\, J\,\sigma(s) - J\,\Pi_H\,\nabla^{g_\theta}_Y\,\sigma(s)\big\}. \tag{4.223}$$

The Levi-Civita connection of $\big(M,\ g_\theta\big)$ and the Tanaka-Webster connection of $(M,\ \theta)$ are related by [cf. (1.61) in [DrTo], p. 37]

$$\nabla^{g_\theta} = \nabla^\theta - (d\,\theta + A)\otimes T + \tau\otimes\theta + 2\,\theta\odot\varphi \tag{4.224}$$

where $\odot$ denotes the symmetric tensor product. In particular

$$\nabla^{g_\theta}_Y Z = \nabla^\theta_Y Z - \big(A(Y,\ Z) + (d\,\theta)(Y,\ Z)\big)T, \quad Y,\ Z \in H(M),$$

hence (4.223) becomes [by $\pi(T) = 0$]

$$\big(\nabla_Y J_Q\big)s = \pi\big(\nabla^\theta_Y\,J\big)\sigma(s) = 0.$$

Together with Lemma 4.175 this proves the first part of Theorem 4.176. Let $\phi : M \to N$ be a CR map and let us set $\lambda = \big(\phi^*\Theta\big)(T) \in C^\infty(M)$. Also let $\mathscr{Z}(\lambda) = \{x \in M\ :\ \lambda(x) = 0\}$ and $U = M \setminus \mathscr{Z}(\lambda)$. As ϕ is CR

$$0 = \Theta_{\phi(x)}\big((d_x\phi)X_x\big) = \big(\phi^*\Theta\big)(X)_x\,, \quad X \in C^\infty\big(H(M)\big), \quad x \in M,$$

and $\dim_{\mathbb{R}} H(M)_x = 2n$ hence $\phi^*\Theta = \lambda\,\theta$. To check the last statement in Theorem 4.176 note that either $\mathscr{Z}(\lambda) = M$ or $\hat\theta$ is a contact form on U. Let us assume that $\phi \in C^\infty\big(\mathscr{F}_\theta,\ \mathscr{F}_\Theta\big)$. Then $(d_x\phi)T_x = \lambda(x)\,T_{N,\,\phi(x)}$ for any $x \in U$ where T_N is the Reeb field of $(N,\ \Theta)$. The Reeb field $\hat T$ of $(U,\ \hat\theta)$ is given by (cf. [DrTo], p. 136)

$$\hat T = \frac{1}{\lambda}\Big\{T - \frac{i}{2\,\lambda}\big(\lambda^\alpha P_\alpha - \lambda^{\overline{\alpha}} P_{\overline{\alpha}}\big)\Big\} \tag{4.225}$$

where $\lambda^\alpha = g^{\alpha\overline{\beta}}\,\overline{P}_\beta(\lambda)$ and $\lambda^{\overline{\alpha}} = \overline{\lambda^\alpha}$. Then

$$\big(\phi^* g_\Theta\big)(T,\ T) = \lambda^2 = g_{\hat\theta}(T,\ T), \quad \big(\phi^* g_\Theta\big)(T,\ X) = 0 = g_{\hat\theta}(T,\ X),$$

$$\big(\phi^* g_\Theta\big)(X,\ Y) = g_{\hat\theta}(X,\ Y), \quad X,\ Y \in H(M),$$

i.e., $\phi^* g_\Theta = g_{\hat\theta}$ on U.

Therefore (by Theorem 4.176 above) unlike holomorphic maps of generalized Hopf manifolds, CR maps of Sasakian manifolds aren't foliated maps in general. Nevertheless if $\phi : M \to N$ is a CR map and $\phi \in C^\infty(\mathscr{F}_\theta\,,\,\mathscr{F}_\Theta)$, then (4.217) is easily seen to hold. Also [by (4.224)] $\nabla^{g_\theta}_T T = 0$ hence $\mathscr{F}_\theta$ is totally geodesic. Next [by (4.219) in Theorem 4.167],

$$K(\phi_t) = K(\phi), \quad |t| < \epsilon, \tag{4.226}$$

for any foliated map CR map $\phi : M \to N$ of compact Sasakian manifolds and any smooth 1-parameter family $\{\phi_t\}_{|t|<\epsilon} \subset C^\infty(\mathscr{F}_\theta\,,\,\mathscr{F}_\Theta)$ such that $\phi_0 = \phi$. Yet ϕ is transversally holomorphic so that [by (4.226)] $t = 0$ is an absolute minimum of $\mathscr{E}$, i.e., ϕ is $(\mathscr{F}, \mathscr{G})$-harmonic and $\mathrm{ind}_B(\phi) = 0$. The last statement in Theorem 4.167 follows from the fact that the total space P^{2n+1} of a principle circle bundle over the complex projective space may be organized as a Sasakian manifold (so that its Reeb flow in transversally Kählerian).

Example 4.177 (*Maps of Heisenberg groups*) Let $\mathbb{H}_n = \mathbb{C}^n \times \mathbb{R}$ be the *Heisenberg group* (cf. e.g., [DrTo], pp. 11–14) with coordinates $(z, t) = (z^1, \cdots, z^n, t)$. Let

$$P_{\overline{\alpha}} = \frac{\partial}{\partial \overline{z}^\alpha} - i\,\frac{\partial}{\partial t}, \quad 1 \le \alpha \le n$$

be the *Lewy operators*. One endows $\mathbb{H}_n$ with the CR structure $T_{1,0}(\mathbb{H}_n)$ spanned by $\{P_\alpha : 1 \le \alpha \le n\}$ (where $P_\alpha = \overline{P_{\overline{\alpha}}}$) and the pseudohermitian structure

$$\theta_0 = d\,t + \sum_{\alpha=1}^{n} \left(z^\alpha\, d\,\overline{z}_\alpha - \overline{z}^\alpha\, d\,z^\alpha\right).$$

Then $G_\theta(P_\alpha\,,\,P_{\overline{\beta}}) = \delta_{\alpha\beta}$ so that $\mathbb{H}_n$ is a strictly pseudoconvex CR manifold of CR dimension n. The Webster metric g_{θ_0} turns out to be a Sasakian metric and actually $\mathbb{H}_n$ is a Sasakian space form (of constant φ-sectional curvature $c = -3$, cf. e.g., [Dra4], p. 195).

Lemma 4.178 *A C^1 map $\Phi = (F^1, \cdots, F^m, f) : \mathbb{H}_n \to \mathbb{H}_m$ is a CR map if and only if F^j, $|F|^2 - i\,f \in \mathrm{CR}^1(\mathbb{H}_n)$ for any $1 \le j \le m$, i.e., F^j and $|F|^2 - i\,f$ are CR functions on $\mathbb{H}_n$.*

Here

$$|F|^2 = \sum_{j=1}^{m} F_j\,\overline{F}_j\,, \quad F = (F_1, \ldots, F_m), \quad F_j = F^j.$$

Let $(w, s) = (w^1, \ldots, w^m, s)$ be the Cartesian coordinates in $\mathbb{H}_m$ and

$$P'_{\overline{j}} \equiv \frac{\partial}{\partial \overline{w}^j} - i\, w^j \frac{\partial}{\partial s}, \quad 1 \le j \le m,$$

the corresponding Lewy operators. Given a CR map $\Phi : \mathbb{H}_n \to \mathbb{H}_m$ one has

$$\Phi_*\Big(\frac{\partial}{\partial t}\Big) = \lambda\Big(\frac{\partial}{\partial s}\Big)^{\Phi} + \frac{\partial F^j}{\partial t}\,(P'_j)^{\Phi} + \frac{\partial \overline{F}^j}{\partial t}\,(P'_{\overline{j}})^{\Phi}, \quad \lambda = f_t - i\,\overline{F}_j\, F^j_t + i\, F_j\, \overline{F}^j_t \in C(\mathbb{H}_n).$$

Therefore, Φ preserves the Reeb flows, i.e., $\Phi \in C^1\big(\mathscr{F}_{\theta_0}, \mathscr{F}_{\Theta_0}\big)$ if and only if $\partial F^j/\partial t = 0$ so that $F^j \in \mathcal{O}\big(\mathbb{C}^n\big)$ for $1 \le j \le m$. Here

$$\Theta_0 = d\,s + i \sum_{j=1}^{m} \big(w^j\, d\,\overline{w}_j - \overline{w}_j\, d\, w^j\big).$$

$\big(\mathbb{H}_n, g_{\theta_0}\big)$ is a noncompact Sasakian manifold so that Theorem 4.167 doesn't apply. Nevertheless

Proposition 4.179 *Any foliated CR map $\Phi \in C^2\big(\mathscr{F}_{\theta_0}, \mathscr{F}_{\Theta_0}\big)$ satisfies* (4.171), *and hence Φ is $\big(\mathscr{F}_{\theta_0}, \mathscr{F}_{\Theta_0}\big)$-harmonic.*

Indeed the Reeb flow $\mathscr{F}_{\theta_0}$ is totally geodesic (in particular $\tau_{\mathscr{F}_{\theta_0}} = 0$. On the other hand,

$$\tau_T(\Phi) = \operatorname{trace}_{g_{\nu(\mathscr{F}_{\theta_0})}} \beta_T(\Phi) = 2\sum_{\alpha=1}^{n} \beta_T(\Phi)(\pi\, P_\alpha, \pi\, P_{\overline{\alpha}}) = 2\sum_{\alpha} \big\{D^{\Phi}_{P_\alpha}\, \Phi_*\, \pi\, P_{\overline{\alpha}} - \Phi_*\, \nabla_{P_\alpha}\, \pi\, P_{\overline{\alpha}}\big\}$$

and [by $T\big(\mathscr{F}_{\theta_0}\big)^{\perp} = H\big(\mathbb{H}_n\big)$ and (4.224)]

$$\nabla_{P_\alpha}\, \pi\, P_{\overline{\alpha}} = \pi\, \nabla^{g_{\theta_0}}_{P_\alpha}\, P_{\overline{\alpha}} = \pi\Big\{\nabla^{\theta_0}_{P_\alpha}\, P_{\overline{\alpha}} - \big(d\,\theta_0\big)\big(P_\alpha, P_{\overline{\alpha}}\big)\frac{\partial}{\partial t}\Big\} = 0.$$

Lemma 4.180 *Let $\mathscr{F}_{\theta_0}$ be the Reeb foliation (tangent to $\partial/\partial t$) of the Heisenberg group $\mathbb{H}_n$ and $z^\alpha = x^\alpha + i\, y^\alpha$ with $1 \le \alpha \le n$. Then $\big(t, x^\alpha, y^\alpha\big) \in \mathscr{F}_{\theta_0}$, i.e., $\big(y^i\big) \equiv \big(x^\alpha, y^\alpha\big)$ are transverse coordinates (here $p = 1$ and $q = 2n$). Consequently, the canonical adapted connection ∇ of $\big(\mathbb{H}_n, \mathscr{F}_{\theta_0}, g_{\theta_0}\big)$ is given by $\Gamma^i_{jk} = 0$.*

Proof Let $P_{\overline{\alpha}} = \frac{1}{2}\big(X_\alpha + i\, Y_\alpha\big)$ be the real and imaginary parts of the Lewy operators. Then $\big\{X_\alpha, Y_\alpha : 1 \le \alpha \le n\big\}$ is a (left invariant) frame of the Levi distribution $H\big(\mathbb{H}_n\big)$. Clearly t is a (global) coordinate along each leaf of $\mathscr{F}_{\theta_0}$ (each maximal integral curve of the Reeb vector). Also

$$\frac{\partial}{\partial x^\alpha} = X_\alpha - 2\, y^\alpha \frac{\partial}{\partial t}, \quad \frac{\partial}{\partial y^\alpha} = Y_\alpha + 2\, x^\alpha \frac{\partial}{\partial t},$$

shows that $\{\pi(\partial/\partial x^\alpha),\ \pi(\partial/\partial y^\alpha)\ :\ 1 \leq \alpha \leq n\}$ is a (global) frame of the transverse bundle $\nu(\mathscr{F}_{\theta_0})$, i.e., $(t,\ x^\alpha,\ y^\alpha)$ is a (globally defined) foliated coordinate system on $(\mathbb{H}_n\ ,\ \mathscr{F}_{\theta_0})$. For instance [by (4.224)],

$$\Gamma^i_{\alpha\beta}\,\pi\frac{\partial}{\partial y^i} = \pi\,\nabla^{g_{\theta_0}}_{X_\alpha} X_\beta - 2\,y^\alpha\,\mathring{\nabla}_{\partial/\partial t}\,\pi\frac{\partial}{\partial x^\beta} = \pi\left\{\nabla^{\theta_0}_{X_\alpha} X_\beta - (d\,\theta_0)(X_\alpha\,,\ X_\beta)\frac{\partial}{\partial t}\right\} = 0$$

(because of $\nabla^{\theta_0}_{P_A} P_B = 0$ for any $A,\ B \in \{0,\ 1,\ \cdots,\ n,\ \overline{1},\ \ldots,\ \overline{n}\}$, cf. e.g., [DrTo], p. 34) yields $\Gamma^i_{\alpha\beta} = 0$ etc.

If $w^j = \xi^j + i\,\eta^j$, then $(s,\ \xi^j,\ \eta^j) \in \mathscr{F}_{\Theta_0}$, i.e., $(Y^u) \equiv (\xi^j,\ \eta^j)$ are transverse coordinates. Then (by (4.180)) $\Gamma'^u_{vw} = 0$ hence

$$D^\Phi_{P_\alpha}\,\Phi_*\,\pi\,P_{\overline{\alpha}} = \left(P_\alpha\,P_{\overline{\alpha}}\,\overline{F}^j\right)\pi'^\Phi\left(P'_{\overline{j}}\right)^\Phi = \left\{\left[P_\alpha\,P_{\overline{\alpha}}\right]\overline{F}^j + P_{\overline{\alpha}}\,P_\alpha\,\overline{F}^j\right\}\pi'^\Phi\left(P'_{\overline{j}}\right)^\Phi = 0$$

due to the commutation relations $\left[P_\alpha\,,\ P_{\overline{\beta}}\right] = -2i\,\delta_{\alpha\beta}\,\partial/\partial\,t$ and the fact that F_j are holomorphic functions. Therefore, $\tau_T(\Phi) = 0$ so that Φ is $(\mathscr{F}_{\theta_0}\,,\ \mathscr{F}_{\Theta_0})$-harmonic, as claimed. □

4.10.2.5 $(\mathscr{F},\ \mathscr{G})$-Harmonic Maps into Spheres

Let $S^m = \left\{x = (x_1\,,\ \ldots,\ x_{m+1}) \in \mathbb{R}^{m+1}\ :\ \sum_{i=1}^{m+1} x_i^2 = 1\right\}$. Drawing inspiration from B. Solomon [Sol], we shall study $(\mathscr{F},\ \mathscr{G}_0)$-harmonic maps $\phi \in C^\infty(\mathscr{F},\ \mathscr{G}_0)$ of a foliated Riemannian manifold $(M,\ \mathscr{F},\ g)$ into S^m thought of as endowed with the trivial foliation by points $\mathscr{G}_0$.

Definition 4.181 We say that $\phi : M \to S^m$ *meets* $S^{m-2} = \{x \in S^m\ :\ x_1 = x_2 = 0\}$ if $\phi(M) \cap S^{m-2} \neq \emptyset$. When ϕ doesn't meet S^{m-2} we say ϕ *links* S^{m-2} if $\phi : M \to S^m \setminus S^{m-2}$ is not null-homotopic. □

We shall prove the following result (containing the result in [Sol] for $\mathscr{F} = \mathscr{F}_0$)

Theorem 4.182 *Let $(M,\ \mathscr{F},\ g)$ be a compact oriented foliated Riemannian manifold with $\mathscr{F}$ transversally oriented and g bundle-like. If $\phi \in C^\infty(\mathscr{F},\ \mathscr{G}_0)$ is a non-constant $(\mathscr{F},\ \mathscr{G}_0)$-harmonic map of M into S^m, then either ϕ meets S^{m-2} or ϕ links S^{m-2}.*

Let $S^m_+ = S^m \cap \mathbb{R}^m_+$ where $\mathbb{R}^m_+ = \{y = (y_1\,,\ \ldots,\ y_m) \in \mathbb{R}^m\ :\ y_m > 0\}$ and let $S^{m-1}_+ \times_f S^1$ be the warped product with warping function $f \in C^\infty(S^{m-1}_+ \times S^1)$ given by $f(y,\ \zeta) = y_m$. Then

$$I : S^{m-1}_+ \times S^1 \to S^m \setminus S^{m-2}\,, \quad I(y,\ \zeta) = (u\,y_m\,,\ v\,y_m\,,\ y'),$$

$$y = (y',\ y_m) \in S^{m-1}_+\,, \quad \zeta = u + i\,v \in S^1 \subset \mathbb{C},$$

is an isometry of $S_+^{m-1} \times_f S^1$ onto $(S^m \setminus S^{m-2}, h_m)$. Here $h_m = \iota^* H$ and H is the Euclidean metric on $\mathbb{R}^{m+1}$ (and $\iota : S^m \to \mathbb{R}^{m+1}$ is the inclusion). Moreover, let

$$p : S_+^{m-1} \times_\gamma \mathbb{R} \to S_+^{m-1} \times_f S^1, \quad p(y, t) = \big(y, \exp(i\, t)\big),$$

be the universal Riemannian covering, where $\gamma \in C^\infty\big(S_+^{m-1}\big)$ is the function $\gamma(y) = y_m$.

The proof of theorem 4.153 is by contradiction, i.e., we may assume that ϕ doesn't meet S^{m-2} and that $\phi : M \to S^m \setminus S^{m-2}$ is null-homotopic. Then ϕ lifts to a $(\mathscr{F}, \mathscr{G}_0)$-harmonic map $\psi : M \to S_+^{m-1} \times_\gamma \mathbb{R}$. Let $F = \pi_1 \circ \psi$ and $u = \pi_2 \circ \psi$ be the components of ψ where

$$\pi_1 : S_+^{m-1} \times \mathbb{R} \to S_+^{m-1}, \quad \pi_2 : S_+^{m-1} \times \mathbb{R} \to \mathbb{R},$$

are the canonical projections. Clearly $\psi \in C^\infty\big(\mathscr{F}, \mathscr{G}_0\big)$ so that $F : M \to S_+^{m-1}$ is a foliated map and $u \in \Omega_B^0(\mathscr{F})$. Throughout $\mathscr{G}_0$ denotes the trivial foliation by points, though on several different manifolds. Let $\{Y_i \,:\, 1 \le i \le q\}$ be a local g-orthonormal frame of the normal bundle $T(\mathscr{F})^\perp$ defined on an open set $U \subseteq M$ [where $q = \operatorname{cod}(\mathscr{F})$]. As $S_+^{m-1} \times_\gamma \mathbb{R}$ is endowed with the (warped product) metric $\pi^* h_{m-1} + (\gamma \circ \pi_1)^2\, d\,t \otimes d\,t$ the transverse energy density is

$$2\, e_T(\psi)_x = \sum_{i=1}^{q} \Big\{ h_{m-1, F(x)}\big((d_x\, F) Y_{i,x}\,,\ (d_x\, F) Y_{i,x}\big) + \gamma\big(F(x)\big)^2\, (d\, u)\big(Y_i\big)_x^2 \Big\} =$$

$$= 2\, e_T(F)_x + \gamma\big(F(x)\big)^2\, \big\| d_B\, u \big\|_x^2$$

for any $x \in U$. Given $\varphi \in \Omega_B^0(\mathscr{F})$ we consider the 1-parameter variation

$$\psi_t = (F,\, u + t\, \varphi) \in C^\infty\big(\mathscr{F}, \mathscr{G}_0\big)\,, \quad |t| < \epsilon\,,$$

so that

$$E_T\big(\psi_t\big) = E_T(F) +$$

$$+ \frac{1}{2} \int_M (\gamma \circ f)^2 \Big\{ \big\| d_B\, u \big\|^2 + 2\, t\, g^*\big(d_B\, u,\, d_B\, \varphi\big) + O\big(r^2\big) \Big\}\, d\, v_g\,.$$

Then [as ψ is $(\mathscr{F}, \mathscr{G}_0)$-harmonic and $\gamma \circ F \in \Omega_B^0(\mathscr{F})$] one has

$$0 = \Big\{ \frac{d}{d\,t}\, E_T\big(\psi_t\big) \Big\}_{t=0} = \big((\gamma \circ F)^2 d_B\, u\,,\ d_B\, \varphi\big)_B$$

so that (as φ is an arbitrary basic function) u must satisfy the transversally elliptic equation

$$\delta_B\big((\gamma \circ F)^2\, d_B\, u\big) = 0. \tag{4.227}$$

Then one may multiply by u in (4.227) and integrate over M to conclude that $d_B\, u = 0$, i.e.,

$$\psi(M) \subset S_+^{m-1} \times \{u_0\}$$

for some $u_0 \in \mathbb{R}$. Let $(U,\, x^\alpha,\, y^i) \in \mathscr{F}$ be a foliated local chart on M. Let $(\mathscr{U},\, X^u)$ be a local coordinate system on $S_+^{m-1} \times \mathbb{R}$ with $X^m = t$ and $\psi(U) \subseteq \mathscr{U}$. Then [again as ψ is $(\mathscr{F}, \mathscr{G}_0)$-harmonic]

$$\Delta_B\, \psi^w + (g_Q)^{ij}\, \frac{\partial\, \psi^u}{\partial\, y^i}\, \frac{\partial\, \psi^v}{\partial\, y^j} \left(\Gamma'^{w}_{uv} \circ \psi\right) = 0, \quad 1 \le w \le m,$$

where $\psi^u = X^u \circ \psi$. Here Γ'^{w}_{uv} are the local coefficients of the Levi-Civita connection of $S_+^{m-1} \times_\gamma \mathbb{R}$. In particular for $w = m$ one obtains

$$\Delta_B\, F^a + (g_Q)^{ij}\, \frac{\partial\, F^b}{\partial\, y^i}\, \frac{\partial\, F^c}{\partial\, y^j} \left(\gamma^a_{bc} \circ F\right) = 0$$

or

$$\Delta_B\, F^a + 2\, e_T(F)\, F^a = 0, \quad 1 \le a \le m-1. \tag{4.228}$$

Here

$$\gamma^c_{ab} = h_{ab}\, X^c\,, \quad h_{ab} = \delta_{ab} + \left(1 - |X|^2\right)^{-1} X_a\, X_b\,.$$

If $F = (F',\, F_m)$ and $F' = (F_1\,,\, \ldots,\, F_{m-1})$, then we may exploit the constraint $|F'|^2 + F_m^2 = 1$ and the identity

$$\Delta_B(u^2) = 2\, u\, \Delta_B\, u + 2\, \|d_B\, u\|^2, \quad u \in \Omega^0_B(\mathscr{F})$$

to show that (4.228) yields

$$\Delta_B\, F_m + 2\, e_T(F) F_m = 0 \tag{4.229}$$

as well. Integration of (4.229) over M and $F_m > 0$ then shows F (and therefore ϕ) is constant, a contradiction.

Since the natural projection $\pi_2 : S_+^{m-1} \times S^1 \to S^1$ is a homotopy equivalence, $\phi : M \to S^m \setminus S^{m-2}$ is null-homotopic if and only if $\pi_2 \circ \left(I^{-1} \circ \phi\right) : M \to S^1$ is null-homotopic. Consequently [as the Bruschlinski group of M and $H^1(M,\, \mathbb{Z})$ are isomorphic],

Corollary 4.183 *Let $(M,\, \mathscr{F},\, g)$ be a compact orientable foliated Riemannian manifold such that*

(i) *$\mathscr{F}$ is a transversally oriented Riemannian foliation,*

(ii) *g is a bundle-like Riemannian metric, and*

(iii) $H^1(M,\, \mathbb{Z}) = 0.$

Then any nonconstant $(\mathscr{F}, \mathscr{G}_0)$*-harmonic map of* $(M, \mathscr{F}, g)$ *into* $(S^m, \mathscr{G}_0, h_m)$ *meets* S^{m-2}.

References

[Rei] B. Reinhart, Foliated manifolds with bundle-like metrics. Ann. Math. **69**, 119–132 (1959)

[Rum] H. Rummler, Quelques notions simples en géométrie Riemanniene et leurs applications aux feuilletages compacts. Comment. Math. Helv. **54**, 224–239 (1979)

[KaTo1] F. Kamber, P. Tondeur, Invariant differential operators and the cohomology of Lie algebra sheaves. Memoirs Am. Math. Soc. **113**, 1–125 (1971)

[KaTo16] F. Kamber, P. Tondeur, De Rham-Hodge theory for Riemannian foliations. Math. Ann. **277**, 415–431 (1987)

[KaTo6] F. Kamber, P. Tondeur, Harmonic foliations, in *Proceedings of NSF Conference on Harmonic Maps*, Tulane (1980). Springer Lecture Notes in Mathematics 949, 87–121 (1982)

[BaWo] Baird, J.C. Wood, *Harmonic Morphisms Between Riemannian Manifolds*, London Mathematical Society Monographs, New Series, vol. 29. (Oxford University Press., Oxford, 2003)

[Vai1] I. Vaisman, Variétés Riemanniennes feuilletés. Czechslovak Math. J. **21**, 46–75 (1971)

[Mol] P. Molino, *Riemannian foliations*, Ed. by Oesterlé & A. Weinstein, Progress in Mathematics, vol. 73. (Birkhäuser, Boston, 1988)

[Mol1] P. Molino, Propriétés cohomologiques et propriétés topologiques des feilletages *à* connexion transverse projectable. Topology **12**, 317–325 (1973)

[KaTo3] F. Kamber, Ph. Tondeur, *Foliated Bundles and Characteristic Classes*, Springer Lecture Notes in Mathematics, vol. 494 (1975)

[KaTo11] F. Kamber, Ph. Tondeur, Foliations and Metrics, in *Proceedings of a Year in Differential Geometry*, University of Maryland, Birkhäuser, Progress in Mathematics, Vol. 32 (1983), pp. 103–152

[Alv1] J. Alvarez, A finiteness theorem for the spectral sequence of a Riemannian foliation. Illinois J. Math., (1) **33**, 79–92 (1989)

[Alv2] J. Alvarez, Duality in the spectral sequence of a Riemannian foliation. Am. J. Math. (6) **111**, 905–926 (1989)

[KoNo] S. Kobayashi, K. Nomizu, *Foundations of Differential Geometry*. (Interscience Publishers, New York, Vol. I, 1963, Vol. II, 1969)

[Ton] P. Tondeur, *Foliations on Riemannian Manifolds* Universitext. (Springer, New York, 1988)

[Ler] J. Lerray, L'anneau d'une representation. C.R. Acad. Sci. Paris **222**, 1366–1368 (1946); Propriétés de l'anneau d'homologie d'une representation. ibid., **222**, 1419-1422 (1946); Sur l'anneau d'homologie de l'espace homogène, quotient d'un groupe clos per un sous-groupe abélien connexe maximum. ibid. **223**, 412–415 (1946); L'anneau spectral et l'anneau filtré d'homologie d'un espace localement compact e d'une application continue. J. Math. Pure Appl. **29**, 1–80, 81–139(1950); L'homologie d'un espace fibré dont la fibre est connexe. ibid. **29**, 169–213 (1950)

[MaLa] S. Mac Lane, *Homology* (Springer, Berlin-Heidelberg-New York, 1994 (fourth printing))

[McCl] J. McCleary, *User's Guide to Spectral Sequences* (Publish or Perish. Inc, Wilmington, Delaware, 1985)

[Jer] R. Jerrard, Curvature of surfaces associated with holomorphic functions. Colloq. Math. **21**, 127–132 (1970)

[JeRu] R. Jerrard, L. Rubel, On the curvature of the level lines of a harmonic function. Proc. Am. Math. Soc. **14**, 29–32 (1963)

[FNS] L. Flatto, D.J. Newman, H.S. Shapiro, The level curves of harmonic functions. Trans. Am. Math. Soc. **123**, 425–436 (1966)

[OrSc] M. Ortel, W. Schneider, Curvature of level curves of harmonic functions. Canad. Math. Bull., (4) **26**, 399–405 (1983)

[KaTo9] F. Kamber, Ph. Tondeur, Duality for Riemannian foliations. Proc. Symp. Pure Math. **40**, 609–618 (1983), Part. 1

[KaTo13] F. Kamber, P. Tondeur, Duality theorems for foliations. Astérisque **116**, 108–116 (1984)

[AlHe1] A. Alaoui, G. Hector, Décomposition de Hodge sur l'espace des feuilles d'un feuilletage Riemannien. C.R. Acad. Sci. Paris **298**, 298–292 (1984)

[AlHe2] A. Alaoui, G. Hector, Décomposition de Hodge basique pour un feuilletage Riemannien. Ann. Inst. Fourier **36**, 207–227 (1986)

[ASH] A. Alaoui, V. Sergiescu, G. Hector, La cohomologie basique d'un feuilletage Riemannien est de dimension finie. Math. Z. **188**, 593–599 (1985)

[Mol0] P. Molino, Géométrie globale des feuilletages Riemanniens. Proc. K. Ned. Akad., Ser. A **85**, 45–76 (1982)

[War] F. Warner, *Foundations of Differentiable Manifolds and Lie Groups* (Heidelberg, New York, Berlin, 1983)

[DrPe] S. Dragomir, D. Perrone, *Levi Harmonic Maps of Contact Riemannian Manifolds*

[Vai] I. Vaisman, The curvature groups of a space form. Ann. Sci. Norm. Sup., Pisa **22**, 331–341 (1968)

[KoSp] K. Kodaira, D.C. Spencer, Multifoliate structures. Ann. Math. **74**, 52–100 (1961)

[Vai0] I. Vaisman, Remarques sur la théorie des formes-jet. C.R. Acad. Sci. Paris **264**, 351–354 (1967)

[GoPe] S.I. Goldberg, N.C. Petridis, The curvature groups on a pseudo-Riemannian manifold. J. Diff. Geometry **9**, 547–555 (1974)

[KaTo7] F. Kamber, P. Tondeur, Infinitesimal automorphisms and second variation of the energy for harmonic foliations. Tohoku Math. J. **34**, 525–538 (1982)

[ToTo] P. Tondeur, G. Toth, On transversal infinitesimal automorphisms for harmonic foliations. Geom. Dedicata. **24**, 229–236 (1987)

[ESH] A. El Kacimi-Alaoui, V. Sergiescu, G. Hector, La cohomologie basique d'un feuilletage Riemannien est de dimension finie. Math. Z. **188**, 593–599 (1985)

[Sch] G. Schwarz, On the de Rham cohomologie of the leaf space of a foliation. Topology **13**, 185–187 (1974)

[Cri] R. Crittenden, Covariant differentiation. Quart. J. Math. Oxford, (2) **13**, 285–298 (1962)

[KTT] F. Kamber, P. Tondeur, G. Toth, Transversal Jacobi fields for harmonic foliations. Michigan Math. J. **34**, 261–266 (1987)

[BaDr] E. Barletta, S. Dragomir, On transversally holomorphic maps of Kählerian foliations. Acta Appl. Math. **54**, 121–134 (1998)

[Hae] A. Haefliger, Variétés feuilletées. Ann. Sc. Norm. Sup., Pisa **16**, 367–397 (1962)

[DrWo] S. Dragomir, J.C. Wood, Sottovarietà minimali ed applicazioni armoniche, Quaderni dell'U.M.I., 35, Pitagora Editrice, Bologna (1989)

[KaAl] A. El Kacimi-Alaoui, Opérateurs transversalment elliptiques sur un feulletage Riemanienne et applications. Compositio Math. **73**, 57–106 (1990)

[DrOr] S. Dragomir, L. Ornea, *Locally conformal Kähler geometry, Progress in Mathematics*, vol. 155 (Birkhäuser, Boston-Basel-Berlin, 1998)

[Chi] Y.J. Chiang, Harmonic maps on V-manifolds. Ann. Global Anal. Geom. **8**, 315–344 (1990)

[NiTo] S. Nishikawa, P. Tondeur, Transversal infinitesimal automorphisms for harmonic Kähler foliations. Tohoku Math. J. **40**, 599–611 (1988)

[Lib] P. Libermann, Sue les structures presque complexes et autres structures infinitésimales réguliers. Bull. Soc. Math. France **83**, 195–224 (1955)

[Vai2] I. Vaisman, Generalized Hopf manifolds. Geom. Dedicata. **13**, 231–255 (1982)

[BrVe] E. Brieskorn, A. Van de Ven, Some complex structures of products of homotopy spheres. Topology **7**, 389–393 (1968)

[Gol] S.I. Goldberg, *Curvature and Homology*

[Tsu1] K. Tsukada, Holomorphic forms and holomorphic vector fields on compact generalized Hopf manifolds. Compositio Math. **93**, 1–22 (1994)

[Tsu2] K. Tsukada, Holomorphic maps of compact generalized Hopf manifolds. Geometriae Dedicata **68**, 61–71 (1997). https://doi.org/10.1023/A:1004949925097

[Lic] A. Lichnerowicz, *Applications harmoniques et variétés kähkeriennes*, Symposia Mathematica, Vol III, Istituto Nazionale di Alta Matematica (Academic Press, London-New York, 1970), pp. 341–402

[Vai3] I. Vaisman, A Schwartz lemma for complex surfaces, in *Global Analysis-Analysis on Manifolds*, Ed. by Th.M. Rassias, Teubner Texte zur Math. 57 (Teubner-Verlag, Leipzig, 1983), pp. 305–323

[DrTom] S. Dragomir, A. Tommasoli, Harmonic maps of foliated Riemannian manifolds. Geometriae Dedicata. https://doi.org/10.1007/s10711-012-9723-3

[Sol] B. Solomon, Harmonic maps to spheres. J. Differ. Geom. **21**, 151–162 (1985)

[KoWo1] J.J. Konderak, R.A. Wolak, Transversally harmonic maps between manifolds with Riemannian foliations. Q.J. Math. **54**, 335–354 (2003)

[KoWo2] J.J. Konderak, R.A. Wolak, Some remarks on transversally harmonic maps. Glasgow J. Math., (1) **50**, 1–16 (2008)

[ChWo] Y.J. Chiang, R. Wolak, Transversally biharmonic maps between foliated Riemannian manifolds. Int. J. Math., (8) **19**, 981–996 (2008)

[PaPa] H.K. Pak, J.H. Park, Transversal harmonic transformations for Riemannian foliations. Ann. Glob. Anal. Geom. (1) **30**, 97–105 (2006)

[JuJu] M.J. Jung, S.D. Jung, On transversally harmonic maps of foliated Riemannian manifolds (2011). arXiv:1109.3932

[KAGo] A. El Kacimi-Alaoui, E.G. Gomez, Applications harmoniques feuilletées. Illinois J. Math., (1) **40**, 115–122 (1996)

[Fug] B. Fuglede, Harnack sets and openness of harmonic morphisms. Math. Ann. **241**, 181–186 (1979)

[PaRi] E. Park, K. Richardson, The basic Laplacian of a Riemannian foliation. Am. J. Math. **118**, 1249–1275 (1996)

[Ura] H. Urakawa, Variational problems over strongly pseudoconvex CR manifolds, in *Differential Geometry, Proceedings of the Symposium in Honour of Prof. Su Buchin on his 90th Birthday*, ed. by C-H. Gu, H-S. Hu, Y-L. Xin, Shanghai, China, 17-23 September 1991(World Scientific, Singapore, 1993), pp. 233–242

[EeLe1] J. Eells, L. Lemaire, A report on harmonic maps. Bull. Lond. Math. Soc. **10**, 1–68 (1978)

[Ish] T. Ishihara, A mapping of Riemannian manifolds which preserves harmonic functions. J. Math. Kyoto Univ. (2) **19**, 215–229 (1979)

[GiTr] D. Gilbarg, N.S. Trudinger, *Elliptic Partial Differential Equations of Second Order* (Springer, New York, 1983)

[ToVa] Ph. Tondeur, L. Vanhecke, Transversally symmetric Riemannian foliations. Tohoku Math. J., (3) **42**, 307–317 (1990)

[Smit] R.T. Smith, The second variation formula for harmonic mappings. Proc. A.M.S., (1) **47**, 229–236 (1975)

[Wel5] R.O. Wells, *Differential Analysis on Complex Manifolds, Graduate Texts in Mathematics*, vol. 65 (Springer, New York, 1980)

[Kon] J.J. Konderak, On sections of fibre bundles which are harmonic maps. Bull. Math. Soc. Sci. Math. Roumanie, (4) **90**, 341–352 (1999)

[Wie] G. Wiegmink, Total bending of vector fields on Riemannian manifolds. Math. Ann., (2) **303**, 325–344 (1995)

[DrTo] S. Dragomir, G. Tomassini, *Differential Geometry and Analysis on CR Manifolds, Progress in Mathematics*, vol. 246 (Birkhäuser, Boston-Basel-Berlin, 2006)

[Dra5] S. Dragomir, Cauchy-Riemann geometry and subelliptic theory. Lecture Notes of Seminario Interdisciplinare di Matematica **7**, 121–162 (2008)

[JoXu] J. Jost $ C.-J. Xu, Subelliptic harmonic maps. Trans. Am. Math, Soc. **350**, 4633–4649 (1998)

[BaDrUr] E. Barletta, S. Dragomir, H. Urakawa, Pseudoharmonic maps from a nondegenerate CR manifold into a Riemannian manifold. Indiana Univ. Math. J. (2) **50**, 719–746 (2001)

[Bla] D.E. Blair, *Contact manifolds in Riemannian geometry*, vol. 509 (Springer, Berlin-Heidelberg-New York, 1976)

[Dra4] S. Dragomir, On pseudohermitian immersions between strictly pseudoconvex CR manifolds. Am. J. Math. (1) **117**, 169–202 (1995)

Chapter 5
Sobolev Inequalities on Submanifolds

5.1 Sobolev Inequalities on Complete Manifolds

The purpose of the present section is to briefly review a few results on complete Riemannian manifolds, with Ricci curvature bounded from below, that support a Sobolev embedding. We follow the presentation by Athanasios, [Ath]. Let (M, g) be a n-dimensional ($n \geq 3$) Riemannian manifold.

Definition 5.1 (M, g) is said to *support a Sobolev embedding* if there exist

$$p, \ q \in \mathbb{R}, \quad m, \ k \in \mathbb{Z}, \quad 1 \leq q < p, \quad 0 \leq m < k, \quad \frac{1}{p} = \frac{1}{q} - \frac{k-m}{n},$$

such that[1]

$$H_k^q(M) \hookrightarrow H_m^p(M). \tag{5.1}$$

□

Here $H_k^p(M)$ denotes the Sobolev space[2] of M, i.e., the completion of

$$C_k^p(M) = \left\{ u \in C^\infty(M) \ : \ \nabla^j u \in L^p(M), \ \ 0 \leq j \leq k \right\}$$

with respect to the norm

$$\|u\|_{H_k^p(M)} = \sum_{j=0}^{k} \|\nabla^j u\|_{L^2(M)} = \sum_{j=0}^{k} \left(\int_M |\nabla^j u|^p \, d\,\mathrm{v}_g \right)^{1/p}.$$

[1] By definition, the relation (5.1) includes the continuity of the inclusion map $H_k^q(M) \to H_m^p(M)$.

[2] Elsewhere in this book, also denoted by $W^{k,p}(M)$.

E. Barletta et al., *Differential Geometry*, Infosys Science Foundation Series,
https://doi.org/10.1007/978-981-96-1631-2_5

As customary $\nabla^k u$ is the k-th covariant derivative of the C^∞ function $u : M \to \mathbb{R}$, i.e., the $(0, k)$-tensor field on M defined recursively, as follows:

$$\nabla^0 u = u,$$

$$(\nabla^1 u)(X) := X(u),$$

$$(\nabla^2 u)(X, Y) = (\nabla_X du)Y = X(Yu)) - (\nabla_X Y)(u),$$

for any $X, Y \in \mathfrak{X}(M)$. If S is a $(0, k)$-tensor field, then its covariant derivative $\nabla_X S$ is the $(0, k)$-tensor field defined by

$$\big(\nabla_X S\big)\big(Y_1, \cdots, Y_k\big) = X\big(S(X_1, \cdots, X_k)\big)+,$$

$$-\sum_{j=1}^{k} S\big(X_1, \cdots, X_{j-1}, \nabla_X X_j, X_{j+1}, \cdots, X_k\big),$$

$$X, X_1, \cdots, X_k \in \mathfrak{X}(M).$$

The covariant derivative ∇S is the $(0, k+1)$-tensor field given by

$$(\nabla S)\big(X_0, X_1, \cdots, X_k\big) := \big(\nabla_{X_0} S\big)\big(X_1, \cdots, X_k\big).$$

The recurrence relation defining the higher order covariant derivatives of u is

$$\nabla^{k+1} u = \nabla\big(\nabla^k u\big), \quad k \geq 0.$$

Let (U, x^i) be a local coordinate system on M, and let $\big(\nabla^k u\big)_{i_1 \cdots i_k}$ be the local components of $\nabla^k u$, i.e.,

$$\big(\nabla^k u\big)_{i_1, \cdots, i_k} = \big(\nabla^k u\big)\Big(\frac{\partial}{\partial x^{i_1}}, \cdots, \frac{\partial}{\partial x^{i_k}}\Big).$$

Let $\big(\nabla^k u\big)^{i_1 \cdots i_k} : U \to \mathbb{R}$ be the C^∞ functions defined by

$$\big(\nabla^k u\big)^{i_1 \cdots i_k} = g^{i_1 j_1} \cdots g^{i_k j_k} \big(\nabla^k u\big)_{j_1, \cdots, j_k}.$$

The pointwise norm of $\nabla^k u$ is the function $\big|\nabla^k u\big| : M \to [0, +\infty)$ defined as follows. Let $x_0 \in M$ and let (U, x^i) be a local coordinate system with $x_0 \in U$. We set

$$\big|\nabla^k u\big|(x_0) = \Big[\big(\nabla^k u\big)_{i_1, \cdots, i_k}(x_0)\big(\nabla^k u\big)^{i_1, \cdots, i_k}(x_0)\Big]^{1/2}.$$

The definition of $|\nabla^k u|(x_0)$ doesn't depend upon the choice of local coordinates about x_0.

5.1.1 Sobolev Inequalities on Euclidean Space

Let $1 \le q < n$ and let us set $q^* = \dfrac{nq}{n-q}$. The simplest Sobolev inequality on $\mathbb{R}^n$ looks like

$$\left(\int_{\mathbb{R}^n} |u|^{q^*} dx\right)^{1/q^*} \le C(n,q) \int_{\mathbb{R}^n} |\nabla u|^q dx\Big)^{1/q}, \quad u \in C_0^\infty(\mathbb{R}^n), \tag{5.2}$$

for some constant $C(n,q) > 0$. The inequality (5.2) implies

$$H_1^q(\mathbb{R}^n) \hookrightarrow H_0^{q^*}(\mathbb{R}^n) := L^{q^*}(\mathbb{R}^n).$$

The Sobolev inequality (5.2) for $q = 1$ was first proved by E. Gagliardo and L. Nirenberg.[3] Let us recall the classical isoperimetric inequality

Theorem 5.2 *For every domain $\Omega \subset \mathbb{R}^n$ with $\partial\Omega \in C^\infty$ and $\overline{\Omega}$ compact,*

$$\left[\mathrm{Vol}(\Omega)\right]^{(n-1)/n} \le K(n,1)\,\mathrm{Area}(\partial\Omega). \tag{5.3}$$

If $b_n = \mathrm{Vol}(\mathbb{B}^n)$, then the optimal constant is $K(n,1) = (1/n)\, b_n^{-1/n}$ and it is achieved when Ω is a ball.

Here $\mathbb{B}^n = \{x \in \mathbb{R}^n : |x| < 1\}$.

Theorem 5.3 *For any $u \in C_0^\infty(\mathbb{R}^n)$,*

$$\left(\int_{\mathbb{R}^n} |u|^{n/(n-1)}\, dx\right)^{(n-1)/n} \le C(n) \int_{\mathbb{R}^n} |\nabla u|\, dx \tag{5.4}$$

for some constant $C(n) > 0$. The optimal constant[4] *is $C(n) = K(n,1) := (1/n)\, b_n^{-1/n}$.*

A proof of Theorem 5.3 relying on Theorem 5.2 was provided in [Ath], pp. 7–8. The proof of (5.2) in the case $1 < q < n$ is due to G. Talenti.[5] The precise statement of Talenti's result is

---

[3] Cf. E. Gagliardo, *Proprietà di alcune classi di funzioni in più variabili*, Ricerche di Mat., Napoli, 7(1958), 102–137; L. Nirenberg, *On elliptic partial differential equations*, Annali della Scuola Normale Superiore di Pisa, (2)13(1959), 115–162.

[4] The same constant that is optimal in the isoperimetric inequality (5.3).

[5] Cf. G. Talenti, *Inequalities in rearrangement invariant function spaces* Nonlinear Analysis, Function Spaces and Applications, Vol. 5, Prometheus, Praha, pp. 177–230, 1995.

Theorem 5.4 *Let $1 < q < n$ and $q^* = nq/(n-q)$. There is a constant $C(n, q) > 0$ such that*

$$\Big(\int_{\mathbb{R}^n} |u|^{q^*}\,dx\Big)^{1/q^*} \le C(n,q)\Big(\int_{\mathbb{R}^n} |\nabla u|^q\,dx\Big)^{1/q} \tag{5.5}$$

for any $u \in C_0^\infty(\mathbb{R}^n)$. The optimal constant is

$$\frac{1}{K(n,q)} = \sqrt{\pi}\,n^{1/q}\left(\frac{n-q}{q-1}\right)^{1-1/q}\left(\frac{\Gamma(n/q)\,\Gamma(1+n-n/q)}{\Gamma(n)\,\Gamma(1+n/2)}\right)^{1/n} \tag{5.6}$$

and is achieved for any

$$u(x) := \left(\alpha + \beta\,|x|^{q/(q-1)}\right)^{1-n/q}, \quad \alpha,\ \beta \in (0, +\infty). \tag{5.7}$$

The proof of Theorem 5.4 provided in [Ath], pp. 9–10, is Talenti's proof (cf. *op. cit.*) using symmetric rearrangements in the sense of Hardy-Littlewood.

5.1.2 Sobolev Inequalities on Complete Riemannian Manifolds with Ricci Curvature Bounded from Below

Through this section, we characterize (by following E. Hebey[6]) complete Riemannian manifolds with Ricci curvature bounded from below that support a Sobolev embedding. The required ingredients are

(i) the Coarea Formula on Riemannian manifolds,
(ii) the calculation of the volume of a ball in geodesic spherical coordinates,
(iii) the Bishop-Gromov volume comparison theorem,
(iv) a covering lemma.

Consequently, one may prove that for every complete Riemannian manifold (M, g) with Ricci curvature bounded from below, all Sobolev embeddings hold on (M, g) if and only if there is a uniform lower bound for the volume of balls, that is independent of their center, i.e.,

$$\inf_{x\in M} \mathrm{Vol}_g\left[B_r(x)\right] > 0 \tag{5.8}$$

for every $r > 0$. It is well known that the embedding

$$H_1^1(M) \hookrightarrow L^{n/(n-1)}(M)$$

[6] Cf. E. Hebey, *Nonlinear Analysis on Manifolds: Sobolev Spaces and Inequalities*, Courant Lecture Notes, 2000; *Sobolev Spaces on Riemannian Manifolds*, Lecture Notes in Mathematics, Springer, 1996.

implies all other Sobolev embeddings, that is

$$\forall\, p, q \in \mathbb{R}, \quad \forall\, m, k \in \mathbb{Z} \quad \text{with,}$$

$$1 \le q < p, \quad 0 \le m < k, \quad \frac{1}{p} = \frac{1}{q} - \frac{k-m}{n},$$

one has

$$H_k^q(M) \hookrightarrow H_m^p(M).$$

Similar arguments show that if

$$\exists\, q_0 \in (1, n) \ : \ H_1^{q_0}(M) \hookrightarrow L^{p_0}(M). \tag{5.9}$$

Then, for all $q \in [q_0, n)$,

$$H_1^q(M) \hookrightarrow L^p(M), \quad \frac{1}{p} = \frac{1}{q} - \frac{1}{n}.$$

Moreover, it can be shown that under the conditions (5.8)–(5.9) the embedding $H_1^q(M) \hookrightarrow L^p(M)$ holds for any exponent $q \in [1, n)$, eventually smaller than q_0, so that condition (5.8) implies all other Sobolev embeddings on (M, g).

5.1.2.1 Coarea Formula

Theorem 5.5 *Let (M, g) be a Riemannian manifold, and let $f \in C_0^\infty(M)$. Then, for every $h \in L^1(M)$,*

$$\int_M h \, d\mathrm{v}_g = \int_{\mathbb{R}} dt \int_{\Gamma(t)} \frac{h}{|\nabla f|} \, dA(t), \tag{5.10}$$

where $\Gamma(t) = f^{-1}(t) = \{p \in M \ : \ f(p) = t\}$ and $dA(t)$ is the induced measure on $\Gamma(t)$.

5.1.2.2 Geodesic Spherical Coordinates

Let (M, g) be a complete Riemannian manifold. For every $p \in M$, let $\mathrm{Cut}(p)$ be the cut locus[7] of p. As well known, $\mathrm{Cut}(p)$ has measure zero.

Theorem 5.6 *Let (M, g) be a complete Riemannian manifold, and let $B_r(p) \subset M$ be a ball such that $B_r(p) \cap \mathrm{Cut}(p) = \emptyset$. For every $v \in T_p(M)$ with $\|v\| = 1$ and for*

[7] Loosely, the set of all points of the manifold where the geodesics issuing at p stop being minimizing.

any orthonormal basis $\{e_1, \cdots, e_{n-1}, v\} \subset T_p(M)$, *let* Y_i *be the Jacobi field along the geodesic* $\gamma(t) = \exp_p(tv)$ *obeying to the initial conditions*

$$Y_i(0) = 0, \quad Y'(0) = e_i, \quad 1 \leq i \leq n-1.$$

Then

$$\mathrm{Vol}_M\big[B_r(p)\big] = \int_{S^{n-1}(1)} \int_0^r J^M(s, v)\, ds\, dv,$$

$$J^M(s, v) := \sqrt{\det\Big[g\big(Y_i(s), Y_j(s)\big)\Big]_{1\leq i,j\leq n-1}},$$

where dv *is the canonical volume form on* $S_p^{n-1}(1) = \{\xi \in T_p(M) : |\xi| = 1\}$.

Cf. e.g., Theorem 3.2.2 in [Ath], pp. 15–19.

5.1.2.3 Bishop-Gromov Volume Comparison

Theorem 5.7 *Let* (M, g) *be a complete Riemannian manifold, and let* $p \in M$ *and* $r > 0$ *such that*

$$\exists\, k \in \mathbb{R}, \quad \forall\, X \in T(M) : \mathrm{Ric}_g(X, X) \geq (n-1)k\, g(X, X) \;\; on \;\; B_r(p).$$

Let $M_{0,k}$ *be a complete simply connected manifold of constant curvature* k. *Then*

(i) $\mathrm{Vol}_M\big[B_r(p)\big] \leq V_k(r)$,

(ii) *The function* $s \mapsto \dfrac{J(s, v)}{J_k(s)}$ *is monotonously decreasing,*

(iii) *The function* $r \mapsto \dfrac{\mathrm{Vol}_M\big[B_r(p)\big]}{V_k(r)}$ *is monotonously decreasing.*

Here $V_k(r)$ is the volume of a ball of radius r in the space-form $M_{0,k}$ and

$$J_k(s) := J^{M_{0,k}}(s, v).$$

Cf. Theorem 3.3.1 in [Ath], pp. 20–26. Theorem 5.7 (the *Bishop-Gromov volume comparison theorem*) provides upper bounds for the volume of balls $B_r(p) \subset M$.

5.1.2.4 A Covering Lemma

Lemma 5.8 *Let* (M, g) *be a complete Riemannian manifold with* $\mathrm{Ric}_g \geq kg$ *for some* $k \in \mathbb{R}$ *and let* $\rho > 0$. *Then there is a sequence* $\{x_\nu\}_{\nu\geq 1} \subset M$ *such that*

(i) $\mu \neq \nu \Longrightarrow B_{\rho/2}(x_\mu) \cap B_{\rho/2}(x_\nu) = \emptyset$.

(ii) *For every $r \geq \rho$, the family $\{B_r(x_\nu)\}_{\nu \geq 1}$ is a uniformly finite covering of M, i.e.,*

(a) $\bigcup_{\nu=1}^{\infty} B_r(x_\nu) = M$.

(b) *There is a positive constant $N = N(n, \rho, r, k)$ and an upper bound of N as a function of n, ρ, r and k, such that each point of M has a neighborhood intersecting at most N of the balls $B_r(x_\nu)$.*

Cf. Lemma 3.4.1 in [Ath], pp. 26–27. The proof of Lemma 5.8 relies on Theorem 5.7.

5.1.3 Poincaré Inequality

Let (M, g) be a n-dimensional ($n \geq 3$) complete Riemannian manifold such that

$$\mathrm{Ric}_g \geq k\, g$$

for some $k \in \mathbb{R}$, and let $\rho > 0$. Through this section, we discuss the validity of the L^1-Poincaré inequality, i.e., the existence of a constant $C = C(n, k, \rho) > 0$ such that for any $0 < r < \rho$ and any $u \in C^\infty(M)$

$$\int_M \left|u(x) - \overline{u}_r(x)\right| d\,\mathrm{v}_g(x) \leq Cr \int_M \left|\nabla u(x)\right| d\,\mathrm{v}_g. \tag{5.11}$$

Here

$$\overline{u}_r(x) := \frac{1}{\mathrm{Vol}_M\big[B_r(x)\big]} \int_{B_r(x)} u(y)\, d\,\mathrm{v}_g(y), \quad x \in M. \tag{5.12}$$

The approach to (5.11) passes through Buser's isoperimetric inequality,[8] that provides a version of (5.11) on a ball $B_r(x)$, namely the existence of a constant $C = C(n, k, \rho) > 0$ such that for any $x \in M$, and any $0 < r < 2\rho$, and any $u \in C^\infty\big(B_r(x)\big)$

$$\int_{B_r(x)} \left|u(y) - \overline{u}_r(y)\right| d\,\mathrm{v}_g(y) \leq C\, r \int_{B_r(x)} |\nabla u(y)|\, d\,\mathrm{v}_g(y). \tag{5.13}$$

The use of (5.13) together with the covering Lemma 5.8 then leads to (5.11).

---

[8] Cf. P. Buser, *A note on the isoperimetric constant*, Ann. Sci. École Norm. Sup., (4)15(1982), 213–230.

5.1.3.1 Buser's Isoperimetric Inequality

Let $x \in M$ and $r > 0$ be fixed. Let $\mathcal{G}$ be the set of all smooth hypersurfaces Γ in $B_r(x)$ such that $\overline{\Gamma} \subset \overline{B_r(x)}$ and there exist open subsets $D_i \subset B_r(x)$ such that

$$B_r(x) \setminus \Gamma = D_1 \cup D_2, \quad D_1 \cap D_2 = \emptyset .$$

Theorem 5.9 *Let (M, g) be a comple Riemannian manifold of dimension n with* $\mathrm{Ric}_g \geq k\, g$ *on M, for some $k < 0$. For every $x \in M$ and any $r > 0$ and any $\Gamma \in \mathcal{G}$, there is a constant $C_B = V_B(n, k, r) > 0$ such that*

$$\min\left\{\mathrm{Vol}_M(D_1),\ \mathrm{Vol}_M(D_2)\right\} \leq C_B\, \mathrm{Area}_g(\Gamma). \tag{5.14}$$

Cf. Theorem 4.1.1 in [Ath], pp. 28–40.

5.1.3.2 L^1-Poincaré Inequality on Balls

Let (M, g) be a n-dimensional complete Riemannian manifold, and let us fix $x \in M$ and $r > 0$.

Definition 5.10 The *isoperimetric constant* is

$$i_\infty = i_\infty\big(B_r(x)\big) := \inf_{\Gamma \in \mathcal{G}} \frac{\mathrm{Area}_g(\Gamma)}{\min\left\{\mathrm{Vol}_g(D_1),\ \mathrm{Vol}_g(D_2)\right\}} \tag{5.15}$$

□

Here Γ is meant to range over all compact $(n-1)$-dimensional C^∞ submanifolds of the ball $B_r(x)$, that divide $B_r(x)$ into two disjoint open submanifolds D_1 and D_2.

Definition 5.11 The *Sobolev constant* is

$$s'_\infty = s'_\infty\big(B_r(x)\big) := \int_{u \in C^\infty(B_r(x))} \frac{\|\nabla u\|_1}{\|u - \overline{u}_r(x)\|_1}. \tag{5.16}$$

□

Here $\|\cdot\|_1$ is the L^1-norm on the ball $B_r(x)$ and

$$\overline{u}_r(x) = \frac{1}{\mathrm{Vol}_g(B_r(x))} \int_{B_r(x)} u(y)\, d\,\mathrm{v}_g(y).$$

We also set

$$s_\infty = s_\infty\big(B_r(x)\big) := \inf_{u\in C^\infty(B_r(x))} \frac{\|\nabla u\|_1}{\inf_{\alpha\in\mathbb{R}} \|u-\alpha\|_1}\,. \tag{5.17}$$

The following result establishes the equivalence between the isoperimetric and Sobolev constants.

Theorem 5.12

$$\frac{i_\infty}{2} \le s'_\infty \le s_\infty \le i_\infty\,. \tag{5.18}$$

Cf. Theorem 4.3.1 in [Ath], pp. 43–45.

5.1.3.3 L^1-Poincaré Inequality on the Whole Manifold

Theorem 5.13 *Let (M, g) be a complete Riemannian manifold such that* $\mathrm{Ric}_g \ge kg$ *for some constant $k\in\mathbb{R}$ and let $R>0$. There is a constant $C = C(n, k, R) > 0$ such that for any $0<r<R$ and any $u\in C^\infty(M)$*

$$\int_M \big|u(x) - \overline{u}_r(x)\big|\, d\,\mathrm{v}_g(x) \le Cr\int_M \big|\nabla u(x)\big|\, d\,\mathrm{v}_g(x). \tag{5.19}$$

Cf. L. Saloff-Coste,[9] and Theorem 4.4.1 in [Ath], pp. 45–48.

5.1.4 Sobolev Inequalities on Complete Riemannian Manifolds

Theorem 5.14 *Let (M, g) be a n-dimensional ($n\ge 3$) complete Riemannian manifold with Ricci curvature bounded from below. The following statements are equivalent:*

(i) *All Sobolev embeddings hold on (M, g).*
(ii) $\inf_{x\in M} \mathrm{Vol}_g\big[B_1(x)\big] > 0$.

Cf. Theorem 5.1.1 in [Ath], p. 49. By the Bishop-Gromov volume comparison theorem the condition (ii) in Theorem 5.14 is equivalent to

$$\inf_{x\in M} \mathrm{Vol}_g\big[B_r(x)\big] > 0$$

for any $r>0$.

[9] L. Saloff-Coste, *Aspects of Sobolev-type inequalities* Cambridge University Press, 2001.

5.1.4.1 The Embedding $H_1^1(M) \hookrightarrow L^{n/(n-1)}(M)$ Implies All Sobolev Embeddings

The following lemma relies on a scaling argument and requires no assumption on the Ricci curvature.

Lemma 5.15 *Let (M, g) be a n-dimensional $(n \geq 3)$ complete Riemannian manifold. Let us assume that the embedding*

$$H_1^1(M) \hookrightarrow L^{n/(n-1)}(M)$$

holds. Then, for any $p, q \in \mathbb{R}$ and any $m, k \in \mathbb{Z}$ with

$$1 \leq q < p, \quad 0 \leq m < k, \quad \frac{1}{p} = \frac{1}{q} - \frac{k-m}{n},$$

the embedding

$$H_k^q(M) \hookrightarrow H_m^p(M)$$

holds.

Cf. Lemma 5.2.1 in [Ath], pp. 50–52.

5.1.4.2 The Case of Compact Riemannian Manifolds

A relatively easier proof of the Sobolev embedding may be provided in the compact case, i.e.,

Theorem 5.16 *Let M be a n-dimensional compact Riemannian manifold. Then*

$$\forall\ p, q \in \mathbb{R}, \quad \forall\ m, k \in \mathbb{Z} \quad \text{with}$$

$$0 \leq m < k, \quad 1 \leq q < n, \quad \frac{1}{p} = \frac{1}{q} - \frac{k-m}{n} \quad :$$

$$H_k^q(M) \hookrightarrow H_m^p(M).$$

Cf. [Ath], pp. 52–54, for a proof of Theorem 5.16.

5.1.4.3 Proof of Theorem 5.14

The proof of Theorem 5.14 relies on the following results.

Theorem 5.17 *Let (M, g) be a n-dimensional complete Riemannian manifold. Let us assume that*

$$H_1^q(M) \hookrightarrow L^p(M)$$

for some $1 \leq q < n$ and $1/p = 1/q - 1/n$. Then, for any $r > 0$, there is a constant $v = v(M, q, r)$ such that

$$\mathrm{Vol}_g\big(B_r(x)\big) \geq v$$

for any $x \in M$.

Cf. Theorem 5.4.1 in [Ath], pp. 55–59.

Theorem 5.18 *Let (M, g) be a n-dimensional complete Riemannian manifold with $\mathrm{Ric}_g \geq kg$ fo some $k \in \mathbb{R}$. Let us assume there is a constant $v > 0$ such that*

$$\mathrm{Vol}_g\big(B_1(x)\big) \geq v, \quad \forall\, x \in M.$$

There exist constants $C = C(n, k, v) > 0$ and $\eta = \eta(n, k, v) \geq v$ such that for any open subset $\Omega \subset M$ with $\partial\Omega \in C^\infty$ and $\overline{\Omega}$ compact

$$\mathrm{Vol}_g(\Omega) \leq \eta \Longrightarrow \Big[\mathrm{Vol}_g(\Omega)\Big]^{(n-1)/n} \leq C\,\mathrm{Area}(\partial\Omega).$$

Cf. Theorem 5.4.3 in [Ath], p. 60.

Corollary 5.19 *Let (M, g) be a n-dimensional complete Riemannian manifold with $\mathrm{Ric}_g \geq kg$ fo some $k \in \mathbb{R}$. Let us assume there is a constant $v > 0$ such that*

$$\mathrm{Vol}_g\big(B_1(x)\big) \geq v, \quad \forall\, x \in M.$$

There exist constants $\delta = \delta(m, k, v) > 0$ and $A = A(n, k, v) > 0$ such that

$$\forall\ x \in M, \quad \forall\ u \in C_0^\infty\big(B_\delta(x)\big)\ :$$

$$\|u\|_{L^n/(n-1)(M)} \leq A\,\|\nabla u\|_{L^1(M)}\,.$$

Cf. Theorem 5.4.4 in [Ath], p. 62.

5.2 Sobolev Inequalities on Manifolds of Negative Curvature

We report on a result by S. Brendle.[10] Let M be a k-dimensional complete, noncompact, Riemannian manifold with nonnegative Ricci curvature.

Definition 5.20 The *asymptotic volume ratio* of M is defined as

$$\theta := \lim_{r\to\infty} \frac{\left|\{p \in M \,:\, d(p,q) < r\}\right|}{\left|\mathbb{B}^k\right| r^k}$$

where $q \in M$ is a fixed point. □

Here $\mathbb{B}^k \subset \mathbb{R}^k$ is the unit ball. By the Bishop-Gromov volume comparison theorem, the limit exists and $\theta \leq 1$. Also the definition on θ doesn't depend upon the choice of $q \in M$. Brendle's main result (cf. *op. cit.*) is a sharp Sobolev inequality on manifolds with nonnegative Ricci curvature.

Theorem 5.21 *Let (M, g) be a n-dimensional, complete, noncompact Riemannian manifold, with nonnegative Ricci curvature. Let $D \subset M$ be a domain and let $f : D \to (0, +\infty)$ be a C^∞ function. Then*

$$\int_D |\nabla f| \, d\,\mathrm{v}_g + \int_{\partial D} f \, d\,\mathrm{v}_g \geq n \left|\mathbb{B}^n\right|^{1/n} \theta^{1/n} \Big(\int_D f^{n/(n-1)} \, d\,\mathrm{v}_g\Big)^{(n-1)/n} \tag{5.20}$$

where θ denotes the asymptotic volume ratio of M.

One may also characterize (cf. S. Brendle, *op. cit.*) equality in the inequality (5.20), as follows.

Theorem 5.22 *Let (M, g) be a n-dimensional, complete, noncompact Riemannian manifold, with nonnegative Ricci curvature. Let $D \subset M$ be a domain. Let us assume that equality is achieved in (5.20) by a C^∞ function $f : D \to (0, +\infty)$ such that $\int_D f^{n/(n-1)} \, d\,\mathrm{v}_g > 0$. Then f is a constant, M is isometric to Euclidean space, and D is a ball.*

Setting $f = 1$ in Theorem 5.21 leads to the following sharp isoperimetric inequality.

Corollary 5.23 *Let M be a n-dimensional, complete, noncompact Riemannian manifold, with nonnegative Ricci curvature. Let $D \subset M$ be a relatively compact domain. Then*

$$|\partial D| \geq n \left|\mathbb{B}^n\right|^{1/n} \theta^{1/n} \, |D|^{(n-1)/n} .$$

[10] Cf. S. Brendle, *Sobolev inequalities in manifolds with nonnegative curvature*, arXiv.2009.13717v5 [math.DG] 28 May 2022.

Brendle's proof pf Theorem 5.21 uses the Alexandrov-Bakelman-Pucci method, cf. X. Cabré.[11]

5.3 Sobolev Inequalities with Euclidean Best Constant

We report on work by M. Ledoux.[12] Let M be a complete Riemannian manifold, of dimension $n \geq 2$. Ledoux is concerned (cf. *op. cit.*) with Sobolev inequalities of the sort

$$\Big(\int_M |f|^p \, d\,\mathrm{v}_g\Big)^{1/p} \leq C\Big(\int_M |\nabla f|^q \, d\,\mathrm{v}_g\Big)^{1/q}, \tag{5.21}$$

$$1 \leq q < n, \quad \frac{1}{p} = \frac{1}{q} - \frac{1}{n},$$

that hold, for some constant $C > 0$, for every $f \in C_0^\infty(M)$. The best constant $C = K(n, q)$ for which (5.21) holds in the Euclidean setting $M = \mathbb{R}^n$ is well known, and is due to G. Talenti.[13] Precisely[14]

$$K(n, 1) = \frac{1}{n\,(\omega_n)^{1/n}}, \quad \omega_n = \mathrm{Vol}\big(\mathbb{B}^n\big), \tag{5.22}$$

$$K(n, q) = \frac{1}{n}\Big[\frac{n(q-1)}{n-q}\Big]^{(q-1)/q}\Big[\frac{\Gamma(n+1)}{n\,\omega_n\,\Gamma(n/q)\,\Gamma(n+1-n/q)}\Big]^{1/n}, \quad q > 1.$$

M. Ledoux looks at the geometry of the complete Riemannian manifolds on which at least one of the Sobolev inequalities (5.21) is satisfied with $C = K(n, q)$. M. Ledoux's main result is (cf. *op. cit.*)

Theorem 5.24 *Let M be a complete n-dimensional Riemannian manifold* ($n \geq 2$) *with nonnegative Ricci curvature. If one of the Sobolev inequalities* (5.21) *holds with $C = K(n, q)$, then M is isometric to $\mathbb{R}^n$.*

The case $q = 1$ [and $p = n/(n-1)$] was previously known, and M. Ledoux generalizes that to $1 < q < n$. It should be observed that when $q = 1$ the proof

[11] X. Cabré, *Elliptic PDEs in probability and geometry. Symmetry and regularity of solutions*, Discrete Cont. Dyn. Systems A, 20(2008), 425–457.

[12] Cf. M. Ledoux, *On manifolds with nonnegative Ricci curvature and Sobolev inequalities*, Communications in Analysis and Geometry, (2)7(1999), 347–353.

[13] Cf. G. Talenti, *Best constants in Sobolev inequality*, Ann. di Matem. Pura ed Appl., 110(1976), 353–372.

[14] As given by Th. Aubin, *Nonlinear Analysis on Manifolds. Monge-Ampére equations*, Springer, 1982, or earlier in this chapter.

of Ledoux's theorem is quite easy. Indeed, when $q = 1$ the Sobolev inequality is equivalent to the isoperimetric inequality

$$\left[\mathrm{Vol}_n(\Omega)\right]^{(n-1)/n} \leq K(n, 1)\,\mathrm{Vol}_{n-1}(\partial\Omega)$$

for any smoothly bounded open set $\Omega \subset M$. If $x_0 \in M$ is fixed and $B(s) = B_s(x_0) \subset M$ is the geodesic ball of center x_0 and radius $s > 0$, then

$$\frac{d}{ds}\mathrm{Vol}_n(B(s)) = \mathrm{Vol}_{n-1}\big(\partial\, B(s)\big).$$

Hence [by the isopetrimetric inequality with $\Omega = B(s)$],

$$V(s)^{(n-1)/n} \leq K(n, 1)\, V'(s),$$

where $V(s) := \mathrm{Vol}_n(B(s))$. Integration yields

$$V(s) \geq \left(\frac{s}{n\,K(n, 1)}\right)^n$$

so that [by (5.22)]

$$V(s) \geq V_0(s) \tag{5.23}$$

where $V_0(s) = \omega_n\, s^n$ is the volume of the Euclidean ball of radius s in $\mathbb{R}^n$. If M has nonnegative Ricci curvature, then one may apply Bishop's comparison theorem[15] to conclude that $V(s) \leq V_0(s)$ so that [by (5.23) and the case of equality] M is isometric to $\mathbb{R}^n$. Therefore, the main interest in Ledoux's theorem is with the case $q > 1$.

The remaining sections of the present chapter are devoted to Sobolev inequalities on submanifolds.

5.4 Hoffman–Spruck Theorem

Let M and $\overline{M}$ be two Riemannian manifolds, respectively, of dimensions m and n, and let $f : M \to \overline{M}$ be an isometric immersion. Through the section, we adopt the following notations.

$\overline{K}$ sectional curvature of $\overline{M}$.

H mean curvature vector of the given immersion f.

$\overline{R}(M)$ injectivity radius of $\overline{M}$ restricted[16] to M.

---

[15] Cf. e.g., I. Chavel, *Riemannian geometry: a modern introduction*, Cambridge University Press, 1993.

[16] The smallest distance from the points of M to the cut locus of $\overline{M}$.

ω_m volume of the unit ball in $\mathbb{R}^m$.

$b \in (0, +\infty) \cup i\mathbb{R}$ positive number or purely imaginary complex number.

The following result is due to D. Hoffman & J. Spruck (cf. [HoSp]).

Theorem 5.25 (Sobolev's inequality) *Let us assume that* $\overline{K} \leq b^2$ *and let* $h : M \to [0, +\infty)$ *be a* C^1 *function such that* $h \equiv 0$ *on* ∂M. *Let* $0 < \alpha < 1$. *Let us assume that*

$$b^2(1-\alpha)^{-2/m}\left[\frac{\mathrm{Vol}\big(\mathrm{supp}(h)\big)}{\omega_m}\right]^{2/m} \leq 1, \tag{5.24}$$

$$2\rho_0 \leq \overline{R}(M), \tag{5.25}$$

where

$$\rho_0 = \begin{cases} \dfrac{1}{b}\arcsin\left\{b(1-\alpha)^{-1/m}\left[\dfrac{\mathrm{Vol}\big(\mathrm{supp}(h)\big)}{\omega_m}\right]^{1/m}\right\} & \text{if } b > 0, \\[2ex] (1-\alpha)^{-1/m}\left[\dfrac{\mathrm{Vol}\big(\mathrm{supp}(h)\big)}{\omega_m}\right]^{1/m} & \text{if } b \in i\mathbb{R}. \end{cases}$$

Then

$$\left(\int_M h^{m/(m-1)}\, dV_M\right)^{(m-1)/m} \leq c(m,\alpha)\int_M [|\nabla h| + h|H|]\, dV_M \tag{5.26}$$

where

$$c(m,\alpha) = \begin{cases} \dfrac{\pi}{2}\, 2^{m-2}\, \alpha^{-1}(1-\alpha)^{-1/m}\, \dfrac{m}{m-1}\, \omega_m^{-1/m} & \text{se } b > 0, \\[2ex] 2^{m-2}\, \alpha^{-1}(1-\alpha)^{-1/m}\, \dfrac{m}{m-1}\, \omega_m^{-1/m} & \text{se } b \in i\,\mathbb{R}. \end{cases}$$

The optimal choice of the parameter $0 < \alpha < 1$ to minimize the constant $c(m,\alpha)$ is

$$\alpha = m/(m+1).$$

When $b > 0$ the condition (5.25) may be replaced by the stronger requirement

$$\overline{R} \geq \pi\, b^{-1}.$$

It is well known (cf. e.g., R. Bishop & R. Crittenden, [BiCr]) that $\overline{R} = \pi b^{-1}$ for any complete and simply connected n-dimensional Riemannian manifold $\overline{M}$ with $\overline{K} = b^2$ [e.g., the n-dimensional sphere of radius $1/b$, i.e., $\overline{M} = S^n(b^{-1}) \subset \mathbb{R}^{n+1}$].

It is equally well known (cf. e.g., D. Gromoll & W. Klingenberg & W. Meyer, [GrKlMe]) that $\overline{R} \geq \pi b^{-1}$ for any complete and simply connected n-dimensional Riemannian manifold $\overline{M}$ with

$$\frac{1}{4} b^2 < \overline{K} \leq b^2 .$$

In general, one may replace (5.25) by a requirement of the form $\overline{R} \geq \epsilon b^{-1}$ if the constant 1 in (5.24) is replaced by a suitable smaller constant.

If $\overline{M}$ is complete and simply connected with b purely imaginary [$b = ia$ with $a \neq 0$ and then $\overline{K} \leq -a^2$] then $\overline{R} = +\infty$ [hence the requirements (5.24)–(5.24) are automatically satisfied].

Theorem 5.26 (Isoperimetric inequality)
Let us assume that M is a compact manifold with boundary $\partial M \neq \emptyset$ and that $\overline{K} \leq b^2$. Moreover, let us assume that

$$b^2(1-\alpha)^{-2/m} \left[\frac{\mathrm{Vol}(M)}{\omega_m}\right]^{2/m} \leq 1, \tag{5.27}$$

$$2\rho_0 \leq \overline{R}(M), \tag{5.28}$$

where

$$\rho_0 = \begin{cases} \dfrac{1}{b} \arcsin\left\{ b(1-\alpha)^{-1/m} \left[\dfrac{\mathrm{Vol}(M)}{\omega_m}\right]^{1/m} \right\} & \text{se } b > 0, \\[2ex] (1-\alpha)^{-1/m} \left[\dfrac{\mathrm{Vol}(M)}{\omega_m}\right]^{1/m} & \text{se } b \in i\mathbb{R}. \end{cases}$$

Then

$$\mathrm{Vol}(M)^{(m-1)/m} \leq c(m) \left[\mathrm{Vol}(\partial M) + \int_M |H| \, d\, V_M \right] \tag{5.29}$$

Theorem 5.26 follows from Theorem 5.25 by approximating the characteristic function of M by a C^1 function $h : M \to [0, +\infty)$ vanishing on ∂M.

Now, let us demonstrate a few geometric applications of Theorem 5.26. Let us assume that $|H| \leq \Lambda$ for some constant $\Lambda > 0$.

Corollary 5.27 *Let us assume that* $\partial M = \emptyset$.

(i) *If* $b > 0$ *and* $\overline{R} \geq \pi b^{-1}$, *then*

$$\operatorname{Vol}(M) \leq (1-\alpha)\omega_m b^{-m} \Longrightarrow \tag{5.30}$$

$$\Longrightarrow \Lambda \geq c(m)^{-1} \operatorname{Vol}(M)^{-1/m} \geq 2^{-(m-3)} \frac{m-1}{m} \alpha \pi b^{-1} .$$

(ii) *If* $b \in i\,\mathbb{R}$, *then*

$$\operatorname{Vol}(M) \leq (1-\alpha)\omega_m \left(\frac{\overline{R}}{2}\right)^m \Longrightarrow$$

$$\Longrightarrow \Lambda \geq c(m)^{-1} \operatorname{Vol}(M)^{-1/m} \geq 2^{-(m-3)} \frac{m-1}{m} \alpha \overline{R}^{-1} .$$

Corollary 5.28 (i) *Let us assume that* $\overline{M}$ *is complete and simply connected and* $\frac{1}{4} b^2 < \overline{K} \leq b^2$ *with* $b > 0$. *Then, for any closed minimal submanifold* M *of* $\overline{M}$

$$\operatorname{Vol}(M) \geq \omega_m b^{-m} . \tag{5.31}$$

(ii) *If* $\overline{M}$ *is complete and simply connected and* $\overline{K} \leq b^2$ *with* $b \in i\,\mathbb{R}$, *then* $\overline{M}$ *doesn't admit closed minimal submanifolds. Moreover, any closed submanifold* M *of* $\overline{M}$ *satisfies*

$$\Lambda \geq c(m)^{-1} \operatorname{Vol}(M)^{-1/m} . \tag{5.32}$$

Covariant Differential Calculus Relative to $f : M \to \overline{M}$

Let $f : M \to \overline{M}$ be a C^∞ map between the Riemannian manifolds M and $\overline{M}$, of dimensions m and n, respectively, equipped with the Riemannian metrics g and $\overline{g}$. Let ∇ and $\overline{\nabla}$ be the Levi-Civita connections of (M, g) and $(\overline{M}, \overline{g})$. Let $f^{-1}T(\overline{M}) \to M$ be the pullback bundle of $T(\overline{M}) \to \overline{M}$ via $f : M \to \overline{M}$. If $X \in \mathfrak{X}(M)$ we denote by $f_*X \in C^\infty(M,\ f^{-1}T\overline{M})$ the section in the pullback bundle defined by

$$\big(f_*X\big)(x) = (d_x f)X_x \in T_{f(x)}(\overline{M}) = \big(f^{-1}T\overline{M}\big)_x, \quad x \in M.$$

Let $f^{-1}\overline{\nabla} = \overline{\nabla}^f$ be the pullback of the connection $\overline{\nabla}$ via f, a connection in the pullback bundle as a vector bundle, i.e., $\overline{\nabla}^f \in \mathcal{C}\big(f^{-1}T\overline{M}\big)$. For any $V \in C^\infty\big(f^{-1}T\overline{M}\big)$, we consider the map

$$\overline{\nabla}^f V : \mathfrak{X}(M) \to C^\infty\big(f^{-1}T\overline{M}\big),$$

$$Y \in \mathfrak{X}(M) \longmapsto \overline{\nabla}^f_Y V.$$

Definition 5.29 $\overline{\nabla}^f_Y V$ is the *gradient* of $V \in C^\infty(f^{-1}T\overline{M})$. □

Definition 5.30 The *divergence*

$$\operatorname{div}_M V \in C^\infty\big(f^{-1}T\overline{M}\big)$$

of $V \in C^\infty\big(f^{-1}T\overline{M}\big)$ is the trace of

$$\overline{\nabla}^f V : \mathfrak{X}(M) \to C^\infty\big(f^{-1}T\overline{M}\big),$$

i.e.,

$$\operatorname{div}_M V = \operatorname{trace}\left(\overline{\nabla}^f V\right).$$

□

Here we follow the terminology in [HoSp]. The last definition is a bit ambiguous, because $\overline{\nabla}^f V$ isn't an endomorphism of $\mathfrak{X}(M)$. Yet, if we consider the pullback

$$f^{-1}\overline{g} =: \overline{g}^f$$

of the metric $\overline{g}$ via f [a Riemannian bundle metric on the vector bundle $f^{-1}T(\overline{M}) \to M$], then one may build the bilinear form

$$(Y, Z) \in \mathfrak{X}(M) \times \mathfrak{X}(M) \longmapsto \overline{g}^f\left(\overline{\nabla}^f_Y V,\ f_* Z\right).$$

Hence, its trace will depend, unlike that of an endomorphism, on a fixed inner product, in the case at hand the Riemannian metric g. Let us write a formula for the local calculation of said trace. Let $x \in M$ be a point, and let $\{E_j : 1 \le j \le m\}$ be a local g-orthonormal frame in $T(M)$, defined on the open subset $U \subset M$ of the point x

$$E_j \in C^\infty\big(U,\ T(M)\big),$$

$$g(E_j,\ E_k) = \delta_{jk}, \quad 1 \le j, k \le m.$$

Next, we set by definition

$$(\mathrm{div}_M V)_x = \sum_{j=1}^{m} \overline{g}^f \left(\overline{\nabla}^f_{E_j} V,\ f_* E_j \right)_x .$$

The definition of $(\mathrm{div}_M V)_x$ doesn't depend upon the choice of local g-orthonormal frame about the point x. If U is also the domain of a local coordinate system $x^j : U \to \mathbb{R}$, as we may assume w.l.o.g., then

$$\mathrm{div}_M V = g^{jk} \overline{g}^f \left(\overline{\nabla}^f_{\partial_j} V,\ f_* \partial_k \right),$$

$$\left[g^{jk} \right] = \left[g_{jk} \right]^{-1}, \quad g_{jk} = g\left(\partial_j,\ \partial_k \right), \quad \partial_j = \partial / \partial x^j,$$

on U.

Definition 5.31 The *second fundamental form* of f is

$$B(X, Y) = \overline{\nabla}^f_X f_* Y - f_* \nabla_X Y, \quad X, Y \in \mathfrak{X}(M).$$

The *tension field* of f is

$$\tau(f) = \mathrm{trace}_g\, B \in C^\infty\left(f^{-1} T\overline{M} \right).$$

□

Lemma 5.32 *Let us assume that* $f : M \to \overline{M}$ *is an isometric immersion and let* $N(f) \to M$ *be the normal bundle of* f*, so that*

$$\left(f^{-1} T\overline{M} \right)_x = [(d_x f) T_x(M)] \oplus N(f)_x, \quad x \in M. \tag{5.33}$$

Let

$$\tan_x : T_{f(x)}(\overline{M}) \to T_x(M), \quad \mathrm{nor}_x : T_{f(x)}(\overline{M}) \to N(f)_x,$$

be the projections associated to the direct sum decomposition (5.33). *If* $V \in C^\infty \left(f^{-1} T\overline{M} \right)$*, then we set* $V^\top = \tan(V)$ *and* $V^\perp = \mathrm{nor}(V)$ *so that* $V = f_* V^\top + V^\perp$.

(i) *One has*

$$\nabla_X Y = \tan\left(\overline{\nabla}^f_X f_* Y \right), \quad B(X, Y) = \mathrm{nor}\left(\overline{\nabla}^f_X f_* Y \right), \tag{5.34}$$

for any $X, Y \in \mathfrak{X}(M)$.

(ii) *If we set*

$$a_\xi X = -\tan\left(\overline{\nabla}^f_X \xi\right), \tag{5.35}$$

$$X \in \mathfrak{X}(M), \quad \xi \in C^\infty\big(N(f)\big),$$

then

$$\overline{g}^f\left(B(X,Y),\,\xi\right) = g\big(a_\xi X,\,Y)\big). \tag{5.36}$$

(iii) *The following identities hold good*

$$\operatorname{div}_M\big(f_* V^\top\big) = \operatorname{div}_M V + \overline{g}^f\left(V,\ \tau(f)\right), \tag{5.37}$$

$$\operatorname{div}_M(\varphi V) = \varphi \operatorname{div}_M V + \overline{g}^f\left(V,\ f_*\nabla\varphi\right) = \tag{5.38}$$

$$= \varphi \operatorname{div}_M V + g\big(V^\top,\ \nabla\varphi\big)$$

for any $\varphi \in C^1(M)$. *Here* $\nabla\varphi \in \mathfrak{X}(M)$ *is the gradient of* φ *with respect to* g, *i.e.,* $g(\nabla\varphi, X) = X(\varphi)$ *for any* $X \in \mathfrak{X}(M)$.

(iv) *If* M *is a manifold with boundary* $\partial M \neq \emptyset$ *and* ν *is the outward unit normal on* ∂M, *then*

$$\int_M \operatorname{div}_M(V)\, d\,\mathrm{v}_M = \tag{5.39}$$

$$= -\int_M \overline{g}^f\left(V,\ \tau(f)\right) d\,\mathrm{v}_M + \int_{\partial M} \overline{g}^f\left(V,\ f_*\,\nu\right) d\,\mathrm{v}_{\partial M}\,.$$

(v) *Let us set*

$$\operatorname{Vol}(M) = \frac{1}{2}\int_M \|df\|^2\, d\,\mathrm{v}_M\,.$$

Let

$$\{f_t\}_{|t|<\epsilon} \subset C^\infty(M, \overline{M})$$

be a smooth 1*-parameter variation of* $f_0 = f$ *and let*

$$V \in C^\infty\big(f^{-1}T\overline{M}\big)$$

be the infinitesimal variation induced by the variation $\{f_t\}_{|t|<\epsilon}$. *In addition to that, let us assume that* $\text{Supp}(V)$ *is a compact subset in* M. *Then*

$$\frac{d}{dt}\left\{\text{Vol}\big(f_t(M)\big)\right\}_{t=0} = \int_M \text{div}_M\big(V\big)\, d\,\text{v}_M\,. \tag{5.40}$$

(vi) *Consequently*

$$\frac{d}{dt}\left\{\text{Vol}\big(f_t(M)\big)\right\}_{t=0} = \tag{5.41}$$

$$= -\int_M \overline{g}^f\big(V,\ \tau(f)\big)\, d\,\text{v}_M + \int_{\partial M} \overline{g}^f\big(V,\ f_*\nu\big)\, d\,\text{v}_{\partial M}$$

so that the formula (5.37) *is an infinitesimal version of the first variation formula* (5.41).

Proof (iii) For any $Y \in \mathfrak{X}(M)$ and any $V \in C^\infty\big(f^{-1}T\overline{M}\big)$,

$$\overline{\nabla}^f_Y V = \overline{\nabla}^f_Y f_* V^\top + \overline{\nabla}^f_Y V^\perp$$

so that

$$\overline{g}^f\big(\overline{\nabla}^f_Y f_* V^\top,\ f_*Y\big) = \overline{g}^f\big(\overline{\nabla}^f_Y V,\ f_*Y\big) - \overline{g}^f\big(\overline{\nabla}^f_Y V^\perp,\ f_*Y\big)$$

hence (locally)

$$\text{div}_M\big(f_*V^\top\big) = \sum_{j=1}^m \overline{g}^f\big(\overline{\nabla}^f_{E_j} f_* V^\top,\ f_*E_j\big) =$$

$$= \text{div}_M V - \sum_{j=1}^m \overline{g}^f\big(\overline{\nabla}^f_{E_j} V^\perp,\ f_*E_j\big) =$$

$$= \text{div}_M V - \sum_{j=1}^m \overline{g}^f\big(f_*\, a_{V^\perp} E_j,\ f_*E_j\big) =$$

(as $f^*\overline{g} = g$)

$$= \text{div}_M V - \sum_{j=1}^m g\big(a_{V^\perp} E_j,\ E_j\big) =$$

[by (5.36)]

$$= \operatorname{div}_M V - \sum_{j=1}^{m} \overline{g}^f \big(B(E_j,\ E_j),\ V^\perp\big) =$$

$$= \operatorname{div}_M V - \overline{g}^f \big(\operatorname{trace}_g B,\ V^\perp\big).$$

The identity (5.37) is proved. Moreover,

$$\overline{\nabla}^f_X\big(\varphi V\big) = X(\varphi)\, V + \varphi\, \overline{\nabla}^f_X V$$

so that

$$\operatorname{div}_M\big(\varphi V\big) = \sum_{j=1}^{m} \overline{g}^f \big(\overline{\nabla}^f_{E_j}(\varphi V),\ f_* E_j\big) =$$

$$= \sum_{j=1}^{m} E_j(\varphi)\, \overline{g}^f\, \big(V,\ f_* E_j\big) + \varphi \operatorname{div}_M V =$$

[as locally $\nabla\varphi = \sum_{j=1}^{m} E_j(\varphi)\, E_j$]

$$= \overline{g}^f \big(V,\ f_* \nabla\varphi\big) + \varphi \operatorname{div}_M V = g\big(V^\top,\ \nabla\varphi\big) + \varphi \operatorname{div}_M V.$$

The identity (5.38) is proved.

(iv) One has

$$\operatorname{div}_M\big(f_* V^\top\big) = \sum_{j=1}^{m} \overline{g}^f \big(\overline{\nabla}^f_{E_j} f_* V^\top,\ f_* E_j\big) =$$

$$= \sum_{j=1}^{m} \overline{g}^f \big(f_* \nabla_{E_j} V^\top,\ f_* E_j\big) =$$

$$= \sum_{j=1}^{m} g\big(\nabla_{E_j} V^\top,\ E_j\big) = \operatorname{div}_g\big(V^\top\big)$$

where $\operatorname{div}_g : \mathfrak{X}(M) \to C^\infty(M)$ is the ordinary divergence of a tangent vector field on M, i.e.,

$$\mathcal{L}_X\, d\,\mathrm{v}_M = \operatorname{div}_g(X)\, d\,\mathrm{v}_M$$

and $\mathcal{L}_X$ is the Lie derivative in the direction $X \in \mathfrak{X}(M)$. By Green's lemma

$$\int_M \operatorname{div}_g(V^\top)\, d\,\mathrm{v}_M = \int_{\partial M} g(V^\top,\, \nu)\, d\,\mathrm{v}_M$$

hence

$$\int_M \operatorname{div}_M(f_* V^\top)\, d\,\mathrm{v}_M = \int_{\partial M} g(V^\top,\, \nu)\, d\,\mathrm{v}_M$$

or [by (5.37)]

$$\int_M \left\{\operatorname{div}_M V + \overline{g}^f(V,\, \tau(f))\right\} d\,\mathrm{v}_M = \int_{\partial M} g(V^\top,\, \nu)\, d\,\mathrm{v}_M\,.$$

The identity (5.39) is proved.

v) Let us set $\tilde{M} = M \times (-\epsilon, \epsilon)$ and let us consider the map $F : \tilde{M} \to \overline{M}$ defined by

$$F(x, t) = f_t(x), \quad x \in M, \quad |t| < \epsilon.$$

Then $F \in C^\infty(\tilde{M},\, \overline{M})$. The infinitesimal variation $V \in C^\infty(f^{-1}T\overline{M})$ induced by the 1-parameter variation $\{f_t\}_{|t|<\epsilon}$ is given by

$$V_x = \left(d_{(x,0)} F\right)\left(\frac{\partial}{\partial t}\right)_{(x,0)}, \quad x \in M.$$

We shall make use of the vector bundle

$$F^{-1}T(\overline{M}) \to M$$

[the pullback of $T(\overline{M}) \to \overline{M}$ via F], of the Riemannian bundle metric

$$F^{-1}\overline{g} = \overline{g}^F$$

on $F^{-1}T(\overline{M}) \to M$, and of the pullback connection

$$F^{-1}\overline{\nabla} = \overline{\nabla}^F \in \mathcal{C}\left(F^{-1}T(\overline{M}),\, \overline{g}^F\right).$$

For any $|t| < \epsilon$, we consider the injection

$$\alpha_t : M \to \tilde{M}, \quad \alpha_t(x) = (x, t), \quad x \in M,$$

so that $f_t = F \circ \alpha_t$. For any $X \in \mathfrak{X}(M)$, we consider the tangent vector field $\tilde{X}$ on $\tilde{M}$ defined by

$$\tilde{X}_{(x,t)} = (d_x \alpha_t) X_x, \quad x \in M, \quad |t| < \epsilon.$$

Moreover,

$$d_x f_t = (d_{(x,t)}F) \circ (d_x \alpha_t)$$

and hence for any $X \in \mathfrak{X}(M)$

$$\left\{(f_t)_* X\right\}_x = (d_x f_t) X_x = (d_{(x,t)}F) \circ (d_x \alpha_t) X_x =$$

$$= (d_{(x,t)}F)\tilde{X}_{(x,t)} = \left\{F_* \tilde{X}\right\}_{(x,t)}$$

or

$$\left(f_t\right)_* X = \left(F_* \tilde{X}\right) \circ \alpha_t, \tag{5.42}$$

$$X \in \mathfrak{X}(M), \quad |t| < \epsilon.$$

Let us observe that

$$\overline{g}_x^{f_t} = \overline{g}_{f_t(x)} = \overline{g}_{F(\alpha_t(x))} = \overline{g}^F_{(x,t)} \, .$$

A common trick, that we adopt here as well, is to express the Hilbert-Schmidt norm $\|df_t\|$ in terms of F as follows:

$$\|df_t\|_x^2 = \sum_{j=1}^{m} \overline{g}^{f_t}\left(\left(f_t\right)_* E_j, \, \left(f_t\right)_* E_j\right)_x =$$

$$= \sum_{j=1}^{m} \overline{g}^F_{(x,t)}\left(\left\{F_* \tilde{E}_j\right\}_{(x,t)}, \, \left\{F_* \tilde{E}_j\right\}_{(x,t)}\right) =$$

$$= \sum_{j=1}^{m} \overline{g}^F\left(F_* \tilde{E}_j, \, F_* \tilde{E}_j\right)_{(x,t)}$$

or

$$\|df_t\|^2 = \sum_{j=1}^{m} \overline{g}^F\left(F_* \tilde{E}_j, \, F_* \tilde{E}_j\right) \circ \alpha_t \tag{5.43}$$

on U. Let us consider now the functions

$$\text{Vol} : (-\epsilon, \epsilon) \to \mathbb{R}, \quad \varphi : \tilde{M} \to \mathbb{R},$$

$$\text{Vol}(t) = \text{Vol}\left[f_t(M)\right], \quad \varphi(x, t) = \frac{1}{2} \|df_t\|_x^2,$$

$$x \in M, \quad |t| < \epsilon,$$

so that

$$\text{Vol}(t) = \int_M \varphi(x, t) \, d\,\mathrm{v}_M(x)$$

and hence

$$\frac{d}{dt}\left\{\text{Vol}\left[f_t(M)\right]\right\}_{t=0} = \text{Vol}'(0) = \int_M \frac{\partial \varphi}{\partial t}(x, 0) \, d\,\mathrm{v}_M(x).$$

We compute separately $(\partial\varphi/\partial t)(x, 0)$ for any $x \in M$ by choosing a local g-orthonormal frame

$$\{E_j : 1 \le j \le m\} \subset C^\infty(U, T(M))$$

defined on an open neighborhood $U \subset M$ of the point x. We may use the formula (5.43) that is

$$\varphi = \frac{1}{2} \sum_{j=1}^{m} \overline{g}^F\left(F_* \tilde{E}_j, \, F_* \tilde{E}_j\right)$$

on $\tilde{U} = U \times (-\epsilon, \epsilon)$. It follows that

$$\frac{\partial \varphi}{\partial t} = \frac{1}{2} \sum_{j=1}^{m} \frac{\partial}{\partial t}\left[\overline{g}^F\left(F_* \tilde{E}_j, \, F_* \tilde{E}_j\right)\right] =$$

[as $\overline{\nabla}^F \overline{g}^F = 0$]

$$= \sum_j \overline{g}^F\left(\overline{\nabla}^F_{\partial/\partial t} F_* \tilde{E}_j, \, F_* \tilde{E}_j\right) =$$

[as the symmetry of the linear connection $\overline{\nabla}$ implies

$$\overline{\nabla}^F_{\partial/\partial t} F_* \tilde{X} = \overline{\nabla}^F_{\tilde{X}} F_* \partial/\partial t + \left[\partial/\partial t, \tilde{X}\right]$$

and $\left[\partial/\partial t, \tilde{X}\right] = 0$ for any $X \in \mathfrak{X}(M)$]

$$= \sum_j \overline{g}^F \left(\overline{\nabla}^F_{\tilde{E}_j} F_* (\partial/\partial t), F_* \tilde{E}_j\right)$$

and hence [as

$$\overline{g}^F_{(x,0)} = \overline{g}^f_x, \quad \left\{\overline{\nabla}^F_{\tilde{X}} F_* (\partial/\partial t)\right\}_{(x,0)} = \left\{\overline{\nabla}^f_X V\right\}_x,$$

for any $x \in M$]

$$\frac{\partial\varphi}{\partial t}(x, 0) = \sum_j \overline{g}^f \left(\overline{\nabla}^f_{E_j} V, f_* E_j\right)_x = \operatorname{div}_M(V)_x$$

so that the formula (5.40) is proved. Q.e.d.

Lemma 5.33 *Let $E : C^\infty(M, \overline{M}) \to \mathbb{R}$ be the Dirichlet energy functional, i.e.,*

$$E(f) = \frac{1}{2} \int_M \|df\|^2 \, d\,\mathrm{v}_M$$

for any $f \in C^\infty\left(M, \tilde{M}\right)$. Then, for any smooth 1-parameter variation

$$\{f_t\}_{|t|<\epsilon} \subset C^\infty(M, \overline{M})$$

whose associated infinitesimal variation $V \in C^\infty(f^{-1} T\overline{M})$ has compact support in M, the following first variation formula holds

$$\frac{d}{dt} \left\{E\left(f_t\right)\right\}_{t=0} = \tag{5.44}$$

$$= -\int_M \overline{g}^f \left(V, \tau(f)\right) d\,\mathrm{v}_M + \int_{\partial M} \overline{g}^f \left(V, f_* \nu\right) d\,\mathrm{v}_{\partial M}.$$

Proof Let us consider the functions

$$\mathcal{E} : (-\epsilon, \epsilon) \to \mathbb{R}, \quad \varphi : \tilde{M} \to \mathbb{R},$$

$$\mathcal{E}(t) = E\big(f_t\big), \quad \varphi(x, t) = \frac{1}{2} \, \|df_t\|_x^2,$$

$$x \in M, \quad |t| < \epsilon.$$

Part of the proof of Lemma 5.32, preceding and up to the derivation of the formula

$$\frac{\partial \varphi}{\partial t} = \sum_{j=1}^{m} \overline{g}^F \big(\overline{\nabla}_{\tilde{E}_j} F_*(\partial/\partial t), \ F_* \tilde{E}_j\big), \tag{5.45}$$

doesn't rely on the assumption that $f : M \to \overline{M}$ be an isometric immersion, hence we may use (5.45) under assumptions more general than those in Lemma 5.33. As

$$\overline{\nabla}^F \overline{g}^F = 0$$

the formula (5.45) leads to

$$\frac{\partial \varphi}{\partial t} = \sum_{j=1}^{m} \Big\{\tilde{E}_j \big(\overline{g}^F \big(F_* \, (\partial/\partial t), \ F_* \tilde{E}_j\big)\big) - \overline{g}^F \big(F_* \, (\partial/\partial t), \ \overline{\nabla}^F_{\tilde{E}_j} F_* \tilde{E}_j\big)\Big\}.$$

For any $|t| < \epsilon$, we consider the tangent vector field $X_t \in \mathfrak{X}(M)$ uniquely determined by

$$g(X_t, \ Y) = \overline{g}^F \big(F_* \, (\partial/\partial t), \ F_* \tilde{Y}\big) \circ \alpha_t \tag{5.46}$$

for any $Y \in \mathfrak{X}(M)$. We wish to compute the ordinary divergence of X_t on the Riemannian manifold (M, g)

$$\operatorname{div}_g\big(X_t\big) = \sum_{j=1}^{m} g\left(\nabla_{E_j} X_t, \ E_j\right) =$$

[as $\nabla g = 0$]

$$= \sum_j \left\{E_j \left(g\big(X_t, \ E_j\big)\right) - g\big(X_t, \ \nabla_{E_j} E_j\big)\right\} =$$

[by the definition (5.46)]

$$= \sum_j \Big\{ E_j\big(\overline{g}^F\big(F_*(\partial/\partial t),\ F_*\tilde{E}_j\big)\circ\alpha_t\big) -$$

$$-\overline{g}^F\big(F_*(\partial/\partial t),\ F_*\widetilde{\nabla_{E_j}E_j}\big)\circ\alpha_t\Big\} =$$

[as

$$\{E_j\big(u\circ\alpha_t\big)\}_x = E_{j,x}\big(u\circ\alpha_t\big) = \big[(d_x\alpha_t)E_{j,x}\big](u) = \tilde{E}_{j,(x,t)}(u)$$

or

$$E_j\big(u\circ\alpha_t\big) = \tilde{E}_j(u)\circ\alpha_t$$

for any $u\in C^\infty(\overline{M})$ and any $|t|>\epsilon$]

$$= \sum_j \big\{\tilde{E}_j\big(\overline{g}^F\big(F_*(\partial/\partial t),\ F_*\tilde{E}_j\big)\big) - \overline{g}^F\big(F_*(\partial/\partial t),\ F_*\widetilde{\nabla_{E_j}E_j}\big)\big\}\circ\alpha_t$$

and hence

$$\sum_{j=1}^m \tilde{E}_j\big(\overline{g}^F\big(F_*(\partial/\partial t),\ F_*\tilde{E}_j\big)\big)\circ\alpha_t = \tag{5.47}$$

$$= \operatorname{div}_g(X_t) + \sum_{j=1}^m \overline{g}^F\big(F_*(\partial/\partial t),\ F_*\widetilde{\nabla_{E_j}E_j}\big)\circ\alpha_t\,.$$

It follows [by (5.47)] that

$$\frac{\partial\varphi}{\partial t} = \operatorname{div}_g\big(X_t\big) - \overline{g}^F\big(F_*(\partial/\partial t),\ \operatorname{trace}_g B(F)\big)\circ\alpha_t \tag{5.48}$$

where the $C^\infty(M)$-bilinear form $B(F)$ is defined by

$$B(F)(X,Y) = \overline{\nabla}^F_{\tilde{X}}F_*\tilde{Y} - F_*\widetilde{\nabla_X Y},$$

$$X,Y\in\mathfrak{X}(M).$$

Let us evaluate (5.48) at the points $(x,0)$ with $x\in M$. We obtain

$$\frac{\partial\varphi}{\partial t}(x,0) = \operatorname{div}_g\big(X_0\big)_x - \overline{g}^f\big(V,\ \operatorname{trace}_g B(f)\big) \tag{5.49}$$

where we have set

$$B(f)(X, Y) = \overline{\nabla}^f_X f_* Y - f_* \nabla_X Y,$$

$$X, Y \in \mathfrak{X}(M).$$

Let us integrate (5.49) along M and take into account that [by Green's lemma]

$$\int_M \operatorname{div}_g(X_0)\, d\,\mathrm{v}_M = \int_{\partial M} g(X_0,\ \nu)\, d\,\mathrm{v}_{\partial M} =$$

[by the definition (5.46) with $t = 0$]

$$= \int_{\partial M} \left\{ \overline{g}^F\left(F_*(\partial/\partial t),\ F_*\nu\right) \circ \alpha_0 \right\} d\,\mathrm{v}_{\partial M} =$$

$$= \int_{\partial M} \overline{g}^f\left(V,\ f_*\nu\right) d\,\mathrm{v}_{\partial M}\,.$$

We obtain

$$\frac{d}{dt}\left\{E(f_t)\right\}_{t=0} = \mathrm{Vol}'(0) = \int_M \frac{\partial\varphi}{\partial t}(x, 0)\, d\,\mathrm{v}_M(x) =$$

$$= \int_{\partial M} \overline{g}^f\left(V,\ f_*\nu\right) d\,\mathrm{v}_{\partial M} - \int_M \overline{g}^f\left(V,\ \tau(f)\right).$$

The formula (5.44) is proved. Q.e.d.

Given a section $V \in C^\infty\left(f^{-1}T\overline{M}\right)$ and an open subset $\mathcal{U} \subset \overline{M}$ such that $f(M) \cap \mathcal{U} \neq \emptyset$ we consider the operator

$$\delta_V : C^1(\mathcal{U}) \to C\left(f^{-1}(\mathcal{U})\right)$$

defined as follows. Let $u \in C^1(\mathcal{U})$ and let us set

$$\left(\delta_V u\right)_x = \operatorname{div}_M\left(u^f\ V\right)_x, \quad \forall\, x \in f^{-1}(\mathcal{U}),$$

where $u^f = u \circ f$. If $f : M \to \overline{M}$ is an isometric immersion, then [by (5.38) with $\varphi = u^f$]

$$\delta_V(u) = \operatorname{div}_M\left(u^f\ V\right) = u^f \operatorname{div}_M(V) + g\left(V^\top,\ \nabla u^f\right)$$

so that, if $\text{div}_M(V) = 0$, then $\delta_V(u)$ is the *directional derivative* of u at the direction $V^\top$.

Let $(\mathcal{U},\ y^\alpha)$ be a local coordinate system on $\overline{M}$ about the point $f(x_0)$, for some $x_0 \in M$. Let

$$V_\alpha = (\partial/\partial y^\alpha)^f = (\partial/\partial y^\alpha) \circ f$$

be the *natural lift* of $\partial/\partial x^\alpha$ so that

$$\{V_\alpha : 1 \le \alpha \le n\} \subset C^\infty\left(f^{-1}(\mathcal{U}),\ f^{-1}T\left(\overline{M}\right)\right)$$

is a local frame in the pullback bundle $f^{-1}T(\overline{M}) \to M$. Given a function $u \in C^1(\mathcal{U})$ we wish to compute $\delta_\alpha(u)$ where $\delta_\alpha \equiv \delta_{V_\alpha}$. We have

$$\delta_\alpha(u) = \text{div}_M\left(u^f V_\alpha\right) = u^f\ \text{div}_M(V_\alpha) + \overline{g}^f\left(V_\alpha,\ f_*\nabla u^f\right).$$

To compute that 0-th-order term, we also fix a g-orthonormal local frame $\{E_j : 1 \le j \le m\} \subset C^\infty(U, T(M))$ defined on an open neighborhood $U \subset M$ of the point x_0 such that $f(U) \subset \mathcal{U}$. Then, for any $X \in \mathfrak{X}(M)$

$$f_*X = X(f^\alpha)\, V_\alpha,$$

$$\overline{\nabla}^f_X V_\alpha = X(f^\beta)\left(\overline{\Gamma}^\gamma_{\beta\alpha} \circ f\right) V_\gamma$$

by the very definition of the pullback connection $f^{-1}\overline{\nabla} = \overline{\nabla}^f$. Here $f^\alpha = y^\alpha \circ f$ and $\overline{\Gamma}^\alpha_{\beta\gamma}$ are the Christoffel symbols of the second species of the Riemannian metric $\overline{g}$

$$\overline{\nabla}_{\partial/\partial y^\beta} \frac{\partial}{\partial y^\gamma} = \overline{\Gamma}^\alpha_{\beta\gamma} \frac{\partial}{\partial y^\alpha}.$$

Consequently

$$\text{div}_M\left(V_\alpha\right) = \sum_{j=1}^m \overline{g}^f\left(\overline{\nabla}^f_{E_j} V_\alpha,\ f_* E_j\right) =$$

$$= \sum_{j=1}^m E_j(f^\beta)\left(\overline{\Gamma}^\gamma_{\beta\alpha} \circ f\right) \overline{g}^f\left(V_\gamma,\ f_* E_j\right) =$$

$$= \sum_{j=1}^m E_j(f^\beta)\, E_j(f^\sigma)\left(\overline{\Gamma}^\gamma_{\beta\alpha} \circ f\right) \overline{g}^f\left(V_\gamma,\ V_\sigma\right) =$$

[as

$$\overline{g}^f(V_\alpha,\, V_\beta) = \overline{g}\left(\frac{\partial}{\partial y^\alpha},\, \frac{\partial}{\partial y^\beta}\right) \circ f = \overline{g}_{\alpha\beta} \circ f$$

for any $1 \leq \alpha, \beta \leq n$]

$$= \sum_{j=1}^{m} E_j(f^\beta)\, E_j(f^\sigma)\, \left(\overline{\Gamma}^\gamma_{\beta\alpha} \circ f\right)\, (\overline{g}_{\gamma\sigma} \circ f) =$$

$$= \frac{1}{2} \sum_{j=1}^{m} E_j(f^\beta)\, E_j(f^\sigma)\, \left(\frac{\overline{g}_{\beta\sigma}}{\partial y^\alpha} \circ f\right).$$

We may assume w.l.o.g. that U is also the domain of a local coordinate system (U, x^i) on M, hence if $E_j = E_j^k\, \partial/\partial x^k$, then

$$\sum_{j=1}^{m} E_j(f^\beta)\, E_j(f^\sigma) = \sum_{j=1}^{m} E_j^k\, E_j^\ell\, \frac{\partial f^\beta}{\partial x^k} \frac{\partial f^\sigma}{\partial x^\ell} =$$

[as the orthonormality of $\{E_j : 1 \leq j \leq m\}$ implies $\sum_{j=1}^m E_j^k\, E_j^\ell = g^{k\ell}$]

$$= g^{k\ell}\, \frac{\partial f^\beta}{\partial x^k} \frac{\partial f^\sigma}{\partial x^\ell}\, .$$

Finally,

$$\text{div}_M(V_\alpha) = \frac{1}{2}\, g^{k\ell}\, \frac{\partial f^\beta}{\partial x^k} \frac{\partial f^\sigma}{\partial x^\ell}\, \left(\overline{g}_{\beta\sigma|\alpha} \circ f\right)$$

where $u_{|\alpha} = \partial u/\partial y^\alpha$ for any $u \in C^1(\mathcal{U})$. As to the first-order term

$$\nabla u^f = g^{ij}\, \frac{\partial u^f}{\partial x^i} \frac{\partial}{\partial x^j} = g^{ij}\, \left(u_{|\gamma} \circ f\right)\, \frac{\partial f^\gamma}{\partial x^i} \frac{\partial}{\partial x^j},$$

$$f_* \nabla u^f = g^{ij}\, (u_{|\gamma} \circ f)\, \frac{\partial f^\gamma}{\partial x^i} \frac{\partial f^\beta}{\partial x j}\, V_\beta,$$

$$\overline{g}^f\left(V_\alpha,\, f_* \nabla u^f\right) = \left(\overline{g}_{\alpha\beta} \circ f\right)\, \frac{\partial f^\beta}{\partial x^i} \frac{\partial f^\gamma}{\partial x^j}\, g^{ij}\, \left(u_{|\gamma} \circ f\right).$$

Finally,

$$\delta_\alpha(u) = \frac{1}{2}\, g^{k\ell}\, \frac{\partial f^\beta}{\partial x^k}\, \frac{\partial f^\sigma}{\partial x^\ell}\, \left(\overline{g}_{\beta\sigma|\alpha} \circ f\right)\, u^f + \tag{5.50}$$

$$+ \left(\overline{g}_{\alpha\beta} \circ f\right) \frac{\partial f^\beta}{\partial x^i}\, \frac{\partial f^\gamma}{\partial x^j}\, g^{ij} \left(\frac{\partial u}{\partial y^\gamma} \circ f\right).$$

Let us examine now the particular case where $\overline{M} = \mathbb{E}^n$ is the n-dimensional Euclidean space ($\overline{g}_{\alpha\beta} = \delta_{\alpha\beta}$). If this is the case

$$\operatorname{div}_M\left(V_\alpha\right) = 0$$

hence [by (5.50)]

$$\delta_\alpha(u) = g^{ij}\, \frac{\partial f^\alpha}{\partial x^i}\, \frac{\partial f^\beta}{\partial x^j} \left(\frac{\partial u}{\partial y^\beta} \circ f\right). \tag{5.51}$$

Let $y^\alpha\, (\partial/\partial y^\alpha) \in \mathfrak{X}\left(\mathbb{E}^n\right)$ be the *radial field* and let

$$V = \left(y^\alpha \circ f\right) V_\alpha \in C^\infty\left(M,\, f^{-1} T(\mathbb{E}^n)\right)$$

be its natural lift. We wish to compute the divergence of V. As

$$g_{ij} = \frac{\partial f^\alpha}{\partial x^i}\, \frac{\partial f^\beta}{\partial x^j}\, \left(\overline{g}_{\alpha\beta} \circ f\right), \quad \overline{g}_{\alpha\beta} = \delta_{\alpha\beta},$$

one has

$$g_{ij} = \sum_{\alpha=1}^{n} \frac{\partial f^\alpha}{\partial x^i}\, \frac{\partial f^\alpha}{\partial x^j}$$

and contracting with g^{ij}

$$\sum_{\alpha=1}^{n} g^{ij}\, \frac{\partial f^\alpha}{\partial x^i}\, \frac{\partial f^\alpha}{\partial x^j} = m.$$

Finally,

$$\operatorname{div}_M\left(V\right) = \operatorname{div}_M\left((y^\alpha \circ f)\, V_\alpha\right) = \delta_\alpha\left(y^\alpha\right) =$$

[by (5.51)]

$$= g^{ij}\, \frac{\partial f^\alpha}{\partial x^i}\, \frac{\partial f^\beta}{\partial x^j} \left(\frac{\partial y^\alpha}{\partial y^\beta} \circ f\right) = m.$$

Let us now go back to the general case of an arbitrary Riemannian manifold $\overline{M}$. Let $U\big(\overline{M}\big) \to \overline{M}$ be the unit tangent bundle, i.e.,

$$U\big(\overline{M}\big)_y = \big\{v \in T_y(\overline{M}) : \overline{g}_y(v, v) = 1\big\}, \quad x \in \overline{M}.$$

Let $\xi \in \overline{M}$ be a fixed point, and let (ρ, x) be the *polar coordinates* on $T_\xi(\overline{M})$, i.e.,

$$\rho : T_\xi\big(\overline{M}\big) \to [0, +\infty), \quad x : T_\xi(\overline{M}) \to U\big(\overline{M}\big)_\xi,$$

$$\rho(v) = \overline{g}_\xi\big(v, v\big)^{1/2}, \quad x(v) = \frac{1}{\rho(v)}\, v, \quad v \in T_\xi(\overline{M}).$$

Let $\exp_\xi : B_{\overline{R}}(0) \to \overline{M}$ be the exponential map, where $\overline{R}$ is the injectivity radius of $\overline{M}$.

Definition 5.34 The *geodesic distance* of ξ to q is

$$r(q) = \rho\left(\exp_\xi^{-1}(q)\right), \quad q \in \mathcal{U} = \exp_\xi\left[B_{\overline{R}}(0)\right].$$

□

Let $u = (\xi,\ \{X_\alpha : 1 \leq \alpha \leq n\})$ be a tangent linear frame on $\overline{M}$ at the point ξ, i.e., the $\mathbb{R}$-linear isomorphism

$$u : \mathbb{R}^n \to T_\xi(\overline{M}), \quad u(e_\alpha) = X_\alpha, \quad 1 \leq \alpha \leq n.$$

Let us choose from the very beginning an orthonormal linear frame, i.e.,

$$\overline{g}_\xi\big(X_\alpha,\ X_\beta\big) = \delta_{\alpha\beta}, \quad 1 \leq \alpha, \beta \leq n.$$

We shall make use of the *normal coordinates* of center ξ, that is we set

$$\chi = \big(y^1, \cdots, y^n\big) : \mathcal{U} \to \mathbb{R}^n,$$

$$\chi = u^{-1} \circ \big(\exp_\xi\big)^{-1}.$$

Said construction may be summed up as the commutative diagram

$$\begin{array}{cccccc}
T_\xi\big(\overline{M}\big) \supset & B_{\overline{R}}(0) & \overset{\exp_\xi}{\longrightarrow} & \mathcal{U} & \subset \overline{M} \\
u \uparrow & u^{-1} \downarrow & & \downarrow \chi & \\
\mathbb{R}^n \quad \supset & \Omega & = & \Omega & \subset \mathbb{R}^n
\end{array}$$

where $\Omega = u^{-1}\left[B_{\overline{R}}(0)\right]$ (an open subset of $\mathbb{R}^n$). Then

$$\left(\frac{\partial}{\partial y^\alpha}\right)_\xi = X_\alpha, \quad \overline{g}_{\alpha\beta}(\xi) = \delta_{\alpha\beta}\,.$$

Let

$$\dot{y}^\alpha : T_\xi(\overline{M}) \to \mathbb{R}$$

be the Cartesian coordinates associated to the local coordinate system $(\mathcal{U},\ y^\alpha)$, i.e.,

$$v = \dot{y}^\alpha(v)\left(\frac{\partial}{\partial y^\alpha}\right)_\xi, \quad v \in T_\xi\big(\overline{M}\big).$$

Definition 5.35 The *radial field* $\mathbf{X} \in C^\infty\big(\mathcal{U},\ T(\overline{M})\big)$ is

$$\mathbf{X}(q) = \dot{y}^\alpha(v)\left(\frac{\partial}{\partial y^\alpha}\right)_q, \quad q \in \mathcal{U},$$

$$v = \big(\exp_\xi\big)^{-1}(q) \in B_{\overline{R}}(0) \subset T_\xi\big(\overline{M}\big).$$

□

Lemma 5.36 *Let* $v \in B_{\overline{R}}(0)$ *and let us set* $q = \exp_\xi v$. *Let*

$$d_v \exp_\xi : T_v\big(T_\xi(\overline{M})\big) \to T_q(\overline{M})$$

be the differential of $\exp_\xi$ *at the point* v. *Then* (i)

$$\big(d_v \exp_\xi\big)\tilde{v} = \mathbf{X}(q) \tag{5.52}$$

where

$$\tilde{v} = \dot{y}^\alpha(v)\left(\frac{\partial}{\partial \dot{y}^\alpha}\right)_v \in T_v\big(T_\xi(\overline{M})\big).$$

Moreover, (ii)

$$r(q)\,\frac{\partial r}{\partial y^\alpha}(q) = \dot{y}^\alpha(v). \tag{5.53}$$

Let us assume that $\xi \in f(M)$. *If* $\hat{\mathbf{X}} \in C^\infty\left(f^{-1}(\mathcal{U}),\ f^{-1}T\overline{M}\right)$ *is the natural lift of the radial field* $\mathbf{X} : \mathcal{U} \to T(\overline{M})$, *then* (iii)

$$\operatorname{div}_M \hat{\mathbf{X}} = \|df\|^2 + \frac{1}{2}\left(\dot{y}^\alpha \circ \exp_\xi^{-1}\right)^f g^{k\ell} \frac{\partial f^\beta}{\partial x^i} \frac{\partial f^\gamma}{\partial x^j} \left(\overline{g}_{\beta\gamma} \circ f\right) \tag{5.54}$$

and $\|df\|^2 = m$.

Proof (i) Let

$$\left(\exp_\xi\right)^\alpha = y^\alpha \circ \exp_\xi$$

be the components of $\exp_\xi$ with respect to the local chart χ. Then

$$\left(d_v \exp_\xi\right)\left(\frac{\partial}{\partial \dot{y}^\beta}\right)_v = \frac{\partial\left(\exp_\xi\right)^\alpha}{\partial \dot{y}^\beta}(v)\left(\frac{\partial}{\partial y^\alpha}\right)_q .$$

To compute the partials

$$\left(\partial\left(\exp_\xi\right)^\alpha / \partial \dot{y}^\beta\right)(v)$$

one should calculate first the local expression of

$$\left(\exp_\xi\right)^\alpha : B_{\overline{R}}(0) \to \mathbb{R}$$

with respect to the (global) chart

$$\dot{\chi} = \left(\dot{y}^1,\ \cdots,\ \dot{y}^n\right) : T_\xi(\overline{M}) \to \mathbb{R}^n$$

that is, for any $\eta = (\eta^1,\ \cdots,\ \eta^n) \in \Omega \subset \mathbb{R}^n$

$$\left[\left(\exp_\xi\right)^\alpha \circ \dot{\chi}^{-1}\right](\eta) = \left(\exp_\xi\right)^\alpha\left[\eta^\beta\left(\frac{\partial}{\partial y^\beta}\right)_\xi\right] =$$

$$= \left(p^\alpha \circ \chi \circ \exp_\xi\right)\left[\eta^\beta\left(\frac{\partial}{\partial y^\beta}\right)_\xi\right] = \left(p^\alpha \circ u^{-1}\right)\left[\eta^\beta\left(\frac{\partial}{\partial y^\beta}\right)_\xi\right] =$$

$$= p^\alpha\left(\eta^\beta e_\beta\right) = \eta^\alpha$$

and hence

$$\frac{\partial\left(\exp_\xi\right)^\alpha}{\partial \dot{y}^\beta}(v) = \frac{\partial\left[\left(\exp_\xi\right)^\alpha \circ \dot{\chi}^{-1}\right]}{\partial \eta^\beta}\left(\dot{\chi}(v)\right) = \delta^\alpha_\beta$$

so that

$$(d_v \exp_\xi)\left(\frac{\partial}{\partial \dot{y}^\beta}\right)_v = \left(\frac{\partial}{\partial y^\beta}\right)_q . \tag{5.55}$$

The formula (5.52) is an immediate consequence of (5.55).

(ii) To compute the partial derivatives $\partial r/\partial y^\beta$ one should determine the local expression of the function $r : \mathcal{U} \to \mathbb{R}$ with respect to the local coordinate system $(\mathcal{U},\ y^\alpha)$, that is

$$\big(r \circ \chi^{-1}\big)(\eta) = \big(\rho \circ \exp_\xi^{-1} \circ \chi^{-1}\big)(\eta) = \big(\rho \circ u\big)(\eta) =$$

$$= g_\xi\big(u(\eta), u(\eta)\big)^{1/2} = \big[\eta^\alpha \eta^\beta g_\xi\big(X_\alpha,\ X_\beta\big)\big]^{1/2} = \|\eta\|$$

where $\|\cdot\|$ is the Euclidean norm on $\mathbb{R}^n$. It follows that

$$\frac{\partial r}{\partial y^\beta}(q) = \frac{\partial(r \circ \chi^{-1})}{\partial \eta^\beta}(\chi(q)) = \frac{\eta^\beta}{\|\eta\|}\bigg|_{\eta=\chi(q)} =$$

[as

$$\eta = \chi(q) = u^{-1}(v) = \dot{y}^\alpha(v) e_\alpha \Longrightarrow \eta^\alpha = \dot{y}^\alpha(v)$$

and hence

$$\|\eta\|^2 = \sum_{\alpha=1}^n \dot{y}^\alpha(v)^2 = g_\xi(v, v)]$$

$$= \frac{\dot{y}^\beta(v)}{g_\xi(v, v)^{1/2}} =$$

[as $r(q) = \rho\big(\exp_\xi^{-1}(q)\big) = \rho(v) = g_\xi(v, v)^{1/2}$]

$$= \frac{\dot{y}^\beta(v)}{r(q)}$$

and the formula (5.53) is proved. Q.e.d.

(iii) If $\xi \in f(M)$, i.e., ξ is actually a point of the given submanifold, let

$$\hat{\mathbf{X}} : f^{-1}(\mathcal{U}) \to f^{-1} T(\overline{M})$$

be the natural lift of the radial field

$$\mathbf{X} : \mathcal{U} \to T(\overline{M}),$$

i.e.,

$$\hat{\mathbf{X}} = \mathbf{X}^f = \mathbf{X} \circ f.$$

Then

$$\mathrm{div}_M\left(\hat{\mathbf{X}}\right) = \mathrm{div}_M\left[\left(\dot{y}^\alpha \circ \exp_\xi^{-1}\right)^f V_\alpha\right] =$$

[where $V_\alpha = \widehat{\partial/\partial y^\alpha} \in C^\infty\big(f^{-1}(\mathcal{U}),\ f^{-1}T\overline{M}\big)$]

$$= \left(\dot{y}^\alpha \circ \exp_\xi^{-1}\right)^f \mathrm{div}\big(V_\alpha\big) + \overline{g}^f\left(V_\alpha,\ f_*\nabla\left(\dot{y}^\alpha \circ \exp_\xi^{-1}\right)^f\right).$$

Moreover, if

$$h^\alpha = \dot{y}^\alpha \circ \exp_\xi^{-1},$$

then

$$\overline{g}^f\left(V_\alpha,\ f_*\nabla\big(h^\alpha\big)^f\right) = \tag{5.56}$$

$$= \left(\overline{g}_{\alpha\beta} \circ f\right) \frac{\partial f^\beta}{\partial x^i} \frac{\partial f^\gamma}{\partial x^j}\, g^{ij} \left({h^\alpha}_{|\gamma} \circ f\right)$$

and the local expression of h^α is

$$\left(h^\alpha \circ \chi^{-1}\right)(\eta) = \left(\dot{y}^\alpha \circ u\right)(\eta) = \eta^\alpha$$

so that

$${h^\alpha}_{|\gamma} = \frac{\partial h^\alpha}{\partial y^\gamma} = \frac{\partial(h^\alpha \circ \chi^{-1})}{\partial \eta^\gamma} \circ \chi = \delta^\alpha_\gamma$$

and the formula (5.56) becomes

$$\overline{g}^f\left(V_\alpha,\ f_*\nabla\big(h^\alpha\big)^f\right) = \left(\overline{g}_{\alpha\beta} \circ f\right) \frac{\partial f^\alpha}{\partial x^i} \frac{\partial f^\beta}{\partial x^j}\, g^{ij} = \|df\|^2 =$$

[as $f^*\overline{g} = g$]

$$= g_{ij}\, g^{ij} = m.$$

The formula (5.54) should be paralleled to the formula (3.7) in [HoSp], p. 721. D. Hoffman & J. Spruck also introduce (cf. *op. cit.*) the matrix

$$\tilde{g}_\gamma^\beta = g^{ij} \frac{\partial f^\beta}{\partial x^i} \frac{\partial f^\sigma}{\partial x^j} \left(\overline{g}_{\gamma\sigma} \circ f\right)$$

so that (5.54) may be rewritten as

$$\operatorname{div}\hat{\mathbf{X}} = m + \left(\dot{y}^\alpha \circ \exp_\xi^{-1}\right)^f \tilde{g}_\gamma^\beta \left(\overline{\Gamma}_{\beta\alpha}^\gamma \circ f\right)$$

together with

$$\operatorname{trace}\left(\tilde{g}\right) = \tilde{g}_\beta^\beta = \|df\|^2 = m.$$

Next, we shall estimate $\operatorname{div}\left(\hat{\mathbf{X}}\right)$ in terms of $r(q)$ [the geodesic distance from ξ to q] and the sectional curvature of $\overline{M}$.

Let $\xi \in \overline{M}$ and $v \in B_{\overline{R}}(0)$ and let us set $q = \exp_\xi(v)$. Let $\mathbf{x} = (1/\rho(v))\, v$ so that $\mathbf{x} \in U(\overline{M})$ and in particular

$$q = \exp_\xi \left[r(q)\, \mathbf{x}\right]$$

is a tautology. Let us consider the geodesic

$$\gamma(t) = \exp_\xi \left(t\mathbf{x}\right)$$

so that $\gamma(0) = \xi$ and $\gamma(r(q)) = q$. To start with we need the following result[17]

Lemma 5.37 *Let $\overline{M}$ be a Riemannian manifold of sectional curvature bounded from above $\overline{K} \le b^2$. If $b^2\, r(q)^2 < \pi^2$, then there are no conjugate points along γ. Consequently, if $Y \in T_q\left(\overline{M}\right)$, then there is $\mathbf{s} \in T_{r(q)\mathbf{x}}\left(T_\xi\left(\overline{M}\right)\right)$ such that*

$$\left(d_{r(q)\mathbf{x}} \exp_\xi\right)(\mathbf{s}) = Y.$$

Let us extend $\mathbf{s}$ *by parallelism along $t \mapsto t\,\mathbf{x}$ in $T_\xi(\overline{M})$ such that to obtain the Jacobi field*

$$V(t) = \left(d_{r(q)\mathbf{x}} \exp_\xi\right)\left[\frac{t}{r(q)}\,\mathbf{s}\right]$$

[17] For a proof, one may see R. Bishop & R. Crittenden, [BiCr], p. 255.

along $\gamma(t)$. *Then*

$$V(0) = 0, \quad V'(0) = \frac{1}{r(q)}\,\mathbf{s}, \quad V(r(q)) = Y,$$

$$\langle \overline{\nabla}_{\dot{\gamma}} V,\ V \rangle_{r(q)} \geq \begin{cases} b \cot\big[b\, r(q)\big] & \text{se } \ b > 0, \\ \\ |b| \coth\big[|b|\, r(q)\big] & \text{se } \ b \in i\mathbb{R}. \end{cases}$$

One has

Lemma 5.38 *Let* $\overline{M}$ *be a Riemannian manifold obeying to the assumptions in Lemma* 5.37 *and let* $\mathbf{X}$ *be the radial field of center* $\xi \in f(M)$. *If* $q \in f(M)$ *is a point such that* $b^2\, r(q)^2 < \pi^2$, *then*

$$\operatorname{div}_M\big(\hat{\mathbf{X}}\big)(q) \geq m\, b\, r(q) \cot\big[b\, r(q)\big]. \tag{5.57}$$

5.5 Brendle's Theorems

We start by reporting on S. Brendles's results[18] on Sobolev inequalities on submanifolds in the Euclidean space.

Theorem 5.39 *Let* $\Sigma \subset \mathbb{R}^{n+m}$ *be a compact n-dimensional submanifold with* $\partial\Sigma = \emptyset$. *Let* $f : \Sigma \to (0, +\infty)$ *be a* C^∞ *function. Then*

$$\int_\Sigma f \left\{ \log f + n + \frac{n}{2}\, \log(4\pi) \right\} d\,\mathrm{v}_g +$$

$$- \int_\Sigma \frac{|\nabla^\Sigma f|^2}{f}\, d\,\mathrm{v}_g - \int_\Sigma f\, |H|^2\, d\,\mathrm{v}_g \leq$$

$$\leq \left(\int_\Sigma f\, d\,\mathrm{v}_g \right) \log \left(\int_\Sigma f\, d\,\mathrm{v}_g \right).$$

[18] S. Brendle, *The logarithmic Sobolev inequality for a submanifold in Euclidean space*, Communications on Pure and Applied Mathematics, Vol. LXXV (2022), 0449–0454.

Here g and H are, respectively, the first fundamental form and mean curvature vector of the given immersion $\Sigma \hookrightarrow \mathbb{R}^{n+m}$. If one sets

$$f = \frac{1}{(4\pi)^{n/2}} \exp\Big(- \frac{1}{4} |x|^2 \Big) \varphi,$$

then Theorem 5.39 yields

Corollary 5.40 *Let $\Sigma \subset \mathbb{R}^{n+m}$ be a compact n-dimensional submanifold with $\partial\Sigma = \emptyset$, and let*

$$d\gamma = \frac{1}{(4\pi)^{n/2}} \exp\Big(- \frac{|x|^2}{4} \Big) d\,\mathrm{v}_g$$

be the Gaussian measure on Σ. Let $\varphi : \Sigma \to (0, +\infty)$ be a C^∞ function. Then

$$\int_\Sigma \varphi\,(\log \varphi)\, d\gamma - \int_\Sigma \frac{|\nabla^\Sigma \varphi|^2}{\varphi}\, d\gamma - \int_\Sigma \varphi \Big| H + \frac{1}{2} x^\perp \Big|^2 d\gamma \leq$$

$$\leq \Big(\int_\Sigma \varphi\, d\gamma \Big) \log \Big(\int_\Sigma \varphi\, d\gamma \Big).$$

The Euclidean logarithmic inequality is well known, and was obtained as a consequence of the isoperimetric inequality, cf. C. Borell.[19] The logarithmic Sobolev inequality plays an important role in Perelman's monotonicity formula for the entropy under Ricci flow.[20] The Gauss measure γ is judged[21] as very natural from a geometric point of view for, in particular, the quantity

$$\gamma(\Sigma) = \frac{1}{(4\pi)^{n/2}} \int_\Sigma e^{-|x|^2/4}\, d\,\mathrm{v}_g$$

is the Gaussian area appearing in Huisken's monotonicity formula for the mean curvature flow.[22] One has $H + \frac{1}{2} x^\perp = 0$ if and only if Σ is a critical point[23] for Gaussian area $\gamma(\Sigma)$.

---

[19] C. Borel, *The Brun-Minkowski inequality in Gauss space*, Invent. Math., (2) 30(1975), 207–216.

[20] Cf. G. Perelman, *The entropy formula for the Ricci flow and its geometric applications*, `arXiv:math/0211159 [math.DG]`.

[21] Cf. S. Brendle, *op. cit.*, p. 450.

[22] Cf. G. Huisken, *Asymptotic behavior for singularities of the mean curvature flow*, J. Differential Geometry, (1)31(1990), 285–299.

[23] Submanifolds with this property are called *self-similar shrinkers* (for they shrink homothetically under the mean curvature flow).

S. Brendle generalized[24] his own previous (Euclidean) result[25] to the Riemannian case, as follows.

Theorem 5.41 *Let M be a complete noncompact Riemannian manifold, of dimension $n + m$, with nonnegative sectional curvature. Let $\Sigma \subset M$ be a n-dimensional compact submanifold, with boundary $\partial\Sigma$. Let $f : \Sigma \to (0, +\infty)$ be a C^∞ function. If $m \geq 2$, then*

$$\int_\Sigma \sqrt{\left|\nabla^\Sigma f\right|^2 + f^2|H|^2}\, d\,\mathrm{v}_g + \int_{\partial\Sigma} f\, d\,\mathrm{v}_g \geq$$

$$\geq n\left[\frac{(n+m)|\mathbb{B}^{n+m}|}{m|\mathbb{B}^m|}\right]^{1/n} \theta^{1/n}\left(\int_\Sigma f^{n/(n-1)}\, d\,\mathrm{v}_g\right)^{(n-1)/n}.$$

Here θ is the asymptotic volume ratio of M and H denotes the mean curvature vector of $\Sigma \hookrightarrow M$. The reader will appreciate the particularly elegant form of the Sobolev inequality for submanifolds of codimension $m = 2$.

Corollary 5.42 *Let M be a complete noncompact Riemannian manifold, of dimension $n + 2$, with nonnegative sectional curvature. Let $\Sigma \subset M$ be a n-dimensional compact submanifold, with boundary $\partial\Sigma$. Let $f : \Sigma \to (0, +\infty)$ be a C^∞ function. Then*

$$\int_\Sigma \sqrt{|\nabla^\Sigma f|^2 + f^2|H|^2}\, d\,\mathrm{v}_g + \int_{\partial\Sigma} f\, d\,\mathrm{v}_g \geq$$

$$\geq n\,|\mathbb{B}^n|^{1/n}\,\theta^{1/n}\left(\int_\Sigma f^{n/(n-1)}\, d\,\mathrm{v}_g\right)^{(n-1)/n}.$$

5.6 Batista–Mirandola Theorem

We report on a result by M. Batista & H. Mirandola[26] for Sobolev inequalities for a submanifold of a Riemannian manifold $(\overline{M}, g)$, where integration of functions defined on the ambient space is performed with respect to a weighted measure $d\mu = e^{-f}\, d\,\mathrm{v}_g$, for some C^∞ function $f : \overline{M} \to \mathbb{R}$ (the *density*).

---

[24] Cf. Theorem 1.4 by S. Brendle, *Sobolev inequalities in manifolds with nonnegative curvature*, `arXiv:2009.13717v5 [math.DG] 28 May 2022`, p. 2.

[25] Cf. S. Brendle, *The isoperimetric inequality for a minimal submanifold in Euclidean space*, J. Amer. Math. Soc., 34(2021), 595–603.

[26] Cf. M. Batista & H. Mirandola, *Sobolev and isoperimetric inequalities for submanifolds in weighted ambient spaces*, `arXiv:1304.2271v2 [math.DG] 12 Apr 2013`.

Let $x : M \to \overline{M}$ be an isometric immersion of a complete Riemanian manifold M, eventually with nonempty boundary ($\partial M \neq \emptyset$) in the Riemannian manifold $(\overline{M}, g)$. Let $f : \overline{M} \to \mathbb{R}$ be a C^∞ function, and let H be the mean curvature vector of the given immersion $x : M \to \overline{M}$. Let $T(M)^\perp \to M$ be the normal bundle of the given immersion, and let $V^\perp$ denote the $T(M)^\perp$-component of $V \in \mathfrak{X}(\overline{M})$. Let us set

$$H_f := H + \left(\overline{\nabla} f\right)^\perp .$$

Definition 5.43 The immersion $x : M \to \overline{M}$ is said to be *f-minimal* if $H_f = 0$. □

Also, if $\Omega \subset \overline{M}$ is a bounded domain, we set

$$\operatorname{vol}_f(\Omega) := \int_\Omega d\mu = \int_\Omega e^{-f(x)}\, d\, \mathrm{v}_g(x). \tag{5.58}$$

By a result of V. Bayle[27] for every smooth 1-parameter variation $\{\Omega_t\}_{|t|<\epsilon}$ of Ω

$$\frac{d}{dt}\left\{\operatorname{vol}_f(\Omega_t)\right\}_{t=0} = \int_\Omega \langle H_f,\, V\rangle\, d\,\mu$$

where V is the normal infinitesimal variation induced by $\{\Omega_t\}_{|t|<\epsilon}$. It may then be argued in favor of the "natural" occurrence of H_f [as springing from the variational problem associated to the functional (5.58)].

Let us look at the following example, associated with work by M. Gromov.[28] Let $\mathbb{E}^n = \left(\mathbb{R}^n,\, g_0\right)$ be the n-dimensional Euclidean spaces, and let us consider the weighted measure $d\mu = \exp(-|x|^2)\, d\, \mathrm{v}_{g_0}(x)$. An isometric immersion $F : M \to \mathbb{R}^n$ is said to be a *self-shrinker* if its mean curvature vector H satisfies $2H = -F^\perp$. This is of course equivalent to saying that F is $(|x|^2/4)$-minimal.

Let $\mathcal{K} : \mathbb{R} \to [0, +\infty)$ be a nonnegative, even, and continuous function, and let h be the solution to the Cauchy problem

$$\begin{cases} h'' + \mathcal{K}\, h = 0, \\ \\ h(0) = 0, \quad h'(0) = 1. \end{cases} \tag{5.59}$$

Let us choose constants

$$r_0 = r_0(\mathcal{K}) > 0, \quad s_0 = s_0(\mathcal{K}) > 0,$$

---

[27] Cf. V. Bayle, *Propriétées de concavité du profil isopérimétrique et applications*, Mathématiques [math]. Université Joseph-Fourier - Grenoble I, 2003, Français, NNT : tel-00004317v1, HAL Id: tel-00004317 https://theses.hal.science/tel-00004317v1.

[28] Cf. M. Gromov, *Isoperimetry of waists and concentration of maps*, Geom. Funct. Anal., (1)13(2003), 178–215.

such that h is increasing on the interval $(0, s_0)$ and h maps $(0, s_0)$ onto the interval $(0, r_0)$. Next, let us assume that the radial curvature of $\overline{M}$ with base point ξ satisfies

$$\left(\overline{K}_{\mathrm{rad}}\right)_\xi \leq \mathcal{K}\left(d_{\overline{M}}(\cdot, \xi)\right) \tag{5.60}$$

for every $\xi \in M$, where $d_{\overline{M}}$ is the distance function associated to the Riemannian metric on $\overline{M}$. The main result by M. Batista & H. Mirandola is (cf. *op. cit.*)

Theorem 5.44 *Let us assume that* $f^* := \sup_M f < \infty$, *and let* $\varphi \in C_0^1(M)$ *be a nonnegative function such that* $\varphi\big|_{\partial M} = 0$. *Moreover, let us assume that there is* $\kappa \in (0, 1)$ *such that*

$$\begin{cases} \overline{J} := \left[\dfrac{e^{f^*}}{(1-\kappa)\omega_m} \operatorname{vol}_f\left(\operatorname{supp}(\varphi)\right)\right]^{1/m} \leq s_0, \\ \\ h^{-1}\left(\overline{J}\right) \leq 2 \operatorname{Inj}_\varphi, \end{cases} \tag{5.61}$$

where $\omega_m = \operatorname{Vol}\left(\mathbb{B}^m\right)$ *and* $\operatorname{Inj}_\varphi$ *is the minimum of the injectivity radius of* $\overline{M}$ *restricted to the points of* $\operatorname{supp}(\varphi)$. *Then there is a constant* $S = S(m, \mathcal{K}) > 0$ *such that*

$$\left(\int_M \varphi^{mp/(m-p)} \, d\mu\right)^{(m-p)/m} \leq$$

$$\leq S\, e^{f^*/m} \int_M \left(|\nabla\varphi| + \varphi\left|H_f - \overline{\nabla} f\right|\right)^p d\,\mu$$

for any $1 \leq p < m$, *where the constant* S *is*

$$S = \frac{2^m m}{\kappa(m-1)} \frac{r_0}{s_0}\left[\frac{1}{(1-k)\omega_m}\right]^{1/m}. \tag{5.62}$$

Theorem 5.44 is an extension (the context of submanifolds of ambient spaces endowed with a weighted measure $d\mu = e^{-f}\, d\,\mathrm{v}_g$) of results by J.H. Michael & L.M. Simon[29] and D. Hoffman & J. Spruck.[30]

---

[29] Cf. J.H. Michael & L.M. Simon, *Sobolev and mean-value inequalities on generalized submanifolds of* $\mathbb{R}^n$, Comm. Pure Appl. Math., 26(1973), 361–379.

[30] Cf. D. Hoffman & J. Spruck, [HoSp], and our presentation earlier in this chapter.

References

[Ath] C. Athanasios, *Sobolev inequalities on complete Riemannian manifolds*, Master Thesis, University of Crete (2012)

[HoSp] D. Hoffman, J. Spruck, Sobolev and isoperimetric inequalities on Riemannian submanifolds. Commun. Pure Appl. Math. **XXVII**, 715–727 (1974)

[BiCr] R. Bishop, R. Crittenden, *Geometry of manifolds* (Academic, New York-London, 1964)

[GrKlMe] D. Gromoll, W. Klingenberg, W. Meyer, *Riemannsche Geometrie im Grossen*, Lecture Notes in Mathematics (Springer, Berlin-Heidelberg-New York, 1968)

Index

E. Barletta et al., *Differential Geometry*, Infosys Science Foundation Series,
https://doi.org/10.1007/978-981-96-1631-2

The manufacturer's authorised representative in the EU is Springer Nature Customer Service Centre GmbH, Europaplatz 3, 69115 Heidelberg, Germany. If you have any concerns regarding our products, please contact ProductSafety@springernature.com

Printed and bound by CPI Group (UK) Ltd, Croydon, CR0 4YY
01/10/2026
02225928-0020